NEUROCUTANEOUS SYNDROMES

HANDBOOK OF CLINICAL NEUROLOGY

Series Editors

MICHAEL J. AMINOFF, FRANÇOIS BOLLER, AND DICK F. SWAAB

VOLUME 132

ELSEVIER

EDINBURGH LONDON NEW YORK OXFORD PHILADELPHIA
ST LOUIS SYDNEY TORONTO 2015

NEUROCUTANEOUS SYNDROMES

Series Editors

MICHAEL J. AMINOFF, FRANÇOIS BOLLER, AND DICK F. SWAAB

Volume Editors

MONICA P. ISLAM AND E. STEVE ROACH

VOLUME 132

3rd Series

ELSEVIER

EDINBURGH LONDON NEW YORK OXFORD PHILADELPHIA
ST LOUIS SYDNEY TORONTO 2015

ELSEVIER B.V.
Radarweg 29, PO Box 211, 1000 AE Amsterdam, Netherlands
The Boulevard, Langford Lane, Kidlington, Oxford OX5 1GB, UK
225 Wyman Street, Waltham, MA 02451, USA

ISBN: 9780444627025

British Library Cataloguing in Publication Data
A catalogue record for this book is available from the British Library

Library of Congress Cataloging in Publication Data
A catalog record for this book is available from the Library of Congress

Commissioning Editor: Mara E. Conner
Development Editor: Michael Parkinson
Project Manager: Sujatha Thirugnana Sambandam
Designer/Design Direction: Alan Studholme

The Publisher's policy is to use **paper manufactured from sustainable forests**

Handbook of Clinical Neurology 3rd Series

Available titles

Vol. 79, The human hypothalamus: basic and clinical aspects, Part I, D.F. Swaab ISBN 9780444513571
Vol. 80, The human hypothalamus: basic and clinical aspects, Part II, D.F. Swaab ISBN 9780444514905
Vol. 81, Pain, F. Cervero and T.S. Jensen, eds. ISBN 9780444519016
Vol. 82, Motor neurone disorders and related diseases, A.A. Eisen and P.J. Shaw, eds. ISBN 9780444518941
Vol. 83, Parkinson's disease and related disorders, Part I, W.C. Koller and E. Melamed, eds. ISBN 9780444519009
Vol. 84, Parkinson's disease and related disorders, Part II, W.C. Koller and E. Melamed, eds. ISBN 9780444528933
Vol. 85, HIV/AIDS and the nervous system, P. Portegies and J. Berger, eds. ISBN 9780444520104
Vol. 86, Myopathies, F.L. Mastaglia and D. Hilton Jones, eds. ISBN 9780444518996
Vol. 87, Malformations of the nervous system, H.B. Sarnat and P. Curatolo, eds. ISBN 9780444518965
Vol. 88, Neuropsychology and behavioural neurology, G. Goldenberg and B.C. Miller, eds. ISBN 9780444518972
Vol. 89, Dementias, C. Duyckaerts and I. Litvan, eds. ISBN 9780444518989
Vol. 90, Disorders of consciousness, G.B. Young and E.F.M. Wijdicks, eds. ISBN 9780444518958
Vol. 91, Neuromuscular junction disorders, A.G. Engel, ed. ISBN 9780444520081
Vol. 92, Stroke – Part I: Basic and epidemiological aspects, M. Fisher, ed. ISBN 9780444520036
Vol. 93, Stroke – Part II: Clinical manifestations and pathogenesis, M. Fisher, ed. ISBN 9780444520043
Vol. 94, Stroke – Part III: Investigations and management, M. Fisher, ed. ISBN 9780444520050
Vol. 95, History of neurology, S. Finger, F. Boller and K.L. Tyler, eds. ISBN 9780444520081
Vol. 96, Bacterial infections of the central nervous system, K.L. Roos and A.R. Tunkel, eds. ISBN 9780444520159
Vol. 97, Headache, G. Nappi and M.A. Moskowitz, eds. ISBN 9780444521392
Vol. 98, Sleep disorders, Part I, P. Montagna and S. Chokroverty, eds. ISBN 9780444520067
Vol. 99, Sleep disorders, Part II, P. Montagna and S. Chokroverty, eds. ISBN 9780444520074
Vol. 100, Hyperkinetic movement disorders, W.J. Weiner and E. Tolosa, eds. ISBN 9780444520142
Vol. 101, Muscular dystrophies, A. Amato and R.C. Griggs, eds. ISBN 9780080450315
Vol. 102, Neuro-ophthalmology, C. Kennard and R.J. Leigh, eds. ISBN 9780444529039
Vol. 103, Ataxic disorders, S.H. Subramony and A. Durr, eds. ISBN 9780444518927
Vol. 104, Neuro-oncology, Part I, W. Grisold and R. Sofietti, eds. ISBN 9780444521385
Vol. 105, Neuro-oncology, Part II, W. Grisold and R. Sofietti, eds. ISBN 9780444535023
Vol. 106, Neurobiology of psychiatric disorders, T. Schlaepfer and C.B. Nemeroff, eds. ISBN 9780444520029
Vol. 107, Epilepsy Part I, H. Stefan and W.H. Theodore, eds. ISBN 9780444528988
Vol. 108, Epilepsy Part II, H. Stefan and W.H. Theodore, eds. ISBN 9780444528995
Vol. 109, Spinal cord injury, J. Verhaagen and J.W. McDonald III, eds. ISBN 9780444521378
Vol. 110, Neurological rehabilitation, M. Barnes and D.C. Good, eds. ISBN 9780444529015
Vol. 111, Pediatric neurology Part I, O. Dulac, M. Lassonde and H.B. Sarnat, eds. ISBN 9780444528919
Vol. 112, Pediatric neurology Part II, O. Dulac, M. Lassonde and H.B. Sarnat, eds. ISBN 9780444529107
Vol. 113, Pediatric neurology Part III, O. Dulac, M. Lassonde and H.B. Sarnat, eds. ISBN 9780444595652
Vol. 114, Neuroparasitology and tropical neurology, H.H. Garcia, H.B. Tanowitz and O.H. Del Brutto, eds. ISBN 9780444534903
Vol. 115, Peripheral nerve disorders, G. Said and C. Krarup, eds. ISBN 9780444529022
Vol. 116, Brain stimulation, A.M. Lozano and M. Hallett, eds. ISBN 9780444534972
Vol. 117, Autonomic nervous system, R.M. Buijs and D.F. Swaab, eds. ISBN 9780444534910
Vol. 118, Ethical and legal issues in neurology, J.L. Bernat and H.R. Beresford, eds. ISBN 9780444535016
Vol. 119, Neurologic aspects of systemic disease Part I, J. Biller and J.M. Ferro, eds. ISBN 9780702040863
Vol. 120, Neurologic aspects of systemic disease Part II, J. Biller and J.M. Ferro, eds. ISBN 9780702040870
Vol. 121, Neurologic aspects of systemic disease Part III, J. Biller and J.M. Ferro, eds. ISBN 9780702040887
Vol. 122, Multiple sclerosis and related disorders, D.S. Goodin, ed. ISBN 9780444520012
Vol. 123, Neurovirology, A.C. Tselis and J. Booss, eds. ISBN 9780444534880

Foreword

The nervous system and the skin have a common embryological origin, and therefore the relation between the two is particularly important. Both originate from cells derived from the embryonic neural crest cell lineage, as pointed out in Chapter 2. Many neurologic diseases affect the skin. Furthermore much pathology affects both. A volume dedicated to neurocutaneous disorders is therefore particularly timely. A previous chapter published in Volume 121 in 2014 covered neurodermatology, presenting the various diseases according to the conditions most often encountered by neurologists such as stroke, epilepsy, and brain tumors. The present volume deals with the most important neurocutaneous disorders in greater detail, with an emphasis on the genetics, clinical aspects, and treatment in each case. Chapter 3 discusses the pathophysiology and manifestation of various acquired neurocutaneous disorders, namely Behçet disease, dermatomyositis, Sjögren syndrome, systemic lupus erythematosus (SLE), scleroderma, Parry–Romberg syndrome and Degos disease. Subsequent chapters illustrate several other conditions with particular emphasis on congenital and genetic disorders.

The encyclopedic coverage provided in the present volume has been guided by the experience and skills of the two volume editors. Monica P. Islam is a pediatric neurologist at the Nationwide Children's Hospital in Columbus, Ohio where the co-editor of this volume, E. Steve Roach, is chief of pediatric neurology and vice-chair of the department of pediatrics. Both are widely recognized experts in the field and together they have made an outstanding team to develop and edit this new volume. They have gathered together a group of international contributors and crafted the text to ensure that the various chapters fit seamlessly together as a valuable synthesis of the literature. We are grateful to them and to all the contributors. We have reviewed and commented on each of the chapters in our capacity as series editors and believe that both clinicians and basic scientists will find the volume appealing. The availability of the volume electronically on Elsevier's Science Direct site should ensure its ready accessibility and facilitate searches for specific information.

As always, it is a pleasure to thank Elsevier, our publisher – and in particular Mica Haley, Michael Parkinson, and Kristi Anderson as well as our new editor, Mara E. Conner – for their assistance in the development and production of this volume.

Michael J. Aminoff
François Boller
Dick F. Swaab

Preface

Prior to the advent of modern imaging and sophisticated laboratory techniques, the concept of neurocutaneous syndromes long served as an extrapolation of visible skin stigmata to underlying structural abnormalities within the nervous system. Historically, neurocutaneous disorders have included congenital disorders manifesting symptoms early in life as well as across the lifespan related to tumors and vascular malformations – not only confined to the skin and neuroaxis but also involving other body systems. Many of these conditions have inheritance patterns that eventually guided the discovery of disease-specific gene mutations. Identification of these mutations and an increasing understanding of expected normal protein function led to an increased appreciation of varying phenotypic expression. Surveillance recommendations for affected individuals – and unaffected family members for some conditions – have evolved with the available knowledge regarding the spectrum of clinical involvement, imaging findings, natural history, preventive measures, and curative intervention.

In this volume, it has been our privilege to compile chapters from authors of varying backgrounds and provide a comprehensive and expert update on neurocutaneous disorders. The introductory chapters examine the foundation for the genetic basis impacting neurocutaneous disorders and examine how genotype and phenotype can inform each other. We have included a chapter summarizing selected acquired chronic neurocutaneous manifestations, noncongenital conditions that feature both neurologic and cutaneous manifestations. For the more traditional neurocutaneous syndromes, the contributors have highlighted the clinical manifestations, laboratory and radiographic evaluations, prognosis, underlying pathophysiology, genetics, surveillance recommendations, interventions, and directions for acquiring future knowledge. We thank these authors, and the patients whose images have been included here, for their willingness to share their knowledge in words and pictures.

As revelations regarding genetics or pathophysiology have occurred, the classification of these conditions in a shared category of neurocutaneous syndromes may border on the obsolete. Advances regarding tumor suppressor genes, treatments targeting angiogenesis, and enzyme replacement therapy (to name but a few) expand our knowledge base and classification schemata beyond the conventional approach to neurocutaneous disorders. Still, the insights available regarding the nervous system on the basis of the examination of the skin remain invaluable to experienced clinicians, novices, and even researchers who are struggling to make a diagnosis or to develop new ideas.

Dr. John Stone wrote, in a memorable graduation address "Gaudeamus Igitur":

For your patients will test you for spleen
… and for the four humors
For they will know the answer
For they have the disease

May we never ignore the answer before us, common or rare, when the recognition of neurologic and abnormal skin or eye findings compels a need for comprehensive, often multidisciplinary care across the lifespan.

Monica P. Islam
E. Steve Roach

Contributors

J.L. Anderson
Department of Neurology, Washington University School of Medicine, St. Louis, MO, USA

S. Asch
Department of Dermatology, University of California San Francisco, San Francisco, CA, USA

F. Ballieux
Center for Vascular Anomalies, Division of Plastic and Reconstructive Surgery, Cliniques Universitaires Saint Luc, Université catholique de Louvain, Brussels, Belgium

B. Ben-Zeev
Pediatric Neurology Unit, Edmond and Lilly Safra Children Hospital, Chaim Sheba Medical Center, Ramat Gan and Sackler School of Medicine, Tel Aviv University, Tel Aviv, Israel

J. Biller
Department of Neurology, Stritch School of Medicine, Loyola University Chicago, Maywood, IL, USA

L.M. Boon
Center for Vascular Anomalies, Division of Plastic and Reconstructive Surgery, Cliniques Universitaires Saint Luc, Université catholique de Louvain, Brussels, Belgium

P. Chittiboina
Surgical Neurology Branch, National Institute of Neurological Disorders and Stroke, National Institutes of Health, Bethesda, MD, USA

M.G. Chung
Division of Pediatric Neurology and Critical Care Medicine, Department of Pediatrics, Nationwide Children's Hospital, Columbus, OH, USA

A.M. Comi
Hunter Nelson Sturge-Weber Center, Kennedy Krieger Institute, Baltimore, MD, USA

K.A. Dies
Department of Neurology, Boston Children's Hospital, Boston, MA, USA

C. Eng
Genomic Medicine Institute, Cleveland Clinic, Cleveland, OH, USA

D.G.R. Evans
University Department of Genomic Medicine, University of Manchester, St Mary's Hospital, Manchester, UK

L. Flores-Sarnat
Departments of Clinical Neurosciences and Paediatrics, University of Calgary and Alberta Children's Hospital Research Institute, Calgary, Alberta, Canada

T. Geller
Department of Child Neurology, St Louis University, St Louis, MO, USA

L.B. Gordon
Department of Pediatrics, Hasbro Children's Hospital and Warren Alpert Medical School of Brown University, Providence, RI; Department of Anesthesia, Boston Children's Hospital and Harvard Medical School, Boston, MA; Progeria Research Foundation, Peabody, MA, USA

D.H. Gutmann
Department of Neurology, Washington University School of Medicine, St. Louis, MO, USA

G.L. Heyer
Division of Pediatric Neurology, Nationwide Children's Hospital and Department of Neurology, Ohio State University, Columbus, OH, USA

M.P. Islam
Department of Child Neurology, Nationwide Children's Hospital, Ohio State University College of Medicine, Columbus, OH, USA

M.L. Levy
Baylor College of Medicine, Houston; Dell Children's Medical Center and Dell Medical School, University of Texas, Austin, TX, USA

R.R. Lonser
Surgical Neurology Branch, National Institute of Neurological Disorders and Stroke, National Institutes of Health, Bethesda, MD and Department of Neurological Surgery, Ohio State University Wexner Medical Center, Columbus, OH, USA

C. Massey
New Braunfels Pediatric Associates, New Braunfels, TX, USA

J.A. McGrath
St John's Institute of Dermatology, King's College London (Guy's Campus), London, UK

J. Mester
Genomic Medicine Institute, Cleveland Clinic, Cleveland, OH, USA

M.J. Narayanan
LSU Health Sciences Center, New Orleans School of Medicine, New Orleans, LA, USA

V. Narayanan
Neurology Research, Barrow Neurological Institute; Center for Rare Childhood Disorders, Translational Genomics Research Institute and Arizona Pediatric Neurology and Neurogenetics Associates, Phoenix, AZ, USA

A. Nissenkorn
Pediatric Neurology Unit and National AT Center, Edmond and Lilly Safra Children Hospital, Chaim Sheba Medical Center, Ramat Gan and Sackler School of Medicine, Tel Aviv University, Tel Aviv, Israel

A. Nouh
Department of Neurology, Hartford Hospital - University of Connecticut, Hartford CT, USA

S. Rangasamy
Neurology Research, Barrow Neurological Institute, Phoenix, AZ, USA

M. Ream
Department of Neurology, Nationwide Children's Hospital, Columbus, OH, USA

E.S. Roach
Department of Child Neurology, Nationwide Children's Hospital, Ohio State University College of Medicine, Columbus, OH, USA

M. Sahin
Department of Neurology, Boston Children's Hospital and Harvard Medical School, Boston, MA, USA

H.B. Sarnat
Departments of Paediatrics, Pathology (Neuropathology) and Clinical Neurosciences, Calgary, Alberta, Canada

R. Schiffmann
Institute of Metabolic Disease, Baylor Research Institute, Dallas, TX, USA

J. Speiser
Section of Dermatopathology, Department of Pathology, Stritch School of Medicine, Loyola University Chicago, Maywood, IL, USA

J.L. Sugarman
Departments of Dermatology and Community and Family Medicine, University of California San Francisco, Santa Rosa, CA, USA

S. Thalakoti
Department of Child Neurology, St Louis University, St Louis, MO, USA

N.J. Ullrich
Department of Neurology, Boston Children's Hospital and Harvard Medical School, Boston, MA, USA

M. Vikkula
Laboratory of Human Molecular Genetics, de Duve Institute, Université catholique de Louvain, Brussels, Belgium

Contents

Section 1

Introductory chapters

Handbook of Clinical Neurology, Vol. 132 (3rd series)
Neurocutaneous Syndromes
M.P. Islam and E.S. Roach, Editors

Chapter 1

Genetics of neurocutaneous disorders: basic principles of inheritance as they apply to neurocutaneous syndromes

KIRA A. DIES[1] AND MUSTAFA SAHIN[1,2]*
[1]*Department of Neurology, Boston Children's Hospital, Boston, MA, USA*
[2]*Harvard Medical School, Boston, MA, USA*

INTRODUCTION

The neurocutaneous syndromes include a heterogeneous group of disorders characterized by abnormalities of both the integument and the central nervous system (CNS). Many of the disorders are familial and believed to arise from a genetic defect in the development of the primitive ectoderm. Within the last decade, there has been an explosion of gene identification efforts through the Human Genome Project and other research consortia, leading to new insights into the causes of neurocutaneous disorders and possible treatments.

Most disorders classified as neurocutaneous syndromes are *single-gene disorders*, meaning they are caused by individual mutant genes and vary widely in inheritance pattern. The human genome contains, by current estimates, about 20 000–30 000 genes. Genes can be thought of as units of genetic information that code for proteins and, in turn, the development and function of the human body. Genes are encoded by deoxyribonucleic acid (DNA) and packaged into chromosomes in the nucleus of each cell. Human beings have 23 pairs of chromosomes, or a total of 46 chromosomes, with one member of each pair inherited from the mother and the other from the father. Of the 23 pairs, 22 are the same in males and females and are called the autosomes. The remaining pair represents the sex chromosomes, XX in females and XY in males. Members of a pair of chromosomes carry matching genes in the same sequence; however, at any specific region, there may be slightly different forms of the genes, called *alleles*. A gene *mutation*, or gene change, causing a neurocutaneous syndrome may be present on only one allele of a pair of chromosomes or on both alleles. This mutation can also be inherited from a parent or arise *de novo*, meaning a new genetic change not previously seen in the family.

Single gene disorders, as is the case for most neurocutaneous disorders, are characterized by their patterns of inheritance in a family. To help determine this pattern of inheritance, it is essential to take a comprehensive family history as part of a patient evaluation. This can be performed by creating a *pedigree*, a graphical representation of a family tree using standard symbols (Fig. 1.1).

A family history can be critical in diagnosis, to determine whether the disorder is hereditary, clarify the pattern of inheritance, and provide information about the natural history of disease and variation within a family. Family history can provide information about *phenotype*, or what we observe physically or clinically, and can help elucidate which gene may be causing the condition and thus the individual's *genotype*, or genetic constitution at a locus. Determining whether a condition is hereditary is important for estimating the risk to other family members and for genetic counseling for recurrence risk for future generations.

AUTOSOMAL DOMINANT INHERITANCE

Autosomal dominant inheritance refers to a mutation on one of the 22 pairs of nuclear chromosomes (i.e. non-sex chromosomes) that leads to syndrome expression when only one copy of the chromosome pair carries the mutant allele. Thus, an individual with an autosomal dominant disorder has one mutant allele and one normal allele, but the mutant allele is sufficient to cause the disorder in an individual. Autosomal dominant pedigrees can be recognized by a genetic condition that is typically seen

*Correspondence to: Mustafa Sahin, Department of Neurology, Boston Children's Hospital, 300 Longwood Avenue, CLS 14073, Boston, MA 02115, USA. Tel: +1-617-919-6258, E-mail: Mustafa.Sahin@childrens.harvard.edu

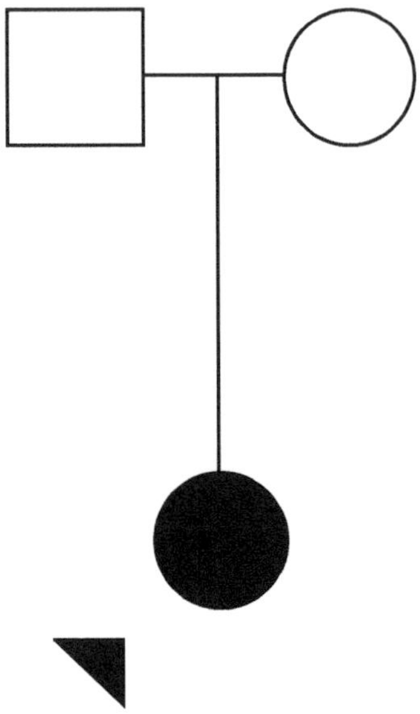

Fig. 1.1. Basic pedigree structure.

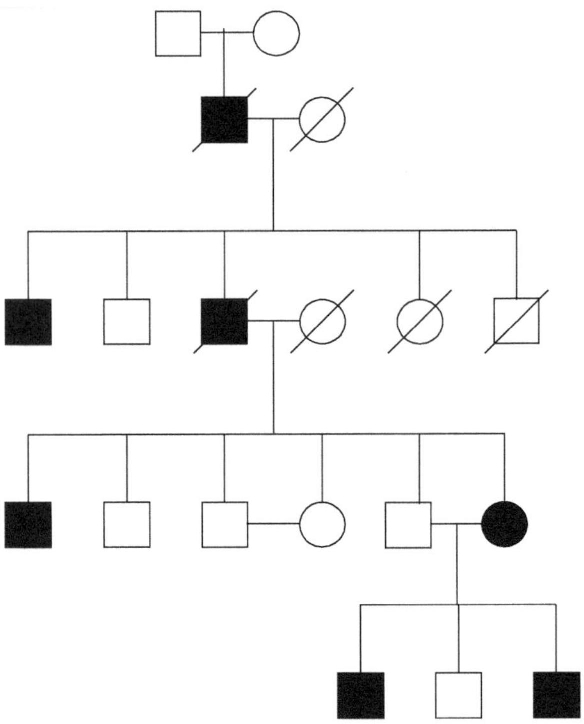

Fig. 1.2. Autosomal dominant inheritance.

in each generation and in which males and females are affected in roughly equal proportions. A typical pedigree of autosomal dominant inheritance is shown in Figure 1.2.

An example of a neurocutaneous disorder with autosomal dominant inheritance is *von Hippel–Lindau disease* (Ch. 10). As with most of the neurocutaneous syndromes, von Hippel–Lindau (VHL) disease affects many organs, including the cerebellum, spinal cord, retina, kidney, pancreas, and epididymis (Maher et al., 2011). It results from an autosomal dominant mutation affecting a tumor suppressor gene, *VHL*. Around 80% of individuals with VHL syndrome have an affected parent, and around 20% have a *de novo* gene mutation. *VHL* mutations are highly *penetrant*, meaning that almost all individuals with a mutation in *VHL* express disease-related symptoms by age 65 years (Maher et al., 1991).

Individuals who carry a known *VHL* mutation have a 50% chance of passing the mutation to their children. Molecular testing is available and detects mutations in almost 100% of probands. This allows for prenatal testing options, including amniocentesis and chorionic villi sampling, as well as preimplantation genetic diagnosis (please refer to "Genetic counseling and reproductive options", below).

Another example of an autosomal dominant neurocutaneous syndrome is *neurofibromatosis*, with two clinically and genetically distinctive types: neurofibromatosis 1 (NF1) and neurofibromatosis 2 (NF2). NF1 is the most prevalent type and is caused by dominant loss-of-function mutations in the *NF1* gene (Ch. 4). The disorder is clinically diagnosed when any two of seven defined features are present (DeBella et al., 2000). Café-au-lait spots are the hallmark of neurofibromatosis and are present in almost 100% of patients. These lesions are found over the body surface, with predilection for the trunk and extremities but sparing the face.

While NF1 is an autosomal dominant disorder, over half the cases are sporadic, representing *de novo* mutations. The *NF1 gene* on chromosome region 17q11.2 encodes for a protein known as neurofibromin. While the diagnosis of NF1 is typically made based on the clinical features, molecular testing for the *NF1* gene mutation is available and can be useful for patients who meet only one of the criteria for clinical diagnosis, those with unusually severe disease, and those seeking prenatal/preimplantation diagnosis.

Notably, another inheritance pattern seen in NF1 is *somatic mosaicism* (see later in this chapter), which is defined as the presence of two or more cell lines that differ genetically but are derived from the same zygote in an individual or tissue. Also called *segmental NF1*, the result of having the mutation in some proportion of the cells is that only one part of the body is affected, for example, café-au-lait macules confined to one side of the body or one of the extremities. This is most likely due to a mutation in the *NF1* gene that arose very early on in development and led to a proportion of affected cells in the same area of the body.

Tuberous sclerosis complex (TSC) is also inherited in an autosomal dominant manner and shows *variable expressivity*, defined as when the severity of disease differs in people who have the same genotype, that is, individuals in the same family sharing the same mutation. TSC has a wide clinical spectrum ranging from severe intellectual disability and intractable epilepsy to normal intelligence and a lack of seizures. A parent with mild symptoms of TSC can transmit the gene mutation to his/her child who can have a severe expression of TSC. Causes of variable expressivity in a family with a shared TSC mutation have not yet been discovered, but may include environmental factors or interactions with other genes.

Spontaneous genetic mutations occur in two-thirds of the cases. TSC is a genetic disorder that exhibits *locus heterogeneity*, meaning a condition caused by mutations at different loci in different families. Molecular studies have identified two foci for TSC: the *TSC1* gene is located on chromosome 9q34, and the *TSC2* gene is on chromosome 16p13. The *TSC1* gene encodes a protein called hamartin while the *TSC2* gene encodes the protein tuberin. Within a cell, these two proteins bind to one another and work together. Consequently, a mutation in either the *TSC1* gene or the *TSC2* gene results in a similar disease expression in patients. The *TSC1* and *TSC2* genes are tumor suppressor genes. The loss of either tuberin or hamartin protein results in the formation of numerous benign tumors (hamartomas) in various organs including skin, brain, heart, kidney, eyes, lungs and bones (Ch. 6).

Hamartin and tuberin are involved in a key pathway in the cell that regulates protein synthesis and cell size (Kwiatkowski et al., 2010). One of the ways cells regulate their growth is by controlling the rate of protein synthesis. A protein called mTOR (mammalian target of rapamycin) was identified as one of the master regulators of cell growth. Once this genetic pathway was understood, researchers began to study the efficacy of mTOR inhibitors for treatment of TSC (Krueger et al., 2010). Everolimus, an mTOR inhibitor, was recently approved by the US Food and Drug Administration (FDA) for the use of two complications associated with TSC, renal angiomyolipoma and subependymal giant cell astrocytoma. Multiple additional clinical trials are currently underway using mTOR inhibitors for other symptoms of TSC including neurocognition and seizures.

Genetic testing for *TSC1* and *TSC2* mutations is available and may be considered when the individual patient does not meet all the clinical criteria. Prenatal testing/preimplantation genetic diagnosis may be offered when a known TSC mutation exists in the family.

AUTOSOMAL RECESSIVE INHERITANCE

Autosomal recessive conditions are expressed only when both chromosomes of a pair carry mutant alleles. An individual with an autosomal recessive condition will have inherited a mutant allele from each parent (unless there is a *de novo* mutation). If both parents are unaffected by the condition they are presumed to be carriers of one mutant allele each and one normal allele each. With each pregnancy, the carrier parents thus have a 25% chance of producing a child with the recessive condition ({½} chance mother × {½} chance father = {¼} or 25%). Autosomal recessive disorders usually occur in one or more siblings but not in an earlier generation. Females and males are affected in approximately equal proportions. A typical autosomal recessive pedigree is shown in Figure 1.3. Of note, the chance that both parents are carriers of a mutant allele of the same gene is increased if the parents are related and share common ancestors, called *consanguinity*.

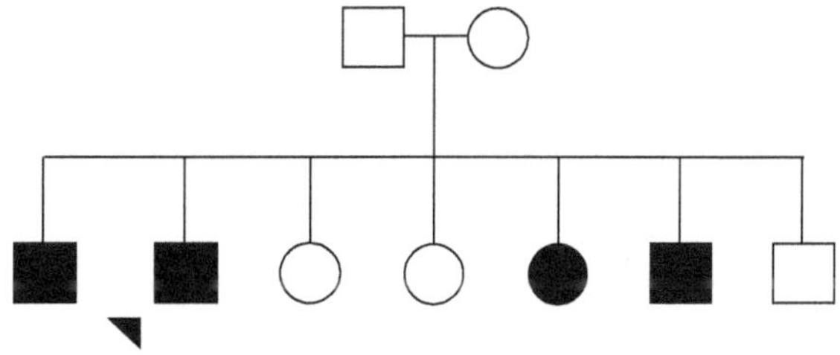

Fig. 1.3. Autosomal recessive inheritance.

Pseudoxanthoma elasticum (PXE) is a neurocutaneous disorder that affects the elastic tissue of the skin, the eye, and the cardiovascular and gastrointestinal systems (Ch. 15). Patients most often present with papules in the skin and/or with angioid streaks of the retina. Diagnosis is made clinically based on characteristic skin and ocular findings and is confirmed by histologic findings from biopsy of skin lesions. PXE is inherited in an autosomal recessive manner. The *ABCC6* gene encodes an ATP-binding cassette protein and is associated with PXE. Molecular genetic testing typically detects a mutation in one allele in almost all affected individuals and in both alleles in close to 90% of affected individuals.

The parents of a child with PXE are presumed to be obligate carriers, each with one mutation. For each pregnancy of parents with a previous child with PXE, each child has a 25% chance of being affected, a 50% chance of being an asymptomatic carrier, and a 25% chance of being unaffected and not a carrier. Once an at-risk sib is known to be unaffected, the chance of his/her being a carrier is {⅔}. Carrier testing for at-risk family members and prenatal testing for pregnancies at increased risk is possible once the mutations have been identified in the family.

An individual with PXE will pass down one mutation to his/her offspring, and thus they will be obligate carriers for a mutant allele causing PXE. Carriers of one mutation are typically asymptomatic. It is possible that carriers of an *ABCC6* mutation may have mild ocular and cutaneous findings, but this is currently under debate in the literature (Bacchelli et al., 1999; Sherer et al., 2001; Martin et al., 2008). Of note, an increased risk for PXE is seen in a few populations including the Afrikaners in South Africa, who display a *founder effect*, defined as reduced genetic variation in an isolated population descending from a small group of ancestors (Le Saux et al., 2002).

X-LINKED INHERITANCE

An X-linked pattern of inheritance occurs in families where a mutation is present on the X chromosome. In contrast to autosomal recessive and dominant conditions, where males and females are typically affected in equal numbers, the incidence of X-linked conditions is much higher in males than females. As females have two X chromosomes, while males have one X chromosome, females have a "spare tire" that may compensate for a mutation affecting only one of the X chromosomes. Females with an X-linked mutation are typically unaffected or more mildly affected than males. Some females may be affected as a result of the principle of *X-inactivation*, in which only one of the X chromosomes in a female's cells is active with the genes transcribed into proteins (Nussbaum et al., 2007). X-inactivation is typically random and established at an early embryonic stage. In rare cases, females may express an X-linked trait due to the particular gene escaping X-inactivation or as a result of *"skewed" inactivation*, where more of the X chromosomes with the mutation are active than expected by chance.

Conditions are considered X-linked dominant if the traits are consistently expressed in carrier females, and X-linked recessive if usually not expressed in a carrier female unless there is skewed inactivation. An X-linked recessive condition is typically expressed in all males who receive it, but only in females who receive two copies of the mutation. An X-linked mutation is passed down from an affected father to all of his daughters and none of his sons (his Y chromosome will go to all of his sons). Females carrying one X-linked mutation have a 50% chance of passing down the affected X with each pregnancy, whether a son or daughter. X-linked dominant and X-linked recessive pedigrees are shown in Figures 1.4 and 1.5, respectively.

Incontinentia pigmenti is an example of X-linked dominant inheritance with *male lethality* in the prenatal period. This rare, heritable, multisystem ectodermal disorder features dermatologic, dental, and ocular abnormalities (Ch. 20). This condition is lethal, causing spontaneous abortion in male pregnancies, thus those born affected are typically females having inherited

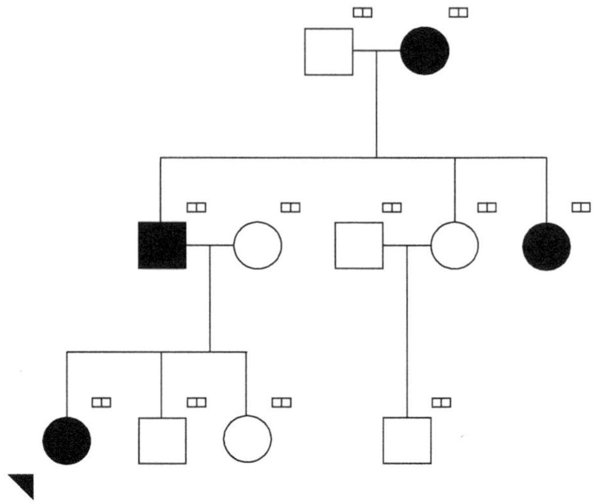

Fig. 1.4. X-linked dominant inheritance.

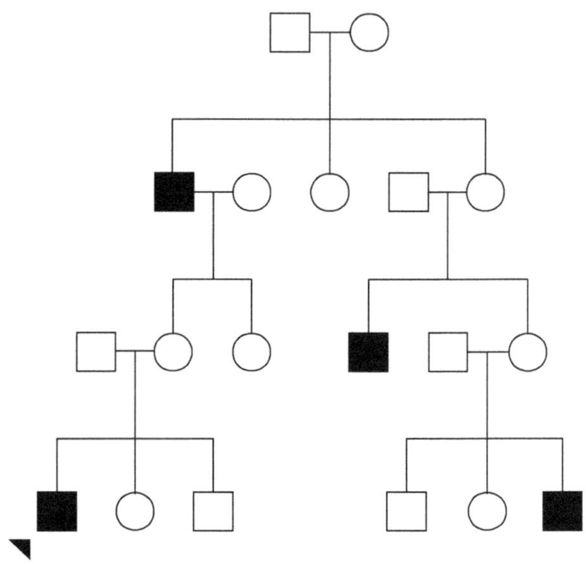

Fig. 1.5. X-linked recessive inheritance.

the mutation from their mother or *de novo*. The phenotype is produced by functional mosaicism caused by random X-inactivation of an X-linked dominant gene (*IKBKG,* previously *NEMO* gene) in females (Smahi et al., 2000). Clinical molecular testing is available, and in 80% of the affected patients a deletion that removes exons 4 through 10 of *IKBKG* gene can be detected.

Fabry disease is inherited in an X-linked recessive manner and is the result of reduced activity of the enzyme α-galactosidase (α-Gal A). There is phenotypic variability depending on the percentage of enzyme activity in males and females (Ch. 17). In males with less than 1% α-Gal A enzyme activity, symptoms typically begin in childhood or adolescence with severe pain in the extremities, vascular cutaneous lesions, sweating abnormalities, characteristic eye opacities, and proteinuria (Deegan et al., 2006). Later in life, heart and renal disease often occurs. Males with more than 1% α-Gal A activity can have either heart disease and proteinuria or renal disease without skin lesions or pain. Females with the mutation typically are asymptomatic or have milder symptoms, with a later age of onset than males. The range of severity in how females are affected is attributed to random X-inactivation.

Fabry disease is caused by the *GLA* gene on the X chromosome. For males, the most efficient and reliable method of diagnosing Fabry disease is measuring α-galactosidase A (α-Gal A) enzyme activity in plasma, isolated leukocytes, and/or cultured cells. In females, measurement of α-Gal A enzyme activity is unreliable and thus molecular testing is more commonly utilized. Nearly 100% of affected males have an identifiable *GLA* mutation. *Enzyme replacement therapy (ERT)* is recommended for males and symptomatic carrier females as early as possible to reduce symptoms; however, the long-term effectiveness of ERT is controversial (Weidemann et al., 2013).

As this disease is inherited in an X-linked recessive manner, the mother of an affected male is most likely a carrier of one mutation, or alternatively the male may have a *de novo* mutation. A carrier female has a 50% chance of transmitting the *GLA* mutation in each pregnancy. An affected male transmits his mutation to all of his daughters and none of his sons. Prenatal genetic testing and carrier testing for family members is available once a mutation is identified in a family.

SOMATIC MOSAICISM

A mutation that occurs during embryonic development may result in somatic mosaicism. Depending on the stage of embryonic development and the lineage of the somatic cell in which the mutation originated, the mutation may only be expressed in isolated tissues or organs. These mutations are not inherited from a parent as they arise during embryogenesis. However, a patient with a mosaic mutation can transmit the mutation to his/her children, depending on whether the mutation has affected the patient's *germline*, or reproductive cells (see next section).

Sturge–Weber syndrome (SWS) is a sporadic vascular disorder and consists of a constellation of symptoms and signs including a facial capillary malformation (port-wine stain), abnormal blood vessels of the brain (leptomeningeal angioma), and abnormal blood vessels of the eye leading to glaucoma (Ch. 11). The sporadic incidence and focal nature of SWS had long suggested the presence of somatic mutations but this was not demonstrated until recently. In 2013, a group of investigators performed whole-genome sequencing from affected and unaffected skin of three patients with SWS (Shirley et al., 2013). They identified a single-nucleotide variant (c.548G→A, p.Arg183Gln) in the *GNAQ* gene. This mutation was confirmed in samples of affected tissue from 88% of a larger cohort of SWS patients as well as 92% of the participants with apparently nonsyndromic port-wine stains. Brain tissue from SWS patients also demonstrated the same change in the *GNAQ* gene. These results strongly suggest that SWS occurs due to mosaic mutations in *GNAQ* and for the first time provides an opportunity to target this syndrome at the molecular level. Clinical genetic testing is not yet available for this condition as it would have to be performed on affected tissues rather than blood; the mutation is likely isolated to brain or other affected tissues and may not be present in a detectable proportion in blood lymphoblasts.

GERMLINE MOSAICISM

In all genetic conditions, there is a theoretical risk of *germline mosaicism*. Germline refers to those cells that can pass DNA from one generation to the next, that is, the eggs from a woman and the sperm from a man. Germline mosaicism means that some sperm or eggs have a gene mutation that may not be present in other tissues of the body (such as the blood, which is typically used for genetic testing). In these cases, the parent with germline mosaicism would not have symptoms of a neurocutaneous disorder, but could pass the mutation on to a proportion of their children. Germline mosaicism is suspected when two or more offspring present with an autosomal dominant condition when there is no family history of the condition. As mutation is a rare event, it is unlikely that multiple affected siblings would be due to multiple new mutations occurring in the same gene in the same family.

Once unaffected parents have one child with a genetic disorder, they will often be offered genetic testing. If no mutations are identified in the parents' blood, it is presumed that the child has a *de novo* mutation or that the child inherited the mutation from a parent with germline mosaicism.

GENETIC COUNSELING AND REPRODUCTIVE OPTIONS

Individuals and families affected with neurocutaneous disorders may benefit from *genetic counseling*. Genetic counseling is the process of helping individuals and families understand and adapt to the medical, psychological and familial implications of genetic contributions to disease (Resta et al., 2006). Genetic counselors provide education about inheritance, recurrence risks, genetic testing, management, prevention, resources, and research.

For individuals who are affected with a genetic neurocutaneous disorder, are carriers, or are at risk of being carriers, a genetic counselor can discuss potential risks to offspring and the availability of prenatal options. If the family has an identified mutation(s), *prenatal diagnosis* may be offered, which involves analysis of DNA extracted from fetal cells obtained by amniocentesis (usually performed at ~15–18 weeks' gestation) or chorionic villus sampling (usually performed at ~10–12 weeks' gestation). This technology can inform the parents if the fetus is a carrier of, or affected or unaffected by a neurocutaneous disorder.

More recently, *preimplantation genetic diagnosis (PGD)* has been developed for parents with a known mutation who are planning a pregnancy. PGD is defined as the testing of embryos or oocytes for known genetic mutations prior to implantation. Embryos or oocytes without the mutation can then be implanted into the mother by *in vitro* pregnancy. PGD is increasingly available, albeit expensive, and provides an alternative to prenatal diagnosis, which is performed after the pregnancy has been established. Long-term follow-up studies of children conceived utilizing PGD to screen for a known neurocutaneous disorder mutation are warranted and are yet to be performed.

CONCLUSION

Neurocutaneous disorders vary widely in genetic inheritance pattern. Genetic testing is available for many of the disorders, and once a mutation is identified, this opens up prenatal testing and preimplantation diagnosis options. Additional research on the genetic causes of neurocutaneous disorders will further elucidate causal molecular pathways and ultimately lead to targeted treatments.

References

Bacchelli B, Quaglino D, Gheduzzi D et al. (1999). Identification of heterozygote carriers in families with a recessive form of pseudoxanthoma elasticum (PXE). Mod Pathol 12: 1112–1123.

DeBella K, Szudek J, Friedman JM (2000). Use of the National Institutes of Health criteria for diagnosis of neurofibromatosis 1 in children. Pediatrics 105: 608–614.

Deegan PB, Baehner AF, Barba Romero MA et al. (2006). Natural history of Fabry disease in females in the Fabry Outcome Survey. J Med Genet 43: 347–352.

Krueger DA, Care MM, Holland K et al. (2010). Everolimus for subependymal giant-cell astrocytomas in tuberous sclerosis. N Engl J Med 363: 1801–1811.

Kwiatkowski DJ, Whittemore VH, Thiele EA (2010). Tuberous Sclerosis Complex: Genes, Clinical Features and Therapeutics. Wiley-VCH, Weinheim, Germany.

Le Saux O, Beck K, Sachsinger C et al. (2002). Evidence for a founder effect for pseudoxanthoma elasticum in the Afrikaner population of South Africa. Hum Genet 111: 331–338.

Maher ER, Iselius L, Yates JR et al. (1991). Von Hippel–Lindau disease: a genetic study. J Med Genet 28: 443–447.

Maher ER, Neumann HP, Richard S (2011). von Hippel–Lindau disease: a clinical and scientific review. Eur J Hum Genet 19: 617–623.

Martin L, Maître F, Bonicel P et al. (2008). Heterozygosity for a single mutation in the ABCC6 gene may closely mimic PXE: consequences of this phenotype overlap for the definition of PXE. Arch Dermatol 144: 301–306.

Nussbaum R, McInnes R, Willard H et al. (2007). Thompson & Thompson Genetics in Medicine. 7th edn. Saunders/Elsevier, Philadelphia.

Resta R, Biesecker B, Bennett R et al. (2006). A new definition of genetic counseling: National Society of Genetic Counselors' task force report. J Genet Couns 15: 77–83.

Sherer DW, Bercovitch L, Lebwohl M (2001). Pseudoxanthoma elasticum: significance of limited phenotypic expression in parents of affected offspring. J Am Acad Dermatol 44: 534–537.

Shirley MD, Tang H, Gallione CJ et al. (2013). Sturge–Weber syndrome and port-wine stains caused by somatic mutation in GNAQ. N Engl J Med 368: 1971–1979.

Smahi A, Courtois G, Vabres P et al. (2000). Genomic rearrangement in NEMO impairs NF-kappaB activation and is a cause of incontinentia pigmenti. The International Incontinentia Pigmenti (IP) Consortium. Nature 405: 466–472.

Weidemann F, Niemann M, Störk S et al. (2013). Long-term outcome of enzyme-replacement therapy in advanced Fabry disease: evidence for disease progression towards serious complications. J Intern Med 274: 331–341.

Handbook of Clinical Neurology, Vol. 132 (3rd series)
Neurocutaneous Syndromes
M.P. Islam and E.S. Roach, Editors

Chapter 2

Phenotype/genotype correlations in epidermal nevus syndrome as a neurocristopathy

LAURA FLORES-SARNAT[1]* AND HARVEY B. SARNAT[2]

[1]*Departments of Clinical Neurosciences and Paediatrics, University of Calgary and Alberta Children's Hospital Research Institute, Calgary, Alberta, Canada*

[2]*Departments of Paediatrics, Pathology (Neuropathology) and Clinical Neurosciences, Calgary, Alberta, Canada*

INTRODUCTION

Phenotype is often a clinical clue to genotype and enables the clinician to choose the genetic study that is most likely to confirm the etiology of a constellation of clinical, imaging, laboratory, and pathologic findings. Phenotype is defined by physical features, some of which may be evident during physical examination while others require special studies of internal organs. *Genotype*, by contrast, is defined by specific mutations, deletions, or duplications in particular genes, by chromosomopathies, or by identification simply of the chromosomal locus.

In many cases phenotype/genotype correlation varies due to considerable individual biological diversity, as can occur with post-zygotic somatic mutations that are the basis of epidermal nevus syndrome (Sugarman and Frieden 2004), tuberous sclerosis complex (TSC) (Curatolo and Bombardieri, 2008) and many other neurocutaneous syndromes. In some genetic disorders with cerebral dysgeneses, the phenotype defines only a syndrome of multiple genetic etiologies, all producing an identical or very similar clinical presentation. An example is holoprosencephaly (HPE), in which at least 11 of distinct genes with loci on different chromosomes can produce a similar cerebral dysgenesis due to failure of prosencephalic cleavage and arrhinencephaly. A phenotype of midfacial hypoplasias occurs, regardless of the gene that is mutated, and the varieties of facial dysmorphism are explained by extension of the genetic rostrocaudal gradient to the embryonic mesencephalon to impair neural crest formation and migration (Sarnat and Flores-Sarnat, 2001; Carstens, 2008). Nevertheless, the phenotype is the starting point for clinicians to suspect genetic disorders and helps guide genetic and metabolic investigations. It also helps indicate other neurologic studies for patient management. For example, in the head, brain US and/or magnetic resonance imaging (MRI) is justified to detect hemimegalencephaly (HME) or other brain and meningeal anomalies (Flores-Sarnat, 2015). This diagnosis can be established even prenatally by fetal ultrasound and MRI during the late second and third trimesters. Neonatal electroencephalography (EEG) is also indicated to detect early epilepsy that often accompanies those anomalies.

The genotype of nearly all neurocutaneous syndromes, and ENS in particular, is somatic mosaicism (Sugarman and Frieden, 2004), in which some progenitor cells are mutated and produce clones with the same mutation, mixed with other nonmutated progenitors producing clones of normal cells, all within the same tissue, whether skin, brain, or other organs. Somatic mutations generally are postzygotic and thus explain the great individual biological variability in distribution, size, and extent of cutaneous lesions, including nevi and capillary hemangiomata. Somatic mutations of the neuroepithelium produce similar variability in the extent and location of central nervous system (CNS) lesions, influenced by timing of the mutation. Somatic mutations occurring late in the 33 mitotic cycles of periventricular neuroepithelium result in HME involving only one lobe or a portion of the telencephalic hemisphere. If the somatic mutation occurs in an earlier mitotic cycle, HME involves the entire hemisphere and even earlier involved cycles can produce "total HME" in which the ipsilateral brainstem and cerebellum are also involved at a time

*Correspondence to: Dr. Laura Flores-Sarnat, Alberta Children's Hospital, 2888 Shaganappi Trail NW, Calgary, Alberta T3B 6A8, Canada. Tel: +1-403-955-7131, Fax: +1-403-955-7609, E-mail: laura.flores-sarnat@albertahealthservices.ca

when the periventricular neuroepithelium and rhombic lip of His are still in continuity (Sarnat et al., 2015). Most neurocutaneous syndromes, including ENS, are not transmitted as Mendelian traits, so that a family history of the disorder is not elicited. TSC and neurofibromatosis 1 and 2 are exceptions, with autosomal dominant inheritance, but some of these cases are novel mutations without either parent being involved.

The somatic mutations in ENS involve several genetic pathways, including mTOR (mammalian target of rapamycin), RAS (rat sarcoma), and PIK3CA (see below). They are also oncogenes that cause or facilitate neoplastic transformation of cells into both benign and malignant tumors. In making phenotype/genotype correlations, it is very important to recognize that even though the keratinocytic and sebaceous forms of ENS are distinctive in their cutaneous appearance and skin pathology, both result from mutations in the same RAS genes (Hafner et al., 2012; Sun et al., 2012; Flores-Sarnat, 2015). They often coexist in the same patient (Solomon and Esterly, 1975; Waltz et al., 1999). These recent genetic data support that epidermal nevus syndrome has subtypes rather than being several different and independent syndromes (Flores-Sarnat, 2015). Epidermal nevi in the scalp are characteristically sebaceous because of the abundance of sebaceous glands and hair follicles in the scalp relative to other regions of skin. RAS pathway mutations and associated fibroblast growth factor receptor 2 (FGFR3) defects also produce vitamin D resistance, which explains a phenotype with osseous lesions and severe rickets (Lim et al., 2014).

Somatic mosaicism should be distinguished from *germline mutations*, in which mutation occurs at the time of conception, so that every cell in the body carries the same genetic defect, though not always expressed equally in cellular or tissue phenotypes.

The nature of epidermal nevi, HME, and dysplasias in other visceral organs in ENS are hamartomatous. *Hamartoma* is defined not only by abnormalities in tissue architecture but by poorly regulated growth and morphology of individual cells. Hamartoma (plural hamartomata) is an antique term derived from the Greek ἁμαρτία, *hamartia*, meaning "error", or "to err". In HME and TSC many megalocytic and dysmorphic neurons are found, and these exhibit abnormal dendrites, axons, and synapses. Balloon cells may also be present (see below). Neuronal maturation in HME may be altered with either delayed or precocious expression of neuron-specific proteins. These lesions thus differ from cortical dysplasias that are primarily neuroblast migratory disturbances, which produce lissencephaly, pachygyria, polymicrogyria and subcortical laminar heterotopia, in which individual neurons and glial cells are relatively normal but their arrangements in the cortical plate and subcortical white matter are abnormal. Involvement of faulty neural crest provides insight into the pathogenesis of many of the systemic lesions, including cutaneous nevi and anomalies in eyes, brain, cardiac and peripheral vascular structures, renal and other viscera (Flores-Sarnat, 2013, 2015).

In this chapter, we address phenotype/genotype correlations of ENS and its increasing recognition and improved understanding of its status as a neurocutaneous disorder and neurocristopathy. Nevertheless, since the term 'Epidermal nevus syndrome' was first introduced by Solomon et al. (1968), it has been subject to unjustified criticism that has led to incongruent terminology. Hence an update of the nomenclature and history is important.

EPIDERMAL NEVUS AND EPIDERMAL NEVUS SYNDROME

Epidermal nevi (EN) are common cutaneous disorders; their incidence is about 1:1000 live births (Solomon and Esterly, 1975). Epidermal nevus syndrome (ENS), by contrast, is less frequent. A detailed review of 133 children with epidermal nevi, with the objective of determining the frequency of associated abnormalities, showed 33% of cases with ENS (Rogers et al., 1989). Nevertheless, it represents an important category among the neurocutaneous syndromes that frequently cause epilepsy, developmental delay, cognitive impairment, and other neurologic features, as well as multisystemic involvement.

EN are histopathologic criteria according to the predominant cell type giving rise to the hamartomatous growth. The cutaneous nevi are classified into morphologic categories of keratinocytic, sebaceous, follicular (hair follicle), apocrine, and eccrine (Solomon and Esterly, 1975; Rogers et al., 1989; Sugarman, 2007). A "smooth muscle hamartoma" has been considered another category of EN, known as Becker nevus (Sugarman, 2007). The two most frequent nevi are the keratinocytic and sebaceous; few lesions are exclusively of one histologic type, however (Solomon and Esterly, 1975; Rogers et al., 1989). Moreover, two or more different nevi can occur in the same patient, mainly keratinocytic and sebaceous nevi in distant or contiguous areas that may show transition; in some patients melanocytic nevi also occur (Marden and Venters, 1966; Solomon et al., 1968; Sugarman and Reed, 1969; Wauschkuhn and Rohde, 1971; Holden and Dekaban, 1972; Lansky et al., 1972; Carlier et al., 1974; Hornstein and Knickenberg, 1974; Solomon and Esterly, 1975; Carey et al., 1986; Eichler et al., 1989; Rogers et al., 1989; David et al., 1990; Tokatli et al., 1997; Olivares et al., 1999; Yu et al., 2000). With the exception of the Becker nevus, epidermal nevi exhibit a linear distribution and therefore the "linear" designation. It is useful to retain linear in the name for clearer comprehension by neurologists and nondermatologists, who may be unaware that epidermal nevi consistently follow this distinctive pattern.

Keratinocytic epidermal nevus (KEN) is also known as linear epidermal nevus (LEN) or verrucous nevus. They appear as skin-colored or slightly hyperpigmented plaques. Initially they may be flat but over time they often become elevated, verrucous, and darker (Lansky et al., 1972; Sugarman, 2007). Acral lesions often have a more warty appearance; in the body folds, lesions are softer and less verrucous. Their distribution and extent are variable: they can be solitary, multiple, large or small; they are common on the trunk or extremities and occur less often on the head (Sugarman, 2007). They may be unilateral or bilateral, stopping abruptly at the dorsal and ventral midline. When they are bilateral and symmetric they are referred to as systematized epidermal nevus or ichthyosis hystrix (Sugarman, 2007). Pathologic involvement is limited to the epidermis, in contrast to sebaceous nevi, which show both dermal and epidermal abnormalities.

Linear sebaceous nevus (LSN) or sebaceous nevus (SN) is also known as Jadassohn nevus or organoid nevus. A typical localization is on the scalp, periauricular area, hemi- or bifacial, or vertically in the midline of the forehead extending to the tip of the nose as a yellowish-orange linear plaque that may reach the upper lip or chin. In the neck and body, the nevus can be either sebaceous or KEN. Histopathologic features of NS are hyperkeratosis and papillomatosis of the epidermis, acanthosis, with increase of sebaceous lobules (or reduced, in small children). Underdeveloped hair follicles result in characteristic alopecia of the scalp. Sebaceous nevi are always present at birth (Sugarman, 2007); those in the scalp and face are more evident. Their natural history, described by Mehregan and Pinkus, includes three stages, in infancy, at puberty, and a third stage complicated by benign or malignant neoplasms (Sugarman, 2007). Infrequently, sebaceous nevi in the scalp and face can be massive in neonates (Nour et al., 2012).

Feuerstein and Mims (1962) and other authors made the accurate observation that sebaceous nevi localized on the head and/or face have a higher association with brain involvement and neurologic manifestations than those on the body (Solomon et al., 1968; Solomon and Esterly, 1975; Zaremba et al., 1978; Clancy et al., 1985; Baker et al., 1987; Pavone et al., 1991; Grebe et al., 1993).

The term epidermal nevus syndrome (ENS) was introduced by Solomon et al. (1968). It encompasses several subtypes, defined by the association of a hamartomatous epidermal nevus with one or more congenital systemic anomalies, mainly skeletal, central nervous system, and ocular (Solomon et al., 1968). In a later review of clinical and histopathologic features of their 60 patients (80% children), Solomon and Esterly (1975) described an even wider spectrum of systemic anomalies. Skeletal anomalies comprised the most frequent complication (70% of patients), followed by involvement of the brain in 50% (33% with epilepsy); ocular anomalies in 33%; six patients had oral lesions and one had Wilms' tumor of the kidney (Solomon and Esterly, 1975). The first report of an infant with ENS, epilepsy, and congenital cardiac anomalies was made by Marden and Venters (1966). The cardiac malformations result from defective cardiac neural crest (Kirby et al., 1983), a specialized part of the rhombencephalic neural crest with unique migration to form multiple cardiac structures, including the septum, valves, and conduction system.

Based on the predominant nevus, several major subtypes of ENS have been delineated (Solomon and Esterly, 1975; Rogers et al., 1989), most of them are characterized by keratinocytic nevus (KN): (1) Linear sebaceous nevus syndrome (LSNS), incorrectly called 'Schimmelpenning syndrome', (2) Keratinocytic epidermal nevus syndrome, (3) CHILD syndrome: Congenital Hemidysplasia with Ichthyosiform nevus and Limb Defect syndrome, (4) Becker nevus syndrome: epidermal nevus with hypertrichosis and 'smooth muscle hamartoma' associated with shoulder, arm, and breast hypoplasia, (5) nevus comedonicus syndrome (KN), (6) ILVEN: Inflammatory Linear Verrucous Epidermal Nevus (KN); and (7) Phakomatosis pigmentokeratotica: nevus sebaceous and melanocytic nevi (another example of coexistence of different nevi in ENS).

"Epidermal nevus syndrome" (ENS), is inclusive for all these categories. Nevertheless, despite their clinical and histopathological overlap, Happle and other dermatologists stated that they are different, independent syndromes (Happle, 1995, 2010). Many authors recognized that these subtypes correspond to the same entity (Solomon et al., 1968; Solomon and Esterly, 1975; Rebollo, 1979; Clancy, 1985; Baker, 1987; Rogers, 1989; Cavenagh, 1993; Waltz, 1999; Sato, 1994) which recently has been confirmed by molecular genetics (Sun, 2012). It is important, however, to distinguish the different phenotypes, in particular the two most frequent: 'epidermal (keratinocytic) nevus syndrome' and 'linear sebaceous nevus syndrome' because they constitute the neurological phenotypes characterized by CNS dysplasias (Flores-Sarnat, 2015). Their distinction provides a basis for classification and guides adequate investigation and patient management, also improving knowledge of epidemiological aspects.

Three other well defined syndromes within the spectrum of ENS with neurologic phenotype are: (1) CLOVES syndrome (congenital lipomatous overgrowth, vascular malformations, epidermal nevi, plus S for scoliosis, skeletal and spinal anomalies) (Sapp et al., 2007; Alomari, 2009); (2) SCALP syndrome (sebaceous nevus syndrome, CNS malformations, aplasia cutis, limbal dermoid, pigmented (melanocytic) nevus) (Lam et al., 2008); (3) Heide's syndrome (hemifacial keratinocytic or sebaceous nevus, hemifacial hyperplasia with lipomatosis and hemimegalencephaly) (Flores-Sarnat, 2002, 2008, 2012). These have been known for many years but have only recently been delineated.

HISTORICAL ASPECTS AND NOMENCLATURE

Epidermal nevi were described in the 19th century using Latin terms. The term *naevus* was used for centuries as synonymous with birthmark. The ancient Romans thought it was caused by a "maternal impression" during pregnancy (Bateman, 1814, cited by Solomon and Esterly, 1975). In 1863, Von Bärensprung indicated that nevi denote disease (Solomon et al., 1968). He reported a patient with a striking unilateral (*nevus unius lateris*) verrucous nevus. The term *ichthyosis hystrix* described more extensive, bilateral nevi. In 1895, Josef Jadassohn, a distinguished pioneer dermatologist, who had a great impact in the development of his specialty (Sultzberger, 1936, 1985; Weyers, 2013), gave the first clinicopathologic description of nevus sebaceous (Jadassohn, 1895). He reported two cases and made a comprehensive review of the literature. He coined the term "talgdrüsen naevi" (sebaceous gland nevi) and also introduced the term organoid nevus to describe these congenital lesions with either an excess or deficiency of sebaceous glands and other appendages of the skin with reduction of hair follicles. Before his description there was no distinction between sebaceous and keratinocytic nevi: the term ichthyosis hystrix was used indiscriminately for both. Jadassohn was also the first to distinguish sebaceous nevus from so-called adenoma sebaceum (a misnomer since it is neither adenomatous nor sebaceous, but a term commonly applied at that time and until recently) in tuberous sclerosis complex. Jadassohn cited authors that had emphasized the coexistence of nevus sebaceous with neurologic manifestations such as a report by Gerhardt in 1871 about a 6-year-old girl with sebaceous nevus in the left side of the nose, upper lip, cheek and retroauricular area, epilepsy and intellectual deficit (Gerhardt, 1871). One of the first reports of the association of nevus sebaceous with congenital ocular anomalies was made by Bögel in 1886.

In 1901 another distinguished dermatologist, Alfred Blaschko, published an extraordinary book in which he portrayed in drawings a system of lines on the human body meticulously documenting a variety of lesions in over 100 patients with diverse skin disorders, including verrucous and sebaceous nevi (Blaschko, 1901). He represented the peculiar and constant patterns as stripes in S- or V-shaped, and linear patterns. Blaschko included personal cases and also acknowledged the work of authors such as Voigt, Sherrington, Brissaud, Bärensprung, Hallopeau, Morrow, Jadassohn and others, some of whom used the term linear nevi. He and Jadassohn postulated an embryonic origin for that linear pattern (now known as "lines of Blaschko").

In the 1920s, Schumacher and Engelhardt reported an asphyxiated neonate with extensive linear sebaceous nevi in the head and bilateral ocular anomalies (Schumacher and Engelhardt, 1926). A few years later Robinson introduced the term "naevus sebaceus (Jadassohn)", citing more than 30 reports in the European literature (Robinson, 1932). He described four additional cases of nevus sebaceous with unilateral localization on the face, forehead, cheek and scalp and preauricular region. This term is still used by many authors. Three years later, van Bogaert described and illustrated a patient with "congenital ichthyosis" in the alopecic scalp, forehead and face, who also had epilepsy and intellectual deficit, clearly corresponding to LSNS (van Bogaert, 1935). In 1941, Pack and Sunderland performed a comprehensive review of the literature and the confusing nomenclature of nevus unius lateris, also adding four more cases (Pack and Sunderland, 1941). Some patients corresponded to sebaceous nevus and in several of them, epilepsy and intellectual deficit were noted.

In 1957, Schimmelpenning published a report of an 18-year-old female with sebaceous nevus in the left parietal area and another nevus in the neck, with the title "Clinical contribution to symptomatology of phakomatosis" (Schimmelpenning, 1957). The patient had been seen in the ophthalmology clinic since her birth, for ocular anomalies. Her milestones were not described; she had generalized epilepsy at 12 years of age. Skin biopsy was not mentioned. Schimmelpenning concluded that the spectrum of symptoms was "within the group of dysontogenetic processes."

The first article that called attention to the association of linear sebaceous nevus with neurologic manifestations was made by Feuerstein and Mims (1962). They reported two unrelated boys under the explicit title "Linear nevus sebaceous with convulsions and mental retardation." Both children had a linear sebaceous nevus in the midline of the forehead extending to the tip of the nose, confirmed by skin biopsy, and both had epilepsy and cognitive deficit but no other systemic anomalies. After this important report, the descriptive term linear sebaceous nevus syndrome has been used by numerous authors, enhancing the recognition of this syndrome (Lansky et al., 1972). Nevertheless, a long and unjustified debate was initiated by other authors against this term, claiming that Schimmelpenning was the first to describe this syndrome, even substituting the eponym Schimmelpenning syndrome (SS). Other authors still recognized the original contribution of Feuerstein and Mims and included them as Schimmelpenning–Feuerstein–Mims syndrome (SFM) (Wauschkuhn and Rohde, 1971; Carlier et al., 1974; Besser, 1976; Neuman et al., 2003; Rijntjes-Jacobs et al., 2010).

In 1983 there was a second report by Schimmelpenning, about the same patient, who paid more attention in her head CT to cranial bones than to brain parenchyma. Schimmelpenning reported dilatation of left lateral ventricle, left temporal calcification, and normal posterior fossa, emphasizing that "the outstanding feature consisting of fibrous bone dysplasia continues to progress into adult life" (Schimmelpenning, 1983). He did not refer to Jadassohn's pioneer descriptions in either of his two papers; nevertheless, in his first article he stated that Jadassohn was the first to describe the nevus sebaceous and to separate if from adenoma sebaceum.

In 2003, Warnke, Schimmelpenning, and colleagues claimed that a wax moulage from 1927 in Kiel, Germany, of an individual, with the legend "Naevus sebaceous" (in the left side of the face, scalp, ear, neck, and shoulder), is the first case of SFM syndrome: "the wax moulage is the first documentation of a patient with SFM syndrome and therefore 30 years older than the original report of Schimmelpenning" (Warnke et al., 2003). This statement is incongruent. More likely that wax moulage was an acknowledgment of Jadassohn's seminal contribution. Warnke and colleagues added: "While epilepsy and the other symptoms are only facultative in cases of SFM syndrome, the linear sebaceous nevus is the hallmark of the syndrome and leads to the diagnosis."

In 2004, Happle stated that Feuerstein and Mims had "rediscovered" the condition described by Schimmelpenning, and he asserted that this syndrome should bear his name (Happle, 2004), again not taking into account that Jadassohn had defined the foundation of the syndrome 60 years before Schimmelpenning (Flores-Sarnat, 2013). Feuerstein and Mims did not attempt to introduce their own names as eponyms of this condition; they employed the accurate term "sebaceous nevus" introduced by Jadassohn and the descriptive term "linear" used by him and other distinguished contemporary dermatologists. Nevertheless, Happle refused to recognize other authors and also failed to distinguish the phenotypical differences between the single patient reported by Schimmelpenning and the children reported by Feuerstein and Mims.

Greene et al. (2007) also endorsed the eponym Schimmelpenning syndrome (SS), attributing to him a variety of vascular anomalies that were absent in the patient seen by Schimmelpenning. They renamed as SS numerous well known cases of LSNS from the literature, beginning with the classic report of Feuerstein and Mims (F-M).

In 2010, Happle stated "An historical account on the delineation of the 'SS' has recently been published," referring to his own laudatory biography of Schimmelpenning (Happle, 2004, 2010), attributing to him brain and other systemic anomalies absent in that patient. Happle also berated Solomon et al. for introducing the encompassing term epidermal nevus syndrome, stating that there are several different epidermal nevus syndromes. His learning objectives were to avoid the terms epidermal nevus syndrome and keratinocytic nevus syndrome (Happle, 2010) that are now well established. This distinction is now clearly artificial.

If Schimmelpenning was not the first to describe the sebaceous nevus and its association with ocular anomalies and neurologic manifestations, why should the LSNS bear his name and why should pioneer authors be ignored? His eponym is inappropriate for many reasons, including ethical considerations. By contrast, the descriptive term linear sebaceous nevus syndrome, with roots in original descriptions in the 19th century, promotes communication and understanding between different specialists. Linear sebaceous nevus syndrome is also the term patients can relate to and understand. As Morrow wisely said in 1898 with respect to terminology of epidermal nevi: "the title of linear naevi … sufficiently identifies the affection, while possessing the advantages of simplicity and conciseness" (Morrow, 1898). The eponym SS, introduced some 75 years after NS was well defined clinically, lacks these advantages.

GENETIC AND EMBRYOLOGIC BASES OF EPIDERMAL NEVUS SYNDROME

The genetics and ontogenetic development of epidermal nevus syndrome are closely intertwined but not identical; they relate to the broader distinction between etiology and pathogenesis. Genetic mutations denote *etiology*; *pathogenesis* is the progression of processes of cellular lineage, maturation, and migration, and eventual morphogenesis of both individual cells and their architecture in forming a tissue or organ. This distinction is particularly important because genotype/phenotype correlations are not always predictable and vary considerably from one tissue to another and between individuals with the same genetic mutation. The terms *somatic mosaicism* and *postzygotic mosaicism* are often used interchangeably; the last refers to all DNA changes acquired during postconceptional life.

Mosaicism is a basic feature of ENS that was inferred clinically by several authors, first in incontinentia pigmenti and later in epidermal nevi (Lenz, 1961; Curth and Warburton, 1965; Happle, 1987; Moss et al., 1993).

One of the earliest clinical examples of epidermal mosaicism are the lines of Blaschko (Blaschko, 1901); they represent the pathways of migration of neural crest cells throughout the body (Sarnat and Flores-Sarnat, 2005, 2015a; Flores-Sarnat, 2013). In 1976, Jackson interpreted the famous lines as a form of human mosaicism (Jackson, 1976). This phenomenon was discovered by Curt Stern in Drosophila; he demonstrated that genetic recombination, normal in meiosis, also can take place in mitosis (Stern, 1936). When it occurs, it results in somatic (body) mosaics.

The discovery that a subgroup of keratinocytic epidermal nevi (KEN) is caused by oncogenic *FGFR3* and *PIK3CA* mutations with mosaicism (Hafner et al., 2006, 2012) was a revelation of these nevi and multiorgan involvement because *FGFR* and *PIK3CA* participate in embryogenesis, angiogenesis, and tissue homeostasis (Hafner et al., 2006). This discovery was followed by the identification of *FGFR3* mutation in keratinocytic epidermal nevus syndrome in a female with epilepsy since infancy (Ousager et al., 2012). However, most patients with KEN have activation of *HRAS* somatic mutations (Hafner et al., 2012). Almost simultaneously, it was demonstrated that nevus sebaceous and linear sebaceous nevus syndrome are also caused by postzygotic *HRAS* and *KRAS* mutations (Groesser et al., 2012; Levinsohn et al., 2013; Sun et al., 2013). The PIK3CA, AKT3 and mTOR genes are involved in cell growth, metabolism, and cancer; while PIK3CA and MTOR have widespread expression, AKT3 is most strongly expressed in the brain (Lee et al., 2012; Poduri et al., 2012). This clarifies why hemimegalencephaly, also a somatic mosaic anomaly caused by *de novo* somatic mutations in the PI3K-AKT3-mTOR pathway (Lee et al., 2012; Poduri et al., 2012, 2013), has a close relationship with LSNS and KENS including Proteus syndrome, caused by mosaic activating mutation in AKT1 (Lindhurst et al., 2011). CLOVES syndrome and Heide's syndrome share a similar association (see below).

The RAS family of genes was first discovered in rats in the 1960s and 1970s in relation to neoplasia (Ras is an abbreviation for rat sarcoma); their functions in normal cellular development were elucidated later (Tanaka et al., 1986; Goodsell, 1999). They are the most common oncogenes in human cancer. Mutations in ras genes can lead to the production of permanently activated Ras proteins (Goodsell, 1999). RAS genes are also involved in cellular growth, differentiation, proliferation, and survival (Goodsell, 1999; Castellano and Santos, 2011). They are expressed in all cell lineages and organs and broadly conserved across species; the highest level is found in brain (Tanaka et al., 1986). In human all RAS proto-oncogenes were detected in almost every normal fetal tissue and in several human tumor cell lines, also with highest concentration in the brain (Fiorucci and Hall, 1988).

Despite identical underlying somatic HRAS and KRAS mutations in NS and KEN, their syndromic forms have distinctive features suggesting that body site and timing determines the extent and distribution of skin involvement (Levinsohn et al., 2013). Sun et al. (2013) confirmed mosaic activating *RAS* mutations in a female with severe neurologic LSNS: Chiari II malformation, epilepsy, cognitive deficit, and middle cerebral infarction, SN in the forehead. They concluded that activating *HRAS* and *KRAS* mutations associated with keratinocytic and sebaceous nevi represent a spectrum of the same entity (Sun et al., 2013), confirming the early interpretation of Solomon and Esterly (1975). They also postulate that the difference between these nevi is related to the extent of the mutation and body site-specific embryological patterns (Sun et al., 2013). Activated RAS genes produce hamartomatous overgrowth and also neoplastic risk in nevus sebaceous (Levinsohn et al., 2013). On this basis, the association of nevi and systemic hamartomas in LSNS and KENS is clarified. The recent discovery that melanocytic nevi are also caused by postzygotic HRAS mutation (Groesser et al., 2013; Luo and Tsao, 2014) explains those cases of ENS in which these three nevi coexist in the same individual.

The fundamental embryonic-pathologic mechanism that explains the development of epidermal nevi and the associated systemic anomalies is a defect in neural crest (Flores-Sarnat, 2013). After Wilhelm His discovered in 1868 that the neural crest was the source of cranial and spinal ganglia, the identification by many investigators of numerous other structures that originate from the neural crest followed (Hall, 2009) [Sarnat and Flores-Sarnat, this issue]. One of the most recent derivatives discovered is the cardiac neural crest, defective in some patients with LSNS (Kirby et al., 1983). The bundle of His was first described by Wilhelm His Jr., a Swiss cardiologist and son of the renowned anatomist with the same name (Roguin, 2006); recently this too has been postulated to be derived from cardiac neural crest (Snider et al., 2007).

The prosencephalon grows within a mesenchymal matrix of local paraxial mesoderm and of neural crest cells (NCC) derived from the posterior diencephalon (i.e. dorsal lamina terminalis) and mesencephalon. Part of this NCC population forms the outer wall of capillaries within the prosencephalic leptomeninges and neuroepithelium itself (Etchevers et al., 2001). Within the meninges, capillaries with pericytes of diencephalic and mesencephalic neural fold origin supply the forebrain, while capillaries with pericytes of mesodermal origin supply the rest of the central nervous system, in a mutually exclusive manner (Etchevers et al., 2001). The two types of head vasculature contact at a few defined points, including the anastomotic vessels of the circle of Willis (Etchevers et al., 2001). Le Douarin et al. (2012) indicate that in addition to its role in building up the face and skull, neural crest is an important signaling center for brain development; the brain and face of vertebrates constitute a "developmental unit" controlled by neural crest (Le Douarin et al., 2012).

The traditional concept that skin and brain have the same ectodermal origin has been updated and changed by molecular genetics because genes are not restricted to a single embryonic germinal layer (Sarnat and Flores-Sarnat, 2005). Neural crest originates from each of the three primary cephalic vesicles of the developing neural tube: prosencephalon, mesencephalon and rhombencephalon. The various phenotypes of ENS may involve more than one site of neural crest origin, so that these disorders cannot easily be classified according to which portion is involved: midline vertical nevi of the nose and forehead from prosencephalic; ocular and lateral facial and scalp nevi from mesencephalic; neck and body (trunk, extremities and, viscera including heart) derives from rhombencephalic neural crest. Moreover, genes are often expressed in gradients in the three axes of not only the neural tube but also the body in general (Sarnat and Flores-Sarnat, 2005). Classification becomes complex because of these embryological patterns. The distinct location of the nevi in the head and neck indicates a difference in embryology and pathogenesis, even though the genetic mutation is the same. Most, if not all the multisystemic anomalies observed in epidermal nevus syndrome can be explained by a defective neural crest (Flores-Sarnat, 2013). Neural crest cells differentiate into their definitive cellular type after reaching their final destination and not during their migration.

The RAS genes are not listed as neural crest genes; however, the present clinical, genetic, and embryological evidence supports this concept, including the manifestations of epidermal nevus syndrome. Developmental "RASopathies" such as DiGeorge syndrome, Costello and others, share characteristic overlapping features, including craniofacial dysmorphism, cardiac malformations, cutaneous, musculoskeletal, and ocular abnormalities with varying degrees of neurologic involvement (Tidyman and Rauen, 2009). These conditions are recognized neurocristopathies, further supporting the concept that RAS genes indeed are expressed in neural crest.

SYSTEMIC ANOMALIES

It is important that neurologists perform careful inspection of the skin, facial or corporal asymmetry and identify the multisystemic complications of ENS and their common pathogenesis as neurocristopathies. It is also important that non-neurologic medical specialists investigate the CNS.

Ocular

Ocular anomalies are characteristic in linear sebaceous nevus syndrome. They correspond to neurocristopathies and are always present at birth. They are usually unilateral, ipsilateral to the scalp or facial nevus sebaceous and the brain dysplasia (Canyigit and Oguz, 2006), often HME. Bilateral involvement can also occur (Alfonso et al., 1987). The most common ocular anomalies are dermoids, lipodermoids, and simple or complex choristomas (Duncan et al., 1998; Singal et al., 2001). Ocular choristomas are congenital overgrowths of normal eye tissue in an abnormal location (Brodsky et al., 1997). Lall described an infant with sebaceous nevus in the scalp, neonatal seizures, and a complex choristoma containing cartilage and lacrimal glands (Lall, 1962). Visual acuity can be severely impaired. Colobomata may affect the retina, choroid, iris, and eyelid (Holden and Dekaban, 1972; Duncan et al., 1998; Canyigit and Oguz, 2006). Other anomalies include microphthalmia and cataracts, ipsilateral to HME (Chen et al., 2006).

Osseous and muscular involvement

Multiple osseous anomalies are frequent in patients with ENS (Solomon et al., 1968; Solomon and Esterly, 1975) that can affect any bony tissue: cranial vault, mandible, scapula, ribs, vertebrae, pelvis, or long bones of the extremities (many derived from neural crest). Cystic lesions in multiple bones were first described by Moynahan and Wolff in a toddler with craniofacial SN, infantile spasms and cerebral asymmetry (1967). Scoliosis and kyphoscoliosis may appear in early infancy (Larrègue et al., 1974; Olivares et al., 1999).

One particular phenotype is the association of ENS, in particular LSNS, with hypophosphatemic vitamin D-resistant rickets, with recurrent pathologic limb fractures, sometimes with severe neurological manifestations including epilepsy (Besser, 1976; Aschinberg et al., 1977; Carey et al., 1986; Oranje et al., 1994; Vidaurri de la Cruz et al., 2004). The first report was made by Sugarman and Reed (1969). Rickets, muscle weakness and bone pain develop at an early age; neurological findings in 86% (Sugarman, 2007). An important feature that I observed in these cases is the presence of melanocytic nevi, also caused by RAS mutations. A recent report of 4 children with association of keratinocytic or sebaceous nevi and a fifth child with giant melanocytic nevus and epidermal nevus with bone dysplasia and hypophosphatemia caused by elevated FGF23 levels, mutations in HRAS and NRAS were demonstrated

(Lim et al., 2014). The authors support that the wide distribution of tissues and multiple segmental skin and bone lesions result from early embryonic somatic mutations.

Cardiovascular

Cardiovascular anomalies such as patent ductus arteriosus, coarctation of the aorta, and septal defects are neurocristopathies; they are found almost exclusively in linear nevus sebaceous syndrome, first described by Marden and Venters (1966). Tripp reported an infant with a scalp and facial SN, infantile spasms, coartaction of the aorta with upper limb hypertension, and a marked flutter fibrillation. He assertively mentioned that cardiac arrhythmia, a rarity in infants, may indicate an abnormality of bundle of his (Tripp, 1971). Indeed, this anomaly also can be attributed to a defective neural crest (Snider et al., 2007; Flores-Sarnat, 2013). Another infant with SN in the scalp and refractory epilepsy had frequent atrial and ventricular premature systole and a wandering pacemaker in the EKG (Lansky et al., 1972). A child with patent ductus arteriosus had chronic occlusion of the left distal internal carotid artery that resulted in ipsilateral cerebral atrophy (Canyigit and Oguz, 2006). Congenital cardiac anomalies are not a feature of any subtype of KEN syndrome.

Congenital hemifacial hyperplasia with lipomatosis

Congenital hemifacial hyperplasia with lipomatosis (CHHL) is also a distinctive mosaic overgrowth disorder that occurs in neurologic phenotypes of ENS, often incompletely recognized. It is characterized by congenital unilateral hyperplasia of multiple osseous and soft facial tissues with an ipsilateral mass of lipomatosis at the level of the zygoma (described as "cheek" or "lower part of the face"). Most affected tissues derive from the neural crest. Recently, oncogenic PIK3CA mutations were demonstrated to be the cause (Maclellan et al., 2014). CHHL can occur as an isolated anomaly but is often associated with neurocutaneous syndromes, in particular sebaceous and keratinocytic ENS (Vles et al., 1985; Donati et al., 1990; Pavone et al., 1991; El-Shanti et al., 1992; Unal et al., 2000; Di Rocco et al., 2001; Aydíngöz et al., 2002; Flores-Sarnat, 2012); tuberous sclerosis complex is an exception. This singular condition is reported under numerous terms: facial hemihypertrophy, true hemifacial hyperplasia, hemihypertrophy of the face, etc. (Flores-Sarnat, 2012). In 1983, Slavin et al. introduced the term congenital infiltrative lipomatosis of the face (CILF), defining the histopathologic and clinical features of the lipomatous mass (Slavin et al., 1983) without addressing the associated hyperplasia of multiple hemifacial structures other than the zygoma, or correlating with reports under the above mentioned terms. Pollock et al. (1985) introduced a more precise term, congenital hemifacial hyperplasia, emphasizing the designation as hyperplasia rather than hypertrophy because the individual cells in different tissues are not enlarged (hypertrophy) and histopathology showed increased numbers of cells (hyperplasia). They correctly suggested the neural crest as the embryological basis. It is important to recognize that congenital hyperplasia of hemifacial structures and the ipsilateral lipomatosis occur simultaneously and correspond to a single entity. *Congenital hemifacial hyperplasia with lipomatosis* is an inclusive term for both, encompassing several specialties. Congenital implies pathogenesis; infiltrative is not necessary in the name of this entity; by definition, lipomatosis infiltrates surrounding structures, and is not encapsulated as it is in lipoma. Even when CHHL is not a neoplastic condition, it cannot be considered benign because of the severe and chronic clinical manifestations.

The widely used terms hemihypertrophy, hemiatrophy, or similar terminology are inaccurate because "half" hypertrophy, hyperplasia, and atrophy do not occur. It is clearer to localize the affected unilateral area as hemifacial, hemicorporal, etc. (e.g. hemifacial hyperplasia). In 1863 Friedreich reported a girl with CHHL under the title "congenital unilateral head hypertrophy" (Flores-Sarnat, 2012, 2013).

Oncogenic *PIK3CA* mutations are related to HME (Lee et al., 2012) and keratinocytic epidermal nevus syndromes (Hafner et al., 2007), further explaining their association. Three neurologic phenotypes of ENS are particularly associated with CHHL: Heide's syndrome, CLOVES syndrome, and Proteus syndrome (see below), usually associated with HME, a natural occurrence due to shared common genetic mutations. Diaz (2000) and Trapp et al. (1981) have addressed the anesthetic challenges in children with CHHL and epilepsy.

BRAIN ANOMALIES IN SEBACEOUS AND EPIDERMAL (KERATINOCYTIC) NEVUS SYNDROMES

Hemimegalencephaly

Hemimegalencephaly (HME) is the most frequent dysgenesis observed in all neurologic phenotypes of ENS. In a review of 109 cases of LSNS, 38.5% had HME (Flores-Sarnat, 2015). Its hamartomatous unilateral overgrowth is explained by mosaicism (Lee et al., 2012). Its extent corresponds to timing in the mitotic cycles of neuroblast proliferation: the earlier the involvement, the more extensive is the resulting dysgenesis (Sarnat et al., 2015). ENS occupies first place amongst neurocutaneous disorders in its association with HME (Sakuta et al., 1991; Flores-Sarnat, 2002). This is the main cause of epilepsy.

Neural crest gives origin to vascular structures throughout the body. In the case of HME a frequent finding is abnormal meningeal and intraparenchymal vessels (Bosemani et al., 2013). This can be explained as a neural crest defect; they are particularly common in the opercular region. It has been confirmed at autopsy (Mollica et al., 1974) and by angiography and MRI (Chalhub et al., 1975; Brihaye et al., 1988). Cerebral vascular malformations have also been reported by other authors (Solomon and Esterly, 1975; Baker et al., 1987). In early reports before CT and MRI, HME could sometimes be inferred from pneumoencephalography or ultrasound studies as asymmetric enlargement of one cerebral hemisphere or one lateral ventricle, but the real frequency is uncertain due to incomplete documentation in many reports.

Other cerebral and meningeal malformations reported in linear sebaceous nevus syndrome

There is a predilection for anomalies to form in the temporal lobe, such as arachnoid cysts, often associated with a sebaceous nevus in the scalp, forehead or face, usually ipsilateral (Holden and Dekaban, 1972; Aizawa et al., 2000; Ousager et al., 2012). Mesencephalic neural crest is the origin of the cranial (but not the spinal) meninges. Temporal lobe hypoplasia (Mullaney and Weatherhead, 1996), dysplasia (Booth and Rollins, 2002), or atrophy (Yu et al., 2000; Özkiris et al., 2003; Rijntjes-Jacobs et al., 2010; Groesser et al., 2012) are probably related to vascular anomalies causing chronic ischemia during development; they can occur without HME (Dobyns and Garg, 1991). Selective temporal lobe enlargement was reported in an infant with infantile spasms (Kwa et al., 1995). Occipital focal cortical dysplasia (Alfonso et al., 1987; Cross, 2005) and pachygyria or polymicrogyria are common (Gurecki et al., 1996). Hypoplasia of corpus callosum, enlarged temporal horn, and leukomalacia of the left temporal lobe were observed in a neonate with refractory seizures (Olivares et al., 1999). Focal temporal cortical dysplasia is a cause of severe epilepsy (Pujol et al., 1987; Loddenkemper et al., 2008). Dandy–Walker malformation is a rare association (Dodge and Dobyns, 1995; Nour et al., 2012).

CLASSIFICATION OF NEUROLOGIC PHENOTYPES IN EPIDERMAL NEVUS SYNDROME

Many authors performed careful reviews of the literature and several tried to categorize neurologic phenotypes for classification (Zaremba et al., 1978; Alfonso et al., 1987; Pujol et al., 1987; Pavone et al., 1991). The early observation by Holden and Dekaban that epilepsy and cognitive deficit (due to HME or other brain anomalies) are more frequent in LSNS than in keratinocytic nevus syndrome (Holden and Dekaban, 1972) is here confirmed. Knowledge of embryology provides a further contribution in delineating genotype/phenotype correlations.

Linear sebaceous nevus syndrome

LSNS (linear nevus sebaceous syndrome) is the most frequent neurologic phenotype; a review of the literature enables us to distinguish three principal subtypes according to the localization of the nevus and associated anomalies. The following classification combines clinical and embryological criteria. More than one site of neural crest origin may be involved in the same patient. The three phenotypes are listed in order of frequency:

1. Mesencephalic neural crest: SN in the scalp and/or face, with ocular anomalies (Alfonso et al., 1987; Brodsky et al., 1997; Duncan et al., 1998; Singal et al., 2001).
2. Prosencephalic neural crest: SN in the midline of the forehead and nose as a vertical line, without ocular involvement (Feuerstein and Mims, 1962; Chalhub et al., 1975; Vigevano et al., 1984).
3. Mesencephalic and rhombencephalic neural crest: midfacial or scalp SN with ocular and cardiac involvement (Marden and Venters, 1966; Tripp, 1971; Diaz, 2000).

SCALP syndrome

SCALP syndrome is another phenotype associated with sebaceous nevus, delineated by Lam et al. (2008). The acronym corresponds to sebaceous nevus syndrome, CNS malformations, aplasia cutis congenita, limbal dermoid, and pigmented nevus (giant congenital melanocytic nevus). Mimouni et al. reported a newborn with coexistence of a left craniofacial sebaceous nevus and ipsilateral HME with a giant melanocytic nevus in a bathing trunk distribution (Mimouni et al., 1986). Somatic mutations in *NRAS* and *BRAF* have been demonstrated in multiple congenital melanocytic nevi, giant congenital melanocytic nevi, and neurocutaneous melanocytosis (Kinsler et al., 2013; Salgado et al., 2015).

Keratinocytic epidermal nevus syndromes

Proteus Syndrome

Proteus syndrome (PS) is a complex hamartomatous disorder consisting of disproportionate, asymmetric overgrowth of any tissue of the body, particularly the skeleton, cerebriform connective tissue nevi, epidermal nevi, vascular malformations, and dysregulated adipose tissue (Cohen, 2005; Biesecker, 2006). Proteus syndrome is a classic neurocristopathy because most of the tissues involved derive from the neural crest. The heart, however, is spared, as in all KEN syndromes. This patchy or mosaic overgrow is not present at birth; it typically starts in infancy, as does the scoliosis (Biesecker, 2006). Linear epidermal nevi (keratinocytic) are characteristic; they can develop in several parts of the body but most frequently in the neck. Proteus syndrome is caused by a somatic activating mutation in AKT1, implicating activation of the PI3K–AKT pathway (Lindhurst et al., 2011). Proteus syndrome was described in the 19th century in accounts of Joseph Merrick, the "Elephant Man" (1862–1898), who was mistakenly diagnosed as having neurofibromatosis; reports from Eve (1883) and Finlayson (1884) correspond to Proteus syndrome with CHHL. Eve mentioned left cerebral enlargement at autopsy, suggesting HME. Another autopsy report in 1907 by Webster described a child with CHHL and HME (Webster, 1907).

Central nervous system involvement occurs in about 40%, with cognitive deficit in about 30% (Turner et al., 2004). The most frequent association is HME (De Lone et al., 1990; Rizzo et al., 1990; Sarnat et al., 1993; Dietrich et al., 1998; Cross, 2005; Demir, 2008; Jackson et al., 2010; Kaduthodil et al., 2012). Epilepsy can be severe (Rizzo et al., 1990); however, it is not investigated in many reports and HME is uncertain because of lack of diagnostic study. The association of Proteus syndrome with HME and CHHL is now explained by common somatic mutations in the PIK/AKT/mTOR pathway (Lindhurst et al., 2011; Lee et al., 2012).

CLOVES syndrome

This constellation of anomalies was delineated by Sapp et al. (2007), with the acronym: CLOVE (congenital lipomatous overgrowth, vascular malformations, epidermal nevi). Later an 'S' was added to emphasize the association with scoliosis and skeletal and spinal anomalies (Alomari, 2009). Large truncal lipomatous masses including variable lymphatic and vascular components are a constant finding. Many structures involved are of neural crest origin, including lymphatic vessels because they are derived from venous endothelium. Initially several patients with CLOVES syndrome were confused with Proteus syndrome (Sapp et al., 2007). Somatic mosaic activating mutations in *PIK3CA* gene were recently found in affected tissue from multiple embryonic lineages (Kurek et al., 2012). The association with HME is explained by a similar mutation (Lee et al., 2012).

Neurologic involvement, epilepsy in particular, and neuroimaging of the brain are not investigated in most reports. A dizygotic twin with CLOVES syndrome detected prenatally had enlarged left cerebral ventricles and ipsilateral fatty overgrowth of the face (Fernandez-Pineda et al., 2010) highly suggestive of HME with ipsilateral CHHL. The only report of CLOVES with HME was made by Gucev et al., in a neonate with hemifacial overgrowth (not shown), bilateral cataracts, and refractory epilepsy (Gucev et al., 2008).

Several patients reported as other phenotypes of ENS probably correspond to CLOVES syndrome; all had HME, epilepsy, and intellectual deficit (Zaremba et al., 1978; Vigevano et al., 1984; Clancy et al., 1985; Pavone et al., 1991). The first case of fetal CLOVES syndrome with HME confirmed by detailed neuropathologic examination was presented by Dr. Robert Hevner at the 2014 meeting of the American Association of Neuropathologists in Portland, Oregon, USA (unpublished). This reinforces the need for neuroimaging in these patients.

Heide's syndrome

This is a distinctive neurologic phenotype of both KENS and LSNS characterized by a typical triad of congenital hemifacial hyperplasia with lipomatosis (CHHL), hemifacial nevus, and ipsilateral hemimegalencephaly (Flores-Sarnat, 2012, 2013). Pavone et al. brought attention to the triad and designated it a "neurologic variant" (Pavone et al., 1991). The first patient reported with this triad was Heide, a 7-year-old German girl who had congenital enlargement of the left side of the face with an ipsilateral verrucous nevus in the face, neck, shoulder and arm. She had epilepsy, starting with infantile spasms and intellectual deficit. In 1944 at almost 10 years of age, she was transferred to Vienna where she was subjected to autopsy and left hemimegalencephaly was demonstrated (Gross and Uiberrak 1955). She was a victim of "active euthanasia" (Ronen et al., 2009). For this entity the name "Heide's syndrome" is medically, historically and ethically justified (Flores-Sarnat, 2012, 2013), not only because she was the first patient described, but also to enhance identification of these patients and the investigation of HME. Epilepsy can be refractory from the neonatal period (Flores-Sarnat, 2002). Several reports in the literature correspond to Heide's syndrome with the triad of CHHL, LEN or KEN and HME (Clancy et al., 1985; Dean et al., 1990; Donati et al., 1990; Trounce et al., 1991; El-Shanti et al., 1992; Cavenagh et al., 1993; Unal et al., 2000; Di Rocco et al., 2001; Flores-Sarnat, 2002).

Epilepsy in keratinocytic epidermal nevus and linear sebaceous nevus syndromes

Macrocephaly is usually the first neurologic manifestation noted at birth, and epilepsy is the second. About one third of children with ENS have CNS involvement, usually epilepsy (Solomon and Esterly, 1975), in most cases due to HME (Vigevano et al., 1989; Pavone et al., 1991; Sakuta et al., 1991; El-Shanti et al., 1992; Demaerel et al., 1994; Wakabayashi et al., 1997; Allison et al., 1998; van de Warrenburg et al., 1998; Herman and Siegel, 2001; Flores-Sarnat, 2002; Loddenkemper et al., 2008). Multiple seizure types are common and epileptic syndromes can be identified (Gurecki et al., 1996; Loddenkemper et al., 2008). Neonatal seizures are usually associated with major unilateral hemispheric malformations, mainly HME; they may be subtle or convulsive, with tonic spasms, myoclonic, or sometimes with facial twitching. The EEG pattern in HME is distinctive and similar in isolated cases or associated with ENS (Paladin et al., 1989). *Infantile spasms* is the most frequent epilepsy of onset in infants, often due to HME (Lall, 1962; Kurokawa et al., 1981; Paladin et al., 1989; Hager et al., 1991; Kwa et al., 1995; Flores-Sarnat, 2002). The neurologic phenotypes of ENS are the 2nd most frequent cause of West syndrome of all neurocutaneous syndromes, following TSC [LFS 2015]. The EEG generally shows bilateral hypsarrhythmia, but asymmetric or unilateral hypsarrhythmic patterns are also observed and a grossly abnormal EEG is reported (Tripp, 1971). An analysis of the frequency and types of epilepsy, age at onset, phenotypic correlations and treatment options is provided in a recent review addressing epilepsy in ENS (Flores-Sarnat, 2015). Therapeutic approaches involve not only antiepileptic medications, but also mTOR inhibitors such as rapamycin and other pharmaceuticals that can alter abnormal genetic and metabolic pathways, and surgical resections of intractable epileptic foci. Early treatment in highly epileptogenic malformations such as HME may be considered to prevent the spread of epileptic foci to the other hemisphere or to normally functioning parts of the cortex and hippocampus.

Most patients with KENS or LSNS with epilepsy also have developmental delay and cognitive impairment of various degrees, but less frequently, development and cognitive functions can be normal (Ousager et al., 2012). Absence of epilepsy in the presence of a cerebral asymmetric anomaly is exceptional (Olivares et al., 1999). Epilepsy with aberrant behavior with marked hyperkinesis, vociferousness, and destructive activity is rare. By contrast with TSC, autism is rare in ENS.

Investigation of patients with neurologic phenotypes of epidermal nevus syndrome

Neuroimaging is essential in the evaluation of neonates or children with sebaceous nevi of the scalp or midline facial area, or keratinocytic nevi in the head, especially in the presence of epilepsy of other neurologic abnormalities. Careful and detailed descriptions of CT and MRI of the brain facilitate the diagnosis of HME or other brain anomalies (Vles et al., 1985; Barkovich and Chuang, 1990; Carter and Hurst, 1990; Sarwar and Schafer, 1998). The first report of linear nevus sebaceous syndrome (LSNS) associated with HME by MRI was by Sarwar and Schafer (1998). In their second report, they described R HME; ulstrasound demonstrated colpocephaly, (selective dilatation of the occipital horns) and a ventriculoperitoneal shunt was placed (Carter and Hurst, 1990). In cases with CHHL, magnetic resonance imaging of the face is indicated to define the extent of lipomatosis and hyperplasia. Positron emission tomography (PET) and single-photon emission computed tomography (SPECT) can provide important functional information and can show the extent of the epileptogenic locations, especially for surgical candidates (Maher et al., 2003; Zhang et al., 2003). Shifting of the midline structures to the opposite side, particularly the occipital lobe, is characteristic of HME (Flores-Sarnat, 2002; Lien et al., 2005). Prenatal 3D sonogram can show ocular and skin lesions of LSNS (Lien et al., 2005; Neis et al., 2006); in both cases, postnatal MRI confirmed hemimegalencephaly. The use of Tc-99 m HMPAO brain SPECT in linear nevus sebaceous syndrome also helps to define brain pathology (Özkiris et al., 2003). Some cases studied by pneumoencephalography (PEG), showing asymmetric enlargement of the left lateral ventricle, suggest HME (Marden and Venters, 1966; Chalhub et al., 1975; Zaremba et al., 1978; Kirby et al., 1983).

Neuropathologic studies

Neuropathologic studies in patients with ENS are sparse. Probably the first autopsy of the brain in a patient with LSNS was provided by Berg and Crome in 1960 in a girl with severe epilepsy and cognitive deficit (Berg and Crome, 1970). The report was inconclusive, probably because they were looking for tuberous sclerosis. There was no macroscopic description of cerebral asymmetry but she had an enlarged left ventricle and a distinctive dysplastic and gliotic area in the left temporal. Neuropathologic studies of surgical specimens for treatment of refractory epilepsy due to HME or focal cortical dysplasia provide valuable information for patient care. Less frequently, there is availability of autopsy tissue.

In a neonate with LSNS and severe epilepsy, HME was confirmed by autopsy (Merks et al., 2003). The cerebral lesions of HME in Proteus syndrome in infancy, including mTOR activation and tau overexpression, were recently described (Sarnat et al., 2012). The histopathologic lesions of HME in ENS are identical to those encountered in isolated HME: architectural disorganization with loss of cortical lamination and convolutions, areas of gyral fusion, and neuronal white matter heterotopia. Not only is the tissue organization very abnormal, but individual neurons themselves are dysmorphic and megalocytic and often show mixed lineage; balloon cells are also found, as in tuberous sclerosis complex (Flores-Sarnat et al., 2003). Leptomeningeal angiomatosis extending into the right temporal lobe discovered at autopsy in a neonate with LSNS and HME (Mollica et al., 1974) is an example of defective neural crest.

Asymmetric enlargement of one cerebral hemisphere, thickened cortex with abnormal gyration: lissencephaly, pachygyria, polymicrogyria; hemilissencephaly, poor distinction of the grey/white matter boundary or serrated junctions are characteristic of HME. The lateral ventricle is straightened and either dilated or collapsed; colpocephaly is characteristic. The contralateral hemisphere is usually normal, though displaced, but scattered dysmorphic neurons can also be seen in the opposite hemisphere, the hippocampus can show granule cell dispersion, and the cingulate gyrus may show an immature cortex with radial microcolumnar architecture (Sarnat et al., 2012). The microtubule-associated protein tau, in abnormal phosphorylated form, is upregulated in HME but expressed differently in the developing brain than in the mature brain, in which tauopathy causes neuronal degeneration and dementia without epilepsy. In the fetus and infant, tau upregulation causes malformation, cytologic dysgenesis, and epilepsy; it has been suggested that this represents an "infantile tauopathy" (Sarnat et al., 2012; Sarnat and Flores-Sarnat, 2015b). The presence of gliosis in HME may also contribute to epileptogenicity through production of inflammatory cytokines. Boer et al. (2007) and Crino (2005) reported that the observed cytomegalic neurons expressed receptors for distinct neurotransmitters, neuropeptides, and growth factors. Analysis of components of the phosphoinositide 3-kinase *PI3K* pathway revealed expression of phospho-S6 ribosomal protein in cytomegalic neurons. Bilateral paracentral cortex agyria with frontal and parietal cortex polymicrogyria and severe dysplasia of midbrain and cerebellum plus neonatal medulloblastoma is a rare finding (Okumura et al., 2012).

It is of paramount importance that neonates and young infants with neurologic phenotypes of LSNS or KENS be followed by a pediatric neurologist, even if they do not yet have epilepsy. The phenotypes reviewed here should be considered in infants with infantile spasms or any refractory epilepsy without an evident etiology. Because of the multisystemic nature of somatic mosaicism, a multidisciplinary approach involving several organ system specialists, neurosurgeons, neuroradiologists, neuropathologists, and neuropsychologists offers the best opportunity for early recognition and treatment of the various manifestations of ENS as a complex disorder.

References

Aizawa K, Nakamura T, Ohyama Y et al. (2000). Renal artery stenosis associated with epidermal nevus syndrome. Nephron 84: 67–70.

Alfonso I, Howard C, Lopez PF et al. (1987). Linear nevus sebaceous syndrome. A review. J Clin Neuroophthalmol 7: 170–177.

Allison MA, Dunn CL, Pederson RC (1998). What syndrome is this? Epidermal nevus syndrome: a neurologic variant with hemimegalencephaly, facial hemihypertrophy and gyral malformation. Pediatr Dermatol 15: 59–61.

Alomari AI (2009). CLOVE(S) syndrome: expanding the acronym. Am J Med Genet 149A: 294.

Aydíngöz U, Emir S, Karlí-Oğuz K et al. (2002). Congenital infiltrating lipomatosis of the face with ipsilateral hemimegalencephaly. Pediatr Radiol 32: 106–109.

Baker RS, Ross PA, Baumann RJ (1987). Neurologic complications of the epidermal nevus syndrome. Arch Neurol 44: 227–232.

Barkovich AJ, Chuang SH (1990). Unilateral megalencephaly: correlation of MR imaging and pathologic characteristics. Am J Neuroradiol 11: 523–531.

Berg JM, Crome L (1970). A possible case of atypical tuberous sclerosis. J Ment Defic Res 4: 24–31.

Besser FS (1976). Linear sebaceous naevi with convulsions and mental retardation (Feuerstein–Mims' syndrome), vitamin-D-resistant rickets. Proc Roy Soc Med 69: 518–520.

Biesecker L (2006). The challenges of Proteus syndrome: diagnosis and management. Eur J Hum Genet 14: 1151–1157.

Blaschko A (1901). Die Nervenverteilung in der Haut ihrerBeziehungzu den Erkrankungen der Haut. Wilhelm Braumuller, Vienna. Quoted by Jackson R (1976). The lines of Blaschko: a review and reconsideration. Br J Dermatol 95: 349–360.

Boer K, Troost D, Spliet WG et al. (2007). A neuropathological study of two autopsy cases of syndromic hemimegalencephaly. Neuropathol Appl Neurobiol 33: 455–470.

Bögel G (1886). Über das subconjunctivale Lipom und eine Kombination desselben mit Ichthyosis hystrix. Arch Ophthalmol 32: 129–150.

Booth TN, Rollins NK (2002). MR imaging of the spine in epidermal nevus syndrome. AJNR Am J Neuroradiol 23: 1607–1610.

Bosemani T, Boltshauser E, Staudt M et al. (2013). Hemimegalencephaly: involvement of cranial nerves and intracranial vessels. Neuropediatrics 44: 1031 [Abstract].

Brihaye J, Brihaye-van Geertruyden M, Retif J et al. (1988). Late occurrence of additional ocular and intracranial pathologies in the linear naevus sebaceous (Feuerstein–Mims) syndrome. Acta Neurochir (Wien) 92: 132–137.

Brodsky MC, Kincannon JM, Nelson-Adesokan P et al. (1997). Oculocerebral dysgenesis in the linear nevus sebaceous syndrome. Ophthalmology 104: 497–503.

Canyigit M, Oguz KK (2006). Epidermal nevus syndrome with internal carotid artery occlusion and intracranial and orbital lipomas. Am J Neuroradiol 27: 1559–1561.

Carey DE, Drezner MK, Hamdan JA et al. (1986). Hypophosphatemic rickets/osteomalacia in linear sebaceous naevus syndrome: a variant of tumor-induced osteomalacia. J Pediatr 109: 994–1000.

Carlier G, Lambotte C, Pierard G (1974). The diversities of phacomatosis (hamartosis). Regarding a case likely to represent a new syndrome. Arch Fr Pediatr 31: 657–668 [in French].

Carstens MH (2008). Neural tube programming and the pathogenesis of craniofacial clefts. Part 1: The neuromeric organization of the head and neck; Part 2: Mesenchyme, pharyngeal arches, developmental fields, and the assembly of the human face. In: HB Sarnat, P Curatolo (Eds.), In: Malformations of the Nervous System. Vol. 87. Elsevier, Edinburgh, pp. 247–276, 277–339.

Carter BS, Hurst DL (1990). Magnetic resonance imaging of the brain in the linear sebaceous nevus syndrome. J Child Neurol 5: 68–69.

Castellano E, Santos E (2011). Ras isoforms: so similar but so different. Genes Cancer 2: 216–231.

Cavenagh EC, Hart BL, Rose D (1993). Association of linear sebaceous nevus syndrome and unilateral megalencephaly. Am J Neuroradiol 14: 405–408.

Chalhub EG, Volpe JJ, Gado MH (1975). Linear nevus sebaceous syndrome associated with porencephaly and nonfunctioning major cerebral venous sinuses. Neurology 25: 857–860.

Chen CC, Lien SH, Hsu ML et al. (2006). Linear nevus sebaceous syndrome. J Med Sci 26: 077–082.

Clancy RR, Kurtz MB, Baker D et al. (1985). Neurologic manifestations of the organoid nevus syndrome. Arch Neurol 42: 236–240.

Cohen Jr MM (2005). Proteus syndrome: an update. Am J Med Genet C Semin Med Genet 137C: 38–52.

Crino PB (2005). Molecular pathogenesis of focal cortical dysplasia and hemimegalencephaly. J Child Neurol 20: 330–336.

Cross JH (2005). Neurocutaneous syndromes and epilepsy – issues in diagnosis and management. Epilepsia 46 (Suppl. 10): 17–23.

Curatolo P, Bombardieri R (2008). Tuberous sclerosis. In: HB Sarnat, P Curatolo (Eds.), Handbook of Clinical Neurology, vol. 87, 3rd series: Malformations of the Nervous System, Elsevier, Edinburgh, pp. 129–151.

Curth HO, Warburton D (1965). The genetics of incontinentia pigmenti. Arch Dermatol 2: 229–351.

David P, Elia M, Garcovich A et al. (1990). A case of epidermal nevus syndrome with carotid malformation. Ital J Neurol Sci 11: 293–296.

De Lone DR, Brown WD, Gentry LR (1999). Proteus syndrome: craniofacial and cerebral MRI. Neuroradiology 41: 840–843.

Dean JCS, Cole GF, Appleton RE et al. (1990). Cranial hemihypertrophy and neurodevelopmental prognosis. J Med Genet 27: 160–164.

Demaerel P, Wilms G, Casaer P (1994). More association of linear sebaceous nevus syndrome and unilateral megalencephaly. Am J Neuroradiol 15: 196–197.

Demir MK (2008). Case 131: Proteus syndrome. Radiology 246: 974–979.

Di Rocco F, Novegno F, Tamburrini G et al. (2001). Hemimegalencephaly involving the cerebellum. Pediatr Neurosurg 35: 274–276.

Diaz JH (2000). Perioperative management of infants with the linear naevus sebaceous syndrome of Jadassohn: a report of two cases. Paediatr Anaesth 10: 669–673.

Dietrich RB, Glidden DE, Roth GM et al. (1998). The Proteus syndrome: CNS manifestations. AJNR Am J Neuroradiol 19: 987–990.

Dobyns WB, Garg BP (1991). Vascular abnormalities in epidermal nevus syndrome. Neurology 41: 276–278.

Dodge NN, Dobyns WB (1995). Agenesis of the corpus callosum and Dandy–Walker malformation associated with hemimegalencephaly in the sebaceous nevus syndrome. Am J Med Genet 56: 147–150.

Donati L, Candiani P, Grappolini S et al. (1990). Congenital infiltrating lipomatosis of the face related to cytomegalovirus infection. Br J Plastic Surg 43: 124–126.

Duncan JL, Golabi M, Fredrick DR et al. (1998). Complex limbal choristomas in linear nevus sebaceous syndrome. Ophthalmology 105: 1459–1465.

Eichler C, Flowers FP, Ross J (1989). Epidermal nevus syndrome: case report and review of clinical manifestations. Pediatr Dermatol 6: 316–320.

El-Shanti H, Bell WE, Waziri M (1992). Epidermal nevus syndrome: subgroup with neuronal migration defects. J Clin Neurol 7: 29–34.

Etchevers HC, Vincent C, Le Douarin NM et al. (2001). The cephalic neural crest provides pericytes and smooth muscle cells to all blood vessels of the face and forebrain. Development 128: 1059–1068.

Eve FS (1883). Congenital hypertrophy of the foot and leg, especially of the skin and subcutaneous tissue of the sole. Pathol Trans (Lond) 34: 298–304.

Fernandez-Pineda I, Fajardo M, Chaudry G et al. (2010). Perinatal clinical and imaging features of CLOVES syndrome. Pediatr Radiol 40: 1436–1439.

Feuerstein RC, Mims LC (1962). Linear nevus sebaceus with convulsions and mental retardation. Am J Dis Child 104: 675–679.

Finlayson J (1884). On the case of a child affected with congenital unilateral hypertrophy and patches of cutaneous congestion. Glasgow Med J 22: 327–337.

Fiorucci G, Hall A (1988). All three human ras genes are expressed in a wide range of tissues. Biochim Biophys Acta 950: 81–83.

Flores-Sarnat L (2002). Hemimegalencephaly: part 1. Genetic, clinical, and imaging aspects. J Child Neurol 17: 373–384.

Flores-Sarnat L (2008). Hemimegalencephaly syndrome. In: HB Sarnat, P Curatolo (Eds.), Handbook of Clinical Neurology, Vol. 87: Malformations of the Nervous System, Elsevier, Edinburgh, pp. 153–176.

Flores-Sarnat L (2012). Congenital infiltrating lipomatosis of the face: recognition and pathogenesis. Neuropediatrics 43: 346–348.

Flores-Sarnat L (2013). Epidermal nevus syndrome. In: O Dulac, M Lassonde, HB Sarnat (Eds.), Handbook of Clinical Neurology, Vol. 111, 3rd series. Pediatric Neurology, Elsevier, Amsterdam, pp. 349–368.

Flores-Sarnat L (2015). Epilepsy in neurological phenotypes of epidermal nevus syndrome. J Pediatr Epilepsy (in press).

Flores-Sarnat L, Sarnat HB, Davila-Gutierrez G et al. (2003). Hemimegalencephaly: part 2. Neuropathology suggests a disorder of cellular lineage. J Child Neurol 18: 776–785.

Gerhardt C (1871). Beobachtungen über neuropathisches Hautpapillom. Jahrb Kinderheilk 4: 270–274. Jahrb f Kinderheilk u phys Erziehung. IL p. 270.

Goodsell DS (1999). The molecular perspective: the ras oncogene. Oncologist 4: 263–264.

Grebe TA, Rimsza ME, Richter SF (1993). Further delineation of the epidermal nevus syndrome: two cases with new findings and literature review. Am J Med Genet 47: 24–30.

Greene AK, Rogers GF, Mulliken JB (2007). Schimmelpenning syndrome: an association with vascular anomalies. Cleft Palate Craniofac J 44: 208–215.

Groesser L, Herschberger E, Ruetten A et al. (2012). Postzygotic HRAS and KRAS mutations cause nevus sebaceous and Schimmelpenning syndrome. Nat Genet 44: 783–787.

Groesser L, Herschberger E, Sagrera A et al. (2013). Phacomatosis pigmentokeratotica is caused by a postzygotic HRAS mutation in a multipotent progenitor cell. J Invest Dermatol 133: 1998–2003.

Gross H, Uiberrak B (1955). Klinisch-anatomische Befunde bei Hemimegalencephalie (Über die Stellung der cerebralen Hyperplasie und des örtlichen Riesenwuchses innerhalb der Phakomatosen). Virch Arch 327: 577–589.

Gucev ZS, Tasic V, Jancevska A et al. (2008). CLOVE syndrome (congenital lipomatous overgrowth, vascular malformations, and epidermal nevi): CNS malformations and seizures may be a component of this disorder. Am J Med Genet A 146A: 2688–2690.

Gurecki PJ, Holden KR, Sahn EE et al. (1996). Developmental neural abnormalities and seizures in epidermal nevus syndrome. Dev Med Child Neurol 38: 716–723.

Hafner C, van Oers JM, Vogt T et al. (2006). Mosaicism of activating FGFR3 mutations in human skin causes epidermal nevi. J Clin Invest 116: 2201.

Hafner C, Lopez-Knowles E, Luis NM et al. (2007). Oncogenic PIK3CA mutations occur in epidermal nevi and seborrheic keratoses with a characteristic mutation pattern. Proc Natl Acad Sci U S A 104: 13450–13454.

Hafner C, Toll A, Gantner S et al. (2012). Keratinocytic epidermal nevi are associated with mosaic RAS mutations. J Med Genet 49: 249–253.

Hager BC, Dyme IZ, Guertin SR et al. (1991). Linear sebaceous nevus syndrome: megalencephaly and heterotopic gray matter. Pediatr Neurol 7: 45–49.

Hall BK (2009). The Neural Crest and Neural Crest Cells in Vertebrate Development and Evolution, Springer, New York.

Happle R (1987). Lethal genes surviving by mosaicism: a possible explanation for sporadic birth defects involving the skin. J Am Acad Dermatol 16: 899–906.

Happle R (1995). Epidermal nevus syndromes. Semin Dermatol 14: 111–121.

Happle R (2004). Gustav Schimmelpenning and the syndrome bearing his name. Dermatology 209: 84–87.

Happle R (2010). The group of epidermal nevus syndromes. Part I. Well defined phenotypes. J Am Acad Dermatol 63: 1–22. quiz 23–24.

Herman TE, Siegel MJ (2001). Hemimegalencephaly and linear nevus sebaceous syndrome. J Perinatol 21: 336–338.

Holden KR, Dekaban AS (1972). Neurological involvement in nevus unis lateris and nevus linearis sebaceus. Neurology 22: 879–884.

Hornstein OP, Knickenberg M (1974). On the Schimmelpenning–Feuerstein–Mims-syndrome (organoid nevus-phakomatosis) [author's translation]. Arch Dermatol Forsch 250: 35–50 [in German].

Jackson R (1976). The lines of Blaschko: a review and reconsideration. Observations of the cause of certain unusual linear conditions of the skin. Br J Dermatol 95: 349–360.

Jackson PE, Huang YT, Fiumara F et al. (2010). Ictal SPECT with MRI fusion localizing a seizure focus in child with hemimegalencephaly and Proteus syndrome. Clin Nucl Med 35: 637–639.

Jadassohn J (1895). Bemerkungen zür Histologie der systematisierten Naevi und über "Talgdrüsen – Naevi". Arch Dermatol Syph (Paris) 33: 355–394.

Kaduthodil MJ, Prasad DS, Lowe AS et al. (2012). Imaging manifestations in Proteus syndrome: an unusual multisystem developmental disorder. Br J Radiol 85: e793–e799.

Kinsler VA, Thomas AC, Ishida M et al. (2013). Multiple congenital melanocytic nevi and neurocutaneous melanosis are caused by postzygotic mutations in codon 61 of NRAS. J Invest Dermatol 133: 2229–2236.

Kirby ML, Gale TF, Stewart DE (1983). Neural crest cells contribute to normal aorticopulmonary septation. Science 220: 1059–1061.

Kurek KC, Luks VL, Ayturk UM et al. (2012). Somatic mosaic activating mutations in PIK3CA cause CLOVES syndrome. Am J Hum Genet 90: 1108–1115.

Kurokawa T, Sasaki K, Hanai T et al. (1981). Linear nevus sebaceous syndrome. Report of a case with Lennox–Gastaut syndrome following infantile spasms. Arch Neurol 38: 375–377.

Kwa VI, Smitt JH, Verbeteen BW et al. (1995). Epidermal nevus syndrome with isolated enlargement of one temporal lobe: a case report. Brain Dev 17: 122–125.

Lall K (1962). Teratoma of conjunctiva (associated with nevus systematicus and epilepsia symptomatica). Acta Ophthalmol 40: 555–558.

Lam J, Dohil MA, Eichenfield LF et al. (2008). SCALP syndrome: sebaceous nevus syndrome, CNS malformations, aplasia cutis congenita, limbal dermoid, and pigmented nevus (giant congenital melanocytic nevus) with neurocutaneous melanosis: a distinct syndromic entity. J Am Acad Dermatol 58: 884–888.

Lansky LL, Funderburk S, Francis E et al. (1972). Linear sebaceous nevus syndrome: a hamartoma variant. Am J Dis Child 123: 587–590.

Le Douarin NM, Couly G, Creuzet SE (2012). The neural crest is a powerful regulator of pre-otic brain development. Dev Biol 366: 74–82.

Lee JH, Huynh M, Silhavy JL et al. (2012). De novo somatic mutations in components of the PI3K-AKT3-mTOR pathway cause hemimegalencephaly. Nat Genet 44: 941–945.

Lenz W (1961). Zur Genetik der Incontinentia Pigmenti. Ann Paediatr 196: 149–165.

Levinsohn J, Tian L, Boyden L et al. (2013). Whole exome sequencing reveals somatic mutations in HRAS and KRAS which cause nevus sebaceus. J Invest Dermatol 133: 827–830.

Lien SH, Hsu ML, Yuh YS et al. (2005). Prenatal three dimensional ultrasound detection of linear nevus sebaceous syndrome. Arch Dis Child Fetal Neonatal Ed 90: F315.

Lim YH, Ovejero D, Sugarman JS (2014). Multilineage somatic activating mutations in HRAS and NRAS cause mosaic cutaneous and skeletal lesions, elevated FGF23 and hypophosphatemia. Hum Mol Genet 23: 397–407.

Lindhurst MJ, Sapp JC, Teer JK et al. (2011). A mosaic activating mutation in AKT1 associated with the Proteus syndrome. N Engl J Med 365: 611–619.

Loddenkemper T, Alexopoulos AV, Kotagal P et al. (2008). Epilepsy surgery in epidermal nevus syndrome variant with hemimegalencephaly and intractable seizures. J Neurol 255: 1829–1831.

Luo S, Tsao H (2014). Epidermal, sebaceous, and melanocytic nevoid proliferations are spectrums of mosaic RASopathies. J Investigative Dermatol 134: 2493–2496.

Maclellan RA, Luks VL, Vivero MP et al. (2014). PIK3CA activating mutations in facial infiltrating lipomatosis. Plast Reconstr Surg 133: 12e–19e.

Maher CO, Cohen-Gadol AA, Raffel C (2003). Cortical resection for epilepsy in children with linear sebaceous nevus syndrome. Pediatr Neurosurg 39: 129–135.

Marden PH, Venters HP (1966). A new neuro-cutaneous syndrome. Am J Dis Child 112: 79–81.

Merks JH, de Vries LS, Zhou XP et al. (2003). PTEN hamartoma tumour syndrome: variability of an entity. J Med Genet 40: e111.

Mimouni F, Han BK, Barnes L et al. (1986). Multiple hamartomas associated with intracranial malformation. Pediatr Dermatol 3: 219–225.

Mollica F, Pavone L, Nuciforo G (1974). Linear sebaceous nevus syndrome in a newborn. Am J Dis Child 128: 868–871.

Morrow PA (1898). Two cases of linear naevus with remarks on its nature and nomenclature. NY Med J 67: 1.

Moss C, Larkins S, Stacey M et al. (1993). Epidermal mosaicism and Blaschko's lines. J Med Genet 30: 752–755.

Mullaney PB, Weatherhead RG (1996). Epidermal nevus syndrome associated with a complex choristoma and a bilateral choroidal osteoma. Arch Ophthalmol 114: 1292–1293.

Neis AE, Johansen KL, Harms RW et al. (2006). Sonographic characteristics of linear nevus sebaceous sequence. Ultrasound Obstet Gynecol 27: 323–324.

Neuman LM, Scheer I, Kunze J et al. (2003). Cerebral manifestations, hemihypertrophy and lymphoedema of one leg in a child with epidermal nevus syndrome (Schimmelpenning–Feuerstein–Mims). Pediatr Radiol 33: 637–640.

Nour I, Abdel-Hady H, Nasef N et al. (2012). Linear nevus sebaceous syndrome in a neonate conceived by intracytoplasmic sperm injection. J Clin Neonatol 1: 143–145.

Okumura A, Lee T, Ikeno M (2012). A severe form of epidermal nevus syndrome associated with brainstem and cerebellar malformations and neonatal medulloblastoma. Brain Dev 34: 881–885.

Olivares JL, Ramos FJ, Carapeto FJ et al. (1999). Epidermal naevus syndrome and hypophosphataemic rickets: description of a patient with central nervous system anomalies and review of the literature. Eur J Pediatr 158: 103–107.

Ousager LB, Bygum A, Hafner C (2012). Identification of a novel S249C FGFR3 mutation in a keratinocytic epidermal nevus syndrome. Br J Dermatol 167: 202–204.

Özkiris A, Vereklioglu CE, Kula M et al. (2003). Tc-99 m HMPAO brain SPECT in linear nevus sebaceous syndrome. Ann Nucl Med 17: 703–706.

Pack GT, Sunderland DA (1941). Nevus unius lateris. Arch Surg 43: 341–372.

Paladin F, Chiron C, Dulac O et al. (1989). Electroencephalographic aspects of hemimegalencephaly. Dev Med Child Neurol 31: 337–383.

Pavone L, Curatolo P, Rizzo R et al. (1991). Epidermal nevus syndrome: a neurologic variant with hemimegalencephaly, gyral malformation, mental retardation, seizures, and facial hemihypertrophy. Neurology 41: 266–271.

Poduri A, Evrony GD, Cai X et al. (2012). Somatic activation of AKT3 causes hemispheric developmental brain malformations. Neuron 74: 41–48.

Poduri A, Evrony GD, Cai X et al. (2013). Somatic mutation, genomic variation, and neurological disease. Science 341: 1237758.

Pollock RA, Newman MH, Burdi AR et al. (1985). Congenital hemifacial hyperplasia: an embryologic hypothesis and case report. Cleft Pal J 22: 173–184.

Pujol RM, Tuneu A, Moragas De et al. (1987). Síndrome del nevus sebáceo lineal. Med Cut I L A 15: 280–284.

Rebollo Alvarez-Amandi M, Berciano Blanco J, Combarros Pascual O et al. (1979). Jadassohn's nevus sebaceus and nevus unius lateris. Report of an observation with multiple dysplasia. Med Clin (Barc) 72(3): 106–110.

Rijntjes-Jacobs EG, Lopriore E, Steggerda SJ et al. (2010). Discordance for Schimmelpenning–Feuerstein–Mims syndrome in monochorionic twins supports the concept of a postzygotic mutation. Am J Med Genet A 152A: 2816–2819.

Rizzo R, Pavone L, Sorge G et al. (1990). Proteus syndrome: report of a case with severe brain impairment and fatal course. J Med Genet 27: 399–402.

Robinson SS (1932). Nevus sebaceous (Jadassohn). Arch Derm Syph 26: 663–670.

Rogers M, McCrossin I, Commens C (1989). Epidermal nevi and the epidermal nevus syndrome. A review of 131 cases. J Am Acad Dermatol 20: 476–488.

Roguin A (2006). Wilhelm His, Jr. (1863–1934). The man behind the bundle. Heart Rhythm 3: 480–483.

Ronen GM, Meaney B, Dan B et al. (2009). From eugenic euthanasia to habilitation of "disabled" children: Andreas Rett's contribution. J Child Neurol 24: 115–127.

Sakuta R, Aikawa H, Takashima S et al. (1991). Epidermal nevus syndrome with hemimegalencephaly. Neuropathological study. Brain Dev 13: 260–265.

Salgado CM, Basu D, Nikiforova M et al. (2015). BRAF mutations are also associated with neurocutaneous melanocytosis and large/giant congenital melanocytic nevi. Pediatr Dev Pathol 18: 1–9.

Sapp JC, Turner JT, van de Kamp JM et al. (2007). Newly delineated syndrome of congenital lipomatous overgrowth, vascular malformations, and epidermal nevi (CLOVE syndrome) in seven patients. Am J Med Genet 143A: 2944–2958.

Sarnat HB, Flores-Sarnat L (2001). Neuropathological research strategies in holoprosencephaly. J Child Neurol 16: 918–931.

Sarnat HB, Flores-Sarnat L (2005). Embryology of the neural crest: its inductive role in neurocutaneous syndromes. J Child Neurol 20: 637–643.

Sarnat H, Flores-Sarnat L (2015a). Neurocutaneous syndrome as embryonic neurocristopathies. J Pediatr Epilepsy (in press).

Sarnat HB, Flores-Sarnat L (2015). Infantile tauopathies: hemimegalencephaly, tuberous sclerosis complex, focal cortical dysplasia 2, ganglioglioma. Brain Dev 37: 553–562.

Sarnat HB, Diadori P, Trevenen CL (1993). Myopathy of the Proteus syndrome: hypothesis of muscular dysgenesis. Neuromusc Disord 3: 293–301.

Sarnat HB, Flores-Sarnat L, Crino PB et al. (2012). Hemimegalencephaly: fetal tauopathy, mTOR activation and neuronal lipidosis. Folia Neuropathol 50: 330–345.

Sarnat HB, Philippart M, Flores-Sarnat L et al. (2015). Timing in neural development: arrest, delay, precociousness, and temporal determination of malformations. Pediatr Neurol 52: 473–486.

Sarwar M, Schafer ME (1998). Brain malformations in linear nevus sebaceous syndrome: an MR study. J Comput Assist Tomogr 2: 338–340.

Schimmelpenning GW (1957). Klinischer Beitrag zur Symptomatologie der Phakomatosen. Fortschr Rontgenstr 87: 716–720.

Schimmelpenning GW (1983). Long-term observation of a case of organoid naevus phakomatosis (Schimmelpenning–Feuerstein–Mims syndrome). Rofo 139: 63–67.

Schumacher E, Engelhardt W (1926). Beitrag zur Genese angeborener Muttermale. Krankheitsforsch 2: 311–326.

Singal A, Dhaliwal U, Bhattacharya SN et al. (2001). Complex ocular choristomas in linear nevus sebaceous syndrome: a report of two cases. J Dermatol 28: 259–264.

Slavin SA, Baker DC, McCarthy JG et al. (1983). Congenital infiltrating lipomatosis of the face: clinicopathologic evaluation and treatment. Plast Reconstr Surg 72: 158–164.

Snider P, Olaopa M, Firulli AB et al. (2007). Cardiovascular development and the colonizing cardiac neural crest lineage. Scient World J 7: 1090–1113.

Solomon LM, Esterly NB (1975). Epidermal and other congenital organic nevi. Curr Probl Pediatr 6: 3–56.

Solomon LM, Fretzin DF, Dewald RL (1968). The epidermal nevus syndrome. Arch Dermatol 97: 273–285.

Stern C (1936). Somatic crossing-over and segregation in Drosophila melanogaster. Genetics 21: 625–730.

Sugarman JL (2007). Epidermal nevus syndromes. Semin Cutan Med Surg 26: 221–230.

Sugarman JL, Frieden IJ (2004). Epidermal nevus syndromes. In: ES Roach, VS Miller (Eds.), Neurocutaneous Diseases. Cambridge University Press, New York, pp. 88–104.

Sugarman GI, Reed WB (1969). Two unusual neurocutaneous disorders with facial cutaneous signs. Arch Neurol 21: 242–247.

Sultzberger M (1936). Josef Jadassohn 1863–1936. Cole HN. An appreciation. KGZ Arch Dermatol Syphilol 33: 1063–1069.

Sultzberger MB (1985). Josef Jadassohn. A personal appreciation. Am J Dermatopathol 7: 31–36.

Sun BK, Saggini A, Sarin KY et al. (2013). Mosaic activating RAS mutations in nevus sebaceus and nevus sebaceus syndrome. J Invest Dermatol 133: 824–827.

Tanaka T, Ida N, Shimoda H et al. (1986). Organ specific expression of ras oncoproteins during growth and development of the rat. Mol Cell Biochem 70: 97–104.

Tidyman WE, Rauen KA (2009). The RASopathies: developmental syndromes of Ras/MAPK pathway dysregulation. Curr Opin Genet Dev 19: 230–236.

Tokatli A, Coskun T, Özalp I (1997). Hypophosphataemic vitamin-D resistant rickets associated with epidermal naevus syndrome. A case report. Turkish J Pediatr 39: 247–251.

Trapp LD, Lee C, Troutman KC et al. (1981). Anesthetic management of the pediatric patient with multiple congenital anomalies including severe hemifacial hypertrophy. Anesth Prog 28: 162–166.

Tripp JH (1971). A new neurocutaneous syndrome (skin, eye, brain, and heart). Proc R Soc Med 64: 23–24.

Trounce JQ, Rutter N, Mellor DH (1991). Hemimegalencephaly: diagnosis and treatment. Dev Med Child Neurol 33: 257–266.

Turner JT, Cohen Jr MM, Biesecker LG (2004). Reassessment of the Proteus syndrome literature: application of diagnostic criteria to published cases. Am J Med Genet A 130A: 111–122.

Unal O, Cirak B, Bekerecioglu M et al. (2000). Congenital infiltrating lipomatosis of the face with cerebral abnormalities. Eur Radiol 10: 1610–1613.

Van Bogaert L (1935). Les dysplasies neuroectodermiques. Rev Neurol 63: 353.

van de Warrenburg BPC, van Gulik S, Renier WO et al. (1998). The linear naevus sebaceus syndrome. Clin Neurol Neurosurg 100: 126–132.

Vidaurri de la Cruz H, Tamayo-Sanchez L, Duran-McKinster C et al. (2004). Epidermal nevus syndromes: clinical findings in 35 patients. Pediatr Dermatol 21: 432–439.

Vigevano F, Aicardi J, Lini M et al. (1984). La sindrome del nevo sebaceo lineare: presentazione di una casistica multicentnca. Boll Lega Ital Epil 45 (46): 59–63.

Vigevano F, Bertini E, Boldrini R et al. (1989). Hemimegalencephaly and intractable epilepsy: benefits of hemispherectomy. Epilepsia 30: 833–843.

Vles JSH, Degraeuwe P, DeCock P et al. (1985). Neuroradiological findings in Jadassohn nevus phakomatosis: a report of 4 cases. Eur J Pediatr 144: 290–294.

Wakabayashi C, Machara T, Shimamura T et al. (1997). Unilateral megalencephaly in linear sebaceous syndrome: a neuroradiological case report. Radiat Med 15: 65–69.

Waltz KM, Helm KF, Billingsley EM (1999). The spectrum of epidermal nevi: a case of verrucous epidermal nevus contiguous with nevus sebaceous. Pediatr Dermatol 16: 211–213.

Warnke PH, Hauschild A, Schimmelpenning GW et al. (2003). The sebaceous nevus as part of the Schimmelpenning–Feuerstein–Mims syndrome – an obvious phacomatosis first documented in 1927. J Cutan Pathol 30: 470–472.

Wauschkuhn J, Rohde B (1971). Systematized sebaceous, pigmented and epithelial nevi with neurologic symptoms. Neuroectodermal Feuerstein–Mims syndrome [in German]. Hautarzt 22: 10.

Webster JHD (1907). A case of unilateral cerebral hyperplasia with co-existent "acromegaly" of the feet and a slight degree of unilateral gigantism. J Path Bact 12: 306–309.

Weyers W (2013). Josef Jadassohn – an appreciation on the occasion of his 150th birthday. Am J Dermatopathol 35: 742–751.

Yu TW, Tsau YK, Young C et al. (2000). Epidermal nevus syndrome with hypermelanosis and chronic hyponatremia. Pediatr Neurol 22: 151–154.

Zaremba J, Wisiawski J, Bidzinski J et al. (1978). Jadassohn's naevus phakomatosis: a report of two cases. J Ment Def Res 22: 91–102.

Zhang W, Simos PG, Isibashi H et al. (2003). Neuroimaging features of epidermal nevus syndrome. Am J Neuroradiol 24: 1468–1470.

Section 2

Neurocutaneous syndromes

Handbook of Clinical Neurology, Vol. 132 (3rd series)
Neurocutaneous Syndromes
M.P. Islam and E.S. Roach, Editors

Chapter 3

Acquired neurocutaneous disorders

AMRE NOUH[1], JODI SPEISER[2], AND JOSÉ BILLER[3]*

[1]*Department of Neurology, Hartford Hospital - University of Connecticut, Hartford CT, USA*

[2]*Section of Dermatopathology, Department of Pathology, Stritch School of Medicine, Loyola University Chicago, Maywood, IL, USA*

[3]*Department of Neurology, Stritch School of Medicine, Loyola University Chicago, Maywood, IL, USA*

INTRODUCTION

A variety of neurologic diseases have cutaneous manifestations. These may precede, coincide with, or follow the neurologic findings. An array of autoimmune, genetic, and environmental factors play a role in expression and severity of the neurologic burden in these conditions. This chapter emphasizes congenital and genetic disorders, but we also discuss the pathophysiology and manifestation of various acquired neurocutaneous disorders with an emphasis Behçet's disease, dermatomyositis, Sjögren syndrome, systemic lupus erythematosus (SLE), scleroderma, Parry–Romberg syndrome and Degos disease.

BEHÇET'S DISEASE

Behçet's disease is a chronic multisystem vasculitic disorder of unknown pathogenesis originally described by the Turkish dermatologist Behçet in the 1930s, with an account of the disease's characteristic recurrent oral aphthous ulcers, genital ulcers, and uveitis (Behçet, 1937). Although observed worldwide, Behçet's disease is most prevalent in regions stretching from the Middle East to the eastern shores of Asia, along the ancient network of trade routes known as the Silk Road that historically ran between China and the eastern shores of the Mediterranean (Davatchi et al., 2010a). Compared to Asia (2.1–19.5/100 000) or Europe (0.3–15.9/100 000), the prevalence of Behçet's disease in Turkey is quite high (20–420/100 000) (Mahr and Maldini, 2014). The mean age of presentation of Behçet's disease is between the third and fourth decades of life. Behçet's disease has a male gender predominance as high as 5:1 and 4:1 in countries such as Egypt and Algeria, with a worldwide average of 1.4:1 (Kural-Seyahi et al., 2003; Davatchi et al., 2010a, b, 2012). Mucocutaneous lesions are the hallmark and the most common findings of Behçet's disease, followed by ophthalmologic, articular, vascular, neurologic, and systemic manifestations (Davatchi et al., 2010a; Saadoun and Wechsler, 2012).

Etiology and pathogenesis

The etiology and pathogenesis of Behçet's disease are unknown. It has been postulated that microbiological antigenic stimuli lead to cellular and humoral mediated inflammatory responses in genetically susceptible individuals (Direskeneli, 2001). Tumor necrosis factor α (TNF-α), a proinflammatory cytokine primarily secreted by mononuclear phagocytic cells, has been implicated in the pathogenesis of Behçet's disease, and its overexpression has been related to the severity of the inflammatory response. In particular TNF (−308A/G, −238A/G, −1031C/T, and −857T/C) polymorphisms have been associated with Behçet's disease susceptibility (Zhang et al., 2013). A prime component of genetic susceptibility to Behçet's disease is within the major histocompatibility complex (MHC) region related to an HLA-B51 subtype present in almost half of patients with Behçet's disease (Direskeneli, 2001; Davatchi et al., 2010b; Piga and Mathieu, 2011). Patients with HLAB27 have less severe clinical manifestations and predominantly anterior uveitis (Al-Araji and Kidd, 2009). Furthermore, the plausible genetic contribution has been supported by a high sibling recurrence risk, particularly among the Turkish population (Gül et al., 2000).

*Correspondence to: José Biller, MD, FACP, FAAN, FANA, FAHA, Professor and Chairman, Department of Neurology, Loyola University Chicago, Stritch School of Medicine, 2160 S. 1st Avenue, Bldg. 105, Room 2700, Maywood, IL 60153, USA. Tel: +1-708-216-2438, Fax: +1-708-216-5617, E-mail: jbiller@lumc.edu

Neurobiological findings of Behçet's disease may include severe meningoencephalitis with necrosis, accumulation of lipid-laden macrophages, gliosis with neuronal apoptosis, perivasculitis, and rarely fibrinoid necrosis of small postcapillary venules (Kidd et al., 1999; Al-Araji and Kidd, 2009). The 2012 Revised International Chapel Hill Consensus Conference Nomenclature classifies Behçet's disease as a "variable vessel vasculitis" due to involvement of both arteries and veins with no predominant vessel type or size affected by the disease (Jennette et al., 2013).

Aphthous oral/genital ulcers and other cutaneous lesions

Painful aphthous ulcers are diagnostic features of Behçet's disease (International Study Group for Behçet's Disease, 1990). Oral, laryngeal and/or pharyngeal ulcers occur in 90–100% of patients, while scrotal, penile, vaginal, and/or vulvar ulcers occur in 57–93% of patients (Zouboulis, 1999). The ulcers are painful, recurrent, often > 1 cm in size, with a yellowish necrotic crust (Fig. 3.1) (Calonje et al., 2012). A direct correlation between ulcer size and severity of oral disease has been postulated. Yet, there is no association with systemic severity (Krause et al., 1999).

Other frequent cutaneous manifestations include erythema nodosum-like lesions and pseudofolliculitis. Additional findings include acneiform papules and pustules, pathergic response to trauma in the form of sterile pustules, furuncles, pyoderma, thrombophlebitis, and Sweet's syndrome-like features (Calonje et al., 2012).

Ocular, vascular, and systemic manifestations

Ocular manifestations of Behçet's disease are often severe and bilateral, and present in approximately half of patients (Tugal-Tutkun et al., 2004; Davatchi et al., 2010b, 2014). These include cataracts, glaucoma, and posterior segment involvement with vasculitis, vitritis, retinitis, panuveitis, retinal edema, cystoid macular degeneration, venous or arterial occlusions, disc edema, and retinal detachment. Uveitis, the most common ocular manifestation (Tugal-Tutkun et al., 2004; Saadoun and Wechsler, 2012), is more severe among men and carries the greatest risk of vision loss (Tugal-Tutkun et al., 2004). Risk factors for vision loss include persistent inflammatory activity, presence of posterior synechiae and hypotony, and elevated intraocular pressure (Kaçmaz et al., 2008).

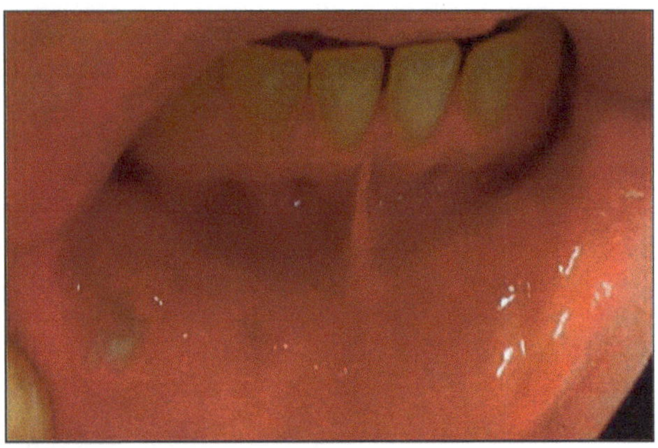

Fig. 3.1. Oral aphthous ulcers in Behçet's disease. (Courtesy of Dr. Iris Aronson and Dr. Amanda Marsch, University of Illinois at Chicago, Department of Dermatology, Chicago, IL, USA.)

Vascular involvement, present in 14–30% of patients, is also more common among men. Venous disease, including superficial phlebitis and deep venous thrombosis, is more common than arterial thrombosis or systemic aneurysms (Sarica-Kucukoglu et al., 2006; Davatchi et al., 2012; Saadoun and Wechsler, 2012).

Systemic manifestations include arthritis, painful bleeding gastrointestinal (GI) ulcerations indistinguishable from Crohn's disease, pleural effusion, pulmonary embolism, endocarditis, myocarditis, lymphadenopathy, hepatosplenomegaly, epididymitis and rarely renal involvement (Davatchi et al., 2010b; Saadoun and Wechsler, 2012).

Neurologic manifestations of Behçet's disease

The frequency of neurologic Behçet's disease (NBD) varies between 3.8% and 31% and may follow or precede the onset of systemic manifestations. Typically, neurologic manifestations present within 3–5.5 years from disease onset (Akman-Demir et al., 1999; Al-Araji et al., 2003; Hirohata et al., 2012). The central nervous system (CNS) is most commonly involved. Peripheral nervous system (PNS) involvement is rare (Al-Araji et al., 2003; Al-Araji and Kidd, 2009; Davatchi et al., 2010a, b). Neurologic findings are more common among men during the third or fourth decades of life (Akman-Demir et al., 1999; Al-Araji et al., 2003). Headache is the most common manifestation, with a prevalence of 66–82.5%. Headache in Behçet's disease may reflect disease activity, cerebral venous sinus thrombosis (CVST), intracranial hypertension, acute meningeal syndrome, uveitis, or coexisting primary headaches. Headaches unrelated to structural disorders typically present within 6 months of disease onset, are often recurrent, and resemble migraines (Kidd, 2003; Saip et al., 2005). Neurologic Behçet's disease is categorized into *primary* or *parenchymal* neurologic dysfunction, *secondary* or *nonparenchymal neurologic dysfunction*, and *mixed forms* (Akman-Demir et al., 1999; Kidd et al., 1999; Al-Araji et al., 2003; Siva et al., 2004; Al-Araji and Kidd, 2009; Hirohata et al., 2012). Parenchymal dysfunction is characterized by focal or diffuse CNS manifestations with meningoencephalitis accounting for two-thirds of cases (Akman-Demir et al., 1999; Al-Araji et al., 2003). Focal forms include brainstem involvement and stroke-like manifestations (Al-Araji et al., 2003). Al-Araji and Kidd stratified neurologic Behçet's disease into *central* (parenchymal and nonparenchymal), *peripheral* (relation to Behçet's disease is uncertain), and *other* (uncommon but recognized symptoms) (Table 3.1).

Meningoencephalitis involves the brainstem in half of the patients. However, diencephalic, cerebral, spinal cord, and multifocal involvement are not uncommon. One-fifth of patients have fever without infection during attacks (Akman-Demir et al., 1999). Affected individuals may present with brainstem syndromes, ophthalmoplegia, other cranial nerve palsies, cerebellar symptoms, myelopathies, and hemiparesis, with or without alterations in sensorium. Onset of symptoms is usually subacute, although it may be acute. Patients may present with stroke-like symptoms, cognitive and behavioral changes, seizures, tumor-like mass lesions, movement disorders, spinal cord involvement, and optic neuropathies. Arterial involvement leading to parenchymal injury is rare (Kidd et al., 1999; Akman-Demir et al., 1999; Al-Araji et al., 2003; Siva et al., 2004; Al-Araji and Kidd, 2009). Patients with spinal cord involvement commonly present with sensory-motor symptoms or sphincter dysfunction. Spinal cord involvement is usually disabling and carries a poor prognosis (Koçer et al., 1999; Moskau et al., 2003; Yesilot et al., 2007; Coulter et al., 2012). Optic neuropathies may be inflammatory or ischemic. They are rare, highly associated with other ocular lesions, and carry poor prognosis (Frigui et al., 2009).

Parenchymal neurologic dysfunction may be acute or chronic progressive (Akman-Demir et al., 1999; Kidd et al., 1999; Hirohata et al., 2012). Acute forms may be self-limiting with infrequent relapses. In contrast, chronic forms frequently relapse and are slowly progressive with poor response to therapy. *HLAB51* and cigarette smoking have been associated with chronic forms of neurologic dysfunction due to Behçet's disease (Hirohata et al., 2012).

Nonparenchymal neurologic dysfunction from Behçet's disease is related to CNS vascular involvement. CVST commonly involves large veins with a reported incidence of 4–25% (Kidd et al., 1999; Al-Araji et al., 2003; Saadoun et al., 2009). Lower rates have been reported in Japan than in Middle Eastern countries (Hirohata et al., 2012). CVST occurs earlier than parenchymal dysfunction and is strongly associated with peripheral large vessel thrombosis (Wechsler et al., 1992; Tunc et al., 2004; Saadoun et al., 2009). Patients typically present with symptoms and signs of raised intracranial pressure (ICP). Hemorrhagic infarctions are infrequent (Wechsler et al., 1992; Tunc et al., 2004). Various prothrombotic states have been documented in 31% of patients, including MTHFR homozygous and heterozygous mutations, hyperhomocysteinemia, anticardiolipin antibodies, lupus anticoagulant, factor V Leiden mutation, prothrombin gene mutation, and protein C deficiency (Tunc et al., 2004). Other series report no clear association with protein C or S activity or antiphospholipid antibodies (Lenk et al., 1998; Alekberova et al., 1999). Individuals with increased ICP without evidence of CVST have idiopathic intracranial hypertension. Cervical artery dissection and intracranial aneurysm formation are rarely reported (Bahar et al., 1993; Nakasu et al., 2001).

Peripheral neuropathy is documented in 3–15% of patients (Akbulut et al., 2007; Hirohata et al., 2012). Neuropathy is mainly axonal in nature and affects predominantly sensory nerves and lacks direct association with disease activity (Birol et al., 2004; Akbulut et al., 2007). Other peripheral nerve syndromes include polyradiculitis, mononeuritis multiplex, and necrotizing myositis (Bakouche and Guillard, 1983; Sarui et al., 2002; Davatchi et al., 2010b).

Table 3.1

Classification of neurologic Behçet's disease*

Central nervous system		Peripheral nervous system/ Neuromuscular	Other
Parenchymal	*Nonparenchymal*	Peripheral neuropathy	Acute meningeal syndrome
Brainstem	CVST	Mononeuritis multiplex	Tumor-like NBD
Diffuse (brainstem plus)	Intracranial hypertension	Myopathy	Optic neuropathy
Spinal cord	Intracranial aneurysms	Myositis	Psychiatric symptoms
Cerebral	Extracranial aneurysms/dissection		
Asymptomatic			

*From Al-Araji and Kidd, 2009.

NBD, neurologic Behçet's disease; CVST, cerebral venous sinus thrombosis.

Neuropsychiatric manifestations occur in 5–25% of patients. Memory impairment and attention deficits are the most prominent cognitive manifestations. Other neuropsychiatric features include personality and behavioral change, disinhibition, apathy, and executive dysfunction (Öktem-Tanör et al., 1999; Akman-Demir et al., 1999; Siva et al., 2004).

Diagnosis and evaluation

Diagnosis of Behçet's disease is primarily clinical. In 1992, the International Study Group for Behçet's Disease (ISG) proposed diagnostic criteria that emphasized the presence of mucocutaneous and ocular features as well as the presence of a positive pathergy test (with a reported sensitivity of 85%, and a specificity of 96%) (ISG, 1990). In 2013, the International Team for the Revision of the International Criteria for Behçet's Disease (ICBD) proposed a validated diagnostic and classification system with a reported sensitivity and specificity of 94.8% and 90.5% respectively. Two points each are assigned to ocular lesions, oral aphthosis, and genital aphthosis, while cutaneous lesions, CNS involvement, vascular manifestations or a positive pathergy test receive 1 point each. A patient scoring 4 points is classified as having Behçet's disease (Table 3.2) (Davatchi et al., 2014). Individuals with neurologic manifestations as the initial presentation require meticulous history taking and careful examination to establish the diagnosis of Behçet's disease based on the aforementioned criteria and the exclusion of secondary causes such as infections, neoplasms, or other mimickers including treatment-induced syndromes.

Neuroimaging

Magnetic resonance imaging (MRI) most commonly demonstrates T2 signal hyperintensities or plaques at the upper brainstem extending to the diencephalon and basal ganglia. More commonly unilateral and present during an acute attack, these lesions are associated with edema. Scattered subcortical lesions involving the pontobulbar junction, basal ganglia, and internal capsule are the second most common imaging findings (Akman-Demir et al., 1999; Koçer et al., 1999; Kidd et al., 1999; Haghighi et al., 2011; Hirohata et al., 2012). Plaques appear hypointense on T1-weighted imaging and may not be evident without gadolinium infusion. Centrally enhancing lesions may be observed (Akman-Demir et al., 1999; Koçer et al., 1999; Al-Arajji et al., 2003). Periventricular perpendicular plaque orientation, as in multiple sclerosis, is not typically observed (Kidd et al., 1999; Al-Arajji et al., 2003). Prominent brainstem atrophy is present in 50–75% of patients. Chronic T1 lesions may appear as "black holes" (Haghighi et al., 2011; Hirohata et al., 2012). Increased plaque burden on serial imaging suggests ongoing disease, and reversibility with treatment can occur (Koçer et al., 1999; Haghighi et al., 2011). Rarely, cerebellar or optic nerve lesions and subacute hemorrhagic lesions are apparent (Koçer et al., 1999). Symptomatic patients have been reported to have normal MRI scans, and T2 hyperintensities have been observed in asymptomatic patients, highlighting the nonspecific nature of these "silent lesions" (Akman-Demir et al., 1999; Kidd et al., 1999; Hirohata et al., 2012). Diffusion-weighted imaging (DWI) findings show hyperintensity with apparent diffusion coefficient (ADC) values in cases of active disease, and hyperintensity on DWI with decreased ADC value in subacute phases (Yildiz et al., 2013). Spinal cord lesions may be longitudinally extensive with a predilection for the posterolateral cervical and thoracic cord (Koçer et al., 1999; Moskau et al., 2003; Yesilot et al., 2007; Coulter et al., 2012). Magnetic resonance venography (MRV) may reveal filling defects within the superior sagittal and transverse sinuses (Wechsler et al., 1992; Tunc et al., 2004; Saadoun et al., 2009). The intracranial vasculature should be thoroughly evaluated in vascular Behçet's disease. MRI perfusion and spectroscopy scans are sometimes helpful in symptomatic patients with normal MRI scans (Alkan et al., 2012; Chuang et al., 2013).

Table 3.2

International Criteria for Behçet's Disease (ICBD) point score system

Sign or symptom	Points
Ocular lesions	2
Genital aphthosis	2
Oral aphthosis	2
Skin lesions	1
Neurologic manifestations	1
Vascular manifestations	1
Positive pathergy test*	1

*Pathergy test is optional and the primary scoring system does not include testing. However, where pathergy testing is conducted 1 extra point may be assigned for a positive result.

Scoring 4 indicates diagnosis (Davatchi et al., 2014).

Laboratory investigations

Elevated erythrocyte sedimentation rate (ESR) values are observed in over half of patients. However, ESR values > 50 mm are only observed in 15% of patients (Davatchi et al., 2010a, b). Abnormal renal sediment, hematuria, and proteinuria occur in approximately 10% of patients (Davatchi et al., 2010b). Cerebrospinal fluid (CSF) studies show neutrophilic or lymphocytic pleocytosis and increased protein content in two-thirds of patients (Kidd et al., 1999; Akman-Demir et al., 1999; Al-Araji et al., 2003; Yesilot et al., 2007; Hirohata et al., 2012). In the acute phase of the disease, significant pleocytosis, protein content elevation, and decreased CSF glucose levels may be observed (Hirohata et al., 2012). Oligoclonal bands or elevated IgG index are typically absent (Akman-Demir et al., 1999; Kidd et al., 1999). Elevated levels of interleukin-6 have been associated with neurologic Behçet's disease and may help differentiate acute nonprogressive and chronic progressive neurologic disease (Hirohata et al., 2012). Elevated CSF pressure warrants further evaluation for possible underlying CVST.

Ancillary investigations

Visual, brainstem, and somatosensory evoked potentials may be useful (Akman-Demir et al., 1999; Kidd et al., 1999; Yesilot et al., 2007). Nerve conduction studies typically demonstrate an axonal pattern of neuropathic involvement (Birol et al., 2004; Akbulut et al., 2007). Neuropsychologic evaluation is recommended for patients with early behavioral or neurocognitive changes. HLA B typing may have value in selected cases.

Management and prognosis

In 2008, the European League Against Rheumatism (EULAR) published treatment recommendations based on a systematic literature review by an expert committee; they addressed nine systemic manifestations (Hatemi et al., 2008). For individuals with neurologic dysfunction, corticosteroids remain the mainstay of therapy. A pulse dose of 1 g methylprednisolone for 3–7 days is recommended during the acute phase, followed by maintenance oral prednisone tapered over 2–3 months. Steroids may also be used for relapses. Other conventional treatments include cyclophosphamide and azathioprine. Due to its neurotoxic potential, ciclosporin is not recommended in patients with Behçet's disease unless it is needed for intraocular inoculation. Likewise, chlorambucil is not advised due to myelotoxicity. These recommendations are supported by level III C/D evidence (Hatemi et al., 2008).

Neurologic involvement carries significant risk of morbidity (Al-Araji et al., 2003; Saadoun and Wechsler, 2012). Other poor prognostic indicators include early age at onset, combined brainstem and other CNS involvement, frequent attacks, steroid dependence, a progressive course, and CSF pleocytosis with elevated protein content at initial study (Siva et al., 1998). Longitudinal studies spanning over two decades found a 10% mortality, primarily due to major vessel disease and neurologic involvement (Kural-Seyahi et al., 2003).

Haghighi and Safari (2010) stratified parenchymal neurologic Behçet's disease into low risk (patients without poor prognostic factors) and high risk (those with poor prognostic factors) groups. In their proposed scheme, first-line drugs include corticosteroids, azathioprine, methotrexate, and cyclophosphamide. Second line drugs include TNF blockers, interferon-α, chlorambucil, and mycophenolate mofetil. Selection of a first or second line agents is ultimately dependent on absence or presence of poor prognostic factors. Other recommendations include management of nonparenchymal brain dysfunction and, as a last resort, targeted therapies such as stem cell transplantation (Haghighi and Safari, 2010).

An analysis of 369 patients who were treated with TNF-α inhibitors such as infliximab in individuals with steroid, ciclosporin, azathioprine, and/or methotrexate refractory neurologic Behçet's disease has shown excellent results, with $>90\%$ improvement and a 6–18-month remission rate of 75% after discontinuation. Furthermore, remission of parenchymal hyperintensities, reduction in CSF IL-6 levels, and amelioration or stabilization of myoclonus, spasticity, depression, or urinary symptoms have been documented (Arida et al., 2011). In parallel, patients with brainstem involvement respond well to corticosteroids, while patients with spinal cord involvement do not (Yesilot et al., 2007; Al-Araji and Kidd, 2009). Relapses in a chronic progressive course have occurred in one-third of patients (Akman-Demir et al., 1999; Kidd et al., 1999; Hirohata et al., 2012).

Thus, identifying patients with chronic neurologic Behçet's disease may have therapeutic and prognostic value because these individuals typically are refractory to conventional treatment. Hirohata and colleagues proposed a preliminary classification system to differentiate between acute and chronic progressive forms based on clinical and ancillary findings (Table 3.3). Patients with CVST can safely and effectively receive anticoagulant therapy (Wechsler et al., 1992; Saadoun et al., 2009). Those with idiopathic increased hypertension should be managed according to established standards (Al-Araji and Kidd, 2009).

DERMATOMYOSITIS

Dermatomyositis (DM) is an acquired idiopathic inflammatory myopathy with distinct cutaneous and histopathologic findings. In contrast to other inflammatory myopathies such as polymyositis or inclusion body myositis, dermatomyositis is the most common inflammatory myopathy across all ages.

The exact incidence of dermatomyositis is not known. Earlier estimates of all inflammatory myopathies range from 0.6 to 1.0 per 100 000 (Dalakas, 1991; Dalakas and Hohlfeld, 2003). More recent, population-based studies found an incidence of DM of 7–9.6 per million (Bendewald et al., 2010; Kuo et al., 2011). DM is more

Table 3.3

Proposed preliminary classification of parenchymal neurologic Behçet's disease*

Acute[†] [§]	Chronic[‡]
1. Fulfilling International Study Group (ISD) criteria for Behçet's disease 2. Acute or subacute onset of neurologic symptoms, including headache, fever, or any focal signs 3. Increase in CSF cell count of > 6.2/mm[3]	1. Fulfilling ISD criteria for Behçet's disease 2. Insidious onset of neuropsychiatric symptoms, including dementia/neurobehavioral changes, ataxia, and dysarthria, with or without other manifestations 3. Presence of one of the following: a. Increase in CSF IL-6 of > 17.0 pg/mL on two different occasions at intervals of at least 2 weeks b. Increase in CSF IL-6 of > 17.0 pg/mL and presence of brainstem atrophy on MRI

*From Hirohata et al., 2012.
†Diagnosis of acute neurologic Behçet's disease requires all items 1–3.
‡Diagnosis of chronic progressive neurologic Behçet's disease requires all items 1–3.
§Exclusion: CNS infections
CNS, central nervous system; CSF, cerebrospinal fluid; MRI, magnetic resonance imaging.

common among women (Dalakas and Hohlfeld, 2003) and substantially higher in association with malignancies (Hill et al., 2001). In addition, dermatomyositis has been associated with scleroderma and mixed connective tissue disease, known as "overlap syndrome" (Amato and Barohn, 2009).

The pathogenesis of dermatomyositis is complex. Autoimmune T cell-mediated binding of antibodies to endothelial antigen, activation of complement, and vascular inflammation leading to ischemia are thought to be the primary mechanism of the characteristic perifascicular atrophy in dermatomyositis (Dalakas and Hohlfeld, 2003). A high expression of interferon-α/β gene inducible protein MxA and abundance of plasmacytoid dendritic CD4+ cells in dermatomyositis challenges the ischemic theory and suggests an innate immune response (Greenberg et al., 2005).

The classic dermatomyositis phenotype consists of characteristic cutaneous manifestations, muscle weakness, and pathologic evidence of typical dermatomyositis features. The clinical presentation is variable and represents a spectrum of cutaneous, myopathic, and systemic involvement.

Cutaneous manifestations

Skin manifestations may be the only initial clinical manifestation (Rockerbie et al., 1989), and most commonly, they precede onset of muscle weakness (Dalakas and Hohlfeld, 2003). Typical skin findings help differentiate dermatomyositis from other inflammatory myopathies. Characteristic skin lesions of dermatomyositis include pink-violet papules and plaques arranged in a symmetric distribution. The lesions range from a mild erythema to a marked poikiloderma with associated telangiectasia, hypopigmentation and/or hyperpigmentation, and superficial atrophy. Besides other characteristic locations, the violaceous hue and associated pruritus are distinguishing characteristics from the cutaneous manifestations of systemic lupus erythematosus (SLE) (Jorizzo and Vleugels, 2012). A brief clinical and histopathologic discussion of the most common cutaneous manifestations of dermatomyositis is summarized below.

Heliotrope rash

A heliotrope rash occurs in 30–60% of patients with dermatomyositis (Kovacs and Kovacs, 1998). It is characterized by pink-violet papules and plaques in a symmetric periorbital distribution with extension to the lateral nasal root and cheeks (Fig. 3.2) (Callen, 1995). It can be transient and is frequently associated with periorbital edema (Dicken, 1991). Differential diagnosis includes SLE, contact or allergic dermatitis, seborrheic dermatitis, scleroderma, photosensitivity reactions, and trichinosis.

Grotton's papules

Grotton's papules present as a pink-violet lichenified papules and plaques in a symmetric distribution over the dorsal interphalangeal joints, sparing the interphalangeal spaces, and are present in 70% of patients with dermatomyositis (Kovacs and Kovacs, 1998) (Fig. 3.3). When they involve the extensor surfaces of the elbows, knees, and knuckles without associated lichenification, they are termed Grotton's sign (Jorizzo and Vleugels, 2012). Differential diagnoses include SLE, psoriasis, atopic dermatitis, allergic or contact dermatitis, cutaneous T-cell lymphoma, and scleroderma.

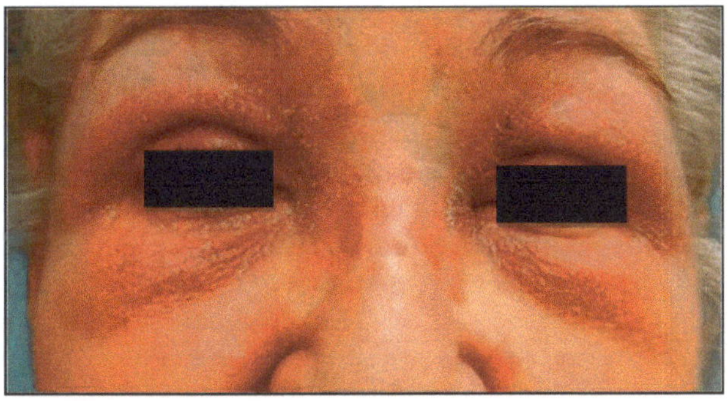

Fig. 3.2. Heliotrope rash in a patient with dermatomyositis. Note the extensive periorbital edema. (Courtesy of Dr. Anthony Peterson, Loyola University Medical Center, Department of Dermatology, Maywood, IL, USA.)

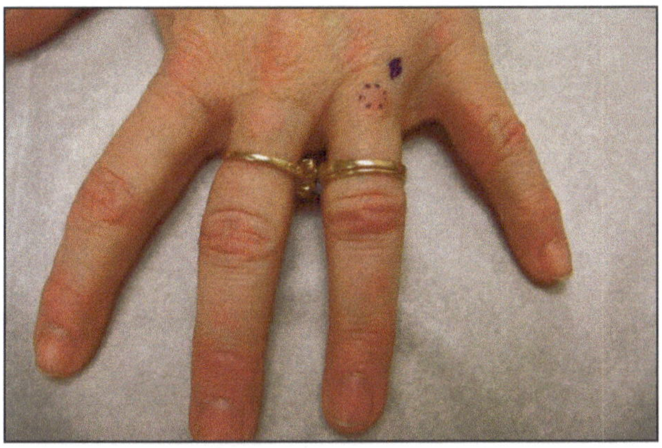

Fig. 3.3. Grotton's pink-violet papules and plaques occurring in a symmetrical distribution over the interphalangeal joints, sparing the interphalangeal spaces. (Courtesy of Dr. Jane Scribner, University of Illinois at Chicago, Department of Dermatology, Chicago, IL, USA.)

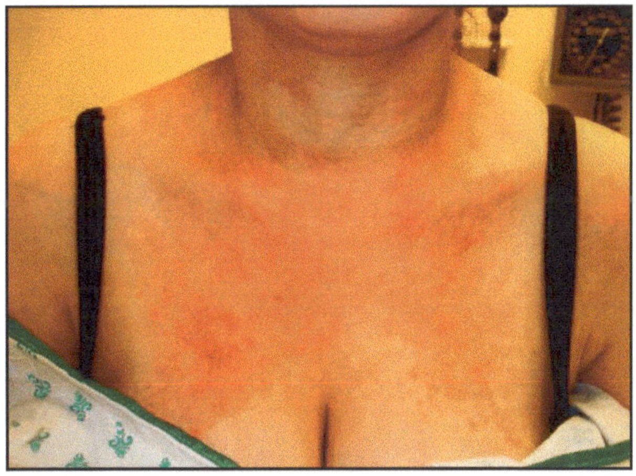

Fig. 3.4. Shawl sign consisting of pink-violet papules and plaques in the "V"-shaped area of the upper chest. (Courtesy of Dr. Iris Aronson, University of Illinois at Chicago, Department of Dermatology, Chicago, IL, USA.)

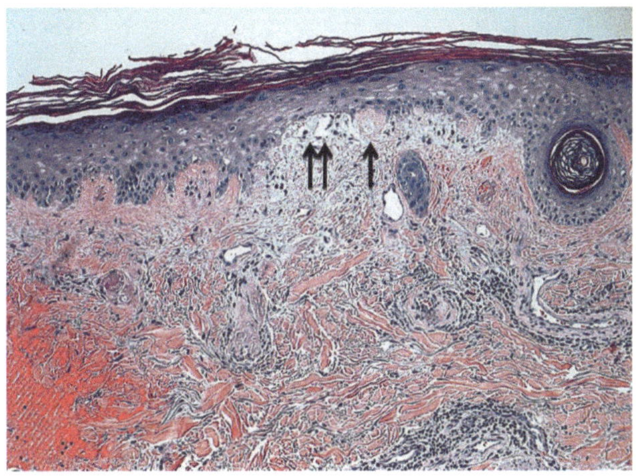

Fig. 3.5. (10 × objective). There is a very mild interface dermatitis with epidermal atrophy, basal cell vacuolization (double arrow), basement membrane thickening (single arrow), dermal mucin accumulation, and a mild superficial perivascular lymphocytic infiltrate. (Courtesy of Dr. Kelli A. Hutchens, Loyola University Medical Center, Department of Pathology, Maywood, IL, USA.)

Photodistributed poikiloderma ("shawl sign")

Photodistributed poikiloderma, colloquially known as the "shawl sign," presents as pink-violet papules and plaques with a symmetric distribution involving the shoulders, arms, and the "V" area of the upper chest and back (Fig. 3.4) (Sontheimer, 1999). Photodistributed poikiloderma typically worsens with exposure to ultraviolet light. Differential diagnosis includes SLE, allergic or contact dermatitis, photosensitivity reaction of any etiology, cutaneous T-cell lymphoma, and scleroderma.

Histopathology

All cutaneous manifestations of dermatomyositis exhibit similar histologic findings that can, at times, be indistinguishable from SLE. Dermatomyositis is classified as mild interface dermatitis with variable associated histologic findings. Lesions of dermatomyositis can include epidermal atrophy with effacement of rete ridges, vacuolar changes of basal keratinocytes, basement membrane thickening, dermal mucin accumulation, and mild to moderate dermal mononuclear inflammatory cell infiltrates in the superficial dermis, sparing the cutaneous adnexae (Fig. 3.5) (Luzar and Calonje, 2012). Direct immunofluorescence demonstrates a granular deposit of IgG, IgM, IgA, and complement at the dermal–epidermal junction in one-third of patients (Fig. 3.6) (Chen et al., 1985).

Other cutaneous manifestations

Other cutaneous findings suggestive of dermatomyositis include cuticular dystrophy, skin calcifications, and

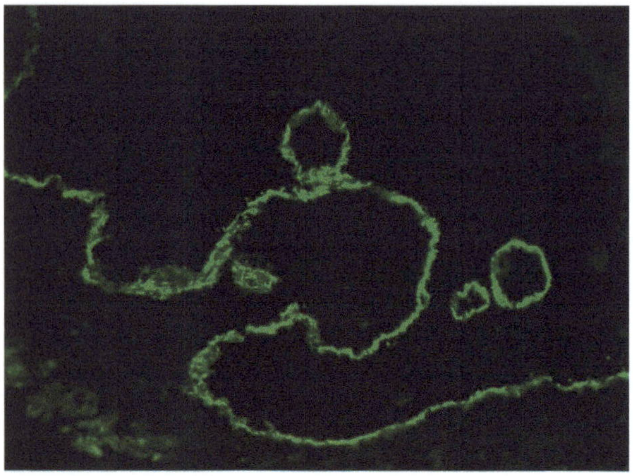

Fig. 3.6. A granular deposit of IgG at the basement membrane on direct immunofluorescence. (Courtesy of Dr. Maria Picken, Loyola University Medical Center, Department of Pathology, Maywood IL, USA.)

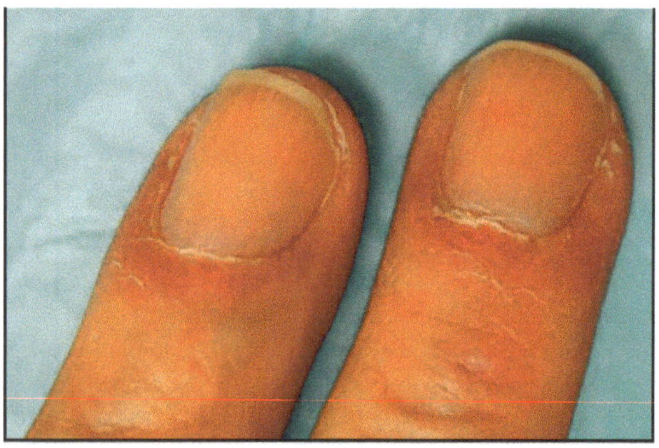

Fig. 3.7. Nail fold telangiectasia. Note the dilated capillary loops and erythema of the nail fold. (Courtesy of Dr. Anthony Peterson, Loyola University Medical Center, Department of Dermatology, Maywood, IL, USA.)

nail fold telangiectasias (Fig. 3.7). Additionally, up to 70% of juvenile dermatomyositis cases and approximately 10% of adult dermatomyositis patients present with calcinosis cutis, usually involving the buttocks, elbows, knees, or traumatized areas (Bowyer et al., 1983; Orlow et al., 1991). Other features such as seborrheic dermatitis, vasculitis, lichen planus-like papules, livedo reticularis, and "mechanic's hand" appearance (due to roughening and cracking of the lateral aspect of the palm and finger creases secondary to hyperkeratosis) are sometimes observed (Parodi et al., 2002).

Dermatomyositis may also present with erythematous scaling hyperkeratotic papules and plaques involving the trunk and limbs that histopathologically resemble pityriasis rubra pilaris. This is known as Wong's dermatomyositis (Wong, 1969), and its most frequent histopathologic feature is characterized by follicular hyperkeratosis with keratotic plugs filling dilated follicular infundibula (Haro et al., 2013).

Cutaneous symptoms of dermatomyositis have a higher impact on the quality of life than psoriasis or atopic dermatitis; with pruritus being a key and potentially treatable factor (Hundley et al., 2006).

Other manifestations

Beyond the characteristic cutaneous findings, subacute onset of proximal greater than distal symmetric limb and girdle muscle weakness, manifested by difficulty rising from a seated position, climbing steps, lifting objects, or combing hair, is common. The severity of muscle weakness varies. Distal muscle weakness can occur late, in contrast to inclusion body myositis which usually involves finger flexors early in the course of the disease. Neck extensor muscle weakness may cause head drop. Facial muscles are typically spared and the extraocular muscles are not affected (Dalakas, 1991; Dalakas and Hohlfeld, 2003). Oropharyngeal weakness resulting in dysphagia occurs in up to 20% of patients (Parodi et al., 2002). In advanced cases, breathing difficulty secondary to respiratory muscle weakness can occur. Muscle stretch reflexes are usually preserved and sensation is spared (Dalakas, 1991; Dalakas and Hohlfeld, 2003). Muscle stretch hyporeflexia or areflexia suggests severe and advanced disease with significant muscle atrophy.

Juvenile dermatomyositis is the most common inflammatory myopathy of childhood with 25% of cases occurring among individuals younger than 4 years of age (Feldman et al., 2008). Irritability, fatigue, decreased sociability, and red flushed facies, also referred to as "misery," are common early abnormalities in juvenile dermatomyositis (Dalakas, 1991; Dalakas and Hohlfeld, 2003). Juvenile dermatomyositis is similar to the adult form of dermatomyositis with more frequent systemic involvement (Dalakas and Hohlfeld, 2003). In contrast to adults, dysphagia or dysphonia is present in approximately 40% of children with juvenile dermatomyositis (Pachman et al., 1998). Muscle weakness is present in approximately 95%, myalgia or arthralgia in up to two-thirds, and one-fourth of children develop contractures. In addition, muscle weakness is both proximal and distal, with frequent muscle atrophy (Feldman et al., 2008). Development of "tip-toe gait" due to ankle contractures is common (Dalakas, 1991).

Atypical presentations

In some instances, classic cutaneous manifestations may not be accompanied by objective clinical weakness or laboratory evidence of muscle breakdown. Patients may have skin manifestations persisting for months to years before developing weakness, or not develop weakness at all; this is referred to as *amyotrophic dermatomyositis* or *dermatomyositis sine myositis* (Euwer and Sontheimer, 1991; Gerami et al., 2006). In contrast, dermatomyositis with classic skin manifestations and laboratory, electromyographic or pathologic evidence of myositis without clinical weakness has been termed *hypomyopathic dermatomyositis*) (Gerami et al., 2006). The term *clinically amyopathic dermatomyositis (CADM)* has been used to describe all these subtypes, estimated to represent approximately 5–20% of all dermatomyositis (Sontheimer, 2002; el-Azhary and Pakzad, 2002; Bendewald et al., 2010).

A systematic review evaluating 291 individuals with CADM found that 13% of patients developed muscle weakness between 15 months and 6 years from the onset of skin disease. The average duration of skin manifestations was approximately 3.7 years (Gerami et al., 2006).

A study conducted at dermatology and rheumatology departments at a tertiary care center found a higher incidence of hypomyopathic dermatomyositis or amyotrophic dermatomyositis in the dermatology as compared to the rheumatology department (28% hypomyopathic dermatomyositis, and 40% amyotrophic dermatomyositis versus 7% HDAM and 1% amyotrophic dermatomyositis respectively) (Klein et al., 2007). Although a higher percentage of patients were seen by dermatology alone as compared to rheumatology (65% versus 29%), these observations highlight the need for careful and skilled dermatologic evaluation to properly identify dermatomyositis suspects in the absence of myopathy.

More rarely, individuals with myopathic features may have no obvious cutaneous lesions and clinically be diagnosed with polymyositis until biopsy findings prove otherwise. The term *dermatomyositis sine dermatitis* has been used to describe these cases (Leteurtre et al., 1994).

Myositis-specific antibodies have been associated with particular dermatomyositis phenotypes. For example, autoantibodies against the histidyl-tRNAsynthetase (also known as anti-Jo-1) associated with myositis, interstitial lung disease, non-erosive arthritis, fever, Raynaud's phenomenon and "mechanic's hands" is known as the antisynthetase syndrome and can be present in both polymyositis and dermatomyositis (Mammen, 2010). Anti-Jo-1 is present in about 80% of patients with antisynthetase syndrome (AnSS) (Dalakas and Hohlfeld, 2003).

Other clinical features

Generalized fatigue is common. Myalgias are present in one-third of affected adults (Dalakas and Hohlfeld, 2003) and arthralgias in up to 40% (Parodi et al., 2002). Cardiac abnormalities are relatively uncommon. The most common abnormalities include subclinical arrhythmias (32.5–72%) or congestive heart failure (3–45%) due to myocarditis, coronary artery disease or myocardial small vessel involvement (Lundberg, 2006). Gastrointestinal symptoms such as ulceration, hemorrhage, or perforation secondary to intestinal vasculopathy, particularly in children, has decreased after the use of immunosuppressive therapy (Feldman et al., 2008). Less frequently, oral manifestations including mucosal ulcers, white patches, and gingival telangiectasias can occur (Tanaka and Geist, 2012). Periungual and capillary nail bed changes are present in > 90% of juvenile dermatomyositis cases (Feldman et al., 2008). Association with pulmonary disease and malignancy is discussed in more detail below.

Pulmonary disease

Pulmonary involvement is common and includes complications of respiratory muscle weakness such as aspiration pneumonia, interstitial lung disease, and malignancies. Interstitial lung disease is characterized by restrictive lung changes with reduced forced vital capacity and decreased diffusion capacity of the lungs for carbon monoxide (DLCO) and is present in 8–20% of patients (Parodi et al., 2002; Gerami et al., 2006; Chen et al., 2009). Interstitial lung disease associated with dermatomyositis most commonly takes the histopathologic and radiographic from of nonspecific interstitial pneumonia (NSIP), with irregular linear opacities with areas of consolidation and ground glass appearance on chest computed tomography (CT). Other forms of interstitial lung disease include interstitial pneumonia with diffuse alveolar damage, bronchiolitis obliterans, organizing pneumonia or idiopathic pulmonary fibrosis. The nonspecific interstitial pneumonia form of interstitial lung disease has a better response to treatment and prognosis as compared to idiopathic pulmonary fibrosis (Douglas et al., 2001).

The most common muscle-specific antibody associated with interstitial lung disease is the anti-Jo-1 antibody (Mammen, 2010). Anti-Jo-1 is present in more than a third of dermatomyositis patients with interstitial lung disease (Douglas et al., 2001; Mammen, 2010). The presence of the anti-Jo-1 antibody, arthralgia, and older age have been reported to be associated with the presence of interstitial lung disease in dermatomyositis (Chen et al., 2009). The presence or absence of Jo-1 antibodies has not been reported to correlate with survival, therefore its prognostic value in this setting is not known (Douglas et al., 2001).

Malignancies

Although all inflammatory myopathies have been associated with some form of cancer, the frequency of malignancy with dermatomyositis is highest (Dalakas and Hohlfeld, 2003). The estimated incidence of malignancy in dermatomyositis is between 10% and 30% (Hill et al., 2001; Parodi et al., 2002; Chen et al., 2010). Geographic variation has shown variability in cancer subtype frequency. Malignancies associated with dermatomyositis in the Scandinavian population are most frequently adenocarcinomas. In particular, ovarian, lung, gastric, colorectal, pancreatic and non-Hodgkin's lymphoma were most common (Hill et al., 2001). In Asian countries, cancers of the nasopharynx, lung, breast, liver and gallbladder, colon, rectum and hematologic malignances were most common. Most malignancies present within the first year of dermatomyositis diagnosis (Hill et al., 2001; Chen et al., 2010). Unlike in adults, dermatomyositis malignancies are unusual in juvenile dermatomyositis (Feldman et al., 2008).

Anti-155/140 antibodies (myositis-specific autoantibodies recognizing 155/140 KDA proteins) have been recently identified in dermatomyositis (Mammen, 2010). These antibodies have a high specificity for dermatomyositis and often correlate with malignancy-associated dermatomyositis (Kaji et al., 2007).

Overlap syndrome

Dermatomyositis and other inflammatory myopathies have been associated with other systemic diseases such as SLE, scleroderma, or inflammatory bowel diseases. In contrast with other inflammatory myopathies, dermatomyositis is the only one that truly overlaps with systemic sclerosis and mixed connective tissue diseases, sharing similar findings such as sclerotic skin changes, microangiopathy, and esophageal hypomotility (Dalakas and Hohlfeld, 2003). In children, overlap myopathies account for up to 10% of all inflammatory myopathies and feature milder forms of myositis, lower creatinine kinase levels, and better response to treatment (Feldman et al., 2008).

Diagnosis and investigation

Proximal limb weakness, stereotypical cutaneous manifestations, elevated muscle enzymes, myopathic features on electromyography, and characteristic histopathologic findings on muscle biopsy represent the 1975 diagnostic criteria proposed by Bohan and Peter, which are still widely used (Bohan and Peter, 1975). However, as previously discussed, not all patients have typical presentation. In addition, criteria for "probable" or "possible" dermatomyositis or polymyositis did not require muscle biopsy (Bohan and Peter, 1975). These criteria have recently been criticized as recent recognition of other subtypes (e.g., CADM), better pathologic characterization, identification of specific and nonspecific myositis-specific antibodies, and use of other diagnostic modalities led to the proposal of new diagnostic criteria emphasizing immunopathologic features (Table 3.4) (Dalakas and Hohlfeld, 2003).

A recent study found these proposed criteria to have a high a sensitivity and specificity (77% and 99%, respectively) compared to the European Neuromuscular Centre criteria (sensitivity 71%, specificity 82%) or the Bohan and Peter criteria (specificity of 29%) (Linklater et al., 2013).

Table 3.4

Dalakas and Hohlfeld diagnostic criteria of dermatomyositis

	Myopathic dermatomyositis		Amyopathic dermatomyositis
Criterion	Definite	Probable	Definite
Myopathic muscle weakness	Yes*	Yes*	No†
Electromyographic findings	Myopathic	Myopathic	Myopathic or nonspecific
Muscle enzymes	High (up to 50 times normal) or normal	High	High (up to 10 times normal) or normal
Muscle biopsy findings	Perifascicular, perimysial or perivascular infiltrates; perifascicular atrophy	Perifascicular, perimysial or perivascular infiltrates; perifascicular atrophy	Nonspecifc or diagnostic for dermatomyositis
Rash or calcinosis	Present	Not detected	Present

*Myopathic muscle weakness, affecting proximal muscles > distal muscles and sparing ocular and facial muscles, is characterized by a subacute onset and rapid progression in patients without family history of neuromuscular disease, exposure to myotoxic drugs or toxins, or signs of biochemical muscle disease.

†Although strength is apparently normal, many patients have new onset of easy fatiguability, myalgia, and reduced endurance. Careful muscle testing may show mild muscle weakness.

Evaluation

For dermatomyositis patients with pulmonary complications, baseline pulmonary function tests and spiral chest CT may be useful. Screening for dysphagia to minimize the risk of aspiration is useful. Identification of myositis-specific antibody phenotypes may help identify individuals with fulminant cutaneous manifestations or those who respond better to treatment. Evaluation of cardiomyopathy or arrhythmia in high-risk patients is prudent. Early identification and treatment of any underlying malignancy is important.

In children, electromyography (EMG) and muscle biopsy are less commonly used, increasing the likelihood of misdiagnosis. An international consensus survey of physicians found that, in addition to proximal muscle weakness, characteristic skin changes and elevated muscle activity enzymes, nail fold capillaroscopy, muscle MRI, and calcinosis were equally as useful as muscle biopsy and EMG findings (Brown et al., 2006).

Laboratory findings

Serum creatine kinase (CK) is the most sensitive assay and may be increased up to 50-fold in individuals with active disease. However, CK levels may be normal in up to 30% of patients and do not always correlate with the severity of muscle weakness (Amato and Barohn, 2009). Serum aldolase, lactate dehydrogenase (LDH), alanine (ALT), and aspartate aminotransferase (AST) and myoglobin may be elevated. Positive ANA titers, although frequent, are nonspecific and only considered "somewhat useful" in the diagnosis of dermatomyositis (Solomon et al., 2002). Myositis-specific antibodies have been associated with specific myositis phenotypes such as the AnSS where the anti-Jo-1 antibody is most common. Other less common anti-aminoacyl-tRNA synthetase antibodies such as the anti-PL-7 have been strongly associated with interstitial lung disease (Hervier et al., 2011). Interestingly, patients with myositis-specific antibodies typically express a single type of autoantibody (Mammen, 2010) and rarely may have more than one (Koreeda et al., 2010). Table 3.5 summarizes serum testing in dermatomyositis.

Electromyographic findings

The EMG typically exhibits a "myopathic" pattern. Findings include increased spontaneous activity, positive sharp waves, complex repetitive discharges, short duration, low amplitude polyphasic motor unit action potentials (MUAPs), and early recruitment. In chronic stages, a mix of small and large MUAPs can be seen (Szmidt-Salkowska et al., 1989; Paganoni and Amato 2013).

Table 3.5

Serum testing in dermatomyositis

CK	Elevated in 70% of patients; up to 50-fold increase in active disease
Aldolase, LDH, AST, ALT	May be elevated; uncertain value in diagnosis of dermatomyositis
ANA	Elevated in 30–60%
ESR and RF	Nonspecific; may be present in overlap syndromes; does not correlate with disease activity
Anti-aminoacyl-tRNA synthetase	(Koreeda et al., 2010; Hervier et al., 2011)
Anti-Jo-1	Anti-Jo-1 positive in one-third of patients; associated with interstitial lung disease and Raynaud's; positive in 80% of patients with antisynthetase syndrome; may correlate with disease activity
Anti-PL-7	Present in < 5%; associated with interstitial lung disease and overlap syndrome
Anti-PL-12	Present in about 20%; strongly associated with interstitial lung disease
Anti-EJ	Present in < 4%; associated with interstitial lung disease and AnSS
Anti-OJ	Present in < 4%; associated with interstitial lung disease and AnSS
Anti-KS	Present in < 4%; associated with interstitial lung disease and AnSS
Anti-Zo	Present in < 4%; associated with interstitial lung disease and ASS
Anti-Mi-2	Positive in 20–30%; associated with fulminant cutaneous manifestations; lower incidence of malignancy and better response to corticosteroid therapy
Anti-SRP antibodies	Positive in 4% of polymoyositis/dermatomyositis; associated with rapidly progressive severe weakness, high CK, myocarditis, and "necrotizing features" on biopsy with absent inflammation and abundant muscle regeneration
Anti-KL6	Associated with interstitial lung disease
Anti-polymyositis-Scl	Present in 15% of dermatomyositis with scleroderma
Anti-U-RNP, Anti-RO and Anti-Ku	Present in some overlap syndromes

MRI of skeletal muscles

In both dermatomyositis and polymyositis, MRI of skeletal muscles may show T2 hyperintensities in affected muscles which may represent inflammation, edema, or fatty/fibrous replacement (Mammen, 2010). Abnormal signal intensity on MRI has been shown to correlate with disease activity, decrease with treatment, and useful in muscle selection for biopsy (Studýnková et al., 2007). This noninvasive modality may aid in diagnosis, especially in children, where invasive modalities may be difficult (Brown et al., 2006; Feldman et al., 2008).

Muscle biopsy

Muscle changes of dermatomyositis are unique and help distinguish dermatomyositis from other inflammatory myopathies. The histopathologic hallmark of dermatomyositis is the combination of perifascicular atrophy with perivascular, perimysial, and endomysial inflammation. Perifascicular atrophy is quite evident in dermatomyositis (Fig. 3.8). Perimysial, endomysial, and perivascular inflammation consist predominantly of lymphocytes, plasma cells, and histiocytes (Fig. 3.9) (Dubowitz et al., 2013).

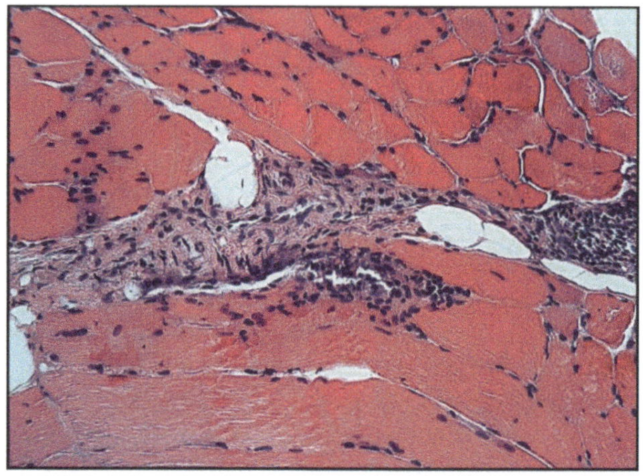

Fig. 3.9. (20 × objective). Muscle biopsy from a patient with dermatomyositis demonstrating perimysial and perivascular inflammation.

Management

Management of cutaneous manifestations consists of photoprotection, topical corticosteroids, and antimalarial agents. Methotrexate is considered second line followed by mycophenolate mofetil or intravenous immunoglobulin (IVIG). Femia and colleagues have summarized the topical and systemic therapies for cutaneous dermatomyositis (Femia et al., 2013).

The mainstay of treatment is immunosuppressive therapy. Corticosteroids is a first-line empiric therapy for inflammatory myopathies, and dermatomyositis has been found to be most responsive. Prednisone doses of 0.75–1.5 mg/kg (up to 100 mg) for 2–4 weeks, titrated over several weeks to alternate day dosing based on disease severity, has been recommended. Adjusting steroid dosing based on CK levels is not recommended; adjustments should be based on objective findings (Dalakas and Hohlfeld, 2003; Amato and Barohn, 2009; Dalakas, 2010).

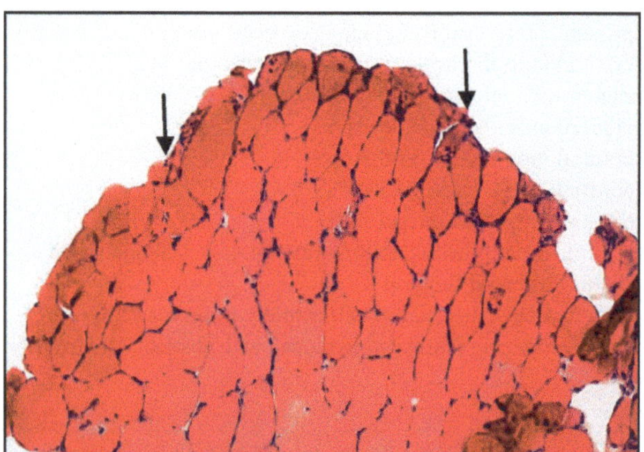

Fig. 3.8. (10 × objective). Perifascicular atrophy of muscle fibers in dermatomyositis (examples highlighted by arrows).

High dose IVIG (2 g/kg monthly) as a second-line therapy in conjunction with prednisone has also been shown to improve muscle weakness (Dalakas et al., 1993). IVIG may be useful in pregnant patients (Mosca et al., 2005; Linardaki et al., 2011). Plasmapheresis has not been shown to be efficacious in controlled trials (Miller et al., 1992). Steroid-sparing agents including methotrexate, azathioprine, mycophenolate mofetil, ciclosporin, cyclophosphamide, rituximab and tumor necrosis factor α blockers have been used. In children, methotrexate can lower steroid requirement by 50% with good efficacy and is used as first-line therapy by some physicians (Feldman et al., 2008). The choice of agent depends on the physician's experience, the side-effect profile of each drug, and patient tolerance. A baseline dose of prednisone is usually required throughout the course of treatment, so monitoring for steroid-related side-effects is prudent.

SJÖGREN SYNDROME

Sjögren syndrome, also known as sicca syndrome, is a chronic systemic autoimmune disease characterized by lymphocytic exocrine glandular infiltration primarily manifesting as dryness of the mouth and eyes with parotid gland enlargement. Sjögren syndrome may present as a primary disease, progressive systemic sclerosis, or in association with other connective tissue disease, secondary Sjögren syndrome (Kassan and Moutsopoulos, 2004; Kivity et al., 2014). The most common cutaneous features of Sjögren syndrome include xerosis and vasculitis (Kassan and Moutsopoulos, 2004).

Beyond xerostomia, xerophthalmia, and keratoconjunctivitis, systemic manifestations reported in over half of patients with Sjögren syndrome include fatigue, musculoskeletal pain, esophageal dysmotility, and hypothyroidism. The most serious concern is an increased risk of malignant lymphoproliferative disease (Kassan and Moutsopoulos, 2004; Peri et al., 2012). Approximately one-third of patients with Sjögren syndrome present with extraglandular systemic manifestations (Vitali et al., 2002). Women between the fifth and sixth decades are more commonly affected, with a reported disease prevalence of 1–3% (Peri et al., 2012).

The pathogenesis of Sjögren syndrome is not fully understood. Recent studies suggest genetically susceptible individuals with strong association to the HLA loci and subclinical immune dysregulation may develop Sjögren syndrome secondary to external immune stimuli including infections, physical stress, and hormone imbalance with autoantibody production (Kivity et al., 2014).

Xerosis

Xerosis, the most common cutaneous finding in Sjögren syndrome (Bernacchi et al., 2004), presents as pruritic, dry, dull skin with a fine white "bran-like" scale most frequently involving the shins, lower flanks, and posterior axillary lines (Fig. 3.10) (Reider and Fritsch, 2012). Histopathologically, xerosis presents as a mildly acanthotic epidermis, overlying compact stratum corneum with parakeratosis and underlying dermis with a mild superficial perivascular lymphocytic infiltrate (Ramdial and Grayson, 2012).

Vasculitis

Vasculitis is a common cutaneous finding in Sjögren syndrome, reported in 5–55% of patients and associated with a greater degree of disease severity (Ramos-Casals et al., 2004; Feist et al., 2009). Clinical presentations include palpable purpura, pigmented purpura, macular hemorrhages, and urticarial lesions. Histopathologically, the vasculitis associated with Sjögren syndrome demonstrates of small vessel leukocytoclastic vasculitis and acute necrotizing vasculitis of medium caliber vessels (Tsokos et al., 1987).

Annular erythema

Erythema multiforme-like annular erythema, a well-documented cutaneous manifestation of Sjögren syndrome (Katayama et al., 1987), is characterized by fixed red papules that can evolve into erythematous plaques with well-defined bright red borders and central dusky discoloration (French and Prins, 2012). Associated histopathologic changes are not well documented.

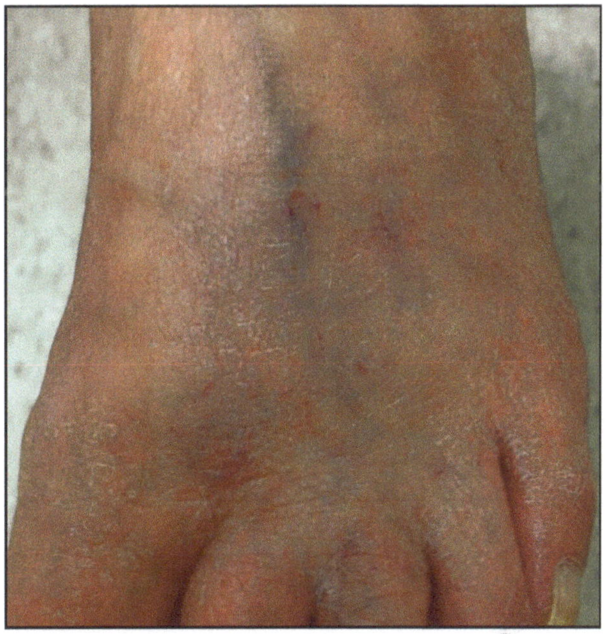

Fig. 3.10. Xerosis on a lower extremity. Note the fine white "bran-like" scale.

Neurologic manifestations

Neurologic involvement in Sjögren syndrome has been reported in 2–67.5% of patients (Delalande et al., 2004; Gøransson et al., 2006; Pavlakis et al., 2011; Morreale et al., 2014). The lack of standardized definition of neurologic involvement accounts for the wide prevalence range. True neurologic sequelae typically occur with primary Sjögren syndrome and may be the first symptoms in over half of patients, or follow disease onset by several years (Delalande et al., 2004). Neurologic manifestations are more common in women in the third or fourth decades of life after average disease duration of 9–12 years (Gøransson et al., 2006; Morreale et al., 2014).

Central nervous system

CNS manifestations are either focal, multifocal, or diffuse in nature and may affect the entire neuroaxis with a relapsing progressive course (Alexander, 1985). Most commonly, the spinal cord, brain, and optic nerves are affected. However, diffuse CNS manifestations are more common than focal involvement (Morreale et al., 2014).

Headaches are one of the most frequent manifestations (Escudero et al., 1994; Morreale et al., 2014). Migraines without aura and chronic tension-type headaches are most frequently reported. Although chronic tension-type headaches are more prevalent among Sjögren syndrome patients as compared to the normal population, they have not been associated with primary Sjögren syndrome-related autoantibodies, MRI or CSF abnormalities, depression or fatigue (Tjensvoll et al., 2013). Conversely, migraine without aura has been related to SSA antibodies and MR spectroscopy abnormalities (Morreale et al., 2014). In addition, late-onset "migraine-like" episodes with prolonged sensorimotor deficits and coexisting neuropsychiatric disease may occur (Escudero et al., 1994). In general, as compared to the normal population, no statistically significant differences in headache prevalence among Sjögren syndrome patients has been observed (Tjensvoll et al., 2013).

Focal brain involvement resulting in hemiparesis, aphasia, dysarthria, cerebellar or brainstem dysfunction, and chorea may present acutely or in a relapsing remitting fashion. In addition, patients with relapsing remitting clinical course often have no sicca symptoms, leading to significant diagnostic delays (Delalande et al., 2004). Partial seizures, generalized seizures, and epilepsia partialis continua have been documented in patients with extensive CNS involvement (Bansal et al., 1987; Delalande et al., 2004).

Acute and chronic myelopathies account for half of all CNS manifestations of Sjögren syndrome, often resembling transverse myelitis, longitudinally extensive transverse myelitis (LETM), progressive myelitis, or Brown-Séquard syndrome (Delalande et al., 2004; Rogers et al., 2004). Over half of these patients have a chronic course, mimicking primary progressive multiple sclerosis, and involving mainly the cervical spinal cord (Delalande et al., 2004; Rogers et al., 2004). Optic nerve involvement may occur in isolation and is rarely the first manifestation of Sjögren syndrome (Kadota et al., 2002). In addition, optic neuritis in association with spinal cord involvement may resemble neuromyelitis optica (NMO) or NMO spectrum disorders (Delalande et al., 2004; Min et al., 2009). Aquaporin-4 antibodies may be present in Sjögren syndrome associated myelopathies, and lesions are usually associated with increased vasogenic edema (Min et al., 2009). Pontine involvement resembling central pontine myelinosis has also been observed (Yoon et al., 2000). Although infrequent, motor neuron disease manifestations have also been described in Sjögren syndrome patients (Lafitte et al., 2001; Delalande et al., 2004). Unusual manifestations include transient monocular blindness, spinal subarachnoid hemorrhage secondary to anterior spinal artery necrotizing vasculitis, meningitis, and meningoencephalitis (Alexander et al., 1981b; Alexander and Alexander, 1983; Rossi and Saddi, 2006; Gerraty et al., 1993).

Neuropsychiatric dysfunction includes primarily affective disturbances and cognitive dysfunction (Malinow et al., 1985; Delalande et al., 2004). Anxiety and depression have significant effects on overall well-being, quality of life, and increased occurrence of headaches, irritability, low mood, impaired memory and concentration (Valtýsdóttir et al., 2000). A psychosomatic pattern of decreased cognitive capacity (Mauch et al., 1994) and cognitive impairment (more commonly a subcortical dysfunctional pattern) not attributable to mood disturbances may occur (Lafitte et al., 2001). Patients often report increased fatigue, depression, impaired physical functioning, and memory dysfunction in the form of loss of thought in the midst of an activity. However, minimal memory, cognitive and psychomotor functioning impairments have been observed on objective testing measures (Epstein et al., 2014). Although the mechanisms behind cognitive dysfunction and memory impairment in Sjögren syndrome are not fully understood, cortical atrophy, smaller hippocampal volume, and elevated *N*-methyl-D-aspartate receptor subtype NR2 (anti-NR2) levels in serum and CSF have been observed (Mauch et al., 1994; Lauvsnes et al., 2013).

PERIPHERAL NERVOUS SYSTEM, AUTONOMIC NERVOUS SYSTEM, NEUROMUSCULAR

Peripheral nervous system involvement occurs in 20–27% of patients with Sjögren syndrome and is mainly characterized by symmetric sensorimotor peripheral neuropathy followed by sensory neuronopathy (ganglionopathy) (Mellgren et al., 1989; Gemignani et al., 1994; Gøransson et al., 2006), mononeuritis multiplex, cranial neuropathies (particularly trigeminal), small fiber neuropathy, radiculopathy, and muscle disease (Gemignani et al., 1994; Delalande et al., 2004; Mori et al., 2005). Neuropathies are typically associated with burning dysesthesias, which may be related to small fiber involvement. Not uncommonly, neuropathy is the presenting symptom of Sjögren syndrome (Mellgren et al., 1989; Gemignani et al., 1994; Mori et al, 2005).

Mori and colleagues classified Sjögren syndrome-associated neuropathies into sensory ataxic, painful sensory neuropathy without sensory ataxia, multiple mononeuropathy, multiple cranial neuropathy, trigeminal neuropathy, autonomic neuropathy, and radiculoneuropathy. Patients with multiple or cranial neuropathies may have an acute to subacute onset, while those with predominantly sensory symptoms often have a chronic progressive course (Mori et al., 2005). Symptoms of small fiber neuropathy include burning dysesthesias, paresthesias, numbness, and allodynia. Often, patients exhibit dysautonomic features in the form of vasomotor symptoms or hyperhidrosis. Patients with small fiber neuropathy tend to be older, have higher rates of xerostomia and arthralgias, and lower prevalence of serum B-cell activation markers (Sène et al., 2013).

The pathogenesis of the neuropathy is mainly due to neuronal vascular or perivascular inflammatory infiltrates, with or without necrosis (Mellgren et al., 1989; Griffin et al., 1990; Delalande et al., 2004; Mori et al., 2005), although alteration of endoneurial small vessels has also been observed (Gemignani et al., 1994). Neuropathies are predominantly axonal and associated with systemic involvement, extraglandular manifestations, such as palpable purpura or vasculitis, and development of lymphoma (Pavlakis et al., 2011). Neuronopathies have been attributed to T-cell lymphocytic infiltration of the dorsal root ganglia (Griffin et al., 1990; Mori et al., 2005).

Autonomic features include both sympathetic and parasympathetic dysfunction. Adie's tonic pupil, orthostatic hypotension, and fixed tachycardia have been reported (Mellgren et al., 1989; Griffin et al., 1990; Mori et al., 2005). Delayed gastric emptying and decreased detrusor muscle tone and contractility are believed to be in part due to autoantibodies reacting with the M3 muscarinic acetylcholine receptor (Kovacs et al., 2002). Of note, autonomic symptoms may have limited associations with other clinical features of Sjögren syndrome and objective autonomic dysfunction parameters (Mandl et al., 2008).

Although muscle involvement (myalgias or mild inflammation) is documented in up to a fourth of patients (Peri et al., 2012), true myositis is unusual and reported in only 3% of patients with Sjögren syndrome (Kraus et al., 1994). Inflammatory myopathies as part of an "overlap syndrome" with interstitial lung disease, or inclusion body myositis with typical clinical features and muscle enzyme level elevations may also occur (Kanellopoulos et al., 2002; Koga et al., 2012).

Clinical evaluation

Meticulous attention is required to exclude other causes of neurologic disease related to treatment or in association with other autoimmune disease such as "overlap syndromes" or SLE. Neurologic manifestations attributed to Sjögren syndrome are typically in those with primary Sjögren syndrome. In patients with suspected Sjögren syndrome, it is vital to rule out other causes of sicca symptoms (e.g., anticholinergic agents, hepatitis C, lymphoma, sarcoidosis, graft-versus-host disease, or prior head and neck radiation) or entities resembling Sjögren syndrome such as IgG4-related systemic disease (Peri et al., 2012). Moreover, the American European Consensus Group revised criteria for classification of Sjögren syndrome requires the presence of ocular and oral symptoms, histopathology, and the presence of autoantibodies. To fulfill the diagnosis, patients must have four of six criteria and either positive autoantibodies or a salivary gland biopsy (Table 3.6) (Vitali et al., 2002).

However, other autoimmune diseases, such as SLE, or infections may share a similar clinical picture and similar neurologic sequelae. For example, patients with HTLV-1 myelopathies may have similar findings of keratoconjunctivitis sicca and salivary gland biopsies consistent with Sjögren syndrome (Williams et al., 2001).

Laboratory investigations

Antinuclear antibodies (ANA) are detected in 40–70% of patients with Sjögren syndrome (Kavanaugh et al., 2000). Antibodies directed against ribonucleoproteins, anti Ro/SSA antibodies and anti La/SSB antibodies are present in 40–90% and in 20–50% of primary Sjögren syndrome patients respectively, but in only 10–50% cases of secondary Sjögren syndrome (Birtane, 2012). However, these antibodies are nonspecific for Sjögren syndrome, and may be present in other autoimmune disease such as SLE (Agmon-Levin et al., 2011). Anti Ro/SSA antibodies are strongly associated with vasculitis, purpura, lymphadenopathy, hematologic manifestations, hyperglobulinemia, and presence of rheumatoid factor (Alexander et al., 1981a; Kavanaugh et al., 2000). Both autoantibodies, particularly anti-Ro/SSA, also correlate with disease severity and extraglandular systemic manifestations (ter Borg et al., 2011). Seronegative patients require salivary gland or lip biopsy to establish the diagnosis. Hypergammaglobulinemia is present in approximately 80% of patients, while IgM-κ cryoglobulins may be present in 20% (Kassan and Moutsopoulos, 2004). ESR may be elevated with CNS disease, although this is a nonspecific finding (Alexander et al., 1981b).

In individuals with neurologic manifestations, ANAs were present in 54–75%, anti-Ro/SSA in 41–63%, anti-La-SSB in 13%, hypergammaglobulinemia in 31–44%, cryoglobulinemia in 25–37%, rheumatoid factor in 33–56%, and a low C4 or total complement in 21% (Alexander et al., 1982; Delalande et al., 2004). Furthermore, the presence of cryoglobulinemia is strongly associated with peripheral neuropathy and mononeuritis multiplex (Delalande et al., 2004). In one series, all patients with neurologic involvement tested positive for the leukocyte antigen (HLA) DQB1*0303 allele (Anaya et al., 2002). In addition, NMO antibodies may be present in patients with NMO-like disease (Min et al., 2009).

CSF abnormalities occur in two-thirds of patients and include mild pleocytosis and increased protein content

Table 3.6

Revised international classification criteria for Sjögren syndrome*

I. Ocular symptoms: a positive response to at least one of the following questions:
  1. Have you had daily, persistent, troublesome dry eyes for more than 3 months?
  2. Do you have a recurrent sensation of sand or gravel in the eyes?
  3. Do you use tear substitutes more than three times a day?

II. Oral symptoms: a positive response to at least one of the following questions:
  1. Have you had a daily feeling of dry mouth for more than 3 months?
  2. Have you had recurrently or persistently swollen salivary glands as an adult?
  3. Do you frequently drink liquids to aid in swallowing dry food?

III. Ocular signs – that is, objective evidence of ocular involvement defined as a positive result for at least one of the following two tests:
  1. Schirmer's test, performed without anesthesia ($\leq$5 mm in 5 minutes)
  2. Rose Bengal score or other ocular dye score ($\geq$4 according to van Bijsterveld's scoring system)

IV. Histopathology: in minor salivary glands (obtained through normal-appearing mucosa) focal lymphocytic sialoadenitis, evaluated by an expert histopathologist, with a focus score $\geq$1, defined as a number of lymphocytic foci (which are adjacent to normal-appearing mucous acini and contain more than 50 lymphocytes) per 4 mm^2 of glandular tissue

V. Salivary gland involvement: objective evidence of salivary gland involvement defined by a positive result for at least one of the following diagnostic tests:
  1. Unstimulated whole salivary flow ($\leq$1.5 mL in 15 minutes)
  2. Parotid sialography showing the presence of diffuse sialectasias (punctate, cavitary, or destructive pattern), without evidence of obstruction in the major ducts
  3. Salivary scintigraphy showing delayed uptake, reduced concentration and/or delayed excretion of tracer

VI. Autoantibodies: presence in serum of the following autoantibodies:
  1. Antibodies to Ro(SSA) or La(SSB) antigens, or both

*From Vitali et al., 2002.

(Alexander, 1985). In one series of 56 patients with CNS involvement, 30% had CSF oligoclonal bands and/or lymphocytic pleocytosis (Delalande et al., 2004). In patients with aseptic meningitis, the CSF exhibits a lymphocytic predominant pleocytosis, increased protein content, but no evidence of intrathecal synthesis of immunoglobulin (Rossi and Saddi, 2006).

Neuroimaging and other ancillary testing

Nonspecific MRI T2 hyperintensities are present in 50–66% of patients (Alexander, 1985; Escudero et al., 1994). Multiple sclerosis-like white matter changes may be observed in up to 40% of patients with CNS involvement. However, involvement of the basal ganglia and rare involvement of the corpus callosum may help differentiate Sjögren syndrome from MS (Delalande et al., 2004). On MR spectroscopy, reduced NAA levels or NAA/Cr ratio in subcortical frontal and basal ganglia white matter may be observed in half of patients (Morreale et al., 2014). Cerebral atrophy and smaller hippocampal volume may be observed in patients with cognitive dysfunction (Mauch et al., 1994; Lauvsnes et al., 2013). Meningeal enhancement may be observed in patients with meningitis or meningoencephalitis (Alexander and Alexander, 1983; Rossi and Saddi, 2006). Spinal cord lesions resembling transverse myelitis typically involve the cervical segment with a third of patients having longitudinally extensive lesions and 40% centromedullary or multifocal involvement (Delalande et al., 2004; Rogers et al., 2004).

Visual evoked potentials may be useful in detecting subclinical optic neuritis. Electromyography and nerve conduction velocities (EMG/NCVs) assist in the diagnosis and characterization of large fiber peripheral neuropathies. Distal sensorimotor axonal neuropathies are the most common EMG/NCVs findings. Pure sensory neuropathies, mononeuritis multiplex, and chronic polyradiculoneuropathy are also observed and at times neuropathies with demyelinating features may be seen (Mellgren et al., 1989; Gemignani et al., 1994; Delalande et al., 2004; Gøransson et al., 2006). Sural nerve biopsy may be required to establish the diagnosis in patients without overt sicca symptoms or patients presenting with neuropathy with suspected primary Sjögren syndrome. Vascular or perivascular inflammatory infiltrates, with or without necrosis (Mellgren et al., 1989; Griffin et al., 1990; Delalande et al., 2004; Mori et al., 2005), are most commonly observed. However, alteration of endoneurial small vessels may occur (Gemignani et al., 1994). In patients with pure sensory neuronopathies, the dorsal root ganglia may show T-cell lymphocytic infiltration (Griffin et al., 1990; Mori et al., 2005).

Patients suspected of having a small fiber neuropathy may require special testing. Abnormal laser evoked potentials, thermal quantitative sensory testing, sympathetic skin reflex or skin biopsy showing low intraepidermal nerve diameter suggests small fiber involvement (Sène et al., 2013). Patients with dysautonomia may require gastric emptying scintigraphy, urodynamic studies, finger skin blood flow testing, measurement of orthostatics, or heart rate variation with deep-breath testing (Kovacs et al., 2002; Mandl et al., 2008). Anti-M3 muscarinic acetylcholine receptor antibodies have been implicated in the pathogenesis of salivary dysfunction in Sjögren syndrome patients (Tsuboi et al., 2013).

Management

Treatment is mainly symptomatic. Typically, patients require a multidisciplinary approach to address glandular and extraglandular manifestations of the disease. Patients with sicca symptoms may require artificial tears, lubricants, temporary occlusion of the puncta, oral moisturizing gels, muscarinic agonists, and routine dental and ophthalmologic evaluations to evaluate and manage potential secondary complications such as dental caries, keratoconjunctivitis, or blepharitis. Corticosteroid-containing drops should be used cautiously as corneal lesions and secondary infections or cataracts may occur with prolonged use (Kassan and Moutsopoulos, 2004; Ramos-Casals et al., 2012; Peri et al., 2012). The management of systemic manifestations is complex, as universally effective therapies are lacking, and systemic therapies have mainly been studied in regards to sicca symptoms. Corticosteroids, antimalarial agents, azathioprine, methotrexate, mycophenolate mofetil, IVIG, plasma exchange (PLEX), antitumor necrosis factor, interferons and B-cell immunomodulators have all been attempted with variable results (Ramos-Casals et al., 2012). Promising trials of rituximab have shown efficacy in treatment of both glandular and extraglandular manifestations of primary Sjögren syndrome (Dass et al., 2008; Mejier et al., 2010; Gottenberg et al., 2013).

For management of neurologic symptoms, systemic therapies are usually required. Unfortunately, randomized controlled trials are lacking. In patients with mild neurologic symptoms, symptomatic management may suffice. Progressive or severe symptoms typically require pulse cyclophosphamide, corticosteroids, and/or adjunct immunosuppressants (Ozgocmen and Gur, 2008).

Treatment of Sjögren syndrome associated myelopathy requires high dose intravenous corticosteroids in combination with immunosuppressive therapy, typically cyclophosphamide or chlorambucil (Wright et al., 1999; Williams et al., 2001). Based on efficacy and side-effect profile, cyclophosphamide may be considered a first-line agent (Rogers et al., 2004). The role of corticosteroids alone is uncertain, as patients may progress despite therapy while patients with multiple sclerosis-like presentations may acutely respond to therapy (Alexander et al., 1981b, 1986).

A favorable response to interferon α in regards to neuropathic symptoms, sicca symptoms, antibody titers, and salivary gland biopsy findings has been observed in cases of sensory ataxic ganglionopathy and sensorimotor neuropathy with demyelinating features (Yamada et al., 2005). In another series of four patients with acute ataxic sensory polyneuropathy with prominent loss of kinesthetic and proprioceptive sensation, all patients received between five and nine sessions of PLEX. However, only two patients (50%) showed a dramatic and sustained effect within 2 weeks of treatment, while no appreciable response was noted in the other patients (Chen et al., 2001). IVIG has been shown to reduce pain and superficial sensory symptoms due to sensory neuropathy (Kizawa et al., 2006; Morozumi et al., 2009). In a long-term study of pure sensory neuropathy of 15 patients (11 with numbness and paresthesia, six with trigeminal neuralgia and four with Adie's tonic pupil), 13 patients received corticosteroids, four cyclophosphamide, and one IVIG. Most patients showed an insidious and indolent course despite treatment with a mean duration of prospective follow-up of 10 years (Font et al., 2003). Rituximab has been shown to decrease the total dose of IVIG by 25% in a patient with IVIG-dependent sensory ataxic neuropathy (Gorson et al., 2007).

Ramos-Casals and colleagues proposed that for CNS manifestations and multineuritis, corticosteroids be considered first line therapy, cyclophosphamide second line, mycophenolate mofetil and azathioprine third line, with rituximab used for refractory cases. In individuals with polyneuropathy or ataxic polyneuropathy, IVIG is considered first line therapy, rituximab second line therapy, and PLEX third line (Ramos-Casals et al., 2012).

SYSTEMIC LUPUS ERYTHEMATOSUS

Systemic lupus erythematosus (SLE) is a systemic autoimmune disorder with widespread multiorgan involvement and a variable clinical course. The worldwide prevalence of SLE is approximately 20–70:100 000, with women carrying a 10-fold greater risk than men, particularly in the reproductive years. Patients of African or Asian heritage have a 2–3 times higher incidence and prevalence than Caucasians (Pons-Estel et al., 2010). Pediatric-onset SLE represents roughly 10–20% of all SLE patients (Kamphuis and Silverman, 2010). The mucocutaneous manifestations of SLE are quite characteristic and represent four of the 11 diagnostic criteria listed by the American College of Rheumatology (ACR) (Hochberg, 1997) as well as four of 11 clinical criteria listed by the Systemic Lupus International Collaboration Clinics (SLICC) classification (Petri et al., 2012).

Cutaneous involvement occurs in 70–80% of all patients with SLE (McCauliffe, 2001). Familiarity with the cutaneous lesions associated with SLE can facilitate early diagnosis and treatment. These cutaneous manifestations are divided into three temporally defined categories: (1) acute cutaneous lupus erythematosus (ACLE), (2) subacute lupus erythematosus (SCLE), and (3) chronic cutaneous lupus erythematosus (CCLE).

Malar rash

Malar rash, colloquially known as the "butterfly rash," is an acute manifestation of SLE, present in 46–65% of patients (Vilá et al., 2004). The rash is characterized by a symmetrical fixed erythematous maculopapular lesion with slight scale occurring over both cheeks and nose, sparing the nasolabial folds (Figs 3.11, 3.12). The rash is typically transient, lasting from days to weeks, associated with occasional symptoms of pain and pruritus (Luzar and Calonje, 2012). Although highly specific (96%) for SLE, it is only moderately sensitive (57%) (Uva et al., 2012), as many other disease states can cause similar cutaneous findings.

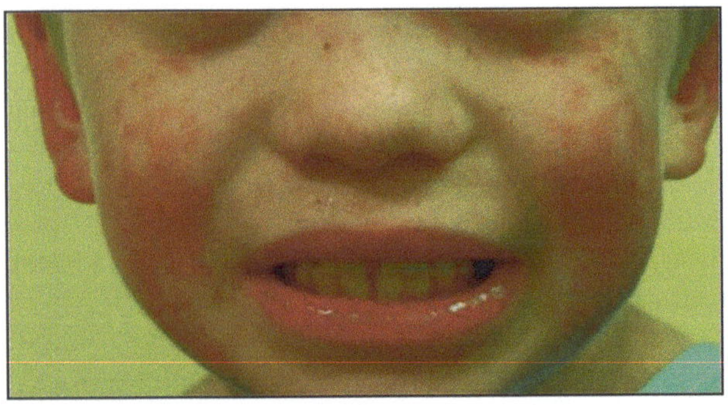

Fig. 3.11. Malar rash in a patient with SLE. Note that the rash does not cross the nasolabial fold. (Courtesy of Dr. Amanda Marsch, University of Illinois at Chicago, Department of Dermatology, Chicago, IL, USA.)

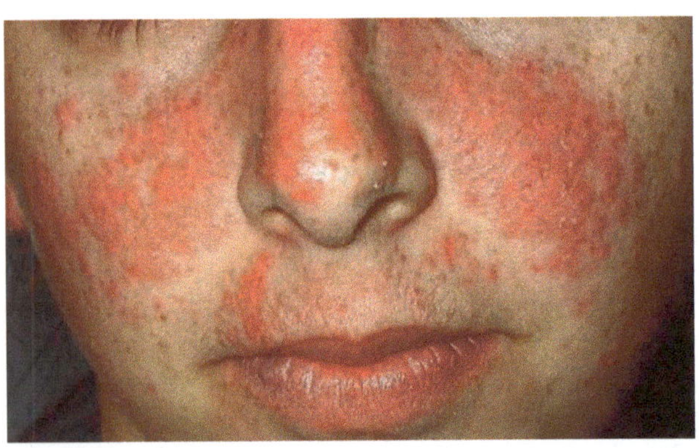

Fig. 3.12. Malar rash characterized by symmetrical fixed erythematous maculopapular rash with slight scale occurring over the bilateral cheeks and nose, with relative sparing of the nasolabial folds.

Differential diagnosis of the malar rash includes other autoimmune diseases, infectious causes, vitamin deficiencies, and some chromosomal disorders. In dermatomyositis, the rash is caused by inflammation of the muscles and skin and involves the nasolabial folds (Sontheimer, 2004). Infectious causes include Lyme disease and erysipelas caused by streptococci. The former can be diagnosed via a careful patient history, biopsy with special stains and/or appropriate microbiological tests, including enzyme-linked immunosorbent assay (ELISA) test confirmed by western blot (Tugwell et al., 1997). The latter causes a malar rash associated with pain that can be diagnosed through biopsy with special stains and/or aerobic culture (O'Connor and Paauw, 2010). The malar rash associated with pellagra can be differentiated by laboratory determination of vitamin levels, particularly vitamin B_3 (Dreizen, 1991). Bloom syndrome, an autosomal recessive mutation in the BLM gene which also causes a malar rash, usually in childhood or adolescence, can be confirmed via genetic studies (Singh et al., 2010). Other cutaneous diseases in the differential diagnosis include rosacea, drug-induced photosensitivities, sunburn, pemphigus erythematosus, and dermatides including atopic, allergic, contact and seborrheic (Lee and Werth, 2012).

Photosensitivity

Photosensitivity is a subacute manifestation of SLE, occurring in about 50% of patients (Watson, 1989). It presents as an erythematous or violaceous maculopapular eruption in sun-exposed areas, most commonly on the face, arms, neck (in "V"-shaped distribution) and hands, sparing the knuckles (Fig. 3.13). The rash, most common amongst Caucasians, is caused by both UVA and UVB light. Similar to the malar rash, it has a high specificity (96%) and moderate sensitivity (43%) for SLE.

Differential diagnosis for lupus-associated photosensitivity includes true photodermatoses such as those of idiopathic, metabolic, genetic, and exogenous etiology, rosacea, Darier's disease, vitiligo, and herpes simplex (Lee and Werth, 2012).

Discoid rash

Discoid rash is a chronic manifestation of SLE occurring in about 20% of patients (Provost, 1994). It is also observed in discoid lupus erythematosus, a separate diagnostic entity in which cutaneous lesions occur

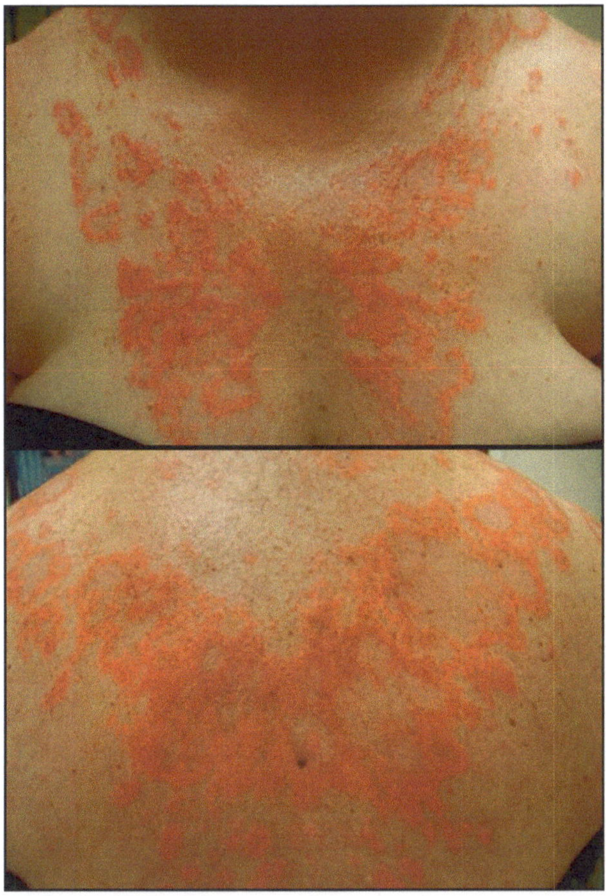

Fig. 3.13. Subacute cutaneous lupus erythematosus with psoriasiform and annular plaques presenting in sun-exposed areas ("V" distribution) on chest and back. (Courtesy of Dr. Anthony Peterson, Loyola University Medical Center, Department of Dermatology, Maywood, IL, USA.)

without systemic symptoms. It develops in sun-exposed areas as thick disc-shaped erythematous plaque(s) with well demarcated scalloped borders, prominent adherent scale, epidermal atrophy, and follicular dilatation and plugging (Fig. 3.14). Removal of the scale is painful, and horny plugs are often seen attached to its undersurface ("carpet tacks" sign) (Luzar and Calonje, 2012). Discoid lesions often heal with marked scarring and can cause hyper- or hypopigmentation in dark-skinned individuals and scarring alopecia on hair-bearing areas (Lee and Werth, 2012) with a sensitivity and specificity of 18% and 59% for SLE respectively (Uva et al., 2012).

Differential diagnosis includes lichen planus/lichen planopilaris, polymorphic light reaction, facial discoid dermatitis, granuloma faciale, sarcoidosis, Jessner's lymphocytic infiltrate, lymphocytoma cutis, lymphoma cutis, dimorphic fungal infections, and dermatophytosis. Single discoid lesions that become hypertrophic or verrucous can mimic verruca vulgaris, actinic keratosis, keratoacanthoma, and squamous cell carcinoma (Lee and Werth, 2012).

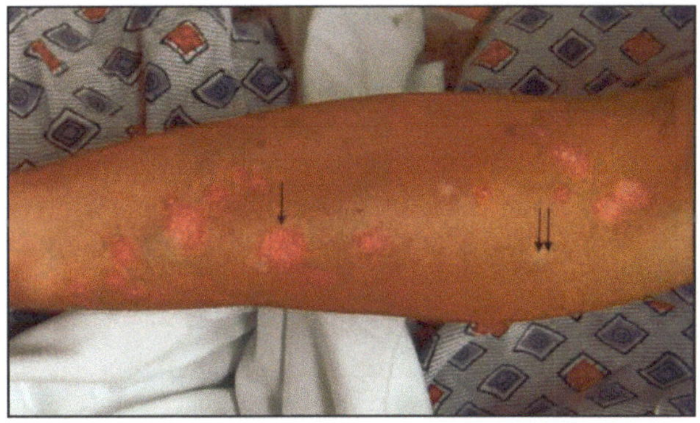

Fig. 3.14. Discoid rash characterized by thick disc-shaped erythematous plaques with well-demarcated scalloped borders and prominent adherent scale (single arrow). There are also intermixed areas of disfiguring scarring and hypopigmentation representing the more progressive form of disease (double arrow). (Courtesy of Dr. Anne Marie Mahoney, Loyola University Medical Center, Department of Dermatology, Maywood, IL, USA.)

Other cutaneous findings

Additional cutaneous findings suggestive of SLE include diffuse non-scarring alopecia, scarring alopecia associated with discoid rash, Raynaud's phenomenon, nail fold telangiectasia and erythema, vasculitis, livedo reticularis, cutaneous signs of antiphospholipid syndrome, livedoid vasculopathy, palmar erythema, and papular and nodular mucinosis (Lee and Werth, 2012). These findings are non-specific but should prompt evaluation for systemic lupus in patients demonstrating cutaneous lupus erythematosus.

Histopathology

All cutaneous manifestations of SLE demonstrate similar histologic findings that in most cases are indistinguishable by light microscopy. Those characteristics that differentiate SLE from SCLE and discoid lupus erythematosus are subtle at best and generally require evaluation by an experienced dermatopathologist as well as appropriate clinical correlation. Therefore, for the scope of this discussion, the most general histologic findings will be discussed.

Lupus erythematosus is generally classified as an interface dermatitis, in which there is a predominant lymphocytic infiltrate present at the dermo-epidermal junction, causing damage to keratinocytes and melanocytes in the form of basal-vacuolar degeneration and pigment incontinence. Additional histopathologic features that suggest a diagnosis of lupus erythematosus include epidermal atrophy, thickening of the basement membrane, dermal mucin deposition, follicular plugging, and superficial and deep perivascular inflammation (Luzar and Calonje, 2012). A helpful adjunct for the definitive diagnosis of lupus erythematosus is evaluation by direct immunofluorescence. Direct immunofluorescence for lupus erythematosus demonstrates a deposition of granular IgG, IgA, IgM, C3 and fibrinogen at the basement membrane (see Fig. 3.6). This direct immunofluorescence finding is colloquially referred to as "a full house."

Neuropsychiatric systemic lupus erythematosus

Neurologic involvement in SLE is quite common and constitutes one of the 11 diagnostic criteria in both classification systems. About half of SLE patients are estimated to experience a neuropsychiatric SLE (NPSLE) manifestation during the span of their illness (Joseph et al., 2007), and 40–50% will develop NPSLE manifestations within the first 2 years of SLE diagnosis (Bertsias and Boumpas, 2010). Patients of Hispanic, African American and Asian heritage tend to have more neurologic manifestations and accrue more damage over time at a faster pace as compared to Caucasians (Pons-Estel et al., 2010). The prevalence of NPSLE in children is similar to adults (Fanouriakis et al., 2013). SLE before age 16 years often presents with neuropsychiatric manifestations (Pons-Estel et al., 2010). NPSLE has been broadly categorized into "primary" or "secondary" based on the underlying pathogenic processes. Direct autoimmune and autoantibody mediated processes involving the CNS or PNS and NPSLE associated with generalized disease activity are referred to as "primary" manifestations. Otherwise, NPSLE resulting from concomitant SLE-related organ dysfunction, complication of immunosuppressive therapy, or as a result of another NPSLE manifestation, are designated as "secondary" NPSLE (Jennekens and Kater, 2002a; Bertsias and Boumpas, 2010; Fanouriakis et al., 2013).

In 1991, the ACR defined NPSLE by developing a standardized nomenclature system, recognizing 19 SLE neuropsychiatric syndromes (Liang et al., 1999) (Table 3.7). This classification serves as a standardized definition for the scope of NPSLE manifestations observed with SLE and includes both major (e.g., cerebrovascular disease, myelopathy, seizures, acute confusional state) and minor (polyneuropathy, headaches, mild psychiatric or cognitive symptoms) categories (Ainiala et al., 2001). Its specificity is increased when the ACR nomenclature excludes minor syndromes (Fanouriakis et al., 2013). However, minor symptoms are the most commonly experienced by patients (Nived et al., 2003).

Pathogenesis and risk factors of NPSLE

The pathogenesis of NPSLE is complex and includes ischemia, thrombosis and inflammation. Ischemia is the leading mechanism of CNS pathology, with vascular occlusions and non-inflammatory cerebral small vessel cerebral angiopathy almost always found in postmortem studies (Jennekens and Kater, 2002b).

Factors contributing to CNS ischemia include autoantibodies, among others (e.g. antiphospholipid antibodies), premature atherosclerosis, small vessel vasculopathy, cerebrovascular thrombosis and embolization, vascular dissections, vasospasm, and vasculitis (Jennekens and Kater, 2002a, b).

Table 3.7

American College of Rheumatology NPSLE nomenclature

Central nervous system	Peripheral nervous system/Neuromuscular
Aseptic meningitis	Acute inflammatory demyelinating polyradiculoneuropathy (Guillain-Barré syndrome)
Cerebrovascular disease	Autonomic nervous system disorders
Demyelinating syndrome	Mononeuropathy, single/multiplex
Headache (including migraine and idiopathic intracranial hypertension)	Cranial Neuropathy
Movement disorder (e.g., chorea)	Plexopathy
Myelopathy	Polyneuropathy
Seizure disorder	Myasthenia gravis
Acute confusional state	
Anxiety disorder	
Cognitive dysfunction	
Mood disorder	
Psychosis	

NPSLE, neuropsychiatric systemic lupus erythematosus.

Less than half of NPSLE manifestations are directly related to active disease (Bertsias and Boumpas, 2010). Other risk factors for NPSLE events include generalized disease activity, concomitant NPSLE involvement, and presence of antiphospholipid antibodies (Bertsias and Boumpas, 2010; Bertsias et al., 2010). The most prominent NPSLE manifestations are discussed individually.

Stroke

The prevalence of stroke in SLE ranges between 2% and 15% (Futrell and Millikan, 1989; Chiu et al., 2012a). Ischemic stroke is most commonly observed among SLE patients (Futrell and Millikan, 1989; Krishnan, 2005), while intracerebral hemorrhage seems to occur in the context of thrombocytopenia (Futrell and Milikan, 1989; Jennekens and Kater, 2002b). Ischemic stroke and TIAs account for over 80% of stroke in SLE, intracerebral hemorrhage for 7–12%, subarachnoid hemorrhage (SAH) for 3–5%, and cerebral venous sinus thrombosis (CVST) for 2% (Bertsias and Boumpas, 2010).

Asymptomatic SLE patients have a greater than seven-fold risk of stroke or coronary heart disease than expected when based on individual vascular risk factors (Esdaile et al., 2001). Patients at high risk are more commonly women < 50 years of age (Mitsias and Levine, 1994; Ward, 1999; Krishnan, 2005; Mikdashi et al., 2007). Stroke customarily occurs within the first few years of SLE onset (Futrell and Millikan, 1989; Mitsias and Levine, 1994). Women with SLE between the ages of 18–44 and 45–64 years have 10 and 2.7 times more frequent events compared to non-SLE women of similar age range, respectively (Ward, 1999). Longitudinal studies found that older age, hypercholesterolemia, and hypertension were the most frequent risk factors associated with stroke in SLE (Mikdashi et al., 2007; Chiu et al., 2012a).

When adjusting for age, gender, or comorbid conditions (hypertension, hypercholesterolemia, diabetes), the risk of stroke in SLE has been reported to be 1.6–2.0 times greater than that of controls (Krishnan, 2005; Chiu et al., 2012a). Population-based studies evaluating hospitalizations for stroke also found an approximately 1.6 to 2 fold higher risk for stroke in all SLE patients compared to controls (Ward, 1999; Krishnan, 2005).

Antiphospholipid antibodies

Antiphospholipid antibodies (APL) are a group of diverse immunoglobulins that react with several anionic plasma proteins. The main components of APL are anticardiolipin antibodies, lupus anticoagulant, and anti-β2-glycoprotein antibodies (Levine et al., 2002; Galli et al., 2003; Meroni el al., 2003; Ruiz-Irastorza et al., 2010). These antibodies bind to antigenic targets such as β2-glycoprotein 1 and prothrombin (Galli et al., 2003). The interaction of these antibodies with target antigens resulting in thrombosis is complex and incompletely understood. Binding promotes activation of endothelial cells, monocytes, platelets, and complement, leading to a procoagulant state, mainly mediated by the increased synthesis of tissue factor and thromboxane A2 (Ruiz-Irastorza et al., 2010). This "state" in turn can lead to arterial or venous thrombi (Levine et al., 2002; Ruiz-Irastorza et al., 2010).

Of interest, brain microvascular endothelium membranes have higher expression of β2-glycoprotein 1 as compared to microvascular endothelium in experimental models. These are recognized by anti-β2-glycoprotein antibodies which bind and promote cell activation and the ensuing proinflammatory cascade. Subsequently, neuronal damage results from ischemia and proinflammatory cytokines that mediate neuronal damage (Meroni et al., 2003). This in part may explain CNS susceptibility to APL.

Antiphospholipid syndrome

Antiphospholipid syndrome (APS) is the most common cause of acquired thrombophilia (Matyja-Bednarczyk et al., 2014), leading to both arterial and venous thrombosis. In 1999, an international panel established consensus diagnostic criteria of APS characterized by evidence of a clinical event (arterial, venous, or small vessel thrombosis in any organ, or related obstetric complication) and specific positive laboratory evidence of antiphospholipid antibodies (Wilson et al., 1999). Individuals with both APS and SLE represent about one-third of all APS patients. Approximately two-thirds of thrombotic events are venous and one-third are arterial (Cervera et al., 2002). The most common arterial manifestations in APS are ischemic stroke and TIA (Cervera et al., 2009; Matyja-Bednarczyk et al., 2014). Although CVST occurs in the context of APL and SLE, it is an uncommon venous thrombotic manifestation (Vidailhet et al., 1990).

Thromboembolic events occur in approximately one-third of APL-positive patients (Galli et al., 2003) and about 40% of patients with SLE have positive antiphospholipid antibodies (Mok et al., 2005). Lupus anticoagulant, present in about one-third of SLE patients (Love and Santoro, 1990), is a stronger risk factor for thrombosis than anticardiolipin antibodies (Horbach et al., 1996; Galli et al., 2003).

Livedo reticularis is a non-specific skin manifestation that occurs in 20–30% of patients with APS (Fig. 3.15). Of interest, it occurs more frequently in SLE-related

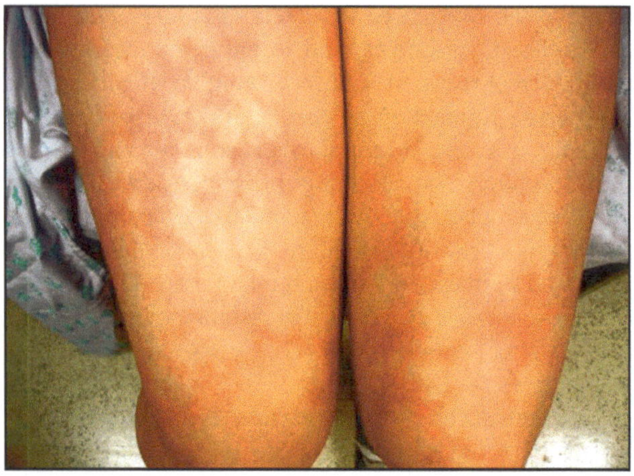

Fig. 3.15. Livedo reticularis. The reticular red skin discoloration surrounding pale central areas, or the "mottled" appearance, of the skin. (Courtesy of Dr. Amanda Marsch, University of Illinois at Chicago, Department of Dermatology, Chicago, IL, USA.)

APS than in primary APS (Cervera et al., 2002, 2009; Matyja-Bednarczyk et al., 2014). The presence of livedo reticularis and stroke is called Sneddon syndrome. A higher proportion of stroke has been associated with APL-positive Sneddon syndrome patients (Caldas and de Carvalho, 2011). The presence of livedo reticularis, history of thrombosis, or fetal loss are tell-tale signs of APS and should be thoroughly investigated in SLE patients.

Premature atherosclerosis

Patients with SLE are at a particularly high risk for premature atherosclerosis. This has been attributed to a chronic inflammatory state, antiphospholipid antibodies, dyslipoproteinemia, renal disease, corticosteroid treatment in addition to other known risk factors prevalent in the general population (Jennekens and Kater, 2002b). A study evaluating the risk of stroke and coronary heart disease in SLE found an eight-fold increased relative risk for stroke and 7.5-fold overall coronary heart disease risk after controlling for traditional Framingham study risk factors (Esdaile et al., 2001). In a case-control study comparing accelerated atherosclerosis in 197 SLE patients and matched controls by measuring carotid plaque burden by ultrasonography, the prevalence of atherosclerosis was more than double. Patients with plaque were more commonly older, postmenopausal, Caucasian, with higher LDL levels, higher systolic blood pressure values, pre-existing coronary artery disease, and pulmonary hypertension. In addition, they had longer disease duration and were less likely to be treated with immunosuppressive agents (Roman et al., 2003). In a recent longitudinal cohort study of women with SLE, a greater mean carotid intimal media thickness (IMT) and presence of carotid artery plaque (defined as focal intimal medial projection greater than 50% than adjacent IMT) at baseline were associated with an increased incidence of stroke, myocardial infarction, percutaneous transluminal angioplasty, coronary artery bypass grafting, and fatal cardiac arrest (Kao et al., 2013).

Cardioembolism

Cardiac involvement in SLE is common. Stroke secondary to cardiac dysfunction leading to cardiac embolism may be related to pump failure, valvular vegetations, or cardiac arrhythmias such as atrial fibrillation. Congestive heart failure may arise secondary to ischemic heart disease, hypertension, renal failure, myocarditis, or valvular heart disease. Studies have shown for every 5–10% reduction in left ventricular ejection fraction, stroke risk increased by 18–58% respectively (Loh et al., 1997; Dries et al., 1997). Ischemic heart disease and congenital heart failure are three and 3.7 times more common in young women with SLE than in unaffected individuals (Ward, 1999). A retrospective follow-up study found a 13% incidence of coronary heart disease events (including angina, non-fatal MI, CHF, and coronary related deaths) over the course of 8.6 years in lupus patients. After controlling for comorbidities, the increased relative risk of 7.5 for coronary heart disease was not fully explained by traditional Framingham risk factors alone (Esdaile et al., 2001).

Valvular heart disease is the most common cardiac manifestation of SLE with a prevalence of 44–62%, and higher amongst those with APL antibodies (Cervera et al., 1992; Kampolis et al., 2014). Libman-Sacks endocarditis, the most common valvular abnormality (Cervera et al., 1992), is characterized by sterile vegetations of fibrin and platelet thrombi on cardiac valves, most commonly the posterior mitral leaflet. Libman-Sacks endocarditis is more commonly asymptomatic; however, it can lead to valvular regurgitation, heart failure, predisposition to infective endocarditis, TIAs, and strokes (Futrell and Millikan, 1989; Lee et al., 2009). A recent longitudinal study evaluating cardiac dysfunction in SLE found that disease progression and SLE associated with APS were associated with valvular disease progression and ventricular diastolic dysfunction. In addition, treatment with anticoagulation, antiplatelets, or immunosuppression did not prevent valvular dysfunction (Kampolis et al., 2014). Long-term use of corticosteroids (greater than 3 years) at a dose of more than 10 mg of prednisone daily may double or triple the risk of major cardiac events (Atzeni et al., 2010). Furthermore, high-dose corticosteroids have been associated with an increased risk of atrial fibrillation (van der Hooft et al., 2006).

Other stroke causes and risk factors

Although systemic vasculitis is present in 10–40% of SLE patients, cerebral vasculitis is an unusual cause of stroke. While cutaneous involvement occurs in 90% of cases, cerebral involvement occurs in only 10% (Cieślik et al., 2008). Multifocal strokes, generalized cerebral involvement, and concomitant cutaneous or retinal vasculitis may suggest underlying CNS vasculitis (Nikolov et al., 2006).

Other traditional vascular risk factors such as arterial hypertension, hyperlipidemia, cigarette smoking, or diabetes mellitus can accelerate stroke risk in SLE patients. These modifiable risk factors can be targeted for lowering the risk. Hypertension is strongly associated with stroke and cardiac disease in SLE (Ward, 1999; Mikdashi et al., 2007).

Seizures

Seizures represent one of the two neurologic disorders included among the 11 criteria of the ACR for classification of SLE. Seizures may result from active SLE or secondary CNS pathology. Epilepsy occurs in up to 22% of SLE patients (Bertsias and Boumpas, 2010). Generalized tonic-clonic and complex partial seizures are the most common seizure types (Appenzeller et al., 2004). Risk factors for development of seizures include disease activity, previous or concomitant cerebrovascular disease, nephritis, and the presence of APL antibodies (Appenzeller et al., 2004; Bertsias and Boumpas, 2010; Toyota et al., 2013). MRI findings in these patients include non-specific multifocal white matter changes and abnormalities in the mesial temporal lobes (Appenzeller et al., 2004; Toyota et al., 2013). In patients with recurrent seizures, interictal epileptic activity is usually present compared to patients with single seizure episodes who more commonly have normal interictal patterns (Appenzeller et al., 2004). Non-specific generalized slowing on EEG is common (Joseph et al., 2007). As carbamazepine and lamotrigine may cause a SLE-like syndrome, alternative antiepileptic drugs (AEDs) such as valproic acid, levetiracetam, or clobazam may be considered (Toyota et al., 2013).

Demyelinating disease and transverse myelitis

Monophasic or recurrent inflammatory CNS attacks resembling multiple sclerosis have been referred to as "lupus sclerosis" (Greenberg, 2009). Lupus sclerosis is far less common than strokes, headaches, or seizures among patients with SLE. Transverse myelitis, with a reported prevalence of 3% in SLE, affects the cervical spine in half of the cases (Mok et al., 1998) and can be the presenting manifestation in one-third of patients (Kovacs et al., 2000). Transverse myelitis is more common among SLE patients with APL antibodies (Lavalle et al., 1990; Campi et al., 1998; Kovacs et al., 2000). Spinal cord lesions are usually longitudinally extensive, and similar to those observed in neuromyelitis optica (NMO), who may also have positive ANA serology (Pittock et al., 2008). A lower occurrence or unilateral presentation of optic neuritis, higher occurrence of thoracic transverse myelitis, peripheral nervous system involvement, lower levels of CSF oligoclonal bands, and improvement after anticoagulation may help distinguish SLE-related transverse myelitis from multiple sclerosis (Ferreira et al., 2005).

Headache

Headaches in patients with SLE are quite common, with a prevalence between 24% and 72% (Hanly et al., 2013). The ACR nomenclature for NPSLE characterized headaches according to the International Headache Society classification, including migraines, tension headache, cluster headache, as well as nonspecific intractable headaches, and headache from intracranial hypertension (IHS, 1988). "Lupus headache" is discussed later in this chapter and is categorized under "nonspecific intractable headache." Moreover, headaches can occur in the context of other SLE manifestations (e.g., CVST). Table 3.8 lists primary headache subtypes as well headaches as a complication of other SLE pathology.

Migraine headaches have been reported in 11–32% of SLE patients, and tension headaches occur in 7–23% of these patients (Mitsikostas et al., 2004; Buse et al., 2013; Hanly et al., 2013). Cluster headache and intracranial hypertension are far less frequent, with one recent study reporting a prevalence of approximately 0.5% and 0.2% respectively (Hanly et al., 2013).

While some studies suggest migraine headaches to be more prevalent in SLE patients (Markus and Hopkinson, 1992; Tjensvoll et al., 2011) and other reports dispute this observation (Mitsikostas et al., 2004; Hanly et al., 2013), there is no established association of migraine headaches with disease activity or presence of anticardiolipin antibodies, lupus anticoagulant (Markus and Hopkinson, 1992; Hanly et al., 2013), anti-β2-glycoprotein 1, antiribosomal P, or anti-NR2 glutamate receptor antibodies (Hanly et al., 2013).

Lupus headache

The term "lupus headache" has been used by clinicians and in earlier literature without precise definition

Table 3.8

Headaches in systemic lupus erythematosus

Primary headaches of the ACR nomenclature for NPSLE	Secondary headaches observed in SLE
Migraine headache with or without aura	Aseptic meningitis
Tension headache	Cerebrovascular disease including ICH, SAH, CVST, subdural hematoma, vasculitis, arterial dissection, and PRES
Cluster headache	Renal dysfunction
Idiopathic intracranial hypertension	Seizures
Unspecified; including "intractable", "severe-persistent" or "lupus" headaches	Use of NSAIDs, side-effects from treatment

ACR, American College of Rheumatology; NPSLE, neuropsychiatric systemic lupus erythematosus; SLE, systemic lupus erythematosus; ICH, intracerebral hemorrhage; SAH, subarachnoid hemorrhage; CVST, cerebral venous sinus thrombosis; PRES, posterior reversible encephalopathy syndrome; NSAIDs, nonsteroidal anti-inflammatory drugs.

(Bombardier et al., 1992). Lupus headache has been conceived as a unique and separate headache subtype characterized by severe, persistent, intractable, and non-narcotic responsive headache (Gladman et al., 2002; Isenberg et al., 2005). Obviously, this is somewhat challenging, as migraine headaches occur frequently in SLE patients and can share similar clinical features (Hanly et al., 2013). Moreover, there is a lack of consensus that headaches in SLE are more prevalent than in the general population (Sfikakis et al., 1998; Jennekens and Kater, 2002a; Mitsikostas et al., 2004; Hanly et al., 2013) or that they are due to underlying specific lupus pathology (Jennekens and Kater, 2002b). In a recent international prospective study encompassing 1732 patients, only a small minority (1.5%) had a "lupus headache" as defined by the aforementioned criteria (Hanly et al., 2013). Although a small statistically significant relationship existed between "lupus headache" and disease activity, previous studies did not find a correlation of headache with disease severity, Raynaud's phenomenon, or the presence of anticardiolipin antibodies (Sfikakis et al., 1998). However, this headache subtype remains incorporated in the Systemic Lupus Erythematosus Disease Activity Index (SLEDAI) and the British Isles Lupus Assessment Group Index (BILAG), two of the most commonly used standardized measures of disease activity (Lam and Petri, 2005).

Cognitive dysfunction

Cognitive dysfunction occurs in SLE patients with and without overt neurologic involvement, with an estimated prevalence between 20% and 60% (Harrison and Ravdin, 2006). Impaired cognitive and functional domains include general intelligence, visuospatial skills, psychomotor speed, verbal learning/memory, and attention/mental flexibility. A recent review suggests at least three of five aforementioned domains are impaired in the majority of patients (Denburg and Denburg, 2003). The mechanism of cognitive dysfunction in SLE is complex, and as yet, there are conflicting data regarding the potential pathogenic role of autoantibodies (Hanly et al., 1993; Denburg et al., 1997; Hanly et al., 1999; McLaurin et al., 2005).

Acute psychotic episodes and chronic cognitive impairment in SLE patients should be differentiated from delirious states and dementia. Overt psychosis is observed in 2.5–3.5% of patients (Bertsias and Boumpas, 2010). Common psychotic symptoms include delusions, paranoia, auditory and visual hallucinations, and depressive psychosis. Most psychotic events occur during active disease, are short lived, and rarely recur (Paholpak et al., 2012). Chronic cognitive dysfunction should be differentiated from vascular cognitive impairment, other dementias, and treatment side-effects.

Posterior reversible encephalopathy syndrome

Posterior reversible encephalopathy syndrome (PRES) is classically associated with severe hypertension, renal failure, exposure to chemotherapeutic or immunosuppressive agents, eclampsia, infections, bone marrow transplants, and other systemic diseases such as hemolytic uremic syndrome (HUS), thrombotic thrombocytopenic purpura (TTP), and SLE (Leroux et al., 2008). The typical presentation of PRES is characterized by encephalopathy, seizures, and bilateral posterior subcortical MRI T2 hyperintensities. Headaches, vomiting, altered sensorium, cortical blindness, and focal weakness are less common. Atypical MRI findings such as anterior location, cortical involvement, unilaterality, hemorrhagic changes, gadolinium enhancement, or irreversibility have been observed (McKinney et al., 2007; Lee et al., 2008). Focal areas of restricted diffusion are documented in 11–25% of patients, and hemorrhages occur in approximately 15% (Bartynski, 2008a). Location of MRI findings in PRES are believed to mirror the degree of vascular sympathetic innervation in a gradient-like fashion from posterior to anterior distribution (McKinney et al., 2007).

The exact pathophysiology of PRES is not fully understood. Failure of cerebral autoregulation leading to endothelial dysfunction and arteriolar vasodilatation or vasoconstriction in response to vascular damage may lead to disruption of the blood–brain barrier and capillary leakage. This has been proposed to account for the potentially reversible "vasogenic" edema in the less tightly packed subcortex, visible on MRI T2 sequences as compared to hypoxic ischemic changes that would cause irreversible "cytotoxic" edema (Stott et al., 2005).

Although not identified as a sole neuropsychiatric manifestation by the ACR criteria, PRES shares many of the neuropsychiatric features listed and has been the initial manifestation of SLE in some instances (Barber et al., 2011; Varaprasad et al., 2011). In addition, the most common associated conditions with PRES may occur in the context of active systemic SLE, as a result of immunosuppressive treatment, or secondary to organ failure. In one series evaluating 46 SLE patients with PRES, 95% were hypertensive, 84% had renal insufficiency, and approximately half had begun receiving immunosuppressive therapy. PRES predominantly occurs in the context of a lupus flare-up (87%) (Leroux et al., 2008). In another series of 13 patients, all had active SLE and hypertension, approximately two-thirds had lupus nephritis, and one-third were on immunosuppressive therapy (cyclophosphamide) (Varaprasad et al., 2011).

One theory suggests that build-up of antigen-antibody complexes due to incorrect T helper cell control over B-cell function observed in SLE results in vascular endothelial injury or activation and potential inflammatory cell attraction (Bartynski, 2008b). However, due to the overwhelming prevalence of arterial hypertension, renal involvement, and immunosuppressive therapy observed in patients with active SLE, establishing a unique pathophysiologic role for lupus in PRES remains challenging. In one series of 22 patients with high SLE disease activity and PRES, approximately half had previous positive antiphospholipid antibodies (Baizabal-Carvallo et al., 2009), which have been previously shown to increase endothelial activation (Meroni et al., 2000). In other series, the prevalence of antiphospholipid antibodies has been much lower (Leroux et al., 2008; Varaprasad et al., 2011). Therefore, a relationship between antiphospholipid antibodies and PRES has been difficult to establish and requires further investigation.

Treatment of PRES involves adequate blood pressure control, removal of offending agents, and use of AEDs to control seizures. Full resolution of neurologic symptoms typically occurs within 2 weeks of onset of symptoms (Stott et al., 2005). Focal MRI abnormalities beyond 2 weeks suggest residual infarctions (Lee et al., 2008). Institution of these measures and management of the SLE flare with corticosteroids or adjunct immunosuppression has led to reversibility and good outcome in the majority of patients (Leroux et al., 2008; Varaprasad et al., 2011). Use of high dose intravenous corticosteroids (methylprednisolone 1 g/day for 3 days) followed by oral prednisone (1–1.5 mg/kg) in 15 cases and adjunct pulses of cyclophosphamide and azathioprine in eight cases rendered approximately 77% of patients free from neurologic sequelae within 7 days of treatment in one series of 22 patients (Baizabal-Carvallo et al., 2009).

Other neurologic manifestations

Polyneuropathy purely attributed to the primary disease process is reported in less than 5% of cases. Neuropathy may result from vasculitic changes affecting the nerve fibers. Chronic demyelinating sensorimotor forms may also occur. Patients with antiphospholipid antibodies and/or lupus nephritis are at increased risk for developing neuropathy (Bertsias and Boumpas, 2010). Rare manifestations including plexopathy, mononeuropathy, cranial neuropathy, autonomic disorders, acute inflammatory demyelinating polyradiculoneuropathy (AIDP), myasthenia gravis, and movement disorders (e.g., chorea) have an estimated prevalence of less than 1% in SLE (Bertsias and Boumpas, 2010).

Evaluation and ancillary testing

A thorough and systematic approach is recommended when evaluating these patients, as neurologic manifestations may be the presenting symptoms in up to half of cases. A systematic evaluation approach is summarized in Figure 3.16. The hallmark of SLE is the presence of SLE-specific antinuclear antibodies (ANA) in the context of clinical signs and symptoms. Autoantibodies may be present for many years prior to clinical disease onset (Arbuckle et al., 2003). ANA-negative lupus is extremely rare (Lam and Petri, 2005). Anti-dsDNA, serum complement levels, cytokine levels, endothelial activation markers and acute phase reactants are associated with disease activity. Few studies have evaluated biomarkers for neurologic involvement. Pro-inflammatory HDL (piHDL) has been found to greatly increase the carotid atherosclerotic burden, while platelet C4d (PC4D) has been associated with occurrence of acute ischemic stroke and mortality (Ahearn et al., 2012).

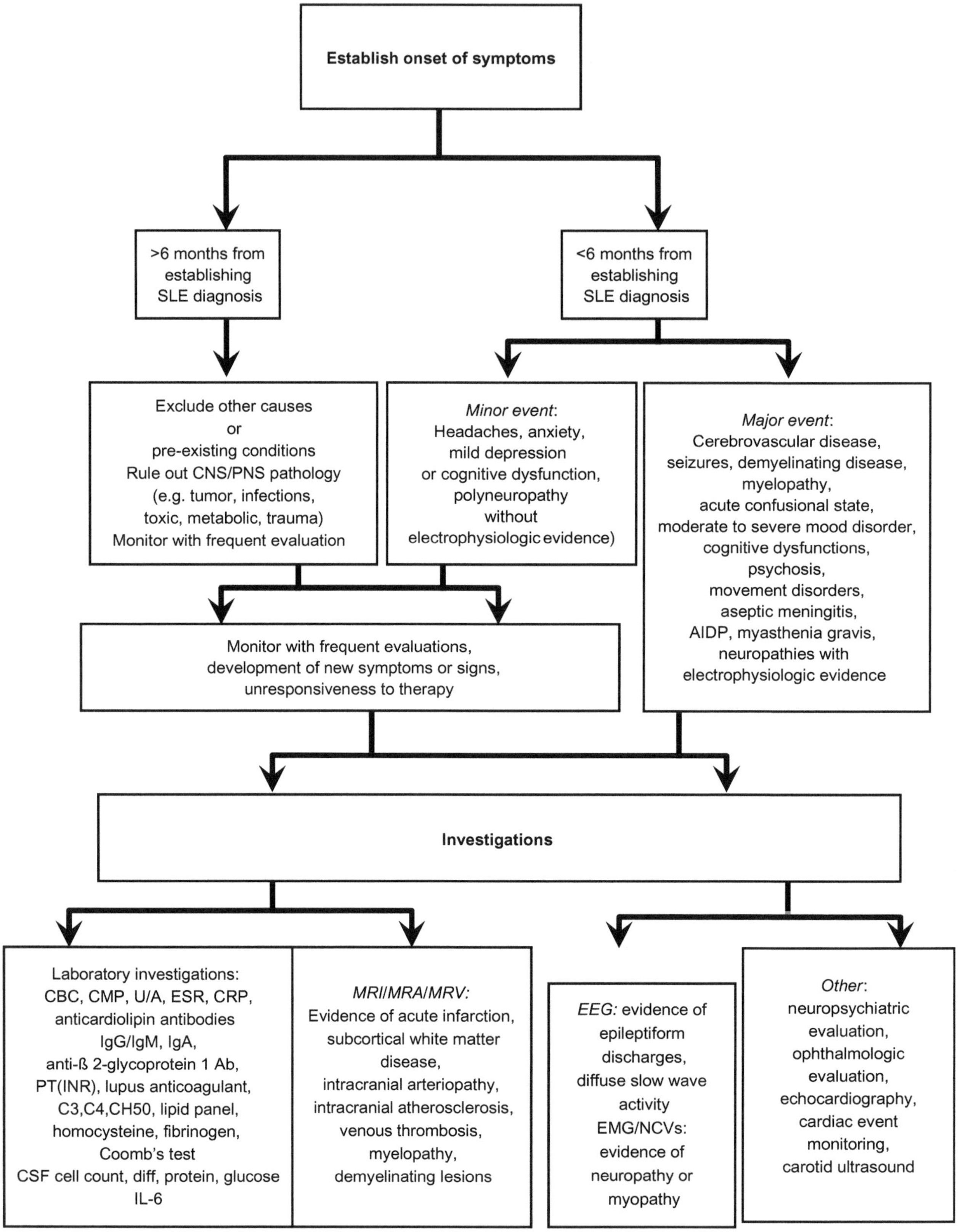

Fig. 3.16. Systematic evaluation of neuropsychiatric features of SLE.

Ancillary symptom-based testing is necessary to evaluate patients with seizures, acute confusional states, and focal neurologic deficits. Abnormal CSF correlates with poor outcomes. Common abnormalities include a lymphocytic pleocytosis, elevated protein levels, and in some patients, the presence of oligoclonal bands and elevated IL-6 levels (Joseph et al., 2007; Bertsias and Boumpas, 2010). Neuropsychiatric testing may be needed to screen for cognitive impairment.

Imaging abnormalities are observed in approximately two-thirds of patients with NPSLE and are associated with disease activity, disease severity, and age (Jung et al., 2010). MRI is the optimal test for NPSLE imaging. Common findings include nonspecific subcortical hyperintense lesions, brain and cortical atrophy, and evidence of infarction (Fig. 3.17) (Jung et al., 2010; Gal et al., 2013). Other imaging modalities may be of some benefit in differentiating acute and chronic NPSLE, such as diffusion tensor imaging (DTI), MR spectroscopy, and functional MRI (Gal et al., 2013). Differentiating demyelinating lesions of multiple sclerosis (both acute and chronic) from SLE is particularly challenging, as the two may be indistinguishable. The presence of T1 black holes and "Dawson's fingers" are more common in multiple sclerosis, but are not pathognomonic. Myelopathy may be longitudinally extensive and span three or more segments but is more commonly located in the thoracic cord and associated with cord swelling and atrophy (Ferreira et al., 2005). Extracranial arterial atherosclerotic disease should be evaluated in individuals with ischemic stroke due to the increased risk of premature atherosclerosis in these patients (Fig. 3.18).

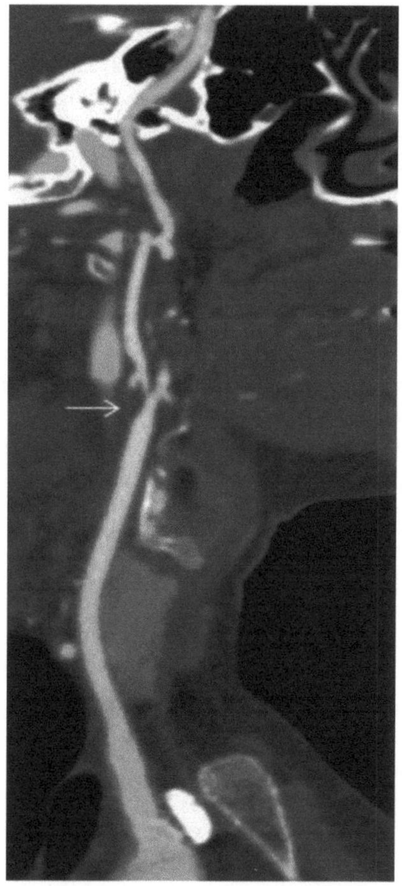

Fig. 3.18. Computed tomography angiography of the same patient showing > 70% stenosis of the proximal right internal carotid artery as well as a noncalcified plaque with adherent thrombus within the lumen (white arrow). (Courtesy of Dr.Christopher Robinson, Loyola University Medical Center Maywood, IL, USA.)

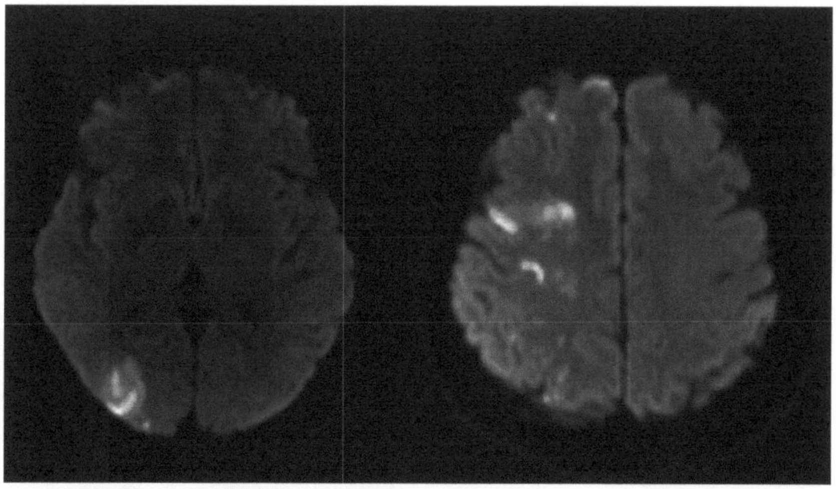

Fig. 3.17. Magnetic resonance diffusion-weighted imaging (DWI) sequence of a 46-year-old woman with SLE and antiphospholipid antibody syndrome who presented with left-sided weakness and confusion. The patient had discontinued warfarin 2 months prior to presentation. (Courtesy of Dr. Christopher Robinson, Loyola University Medical Center, Maywood, IL, USA.)

Management

Although sufficient controlled trials are not available, the European League Against Rheumatism (EULAR) has developed recommendations for the management of NPSLE manifestations based on a systematic review (Bertsias et al., 2010). The mainstay of treatment includes immunosuppressive therapy to treat the primary disease. Antimalarials are the mainstay of immunosuppression in SLE. Most data regarding organ-specific treatment targets renal disease. Belimumab, a monoclonal antibody-targeted drug, has been shown to be superior to standard care; however, data regarding efficacy in NPSLE are lacking because patients with CNS disease have not been studied (Belmont, 2013). Other novel treatments include TNF-α antagonists, rituximab, abatacept, and tocilizumab.

Headaches without obvious etiology are classified according to the International Headache Society criteria and managed accordingly (Mitsikostas et al., 2004). Ultimately, the majority of headaches abate with time irrespective of SLE treatment (Hanly et al., 2013). Symptomatic management of seizures with medication and further evaluation of seizure etiology may be necessary. Patients with acute psychosis should be evaluated for the possibility of steroid-induced psychosis rather than a primary SLE-related pathogenesis. Most patients with mild-to-moderate cognitive impairment have a benign course. Low dose antidopaminergic agents may improve chorea and parkinsonism (Baizabal-Carvallo et al., 2013). Treatment of ischemic stroke depends on the underlying mechanism. The use of platelet antiaggregants is reasonable in SLE patients without antiphospholipid antibodies. The mainstay of secondary thromboprophylaxis in patients with APL antibodies is anticoagulation, with hydroxychloroquine as a potential additional treatment in patients with APL syndrome (Ruiz-Irastorza et al., 2010). Myelopathy may respond to acute treatment with high-dose corticosteroids and adjunct cyclophosphamide treatment. Other treatments with limited data include the use of IVIG and PLEX (Bertsias et al., 2010).

SCLERODERMA

Scleroderma, or progressive systemic sclerosis, is a rare multisystem connective tissue disease of unknown origin characterized by inflammation, vascular injury, and fibrosis. The characteristic skin lesions of scleroderma consist of circumscribed or diffuse hard, smooth, ivory-colored areas that are immobile and give the appearance of hidebound skin (James et al., 2005). In the absence of associated systemic symptoms, localized scleroderma is also termed *morphea* or *linear scleroderma*. If associated with other systemic symptoms, the scleroderma is termed either *limited systemic scleroderma*, associated with calcinosis, Raynaud's phenomenon, esophageal dysmotility, sclerodactyly and telangiectasia, known as the CREST syndrome, or *diffuse systemic sclerosis* (Connolly, 2012). A brief clinical and histopathologic discussion of common cutaneous manifestations of localized and systemic scleroderma follows.

Localized scleroderma: morphea

Morphea is an inflammatory disease of the dermis and subcutaneous tissue. The disorder is more common among women and has no age predilection. Moreover, it is characterized by asymmetric distribution of slightly elevated, erythematous or violaceous edematous plaques that undergo centrifugal expansion, with central scarring and postinflammatory pigmentation as the lesions age (Fig. 3.19). It commonly develops on the trunk, although the extremities can also be affected (James et al., 2005). There are several variants of morphea, including the most common form, plaque type, as well as nodular, pansclerotic, guttate, superficial (atrophoderma of Pasini and Pierini), deep (morphea profunda), and bullous.

Localized scleroderma: linear morphea

Linear scleroderma is an inflammatory disease of the dermis and subcutaneous tissue. In contrast to morphea, linear scleroderma can affect the underlying bones and muscles. Presenting in the first decade of life, it involves the extremities either as a linear erythematous streak or as coalescing morpheaform plaques extending longitudinally to form a scar-like band (Fig. 3.20). These bands often form along the lines of Blaschko and can affect mobility of the adjacent underlying joint or limb (Connolly, 2012). *En coup de sabre* refers to a type of linear morphea that involves unilaterally the scalp and forehead; it is most prevalent among children (Amaral

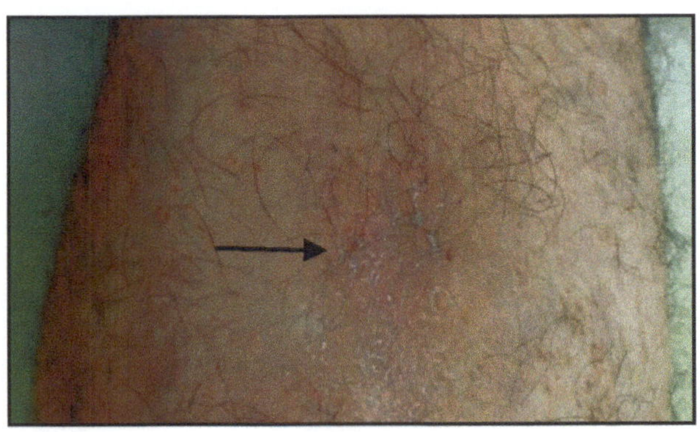

Fig. 3.19. Morphea. Note the slightly elevated, erythematous plaque (arrow).

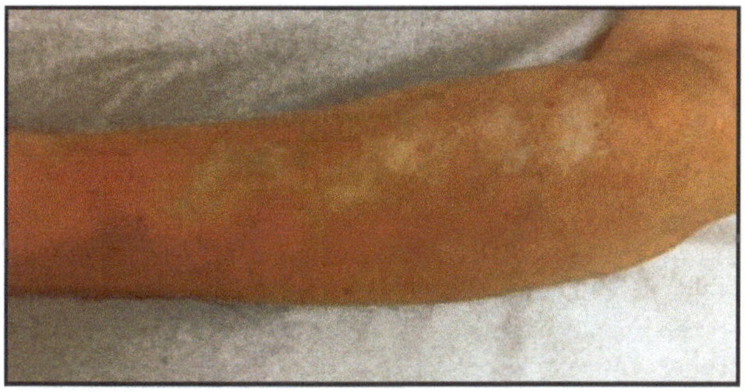

Fig. 3.20. Linear morphea. Note the linear white coalescing plaques extending longitudinally to form a scar-like band.

et al., 2012). Parry–Romberg syndrome, a rare disorder characterized by progressive hemifacial atrophy, is discussed in detail separately.

Systemic scleroderma: morpheaform lesions

Both types of systemic scleroderma demonstrate tightness and thickening of the skin with nonpitting induration, similar to the lesions described in localized morphea. However, in systemic scleroderma, the skin changes occur in a bilateral and symmetric fashion. The skin appears smooth and shiny, with a wax-like appearance. The facial skin can become taught and expressionless, with thinned lips, perioral radiating furrows, tightened lower eyelids, and a smooth forehead (Connolly, 2012). Systemic scleroderma is further subdivided into (1) localized systemic scleroderma when cutaneous changes are limited to the distal extremities and face along with associated organ involvement, and (2) diffuse systemic scleroderma when the cutaneous changes expand to the proximal extremities and trunk. The American Academy of Rheumatology has developed criteria that are 97% sensitive and 98% specific for the diagnosis of both limited and diffuse systemic scleroderma. The sole major criterion includes proximal scleroderma (cutaneous changes confined to skin proximal to the metacarpophalangeal or metatarsophalangeal joints). Minor criteria include sclerodactyly (cutaneous changes limited to the fingers and toes), digital pitting scars, and loss of fingertip pulp. A diagnosis requires one major and at least two minor criteria (Subcommittee for Scleroderma Criteria of the American Rheumatism Association Diagnostic and Therapeutic Criteria Committee, 1980).

Histopathology

Cutaneous changes of localized and systemic scleroderma are similar histologically. With low magnification, punch biopsies commonly demonstrate a "box-car sign," or rigid rectangular shape. At higher power, there is a largely unremarkable epidermis and papillary dermis with thick, hyalinized collagen bundles occupying the reticular dermis, extending to subcutaneous septa with subsequent fibrosis of the lobule (Connolly, 2012). There is a marked lack of adnexal structures with eccrine duct remnants. Mucin and sparse mixed inflammation can be present (Fig. 3.21).

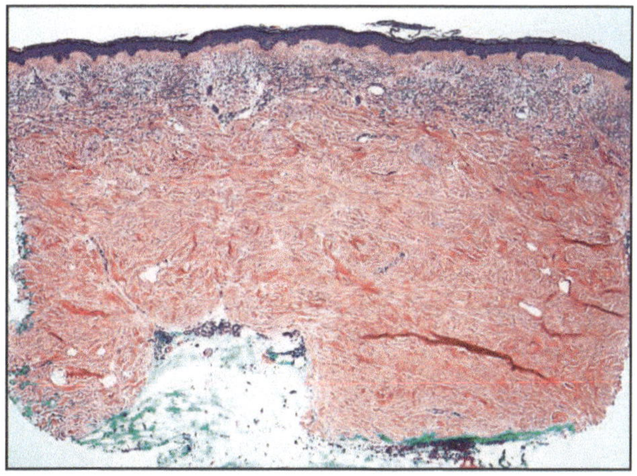

Fig. 3.21. (2× objective). Localized and systemic scleroderma demonstrate an expansion of the reticular dermis and subcutaneous tissue with thick, collagen bundles. Notice the conspicuous decrease in adnexal structures. (Courtesy of Dr. Kelli A. Hutchens, Loyola University Medical Center, Department of Pathology, Maywood, IL, USA.)

Systemic scleroderma: other associated cutaneous findings

Raynaud's phenomenon

Raynaud's phenomenon presents as white, blue, and red color changes in the fingers and toes secondary to pallor, cyanosis, and reactive hyperemia, usually in response to cold (Figs 3.22 and 3.23) (James et al., 2005). It is commonly observed in patients with systemic scleroderma and is often the first presenting sign. Raynaud's is present in 99% of patients with limited systemic scleroderma, and in 90% of those with diffuse systemic scleroderma (Hawk and English, 2001; Hochberg et al., 2003).

Mat telangiectasias and nail fold capillary abnormalities

Mat telangiectasias present as matted or squared-off macules with smooth borders that typically involve the face, neck, and arms (Connolly, 2012). They are present in 90% of limited systemic scleroderma, and 60% of diffuse systemic scleroderma patients (Hochberg et al., 2003). Nail fold capillary abnormalities are present in 90% of systemic scleroderma patients and present as

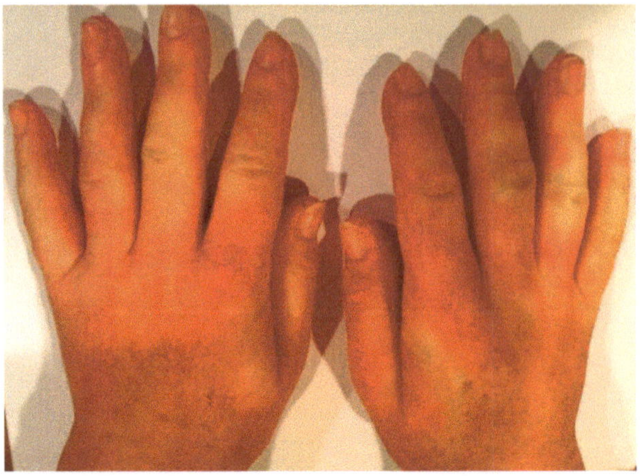

Fig. 3.22. Raynaud's phenomenon. Notice the ill–defined red and white color changes on the proximal and distal fingers.

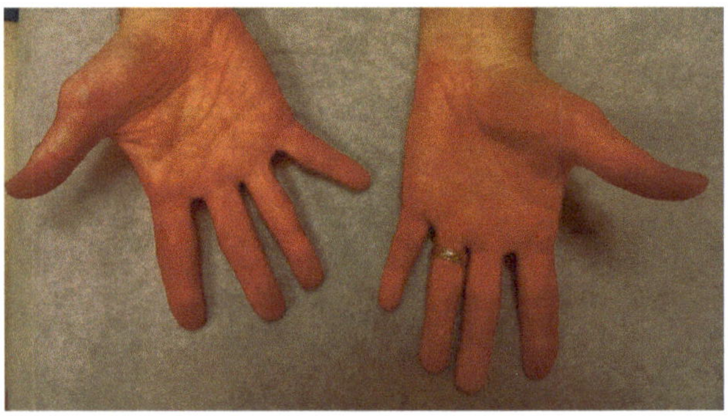

Fig. 3.23. Raynaud's phenomenon.

symmetrically dilated capillaries adjacent to avascular regions (James et al., 2005).

Dystrophic calcinosis cutis

Dystrophic calcinosis cutis occurs in 40% of limited systemic scleroderma patients and in 20% of individuals with diffuse systemic scleroderma (Hochberg et al., 2003). It is characterized by dermal or subcutaneous depositions of calcium (Fig. 3.24) that are generally palpable on physical examination. Associated symptoms may include xerosis and pruritus.

Pathophysiology

The pathogenesis of scleroderma is complex and not fully understood. The proposed explanations involve autoimmune-mediated perivascular inflammation, resulting in early intimal thickening and fibrotic vessel narrowing. In the early phase, transforming growth factor-β (TGF-β) induces fibrosis followed by connective tissue growth factor (CTGF), which acts to maintain fibrosis (Takehara, 2003; Katsumoto et al., 2011).

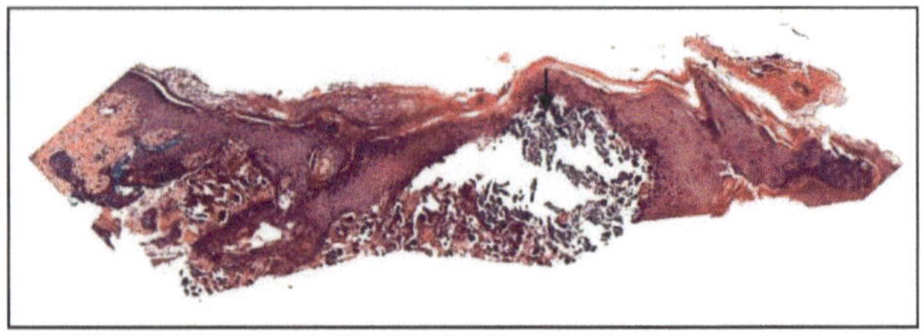

Fig. 3.24. Calcinosis cutis. There is a subepidermal collection of amorphous basophilic material representing calcium deposits (arrow). (Courtesy of Dr. Kelli A. Hutchens, Loyola University Medical Center, Department of Pathology, Maywood, IL, USA.)

Neurologic manifestations of scleroderma

Although CNS, PNS, and autonomic nervous system manifestations may occur in scleroderma, a different pattern of involvement is observed depending on the presence or absence of systemic involvement. In linear scleroderma, seizures and headaches are most commonly reported while peripheral nervous system manifestations, particularly myopathies, and autonomic nervous system manifestations dominate systemic scleroderma (Amaral et al., 2013). Skin lesion severity does not correlate well with CNS involvement, and a minority of patients present with neurologic symptoms as the initial disease manifestation (Amaral et al., 2012).

CNS manifestations

Epilepsy occurs in approximately 8% of patients with systemic scleroderma, and in approximately 73% of those with linear scleroderma or Parry–Romberg syndrome (Kister at al., 2008; Araújo et al., 2013). Generalized tonic-clonic seizures were the only seizure type observed in one study evaluating epilepsy in systemic scleroderma (Araújo et al., 2013). Interictal epileptiform patterns are typically characterized by bilateral synchronous spike-and-wave discharges. The absence of structural MRI findings and the exclusion of underlying metabolic abnormalities support these findings. In addition, older age and higher activity scores are associated with epilepsy (Araújo et al., 2013). Localization-related epilepsy is most prevalent among patients with linear scleroderma. MRI often shows subcortical brain lesions or cortical atrophy. Lesions may be quite epileptogenic and epilepsy may be severe, with EEG abnormalities present in more than two-thirds of patients and refractory epilepsy in one-third of patients (Kister at al., 2008).

Headaches account for approximately 18% of CNS manifestations in scleroderma and include both migraines and tension-type cephalgias (Amaral et al., 2013). They occur more frequently in linear scleroderma and Parry–Romberg syndrome and should be appropriately evaluated to exclude secondary pathologies, as patients with scleroderma and CNS vasculitis or with linear scleroderma and ipsilateral brain cavernous malformation presenting with headaches as an initial manifestation have been documented (Pathak and Gabor, 1991; Fain et al., 2011).

Optic neuropathy and spinal cord involvement are rare. Bilateral optic neuritis in a patient with CREST syndrome and unilateral optic atrophy have been reported (Boschi et al., 1992; Das et al., 2002). Likewise, spinal cord involvement secondary to external compression by ectopic calcinosis and transverse myelitis in systemic scleroderma may occur (Lima et al., 2004; Torabi et al., 2004). Stroke is unusual in patients with scleroderma and may be associated with prominent renal or pulmonary involvement with malignant hypertension secondary to vasculitis or of unknown etiology (Pathak and Gabor, 1991; Das et al., 2002).

Other CNS manifestations include spinal cord involvement (Amaral et al., 2013), leukoencephalopathy, cerebellar ataxia or degeneration, PRES, cognitive impairments, and behavioral disorders. The most common psychiatric manifestations are depression followed by anxiety (Legendre et al., 2005; Amaral et al., 2013). Pain is a major factor and a potential determinant of physical function and social adjustment in depression associated with scleroderma (Benrud-Larson et al., 2002).

Peripheral nervous system/autonomic nervous system, and muscle manifestations

Peripheral neuropathy, autonomic neuropathy, and myopathies are the most common neurologic manifestations. Peripheral nervous system manifestations occur in 5–16% of patients and include carpal tunnel syndrome, trigeminal neuropathies, mononeuritis multiplex, plexopathy, lumbosacral radiculopathy, peripheral sensorimotor polyneuropathy, and myopathy (Lee et al., 1984a; Hietaharju et al., 1993). Cranial nerve abnormalities occur in less than 10% of patients; trigeminal neuropathy is the most common (Amaral et al., 2013). Facial neuropathy and extraocular movement disorders may be related to cutaneous involvement (Amaral et al., 2012). Sensory trigeminal neuropathy and masticatory spasms may represent primary peripheral nervous system involvement (Teasdall et al., 1980; Kim et al., 2000; Amaral et al., 2013). Common entrapment neuropathies include median and ulnar neuropathy. Typically, sensorimotor peripheral nerve involvement follows a topographical pattern in relation to overlying tissue involvement and vascular tone. However, electrophysiologic findings may not be fully explained by external compression, suggesting an ischemic microvascular component (Schady et al., 1991; Lori et al., 1995). Furthermore, asymptomatic peripheral nervous system involvement is not uncommon, suggesting a higher prevalence of peripheral nervous involvement in scleroderma than reported (Raszewa et al., 1991; Schady et al., 1991).

The prevalence of autonomic nervous system dysfunction in scleroderma varies from 14% to 79% (Amaral et al., 2013). Systemic autonomic dysfunction involves cardiovascular, gastrointestinal, bladder, skin, and pupillary dysfunction. Sympathetic overdrive and parasympathetic underactivity account for heart rate and blood pressure changes, gastrointestinal motility disorders, anorectal disorders, abnormal skin responses, pupillary basal diameter and latency abnormalities even prior to onset of organ fibrosis (Amaral et al., 2013). Abnormal sympathetic skin responses are reported in two-thirds of patients and more common in linear scleroderma. Sympathetic skin responses do not correlate with disease severity, localization, or duration or symptoms of the autonomic system disorder (Raszewa et al., 1991). Abnormal sympathetic skin responses have also been observed in patients with Raynaud's phenomenon (Schady et al., 1991).

Myopathy occurs in 14% of patients with scleroderma (Mimura et al., 2005; Amaral et al., 2013). Myopathy is more common among men, and develops at a median of 6 years from disease onset. These patients often present at a younger age than nonmyopathic scleroderma patients. Almost all patients have proximal muscle weakness and two-thirds have abnormal EMG findings and abnormal histopathology. Anticentromere antibodies are commonly negative (Mimura et al., 2005).

Although myopathy associated with scleroderma is typically mild, some patients may develop a severe and debilitating course (Olsen et al., 1996). Careful consideration of the existence of "overlap syndrome" is required, as inflammatory myopathy may coexist. In particular, dermatomyositis only *truly* overlaps with scleroderma as it shares similar findings such as sclerotic skin changes, microangiopathy, and esophageal hypomotility (Dalakas and Hohlfeld, 2003). The term "scleromyositis" has been used to describe a unique entity with distinct clinical and immunologic features from systemic scleroderma associated with myositis (Jablonska and Blaszcyk, 1998). Patients with scleromyositis typically have a milder disease course with infrequent visceral involvement, but more common association with interstitial lung disease, scleroderma-like changes of the face and hands, arthralgia, myalgia, myositis, Raynaud's disease, and calcinosis. Although not disease-specific, the polymyositis/Scl antibody is present in about 90% and the HLA-DR3 in almost all patients with scleromyositis (Jablonska and Blaszcyk, 1998).

Evaluation

Typically, most patients with systemic scleroderma will exhibit clinically characteristic skin and nail fold capillary changes, Raynaud's phenomenon, antinuclear antibodies, and some degree of systemic involvement. Atypical cases have a combination of any of the aforementioned features without skin changes (Barnett et al., 1988). Over 90% of scleroderma patients have positive antinuclear antibodies. Anti-Ro/SSA and anti-La/-SSB antibodies may be present in patients with primary scleroderma. Anticentromere antibodies are highly specific and virtually diagnostic of limited cutaneous systemic sclerosis and the CREST syndrome, although they have a sensitivity of only 65%. The anti-Scl-70 antibody has a specificity of almost 100% and diagnoses scleroderma in a patient with Raynaud's or CREST syndrome, although it has poor sensitivity (Birtane, 2012).

Neuroimaging is required to evaluate patients with CNS involvement. Typically, neurologic lesions are ipsilateral to skin lesions. Findings include intracranial calcifications, cortical atrophy, and subcortical T2 hyperintensity abnormalities which may enhance following gadolinium administration. Skin or skull atrophy may be observed (Fig. 3.25). Other findings include depression of the skull cortex, encephalomalacia, hemorrhage, spinal cord lesions, and rarely vascular abnormalities, such as aneurysms (Kister et al., 2008; Amaral et al., 2012, 2013). EEG should be considered in patients with epileptic seizures.

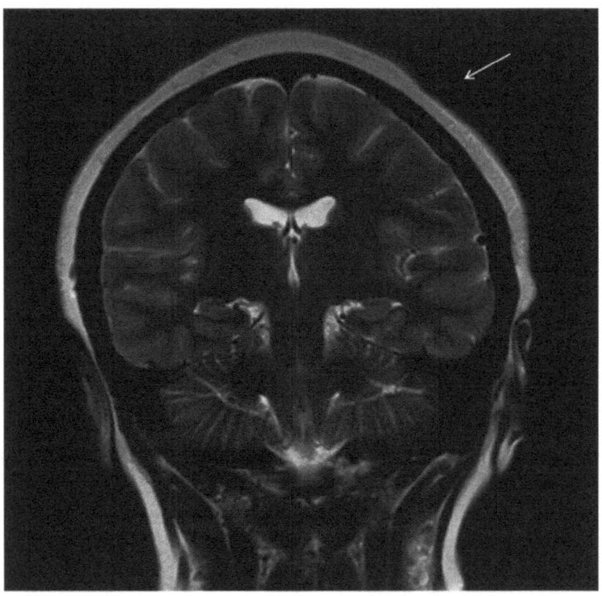

Fig. 3.25. Coronal T2-weighted image showing skin atrophy over the left frontal bone (white arrow).

Peripheral neuropathy requires evaluation with appropriate electrodiagnostic studies. Quantitative sensory testing may yield evidence of increased cold or vibration thresholds, suggesting small-fiber non-length-dependent involvement (Poncelet and Connolly, 2003). Blink reflex analysis may be abnormal in patients with trigeminal neuropathy. Autonomic testing such as sympathetic skin responses, heart rate variability, or sweat-spot testing may be required for evaluation of sympathetic dysfunction (Amaral et al., 2013).

Serum CK levels may be elevated in myositis. EMG findings include low voltage and/or short duration during muscle contraction, fibrillations or sharp waves. Muscle biopsy may show evidence of myofibrillary necrosis and cellular infiltration (Mimura et al., 2005). Other studies such as MR imaging, P-31 magnetic resonance spectroscopy and near infrared spectroscopy have been used to evaluate muscular involvement in scleroderma (Olsen et al., 1996). Paraspinal and intraspinal calcifications may be present (Amaral et al., 2013). Antibodies of extractable nuclear antigens (ENA) such as histidyl-tRNA synthetase (Jo-1), and polymyositis1 may be of use in overlap disease and patients with myopathy (Birtane, 2012). U1RNP is present in cases of scleroderma with overlap, *termed mixed connective tissue disease* (Jablonska and Blaszcyk, 1998). Other antibodies in association with myopathy are reviewed in the section on dermatomyositis.

Management and prognosis

At present, there are no established treatment guidelines or randomized controlled trials to aid the management of the neurologic manifestations of scleroderma. Epilepsy is managed with standard AEDs, with the majority of patients having mild forms and many achieving seizure control with monotherapy. However, up to a third of patients with linear scleroderma or Parry–Romberg syndrome may have refractory epilepsy (Kister et al., 2008; Amaral et al., 2013). The majority of patients with refractory epilepsy respond over time to adjunct immunosuppression with corticosteroids; however, epilepsy surgery may be required (Amaral et al., 2013).

Other forms of CNS involvement are managed with corticosteroids. Adjunct therapies with methotrexate, mycophenolate mofetil, *D*-penicillamine, and azathioprine have been used with variable success (Amaral et al., 2012, 2013). IVIG has not been as successful in treating CNS symptoms (Sathornsumetee et al., 2005; Paprocka et al., 2006). Surgery may be required for compressive myelopathy, radiculopathy, or entrapment peripheral nerve syndromes. Variable responses to corticosteroids and immunosuppression have been observed in the treatment of peripheral and cranial neuropathies.

However, some patients with myopathy respond (Amaral et al., 2013). Pain control is an important component in scleroderma management, as almost half of patients with systemic scleroderma report mid-level pain that impacts functionality and overall quality of life (Benrud-Larson et al., 2002).

PARRY–ROMBERG SYNDROME

Parry Romberg syndrome, first described by Parry (1825) and later by Romberg (1846), was once known as "trophoneurosis of the face." In 1871, Eulenberg coined the term "progressive facial hemiatrophy," which is currently used to describe Parry–Romberg syndrome (Kumar et al., 2009). Progressive facial hemiatrophy is a rare disorder occurring more often in women. Median age of onset is 7–10 years (Stone, 2003; Chiu et al., 2012b), although later occurrence has been described (Stone, 2003; Mendonça et al., 2005). Prominent facial atrophy with enophthalmos is quite common, while ipsilateral extremity or bilateral facial involvement is unusual (Bellusci et al., 2002).

Considered to be a severe variant of scleroderma, Parry–Romberg syndrome presents with progressive loss of subcutaneous fat, hyperpigmentation or hypopigmentation, and mild sclerosis generally affecting one side of the face along the trigeminal nerve distribution (Connolly, 2012). This is referred to as *en coup de sabre* (ECDS), as it resembles the scar of a wound inflicted by a saber. Clinically and histopathologically, it is indistinguishable from linear morphea (Amaral et al., 2012). It has been suggested that Parry–Romberg syndrome may have two separate clinical subtypes, one similar to linear scleroderma (ECDS), involving cutaneous structures, while the second only involves subcutaneous structures primarily within the cheek area (Sommer et al., 2006). Overlap between these variants is common and both have been described under the rubric of "craniofacial linear scleroderma" (Kister et al., 2008).

Etiology and pathophysiology

The cause of Parry–Romberg syndrome is not established. The hemiatrophy may evolve for 2–20 years before stabilizing (Duymaz et al., 2009). The role of autonomic denervation was proposed more than five decades ago, after cervical sympathectomy led to progressive facial hemiatrophy in a rat model (Moss and Crikelair, 1960). Autonomic testing has yielded conflicting results on the presence (Lonchampt et al., 1995; Drummond et al., 2006; Guo et al., 2011) or absence (Larner and Bennison, 1993) of autonomic dysregulation.

Vasculitic cranial nerve involvement has been proposed to account for trigeminal neuralgia and MRI findings in Parry–Romberg syndrome (Pensler et al., 1990; Cory et al., 1997; Moko et al., 2003). Facial trauma has been reported by more than a fourth of these patients, although the exact temporal correlation has not been established (Stone, 2003). Failure of neural crest migration has been proposed to have a role in Parry–Romberg syndrome associated with intracranial vascular abnormalities (Schievink et al., 1996; Pichiecchio et al., 2002; Qureshi et al., 2010). Other proposed mechanisms include genetic inheritance, inflammation, alterations in fat metabolism, viral infection and autoimmunity (Wolf and Verity, 1974; Larner and Bennison, 1993, Bellusci et al., 2002; Kister et al., 2008; Kumar et al., 2009; Lee et al., 2015).

Neurologic manifestations

The prevalence of neurologic complications associated with Parry–Romberg syndrome is unknown. Retrospective studies estimate that 19–39% of children with ECDS and Parry–Romberg syndrome have neurologic symptoms (Zulian et al., 2005; Zulian et al, 2006; Christen-Zaech et al., 2008; Chiu et al., 2012b). Neurologic involvement includes epilepsy, migraines, facial pain, masticatory spasms, cognitive impairment, cranial nerve palsies, focal motor or sensory deficits, as well as intracranial abnormalities usually ipsilateral to the facial hemiatrophy (Woolfenden et al., 1998; Stone, 2003; Chiu et al., 2012b) and rarely contralateral (Guo et al., 2011). In an Internet patient response-based global survey of 205 patients with Parry–Romberg syndrome, migraine occurred in 53%, facial pain in 46%, and epilepsy in 11% (Stone, 2003). Retrospective reviews suggest that epilepsy is the most frequent complication of Parry–Romberg syndrome (Kister et al., 2008; Chiu et al., 2012b) with a third of patients suffering from intractable epilepsy (Kister et al., 2008).

Migraine headaches with or without aura are more prevalent in Parry–Romberg syndrome than in the general population (Cory et al., 1997; Woolfenden et al., 1998; Pichiecchio et al., 2002; Stone, 2003; Drummond et al., 2006). Mild auras in the form of focal paresthesias (Pichiecchio et al., 2002), recurrent motor deficits (Sagild and Alving, 1985; Woolfenden et al., 1998), and status migrainosus have been described (Menascu et al., 2009).

Facial pain in association with headache is common (Drummond et al., 2006; Kumar et al., 2009). It occurs within years after development of Parry–Romberg syndrome, and is characterized as sharp, stabbing or burning and often aggravated by touching the affected area or chewing, thus resembling trigeminal neuralgia (Drummond et al., 2006; Kumar et al., 2009; Viana et al., 2011).

Unusual neurologic manifestations include Adie's tonic pupil (Aynaci et al., 2001), strokes (Menni et al., 1997; Tomizawa et al., 2014) and Rasmussen's encephalitis (Shah et al., 2003; Longo et al., 2011). Brainstem involvement characterized by monoplegic ataxia secondary to ipsilateral pontine involvement (Lee et al., 2015) and dysarthria, dysphagia, and oculomotor palsy due to pontomesencephalic involvement have been described (Sathornsumetee et al., 2005).

Neuroimaging findings

Intracranial abnormalities may be present in two-thirds of patients with Parry–Romberg syndrome (Chiu et al., 2012b). Common MRI findings include white matter hyperintense lesions on T2-weighted images (49%), intracranial calcifications (25%), and ipsilateral cerebral atrophy (17%) (Chiu et al., 2012b). Multiple or diffuse lesions are present in over half of the patients (Kister et al., 2008). Other findings include leptomeningeal enhancement and cortical thickening (Cory et al., 1997; Chiu et al., 2012b). Intracranial vascular abnormalities (Miedziak et al., 1998), aneurysms (Schievink et al., 1996; Pichiecchio et al., 2002; Kuechler et al., 2011), reversible constrictive changes (Woolfenden et al., 1998), and hypoplastic vessels (Qureshi et al., 2010) have been reported. MRA findings suggestive of vasculitis have been found in patients with craniofacial linear scleroderma (Kister et al., 2008).

Monoventricular enlargement, meningocortical dysmorphia, and white-matter changes have been postulated to be secondary to localized meningoencephalitis with vascular involvement (Terstegge et al., 1994). While intracranial abnormalities are common, their presence do not correlate with neurologic symptoms, as neurologically asymptomatic patients may have intracranial abnormalities and vice versa (Moko et al., 2003; Chiu et al., 2012b).

Diagnosis and management

Parry–Romberg syndrome is diagnosed by history and characteristic clinical findings. Differentiating between Parry–Romberg syndrome and localized scleroderma is challenging, as some patients initially diagnosed with localized scleroderma have been known to later develop progressive facial hemiatrophy (Kister et al., 2008). Preservation of elastic tissue on histologic studies may help differentiate between these entities (Terstegge et al., 1994).

Variable reports include evaluation of rheumatologic and autoimmune serologies such as serum ANA titers, ds-DNA, rheumatoid factor and CSF studies for presence of pleocytosis and oligoclonal bands. Electroencephalography (EEG), brain MRI and angiography should be considered in symptomatic patients. MRI spectroscopy may detect early CNS involvement in patients with normal MRI studies (Kister et al., 2008).

Due to the complexity of Parry–Romberg syndrome, a multidisciplinary approach by dermatology, neurology, ophthalmology, and rheumatology seems appropriate. Contrast enhanced MRI should be considered as well as serum or CSF studies in patients with suspected inflammatory or associated autoimmune disease (Kister et al., 2008). Treatment may include surgical intervention for cosmetic purposes and medical management of pertinent neurologic, dermatologic, or rheumatologic manifestations. Cosmetic surgery involves the use of fat grafts for mild deformity and cartilage or bone grafts for more advanced disease (Guerrerosantos et al., 2007). Facial pain is typically managed as in cases of trigeminal neuralgia (Kumar et al., 2009). Successful use of occipital nerve has been reported (Viana et al., 2011). Patients with cerebral involvement, evidence of leptomeningeal enhancement, angiographic findings suggestive of vasculitis, or fleeting neurologic deficits with hyperintense lesions on MRI may require corticosteroids and other immunosuppressive therapy (Korkmaz et al., 2005; Kister et al., 2008; Lee et al., 2015).

DEGOS DISEASE (MALIGNANT ATROPHIC PAPULOSIS)

First described by Kohlmeier and Degos, malignant atrophic papulosis or Degos disease is a rare thrombo-obliterative vasculopathy of undetermined origin (Kohlmeier, 1941; Degos et al., 1942). Skin manifestations are often the initial manifestation of Degos disease and present with equal gender distribution, usually in the second to fourth decade of life. Skin lesions appear as crops of small (2–5 mm) rounded edematous and erythematous papules occurring most commonly on the trunk or extremities, sparing the palms and soles. They may also involve the bulbar conjunctiva, oral mucosa, and genitalia. Within weeks, the papules typically develop an area of central depression that becomes a porcelain white scar with a livid red rim of telangiectasia (James et al., 2011; Bolognia et al., 2012). Histopathologically, the lesions demonstrate epidermal atrophy, hyperkeratosis, and an underlying wedge-shaped area of ischemia and necrosis in the dermis with a prominent lymphocytic infiltrate (Calonje et al., 2012).

While the pathophysiology of Degos disease is incompletely understood, a dominantly inherited predisposition has been suggested (Kisch and Bruynzeel, 1984; Katz et al., 1997). Other proposed mechanisms are vasculitis, coagulopathy, and endothelial cell dysfunction (Theodoridis et al., 2013).

Systemic and central nervous system manifestations

Beyond cutaneous manifestations, the gastrointestinal tract is most commonly involved and may present with bleeding or perforation. Systemic symptoms may develop suddenly, years after the skin lesions and less frequently prior to appearance the dermatologic manifestations (Theodoridis et al., 2013). Approximately 70% of patients develop systemic manifestations within 1 year of disease onset (Theodoridis et al., 2014). The heart, lungs, eyes, and kidneys may be compromised. CNS involvement has been reported in 20% of patients (Subbiah et al.,1996). Systemic involvement, particularly gastrointestinal has serious implications, with a 2 year fatality rate of over 50% (Degos, 1979; Loewe et al., 2005; Theodoridis et al., 2014). Degos disease with systemic involvement has been termed "malignant," and comprises 75% of cases, in contrast to the more self-isolated cutaneous variant (Heymann, 2009; Theodoridis et al., 2013).

Neurologic manifestations include thrombosis of cerebral arteries and venous sinuses, leading to multiple infarctions, intracranial hemorrhage, encephalitis, meningitis, seizures, and radiculitis. Myelopathy, ophthalmoplegia, cranial neuropathy, and myopathy are less common (Olmos and Laugier, 1977; McFarland et al., 1978; Lee et al., 1984b; Subbiah et al., 1996; Amato et al., 2005). When CNS lesions are present, gastrointestinal symptoms are usually mild or absent (Dastur et al., 1981). Arterial fibrous subendothelial plaques, meningeal fibrosis, and superimposed inflammation may be underlying pathologic causes of CNS involvement (Amato et al., 2005).

EVALUATION AND MANAGEMENT

Degos disease is potentially life-threatening, and differentiating it from SLE, granulomatosis with polyangiitis, and other connective tissue diseases may be challenging (Scheinfeld, 2007; Su et al., 2015). Identifying the characteristic cutaneous lesions is key, because there are no specific laboratory tests for the diagnosis of Degos disease. Patients may have antiphospholipid antibodies, but these are of no specific diagnostic value (Levine and Welch, 1987). MRI findings may show ischemic changes, subdural or intracerebral hemorrhage, focal areas of abnormal signal intensities, subdural fluid accumulation, and ependymal or dural enhancement (Subbiah et al.,1996; Amato et al., 2005; Slaviero et al., 2009). Intracranial vasculitis affecting medium sized arteries has been observed (Petit et al., 1982). Elevated CSF protein content and generalized nonspecific slowing on EEG may occur in patients with definite neurologic involvement (Subbiah et al.,1996).

There is no specific treatment for Degos disease. Platelet antiaggregants or anticoagulants are often used in patients with both cutaneous and systemic variants and are considered first line therapy in newly diagnosed patients (Theodoridis et al., 2013). Other proposed treatments include corticosteroids and immunosuppressants. Successful treatment with IVIG and combination therapy of cyclophosphamide with corticosteroids have been reported (Kameda et al., 2012; Guo et al., 2014). However, extra caution is required, as the use of corticosteroids may be associated with increased risk of infection and intestinal perforation.

Albeit a rare disease, systemic manifestations of Degos disease are common and CNS involvement is potentially life-threatining. Early identification and differentiation between cutaneous and malignant forms is key.

References

Agmon-Levin N, Lian Z, Shoenfeld Y (2011). Explosion of autoimmune diseases and the mosaic of old and novel factors. Cell Mol Immunol 8: 189–192.

Ahearn JM, Liu CC, Kao AH et al. (2012). Biomarkers for systemic lupus erythematosus. Transl Res 159: 326–342.

Ainiala H, Hietaharju A, Loukkola J et al. (2001). Validity of the new American College of Rheumatology criteria for neuropsychiatric lupus syndromes: a population-based evaluation. Arthritis Care Res 45: 419–423.

Akbulut L, Gur G, Bodur H et al. (2007). Peripheral neuropathy in Behçet disease: an electroneurophysiological study. Clin Rheumatol 26: 1240–1244.

Akman-Demir G, Serdaroglu P, Taşçi B (1999). Clinical patterns of neurological involvement in Behçet's disease: evaluation of 200 patients. Brain 122: 2171–2182.

Al-Araji A, Kidd DP (2009). Neuro-Behçet's disease: epidemiology, clinical characteristics and management. Lancet Neurol 8: 192–204.

Al-Araji A, Sharquie K, Al-Rawi Z (2003). Prevalence and patterns of neurological involvement in Behçet's disease: a prospective study from Iraq. J Neurol Neurosurg Psychiatry 74: 608–613.

Alekberova ZS, Prokaeva TB, Reshetniak TM et al. (1999). Antiphospholipid antibodies in Behcet's disease. Klin Med 78: 37–38.

Alexander EL (1985). Central nervous system (CNS) manifestations of primary Sjögren's syndrome: an overview. Scand J Rheumatol Suppl 61: 161–165.

Alexander EL, Alexander GE (1983). Aseptic meningoencephalitis in primary Sjögren's syndrome. Neurology 33: 593.

Alexander EL, Hirsch TJ, Arnett FC et al. (1981a). Ro (SSA) and La (SSB) antibodies in the clinical spectrum of Sjögren's syndrome. J Rheumatol 9: 239–246.

Alexander GE, Provost TT, Stevens MB et al. (1981b). Sjögren syndrome. Central nervous system manifestations. Neurology 31: 1391.

Alexander EL, Provost TT, Stevens MB et al. (1982). Neurologic complications of primary Sjögren's syndrome. Medicine 61: 247–257.

Alexander EL, Malinow K, Lejewski JE et al. (1986). Primary Sjögren's syndrome with central nervous system disease mimicking multiple sclerosis. Ann Intern Med 104: 323–330.

Alkan A, Goktan A, Karincaoglu Y et al. (2012). Brain perfusion MRI findings in patients with Behçet's disease. Scientific World Journal 2012: 261502.

Amaral TN, Neto JFM, Lapa AT et al. (2012). Neurologic involvement in scleroderma en coup de sabre. Autoimmune Dis 2012: 719685.

Amaral TN, Peres FA, Lapa AT et al. (2013). Neurologic involvement in scleroderma: a systematic review. Semin Arthritis Rheum 43: 335–347.

Amato AA, Barohn RJ (2009). Evaluation and treatment of inflammatory myopathies. J Neurol Neurosurg Psychiatry 80: 1060–1068.

Amato C, Ferri R, Elia M et al. (2005). Nervous system involvement in Degos disease. Am J Neuroradiol 26: 646–649.

Anaya JM, Villa LA, Restrepo L et al. (2002). Central nervous system compromise in primary Sjögren's syndrome. JCR J Clin Rheumatol 8: 189–196.

Appenzeller S, Cendes F, Costallat LT (2004). Epileptic seizures in systemic lupus erythematosus. Neurology 63: 1808–1812.

Araújo FG, Amaral TN, Appenzeller S et al. (2013). Seizures in systemic sclerosis. Rheumatol Int 33: 1721–1724.

Arbuckle MR, McClain MT, Rubertone MV et al. (2003). Development of autoantibodies before the clinical onset of systemic lupus erythematosus. New Engl J Med 349: 1526–1533.

Arida A, Fragiadaki K, Giavri E et al. (2011). Anti-TNF agents for Behcet's disease: analysis of published data on 369 patients. Semin Arthritis Rheum 41: 61–70.

Atzeni F, Turiel M, Caporali R et al. (2010). The effect of pharmacological therapy on the cardiovascular system of patients with systemic rheumatic diseases. Autoimmun Rev 9: 835–839.

Aynaci FM, Şen Y, Erdöl H et al. (2001). Parry–Romberg syndrome associated with Adie's pupil and radiologic findings. Pediatr Neurol 25: 416–418.

Bahar S, Coban O, Gürvit IH et al. (1993). Spontaneous dissection of the extracranial vertebral artery with spinal subarachnoid haemorrhage in a patient with Behçet's disease. Neuroradiology 35: 352–354.

Baizabal-Carvallo JF, Barragán-Campos HM, Padilla-Aranda HJ et al. (2009). Posterior reversible encephalopathy syndrome as a complication of acute lupus activity. Clin Neurol Neurosurg 111: 359–363.

Baizabal-Carvallo JF, Bonnet C, Jankovic J (2013). Movement disorders in systemic lupus erythematosus and the antiphospholipid syndrome. J Neural Transm 120: 1579–1589.

Bakouche P, Guillard A (1983). Polyradiculoneuritis in a developmental flare–up of Behcet's disease. Rev Neurol 140: 520–522.

Bansal SK, Sawhney IMS, Chopra JS (1987). Epilepsia partialis continua in Sjögren's syndrome. Epilepsia 28: 362–363.

Barber CE, Leclerc R, Gladman DD et al. (2011). Posterior reversible encephalopathy syndrome: an emerging disease manifestation in systemic lupus erythematosus. Semin Arthritis Rheum 41: 353–363.

Barnett AJ, Miller M, Littlejohn GO (1988). The diagnosis and classification of scleroderma (systemic sclerosis). Postgrad Med J 64: 121–125.

Bartynski WS (2008a). Posterior reversible encephalopathy syndrome. Part 1: fundamental imaging and clinical features. Am J Neuroradiol 29: 1036–1042.

Bartynski WS (2008b). Posterior reversible encephalopathy syndrome. Part 2: controversies surrounding pathophysiology of vasogenic edema. Am J Neuroradiol 29: 1043–1049.

Behçet H (1937). Uber residivierende aphtose durch ein Virusverursachtes Geschwure am Mund am Auge und an der Genitalien. Dermatol Wochenschr 105: 1152–1157.

Bellusci C, Liguori R, Pazzaglia A et al. (2002). Bilateral Parry–Romberg syndrome associated with retinal vasculitis. Eur J Ophthalmol 13: 803–806.

Belmont HM (2013). Treatment of systemic lupus erythematosus – 2013 update. Bull Hosp Jt Dis 71: 208–213.

Bendewald MJ, Wetter DA, Li X et al. (2010). Incidence of dermatomyositis and clinically amyopathic dermatomyositis: a population-based study in Olmsted County, Minnesota. Arch Dermatol 146: 26.

Benrud-Larson LM, Haythornthwaite JA, Heinberg LJ et al. (2002). The impact of pain and symptoms of depression in scleroderma. Pain 95: 267–275.

Bernacchi E, Amato L, Parodi A et al. (2004). Sjögren's syndrome: a retrospective review of the cutaneous features of 93 patients by the Italian Group of Immunodermatology. Clin Exp Rheumatol 22: 55–62.

Bertsias GK, Boumpas DT (2010). Pathogenesis diagnosis and management of neuropsychiatric SLE manifestations. Nat Rev Rheumatol 6: 358–367.

Bertsias GK, Ioannidis JPA, Aringer M et al. (2010). EULAR recommendations for the management of systemic lupus erythematosus with neuropsychiatric manifestations: report of a task force of the EULAR standing committee for clinical affairs. Ann Rheum Dis 69: 2074–2082.

Birol A, Ulkatan S, Kocak M et al. (2004). Peripheral neuropathy in Behçet's disease. J Dermatol 31: 455–459.

Birtane M (2012). Diagnostic role of anti-nuclear antibodies in rheumatic diseases. Turkish J Rheumatol 27: 79–89.

Bohan A, Peter JB (1975). Polymyositis and dermatomyositis. N Engl J Med 292: 344–347.

Bolognia JL, Jorizzo JL, Schaffer JV (2012). Systemic sclerosis (scleroderma) and related disorders. In: JL Bolognia, JL Jorizzo, JV Schaffer (Eds.), Dermatology. Elsevier, Oxford, ch. 43.

Bombardier C, Gladman DD, Urowitz MB et al. (1992). Derivation of the SLEDAI. A disease activity index for lupus patients. Arthritis Rheum 35: 630–640.

Boschi A, Snyers B, Lambert M (1992). Bilateral optic neuropathy associated with the CREST variant of scleroderma. Eur J Ophthalmol 3: 219–222.

Bowyer SL, Blane CE, Sullivan DB et al. (1983). Childhood dermatomyositis. J Pediatr 103: 882–888.

Brown VE, Pilkington CA, Feldman BM et al. (2006). An international consensus survey of the diagnostic criteria for juvenile dermatomyositis (JDM). Rheumatology 45: 990–993.

Buse DC, Loder EW, Gorman JA et al. (2013). Sex differences in the prevalence symptoms and associated features of migraine probable migraine and other severe headache: results of the American Migraine Prevalence and Prevention (AMPP) study. Headache 53: 1278–1299.

Caldas CAM, de Carvalho JF (2011). Primary antiphospholipid syndrome with and without Sneddon's syndrome. Rheumatol Int 31: 197–200.

Callen JP (1995). Dermatomyositis. In: JP Callen (Ed.), Dermatological Signs of Internal Disease, 2nd edn. Saunders, Oxford, pp. 13–20.

Calonje E, Brenn T, Lazar A (2012). The vasculopathic reaction pattern. In: E Calonje, T Brenn, A Lazar et al. (Eds.), McKee's Pathology of the Skin, 4th edn. Elsevier, Oxford, ch. 8.

Campi A, Filippi M, Comi G et al. (1998). Recurrent acute transverse myelopathy associated with anticardiolipin antibodies. Am J Neuroradiol 19: 781–786.

Cervera R, Font J, Pare C et al. (1992). Cardiac disease in systemic lupus erythematosus: prospective study of 70 patients. Ann Rheum Dis 51: 156–159.

Cervera R, Piette JC, Font J et al. (2002). Antiphospholipid syndrome: clinical and immunologic manifestations and patterns of disease expression in a cohort of 1,000 patients. Arthritis Rheum 464: 1019–1027.

Cervera R, Boffa MC, Khamashta MA et al. (2009). The Euro-Phospholipid project: epidemiology of the antiphospholipid syndrome in Europe. Lupus 18: 889–893.

Chen Z, Maize JC, Silver RM et al. (1985). Direct and indirect immunofluorescent findings in dermatomyositis. J Cutan Pathol 12: 18–27.

Chen WH, Yeh JH, Chiu HC (2001). Plasmapheresis in the treatment of ataxic sensory neuropathy associated with Sjögren's syndrome. Eur Neurol 45: 270–274.

Chen IJ, Wu YJJ, Lin CW et al. (2009). Interstitial lung disease in polymyositis and dermatomyositis. Clin Rheumatol 28: 639–646.

Chen YJ, Wu CY, Huang YL et al. (2010). Cancer risks of dermatomyositis and polymyositis: a nationwide cohort study in Taiwan. Arthritis Res Ther 12: R70.

Chiu CC, Huang CC, Chan WL et al. (2012a). Increased risk of ischemic stroke in patients with systemic lupus erythematosus: a nationwide population-based study. Intern Med 51: 17–21.

Chiu YE, Vora S, Kwon EKM et al. (2012b). A significant proportion of children with morphea en coup de sabre and Parry–Romberg syndrome have neuroimaging findings. Pediatr Dermatol 29: 738–748.

Christen-Zaech S, Hakim MD, Afsar FS et al. (2008). Pediatric morphea (localized scleroderma): review of 136 patients. J Am Acad Dermatol 59: 385–396.

Chuang YW, Hsu CC, Huang YF et al. (2013). Brain perfusion SPECT in patients with Behçet's disease. J Neuroradiol 40: 288–293.

Cieślik P, Hrycek A, Kłuciński P (2008). Vasculopathy and vasculitis in systemic lupus erythematosus. Pol Arch Med Wewn 118: 57–63.

Connolly MK (2012). Systemic sclerosis (scleroderma). and related disorders. In: JL Bolognia, JL Jorizzo, JV Schaffer (Eds.), Dermatology. Elsevier, Oxford, ch. 43.

Cory RC, Clayman DA, Faillace WJ et al. (1997). Clinical and radiologic findings in progressive facial hemiatrophy (Parry–Romberg syndrome). Am J Neuroradiol 18: 751–757.

Coulter I, Huda S, Baborie A et al. (2012). Longitudinally extensive transverse myelitis as the sole presentation of neuro-Behçet's disease responding to infliximab. J Spinal Cord Med 35: 122–124.

Dalakas MC (1991). Polymyositis dermatomyositis and inclusion-body myositis. N Engl J Med 325: 1487–1498.

Dalakas MC (2010). Immunotherapy of myositis: issues concerns and future prospects. Nat Rev Rheumatol 6: 129–137.

Dalakas MC, Hohlfeld R (2003). Polymyositis and dermatomyositis. Lancet 362: 971–982.

Dalakas MC, Illa I, Dambrosia JM et al. (1993). A controlled trial of high-dose intravenous immune globulin infusions as treatment for dermatomyositis. N Engl J Med 329: 1993–2000.

Das CP, Prabhakar S, Lal V et al. (2002). Scleroderma stroke optic neuropathy: a rare association. Neurol India 50: 504.

Dass S, Bowman SJ, Vital EM et al. (2008). Reduction of fatigue in Sjögren syndrome with rituximab: results of a randomised double-blind placebo-controlled pilot study. Ann Rheum Dis 67: 1541–1544.

Dastur DK, Singhal BS, Shroff HJ (1981). CNS involvement in malignant atrophic papulosis (Kohlmeier–Degos disease): vasculopathy and coagulopathy. J Neurol Neurosurg Psychiatry 44: 156–160.

Davatchi F, Shahram F, Chams-Davatchi C et al. (2010a). Behçet's disease: from East to West. Clin Rheumatol 29: 823–833.

Davatchi F, Shahram F, Chams-Davatchi C et al. (2010b). Behçet's disease in Iran: analysis of 6500 cases. Int J Rheum Dis 13: 367–373.

Davatchi F, Shahram F, Chams-Davatchi C et al. (2012). Behçet's disease: is there a gender influence on clinical manifestations? Int J Rheum Dis 15: 306–314.

Davatchi F, Assaad-Khalil S, Calamia KT et al. (2014). The International Criteria for Behçet's Disease (ICBD): a collaborative study of 27 countries on the sensitivity and specificity of the new criteria. J Eur Acad Dermatol Venereol 28: 338–347.

Degos R (1979). Malignant atrophic papulosis. Br J Dermatol 100: 21–35.

Degos R, Delorot J, Tricot R (1942). Dermatite paplulosquameuse atrophinate. Bull Soc Fr Dermatol Syphiligr 49: 148–150.

Delalande S, de Seze J, Fauchais AL et al. (2004). Neurologic manifestations in primary Sjögren syndrome: a study of 82 patients. Medicine 83 (5): 280–291.

Denburg SD, Denburg JA (2003). Cognitive dysfunction and antiphospholipid antibodies in systemic lupus erythematosus. Lupus 12: 883–890.

Denburg SD, Carbotte RM, Ginsberg JS et al. (1997). The relationship of antiphospholipid antibodies to cognitive function in patients with systemic lupus erythematosus. J Int Neuropsychol Soc 3: 377–386.

Dicken CH (1991). Periorbital edema: an important physical finding in dermatomyositis. Cutis 48: 116–117.

Direskeneli H (2001). Behçet's disease: infectious aetiology new autoantigens and HLA–B51. Ann Rheum Dis 60: 996–1002.

Douglas WW, Tazelaar HD, Hartman TE et al. (2001). Polymyositis–dermatomyositis-associated interstitial lung disease. Am J Respir Crit Care Med 164 (7): 1182–1185.

Dreizen S (1991). The butterfly rash and the malar flush. What diseases do these signs reflect? Postgrad Med 89: 225–228, 233–234.

Dries DL, Rosenberg YD, Waclawiw MA et al. (1997). Ejection fraction and risk of thromboembolic events in patients with systolic dysfunction and sinus rhythm: evidence for gender differences in the studies of left ventricular dysfunction trials. J Am Coll Cardiol 29: 1074–1080.

Drummond PD, Hassard S, Finch PM (2006). Trigeminal neuralgia migraine and sympathetic hyperactivity in a patient with Parry–Romberg syndrome. Cephalalgia 26: 1146–1149.

Dubowitz V, Sewry CA, Oldfors A et al. (2013). Inflammatory myopathies. In: V Dubowitz, CA Sewry, A Oldfors et al. (Eds.), Muscle Biopsy: A Practical Approach. Elsevier, Oxford, ch. 22.

Duymaz A, Karabekmez FE, Keskin M et al. (2009). Parry–Romberg syndrome: facial atrophy and its relationship with other regions of the body. Ann Plast Surg 63: 457–461.

El-Azhary RA, Pakzad SY (2002). Amyopathic dermatomyositis: retrospective review of 37 cases. J Am Acad Dermatol 46: 560–565.

Epstein LC, Masse G, Harmatz JS et al. (2014). Characterization of cognitive dysfunction in Sjögren's syndrome patients. Clin Rheumatol 33: 511–521.

Escudero D, Latorre P, Codina M et al. (1994). Central nervous system disease in Sjögren's syndrome. Ann Med Interne 146: 239–242.

Esdaile JM, Abrahamowicz M, Grodzicky T et al. (2001). Traditional Framingham risk factors fail to fully account for accelerated atherosclerosis in systemic lupus erythematosus. Arthritis Rheum 44: 2331–2337.

Euwer RL, Sontheimer RD (1991). Amyopathic dermatomyositis (dermatomyositis sine myositis): presentation of six new cases and review of the literature. J Am Acad Dermatol 24: 959–966.

Fain ET, Mannion M, Pope E et al. (2011). Brain cavernomas associated with en coup de sabre linear scleroderma: two case reports. Pediatr Rheumatol 9: 18.

Fanouriakis A, Boumpas DT, Bertsias GK (2013). Pathogenesis and treatment of CNS lupus. Curr Opin Rheumatol 25: 577–583.

Feist E, Hermann KG, Dankof A (2009). Citation from translated abstract [Vasculopathy in Sjögren's syndrome]. Z Rheumatol 68: 305–311.

Feldman BM, Rider LG, Reed AM et al. (2008). Juvenile dermatomyositis and other idiopathic inflammatory myopathies of childhood. Lancet 371: 2201–2212.

Femia AN, Vleugels RA, Callen JP (2013). Cutaneous dermatomyositis: an updated review of treatment options and internal associations. Am J Clin Dermatol 14: 291–313.

Ferreira S, D'Cruz DP, Hughes GRV (2005). Multiple sclerosis neuropsychiatric lupus and antiphospholipid syndrome: where do we stand? Rheumatology 44: 434–442.

Font J, Ramos-Casals M, de la Red G et al. (2003). Pure sensory neuropathy in primary Sjögren's syndrome. Longterm prospective followup and review of the literature. J Rheumatol 30: 1552–1557.

French LE, Prins C (2012). Erythema multiforme, Stevens–Johnson syndrome and toxic epidermal necrolysis. In: JL Bolognia, JL Jorizzo, JV Schaffer (Eds.), Dermatology. Elsevier, Oxford, ch. 20.

Frigui M, Kechaou M, Jemal M et al. (2009). Optic neuropathy in Behçet's disease: a series of 18 patients. Rev Med Interne 30: 486–491.

Futrell N, Millikan C (1989). Frequency etiology and prevention of stroke in patients with systemic lupus erythematosus. Stroke 20: 583–591.

Gal Y, Twig G, Mozes O et al. (2013). Central nervous system involvement in systemic lupus erythematosus: an imaging challenge. Israel Med Assoc J IMAJ 15: 382–386.

Galli M, Luciani D, Bertolini G et al. (2003). Lupus anticoagulants are stronger risk factors for thrombosis than anticardiolipin antibodies in the antiphospholipid syndrome: a systematic review of the literature. Blood 101: 1827–1832.

Gemignani F, Marbini A, Pavesi G et al. (1994). Peripheral neuropathy associated with primary Sjögren's syndrome. J Neurol Neurosurg Psychiatry 57: 983–986.

Gerami P, Schope JM, McDonald L et al. (2006). A systematic review of adult-onset clinically amyopathic dermatomyositis (dermatomyositis sine myositis): a missing link within the spectrum of the idiopathic inflammatory myopathies. J Am Acad Dermatol 54: 597–613.

Gerraty RP, McKelvie PA, Byrne E (1993). Aseptic meningoencephalitis in primary Sjögren's syndrome. Acta Neurol Scand 88 (4): 309–311.

Gladman DD, Ibañez D, Urowitz MB (2002). Systemic lupus erythematosus disease activity index 2000. J Rheumatol 29: 288–291.

Gøransson LG, Herigstad A, Tjensvoll AB et al. (2006). Peripheral neuropathy in primary Sjögren syndrome: a population-based study. Arch Neurol 63: 1612–1615.

Gorson KC, Natarajan N, Ropper AH et al. (2007). Rituximab treatment in patients with IVIg-dependent immune polyneuropathy: a prospective pilot trial. Muscle Nerve 35: 66–69.

Gottenberg JE, Cinquetti G, Larroche C et al. (2013). Efficacy of rituximab in systemic manifestations of primary Sjögren's syndrome: results in 78 patients of the AutoImmune and Rituximab registry. Ann Rheum Dis 72: 1026–1031.

Greenberg BM (2009). The neurologic manifestations of systemic lupus erythematosus. Neurologist 15: 115–121.

Greenberg SA, Pinkus JL, Pinkus GS et al. (2005). Interferon-α/β-mediated innate immune mechanisms in dermatomyositis. Ann Neurol 57: 664–678.

Griffin JW, Cornblath DR, Alexander E et al. (1990). Ataxic sensory neuropathy and dorsal root ganglionitis associated with Sjögren's syndrome. Ann Neurol 27: 304–315.

Guerrerosantos J, Guerrerosantos F, Orozco J (2007). Classification and treatment of facial tissue atrophy in Parry–Romberg disease. Aesthetic Plast Surg 31: 424–434.

Gül A, Inanç M, Öcal L et al. (2000). Familial aggregation of Behçet's disease in Turkey. Ann Rheum Dis 59: 622–625.

Guo ZN, Zhang HL, Zhou HW et al. (2011). Progressive facial hemiatrophy revisited: a role for sympathetic dysfunction. Arch Neurol 68: 1195–1197.

Guo YF, Pan WH, Cheng RH et al. (2014). Successful treatment of neurological malignant atrophic papulosis in child by corticosteroid combined with intravenous immunoglobulin. CNS Neurosci Ther 20: 88–91.

Haghighi AB, Safari A (2010). Proposing an algorithm for treatment of different manifestations of neuro-Behcet's disease. Clin Rheumatol 29: 683–686.

Haghighi AB, Sarhadi S, Farahangiz S (2011). MRI findings of neuro-Behçet's disease. Clin Rheumatol 30: 765–770.

Hanly JG, Walsh NM, Fisk JD et al. (1993). Cognitive impairment and autoantlbodies in systemic lupus erythematosus. Rheumatology 32: 291–296.

Hanly JG, Hong C, Smith S et al. (1999). A prospective analysis of cognitive function and anticardiolipin antibodies in systemic lupus erythematosus. Arthritis Rheum 42: 728–734.

Hanly JG, Urowitz MB, O'Keeffe AG et al. (2013). Headache in systemic lupus erythematosus: results from a prospective international inception cohort study. Arthritis Rheum 65: 2887–2897.

Haro R, Revelles JM, Fariña Mdel C et al. (2013). Wong's dermatomyositis: a new case and review of the literature. Int J Dermatol 52: 466–470.

Harrison MJ, Ravdin LD (2006). Cognitive dysfunction in the patient with antiphospholipid antibodies: considerations for cause and treatment. Curr Rheumatol Rep 8: 89–94.

Hatemi G, Silman A, Bang D et al. (2008). EULAR recommendations for the management of Behçet disease. Ann Rheum Dis 67: 1656–1662.

Hawk A, English 3rd JC (2001). Localized and systemic scleroderma. Semin Cutan Med Surg 20: 27–37.

Hervier B, Uzunhan Y, Hachulla E et al. (2011). Antisynthetase syndrome positive for anti-threonyl-tRNA synthetase (anti-PL7). antibodies. Eur Resp J 37: 714–717.

Heymann WR (2009). Degos disease: considerations for reclassification. J Am Acad Dermatol 61: 505–506.

Hietaharju A, Jääskeläinen S, Kalimo H et al. (1993). Peripheral neuromuscular manifestations in systemic sclerosis (scleroderma). Muscle Nerve 16: 1204–1212.

Hill CL, Zhang Y, Sigurgeirsson B et al. (2001). Frequency of specific cancer types in dermatomyositis and polymyositis: a population-based study. Lancet 357: 96–100.

Hirohata S, Kikuchi H, Sawada T et al. (2012). Clinical characteristics of neuro-Behçet's disease in Japan: a multicenter retrospective analysis. Mod Rheumatol 22: 405–413.

Hochberg MC (1997). Updating the American College of Rheumatology revised criteria for the classification of systemic lupus erythematosus. Arthritis Rheum 40: 1725.

Hochberg MC, Silman AJ, Smolen JS et al. (Eds.), (2003). Rheumatology, 3rd edn. Mosby/Elsevier, Edinburgh.

Horbach DA, Van Oort E, Donders RC et al. (1996). Lupus anticoagulant is the strongest risk factor for both venous and arterial thrombosis in patients with systemic lupus erythematosus. Comparison between different assays for the detection of antiphospholipid antibodies. Thromb Haemost 76: 916–924.

Hundley JL, Carroll CL, Lang W et al. (2006). Cutaneous symptoms of dermatomyositis significantly impact patients' quality of life. J Am Acad Dermatol 54: 217–220.

International Headache Society Headache Classification Committee (1988). Classification and diagnostic criteria for headache disorders cranial neuralgias and facial pain. Cephalalgia 8 (Suppl 7): 1–96.

International Study Group for Behçet's Disease (1990). Criteria for diagnosis of Behçet's disease. International Study Group for Behçet's Disease. Lancet 335: 1078–1080.

Isenberg DA, Rahman A, Allen E et al. (2005). BILAG 2004 Development and initial validation of an updated version of the British Isles Lupus Assessment Group's disease activity index for patients with systemic lupus erythematosus. Rheumatology 44: 902–906.

Jablonska S, Blaszcyk M (1998). Scleromyositis: a scleroderma/polymyositis overlap syndrome. Clin Rheumatol 17: 465–467.

James W, Berger T, Elston D (2005). Andrews' Diseases of the Skin: Clinical Dermatology. 10th edn. Saunders, Oxford, pp. 169–172.

James WD, Berger TG, Elston DM (2011). Cutaneous vascular disease. In: WD James, TG Berger, DM Elston (Eds.), Elsevier, Oxford, ch. 35.

Jennekens FGI, Kater L (2002a). The central nervous system in systemic lupus erythematosus. Part 1. Clinical syndromes: a literature investigation. Rheumatology 41: 605–618.

Jennekens FGI, Kater L (2002b). The central nervous system in systemic lupus erythematosus. Part 2. Pathogenetic mechanisms of clinical syndromes: a literature investigation. Rheumatology 41: 619–630.

Jennette JC, Falk RJ, Bacon PA et al. (2013). 2012 Revised International Chapel Hill Consensus Conference nomenclature of vasculitides. Arthritis Rheum 65: 1–11.

Jorizzo JL, Vleugels RA (2012). Dermatomyositis. In: JL Bolognia, JL Jorizzo, JV Schaffer (Eds.), Dermatology. Elsevier, Oxford, ch. 42.

Joseph FG, Lammie GA, Scolding NJ (2007). CNS lupus. A study of 41 patients. Neurology 69: 644–654.

Jung RE, Segall JM, Grazioplene RG et al. (2010). Cortical thickness and subcortical gray matter reductions in neuropsychiatric systemic lupus erythematosus. PLoS One 5, e9302.

Kassan SS, Moutsopoulos HM (2004). Clinical manifestations and early diagnosis of Sjögren syndrome. Arch Intern Med 164 (12): 1275–1284.

Kaçmaz RO, Kempen JH, Newcomb C et al. (2008). Ocular inflammation in Behçet disease: incidence of ocular complications and of loss of visual acuity. Am J Ophthalmol 146: 828–836.

Kadota Y, Tokumaru A, Kamakura K et al. (2002). Primary Sjögren's syndrome initially manifested by optic neuritis: MRI findings. Neuroradiology 44: 338–341.

Kaji K, Fujimoto M, Hasegawa M et al. (2007). Identification of a novel autoantibody reactive with 155 and 140 kDa nuclear proteins in patients with dermatomyositis: an association with malignancy. Rheumatology 46: 25–28.

Kameda T, Dobashi H, Yoneda K et al. (2012). A case of Degos disease successfully treated with corticosteroid combined with cyclophosphamide. Rheumatol Int 32: 2169–2173.

Kamphuis S, Silverman ED (2010). Prevalence and burden of pediatric-onset systemic lupus erythematosus. Nat Rev Rheumatol 6: 538–546.

Kampolis C, Tektonidou M, Moyssakis I et al. (2014). Evolution of cardiac dysfunction in patients with antiphospholipid antibodies and/or antiphospholipid syndrome: a 10-year follow-up study. Semin Arthritis Rheum 43: 558–565.

Kanellopoulos P, Baltoyiannis C, Tzioufas AG (2002). Primary Sjögren's syndrome associated with inclusion body myositis. Rheumatology 41: 440–444.

Kao AH, Lertratanakul A, Elliott JR et al. (2013). Relation of carotid intima-media thickness and plaque with incident cardiovascular events in women with systemic lupus erythematosus. Am J Cardiol 112: 1025–1032.

Katayama I, Kotobuki Y, Kiyohara E et al. (1987). Vasculitis in primary Sjögren's syndrome. Histologic classification and clinical presentation. Am J Clin Pathol 88: 26–31.

Katsumoto TR, Whitfield ML, Connolly MK (2011). The pathogenesis of systemic sclerosis. Annu Rev Pathol 6: 509–537.

Katz SK, Mudd LJ, Roenigk Jr HH (1997). Malignant atrophic papulosis (Degos' disease) involving three generations of a family. J Am Acad Dermatol 37: 480–484.

Kavanaugh A, Tomar R, Reveille J et al. (2000). Guidelines for clinical use of the antinuclear antibody test and tests for specific autoantibodies to nuclear antigens. Arch Pathol Lab Med 124: 71–81.

Kidd D (2003). The prevalence of headache in Behçet's syndrome. In: C Zouboulis (Ed.), Adamantiades–Behçet's Disease. Springer, US, pp. 377–379.

Kidd D, Steuer A, Denman AM et al. (1999). Neurological complications in Behçet's syndrome. Brain 122: 2183–2194.

Kim HJ, Jeon BS, Lee KW (2000). Hemimasticatory spasm associated with localized scleroderma and facial hemiatrophy. Arch Neurol 57: 576–580.

Kisch LS, Bruynzeel DP (1984). Six cases of malignant atrophic papulosis (Degos' disease) occurring in one family. Br J Dermatol 111: 469–471.

Kister I, Inglese M, Laxer RM et al. (2008). Neurologic manifestations of localized scleroderma. A case report and literature review. Neurology 71: 1538–1545.

Kivity S, Arango MT, Ehrenfeld M et al. (2014). Infection and autoimmunity in Sjögren's syndrome: a clinical study and comprehensive review. J Autoimmun 51: 17–22.

Kizawa M, Mori K, Iijima M et al. (2006). Intravenous immunoglobulin treatment in painful sensory neuropathy without sensory ataxia associated with Sjögren's syndrome. J Neurol Neurosurg Psychiatry 77: 967–969.

Klein RQ, Teal V, Taylor L et al. (2007). Number characteristics and classification of patients with dermatomyositis seen by dermatology and rheumatology departments at a large tertiary medical center. J Am Acad Dermatol 57: 937–943.

Koçer N, Islak C, Siva A et al. (1999). CNS involvement in neuro-Behçet syndrome: an MR study. Am J Neuroradiol 20: 1015–1024.

Koga T, Kouhisa Y, Nakamura H et al. (2012). A case of primary Sjögren's syndrome complicated with inflammatory myopathy and interstitial lung disease. Rheumatol Int 32: 3647–3649.

Kohlmeier W (1941). Multiple Hautnekrosen bei thromboangitis obliterans. Arch Dermatol Syph 181: 783–792.

Koreeda Y, Higashimoto I, Yamamoto M et al. (2010). Clinical and pathological findings of interstitial lung disease patients with anti-aminoacyl-tRNA synthetase autoantibodies. Int Med 49: 361–369.

Korkmaz C, Adapinar B, Uysal S (2005). Beneficial effect of immunosuppressive drugs on Parry–Romberg syndrome: a case report and review of the literature. South Med J 98: 939–941.

Kovacs SO, Kovacs SC (1998). Dermatomyositis. J Am Acad Dermatol 39: 899–920.

Kovacs B, Lafferty TL, Brent LH et al. (2000). Transverse myelopathy in systemic lupus erythematosus: an analysis of 14 cases and review of the literature. Ann Rheum Dis 59: 120–124.

Kovacs L, Papos M, Takacs R et al. (2002). Autonomic nervous system dysfunction involving the gastrointestinal and the urinary tracts in primary Sjögren's syndrome. Clin Exp Rheumatol 21: 697–703.

Kraus A, Cifuentes M, Villa AR et al. (1994). Myositis in primary Sjögren's syndrome. Report of 3 cases. J Rheumatol 21: 649–653.

Krause I, Rosen Y, Kaplan I et al. (1999). Recurrent aphthous stomatitis in Behçet's disease: clinical features and correlation with systemic disease expression and severity. J Oral Pathol Med 28: 193–196.

Krishnan E (2005). Stroke subtypes among young patients with systemic lupus erythematosus. Am J Med 118: 1415.e1–1415.e7.

Kuechler D, Kaliaperumal C, Hassan A et al. (2011). Rare disease: aneurysmal subarachnoid haemorrhage in Parry–Rhomberg syndrome. BMJ Case Rep 2011.

Kumar AA, Kumar RA, Shantha GPS et al. (2009). Progressive hemi facial atrophy – Parry Romberg syndrome presenting as severe facial pain in a young man: a case report. Cases J 2: 6776.

Kuo CF, See LC, Yu KH et al. (2011). Incidence cancer risk and mortality of dermatomyositis and polymyositis in Taiwan: a nationwide population study. Br J Dermatol 165: 1273–1279.

Kural-Seyahi E, Fresko I, Seyahi N et al. (2003). The long-term mortality and morbidity of Behçet syndrome: a 2-decade outcome survey of 387 patients followed at a dedicated center. Medicine 82: 60–76.

Lafitte C, Amoura Z, Cacoub P et al. (2001). Neurological complications of primary Sjögren's syndrome. J Neurol 248: 577–584.

Lam GKW, Petri M (2005). Assessment of systemic lupus erythematosus. Clin Exp Rheumatol 23: S120.

Larner AJ, Bennison DP (1993). Some observations on the aetiology of progressive hemifacial atrophy ("Parry–Romberg syndrome"). J Neurol Neurosurg Psychiatry 56: 1035.

Lauvsnes MB, Maroni SS, Appenzeller S et al. (2013). Memory dysfunction in primary Sjögren's syndrome is associated with anti-NR2 antibodies. Arthritis Rheum 65: 3209–3217.

Lavalle C, Pizarro S, Drenkard C et al. (1990). Transverse myelitis: a manifestation of systemic lupus erythematosus strongly associated with antiphospholipid antibodies. J Rheumatol 17: 34–37.

Lee LA, Werth VP (2012). Lupus erythematosus. In: JL Bolognia, JL Jorizzo, JV Schaffer (Eds.), Dermatology. Elsevier, Oxford, ch. 41.

Lee P, Bruni J, Sukenik S (1984a). Neurological manifestations in systemic sclerosis (scleroderma). J Rheumatol 11: 480–483.

Lee DA, Su WP, Liesegang TJ (1984b). Ophthalmic changes of Degos' disease (malignant atrophic papulosis). Ophthalmology 91: 295–299.

Lee VH, Wijdicks EF, Manno EM et al. (2008). Clinical spectrum of reversible posterior leukoencephalopathy syndrome. Arch Neurol 65: 205.

Lee JL, Naguwa SM, Cheema GS et al. (2009). Revisiting Libman–Sacks endocarditis: a historical review and update. Clin Rev Allergy Immunol 36: 126–130.

Lee YJ, Chung KY, Kang HC et al. (2015). Parry–Romberg syndrome with ipsilateral hemipons involvement presenting as monoplegic ataxia. Korean J Pediatr 58(9): 354–357.

Legendre C, Allanore Y, Ferrand I et al. (2005). Evaluation of depression and anxiety in patients with systemic sclerosis. Joint Bone Spine 72: 408–411.

Lenk MD, Özet MD, Alli MD et al. (1998). Protein C and protein S activities in Behçet's disease as risk factors of thrombosis. Int J Dermatol 37: 124–125.

Leroux G, Sellam J, Costedoat-Chalumeau N et al. (2008). Posterior reversible encephalopathy syndrome during systemic lupus erythematosus: four new cases and review of the literature. Lupus 17: 139–147.

Leteurtre E, Hachulla E, Janin A et al. (1994). Vascular manifestations of dermatomyositis and polymyositis. Clinical capillaroscopic and histological aspects. Rev Med Interne 15: 800–807.

Levine SR, Welch KMA (1987). The spectrum of neurologic disease associated with antiphospholipid antibodies: lupus anticoagulants and anticardiolipin antibodies. Arch Neurol 44: 876.

Levine JS, Branch DW, Rauch J (2002). The antiphospholipid syndrome. N Engl J Med 346: 752–763.

Liang MH, Corzillius M, Bae SC et al. (1999). The American College of Rheumatology nomenclature and case definitions for neuropsychiatric lupus syndromes. Arthritis Rheum 42: 599–608.

Lima IV, Galrao LA, Maia TS et al. (2004). Spinal cord compression by ectopic calcinosis in scleroderma. Clin Exp Rheumatol 23: 704–706.

Linardaki G, Cherouvim E, Goni G et al. (2011). Intravenous immunoglobulin treatment for pregnancy-associated dermatomyositis. Rheumatol Int 31: 113–115.

Linklater H, Pipitone N, Rose MR et al. (2013). Classifying idiopathic inflammatory myopathies: comparing the performance of six existing criteria. Clin Exp Rheumatol 31: 767–769.

Loewe R, Palatin M, Petzelbauer P (2005). Degos disease with an inconspicuous clinical course. J Eur Acad Dermatol Venereol 19: 477–480.

Loh E, Sutton MS, Wun CC et al. (1997). Ventricular dysfunction and the risk of stroke after myocardial infarction. N Engl J Med 336: 251–257.

Lonchampt P, Emile J, Pelier-Cady MC et al. (1995). Central sympathetic dysregulation and immunological abnormalities in a case of progressive facial hemiatrophy (Parry–Romberg disease). Clin Auton Res 5: 199–204.

Longo D, Paonessa A, Specchio N et al. (2011). Parry-Romberg syndrome and Rasmussen encephalitis: possible association clinical and neuroimaging features. J Neuroimaging 21: 188–193.

Lori S, Matucci-Cerinic M, Casale R et al. (1995). Peripheral nervous system involvement in systemic sclerosis: the median nerve as target structure. Clin Exp Rheumatol 14: 601–605.

Love PE, Santoro SA (1990). Antiphospholipid antibodies: anticardiolipin and the lupus anticoagulant in systemic lupus erythematosus (SLE) and in non-SLE disorders, prevalence and clinical significance. Ann Intern Med 112: 682–698.

Lundberg IE (2006). The heart in dermatomyositis and polymyositis. Rheumatology 45 (Suppl 4): iv18–iv21.

Luzar B, Calonje E (2012). Idiopathic connective tissue disorders. In: E Calonje, T Brenn, A Lazar (Eds.), McKee's Pathology of the Skin, 4th edn. Elsevier, Oxford, pp. 711–759, ch. 17.

Mauch E, Völk C, Kratzsch G et al. (1994). Neurological and neuropsychiatric dysfunction in primary Sjögren's syndrome. Acta Neurol Scand 89 (1): 31–35.

McCauliffe DP (2001). Cutaneous lupus erythematosus. Semin Cuta Med Surg 20: 14–26.

McFarland HR, Wood WG, Drowns BV et al. (1978). Papulosis atrophicans maligna (Köhlmeier-Degos disease): a disseminated occlusive vasculopathy. Ann Neurol 3: 388–392.

McKinney AM, Short J, Truwit CL et al. (2007). Posterior reversible encephalopathy syndrome: incidence of atypical regions of involvement and imaging findings. Am J Roentgenol 189: 904–912.

McLaurin EY, Holliday SL, Williams P et al. (2005). Predictors of cognitive dysfunction in patients with systemic lupus erythematosus. Neurology 64: 297–303.

Mahr A, Maldini C (2014). Epidemiology of Behçet's disease. Rev Med Interne 35: 81–89.

Malinow KL, Molina R, Gordon B et al. (1985). Neuropsychiatric dysfunction in primary Sjögren's syndrome. Ann Intern Med 103: 344–350.

Mammen AL (2010). Dermatomyositis and polymyositis. Ann N Y Acad Sci 1184: 134–153.

Mandl T, Granberg V, Apelqvist J et al. (2008). Autonomic nervous symptoms in primary Sjögren's syndrome. Rheumatology 47: 914–919.

Markus HS, Hopkinson N (1992). Migraine and headache in systemic lupus erythematosus and their relationship with antibodies against phospholipids. J Neurol 239: 39–42.

Matyja-Bednarczyk A, Swadźba J, Iwaniec T et al. (2014). Risk factors for arterial thrombosis in antiphospholipid syndrome. Thromb Res 133: 173–176.

Mauch E, Völk C, Kratzsch G et al. (1994). Neurological and neuropsychiatric dysfunction in primary Sjögren's syndrome. Acta Neurol Scand 89 (1): 31–35.

Meijer JM, Meiners PM, Vissink A et al. (2010). Effectiveness of rituximab treatment in primary Sjögren's syndrome: a randomized double-blind placebo-controlled trial. Arthritis Rheum 62: 960–968.

Mellgren SI, Conn DL, Stevens JC et al. (1989). Peripheral neuropathy in primary Sjögren's syndrome. Neurology 39: 390.

Menascu S, Padeh S, Hoffman C et al. (2009). Parry–Romberg syndrome presenting as status migrainosus. Pediatr Neurol 40: 321–323.

Mendonça J, Viana SL, Freitas F et al. (2005). Late-onset progressive facial hemiatrophy (Parry–Romberg syndrome). J Postgrad Med 51: 135.

Menni S, Marzano AV, Passoni E (1997). Neurologic abnormalities in two patients with facial hemiatrophy and sclerosis coexisting with morphea. Pediatr Dermatol 14 (2): 113–116.

Meroni PL, Raschi E, Camera M et al. (2000). Endothelial activation by aPL: a potential pathogenetic mechanism for the clinical manifestations of the syndrome. J Autoimmunity 15: 237–240.

Meroni PL, Tincani A, Sepp N et al. (2003). Endothelium and the brain in CNS lupus. Lupus 12: 919–928.

Miedziak AI, Stefanyszyn M, Flanagan J et al. (1998). Parry–Romberg syndrome associated with intracranial vascular malformations. Arch Ophthalmol 116: 1235–1237.

Mikdashi J, Handwerger B, Langenberg P et al. (2007). Baseline disease activity hyperlipidemia and hypertension are predictive factors for ischemic stroke and stroke severity in systemic lupus erythematosus. Stroke 38: 281–285.

Miller FW, Leitman SF, Cronin ME et al. (1992). Controlled trial of plasma exchange and leukapheresis in polymyositis and dermatomyositis. N Engl J Med 326: 1380–1384.

Mimura Y, Ihn H, Jinnin M et al. (2005). Clinical and laboratory features of scleroderma patients developing skeletal myopathy. Clin Rheumatol 24: 99–102.

Min JH, Kim HJ, Kim BJ et al. (2009). Brain abnormalities in Sjögren syndrome with recurrent CNS manifestations: association with neuromyelitis optica. Mult Scler 15: 1069–1076.

Mitsias P, Levine SR (1994). Large cerebral vessel occlusive disease in systemic lupus erythematosus. Neurology 44: 385–393.

Mitsikostas DD, Sfikakis PP, Goadsby PJ (2004). A meta-analysis for headache in systemic lupus erythematosus: the evidence and the myth. Brain 127: 1200–1209.

Mok CC, Lau CS, Chan EY et al. (1998). Acute transverse myelopathy in systemic lupus erythematosus: clinical presentation treatment and outcome. J Rheumatol 25: 467–473.

Mok CC, Tang SS, To CH et al. (2005). Incidence and risk factors of thromboembolism in systemic lupus erythematosus: a comparison of three ethnic groups. Arthritis Rheum 52: 2774–2782.

Moko SB, Mistry Y, Blandin de Chalain TM (2003). Parry–Romberg syndrome: intracranial MRI appearances. J Cranio-Maxillofacial Surgery 31: 321–324.

Mori K, Iijima M, Koike H et al. (2005). The wide spectrum of clinical manifestations in Sjögren's syndrome-associated neuropathy. Brain 128: 2518–2534.

Morozumi S, Kawagashira Y, Iijima M et al. (2009). Intravenous immunoglobulin treatment for painful sensory neuropathy associated with Sjögren's syndrome. J Neurol Sci 279: 57–61.

Morreale M, Marchione P, Giacomini P et al. (2014). Neurological involvement in primary Sjögren syndrome: a focus on central nervous system. PLoS One 9, e84605.

Mosca M, Strigini F, Carmignani A et al. (2005). Pregnant patient with dermatomyositis successfully treated with intravenous immunoglobulin therapy. Arthritis Care Res 53: 119–121.

Moskau S, Urbach H, Hartmann A et al. (2003). Multifocal myelitis in Behçet's disease. Neurology 60: 517.

Moss ML, Crikelair GF (1960). Progressive facial hemiatrophy following cervical sympathectomy in the rat. Arch Oral Biol 1: 254–258.

Nakasu S, Kaneko M, Matsuda M (2001). Cerebral aneurysms associated with Behçet's disease: a case report. J Neurol Neurosurg Psychiatry 70: 682–684.

Nikolov NP, Smith JA, Patronas NJ et al. (2006). Diagnosis and treatment of vasculitis of the central nervous system in a patient with systemic lupus erythematosus. Nat Clin Pract Rheumatol 2: 627–633.

Nived O, Sturfelt G, Liang MH et al. (2003). The ACR nomenclature for CNS lupus revisited. Lupus 12: 872–876.

O'Connor K, Paauw D (2010). Erysipelas: rare but important cause of malar rash. Am J Med 23: 414–416.

Öktem-Tanör Ö, Baykan-Kurt B, Gürvit IH et al. (1999). Neuropsychological follow-up of 12 patients with neuro-Behçet disease. J Neurol 246: 113–119.

Olmos L, Laugier P (1977). Ultrastructure de la maladie de Degos: apport d'un nouveau cas et revue de la littérature. Ann Dermatol Venereol 104: 280–293.

Olsen NJ, King LE, Park JH (1996). Muscle abnormalities in scleroderma. Rheum Dis Clin North Am 22 (4): 783–796.

Orlow SJ, Watsky KL, Bolognia JL (1991). Skin and bones II. J Am Acad Dermatol 25: 447–462.

Ozgocmen S, Gur A (2008). Treatment of central nervous system involvement associated with primary Sjögren's syndrome. Curr Pharmaceut Des 14: 1270–1273.

Pachman LM, Hayford JR, Chung A et al. (1998). Juvenile dermatomyositis at diagnosis: clinical characteristics of 79 children. J Rheumatol 25: 1198–1204.

Paganoni S, Amato A (2013). Electrodiagnostic evaluation of myopathies. Phys Med Rehabil Clin N Am 24: 193–207.

Paholpak P, Rangseekajee P, Foocharoen C (2012). Characteristics treatments and outcome of psychosis in Thai SLE patients. J Psychosom Res 73: 448–451.

Paprocka J, Jamroz E, Adamek D et al. (2006). Difficulties in differentiation of Parry–Romberg syndrome, unilateral facial sclerodermia and Rasmussen syndrome. Child Nerv Syst 22: 409–415.

Parodi A, Caproni M, Marzano AV et al. (2002). Dermatomyositis in 132 patients with different clinical subtypes: cutaneous signs constitutional symptoms and circulating antibodies. Acta Derm Venereol 82: 48–51.

Parry CH (1825). Collections from Unpublished Papers. Unterwood, London.

Pathak R, Gabor AJ (1991). Scleroderma and central nervous system vasculitis. Stroke 22: 410–413.

Pavlakis PP, Alexopoulos H, Kosmidis ML et al. (2011). Peripheral neuropathies in Sjögren syndrome: a new reappraisal. J Neurol Neurosurg Psychiatry 82: 798–802.

Pensler JM, Murphy GF, Mulliken JB (1990). Clinical and ultrastructural studies of Romberg's hemifacial atrophy. Plast Reconstruct Surg 85: 669–674.

Peri Y, Agmon-Levin N, Theodor E et al. (2012). Sjögren's syndrome, the old and the new. Best Pract Res Clin Rheumatol 26: 105–117.

Petit WA, Soso MJ, Higman H (1982). Degos disease: neurologic complications and cerebral angiography. Neurology 32: 1305–1309.

Petri M, Orbai AM, Alarcón GS et al. (2012). Derivation and validation of the Systemic Lupus International Collaborating Clinics classification criteria for systemic lupus erythematosus. Arthritis Rheum 64: 2677–2686.

Pichiecchio A, Uggetti C, Egitto MG et al. (2002). Parry–Romberg syndrome with migraine and intracranial aneurysm. Neurology 59: 606–608.

Piga M, Mathieu A (2011). Genetic susceptibility to Behçet's disease: role of genes belonging to the MHC region. Rheumatology 50: 299–310.

Pittock SJ, Lennon VA, de Seze J et al. (2008). Neuromyelitis optica and non organ-specific autoimmunity. Arch Neurol 65: 78–83.

Poncelet AN, Connolly MK (2003). Peripheral neuropathy in scleroderma. Muscle Nerve 28: 330–335.

Pons-Estel GJ, Alarcón GS, Scofield L et al. (2010). Understanding the epidemiology and progression of systemic lupus erythematosus. Sem Arthritis Rheum 39: 257–268.

Provost T (1994). The relationship between discoid and systemic lupus erythematosus. Arch Dermatol 130: 1308–1310.

Qureshi UA, Wani NA, Altaf U (2010). Parry–Romberg syndrome associated with unusual intracranial vascular malformations and Phthisis bulbi. J Neurol Sci 291: 107–109.

Ramdial PK, Grayson W (2012). Human immunodeficiency virus (HIV) and acquired immunodeficiency syndrome (AIDS)-associated cutaneous diseases. In: E Calonje, T Brenn, A Lazar et al. (Eds.), McKee's Pathology of the Skin, 4th edn. Elsevier, Oxford, Chapter 19.

Ramos-Casals M, Anaya JM, Garcia-Carrasco M et al. (2004). Cutaneous vasculitis in primary Sjögren syndrome: classification and clinical significance of 52 patients. Medicine (Baltimore) 83: 96–106.

Ramos-Casals M, Brito-Zerón P, Sisó-Almirall A et al. (2012). Topical and systemic medications for the treatment of primary Sjögren's syndrome. Nat Rev Rheumatol 8: 399–411.

Raszewa M, Hausmanowa-Petrusewicz I, Błaszczyk M et al. (1991). Sympathetic skin response in scleroderma. Electromyog Clin Neurophysiol 31: 467–472.

Reider N, Fritsch PO (2012). Other eczematous eruptions. In: JL Bolognia, JL Jorizzo, JV Schaffer (Eds.), Dermatology. Elsevier, Oxford, ch. 13.

Rockerbie NR, Woo TY, Callen JP et al. (1989). Cutaneous changes of dermatomyositis precede muscle weakness. J Am Acad Dermatol 20: 629–632.

Rogers SJ, Williams CS, Román GC (2004). Myelopathy in Sjögren's syndrome. Drugs 64: 123–132.

Roman MJ, Shanker BA, Davis A et al. (2003). Prevalence and correlates of accelerated atherosclerosis in systemic lupus erythematosus. N Engl J Med 349: 2399–2406.

Romberg MH (1846). Trophoneurosen. Forstner, Klinische Ergebnisse Berlin, pp. 75–81.

Rossi R, Saddi MV (2006). Subacute aseptic meningitis as neurological manifestation of primary Sjögren's syndrome. Clin Neurol Neurosurg 108: 688–691.

Ruiz-Irastorza G, Crowther M, Branch W et al. (2010). Antiphospholipid syndrome. Lancet 376: 1498–1509.

Saadoun D, Wechsler B (2012). Behçet's disease. Orphanet J Rare Dis 7: 20.

Saadoun D, Wechsler B, Resche-Rigon M et al. (2009). Cerebral venous thrombosis in Behçet's disease. Arthritis Care Res 61 (4): 518–526.

Sagild JC, Alving J (1985). Hemiplegic migraine and progressive hemifacial atrophy. Ann Neurol 17: 620.

Saip S, Siva A, Altintas A et al. (2005). Headache in Behçet's syndrome. Headache 45: 911–919.

Sarica-Kucukoglu R, Akdag-Kose A, Kayabalı M et al. (2006). Vascular involvement in Behçet's disease: a retrospective analysis of 2319 cases. Int J Dermatol 45: 919–921.

Sarui H, Maruyama T, Ito I et al. (2002). Necrotising myositis in Behçet's disease: characteristic features on magnetic resonance imaging and a review of the literature. Ann Rheum Dis 61: 751–752.

Sathornsumetee S, Schanberg L, Rabinovich E et al. (2005). Parry–Romberg syndrome with fatal brain stem involvement. J Pediatr 146: 429–431.

Schady W, Sheard A, Hassell A et al. (1991). Peripheral nerve dysfunction in scleroderma. QJM 80: 661–675.

Scheinfeld N (2007). Malignant atrophic papulosis. Clin Exp Dermatol 32: 483–487.

Schievink WI, Mellinger JF, Atkinson JL (1996). Progressive intracranial aneurysmal disease in a child with progressive hemifacial atrophy (Parry–Romberg disease): case report. Neurosurgery 38: 1237–1241.

Sène D, Cacoub P, Authier FJ et al. (2013). Sjögren syndrome-associated small fiber neuropathy: characterization from a prospective series of 40 cases. Medicine 92: e10–e18.

Sfikakis PP, Mitsikostas DD, Manoussakis MN et al. (1998). Headache in systemic lupus erythematosus: a controlled study. Rheumatology 37: 300–303.

Shah JR, Juhasz C, Kupsky WJ et al. (2003). Rasmussen encephalitis associated with Parry–Romberg syndrome. Neurology 61: 395–397.

Singh A, Ambujam S, Uma AN (2010). The cytogenetics of Bloom's syndrome. J Pediatr Neurosci 5: 171–172.

Siva A, Kantarcy O, Saip S et al. (1998). Neuro–Behcet syndrome: prognostic factors and survival. J Neurol 245: 362.

Siva A, Altintas A, Saip S (2004). Behçet's syndrome and the nervous system. Curr Opin Neurol 17: 347–357.

Slaviero F, Annes RD, Frighetto L et al. (2009). Köhlmeier–Degos disease (malignant atrophic papulosis) and neurologic involvement. Arq Neuropsiquiatr 67: 692–694.

Solomon DH, Kavanaugh AJ, Schur PH (2002). Evidence-based guidelines for the use of immunologic tests: antinuclear antibody testing. Arthritis Care Res 47 (4): 434–444.

Sommer A, Gambichler T, Bacharach-Buhles M et al. (2006). Clinical and serological characteristics of progressive facial hemiatrophy: a case series of 12 patients. J Am Acad Dermatol 54: 227–233.

Sontheimer RD (1999). Dermatomyositis. In: IM Freedberg, AZ Eisen, K Wolff et al. (Eds.), Fitzpatrick's Dermatology in General Medicine, 5th edn. McGraw-Hill, New York, pp. 2009–2020.

Sontheimer RD (2002). Would a new name hasten the acceptance of amyopathic dermatomyositis (dermatomyositis sine myositis) as a distinctive subset within the idiopathic inflammatory dermatomyopathies spectrum of clinical illness? J Am Acad Dermatol 46: 626–636.

Sontheimer RD (2004). Skin manifestations of systemic autoimmune connective tissue disease: diagnostics and therapeutics. Best Pract Res Clin Rheumatol 18: 429–462.

Stone J (2003). Parry–Romberg syndrome. A global survey of 205 patients using the Internet. Neurology 61: 674–676.

Stott VL, Hurrell MA, Anderson TJ (2005). Reversible posterior leukoencephalopathy syndrome: a misnomer reviewed. Int Med J 35: 83–90.

Studýnková JT, Charvat F, Jarošová K et al. (2007). The role of MRI in the assessment of polymyositis and dermatomyositis. Rheumatology 46: 1174–1179.

Su Z, Lu Y, Ge Y et al. (2015). Central nervous system involvement in systemic malignant atrophic papulosis (Degos disease): a case report. Int J Dermatol 54: 699–703.

Subbiah P, Wijdicks E, Muenter M et al. (1996). Skin lesion with a fatal neurologic outcome (Degos' disease). Neurology 46: 636–640.

Subcommittee for Scleroderma Criteria of the American Rheumatism Association (1980). Diagnostic and Therapeutic Criteria Committee. Preliminary criteria for the classification of systemic sclerosis (scleroderma). Arthritis Rheum 23: 581–590.

Szmidt-Salkowska E, Rowinska-Marcinska K, Lovelace RE (1989). EMG dynamics in polymyositis and dermatomyositis in adults. Electromyog Clin Neurophysiol 29: 399.

Takehara K (2003). Hypothesis: pathogenesis of systemic sclerosis. J Rheumatol 30: 755–759.

Tanaka TI, Geist SM (2012). Dermatomyositis: a contemporary review for oral health care providers. Oral Surg Oral Med Oral Pathol Oral Radiol 114: e1–e8.

Teasdall RD, Frayha RA, Shulman LE (1980). Cranial nerve involvement in systemic sclerosis (scleroderma): a report of 10 cases. Medicine 59: 149–159.

ter Borg EJ, Risselada AP, Kelder JC (2011). Relation of systemic autoantibodies to the number of extraglandular manifestations in primary Sjögren's syndrome: a retrospective analysis of 65 patients in the Netherlands. Semin Arthritis Rheum 40: 547–551.

Terstegge K, Kunath B, Felber S et al. (1994). MR of brain involvement in progressive facial hemiatrophy (Romberg disease): reconsideration of a syndrome. Am J Neuroradiol 15: 145–150.

Theodoridis A, Makrantonaki E, Zouboulis CC (2013). Malignant atrophic papulosis (Köhlmeier–Degos disease) – a review. Orphanet J Rare Dis 8: 10.

Theodoridis A, Konstantinidou A, Makrantonaki E et al. (2014). Malignant and benign forms of atrophic papulosis (Köhlmeier–Degos disease): systemic involvement determines the prognosis. Br J Dermatol 170: 110–115.

Tjensvoll AB, Harboe E, Gøransson LG et al. (2011). Migraine is frequent in patients with systemic lupus erythematosus: a case-control study. Cephalalgia 31: 401–408.

Tjensvoll AB, Harboe E, Gøransson LG et al. (2013). Headache in primary Sjøgren's syndrome: a population-based retrospective cohort study. Eur J Neurol 20: 558–563.

Tomizawa Y, Tanaka R, Sekiguchi K et al. (2014). Cerebral infarction in a case of Parry–Romberg syndrome. J Stroke Cerebrovasc Dis 23: 393–394.

Torabi AM, Patel RK, Wolfe GI et al. (2004). Transverse myelitis in systemic sclerosis. Arch Neurol 61: 126–128.

Toyota T, Akamatsu N, Tanaka A et al. (2013). Mesial temporal lobe epilepsy as a neuropsychiatric syndrome of systemic lupus erythematosus. Epilepsia 54: e33–e36.

Tsokos M, Lazarou SA, Moutsopoulos HM (1987). Vasculitis in primary Sjögren's syndrome. Histologic classification and clinical presentation. Am J Clin Pathol 88: 26–31.

Tsuboi H, Iizuka M, Asashima H et al. (2013). Anti-M3 muscarinic acetylcholine receptor antibodies and Sjögren's syndrome. Nihon Rinsho Meneki Gakkai Kaishi 36: 77–85.

Tugal-Tutkun I, Onal S, Altan-Yaycioglu R et al. (2004). Uveitis in Behçet disease: an analysis of 880 patients. Am J Ophthalmol 138: 373–380.

Tugwell P, Dennis DT, Weinstein A et al. (1997). Laboratory evaluation in the diagnosis of Lyme disease. 127: 1109–1123.

Tunc R, Saip S, Siva A et al. (2004). Cerebral venous thrombosis is associated with major vessel disease in Behçet's syndrome. Ann Rheum Dis 63: 1693–1694.

Uva L, Miguel D, Pinheiro C et al. (2012). Cutaneous manifestations of SLE. Autoimmune Dis 2012: 834291.

Valtýsdóttir ST, Gudbjörnsson B, Lindqvist U et al. (2000). Anxiety and depression in patients with primary Sjögren's syndrome. J Rheumatol 27: 165–169.

van der Hooft CS, Heeringa J, Brusselle GG et al. (2006). Corticosteroids and the risk of atrial fibrillation. Arch Intern Med 166: 1016.

Varaprasad IR, Agrawal S, Prabu VN et al. (2011). Posterior reversible encephalopathy syndrome in systemic lupus erythematosus. J Rheumatol 38: 1607–1611.

Viana M, Glastonbury CM, Sprenger T et al. (2011). Trigeminal neuropathic pain in a patient with progressive facial hemiatrophy (Parry–Romberg syndrome). Arch Neurol 68: 938–943.

Vidailhet M, Piette JC, Wechsler B et al. (1990). Cerebral venous thrombosis in systemic lupus erythematosus. Stroke 21: 1226–1231.

Vilá L, Alarcón G, McGwin G et al. (2004). Early clinical manifestations disease activity and damage of systemic lupus erythematosus among two distinct US Hispanic subpopulations. Rheumatology (Oxford) 43: 358–363.

Vitali C, Bombardieri S, Jonsson R et al. (2002). Classification criteria for Sjögren's syndrome: a revised version of the European criteria proposed by the American–European Consensus Group. Ann Rheum Dis 61: 554–558.

Ward MM (1999). Premature morbidity from cardiovascular and cerebrovascular diseases in women with systemic lupus erythematosus. Arthritis Rheum 42: 338–346.

Watson R (1989). Cutaneous lesions in systemic lupus erythematosus. Med Clin North Am 73: 1091–1111.

Wechsler B, Vidailhet M, Piette JC et al. (1992). Cerebral venous thrombosis in Behçet's disease: clinical study and long-term follow-up of 25 cases. Neurology 42: 614–618.

Williams CS, Butler E, Román GC (2001). Treatment of myelopathy in Sjögren syndrome with a combination of prednisone and cyclophosphamide. Arch Neurol 58: 815–819.

Wilson WA, Gharavi AE, Koike T et al. (1999). International consensus statement on preliminary classification criteria for definite antiphospholipid syndrome. Arthritis Rheum 42: 1309–1311.

Wolf SM, Verity MA (1974). Neurological complications of progressive facial hemiatrophy. J Neurol Neurosurg Psychiatry 37: 997–1004.

Wong KO (1969). Dermatomyositis: a clinical of twenty-three cases in Hong Kong. Br J Dermatol 81: 544–547.

Woolfenden AR, Tong DC, Norbash AM et al. (1998). Progressive facial hemiatrophy. Abnormality of intracranial vasculature. Neurology 50: 1915–1917.

Wright RA, O'Duffy JD, Rodriguez M (1999). Improvement of myelopathy in Sjögren's syndrome with chlorambucil and prednisone therapy. Neurology 52: 386–388.

Yamada S, Mori K, Matsuo K et al. (2005). Interferon alfa treatment for Sjögren's syndrome associated neuropathy. J Neurol Neurosurg Psychiatry 76: 576–578.

Yesilot N, Mutlu M, Gungor O et al. (2007). Clinical characteristics and course of spinal cord involvement in Behçet's disease. Eur J Neurol 14: 729–737.

Yildiz S, Boyaci N, Dokumaci DS et al. (2013). Different diffusion-weighted MRI findings in brainstem neuro-Behçet's disease. BMJ Case Reports, bcr. 2013200738.

Yoon KH, Fong KY, Koh DR et al. (2000). Central pontine myelinolysis a rare manifestation of CNS Sjögren's syndrome. Lupus 9: 471–473.

Zhang M, Xu WD, Wen PF et al. (2013). Polymorphisms in the tumor necrosis factor gene and susceptibility to Behçet's disease: an updated meta-analysis. Mol Vis 19: 1913–1924.

Zouboulis CC (1999). Epidemiology of Adamantiades–Behçet's disease. Ann Med Interne (Paris) 150: 488–498.

Zulian F, Vallongo C, Woo P et al. (2005). Localized scleroderma in childhood is not just a skin disease. Arthritis Rheum 52: 2873–2881.

Zulian F, Athreya BH, Laxer R et al. (2006). Juvenile localized scleroderma: clinical and epidemiological features in 750 children. An international study. Rheumatology 45: 614–620.

Handbook of Clinical Neurology, Vol. 132 (3rd series)
Neurocutaneous Syndromes
M.P. Islam and E.S. Roach, Editors

Chapter 4

Neurofibromatosis type 1

JACQUELINE L. ANDERSON AND DAVID H. GUTMANN*
Department of Neurology, Washington University School of Medicine, St. Louis, MO, USA

INTRODUCTION

History

Neurofibromatosis type 1 (NF1), previously known as von Recklinghausen disease, is a common neurogenetic condition affecting 1:2500 people worldwide. NF1 probably existed in ancient times, with art and literature from the 3rd century BCE documenting descriptions consistent with the disease (Zanca, 1980). In 1849, an Irish surgeon named Robert W. Smith differentiated patients with traumatic neuromas from those with cases of multiple, idiopathic neuromas (Smith, 1849). However, it was not until 1882 that the disease entity was fully recognized: the German pathologist Frederick von Recklinghausen first published a classic monograph, in which he described the disease as well as the pathologic basis of neurofibromas (von Recklinghausen, 1882). Iris hamartomas, or Lisch nodules, were first described in patients with NF1 by the Austrian ophthalmologist Karl Lisch in 1937 (Lisch, 1937). Later, Frank Crowe and his colleagues (1956) were the first to recognize NF1 as a hereditary disease, affecting 50% of offspring. In 1964, Dr. Crowe then described skinfold freckling (Crowe, 1964). With the recognition that NF1 was a genetic condition, the US National Institutes of Health (NIH) convened a consensus development conference to establish consistent diagnostic criteria to enable the identification of people with NF1 (National Institutes of Health Consensus Development Conference, 1988). This landmark conference laid the foundations for the genetic analysis of families with NF1, culminating in the discovery of the *NF1* gene in 1990 (Viskochil et al., 1990; Wallace et al., 1990).

Epidemiology

The prevalence of NF1 is approximately 1:2500 to 1:3500 in individuals, regardless of ethnic and racial background (Huson et al., 1989; Rasmussen and Friedman, 2000; Johnson et al., 2013). While NF1 is an autosomal dominant condition, only 50% of people have an affected family member with NF1 (familial cases). As such, 50% of patients will be the first person in their family with NF1, arising from a sporadic *NF1* gene mutation (De Luca et al., 2004; Evans et al., 2010). Life expectancy is reduced by 8–15 years relative to the general population, with malignancy constituting the major reason for death prior to the age of 30 (Rasmussen et al., 2001; Evans et al., 2011). With the establishment of an online worldwide registry for patients with NF1, new insights into the epidemiology of this common condition will likely emerge (Johnson et al., 2013).

CLINICAL MANIFESTATIONS

Hallmark signs and symptoms

The diagnostic criteria for NF1 were first established by the NIH Consensus Development panel in 1987 (National Institutes of Health Consensus Development Conference, 1988) and updated in 1997 (Gutmann et al., 1997). To make the diagnosis of NF1, two of the following clinical features are required:

- six or more café-au-lait macules with diameters greater than 5 mm in a prepubertal patient and greater than 15 mm in a postpubertal patient
- two or more neurofibromas or one plexiform neurofibroma
- skinfold (axillary or inguinal) freckling
- optic pathway tumor
- two or more iris hamartomas
- characteristic bony lesion
- first-degree relative with neurofibromatosis type 1.

In most cases, the diagnosis of NF1 can be made on clinical grounds; however, only in rare circumstances is it

*Correspondence to: David H. Gutmann, MD, PhD, Department of Neurology, Washington University School of Medicine, Box 8111, 660 S. Euclid Avenue, St. Louis, MO 63110, USA. Tel: +1-314-362-7149, Fax: +1-314-362-9462, E-mail: gutmannd@neuro.wustl.edu

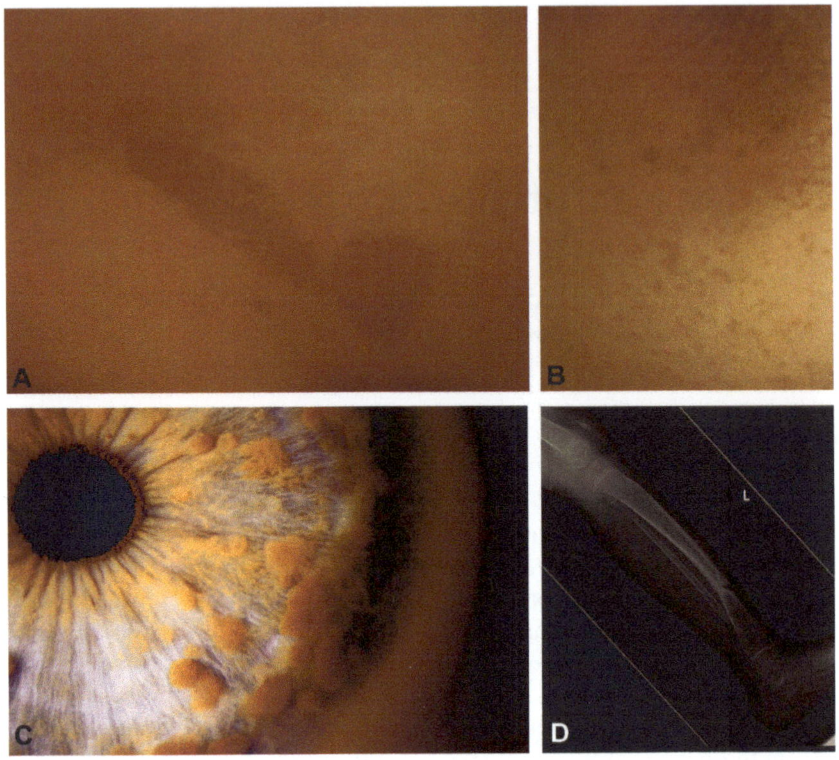

Fig. 4.1. Non-neoplastic features of NF1. (**A**) Typical café-au-lait macule in a child with NF1. (**B**) Skinfold freckling in the axilla of an adult with NF1. (**C**) Lisch nodules in an adult with NF1. (**D**) Tibial pseudarthrosis and fracture in a child with NF1.

necessary to pursue genetic testing. When employed, *NF1* mutation analysis is 95% sensitive (Messiaen et al., 2000; Valero et al., 2011). The features that are typically evident from birth or early infancy include a positive family history and café-au-lait macules. Café-au-lait macules grow in size and number during the first 2 years of life (Fig. 4.1A). Skinfold freckling, most commonly observed in the axillary and inguinal regions, begins to appear in early childhood, most commonly between 5 and 8 years of age (Fig. 4.1B). Optic pathway gliomas develop almost entirely in the pediatric population, usually prior to the age of 7 years old, with a median age at presentation of 4 years (Listernick et al., 1994). Lisch nodules appear as a function of age, such that 30–50% harbor these iris hamartomas by age 6 years, and 92% are present by adulthood (Fig. 4.1C) (Nichols et al., 2003). Characteristic bony abnormalities, such as long bone pseudarthrosis and sphenoid wing dysplasia, when present, are seen in early infancy (Fig. 4.1D). Dermal neurofibromas typically appear in the peripubertal years, and increase in number over the ensuing years. Plexiform neurofibromas are considered congenital, but may not cause problems until later during development or in adulthood.

In addition to the classic features of NF1, people with NF1 are prone to developing aqueductal stenosis, pheochromocytoma, learning and intellectual disabilities, attention deficit, scoliosis, seizures, and vasculopathy as well as other types of tumors and malignancies (e.g., breast cancer and malignant brain tumors).

Cutaneous manifestations

Café-au-lait macules occur in at least 95% of patients with NF1 (Johnson et al., 2013). A child with NF1 usually has at least one café-au-lait macule present at birth, and there will be an increase in number of macules as well as size of the existing macules over the first 1–2 years of life (Nunley et al., 2009). These macules range in color from light to dark brown, depending on the background skin pigmentation. Typically, café-au-lait macules are homogeneous in color with smooth borders. Pathologic examination of these lesions reveals an increased number of macromelanosomes (Slater et al., 1986).

Skinfold freckling is present in 50% of children with NF1 by 10 years of age (Huson et al., 1988; DeBella et al., 2000). The freckles are typically 1–3 mm in diameter, and occur in symmetric clusters in the intertriginous areas of the axillary and inguinal regions as well as under the chin and breasts in women.

Neurofibromas and malignant peripheral nerve sheath tumors

Neurofibromas are the most common tumor type in NF1, affecting 40–60% of patients with NF1 (Friedman and Birch, 1997; McGaughran et al., 1999). Neurofibromas are benign tumors of peripheral nerve sheath cells (WHO grade I) and can occur throughout the peripheral nervous system. Dermal neurofibromas arise from a single peripheral nerve, whereas plexiform neurofibromas arise from a bundle of fascicles or a larger nerve plexus (sacral or brachial plexus).

Cutaneous, localized neurofibromas appear on the surface and can be pedunculated, subcutaneous, or sessile (Fig. 4.2A). They may show slight overlying skin discoloration, sometimes initially appearing as raised erythematous areas. Dermal neurofibromas first appear around the time of puberty, and they typically increase in number with age. While these tumors are benign and do not transform into malignant cancers (Boyd et al., 2009; Jouhilahti et al., 2011), they are frequently associated with significant cosmetic impact or cause irritation because of rubbing or clothing irritation.

Between 30% and 50% of patients with NF1 have plexiform neurofibromas (Waggoner et al., 2000; Mautner et al., 2008). Plexiform neurofibromas are clinically distinct from localized neurofibromas in that they have potential for malignant transformation. Cutaneous plexiform neurofibromas are characterized by overlying skin hyperpigmentation and a thickened dermis, and have been described as "a bag of worms" on palpation (Fig. 4.2B). Internal plexiform neurofibromas can appear as extensive tumors on imaging studies (Fig. 4.2C). Plexiform neurofibromas are most likely congenital, and usually grow most rapidly during the first decade of life. Although the majority of plexiform neurofibromas remain benign, there is still considerable

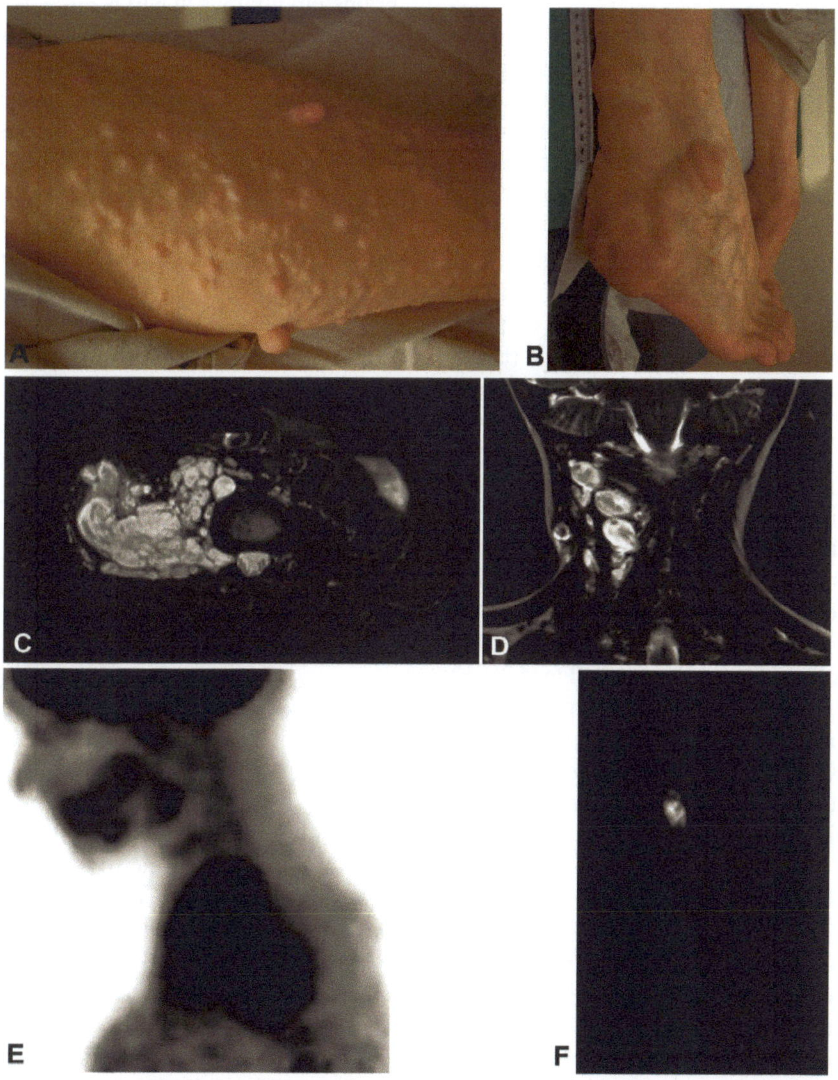

Fig. 4.2. NF1-associated peripheral nerve sheath tumors. (**A**) Dermal neurofibromas on the arm of an adult with NF1. (**B**) Plexiform neurofibroma on the foot of an adult with NF1. (**C**) Internal plexiform neurofibroma in the abdomen/pelvis of an adult with NF1. (**D**) Neck plexiform neurofibroma in an adolescent with NF1. (**E, F**) Positron emission tomography reveals malignant peripheral nerve sheath tumors in the neck (**E**) and leg (**F**) of two different adults with NF1.

morbidity associated with them, including disfigurement and local invasion of neighboring structures (e.g., bone), leading to pain and bony deformities (stimulation of bone growth or bony erosion) as well as rare instances of internal organ, trachea, or vascular compression (Prada et al., 2012) (Fig. 4.2D).

Spinal neurofibromas may cause neurologic symptoms by compressing the spinal cord or spinal roots within the foraminal spaces. Symptoms may include pain, numbness, weakness, or bowel/bladder dysfunction. When arising from the nerve root, the tumor grows in a dumbbell-shaped pattern as it passes through the foramen.

On pathologic examination, neurofibromas consist of neoplastic Schwann cell progenitors growing within a microenvironment of non-neoplastic perineural cells, fibroblasts, mast cells, and collagen (Woodruff, 1999; Jouhilahti et al., 2011).

Although uncommon, new onset of pain or a neurologic deficit in a person with an NF1-associated plexiform neurofibroma should warrant prompt evaluation to exclude a malignant peripheral nerve sheath tumor (MPNST) (Korf, 1999; King et al., 2000). MPNSTs are high-grade spindle-cell sarcomas, found in 8–13% of patients with NF1. Unlike their sporadic counterparts, which typically appear in the 50s and 60s, the mean age at presentation of NF1-associated MPNST is in the mid-30s (Evans et al., 2002). Whereas 5–10% of plexiform neurofibromas transform into MPNSTs (Evans et al., 2002), these cancers can also arise *de novo* in the absence of a known plexiform neurofibroma (Woodruff, 1999).

MRI is not adequate for detecting malignant transformation. For this reason, most clinicians employ 18-FDG-positron emission tomography (PET), which has been shown to be both a sensitive and specific diagnostic test (Mautner et al., 2007; Ferner et al., 2008; Derlin et al., 2013). Standard uptake values greater than 4.0 should raise suspicion for a malignancy (Fig. 4.2E, F). MPNST frequently metastasize, most commonly to the lungs and bone (Ducatman et al., 1986). Unfortunately, the prognosis for NF1-associated MPNST is poor, even after treatment, with overall survival typically less than 5 years (Porter et al., 2009).

Brain tumors

Within the central nervous system, the majority of tumors arising in pediatric patients with NF1 are World Health Organization (WHO) grade I pilocytic astrocytomas. The optic pathway glioma (OPG) is the most common brain tumor associated with NF1, with as many as 15–20% of children with NF1 harboring an optic pathway tumor (Lewis et al., 1984; Listernick et al., 1994). These tumors can occur anywhere along the optic pathway, including the optic nerves, chiasm, and postchiasmatic tracts and radiations (Fig. 4.3A–C) (Listernick et al., 1989, 1994, 1995). Up to half of optic pathway gliomas become symptomatic, but typically only one-third of children with NF1-OPG require therapeutic intervention (Lewis et al., 1984; Listernick et al., 1994; Fisher et al., 2012; de Blank et al., 2013). The decision to treat should be based on increasing visual loss (Listernick et al., 1997, 2007; Avery et al., 2012). Other signs and symptoms may include color vision changes, subacute progressive proptosis, strabismus, papilledema, and optic nerve atrophy. When locally invasive into the hypothalamus, precocious puberty may ensue (Habiby et al., 1995).

Low-grade gliomas may also be found in the brainstem, and these tumors typically exhibit an indolent course (Pollack et al., 1996). The lifetime incidence of brainstem gliomas in NF1 is ~4%, with presentation typically before the age of 10 years (Molloy et al., 1995; Ullrich et al., 2007).

While rare, adults with NF1 have a 50-fold increased risk of developing malignant gliomas, typically glioblastoma (GBM) (Matsui et al., 1993; Gutmann et al., 2002). These cancers appear earlier in life than those observed in the general population; however, the clinical presentation, pathology, and outcomes are similar to sporadically occurring GBM.

Other tumors

Individuals with NF1 are also at risk for developing other cancers. Of these, pheochromocytomas occur with increased frequency in people with NF1 (0.1–13%) (Walther et al., 1999; Vlenterie et al., 2013). In addition, there is also an increased incidence of NF1-associated leukemia (juvenile chronic myeloid leukemia and myelodysplastic syndromes), gastrointestinal stromal tumors, rhabdomyosarcoma, and early-onset breast cancer (Matsui et al., 1993; Stiller et al., 1994; Side et al., 1997, 1998; Gutmann and Gurney, 1999; Sung et al., 2004; Sharif et al., 2007; Vlenterie et al., 2013).

Neurologic manifestations

In addition to tumor-related clinical problems, children with NF1 are also prone to exhibit learning disabilities, cognitive delays, and attention deficits. Compared to the general population, the mean IQ of children with NF1 is 85 (Hyman et al., 2005). However, mental retardation ($IQ < 70$) is rare in children with NF1. When examined across several studies, the frequency of learning disabilities in children with NF1 is estimated to be between 30% and 65% (North et al., 1997). The most commonly affected intellectual domains include verbal learning, visuospatial and visual perceptual processing

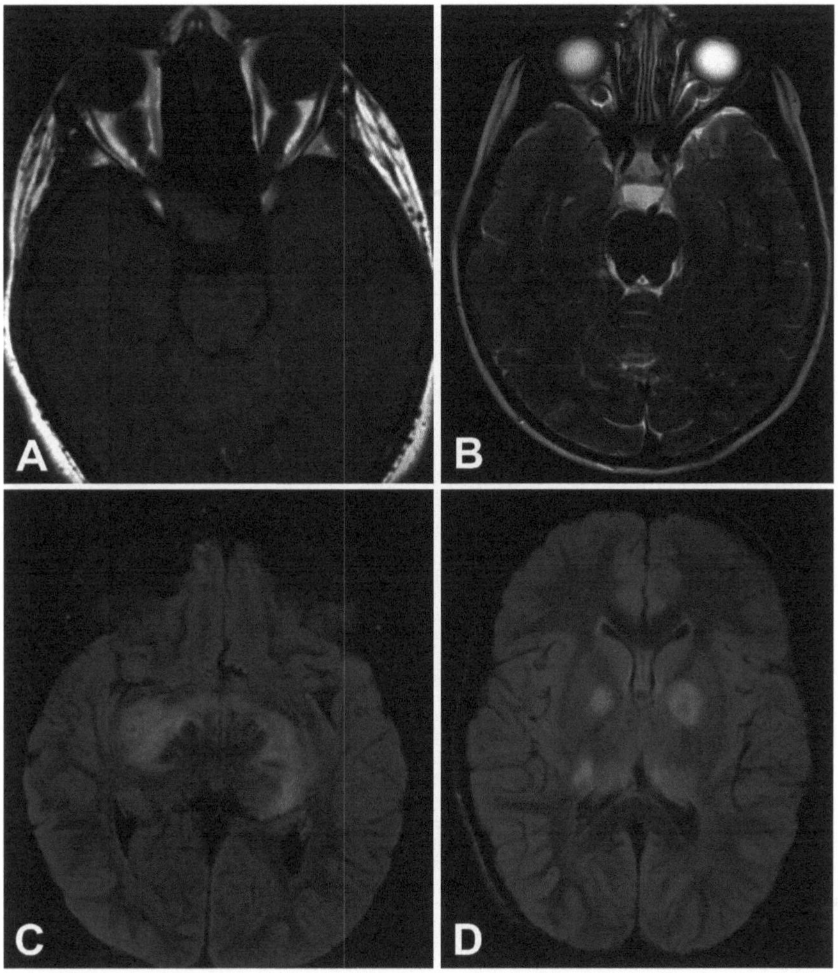

Fig. 4.3. NF1-associated brain tumors. (**A**) Right optic nerve and chiasmal optic pathway glioma in a child with NF1. (**B**) Bilateral optic nerve gliomas with gadolinium enhancement in a child with NF1. (**C**) Postchiasmal optic radiation glioma in a child with NF1. (**D**) T2-hyperintensities found in young adults and children with NF1 can be difficult to distinguish from low-grade gliomas on MRI. These non-neoplastic T2-hyperintensities are typically found in the basal ganglia, brainstem, cerebellum, and optic radiations.

(Dilts et al., 1996; Hyman et al., 2006). In addition, children with NF1 have an increased prevalence of attention deficits (Mautner et al., 2002; Pride et al., 2012; Isenberg et al., 2013), sleep disturbances (Licis et al., 2013), motor delays (Soucy et al., 2012; Wessel et al., 2013), autism spectrum disorders (Garg et al., 2013; Walsh et al., 2013), and impaired social functioning (Huijbregts et al., 2010; Lehtonen et al., 2013), each of which can impact on overall scholastic performance.

Seizures occur in 4–9% of patients with NF1 (Riccardi, 1981; Kulkantrakorn and Geller, 1998; Hsieh et al., 2011). Relative to the general population, seizures in people with NF1 are more often focal and related to a brain tumor. Moreover, individuals with NF1 and seizures frequently require more aggressive therapy than those without NF1, and some patients with NF1-related epilepsy should be considered for surgery when appropriate (Ostendorf et al., 2013).

With the increase in availability of magnetic resonance imaging (MRI), benign abnormalities have been uncovered on neuroimaging of pediatric NF1 patients. More than half of all children with NF1 harbor T2-high signal intensity lesions on brain MRI (Gill et al., 2006). The most common locations are the brainstem, thalamus, cerebellum, and basal ganglia (Fig. 4.3D). These abnormalities are typically well-circumscribed and nonenhancing; the presence of mass effect (architectural distortion), diffuse parenchymal infiltration, or contrast enhancement should warrant further investigation for an underlying brain tumor. While the precise etiology of these lesions remains unknown, one study revealed that these abnormalities may represent vacuolar or spongiotic changes (DiPaolo et al., 1995). In most cases, the lesions disappear by late adolescence or early adulthood (Gill et al., 2006).

Orthopedic manifestations

Tibial pseudarthrosis and sphenoid wing dysplasia are both relatively specific to children with NF1, and typically are detected in early infancy. Sphenoid wing dysplasia usually presents as an orbital abnormality. Orbital dysplasia may result from an associated plexiform neurofibroma.

Long bone dysplasia manifests as cortical thinning and bowing, which may lead to a pathologic fracture. Repetitive cycling of fracture with incomplete healing leads to the development of a pseudarthrosis ("false joint"). In certain situations, a pathologic fracture may indicate bony erosion from a plexiform neurofibroma, but also may be secondary to a nonossifying cyst or osteopenia, both of which occur more frequently in NF1 (Dulai et al., 2007; Stevenson et al., 2007; Brunetti-Pierri et al., 2008; Elefteriou et al., 2009; Petramala et al., 2012). Vertebral anomalies are also associated with NF1, and may appear as benign scalloping of the vertebral body. Scoliosis is common in NF1 and is most commonly lower cervical or upper thoracic. In rare instances, the scoliosis may be dystrophic, leading to significant disfigurement.

Vascular manifestations

The two most common vascular changes associated with NF1 are hypertension and vascular dysplasia. Most cases of NF1-associated hypertension are primary hypertension, but secondary causes include pheochromocytoma and renal vascular dysplasia (renal artery stenosis). NF1-associated vascular dysplasia more commonly affects arteries (Salyer and Salyer, 1974). Dysplasia of the intracranial vessels may cause moyamoya syndrome, which may lead to ischemic stroke (Cairns and North, 2008), whereas vascular dysplasia in adults typically causes hemorrhage and arterial dissection (Friedman et al., 2002). Cerebral vasculopathy has been associated with prior cranial radiation therapy in individuals with NF1.

Variants

Segmental NF1 is a clinical variant of NF1 in which only a single region of the body harbors the manifestations of NF1 (café-au-lait macules, skinfold freckling, neurofibromas). Segmental NF1 results from a somatic mutation in the *NF1* gene during early embryonic development, leading to NF1 restricted to one portion of the child's body. However, if the gonads are involved, a parent with segmental NF1 may have children with generalized, not segmental, NF1 (Ruggieri, 2001).

GENETICS

Molecular basis

NF1 is an autosomal dominant disorder that exhibits complete penetrance. In this regard, there are no carriers of NF1. The *NF1* gene is located on the long arm of chromosome 17 in humans, and forms an 11-13 kb mRNA containing at least 60 common and three alternatively spliced exons (Fig. 4.4A). The encoded protein, termed neurofibromin, is 220–250 kDa and is abundantly expressed in neurons, oligodendrocytes, and Schwann cells. Neurofibromin functions primarily as a GTPase-activating protein (GAP), and inhibits RAS activity by accelerating the conversion of GTP-bound active RAS to its inactive GDP-bound state (Buchberg et al., 1990; Xu et al., 1990; Basu et al., 1992; Cichowski and Jacks, 2001). As a proto-oncogene, RAS promotes cell division and proliferation (Pylayeva-Gupta et al., 2011). In NF1-associated tumors, loss of neurofibromin expression, due to bi-allelic *NF1* gene inactivation, is associated with high levels of active RAS. Depending on the cell type, RAS hyperactivation leads to increased signaling through the RAS downstream pathway intermediates, AKT/mTOR and RAF/MEK (Fig. 4.4B) (Basu et al., 1992; DeClue et al., 1992; Gutmann et al., 1994; Bollag et al., 1996; Kimura et al., 2002; Dasgupta et al., 2005b; Jessen et al., 2013). Each of these RAS

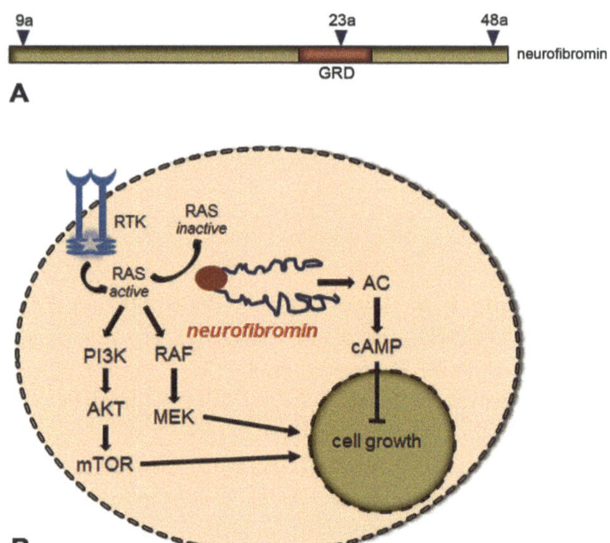

Fig. 4.4. *NF1* gene structure and function. (**A**) The structure of the *NF1* gene product (neurofibromin) with the alternatively spliced exons (9a, 23a, 48a) labeled. The GRD denotes the GAP-related domain. (**B**) Neurofibromin negatively regulates RAS activity and downstream RAS effector (PI3K/AKT/mTOR and RAF/MEK) signaling as well as positively controls cyclic AMP (cAMP) production. In *NF1*-deficient tumor cells, increased RAS function and reduced cAMP levels promote cell growth.

downstream effectors has been investigated as potential rational therapies for NF1-associated tumors. In addition, neurofibromin is also a positive regulator of intracellular cyclic AMP (cAMP) production (Tong et al., 2002; Dasgupta et al., 2003), which in neurons is responsible for maintaining neuronal viability in the setting of optic glioma (Brown et al., 2010).

Animal models

Over the past decade, numerous laboratories have developed accurate genetically engineered mouse (GEM) models of NF1-associated cognitive deficits (Silva et al., 1997; Costa et al., 2002; Li et al., 2005; Cui et al., 2008; Shilyansky et al., 2010), skeletal abnormalities (Wang et al., 2011; Zhang et al., 2011; El-Hoss et al., 2012; El Khassawna et al., 2012), optic glioma (Bajenaru et al., 2003; Dasgupta et al., 2005a; Zhu et al., 2005b), malignant glioma (Zhu et al., 2005a; Kwon et al., 2008), cutaneous neurofibroma (Zhu et al., 2002; Mayes et al., 2011; Wu et al., 2008), MPNST (Cichowski et al., 1999; Vogel et al., 1999), myeloid leukemia (Le et al., 2004), and pheochromocytoma (Tischler et al., 1995). These preclinical models have led to a better understanding of the cellular and molecular bases that underlie the clinical features in children and adults with NF1, and have generated several promising new treatments for NF1-associated tumors and cognitive problems (Gutmann et al., 2013; Lin and Gutmann, 2013).

MANAGEMENT AND TREATMENT

The mainstay of the management of NF1 is anticipatory guidance. Genetic counseling as well as the evaluation of first-degree family members is important. At every office visit, monitoring for macrocephaly, growth failure, precocious puberty, hypertension, developmental delays, learning disabilities, and scoliosis should occur. At each age, there are different problems that may develop, necessitating a focused and age-appropriate evaluation for children and adults. Annual ophthalmologic examinations by an ophthalmologist expert in NF1 should be performed until the age of 12 years to screen for optic pathway gliomas (Listernick et al., 2007).

Children with developmental delay should be referred for appropriate therapies. As such, a concern for intellectual disability or learning disabilities should prompt neuropsychological evaluation. When appropriate, treatment of ADHD with stimulant medications should be considered (Mautner et al., 2002). In all cases, the management of neurocognitive disabilities requires teacher engagement and educational adaptations as indicated.

Surveillance neuroimaging in asymptomatic patients as a screening test for optic glioma pathways is not recommended (King et al., 2003; Segal et al., 2010). However, the development of visual loss, or other concerning symptoms such as precocious puberty, should warrant prompt brain MRI. If an optic pathway glioma is identified on neuroimaging, repeat ophthalmologic examinations should be performed every 3 months for the first year (Listernick et al., 2007). A two-line decrement in visual acuity should prompt treatment, typically with carboplatin and vincristine. Of patients with NF1-associated OPG causing visual impairment who received chemotherapy, 32% had improved visual acuity on follow-up, 40% had stable visual acuity, and 28% had worsened visual acuity (Fisher et al., 2012). Surgery for optic pathway glioma is indicated only in cases of intraorbital tumor causing proptosis and a blind eye. Radiation is not employed, because of increased risk for secondary high-grade CNS gliomas (Sharif et al., 2006).

Cutaneous neurofibromas may be treated with surgery and, occasionally, with CO_2 laser therapy or electrodessication (Levine et al., 2008). In certain instances, plexiform neurofibromas may benefit from surgical debulking, although there is a high risk of iatrogenic injury to associated nerves and surrounding soft tissue as well as hemorrhage due to the significant degree of tumor vascularity. Currently, there are several chemotherapeutic trials underway aimed at halting plexiform neurofibroma growth (Robertson et al., 2012).

The management of MPNSTs involves the coordinated involvement of surgical oncologists, medical oncologists, and radiation oncologists. Small biopsies are notoriously inaccurate for diagnosing MPNST: for this reason, when clinical symptoms or 18-FDG-PET imaging suggests the possibility of malignancy, open biopsy or wide surgical excision is recommended (Ducatman et al., 1986; Ferner and Gutmann, 2002). Treatment following surgical excision entails local radiation and chemotherapy. While radiation therapy delays the time to tumor recurrence, it does not improve long-term survival (Ferner and Gutmann, 2002). Chemotherapy for MPNST has sometimes entailed the use of doxorubicin and ifosfamide; however, there is no current effective chemotherapy for these cancers (Moretti et al., 2011). In addition to local recurrence, these malignancies are prone to metastasis to the lungs and bone. Even with treatment, most patients with NF1-associated MPNST die within 5 years of diagnosis (Porter et al., 2009).

RECENT ADVANCES

Advances from neuroimaging

One of the major areas of focus is the identification of prognostic factors that provide risk assessment for people with NF1-associated medical problems. Recent evidence suggests that favorable radiographic outcomes after chemotherapy for NF1-OPG do not correlate with visual acuity outcomes; rather, the location of the tumor, irrespective of radiographic response, was the single most consistent prognostic indicator (Fisher et al., 2012). In this study, tumors in the postchiasmal optic radiations were most likely to lead to visual loss.

Other studies have focused on anatomic and diffusion-based abnormalities. While optic nerve tortuosity is frequently observed in children with NF1 patients, this radiographic feature has little predictive value in identifying optic gliomas (Ji et al., 2013). Similarly, fractional anisotropy has been explored as an easily quantifiable prognostic indicator for vision loss in NF1-OPG (de Blank et al., 2013).

The future of precision medicine

In 2005, the US Department of Defense established the Neurofibromatosis Clinical Trials Consortium (NFCTC) in order to efficiently deploy resources to critically evaluate the most promising experimental agents in a nationwide testing cohort. These efforts are likely to lead to a therapeutic paradigm shift from the current model of varied treatments to one of targeted and informed use of biologically targeted agents.

With the availability of accurate preclinical mouse models, an efficient clinical trials consortium, and a detailed understanding of neurofibromin function, we are uniquely poised to develop treatments tailored to specific features and subgroups of people with NF1-associated medical problems. For example, rapamycin, which inhibits RAS-dependent mammalian target of rapamycin (mTOR) function, first shown to inhibit the growth of optic glioma in mice (Hegedus et al., 2008), is now in clinical trial for NF1-associated glioma. Similarly, imatinib, which targets the c-kit signaling pathway deregulated in mouse plexiform neurofibromas (Yang et al., 2008), has been investigated in early clinical trials for people with NF1-associated plexiform neurofibroma (Robertson et al., 2012). Finally, based on exciting findings in *Nf1* mouse models of learning and memory defects (Li et al., 2005), lovastatin, a nonselective RAS inhibitor, has been evaluated in children with NF1-associated cognitive problems (Krab et al., 2008; van der Vaart et al., 2013). Additional promising agents are also now in human clinical trials.

Future therapies will also begin to consider cell type-specific growth control pathways downstream of RAS as well as the contribution of non-neoplastic cells present in the tumor microenvironment. As we envision the possibility of personalized treatments for NF1, it will be critical to employ various converging approaches, including registry-based epidemiologic data, *NF1* genetic/genomic sequencing, and patient-derived cell types, to inform novel therapeutic strategies targeted against NF1-associated clinical problems arising in a specific individual with NF1.

References

Avery RA, Ferner RE, Listernick R et al. (2012). Visual acuity in children with low grade gliomas of the visual pathway: implications for patient care and clinical research. J Neurooncol 110: 1–7.

Bajenaru ML, Hernandez MR, Perry A et al. (2003). Optic nerve glioma in mice requires astrocyte Nf1 gene inactivation and Nf1 brain heterozygosity. Cancer Res 63: 8573–8577.

Basu TN, Gutmann DH, Fletcher JA et al. (1992). Aberrant regulation of ras proteins in malignant tumour cells from type 1 neurofibromatosis patients. Nature 356: 713–715.

Bollag G, Clapp DW, Shih S et al. (1996). Loss of NF1 results in activation of the Ras signaling pathway and leads to aberrant growth in haematopoietic cells. Nat Genet 12: 144–148.

Boyd KP, Korf BR, Theos A (2009). Neurofibromatosis type 1. J Am Acad Dematol 61: 1–14.

Brown JA, Gianino SM, Gutmann DH (2010). Defective cAMP generation underlies the sensitivity of CNS neurons to neurofibromatosis-1 heterozygosity. J Neurosci 30: 5579–5589.

Brunetti-Pierri N, Doty SB, Hicks J et al. (2008). Generalized metabolic bone disease in neurofibromatosis type 1. Mol Genet Metab 94: 105–111.

Buchberg AM, Cleveland LS, Jenkins NA et al. (1990). Sequence homology shared by neurofibromatosis type-1 gene and IRA-1 and IRA-2 negative regulators of the RAS cyclic AMP pathway. Nature 347: 291–294.

Cairns AG, North KN (2008). Cerebrovascular dysplasia in neurofibromatosis type 1. J Neurol Neurosurg Psychiatry 79: 1165–1170.

Cichowski K, Jacks T (2001). NF1 tumor suppressor gene function: narrowing the GAP. Cell 104: 593–604.

Cichowski K, Shih TS, Schmitt E et al. (1999). Mouse models of tumor development in neurofibromatosis type 1. Science 286: 2172–2176.

Costa RM, Federov NB, Kogan JH et al. (2002). Mechanism for the learning deficits in a mouse model of neurofibromatosis type 1. Nature 415: 526–530.

Crowe F (1964). Axillary freckling as a diagnostic aid in neurofibromatosis. Ann Intern Med 61: 1142–1143.

Crowe F, Schull W, Neel J (1956). A Clinical, Pathological and Genetic Study of Multiple Neurofibromatosis, Charles C Thomas, Springfield, IL.

Cui Y, Costa RM, Murphy GG et al. (2008). Neurofibromin regulation of ERK signaling modulates GABA release and learning. Cell 135: 549–560.

Dasgupta B, Dugan LL, Gutmann DH (2003). The neurofibromatosis 1 gene product neurofibromin regulates pituitary adenylate cyclase-activating polypeptide-mediated signaling in astrocytes. J Neurosci 23: 8949–8954.

Dasgupta B, Li W, Perry A et al. (2005a). Glioma formation in neurofibromatosis 1 reflects preferential activation of K-RAS in astrocytes. Cancer Res 65: 236–245.

Dasgupta B, Yi Y, Chen DY et al. (2005b). Proteomic analysis reveals hyperactivation of the mammalian target of rapamycin pathway in neurofibromatosis 1-associated human and mouse brain tumors. Cancer Res 65: 2755–2760.

de Blank PM, Berman JI, Liu GT et al. (2013). Fractional anisotropy of the optic radiations is associated with visual acuity loss in optic pathway gliomas of neurofibromatosis type 1. Neuro Oncol 15: 1088–1095.

De Luca A, Schirinzi A, Buccino A et al. (2004). Novel and recurrent mutations in the NF1 gene in Italian patients with neurofibromatosis type 1. Hum Mutat 23: 629.

DeBella K, Szudek J, Friedman JM (2000). Use of the National Institutes of Health criteria for diagnosis of neurofibromatosis 1 in children. Pediatrics 105: 608–614.

DeClue JE, Papageorge AG, Fletcher JA et al. (1992). Abnormal regulation of mammalian p21ras contributes to malignant tumor growth in von Recklinghausen (type 1) neurofibromatosis. Cell 69: 265–273.

Derlin T, Tornquist K, Münster S et al. (2013). Comparative effectiveness of 18 F-FDG PET/CT versus whole-body MRI for detection of malignant peripheral nerve sheath tumors in neurofibromatosis type 1. Clin Nucl Med 38: e19–e25.

Dilts CV, Carey JC, Kircher JC et al. (1996). Children and adolescents with neurofibromatosis 1: a behavioral phenotype. J Dev Behav Pediatr 17: 229–239.

DiPaolo DP, Zimmerman RA, Rorke LB et al. (1995). Neurofibromatosis type 1: pathologic substrate of high-signal-intensity foci in the brain. Radioloy 195: 721–724.

Ducatman BS, Scheithauer BW, Piepgras DG et al. (1986). Malignant peripheral nerve sheath tumors. A clinicopathologic study of 120 cases. Cancer 57: 2006–2021.

Dulai S, Briody J, Schindeler A et al. (2007). Decreased bone mineral density in neurofibromatosis type 1: results from a pediatric cohort. J Pediatr Orthop 27: 472–475.

El Khassawna T, Toben D, Kolanczyk M et al. (2012). Deterioration of fracture healing in the mouse model of NF1 long bone dysplasia. Bone 51: 651–660.

Elefteriou F, Kolanczyk M, Schindeler A et al. (2009). Skeletal abnormalities in neurofibromatosis type 1: approaches to therapeutic options. Am J Med Genet A 149A: 2327–2338.

El-Hoss J, Sullivan K, Cheng T et al. (2012). A murine model of neurofibromatosis type 1 tibial pseudarthrosis featuring proliferative fibrous tissue and osteoclast-like cells. J Bone Miner Res 27: 68–78.

Evans DG, Baser ME, McGaughran J et al. (2002). Malignant peripheral nerve sheath tumors in neurofibromatosis 1. J Med Genet 39: 311–314.

Evans DG, Howard E, Giblin C et al. (2010). Birth incidence and prevalence of tumor-prone syndromes: estimates from a UK family genetic register service. Am J Med Genet A 152A: 327–332.

Evans DG, O'Hara C, Wilding A et al. (2011). Mortality in neurofibromatosis 1: in North West England: an assessment of actuarial survival in a region of the UK since 1989. Eur J Hum Genet 19: 1187–1191.

Ferner RE, Gutmann DH (2002). International consensus statement on malignant peripheral nerve sheath tumors in neurofibromatosis. Cancer Res 62: 1573–1577.

Ferner RE, Golding JF, Smith M et al. (2008). [18 F]2-fluoro-2-deoxy-D-glucose positron emission tomography (FDG PET) as a diagnostic tool for neurofibromatosis 1 (NF1) associated malignant peripheral nerve sheath tumours (MPNSTs): a long-term clinical study. Ann Oncol 19: 390–394.

Fisher MJ, Loquidice M, Gutmann DH et al. (2012). Visual outcomes in children with neurofibromatosis type 1-associated optic pathway glioma following chemotherapy: a multicenter retrospective analysis. Neuo Oncol 14: 790–797.

Friedman JM, Birch PH (1997). Type 1 neurofibromatosis: a descriptive analysis of the disorder in 1,728 patients. Am J Med Genet 70: 138–143.

Friedman JM, Arbiser J, Epstein JA et al. (2002). Cardiovascular disease in neurofibromatosis 1: report of the NF1 Cardiovascular Task Force. Genet Med 4: 105–111.

Garg S, Lehtonen A, Huson SM et al. (2013). Autism and other psychiatric comorbidity in neurofibromatosis type 1: evidence from a population-based study. Dev Med Child Neurol 55: 139–145.

Gill DS, Hyman SL, Steinberg A et al. (2006). Age-related findings on MRI in neurofibromatosis type 1. Pediatr Radiol 36: 1048–1056.

Gutmann DH, Gurney J (1999). Other malignancy. In: JM Friedman, DH Gutmann (Eds.), Neurofibromatosis: Phenotype, Natural History, and Pathogenesis. Johns Hopkins University Press, Baltimore, pp. 231–249.

Gutmann DH, Cole JL, Stone WJ et al. (1994). Loss of neurofibromin in adrenal gland tumors from patients with neurofibromatosis type I. Genes Chromosomes Cancer 10: 55–58.

Gutmann DH, Aylsworth A, Carey JC et al. (1997). The diagnostic evaluation and multidisciplinary management of neurofibromatosis 1 and 2. JAMA 278: 51.

Gutmann DH, Rasmussen SA, Wolkenstein P et al. (2002). Gliomas presenting after age 10 in individuals with neurofibromatosis type 1 (NF1). Neurology 59: 759–761.

Gutmann DH, Blakeley JO, Korf BR et al. (2013). Optimizing biologically targeted clinical trials for neurofibromatosis. Expert Opin Investig Drugs 22: 443–462.

Habiby R, Silverman B, Listernick R et al. (1995). Precocious puberty in children with neurofibromatosis type 1. J Pediatr 126: 364.

Hegedus B, Banerjee D, Yeh TH et al. (2008). Preclinical cancer therapy in a mouse model of neurofibromatosis-1 optic glioma. Cancer Res 68: 1520–1528.

Hsieh HY, Fung HC, Wang CJ et al. (2011). Epileptic seizures in neurofibromatosis type 1 are related to intracranial tumors but not to neurofibromatosis bright objects. Seizure 20: 606–611.

Huijbregts S, Jahja R, De Sonneville L et al. (2010). Social information processing in children and adolescents with neurofibromatosis type 1. Dev Med Child Neurol 52: 620–625.

Huson SM, Harper PS, Compston DA (1988). Von Recklinghausen neurofibromatosis. A clinical and population study in south-east Wales. Brain 111: 1355–1381.

Huson SM, Compston DA, Clark P et al. (1989). A genetic study of von Recklinghausen neurofibromatosis in south east Wales. 1. Prevalence, fitness, mutation rate, and effect of parental transmission of severity. J Med Genet 26: 704–711.

Hyman S, Shores A, North K (2005). The nature and frequency of cognitive deficits in children with neurofibromatosis type 1. Neurology 65: 1037–1044.

Hyman SL, Arthur Shores E, North KN (2006). Learning disabilities in children with neurofibromatosis type 1: subtypes, cognitive profile, and attention-deficit-hyperactivity disorder. Dev Med Child Neurol 48: 973–977.

Isenberg JC, Templer A, Gao F et al. (2013). Attention skills in children with neurofibromatosis type 1. J Child Neurol 28: 45–49.

Jessen WJ, Miller SJ, Jousma E et al. (2013). MEK inhibition exhibits efficacy in human and mouse neurofibromatosis tumors. J Clin Invest 123: 340–347.

Ji J, Shimony J, Gao F et al. (2013). Optic nerve tortuosity in children with neurofibromatosis type 1. Pediatr Radiol 43: 1336–1343.

Johnson KJ, Hussain I, Williams K et al. (2013). Development of an international internet-based neurofibromatosis type 1 patient registry. Contemp Clin Trials 34: 305–311.

Jouhilahti EM, Peltonen S, Callens T et al. (2011). The development of cutaneous neuofibromas. Am J Pathol 178: 500–505.

Kimura N, Watanabe T, Fukase M et al. (2002). Neurofibromin and NF1 gene analysis in composite pheochromocytoma and tumors associated with von Recklinghausen's disease. Mod Pathol 15: 183–188.

King AA, Debaun MR, Riccardi VM et al. (2000). Malignant peripheral nerve sheath tumors in neurofibromatosis 1. Am J Med Genet 93: 388–392.

King A, Listernick R, Charrow J et al. (2003). Optic pathway gliomas in neurofibromatosis type 1: the effect of presenting symptoms on outcome. Am J Med Genet A 122A: 95–99.

Korf BR (1999). Plexiform neurofibromas. Am J Med Genet 89: 31–37.

Krab LC, de Goede-Bolder A, Aarsen FK et al. (2008). Effect of simvastatin on cognitive functioning in children with neurofibromatosis type 1: a randomized controlled trial. JAMA 300: 287–294.

Kulkantrakorn K, Geller TJ (1998). Seizures in neurofibromatosis 1. Pediatr Neurol 19: 347–350.

Kwon CH, Zhao D, Chen J et al. (2008). Pten haploinsufficiency accelerates formation of high-grade astrocytomas. Cancer Res 68: 3286–3294.

Le DT, Kong N, Zhu Y et al. (2004). Somatic inactivation of Nf1 in hematopoietic cells results in a progressive myeloproliferative disorder. Blood 103: 4243–4250.

Lehtonen A, Howie E, Trump D et al. (2013). Behaviour in children with neurofibromatosis type 1: cognition, executive function, attention, emotion, and social competence. Dev Med Child Neurol 55: 111–125.

Levine SM, Levine E, Taub PJ et al. (2008). Electrosurgical excision technique for the treatment of multiple cutaneous lesions in neurofibromatosis type I. J Plast Reconstr Aesthet Surg 61: 958–962.

Lewis RA, Gerson LP, Axelson KA et al. (1984). Von Recklinghausen NF1: incidence of optic gliomas. Ophthalmology 91: 929.

Li W, Cui Y, Kushner SA et al. (2005). The HMG-CoA reductase inhibitor lovastatin reverses the learning and attention deficits in a mouse model of neurofibromatosis type 1. Curr Biol 15: 1961–1967.

Licis AK, Vallorani A, Gao F et al. (2013). Prevalence of sleep disturbances in children with neurofibromatosis type 1. J Child Neurol 28: 1400–1405.

Lin A, Gutmann DH (2013). Advances in the treatment of neurofibromatosis-associated tumors. Nat Rev Clin Oncol 10: 616–624.

Lisch K (1937). Ueber Beteiligung der Augen, insbesondere das Vorkommen von Irisknotchen bei der Neurofibromatose (Recklinghausen). Z Augenheilkd 93: 137–143.

Listernick R, Charrow J, Greenwald MJ et al. (1989). Optic gliomas in children with neurofibromatosis type 1. J Pediatr 114: 788–792.

Listernick R, Charrow J, Greenwald M et al. (1994). Natural history of optic pathway tumors in children with neurofibromatosis type 1: a longitudinal study. J Pediatr 125: 63–66.

Listernick R, Darling C, Greenwald M et al. (1995). Optic pathway tumors in children: the effect of neurofibromatosis type 1 on clinical manifestations and natural history. J Pediatr 127: 718–722.

Listernick R, Louis DN, Packer RJ et al. (1997). Optic pathway gliomas in children with neurofibromatosis 1: consensus statement from the NF1 Optic Pathway Glioma Task Force. Ann Neurol 41: 143–149.

Listernick R, Ferner RE, Liu GT et al. (2007). Optic pathway gliomas in neurofibromatosis-1: controversies and recommendations. Ann Neurol 61: 189–198.

Matsui I, Tanimura M, Kobayashi N et al. (1993). Neurofibromatosis type 1 and childhood cancer. Cancer 72: 2746–2754.

Mautner VF, Kluwe L, Thakker SD et al. (2002). Treatment of ADHD in neurofibromatosis type 1. Dev Med Child Neurol 44: 164–170.

Mautner V, Brenner W, Fünsterer C et al. (2007). Clinical relevance of positron emission tomography and magnetic resonance imaging in the progression of internal plexiform neurofibroma in NF1. Anticancer Res 27: 1819–1822.

Mautner VF, Asuagbor FA, Dombi E et al. (2008). Assessment of benign tumor burden by whole-body MRI in patients with neurofibromatosis 1. Neuro Oncol 10: 593–598.

Mayes DA, Rizvi TA, Cancelas JA et al. (2011). Perinatal or adult Nf1 inactivation using tamoxifen-inducible PlpCre each cause neurofibroma formation. Cancer Res 71: 4675–4685.

McGaughran JM, Harris DI, Donnai D et al. (1999). A clinical study of type 1 neurofibromatosis in north west England. J Med Genet 36: 197–203.

Messiaen LM, Callens T, Mortier G et al. (2000). Exhaustive mutation analysis of the NF1 gene allows identification of 95% of mutations and reveals a high frequency of unusual splicing defects. Hum Mutat 15: 541–555.

Molloy PT, Bilaniuk LT, Vaughan SN et al. (1995). Brainstem tumors in patients with neurofibromatosis type 1: a distinct clinical entity. Neurology 45: 1897–1902.

Moretti VM, Crawford EA, Staddon AP et al. (2011). Early outcomes for malignant peripheral nerve sheath tumor treated with chemotherapy. Am J Clin Oncol 34: 417–421.

National Institutes of Health Consensus Development Conference (1988). Neurofibromatosis Conference Statement. Arch Neurol 45: 575–578.

Nichols JC, Amato JE, Chung SM (2003). Characteristics of Lisch nodules in patients with neurofibromatosis type 1. J Pediatr Ophthalmol Strabismus 40: 293–296.

North KN, Riccardi V, Samango-Sprouse C et al. (1997). Cognitive function and academic performance in neurofibromatosis 1: consensus statement from the NF1 Cognitive Disorders Task Force. Neurology 48: 1121–1127.

Nunley KS, Gao F, Albers AC et al. (2009). Predictive value of café au lait macules at initial consultation in the diagnosis of neurofibromatosis type 1. Arch Dermatol 145: 883–887.

Ostendorf AP, Gutmann DH, Weisenberg JL (2013). Epilepsy in individuals with neurofibromatosis type 1. Epilepsia 54: 1810–1814.

Petramala L, Giustini S, Zinnamosca L et al. (2012). Bone mineral metabolism in patients with neurofibromatosis type 1 (von Recklingausen disease). Arch Dermatol Res 304: 325–331.

Pollack IF, Shultz B, Mulvihill JJ (1996). The management of brainstem gliomas in patients with neurofibromatosis 1. Neurology 46: 1652–1660.

Porter DE, Prasad V, Foster L et al. (2009). Survival in malignant peripheral nerve sheath tumours: a comparison between sporadic and neurofibromatosis type 1-associated tumours. Sarcoma 2009: 756395.

Prada CE, Rangwala FA, Martin LJ et al. (2012). Pediatric plexiform neurofibromas: impact on morbidity and mortality in neurofibromatosis type 1. J Pediatr 160: 461–467.

Pride NA, Payne JM, North KN (2012). The impact of ADHD on the cognitive and academic functioning of children with NF1. Dev Neuropsychol 37: 590–600.

Pylayeva-Gupta Y, Grabocka E, Bar-Sagi D (2011). RAS oncogenes: weaving a tumorigenic web. Nat Rev Cancer 11: 761–774.

Rasmussen SA, Friedman JM (2000). NF1 gene and neurofibromatosis 1. Am J Epidemiol 151: 33–40.

Rasmussen SA, Yang Q, Friedman JM (2001). Mortality in neurofibromatosis 1: an analysis using U.S. death certificates. Am J Hum Genet 68: 1110–1118.

Riccardi V (1981). Von Recklinghausen neurofibromatosis. N Engl J Med 305: 1617–1627.

Robertson KA, Nalepa G, Yang FC et al. (2012). Imatinib mesylate for plexiform neurofibromas in patients with neurofibromatosis type 1: a phase 2 trial. Lancet Oncol 13: 1218–1224.

Ruggieri M (2001). Mosaic (segmental) neurofibromatosis type 1 (NF1) and type 2 (NF2): no longer neurofibromatosis type 5 (NF5). Am J Med Genet 101: 178–180.

Salyer WR, Salyer DC (1974). The vascular lesions of neurofibromatosis. Angiology 25: 510–519.

Segal L, Darvish-Zargar M, Dilenge ME et al. (2010). Optic pathway gliomas in patients with neurofibromatosis type 1: follow-up of 44 patients. J AAPOS 14: 155–158.

Sharif S, Ferner R, Birch JM et al. (2006). Second primary tumors in neurofibromatosis 1 patients treated for optic glioma: substantial risks after radiotherapy. J Clin Oncol 24: 2570–2575.

Sharif S, Moran A, Huson SM et al. (2007). Women with neurofibromatosis 1 are at a moderately increased risk of developing breast cancer and should be considered for early screening. J Med Genet 44: 481–484.

Shilyansky C, Lee YS, Silva AJ (2010). Molecular and cellular mechanisms of learning disabilities: a focus on NF1. Annu Rev Neurosci 33: 221–243.

Side L, Taylor B, Cayouette M et al. (1997). Homozygous inactivation of the NF1 gene in bone marrow cells from children with neurofibromatosis type 1 and malignant myeloid disorders. N Engl J Med 336: 1713–1720.

Side LE, Emanuel PD, Tyalor B et al. (1998). Mutations of the NF1 gene in children with juvenile myelomonocytic leukemia without clinical evidence of neurofibromatosis type 1. Blood 92: 267.

Silva AJ, Frankland PW, Marowitz Z et al. (1997). A mouse model for the learning and memory deficits associated with neurofibromatosis type I. Nat Genet 15: 281–284.

Slater C, Hayes M, Saxe N et al. (1986). Macromelanosomes in the early diagnosis of neurofibromatosis. Am J Dermatopathol 8: 284–289.

Smith R (1849). A Treatise on the Pathology, Diagnosis and Treatment of Neuroma. Hodges and Smith, Dublin.

Soucy EA, Gao F, Gutmann DH et al. (2012). Developmental delays in children with neurofibromatosis type 1. J Child Neurol 27: 641–644.

Stevenson DA, Moyer-Mileur LJ, Murray M et al. (2007). Bone mineral density in children and adolescents with neurofibromatosis type 1. J Pediatr 150: 83–88.

Stiller CA, Chessells JM, Fitchett M (1994). Neurofibromatosis and childhood leukaemia/lymphoma: a population-based UKCCSG study. Br J Cancer 70: 969–972.

Sung L, Anderson JR, Arndt C et al. (2004). Neurofibromatosis in children with rhabdomyosarcoma: a report from the Intergroup Rhabdomyosarcoma study IV. J Pediatr 144: 666–668.

Tischler AS, Shih TS, Williams BO et al. (1995). Characterization of pheochromocytomas in a mouse strain with a targeted disruptive mutation of the neurofibromatosis gene Nf1. Endocr Pathol 6: 323–335.

Tong J, Hannan F, Zhu Y et al. (2002). Neurofibromin regulates G protein-stimulated adenylyl cyclase activity. Nat Neurosci 5: 95–96.

Ullrich NJ, Raja AI, Irons MB et al. (2007). Brainstem lesions in neurofibromatosis type 1. Neurosurgery 61: 762–767.

Valero MC, Martín Y, Hernández-Imaz E et al. (2011). A highly sensitive genetic protocol to detect NF1 mutations. J Mol Diagn 13: 113–122.

van der Vaart T, Plasschaert E, Rietman AB et al. (2013). Simvastatin for cognitive deficits and behavioural problems in patients with neurofibromatosis type 1 (NF1-SIMCODA): a randomized, placebo-controlled trial. Lancet Neurol 12: 1076–1083.

Viskochil D, Buchberg AM, Xu G et al. (1990). Deletions and a translocation interrupt a cloned gene at the neurofibromatosis type 1 locus. Cell 62: 187.

Vlenterie M, Flucke U, Hofbauer LC (2013). Pheochromoctyoma and gastrointestinal stromal tumors in patients with neurofibromatosis type 1. Am J Med 126: 174–180.

Vogel KS, Klesse LJ, Velasco-Miguel S et al. (1999). Mouse tumor model for neurofibromatosis type 1. Science 286: 2176–2179.

Von Recklinghausen F (1882). Über die multiplen Fibrome der Haut und ihre Beziehung zu den multiplen Neuromen. Hirschwald, Berlin.

Waggoner DJ, Towbin J, Gottesman G et al. (2000). Clinic-based study of plexiform neurofibromas in neurofibromatosis 1. Am J Med Genet 92: 132–135.

Wallace MR, Marchuk DA, Andersen LB et al. (1990). Type 1 neurofibromatosis gene: identification of a large transcript disrupted in three NF1 patients. Science 249: 181.

Walsh KS, Vélez JI, Kardel PG et al. (2013). Symptomatology of autism spectrum disorder in a population with neurofibromatosis type 1. Dev Med Child Neurol 55: 131–138.

Walther MM, Herring J, Enquist E et al. (1999). von Recklinghausen's disease and pheochromoctyomas. J Urol 162: 1582–1586.

Wang W, Nyman JS, Ono K et al. (2011). Mice lacking Nf1 in osteochondroprogenitor cells display skeletal dysplasia similar to patients with neurofibromatosis type I. Hum Mol Genet 20: 3910–3924.

Wessel LE, Gao F, Gutmann DH et al. (2013). Longitudinal analysis of developmental delays in children with neurofibromatosis type 1. J Child Neurol 28: 1689–1693.

Woodruff JM (1999). Pathology of tumors of the peripheral nerve sheath in type neurofibromatosis. Am J Med Genet 89: 38.

Wu J, Williams JP, Rizvi TA et al. (2008). Plexiform and dermal neurofibromas and pigmentation are caused by Nf1 loss in desert hedgehog-expressing cells. Cancer Cell 13: 105–116.

Xu GF, O'Connell P, Viskochil D et al. (1990). The neurofibromatosis type 1 gene encodes a protein related to GAP. Cell 62: 599–608.

Yang FC, Ingram DA, Chen S et al. (2008). Nf1-dependent tumors require a microenvironment containing Nf1+/−− and c-kit-dependent bone marrow. Cell 135: 437–438.

Zanca A (1980). Antique illustrations of NF1. Int J Dermatol 19: 55.

Zhang W, Rhodes SD, Zhao L et al. (2011). Primary osteopathy of vertebrae in a neurofibromatosis type 1 murine model. Bone 48: 1378–1387.

Zhu Y, Ghosh P, Charnay P et al. (2002). Neurofibromas in NF1: Schwann cell origin and role of tumor environment. Science 296: 920–922.

Zhu Y, Guignard F, Zhao D et al. (2005a). Early inactivation of p53 tumor suppressor gene cooperating with NF1 loss induces malignant astrocytoma. Cancer Cell 8: 119–130.

Zhu Y, Harada T, Liu L et al. (2005b). Inactivation of NF1 in CNS causes increased glial progenitor proliferation and optic glioma formation. Development 132: 5577–5588.

Handbook of Clinical Neurology, Vol. 132 (3rd series)
Neurocutaneous Syndromes
M.P. Islam and E.S. Roach, Editors

Chapter 5

Neurofibromatosis type 2

D.G.R. EVANS*
University Department of Genomic Medicine, University of Manchester, St Mary's Hospital, Manchester, UK

INTRODUCTION

Type 2 neurofibromatosis (NF2) was probably first described by Wishart in 1822. He described a man with multiple intracranial tumors including what Wishart described as bilateral seventh nerve neuromas, but which Worster-Drought et al. (1937) later suggested were probably eighth nerve tumors (previously called acoustic neuromas but now more accurately labeled vestibular schwannoma). However, following reports of patients with type 1 neurofibromatosis (NF1) by von Recklinghausen (1882), various reports of NF2 cases around the turn of the century were lumped together with von Recklinghausen's disease. Probably the most responsible for the prolonged co-classification of NF1 and NF2 was none other than Harvey Cushing (1917).

Although many reports emphasized the lack of skin tumors or café-au-lait patches in patients and families with bilateral vestibular schwannomas (Gardner and Frazier, 1930; Worster-Drought et al., 1937; Young et al., 1971), the final separation of NF1 and NF2 and their delineation only came in 1987. In that year the gene for NF1 was localized to chromosome 17 (Seizinger et al., 1987) and NF2 to chromosome 22 (Rouleau et al., 1987) by genetic linkage analysis. As a result of this and the increasing clinical evidence to implicate two distinct disorders (Kanter et al., 1980), the US National Institutes of Health (NIH) Consensus Statement published that year (NIH Consensus Development Conference Statement, 1987) formally separated them. The still widely held belief that vestibular schwannomas occur as part of NF1 has now been refuted by large population-based studies of the more common type 1 disease (Huson et al., 1988; McGaughran et al., 1999). Previous reports of vestibular schwannomas in NF1 clearly were contaminated with NF2 cases (Crowe et al., 1956).

CLINICAL MANIFESTATIONS

NF2 is an autosomal dominantly inherited disorder which predisposes affected individuals to the development of vestibular schwannomas (usually bilateral) (Fig. 5.1), schwannomas of the other cranial, spinal and peripheral nerves, meningiomas both intracranial (including on the optic nerve) and intraspinal, and some low grade CNS malignancies (ependymomas, and rarely low grade gliomas). Three large clinical studies now have confirmed the clinical picture (Kanter et al., 1980; Evans et al., 1992a; Parry et al., 1994).

Our own published diagnostic criteria for NF2 (Evans et al., 1992b) are shown in Table 5.1. The original NIH criteria have been expanded to include patients with no family history who have multiple schwannomas and or meningiomas, but who have not yet developed bilateral eighth nerve tumors. These criteria have been shown to be more sensitive and no less specific than the original NIH criteria which acted as the template (Baser et al., 2002a). Individuals may present with cranial meningiomas or a spinal tumor long before the appearance of a vestibular schwannoma (Evans et al., 1992a, 1999a). As at least 50–60% of cases represent new dominant mutations (Evans et al., 1992c) the criteria are more inclusive, but are still extremely unlikely to include chance associations (see below under genetics).

The results of the three major clinical studies (Kanter et al., 1980; Evans et al., 1992a; Parry et al., 1994) are shown in Table 5.2. The majority of individuals with NF2 present with hearing loss, which is usually unilateral at onset. This may be accompanied or preceded by tinnitus. Occasionally vestibular schwannomas may cause dizziness or imbalance as the first symptom. A significant number of patients present initially with a cranial meningioma, spinal tumor, or cutaneous tumor (20–30%). Indeed the more severe multitumor disease

*Correspondence to: Professor D.G.R. Evans, Consultant Clinical Geneticist, Genomic Medicine, MAHSC, St Mary's Hospital, Oxford Road, Manchester, M13 9WL, UK. Tel: +44-(0)161-276-6206, Fax: +44-(0)161-276-6145, E-mail: gareth.evans@cmft.nhs.uk

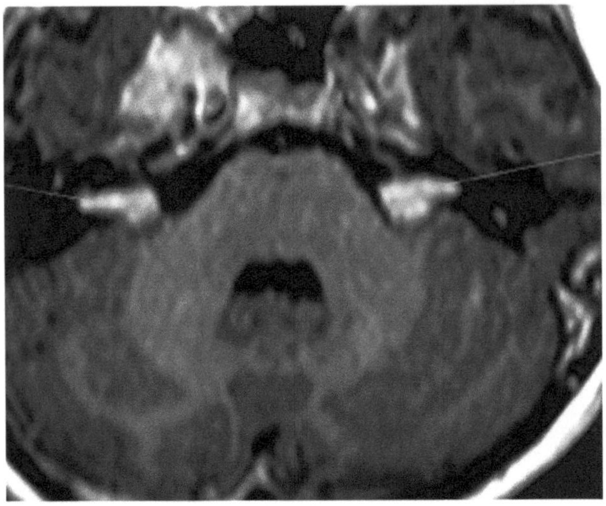

Fig. 5.1. Magnetic resonance image showing bilateral enhancing masses (vestibular schwannomas) in the cerebellopontine angle.

Table 5.1

Manchester diagnostic criteria for neurofibromatosis type 2

Bilateral vestibular schwannomas OR family history of NF2 PLUS
- Unilateral acoustic OR
- Any two of: meningioma, glioma, neurofibroma, schwannoma, posterior subcapsular lenticular opacities

Additional criteria
- Also cerebral calcification

Unilateral vestibular schwannomas + any two of: meningioma, glioma, neurofibroma, schwannoma, cataract, cerebral calcification

Multiple meningioma (two or more) + unilateral acoustic or any two of: glioma, neurofibroma, schwannoma, cataract, cerebral calcification

Table 5.2

Age at onset, frequency of tumour, and other manifestations of neurofibromatosis type 2

	Series		
	Kanter et al., 1980	Evans et al., 1992a	Parry et al., 1994
Number of cases	73	120	63
Number of families	17	75	32
Isolated cases	0	45	17
Age at onset (years)	20.4 (of 59)	22.2	20.3
Meningiomas	18%	45%	49%
Spinal tumors	?	25.8%	67%
Skin tumors	32% (73)	68% (of 100)	67%
>10 skin tumors	?	10% (100)	?
Café-au-lait	42% (31)	43% (100)	47%
Cataracts	–	38% (90)	81%
Astrocytoma	?	4.1%	1.6%
Ependymoma	?	2.5%	3.2%
Optic sheath meningioma	?	4.1%	4.8%

is often manifested first in early childhood with a non-eighth cranial nerve tumor. Pediatric presentation is thus quite different to that in adulthood with vestibular schwannoma accounting for as little as 15–30% of initial symptoms (Evans et al., 1999b).

There is also a tendency for mononeuropathy, particularly affecting the facial nerve to cause a Bell's-like palsy which often does not fully recover. Facial palsy may occur years before a vestibular schwannoma. Some children present with a polio-like illness with wasting of muscle groups in a lower limb, which again does not fully recover. About 3–5% of adults develop a more generalized polyneuropathy, often associated with an "onion bulb" appearance on nerve biopsy (Thomas et al., 1990; Evans et al., 1992a). The pathogensis of this is related to an important function of the NF2 protein merlin (Schulz et al., 2013). This neuropathy can progress, leading to severe muscle wasting and even death.

Ophthalmic features are also prominent in NF2. Between 60% and 80% of patients with the disease have evidence of cataract (Kaiser-Kupfer et al., 1989; Parry et al., 1994). These are usually presenile posterior subcapsular lenticular opacities that rarely require removal. However, childhood cortical wedge opacities may be present from near birth (Evans et al., 1999b). Children may present with ophthalmic features. Optic nerve meningiomas can cause visual loss in the first years of life and extensive retinal hamartomas can also affect vision. Both problems have in my experience led to the eye being removed in the first few years of life due to misdiagnosis as retinoblastoma.

The skin is a useful aid to diagnosis, but features in NF2 are much more subtle than NF1. About 70% of NF2 patients have skin tumors, but only 10% have more than 10 (Evans et al., 1992a). These tumors appear to be of at least three different types. The most frequent is a plaque-like lesion (Fig. 5.2), which is intracutaneous, slightly raised, more pigmented than surrounding skin, and often contains excess hair. More deep-seated subcutaneous nodular tumors (Fig. 5.3) can often be felt,

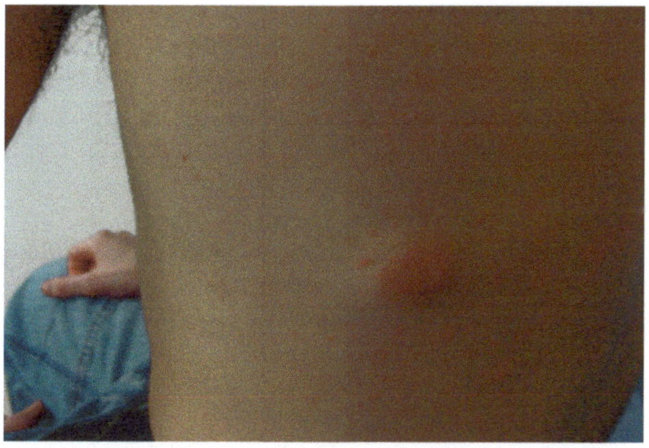

Fig. 5.2. Plaque-like lesions on the abdomen in a patient with NF2. These are raised, often slightly pigmented lesions, which are also frequently hairy.

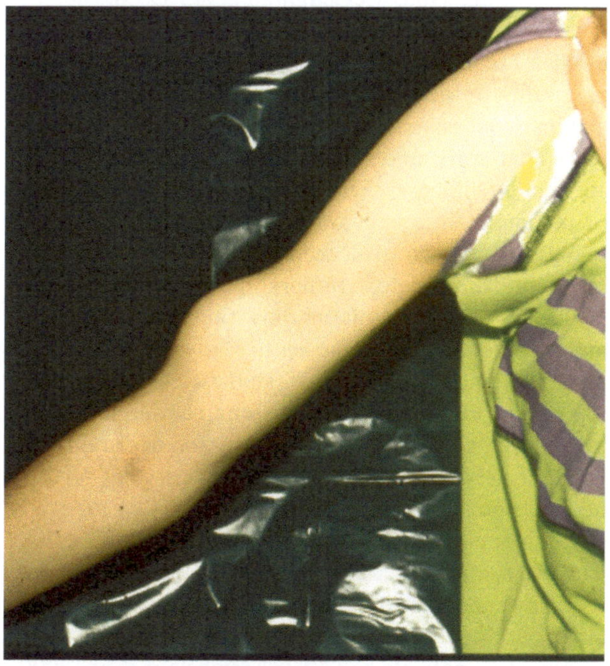

Fig. 5.3. Subcutaneous schwannoma on a major nerve in the upper arm in a patient with NF2.

sometimes on major peripheral nerves. These can manifest as a fusiform swelling of the nerve with thickened nerve either side. There are also occasional intracutaneous tumors similar to those in NF1. The great majority of these tumors are schwannomas, but occasional definite neurofibromas do occur. I have seen only four patients (out of over 450) who had a plexiform tumor similar to those occurring in NF1.

Even with improvements in microsurgery and with use of radiation therapy, the great majority of individuals with NF2 become completely deaf. The tumors in NF2 are more difficult to treat as they are often multifocal, appearing "like a bunch of grapes" (Martuza and Ojemann, 1982; Evans et al., 1993, 2005), around the vestibular nerve in particular. A vestibular schwannoma due to NF2 is histologically more lobular and less vascular than its sporadic counterparts (Sobel and Wang, 1993). Patients may also be severely disabled by a combination of poor balance, visual problems, and weakness due to spinal tumors. Indeed many become wheelchair bound in early adulthood. Loss of facial nerve function is one of the most feared aspects of the condition for many sufferers, although in good surgical hands this is now much less common (Evans et al., 1993, 2005; Samii et al., 1997; Slattery et al., 1998). Mortality in NF2 was on average 15 years after diagnosis with a mean actuarial survival of 60 years pre 1990 (Evans et al., 1992a). Many patients with multitumor disease were dying in their twenties and thirties. This has improved recently but NF2 patients still have significantly reduced survival compared to the general population (Wilding et al., 2012) and survival is worst in those with constitutional truncating mutations (Baser et al., 2002b).

Radiographic findings

NF2 can be diagnosed if the criteria in Table 5.1 are fulfilled. The most precise means of diagnosis is magnetic resonance imaging (MRI) with gadolinium enhancement; this should include a complete spinal as well as cranial scan (Evans et al., 2005). MRI can now detect tumors as small as 1–2 mm in diameter on cranial and spinal nerve roots, but many tumors now being identified on spinal and cranial scans never lead to symptoms. Studies published before widespread use of spinal scanning suggested that only 25–30% of NF2 patients had symptomatic spinal tumors (Evans et al., 1992a; Parry et al., 1994). However, full spinal imaging will detect evidence of spinal tumors in 80–90% of individuals with NF2 (Mautner et al., 1996). There is also increasing awareness that intramedullary tumors in the upper cervical spine and brainstem are often associated with syringomyelia. On biopsy these tumors are usually low grade ependymomas (Hagel et al., 2012). Although these initially can be very worrying for the radiologist or treating clinician, the great majority do not progress.

An MRI can also detect schwannomas on other cranial nerves. These occur most commonly on the fifth and 12th cranial nerves (Evans et al., 1992b), but any cranial nerve except the olfactory can be affected in NF2. Nonetheless it is rare for cranial nerve schwannomas other than vestibular schwannomas to grow to a size where removal is necessitated. Meningiomas can easily be detected on MRI as enhanced areas on the meninges around the spinal cord, brain, or optic nerves. These can form confluent areas on scan or "meningioma en plaque". In contrast to vestibular schwannomas, which grow on average 2 mm per year, meningiomas can grow very rapidly.

There are several groups of individuals who should be considered at risk and investigated further. These include those with a family history of NF2, patients under 30 years presenting with a unilateral vestibular schwannoma or meningioma, patients with multiple spinal tumors (schwannomas or meningiomas) or cutaneous schwannomas (Evans et al., 1992b, 1999a, 2012; Gutmann et al., 1997). MRI scanning is vital in their further assessment and full recommendations on following this group have been published (Evans et al., 2012).

An important goal is the development of more reproducible results on MRI methods to determine tumor growth. Most authors who assess the effect of either conservative (Bederson et al., 1991) or radiosurgical (Kondziolka et al., 1998) management of vestibular schwannomas rely on fairly basic measurements such as cross-sectional diameters. There is great variation from scan to scan and observer to observer with these limited measurement techniques. Use of high technology MRI with volumetric analysis is the only way in which the efficacy of nonsurgical treatment (radiation treatment, drug therapy) can be compared usefully with observation alone (Blakeley et al., 2012).

Although computed tomography (CT) scans have poor sensitivity in detecting small vestibular schwannomas (Evans et al., 1992a, b), CT can identify another feature of NF2. A proportion of NF2 patients have intracranial calcification from a young age (Evans et al., 1992a). However, this sign is not useful enough to supplement MRI with CT. Repeated CT scans should probably be avoided in individuals with NF2 due to their predisposition to tumor formation.

PATHOLOGY

The main tumor sites, their frequencies, and their pathologies are presented in Table 5.2. As stated before, schwannomas can occur in all locations where there are nerves with Schwann cells. The predilection for the superior vestibular branch of the eighth cranial nerve remains unexplained. Schwannomas are encapsulated tumors of pure Schwann cells that grow around the nerve. They may contain blood vessels and have areas of sheets in intertwining fascicles (Antoni A) and looser arrangements called Antoni B (Sobel and Wang, 1993). The tumors also stain for S-100 protein and vimentin. In NF2 these tumors can be multifocal and have a more lobular architecture than sporadic tumors (Sobel and Wang, 1993). Spontaneous malignant transformation of these tumors to malignant peripheral nerve sheath tumors occurs (Higami et al., 1998), but is more than 10 times as likely to happen after radiation treatment (Baser et al., 2000). The background rate of 0.5% for central nervous system (CNS) malignancy in NF2 is very much less than for NF1 (McGaughran et al., 1999). A small proportion of nerve-related tumors in NF2 is pathologically delineated as neurofibroma. In these tumors there is an admixture of cell types including Schwann cells, fibroblasts, and mast cells, and the tumor usually contains identifiable axons. Neurofibromas occur mainly in the skin (where they still are outnumbered by schwannomas by a factor of 5–10) but also develop on the spinal nerve roots; they are not, however, seen in the cranium. Occasionally tumors show features of both schwannoma and neurofibroma, particularly on spinal nerve roots. In contradistinction to NF2, schwannomas and meningiomas do not occur in excess in NF1 (Huson et al., 1988; McGaughran et al., 1999).

Meningiomas are the second most characteristic tumor of NF2. The most typical sites for meningiomas are supratentorially in the falx and around the frontal, temporal and parietal regions. Meningiomas also occur around the spinal cord, and these can be difficult to remove surgically. Although there are different histologic types of meningioma (meningothelial, fibroblastic and transitional), there is no evidence for a subdivision into NF2 and non-NF2 related tumors clinically (Antinheimo et al., 1997). Collision tumors consisting of a schwannoma and meningioma can occur particularly in the cerebellopontine angle. A number of studies have determined the proportion of meningiomas and schwannomas due to NF2. In one series of spinal schwannomas and neurofibromas (Halliday et al., 1991), all the cases of NF1 had neurofibromas, whereas only a proportion of the spinal schwannomas were due to NF2 and one NF2 patient had a mixed tumor. Another study of all meningiomas and schwannomas from an 11-year period in the Helsinki area found that 3% of schwannoma patients and 1% of meningioma patients had NF2 (Antinheimo et al., 2000). However, a further 2% of schwannoma patients and 4% of meningioma patients had multiple tumors without fulfilling diagnostic criteria for NF2. The great majority of NF2 patients do not present with an isolated tumor, and the risk of NF2 in a person with an isolated vestibular schwannoma is small (Evans et al., 1999b), but as many as 10% of children with an apparently isolated meningioma go on to develop NF2 (Evans et al., 1999a).

Low grade ependymomas and gliomas are now being increasingly recognized in individuals with NF2 (Fig. 5.4). These are indolent tumors and rarely metastasize within the CNS. Their prime location is in the cervical spine and brainstem (Hagel et al., 2012). Malignant progression is sometimes associated with radiotherapy treatment (Baser et al., 2000).

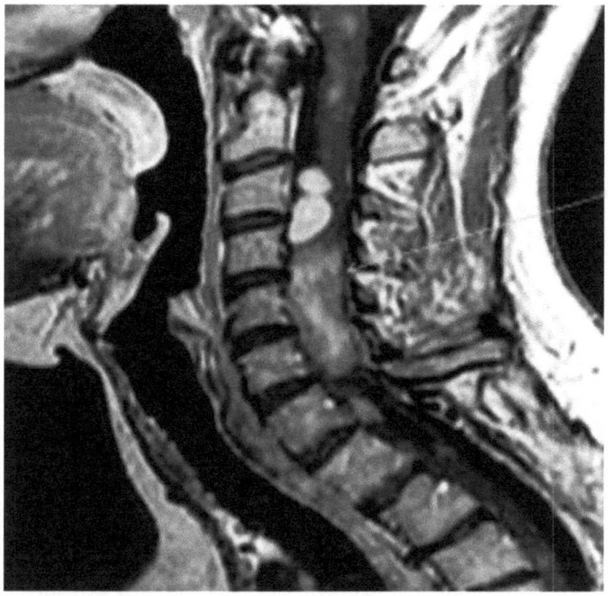

Fig. 5.4. Ependymoma.

GENETICS

Type 2 neurofibromatosis is an autosomal dominantly inherited disorder due to mutations in a single gene on chromosome 22. Individuals who inherit a mutation of the *NF2* gene inevitably develop schwannomas, particularly of the eighth cranial nerve, but also other cranial, spinal and cutaneous nerves. The great majority of nerve tumors are schwannomas, but cutaneous and spinal neurofibromas do occur. Around 60% of individuals will also develop a meningioma. All these tumors occur by inactivation of the second normal copy of the *NF2* gene in a single cell, although further events may be necessary for tumor progression. As expected with autosomal dominant inheritance, half of the offspring of individuals with full blown constitutional NF2 are affected.

NF2 affects about 1 in 35–40 000 people (Evans et al., 1992c, 2010). However, because many people do not develop features of the condition until the third decade or later – and many others die before this time – the actual diagnostic prevalence was only 1 in 200 000 in 1992. This has increased to 1 in 56 000 in 2010. The annual incidence rate is 1 per 2 355 000 representing about one new case per year for each health region in the UK (Evans et al., 1992c), or 100 cases per year in the USA. Individuals with NF2 have reduced genetic fitness and this is more marked in males, who may delay having families until their disease has progressed (Evans et al., 1992c). This reduction of potential gene carriers to pass on the condition is counterbalanced by the high new mutation rate, with 50–60% of patients having no affected parent.

The gene has a high degree of penetrance and is nearly always expressed by the late fifties (Evans et al., 1992c). The initial clue to the whereabouts of the gene came with the discovery of chromosome 22 abnormalities on cytogenetic analysis of meningiomas (Zang, 1982), a location that was confirmed at the molecular level in several different tumors from an NF2 patient with loss of constitutional heterozygosity for chromosome 22q markers (Seizinger et al., 1986). This provided strong evidence that the NF2 gene was a tumor suppressor, with both copies of the gene needing to be inactivated to initiate tumorigenesis (Knudson's 1971 "two hit" hypothesis of tumorigenesis). NF2 is perhaps the best example of this mechanism, other than retinoblastoma, for which the hypothesis was made. Every Schwann cell in NF2 already carries a mutant copy of the gene and if the other copy is mutated or deleted then a tumor ensues. There is little evidence for the involvement of other genes in the formation of schwannomas. Indeed while many studies have shown direct evidence of NF2 mutation or deletion in about 60% of schwannomas, at the biological level virtually 100% show loss of NF2 protein (Huynh et al., 1997).

After the confirmation of a site of tumor loss, linkage studies confirmed that all affected members of the large Pennsylvanian family first reported in 1930 by Gardner and Frazier carried the same copy of chromosome 22 (Rouleau et al., 1987). Further studies have shown no evidence for another gene to account for the NF2 phenotype, and NF2 is almost certainly caused by a single gene on 22q (Narod et al., 1992; Ruttledge et al., 1993). The *NF2* gene was further localized by the discovery of germ line deletions in NF2 families, one of which involved the neurofilament heavy chain gene (Watson et al., 1993). The gene itself was eventually isolated by the simultaneous discovery of constitutional and tumor deletions in a gene coding for a cell membrane-related protein termed merlin (Trofatter et al., 1993) or schwannomin (Rouleau et al., 1993).

Many studies have now confirmed this locus as the *NF2* gene. Standard mutation techniques such as SSCP (single strand confirmation polymorphism) or DGGE (denaturing gradient gel electrophoresis) detect between 35% and 65% of causative mutations (Bourn et al., 1994a; MacCollin et al., 1994; Mérel et al., 1995; Evans et al., 1998a). The majority of these mutations are truncating mutations, leading to a smaller and probably ineffective protein product. Currently full sequencing of the coding exons and intron exon boundaries is the best method, with next generation sequencing offering a valuable rapid approach that can detect mosaicism.

There is now clear evidence of a genotype/phenotype correlation with most missense mutations, which will give rise to a complete protein product, or deletions giving no protein product having a mild phenotype (Parry et al., 1996; Ruttledge et al., 1996; Evans et al., 1998a). Splice site mutations that may splice out a sequence of amino acids from the protein also appear to cause a mild phenotype, but this is more variable (Kluwe et al., 1998). The more severe phenotype with truncating mutations, such as nonsense and frameshifting changes, may be due to a dominant negative effect with mutant protein dimerizing with the normal product leaving less wild type protein for tumor suppression. Recent evidence also suggests that up to 33% of new cases of NF2 with no family history of the disease carry the mutation in only a small proportion of their cells (Evans et al., 1998b, 2007; Kluwe and Mautner, 1998). In these mosaic individuals, the *NF2* gene mutation occurs after conception. If the mutation is carried by < 10% of lymphocytes, it is not detectable by standard mutation detection techniques (including sequencing), although the percentage of cells affected can vary from tissue to tissue (Bourn et al., 1994b). A higher proportion of affected neural crest cells may increase the severity of the disease, but, in general, mosaic individuals are less severely affected than would be predicted from their mutation because fewer cells receive a second hit (Evans et al., 1998b, 2013). In sporadic NF2 patients in whom a mutation is not detected in blood DNA, mutations can be identified in their tumors. An identical mutation in two tumors from the same individual is virtually certain to represent the underlying defect (Evans et al., 1998b). A further cause of the low detection rate for mutations using standard techniques is large deletions and rearrangements at the NF2 locus (Zucman-Rossi et al., 1998). Some of these can be detected by cytogenetic studies using FISH where the signal from one chromosome 22 is missing in each cell. However, others are too small for this type of detection, and techniques such as Southern blotting and PCR dosage studies need to be employed. Mosaic patients are unlikely to be detected with these methods (Zucman-Rossi et al., 1998; Evans et al., 2007).

Predictive diagnosis using markers flanking the *NF2* gene has been possible for 20 years in the vast majority of families containing two or more living affected individuals by linkage analysis (Ruttledge et al., 1996; Evans et al., 1995a). Once a mutation has been identified in an affected individual, a 100% specific test is available for that family. However, blood mutation analysis may not reveal the causative mutation. As such, tumor analysis is a vital step in sporadic cases so that a mosaic mutation can be identified (Evans et al., 2007). Mutation analysis can be undertaken on the child of a sporadic NF2 patient (Evans et al., 2012). By combining this with a cumulative age at onset curve (Evans et al., 1992b; updated in Fig. 5.5), the risk to an unaffected 30-year-old with a

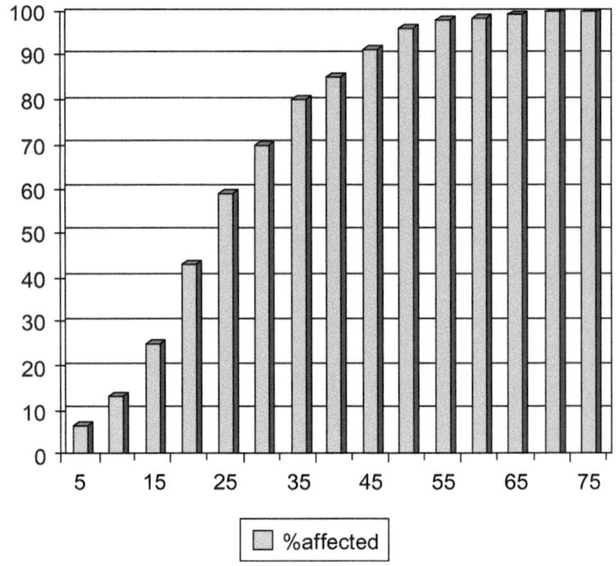

Fig. 5.5. Cumulative age at onset of symptoms in patients with NF2.

normal scan and favorable DNA result would be infinitesimally small. Complete confidence can still be attained only with the identification of family-specific mutations.

Age at onset curves aid genetic counseling. For example, for an asymptomatic at-risk 20-year-old individual, their risk of having inherited NF2, prior to screening, will have dropped to 25%. As more families are followed prospectively a curve for age at diagnosis, with screening, will become available. At-risk individuals can be discharged if they have been shown not to have inherited the mutated *NF2* gene. Uptake of presymptomatic genetic tests for NF2 are high (Evans et al., 1997a), with around 90% of asymptomatic individuals at 50% risk opting for genetic tests.

The main diagnostic dilemma with NF2 is patients with isolated multiple noncranial schwannomas. Some of these patients may well go on to develop NF2 (Evans et al., 1997b) and all require a cranial MRI scan. Some patients prove to be mosaic for an *NF2* mutation (Jacoby et al., 1997; Evans et al., 1998b). A small group of patients with tumors largely confined to the peripheral nerves and spine, with sparing of the eighth nerve do not to have an identifiable *NF2* mutation (Jacoby et al., 1997). These individuals have schwannomatosis and may pass the condition on to their children, and in families where this occurs there is still linkage to the NF2 locus (Evans et al., 1997b; Jacoby et al., 1997). Two genes centromeric to NF2 on chromosome 22 have now been found and account for about 70% of familial and 20% of sporadic schwannomatosis patients: *SMARCB1* (Hulsebos et al., 2007; Smith et al., 2014) and *LZTR1* (Piotrowski et al., 2014; Smith et al., 2015).

Patients with unilateral vestibular schwannoma and other schwannomas may have a *LZTR1* mutation (Smith et al., 2015). Patients may also present with multiple meningiomas with no other features of NF2. The key feature in both these related conditions is the absence of other NF2-related tumors such as intracutaneous schwannoma and ependymoma or cataracts.

Modifying factors

There is evidence that NF2 may be worse if it is inherited from the mother (Kanter et al., 1980; Evans et al., 1992c). However, this could also be due to decreased genetic fitness in males, with only the less severely affected men having children (Baser et al., 2001). As there is no evidence of genomic imprinting on chromosome 22q, the previous reports of a maternal gene effect may be spurious. General worsening of the disease with successive generations ("anticipation") has also been suggested, but objective evidence for this is lacking. Some authors strongly believe that females are more severely affected than males and women become worse during pregnancy (Allen et al., 1974). However, a more complete analysis suggests that this effect applies mainly to meningiomas rather than schwannomas (Evans et al., 1995b).

On the whole, the disease course of NF2 does breed true in individual families. Some families have a mild disease course with late onset and few if any CNS tumors other than vestibular schwannomas. Other families have a more virulent course with early onset and death due to multitumor disease. The evidence of a genotype/phenotype correlation is encouraging and may provide insight into the disease process. Nonetheless, much will depend on stochastic events such as loss of the second NF2 allele, because the disease course is not identical even in monozygotic twins (Baser et al., 1996).

MANAGEMENT ISSUES

We have previously published protocols for screening individuals at risk for NF2 (Evans et al., 1992b, 2005, 2012). The potential for early onset symptoms (10% of cases symptomatic before 10 years of age) warrants earlier screening than generally suggested. The presence of congenital or very early cataracts necessitates detailed early ophthalmologic screening, and slit lamp evaluation and ophthalmoscopy should be done at least every 5 years. Screening for vestibular schwannomas should begin around 10–12 years of age with audiologic testing including auditory brainstem responses (ABRs) which should be performed annually. ABRs have been shown to be more useful in screening for vestibular schwannomas in NF2 than other audiologic tests (NIH Conference Statement, 1990). MRI scans are the imaging technique of choice, because they detect smaller tumors than CT and avoid the risk of radiation. Cranial MRI should be carried out annually in individuals with tumors and every 3 years in those at risk. A baseline scan should be performed aged 14 years. A normal MRI scan at 16–18 years approximately halves the risk of having inherited NF2 and will facilitate reproductive decision making. A normal scan at 30 years will make inheritance very unlikely except in late onset families and could justify cessation of formal screening. When onset has been late, screening should be continued appropriately. In view of the increasing availability of MRI, it may well replace audiologic testing altogether as a screening tool. The baseline scan should include full spinal imaging, but because the majority of asymptomatic spinal tumors would not be surgically removed, further spinal imaging should be performed as indicated by signs and symptoms. Clinical review including an annual neurologic examination should accompany any screening. Although annual review could be omitted in very mild families in the first 10 years of life, it is useful for there to be a point of contact for all at-risk individuals.

The surgical management of individuals with NF2 is complex and probably should be confined to specialized centers with experienced otolaryngologists and neurosurgeons. Data from our own group and elsewhere (Evans et al., 1993, 2005; Samii et al., 1997; Slattery et al., 1998) indicate that individuals treated at specialist centers have a better outcome. As NF2 is a relatively rare condition, in a country the size of the UK, perhaps only two or three such centers would be needed (perhaps 10 in the USA). Otolaryngologists, neurosurgeons and clinical geneticists working together could coordinate the care and follow-up of affected individuals and their at-risk relatives. Timing of surgery can be critical, and the decision to operate early or wait until symptoms develop is crucial. Availability of new techniques such as brainstem implants is only available in a few centers. The cloning of the *NF2* gene may provide insights into pathogenesis and suggest possible preventative treatments. Specialist centers with vast experience in NF2 management would ensure clinical application of such developments, with careful appraisal at the earliest opportunity.

Drug therapy

One of the most significant breakthroughs in NF2 was the publication of a small pragmatic study of the use of the VEGF antibody bevacizumab in treatment of growing vestibular schwannomas (Plotkin et al., 2009). This study and a larger follow-up study showed that 2 weekly infusions of bevacizumab shrinks around 40–50% of growing vestibular tumors and controls growth in around 70%. There is also evidence of hearing improvement in a minority (Plotkin et al., 2012). There is not convincing evidence of the drug in treating NF2-related meningioma disease.

THE FUTURE

The much awaited step in NF2 was the localization and cloning of the gene. This achieved, precise diagnosis is possible if a specific mutation can be found, and this may also allow prediction of disease course by genotype/phenotype studies. This will not be as important in families where the extent of the disease is already known, but for new mutations insight into the likely speed of tumor progression and risk of other tumors would be very helpful. Although DNA predictive testing is now available with flanking markers, there is currently little demand for prenatal or preimplantation diagnoses.

The real hope was that the discovery of the gene and its transcriptional protein product merlin/schwannomin would lead to the development of somatic gene therapy. The prospects for NF2 appeared promising, because of the lack of variation in individuals with the same mutation and in the paucity of involvement of other genes in the tumors themselves. Replacement of the tumor suppressor product in the tumors through viral vectors or direct recombination of the *NF2* gene, although requiring great advances in our knowledge, could be very rewarding. Nonetheless many trials of agents utilizing knowledge of the NF2 pathway such as sorafenib, nilotinib, erlotinib, lapatinib and rapamycin have been initiated and it is hoped there will soon be additional agents with proven efficacy available (Blakeley et al., 2012).

References

Allen J, Eldridge R, Koerber T (1974). Acoustic neuroma in the last months of pregnancy. Am J Obstet Gynecol 119: 516–520.

Antinheimo J, Haapasalo H, Haltia M et al. (1997). Proliferation potential and histological features in neurofibromatosis 2-associated and sporadic meningiomas. J Neurosurg 87: 610–614.

Antinheimo J, Sankila R, Carpen O et al. (2000). Population based analysis of sporadic and NF2-associated meningiomas and schwannomas. Neurology 54: 71–76.

Baser ME, Ragge N, Riccardi V et al. (1996). Phenotypic variability in monozygotic twins with neurofibromatosis 2. Am J Med Genet 64: 563–567.

Baser ME, Evans DGR, Jackler RK et al. (2000). Malignant peripheral nerve sheath tumors radiotherapy and neurofibromatosis 2. Br J Cancer 82: 998.

Baser M, Friedman J, Evans DGR et al. (2001). Maternal gene effect in neurofibromatosis 2: fact or artefact? J Med Genet 38: 783–784.

Baser ME, Friedman JM, Wallace AJ et al. (2002a). Evaluation of diagnostic criteria for neurofibromatosis 2. Neurology 59: 1759–1765.

Baser ME, Friedman JM, Aeschilman D et al. (2002b). Predictors of the risk of mortality in neurofibromatosis 2. Am J Hum Genet 71: 715–723.

Bederson JB, von Ammon K, Wichmann WW et al. (1991). Conservative treatment of patients with acoustic tumours. Neurosurgery 28: 646–651.

Blakeley JO, Evans DG, Adler J et al. (2012). Consensus recommendations for current treatments and accelerating clinical trials for patients with neurofibromatosis type 2. Am J Med Genet A 158A: 24–41.

Bourn D, Carter SA, Mason S et al. (1994a). Germline mutations in the neurofibromatosis type 2 tumour suppressor gene. Hum Mol Genet 5: 813–816.

Bourn D, Carter SA, Evans DGR et al. (1994b). A mutation in the neurofibromatosis type 2 tumor-suppressor gene giving rise to widely different clinical phenotypes in two unrelated individuals. Am J Hum Genet 55: 69–73.

Crowe FW, Schull WJ, Neal JV (1956). A Clinical, Pathological and Genetic Study of Multiple Neurofibromatosis. Charles C Thomas, Springfield, Illinois.

Cushing H (1917). Tumours of the Nervus Acusticus and the Syndrome of the Cerebello-pontine Angle. WB Saunders, Philadelphia.

Evans DG, Huson S, Donnai D et al. (1992a). A clinical study of type 2 neurofibromatosis. Q J Med 84: 603–618.

Evans DG, Huson SM, Donnai D et al. (1992b). A genetic study of type 2 neurofibromatosis: II. Guidelines for genetic counselling. J Med Genet 29: 847–852.

Evans DG, Huson S, Donnai D et al. (1992c). A genetic study of type 2 neurofibromatosis in the north west of England and the UK: prevalence mutation rate fitness and confirmation of maternal gene effect. J Med Genet 29: 841–846.

Evans DG, Ramsden R, Huson SM et al. (1993). Type 2 neurofibromatosis: the need for supraregional care? J Laryngol Otol 107: 401–406.

Evans DG, Bourn D, Wallace A et al. (1995a). Diagnostic issues in a family with late onset type 2 neurofibromatosis. J Med Genet 32: 470–474.

Evans DG, Blair V, Strachan T et al. (1995b). Variation in expression of the gene for type 2 neurofibromatosis: absence of a gender effect on vestibular schwannoma. J Laryngol Otol 109: 830–835.

Evans DG, Maher ER, Macleod R et al. (1997a). Uptake of genetic testing for cancer predisposition. J Med Genet 34: 746–748.

Evans DGR, Mason S, Huson S et al. (1997b). Spinal and cutaneous schwannpmatosis is a variant form of type 2 neurofibromatosis; a clinical and molecular study. J Neurosurg Neurol Psychiatry 62: 361–366.

Evans DG, Trueman L, Wallace A et al. (1998a). Genotype/phenotype correlations in type 2 neurofibromatosis (NF2): evidence for more severe disease associated with truncating mutations. J Med Genet 35: 470–474.

Evans DG, Trueman L, Wallace A et al. (1998b). Mosaicism in classical NF2: a common mechanism for sporadic disease in tumour prone syndromes? Am J Hum Genet 63: 727–736.

Evans DG, Birch JM, Ramsden R (1999a). Paediatric presentation of type 2 neurofibromatosis. Arch Dis Child 81: 496–499.

Evans DG, Wallace A, Ramsden RT et al. (1999b). Probability of bilateral disease in people presenting with a unilateral vestibular schwannoma. J Neurosurg Neurol Psychiatry 66: 764–767.

Evans DG, Baser ME, O'Reilly B et al. (2005). Management of the patient and family with neurofibromatosis 2: a consensus conference statement. Br J Neurosurg 19: 5–12.

Evans DG, Howard E, Giblin C et al. (2010). Birth incidence and prevalence of tumour prone syndromes: estimates from a UK genetic family register service. Am J Med Genet 152A: 327–332.

Evans DGR, Ramsden RT, Shenton A et al. (2007). Mosaicism in NF2 an update of risk based on uni/bilaterality of vestibular schwannoma at presentation and sensitive mutation analysis including MLPA. J Med Genet 44: 424–428.

Evans DG, Raymond F, Barwell J et al. (2012). Genetic testing and screening of individuals at risk of NF2. Clin Genet 82: 416–424.

Evans D, Bowers N, Huson S et al. (2013). Wallace A mutation type and position varies between mosaic and inherited NF2 and correlates with disease severity. Clin Genet 83: 594–595.

Gardner WJ, Frazier CH (1930). Bilateral acoustic neurofibromas: a clinical study and field survey of a family of five generations with bilateral deafness in thirty eight members. Arch Neurol Psychiatry 23: 266–302.

Gutmann DH, Aylesworth A, Carey JC et al. (1997). The diagnostic evaluation and multi-disciplinary management of neurofibromatosis 1 and neurofibromatosis 2. JAMA 278: 51–57.

Hagel C, Stemmer-Rachamimov AO, Bornemann A et al. (2012). Clinical presentation immunohistochemistry and electron microscopy indicate neurofibromatosis type 2-associated gliomas to be spinal ependymomas. Neuropathology 32: 611–616.

Halliday AL, Sobel RA, Martuza RL (1991). Benign spinal nerve sheath tumours their occurrence sporadically and in neurofibromatosis types 1 and 2. J Neurosurg 74: 248–253.

Higami Y, Shimokawa I, Kishikawa M et al. (1998). Malignant peripheral nerve sheath tumors developing mutifocally in the central nervous system in a patient with neurofibromatosis 2. Clin Neuropathol 17: 115–120.

Hulsebos TJ, Plomp AS, Wolterman RA et al. (2007). Germline mutation of INI1/SMARCB1 in familial schwannomatosis. Am J Hum Genet 80: 805–810.

Huson SM, Harper PS, Compston DAS (1988). Von Recklinghausen neurofibromatosis: a clinical and population based study in southeast Wales. Brain 111: 1355–1381.

Huynh DP, Mautner V, Baser ME et al. (1997). Immunohistochemical detection of schwannomin and neurofibromin in vestibular schwannomas ependymomas and meningiomas. J Neuropathol Exp Neurol 56: 382–390.

Jacoby L, Jones D, Davis K et al. (1997). Molecular analysis of the NF2 tumor suppressor in schwannomatosis. Am J Hum Genet 61: 1293–1302.

Kaiser-Kupfer MI, Freidlin V, Datiles MB et al. (1989). The association of posterior capsular lens opacities with bilateral acoustic neuromas in patients with neurofibromatosis type 2. Arch Ophthalmol 107: 541–544.

Kanter WR, Eldridge R, Fabricant R et al. (1980). Central neurofibromatosis with bilateral acoustic neuroma: genetic clinical and biochemical distinctions from peripheral neurofibromatosis. Neurology 30: 851–859.

Kluwe L, Mautner VF (1998). Mosaicism in sporadic neurofibromatosis 2 patients. Hum Mol Genet 7: 2051–2055.

Kluwe L, MacCollin M, Tatagiba M et al. (1998). Phenotypic variability associated with 14 splice-site mutations in the NF2 gene. Am J Med Genet 77: 228–233.

Knudson Jr AG (1971). Mutation and cancer: statistical study of retinoblastoma. Proc Natl Acad Sci 68: 820–823.

Kondziolka D, Lunsford LD, McLaughlin MR et al. (1998). Long-term outcomes after radiosurgery for acoustic neuroma. N Engl J Med 339: 1426–1433.

MacCollin M, Ramesh V, Jacoby LB et al. (1994). Mutational analysis of patients with neurofibromatosis 2. Am J Hum Genet 55: 314–320.

Martuza RL, Ojemann RG (1982). Bilateral acoustic neuromas: clinical aspects pathogenesis and treatment. Neurosurgery 10: 1–12.

Mautner VF, Lindenau M, Baser ME et al. (1996). The neuroimaging and clinical spectrum of neurofibromatosis 2. Neurosurgery 38: 880–885.

McGaughran JM, Harris DI, Donnai D et al. (1999). A clinical study of type 1 neurofibromatosis in north west England. J Med Genet 36: 197–203.

Mérel P, Hoang-Xuan K, Sanson M et al. (1995). Screening for germline mutations in the NF2 gene. Genes Chromosomes Cancer 12: 117–127.

Narod SA, Parry DM, Parboosingh J et al. (1992). Neurofibromatosis type 2 appears to be a genetically homogenous disease. Am J Hum Genet 51: 486–496.

National Institutes of Health Consensus Development Conference Statement on Neurofibromatosis (1987). Neurofibromatosis Research newsletter 3: 3–6.

NIH Conference Statement (1990). Neurofibromatosis 1 (von Recklinghausen disease). and neurofibromatosis (bilateral acoustic neurofibromatosis). Ann Intern Med 113: 39–52.

Parry DM, Eldridge R, Kaiser-Kupfer MI et al. (1994). Neurofibromatosis 2 (NF2): clinical characteristics of 63 affected individuals and clinical evidence for heterogeneity. Am J Med Genet 52: 450–451.

Parry DM, MacCollin M, Kaiser-Kupfer MI et al. (1996). Germ-line mutations in the neurofibromatosis 2 gene: correlations with disease severity and retinal abnormalities. Am J Hum Genet 59: 529–539.

Piotrowski A, Xie J, Liu YF et al. (2014). Germline loss-of-function mutations in LZTR1 predispose to an inherited disorder of multiple schwannomas. Nat Genet 46: 182–187.

Plotkin SR, Stemmer-Rachamimov AO, Barker FG et al. (2009). Hearing improvement after bevacizumab in patients with neurofibromatosis type 2. N Engl J Med 361: 358–367.

Plotkin SR, Merker VL, Halpin C et al. (2012). Bevacizumab for progressive vestibular schwannoma in neurofibromatosis type 2: a retrospective review of 31 patients. Otol Neurotol 33: 1046–1052.

Rouleau G, Seizinger BR, Wertelecki W et al. (1987). Genetic linkage of bilateral acoustic neurofibromatosis to a DNA marker on chromosome 22. Nature 329: 246–248.

Rouleau GA, Merel P, Lutchman M et al. (1993). Alteration in a new gene encoding a putative membrane-organizing protein causes neuro-fibromatosis type 2. Nature 363: 515–521.

Ruttledge MH, Narod SA, Dumanski JP (1993). Presymptomatic diagnosis for neurofibromatosis 2 with chromosome 22 markers. Neurology 43: 1753–1760.

Ruttledge MH, Andermann AA, Phelan CM et al. (1996). Type of mutation in the neurofibromatosis type 2 gene (NF2) frequently determines severity of disease. Am J Hum Genet 59: 331–342.

Samii M, Matthies C, Tatagiba M (1997). Management of vestibular schwannomas (acoustic neuromas): auditory and facial nerve function after resection of 120 vestibular schwannomas in patients with neurofibromatosis 2. Neurosurgery 40: 696–706.

Schulz A, Baader SL, Niwa-Kawakita M et al. (2013). Merlin isoform 2 in neurofibromatosis type 2-associated polyneuropathy. Nat Neurosci 16: 426–433.

Seizinger BR, Martuza RL, Gusella JF (1986). Loss of genes on chromosome 22 in tumorigenesis of human acoustic neuroma. Nature 322: 644–647.

Seizinger BR, Rouleau GA, Ozelius LG et al. (1987). Genetic linkage of von Recklinghausen neurofibromatosis to the nerve growth factor receptor gene. Cell 49: 589–594.

Slattery WH, Brackmann DE, Hitselberger W (1998). Hearing preservation in neurofibromatosis type 2. Am J Otol 19: 638–643.

Smith MJ, Wallace AJ, Bowers NL et al. (2014). SMARCB1 mutations in schwannomatosis and genotype correlations with rhabdoid tumors. Cancer Genet 9: 373–378.

Smith MJ, Isidor B, Beetz C et al. (2015). Mutations in LZTR1 add to the complex heterogeneity of schwannomatosis. Neurology 84: 141–147.

Sobel RA, Wang Y (1993). Vestibular (acoustic) schwannomas: histological features in neurofibromatosis 2 and unilateral cases. J Neuropathol Exp Neurol 52: 106–113.

Thomas PK, King RHM, Chiang TR et al. (1990). Neurofibromatous neuropathy. Muscle Nerve 13: 93–101.

Trofatter JA, MacCollin MM, Rutter JL et al. (1993). A novel moesin- ezrin- radixin-like gene is a candidate for the neurofibromatosis 2 tumor suppressor. Cell 72: 791–800.

von Recklinghausen F (1882). Ueber die multiplen Fibroma der haut und ihre Beziehung zu den multiplen Neuromen, A Hirschwald, Berlin.

Watson C, Gaunt L, Evans G et al. (1993). A germinal interstitial deletion maps the locus for type 2 neurofobromatosis to the interval between D22S56 and LIF (leukaemia inhibitory factor). Hum Mol Genet 2: 1215–1220.

Wilding A, Ingham SL, Lalloo F et al. (2012). Life expectancy in hereditary cancer predisposing diseases: an observational study. J Med Genet 49: 264–269.

Wishart JH (1822). Case of tumours in the skull dura mater and brain. Edinburgh Medical and Surgical Journal 18: 393–397.

Worster-Drought C, Dickson WEC, McMenemy WH (1937). Multiple meningeal and perineural tumours with analogous changes in the glia and ependyma (neurofibromatosis). Brain 60: 88–117.

Young PF, Eldridge R, Nager GT et al. (1971). Hereditary bilateral acoustic neuroma (central neurofibromatosis). Birth Defects Orig Artic Ser 7: 73–86.

Zang KD (1982). Cytological and cytogenetic studies on human meningioma. Cancer Genet Cytogenet 6: 249–274.

Zucman-Rossi J, Legoix P, Sarkissian H et al. (1998). NF2 gene in neurofibromatosis type 2 patients. Hum Mol Genet 7: 2095–2101.

Handbook of Clinical Neurology, Vol. 132 (3rd series)
Neurocutaneous Syndromes
M.P. Islam and E.S. Roach, Editors

Chapter 6

Tuberous sclerosis complex

MONICA P. ISLAM* AND E. STEVE ROACH

Department of Child Neurology, Nationwide Children's Hospital, Ohio State University College of Medicine, Columbus, OH, USA

INTRODUCTION

With multisystem involvement beyond the brain and the skin, tuberous sclerosis complex (TSC) is an important neurocutaneous syndrome to diagnose because of its phenotypic variability and the often serious nature of its manifestations. Disordered cellular differentiation and proliferation, primarily in the form of hamartomas, affect the skin, brain, kidney, heart, lungs and eyes. Although TSC was first recognized over a century and a half ago, the past few decades have seen an exponential increase in our understanding of TSC and greatly expanded our ability to diagnose and manage its complications.

PREVALENCE OF TUBEROUS SCLEROSIS COMPLEX

Prior to the advent of modern imaging techniques, the recognition of milder phenotypes of TSC, and the adoption of standardized diagnostic criteria, the incidence of TSC was estimated to be less than 1/100 000. More recent population based studies indicate that TSC is far more common than these earlier estimates. Sampson and colleagues estimate the prevalence of TSC to be 3.7 per 100 000 people in western Scotland (Sampson et al., 1989a), and Webb et al. (1996) found a similar 3.8 per 100 000 individuals in Wessex. O'Callaghan and colleagues, however, used a capture-recapture analysis of the Wessex data and concluded that the original survey failed to identify about half of the prevalent cases, suggesting that the true prevalence was closer to 8–9 per 100 000 individuals (O'Callaghan et al., 1998).

HISTORY

In 1862, von Recklinghausen described concomitant cardiac and cerebral manifestations found at autopsy in a newborn. This is regarded as the earliest report of tuberous sclerosis, although there had been descriptions by Pierre François Olive Rayer in a dermatologic atlas in 1835 of "végétations vasculaires" on the face, later named "adenoma sebaceum" by John Pringle (Pringle, 1890). Describing two individuals with epilepsy and cortical hamartomas at autopsy, Désiré Magloire Bourneville coined the phrase "sclérose tubéreuse des circonvolutions cérébrales" (Bourneville, 1880). He also described kidney findings in one individual. Five years later, Balzer and Menetrier recognized the familial nature of TSC, documenting TSC in a mother and her daughter (Balzer and Menetrier, 1885). Heinrich Vogt emphasized the clinical findings of epilepsy, idiocy, and adenoma sebaceum, creating crude early diagnostic criteria that became known as "Vogt's triad" (Vogt, 1908). A clinician who focused narrowly on the Vogt triad would have failed to identify many of the individuals who are now routinely diagnosed with TSC, and it was not until the 1979 first edition of Manuel Rodriguez Gómez's monograph on TSC that the highly variable phenotype of TSC was recognized (Gómez, 1979). The advent of computed tomography (CT) and magnetic resonance imaging (MRI) made it easier to confirm a TSC diagnosis in less severely affected individuals and has led to earlier recognition of some TSC complications.

Comparison of TSC status and ABO blood grouping in multigenerational families with TSC led to the identification of a TSC candidate gene region on the distal long arm of chromosome 9 (Fryer et al., 1987). It quickly became apparent that some large families harboring TSC did not share the linkage to chromosome 9, suggesting the existence of one or more additional TSC genes (Sampson et al., 1989b; Northrup et al., 1992). The second gene was linked to chromosome 16p13, the region adjacent to the polycystic kidney disease gene, after Kandt and colleagues noted the occurrence of polycystic kidney disease in a small subset of TSC patients

*Correspondence to: Monica P. Islam, MD, Department of Neurology, 700 Children's Drive, Columbus, OH 43205, USA. Tel: +1-614-722-4689, E-mail: Monica.Islam@nationwidechildrens.org

(Kandt et al., 1992). This gene was later confirmed as the *TSC2* gene coding for *tuberin* (European Chromosome 16 Tuberous Sclerosis Consortium, 1993; Wienecke et al., 1995). A 9q34 locus was subsequently identified as the *TSC1* gene (van Slegtenhorst et al., 1997) coding for the gene product *hamartin*. Our understanding of the molecular function of tuberin and hamartin moved closer to reality with the discovery that the two gene products form a physical complex *in vivo* (van Slegtenhorst et al., 1998) to form a key component of the mTOR (mammalian target of rapamycin) pathway (Tee et al., 2002).

ALTERNATE TERMINOLOGY

The term *tuberous sclerosis*, "tuber" (swelling) + "skleros" (hard), derives from Bourneville's "sclérose tubéreuse" and now applies not only to the cerebral lesions but also describes the disease as a whole. The term *tuberous sclerosis complex* (TSC) was popularized in part to avoid confusion with the abbreviation TS, which was already widely used for Tourette syndrome. Bourneville's disease was a commonly utilized early eponym; others have included Pringle's disease and Bourneville–Pringle disease. "Epiloia" was another early designation, an acronym combining the hallmark epilepsy, low IQ, and adenoma sebaceum.

It is now recognized that the mTOR pathway is highly preserved in nonmammalian species, so there has been a recent effort to designate mTOR as the *mechanistic* target of rapamycin instead of the *mammalian* target of rapamycin.

CLINICAL MANIFESTATIONS

The earliest signs of tuberous sclerosis complex can manifest in the form of cardiac rhabdomyomas on prenatal ultrasound. Otherwise, a common time for diagnosis is early infancy with the onset of infantile spasms and concomitant recognition of skin findings. Multiple organ systems can be affected, and early diagnosis allows recognition of current or future involvement that may otherwise go unrecognized. Occasionally parents with multiorgan involvement are diagnosed with TSC only after the diagnosis is made in their child.

Skin manifestations

In relative order of appearance, hallmark skin findings include hypomelanotic macules ("ash leaf" lesions), facial angiofibromas, shagreen patch, and subungual fibromas. While these skin findings represent hamartomas and can increase in number, they do not undergo malignant transformation.

With a frequency of greater than 90%, hypomelanotic macules ("ash leaf" spots) are the single most consistent physical sign of TSC (Fig. 6.1A), but these lesions are not pathognomonic (Northrup and Krueger, 2013). In a prospective study of 423 healthy Caucasian individuals, only 20 individuals had one or more hypomelanotic macules, only four of these 20 had more than one lesion, and none of them had more than three lesions (Vanderhooft et al., 1996). Based upon this study, the diagnostic criteria have specified a minimum of three hypomelanotic macules to constitute a major diagnostic criterion (Roach et al., 1998). The updated 2012 TSC diagnostic criteria also stipulate a threshold 5 mm diameter (Northrup and Krueger, 2013).

Classically ovoid or leaf-shaped in appearance (thus the name ash leaf spot), hypomelanotic macules can assume any shape. These lesions are typically present in early infancy (Gold and Freeman, 1965) but may be difficult to appreciate, especially in a fair-skinned child. Inspection of the entire exposed skin with the aid of an ultraviolet light is important in an individual with suspected TSC. Clusters of smaller hypomelanotic macules have been termed "confetti macules" and also appear in the first decade of life and asymmetrically. Furthermore, poliosis (hypomelanosis of hair in patches) occurs in many individuals with TSC, and this finding can be used toward the count of hypomelanotic findings (Northrup and Krueger, 2013).

Facial angiofibromas (Fig. 6.1B) appear in about 75% of affected individuals. The 2012 TSC diagnostic criteria stipulate that these lesions should number at least three to constitute a major criterion (Northrup and Krueger, 2013). These lesions present in young children as flat erythematous macules reflecting underlying blood vessels and develop into small, raised, hardened structures from underlying fibrous overgrowth. Similar histologically and very specific for TSC (even if present singly), a fibrous cephalic plaque may be present on the forehead (commonly referred to as a forehead plaque) but also elsewhere on the face or scalp.

Unlike hypomelanotic macules, shagreen patches (Northrup and Krueger, 2013) are raised and often large (Fig. 6.1C). Classically they appear over the lower back with a bumpy texture resembling orange peel (*peau d'orange*). These manifest in the first decade and are very specific to TSC in this location. They occur in about half the affected individuals.

Ungual fibromas are another major manifestation in the 2012 TSC diagnostic criteria, but these lesions sometimes develop after trauma to a digit, so the 2012 TSC diagnostic criteria stipulate that there must be two or more ungual fibromas if this finding is to count toward a diagnosis (Northrup and Krueger, 2013). Ungual fibromas seldom manifest before the second decade, making it relatively unhelpful for making a diagnosis at a young

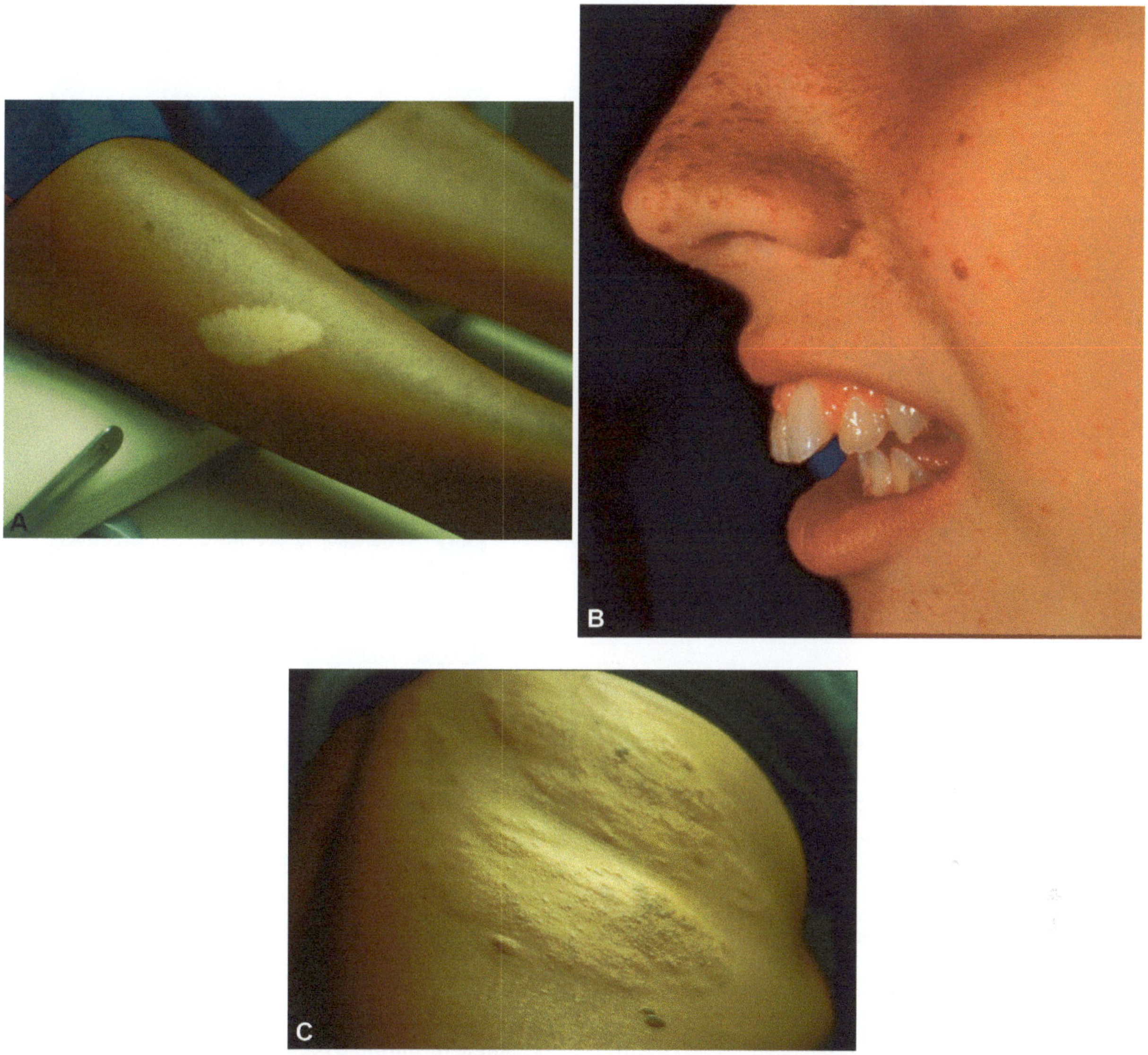

Fig. 6.1. (**A**) Hypomelanotic macule in a woman with tuberous sclerosis complex. (**B**) Facial angiofibromas in an individual with tuberous sclerosis complex. (**C**) Shagreen lesion of the right flank region. (Images (A) and (C) are reproduced with permission from Roach ES (1988). Semin Neurol; 8: 83–96 © Georg Thieme Verlag KG.)

age when the seizures often begin. Up to 80% of adults with TSC have one or more ungual fibromas.

Ophthalmic manifestations

Retinal lesions of TSC consist of astrocytic hamartomas (Fig. 6.2) and achromic retinal lesions. The likelihood of identifying a retinal lesion due to TSC varies with the setting of the examination and with the cooperativeness of the patient. In one report, 87% of individuals with TSC exhibited retinal lesions (Kiribuchi et al., 1986). In most series, the likelihood of retinal abnormalities due to TSC is lower. Retinal hamartomas occur in individuals with other disorders, so they are suggestive but not pathognomonic of TSC (Northrup and Krueger, 2013).

Aronow and colleagues found that retinal astrocytic hamartomas were bilateral in 43% and multiple in 40% of TSC patients (Aronow et al., 2012). Achromic patches occurred in 12% of the patients, half of whom had multiple achromic patches. Individuals with retinal findings were more likely to have a TSC2 mutation than a TSC1 mutation (Aronow et al., 2012).

Most TSC retinal lesions remain asymptomatic. Occasional patients develop acute visual loss following retinal detachment or vitreous hemorrhage (Nyboer et al., 1976). The authors have seen one patient whose vision failed to develop in one eye because of a large hamartoma in the macular region and another child with TSC who lost vision due to severe papilledema related to a giant cell tumor. For the most part, treatment of the

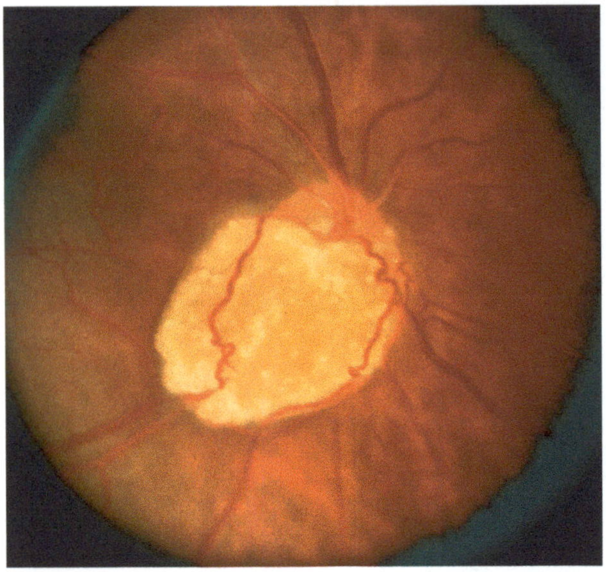

Fig. 6.2. Retinal astrocytic hamartoma in an individual with tuberous sclerosis complex. (Reproduced with permission from Roach, 1992.)

retinal lesions and repeated ophthalmologic examinations are unnecessary.

Pigmentary iris defects have been documented in individuals with TSC (Gutman et al., 1982) but appear to be uncommon compared to the retinal lesions. These lesions remain asymptomatic, and they are too uncommon to be of much use in establishing a diagnosis of TSC.

Oral lesions

Dental enamel pits are occasionally present in normal individuals, but they are present in most adults with TSC (Mlynarczyk, 1991; Sampson et al., 1992). Given the low specificity of dental lesions, dental pits are classified as a minor diagnostic feature (Northrup and Krueger, 2013). Fibromas of the gingiva, buccal mucosa, and labial mucosa also are minor features. Both dental pits and intraoral fibromas must be multiple to count toward diagnostic criteria.

Renal manifestations

Renal angiomyolipomas (AMLs) occur in about two-thirds of TSC patients and are the leading cause of death in adults with TSC (Shepherd et al., 1991). AMLs are histologically benign tumors that contain variable amounts of vascular tissue, smooth muscle, and fat (Monaghan et al., 1981). The prevalence and size of these renal tumors increase with age (Ewalt et al., 1998). Multiple and bilateral AMLs are typical (van Baal et al., 1989).

Tumors larger than 4 cm in diameter (Fig. 6.3) are more likely to become symptomatic (Oesterling et al., 1986; Steiner et al., 1993). Hematuria, flank or abdominal pain, and retroperitoneal hemorrhage are typical presentations. As the tumor increases in size, it becomes more vascular and may develop multiple tortuous vessels and aneurysms that are prone to rupture (Ewalt et al., 2005). Multiple bilateral tumors can also lead to renal failure and arterial hypertension.

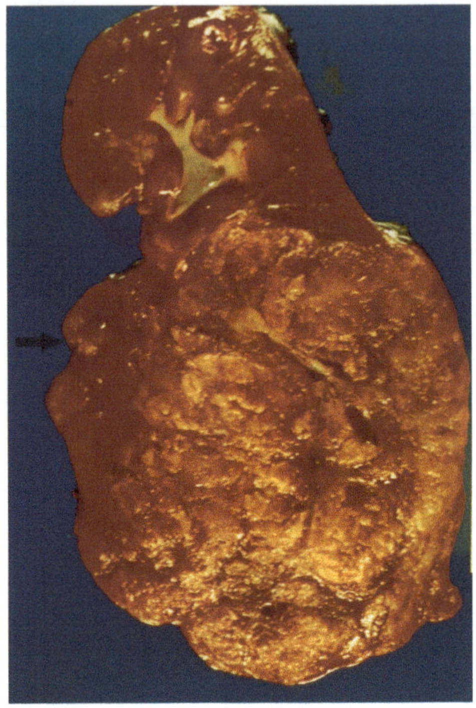

Fig. 6.3. Longitudinally sectioned kidney reveals a large angiomyolipoma in the lower pole. Several small angiomyolipomas are also evident in the renal cortex (arrow). (Reproduced from Weiner et al., 1998.)

Single or multiple renal cysts are common in individuals with TSC, and perhaps 1–2% of TSC patients have adult polycystic kidney disease (Webb et al., 1993). The co-occurrence of TSC and polycystic kidney disease was the clue that led to the discovery of the TSC2 gene adjacent to the gene for adult polycystic kidney disease on chromosome 16 (Kandt et al., 1992).

Cardiovascular manifestations

One or more cardiac rhabdomyomas are demonstrable in two-thirds of newborns with TSC (Gibbs, 1985), and the presence of a single cardiac rhabdomyoma in a newborn imparts a 75–80% risk of TSC. Cardiac rhabdomyomas may be detected on prenatal ultrasonography, providing the earliest clue to the presence of TSC (Crawford et al., 1983). Over time, these tumors typically involute (Smythe et al., 1990; DiMario et al., 1996), and only about a third of older individuals undergoing autopsy have a cardiac rhabdomyoma (Nixon et al., 1989).

Most individuals with a cardiac rhabdomyoma remain asymptomatic, although one child's rhabdomyoma enlarged rapidly during corticotropin administration for infantile spasms (Hishitani et al., 1997). Those who develop symptoms typically exhibit cardiac failure immediately after birth due to obstruction of blood flow by an intraluminal tumor. Some older patients develop an arrhythmia, most commonly Wolff–Parkinson–White syndrome (Nir et al., 1995). One child developed a heart block after starting carbamazepine for epilepsy (Weig and Pollack, 1993). Occasional individuals with cardiac rhabdomyoma experience a cerebral thromboembolism (Kandt et al., 1985; Konkol et al., 1986). Intracranial arteriopathy or aneurysm is infrequently documented in individuals with TSC (Boronat et al., 2013).

Asymptomatic individuals with a cardiac rhabdomyoma need no treatment. Surgical removal of a symptomatic intraluminal tumor may be beneficial, but the surgery may be difficult in babies with unstable cardiac function. Recently, mTOR inhibitors have been used successfully in a few babies with cardiac dysfunction due to a rhabdomyoma (Goyer et al., 2015).

Pulmonary manifestations

Although lymphangioleiomyomatosis (LAM) presents late relative to other signs and symptoms, pulmonary involvement is associated with high morbidity and mortality (Cudzilo et al., 2013). About 30% of women with TSC develop pulmonary involvement, a number that is considerably higher than once appreciated. Symptomatic TSC-related pulmonary disease is much less common among males (Adriaensen et al., 2011). Multifocal multinodular pneumocyte hyperplasia (MMPH) occurs less consistently that LAM.

Many individuals remain asymptomatic despite documented pulmonary lesions on computed tomography, and symptoms before the third or fourth decade are relatively uncommon (Dawson, 1954). Symptoms include dyspnea on exertion, cough, and hemoptysis. Spontaneous pneumothorax may trigger acute pulmonary dysfunction. The authors have encountered individuals presenting with pulmonary failure prior to the diagnosis of TSC. Of individuals with TSC who develop pulmonary symptoms, 10% die within 10 years of symptom onset from pulmonary complications (Johnson et al., 2004).

Better understanding of the role of the mTOR pathway has led to treatment options with sirolimus and other mTOR inhibitors (Henske and McCormack, 2012).

Neurologic manifestations

The principal neurologic manifestations of TSC are epilepsy, autism spectrum disorders, and intellectual disability. However, individuals with little or no neurologic impairment are common. Other manifestations include giant cell astrocytoma, focal neurologic lesions, and various behavioral disturbances. The severity of neurologic dysfunction loosely correlates with the number and size of the cortical lesions shown on neuroimaging studies (Roach et al., 1987; Goodman et al., 1997), but the factors that determine this lesion burden are not well understood.

Seizures occur in over 90% of individuals with TSC. Infantile spasms are particularly common during the first year of life and may be the first indication that a child has TSC. Generalized tonic-clonic seizures and complex partial seizures predominate in older individuals. However, virtually every seizure type has been documented in individuals with TSC, including focal motor seizures, tonic, atonic, myoclonic, and atypical absence seizures. The choice of antiepileptic medications should be tailored to the patient's age and seizure type.

Intellectual disability strongly correlates with the occurrence of uncontrolled seizures, although some individuals with seizures continue to have a normal intellect. Some of the deficits in these children are probably caused by uncontrolled seizures, but there is also MRI evidence that patients with severe cerebral disruption from TSC tend to have a much greater chance of both uncontrolled seizures and intellectual disability, whereas patients with only a few small cerebral lesions typically have better intellectual function and seizure control (Roach et al., 1987; Jambaque et al., 1991). In addition, patients with infantile spasms have more cerebral lesions than children with other seizure types (Shepherd et al., 1995). The eventual neurologic function is probably determined by a combination of the severity of the seizures and the number, size, and location of the cerebral lesions (Jambaque et al., 1991).

At least half of TSC patients experience intellectual disability (Winterkorn et al., 2007), although the degree of intellectual dysfunction ranges from borderline to profound intellectual dysfunction. Some children seem to develop normally until the onset of seizures, then lose function if the seizures are not controlled. Children whose seizures begin early in life and fail to respond to treatment are more likely to experience cognitive impairment (Shepherd and Stephenson, 1992). Individuals who are cognitively normal by school age are likely to remain so (Webb et al., 1991).

Autism occurs in about a quarter of individuals with TSC, but this increases to about half of the TSC patients with intellectual disability (Hunt and Shepherd, 1993). In addition, many individuals with TSC experience a broad range of subtle to severe behavioral, psychiatric, learning, and psychosocial difficulties, such as aggression, sleep disturbance, attention deficit disorder, and frank psychosis (Hunt and Dennis, 1987; Curatolo et al.,

1991). To address the sometimes difficult task of recognizing and characterizing these multiple areas of neuropsychiatric dysfunction, de Vries and colleagues developed the concept of TSC-associated neuropsychiatric disorders (TAND) and with it the TAND Checklist, a screening tool designed to be administered annually (de Vries et al., 2015; Leclezio et al., 2015).

Subependymal giant cell astrocytomas (SEGAs) develop in 5–15% of TSC patients (Northrup and Krueger, 2013; Roth et al., 2013). These tumors are typically found in the anterior portion of the lateral ventricles and are sometimes bilateral. SEGAs are histologically benign and usually grow slowly (Torres et al., 1998). These tumors may be heralded by a new focal neurologic deficit, an unexplained behavior change, or loss of seizure control. Acute symptoms may result from sudden obstruction of the ventricular system by the intraventricular tumor or by hemorrhage within the tumor itself (Waga et al., 1977).

NEURORADIOGRAPHIC FINDINGS

Neuroimaging abnormalities often provide evidence that confirms the diagnosis of TSC. Imaging studies are also important when monitoring for TSC-related complications. Brain lesions of TSC include subependymal nodules, cortical tubers, and giant cell astrocytomas.

Calcified subependymal nodules bulge into the cerebral ventricles from the ependymal area, most commonly involving the anterior half of the lateral ventricles. These lesions occur in about 80% of TSC patients and are a useful diagnostic sign when visible on CT (Fig. 6.4). However, subependymal nodules sometimes contain little calcium at birth, making smaller lesions difficult to visualize with CT until the calcium becomes more prominent (Datta et al., 2008). Noncalcified subependymal nodules appear as a high-signal lesion on T2-weighted MRI sequences. Large densely calcified nodules exhibit a central signal void on MRI.

Larger cortical tubers and areas of focal cortical and subcortical dysplasia are sometimes apparent as hypodensities on CT, but these lesions are more obvious on MRI (Fig. 6.5). Individuals with larger and more numerous cerebral cortical lesions on MRI tend to have more severe epilepsy and cognitive impairment (Roach et al., 1987; Goodman et al., 1997), although this pattern is not consistent enough to reliably predict an individual patient's prognosis. A linear tract perpendicular to the

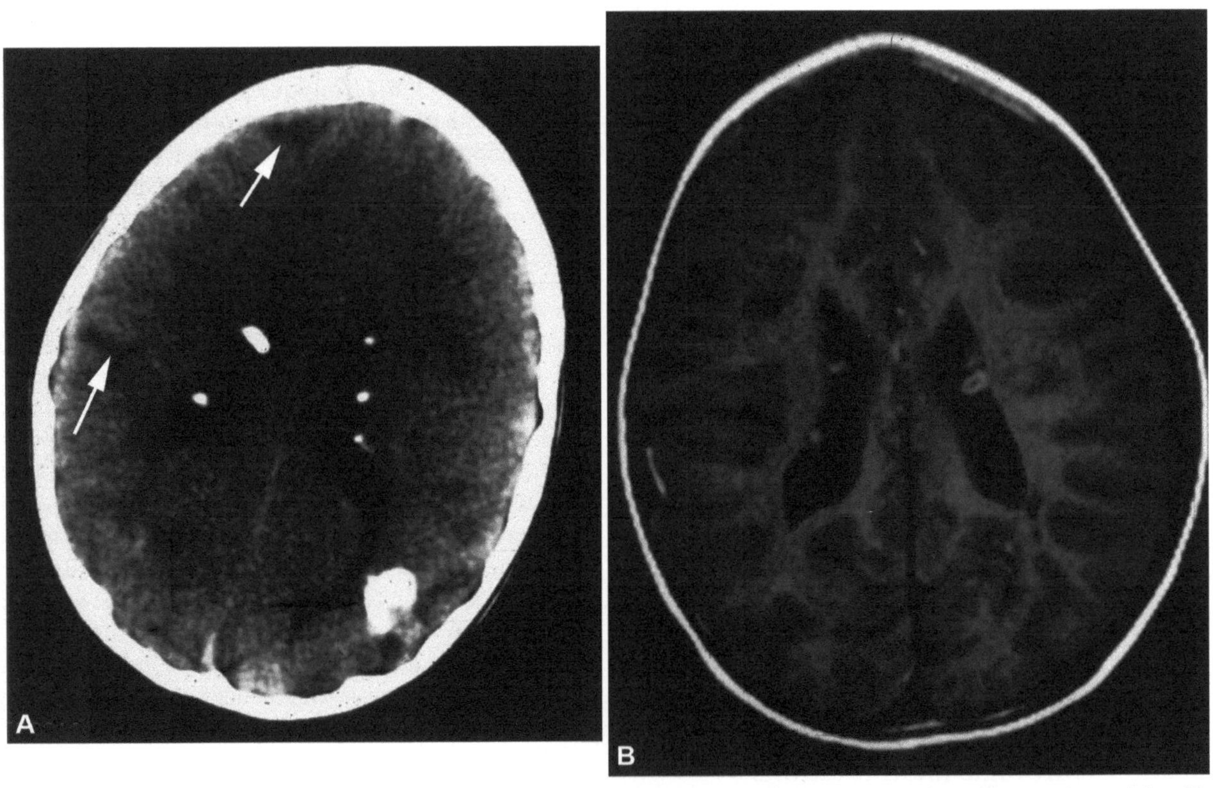

Fig. 6.4. (**A**) Computed tomography reveals typical calcified subependymal nodules protruding into the lateral ventricles. Note also a large calcified cortical lesion in the left posterior hemisphere (black arrow) and scattered cortical hypodense cortical tubers (white arrows). (**B**) T1-weighted magnetic resonance imaging (of a different child) shows subependymal nodules with central calcification. (Image (A) reproduced with permission from Roach et al., 1991b.)

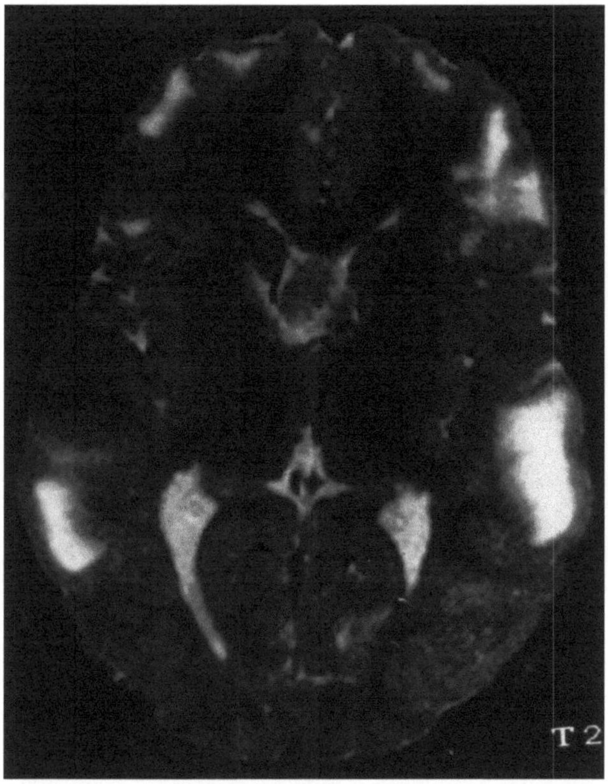

Fig. 6.5. T2-weighted magnetic resonance image shows numerous cortical and subcortical lesions of tuberous sclerosis complex.

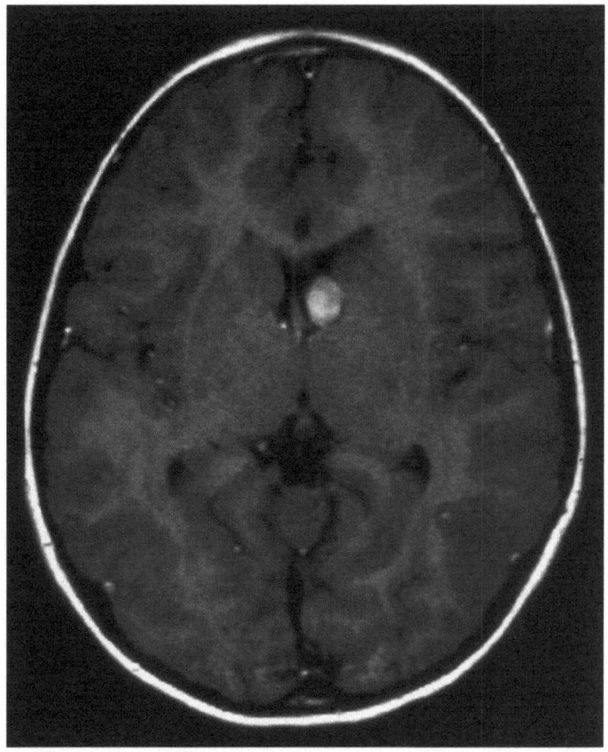

Fig. 6.6. T1-weighted magnetic resonance imaging reveals a contrast-enhancing subependymal giant cell astrocytoma protruding into the anterior horn of the left lateral ventricle.

cerebral cortex beneath a cortical lesion is sometimes apparent on MRI, likely a residual lesion from abnormal neuronal migration (Roach et al., 1991b). Cerebellar lesions are found with MRI in about one-fourth of TSC patients (Vaughn et al., 2013).

SEGAs develop in 5–15% of individuals with TSC, most often arising during childhood (Roth et al., 2013). They may be present at birth (Tien et al., 1990; Hahn et al., 1991), and are occasionally apparent even with prenatal ultrasonography (Mirkin et al., 1999). However, it is unusual for a SEGA to appear *de novo* after the age of 20 years. SEGAs may be unilateral or bilateral, and typically protrude into the lateral ventricles near the foramen of Monro (Fig. 6.6). This lesion localization explains the common occurrence of secondary obstructive hydrocephalus in individuals with a SEGA. Because SEGAs often develop from a calcified subependymal nodule, they often contain calcium deposits in a portion of the lesion. Many SEGAs exhibit contrast enhancement on either CT or MRI (Roth et al., 2013), but contrast enhancement does not reliably predict subsequent tumor enlargement.

A single cortical lesion may be responsible for most of the seizures even when there are numerous areas of abnormality, and identifying the one or two most epileptogenic regions could allow the patient to be considered for epilepsy surgery. Interictal positron emission tomography (PET) imaging with 2-deoxy-2-(18 F) fluoro-D-glucose (FDG) may depict multiple areas of abnormal hypometabolism, but the tracer α-[11C]-methyl-L-tryptophan (AMT) sometimes pinpoints a uniquely epileptogenic region on the basis of tryptophan metabolism (Chugani et al., 2013).

NEUROPATHOLOGY

Several types of brain lesions characterize TSC, but like the clinical manifestations, the neuropathologic pattern is quite variable. Most patients have multiple cortical tubers (Fig. 6.7), although some individuals have none. Histologically the tuber is a hamartoma, composed of neuronal and glial elements with disorganized lamination (Huttenlocher and Heydemann, 1984; Grajkowska et al., 2010). Both the number and the size of the tubers are highly variable, even among affected members of the same family.

Altered neuronal migration during neocortical development leads to the formation of cortical dysplasias that are readily identifiable on imaging. Sometimes islands of heterotopic gray matter or areas with little myelin

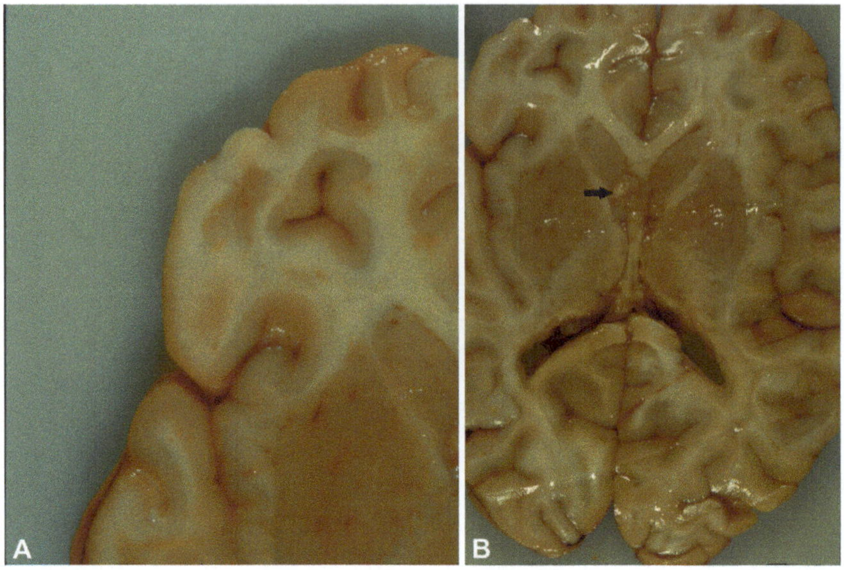

Fig. 6.7. **(A)** Brain section shows a cortical tuber in the left frontal region. **(B)** Subependymal nodules protruding into the ventricles (arrow). (Reproduced with permission from Roach, 1992.)

formation underlie these obvious cortical lesions. The appearance of the cortex aside from cortical defects and tubers is generally normal, but subtle changes in the organizational pattern of the neurons and gliosis can be demonstrated even in parts of the brain that appear to be normal (Marcotte et al., 2012).

Subependymal nodules, as the name implies, occur in the subependymal region of the ventricles (Fig. 6.7B). They are most likely to occur in the anterior half of the lateral ventricles. These nodules are made up of glial and vascular moieties and over time accumulate the calcium deposits that identify them radiographically. Some of the lesions contain large cells similar to those of the less common giant cell tumor (Reagan, 1988), and the giant cell tumor in most instances arises from a subependymal nodule.

DIFFERENTIAL DIAGNOSIS

When the typical manifestations are present, the diagnosis of TSC is not difficult. However, given the phenotypic variability of TSC, the diagnosis can be challenging in individuals with more subtle findings. Additionally, some manifestations of TSC (e.g. facial angiofibromas, renal AMLs, or ungual fibromas) may not be apparent in younger children at the time they begin to experience seizures or other manifestations of TSC, sometimes making the diagnosis difficult in these individuals. Other individuals have relatively subtle manifestations even when older, often due to mosaicism for *TSC1* or *TSC2*. The clinical diagnostic criteria (Table 6.1) often confirm the diagnosis in these individuals, while the commercial gene tests do not.

What often solidifies the diagnosis of TSC is a pattern of clinical manifestations involving multiple organs. When these lesions occur as entities in isolation, however, one must distinguish a mild form of TSC from other diagnoses (Norio et al., 1996). Facial angiofibromas with onset later in childhood or adulthood may be indicators of multiple endocrine neoplasia type 1 (MEN1) or Birt–Hogg–Dubé syndrome (Darling et al., 1997; Pack et al., 1998). Confetti skin lesions in adults may be evidence of chronic sun exposure. Sporadic LAM (rather than TSC-related LAM) can occur via two somatic mutations in the *TSC2* gene; one-third of these individuals may also develop angiomyolipomas. For this reason the presence of these two features alone is not sufficient to make a diagnosis of TSC.

GENETICS AND MOLECULAR BASIS

TSC is inherited as an autosomal dominant trait, but over half of the affected individuals arise via a spontaneous mutation (Roach et al., 1991a). The rate of spontaneous mutation varies from 66% to 75%, depending in part on the completeness of investigation of the extended family (Fleury et al., 1980; Osborne et al., 1991). The clinical manifestations of TSC are highly variable, even among affected family members who share the same mutation. Phenotypic heterogeneity is best illustrated by the sometimes dramatic discordance between homozygous twins with TSC (Gomez et al., 1982).

TSC results from a mutation of either *TSC1* on chromosome 9q34 (van Slegtenhorst et al., 1997) or *TSC2* on chromosome 16p13 (European Chromosome 16 Tuberous Sclerosis Consortium, 1993). Tuberin and

Table 6.1

Diagnostic criteria for tuberous sclerosis complex 2012*

A. Genetic diagnostic criteria

The identification of a pathogenic mutation in either TSC1 or TSC2 is sufficient to make a definite diagnosis of tuberous sclerosis complex (TSC). A pathogenic mutation is defined as a mutation that clearly inactivates the function of the TSC1 or TSC2 proteins (e.g., out-of-frame indel or nonsense mutation), prevents protein synthesis (e.g., large genomic deletion), or is a missense mutation whose effect on protein function has been established. Other TSC1 or TSC2 variants whose effect on function is less certain do not meet these criteria and are insufficient to support a definite diagnosis of TSC. Note that 10–25% of TSC patients have no mutation identified by conventional genetic testing, so a normal result does not exclude TSC or affect the use of clinical diagnostic criteria to diagnose TSC.

B. Clinical diagnostic criteria

Major features
1. Hypomelanotic macules (>3, at least 5 mm diameter)
2. Angiofibromas (>3) or fibrous cephalic plaque
3. Ungual fibromas (>2)
4. Shagreen patch
5. Multiple retinal hamartomas
6. Cortical dysplasias†
7. Subependymal nodules
8. Subependymal giant cell astrocytoma
9. Cardiac rhabdomyoma
10. Lymphangioleiomyomatosis (LAM)‡
11. Angiomyolipomas (>2)‡

Minor features
1. "Confetti" skin lesions
2. Dental enamel pits (>3)
3. Intraoral fibromas (>2)
4. Retinal achromic patch
5. Multiple renal cysts
6. Nonrenal hamartomas

Definite diagnosis: Two major features or one major feature with > 2 minor features

Possible diagnosis: Either one major feature or > 2 minor features

†Includes tubers and cerebral white matter radial migration lines.
‡A combination of the two major clinical features (LAM and angiomyolipomas) without other features does not meet criteria for a definite diagnosis.
*Adapted from Northrup and Krueger, 2013.

hamartin form a physical complex *in vivo* (van Slegtenhorst et al., 1998; Johnson et al., 1999) to form a key inhibitory component of the mTOR pathway (Tee et al., 2002). The tuberin–hamartin complex inhibits the GTPase ras homolog enriched in brain (Rheb) (Zhang et al., 2003). Rheb modulates the mTOR complex 1 (mTORC1), which regulates cell growth and metabolism (Tee et al., 2002).

Sporadic mutations are more likely to occur in *TSC2* than in *TSC1* (Jones et al., 1999; Dabora et al., 2001; Au et al., 2007), possibly because of the larger coding region of *TSC2*. Among familial TSC cases, the two genes are more equally affected (Dabora et al., 2001). *TSC1* mutations often result from single-base substitutions that result in nonsense mutations, while *TSC2* mutations are more likely to involve a large deletion. There is considerable overlap in the clinical severity resulting from the two gene mutations, but a *TSC2* mutation generally tends to cause more severe dysfunction than a *TSC1* mutation (Jones et al., 1999; Dabora et al., 2001; Au et al., 2007).

Commercial mutational analysis will detect a mutation of one of the two TSC genes in about 90% of individuals who meet the clinical diagnostic criteria. Somatic mosaicism may be responsible for the failure to detect a mutation in the remaining individuals (Kwiatkowska et al., 1999). Genotype-phenotype comparisons also suggest the existence of mosaicism; individuals who meet the TSC diagnostic criteria whose mutation is not identified tend to have less severe clinical manifestations than those with a defined mutation (Au et al., 2007). In contrast, somatic mosaicism for TSC is not present among completely unaffected parents of children with TSC. Roberts and colleagues found no mosaic individuals among 54 parents of 30 affected children (Roberts et al., 2004). Germline mosaicism is responsible for the occasional occurrence of siblings with TSC whose parents exhibit neither clinical signs of the disease nor a disease-causing mutation (Yates et al., 1997; Rose et al., 1999).

TREATMENT AND MANAGEMENT

Optimal management of TSC includes a thorough initial evaluation to identify existing concerns and establish a baseline followed by a program designed to identify treatable complications of the disease at an early stage (Krueger and Northrup, 2013). To justify surveillance testing, a complication should be relatively common, clinically significant, and more easily treated when identified earlier. For example, SEGAs typically occur in children and adolescents, and it is easier to manage SEGAs before they become large (Torres et al., 1998). So the 2012 consensus panel recommended a cranial MRI every 1–3 years in asymptomatic TSC patients younger than age 25 years. Individuals with a SEGA by young adulthood should continue to be monitored (Roth et al., 2013). Similarly, each individual should undergo a kidney MRI every 1–3 years to identify clinically significant angiomyolipomas (Krueger and Northrup, 2013).

Accurate blood pressure monitoring tracks the potential for secondary hypertension. Glomerular filtration rate should be periodically calculated.

Even in the absence of cardiac signs and symptoms, newly diagnosed individuals who are less than 3 years of age may benefit from an echocardiogram to evaluate for rhabdomyomas and electrocardiogram to evaluate for arrhythmia. In adults, for whom the presence of significant rhabdomyomas would be low, a baseline electrocardiogram is reasonable.

Given the frequency of early-life seizures and infantile spasms, families should be educated about seizure recognition. A baseline electroencephalogram (EEG) is recommended at the time of diagnosis, with possible longer term EEG monitoring if the initial study is suspicious for subclinical or subtle seizures. The medication of choice for TSC-related infantile spasms is currently vigabatrin (Thiele, 2004), and some physicians believe that vigabatrin is also more effective for other seizure types in individuals with TSC (Greiner et al., 2012). Although many individuals with TSC-related epilepsy remain on antiepileptic medications throughout life, some patients with relatively normal cognitive function who have achieved prolonged seizure freedom manage to stop medications without a return of seizures (Sparagana et al., 2003).

Even when there are multiple cortical tubers, a single dysplastic region may be responsible for all or most of the seizures at a given time – and therefore amenable to surgical resection for treatment (Bebin et al., 1993). However, pinpointing the one tuber that is responsible for the seizures can be challenging. Anecdotal observations suggest that mTOR inhibitors sometimes improve epilepsy related to TSC. This is now under investigation (Krueger et al., 2013).

Surgical resection may be necessary for an acutely symptomatic SEGA, although cerebrospinal fluid diversion might be a temporary option in an individual with acute deterioration due to hydrocephalus. For asymptomatic SEGAs that are increasing size, the location and presence of TSC comorbidities can guide choice of surgical resection versus use of an mTOR inhibitor. Soon after the recognition of the importance of hamartin and tuberin in the mTOR pathway, Franz et al. demonstrated that SEGAs became smaller in TSC patients receiving rapamycin (Franz et al., 2006). Subsequently, the EXIST-1 trial demonstrated that the mTOR inhibitor everolimus effectively decreased the volume of SEGAs and also reduced the frequency of seizures in these individuals (Krueger et al., 2010). Whether an mTOR inhibitor improves epilepsy or cognition in the absence of a tumor is now being investigated. If non-neoplastic TSC-related symptoms improve with mTOR inhibition, medical therapy of TSC-related neoplasms is likely to become standard practice.

The EXIST-2 trial demonstrated a significant reduction in the size of renal AMLs in response to everolimus (Franz, 2013), and the administration of an mTOR inhibitor is now the treatment of choice for larger renal AMLs (Bissler et al., 2008). Endovascular occlusion of the arteries supplying acutely symptomatic renal tumors can help to prevent loss of the kidney (Ewalt et al., 2005). Chronic renal dialysis or kidney transplantation is sometimes necessary in the face of late stage renal failure or severe polycystic renal disease.

LAM occurs more often in individuals with TSC than previously suspected and carries high morbidity and mortality. Women with TSC should be evaluated at 18 years of age with baseline pulmonary function testing, a 6 minute walk test and high-resolution chest CT (preferentially with a low-radiation protocol). Since estrogen and smoking independently may exacerbate LAM, counseling regarding these issues should occur in adolescence into adulthood (Krueger and Northrup, 2013). Treatment with an mTOR inhibitor improves pulmonary function, but the benefit is short-lived if the treatment is halted (Bissler et al., 2008; Franz 2013).

Baseline developmental or neurocognitive testing helps identify TAND and areas of concern can guide further testing as well as interventions. Some school-aged children may benefit from an individualized education plan (IEP).

A pedigree analysis should be done in an effort to identify undiagnosed family members. Gene testing is important for counseling or in situations in which the diagnosis cannot be confirmed clinically. As a child with TSC approaches reproductive age, genetic testing and counseling should be offered. A good source of information for families is the website of the Tuberous Sclerosis Alliance (http:/www.tsalliance.org/).

References

Adriaensen ME, Schaefer-Prokop CM, Duyndam DA et al. (2011). Radiological evidence of lymphangioleiomyomatosis in female and male patients with tuberous sclerosis complex. Clin Radiol 66: 625–628.

Aronow ME, Nakagawa JA, Gupta A et al. (2012). Tuberous sclerosis complex: genotype/phenotype correlation of retinal findings. Ophthalmology 119: 1917–1923.

Au KS, Williams AT, Roach ES et al. (2007). Genotype/phenotype correlation in 325 individuals referred for a diagnosis of tuberous sclerosis complex in the United States. Genet Med 9: 88–100.

Balzer F, Menetrier P (1885). Etude sur un cas d'adénomes sébacés de la face et du cuir chevelu. Arch Physiol Norm Pathol (série III) 6: 564–576.

Bebin EM, Kelly PJ, Gomez MR (1993). Surgical treatment for epilepsy in cerebral tuberous sclerosis. Epilepsia 34: 651–657.

Bissler JJ, McCormack FX, Young LR et al. (2008). Sirolimus for angiomyolipoma in tuberous sclerosis complex or lymphangioleiomyomatosis. N Engl J Med 358: 140–151.

Boronat S, Shaaya EA, Auladell M et al. (2013). Intracranial arteriopathy in tuberous sclerosis complex. J Child Neurol 29: 912–919.

Bourneville DM (1880). Sclérose tubéreuse des circonvolution cérébrales: idiotie et épilepsie hemiplégique. Archives de Neurologie Paris 1: 81–91.

Chugani HT, Luat AF, Kumar A et al. (2013). alpha-[11C]-Methyl-L-tryptophan – PET in 191 patients with tuberous sclerosis complex. Neurology 81: 674–680.

Crawford DC, Garrett C, Tynan M (1983). Cardiac rhabdomyomata as a marker for the antenatal detection of tuberous sclerosis. J Med Genet 20: 303–304.

Cudzilo CJ, Szczesniak RD, Brody AS et al. (2013). Lymphangioleiomyomatosis screening in women with tuberous sclerosis. Chest 144: 578–585.

Curatolo P, Cusmai R, Cortesi F et al. (1991). Neuropsychiatric aspects of tuberous sclerosis. Ann N Y Acad Sci 615: 8–16.

Dabora SL, Jozwiak S, Franz DN et al. (2001). Mutational analysis in a cohort of 224 tuberous sclerosis patients indicates increased severity of TSC2 compared with TSC1 disease in multiple organs. Am J Hum Genet 68: 64–80.

Darling TN, Skarulis MC, Steinberg SM et al. (1997). Multiple facial angiofibromas and collagenomas in patients with multiple endocrine neoplasia type 1. Arch Dermatol 133: 853–857.

Datta AN, Hahn CD, Sahin M (2008). Clinical presentation and diagnosis of tuberous sclerosis complex in infancy. J Child Neurol 23: 268–273.

Dawson J (1954). Pulmonary tuberous sclerosis and its relationship to other forms of the disease. Q J Med 23: 113–145.

de Vries PJ, Whittemore VH, Leclezio L et al. (2015). Tuberous sclerosis associated neuropsychiatric disorders (TAND) and the TAND Checklist. Pediatr Neurol 52: 25–35.

DiMario FJ, Diana D, Leopold H et al. (1996). Evolution of cardiac rhabdomyoma in tuberous sclerosis complex. Clin Pediatr 12: 615–619.

European Chromosome 16 Tuberous Sclerosis Consortium (1993). Identification and characterization of the tuberous sclerosis gene on chromosome 16. Cell 75: 1305–1315.

Ewalt DH, Sheffield E, Sparagana SP et al. (1998). Renal lesion growth in children with tuberous sclerosis complex. J Urol 160: 141–145.

Ewalt DH, Diamond N, Rees C et al. (2005). Long-term outcome of transcatheter embolization of renal angiomyolipomas due to tuberous sclerosis complex. J Urol 174: 1764–1766.

Fleury P, De Groot WP, Delleman JW et al. (1980). Tuberous sclerosis: the incidence of sporadic cases versus familial cases. Brain Dev 2: 107–117.

Franz DN (2013). Everolimus in the treatment of subependymal giant cell astrocytomas, angiomyolipomas and pulmonary and skin lesions associated with tuberous sclerosis complex. Biologics 7: 211–221.

Franz DN, Leonard J, Tudor C et al. (2006). Rapamycin causes regression of astrocytomas in tuberous sclerosis complex. Ann Neurol 59: 490–498.

Fryer AED, Chalmers A, Connor JM et al. (1987). Evidence that the gene for tuberous sclerosis is on chromosome 9. Lancet 1: 659–661.

Gibbs JL (1985). The heart and tuberous sclerosis. An echocardiographic and electrocardiographic study. Br Heart J 54: 596–599.

Gold AP, Freeman JM (1965). Depigmented nevi: the earliest sign of tuberous sclerosis. Pediatrics 35: 1003–1005.

Gomez MR (Ed.), (1979). Tuberous Sclerosis, 1st edn. Raven Press, New York.

Gomez MR, Kuntz NL, Westmoreland BF (1982). Tuberous sclerosis early onset of seizures and mental subnormality: study of discordant homozygous twins. Neurology 32: 604–611.

Goodman M, Lamm SH, Engel A et al. (1997). Cortical tuber count: a biomarker indicating cerebral severity of tuberous sclerosis complex. J Child Neurol 11: 85–90.

Goyer I, Dahdah N, Major P (2015). Use of mTOR Inhibitor everolimus in three neonates for treatment of tumors associated with tuberous sclerosis complex. Pediatr Neurol 52: 450–453.

Grajkowska W, Kotulska K, Jurkiewicz E et al. (2010). Brain lesions in tuberous sclerosis complex. Review. Folia Neuropathol 48: 139–149.

Greiner HM, Lynch ER, Fordyce S et al. (2012). Vigabatrin for childhood partial-onset epilepsies. Pediatr Neurol 46: 83–88.

Gutman I, Dunn D, Behrens M et al. (1982). Hypopigmented iris spot. An early sign of tuberous sclerosis. Ophthalmology 89: 1155–1159.

Hahn JS, Bejar R, Gladson CL (1991). Neonatal subependymal giant cell astrocytoma associated with tuberous sclerosis: MRI CT and ultrasound correlation. Neurology 41: 124–128.

Henske EP, McCormack FX (2012). Lymphangioleiomyomatosis – a wolf in sheep's clothing. J Clin Invest 122: 3807–3816.

Hishitani T, Hoshino K, Ogawa K et al. (1997). Rapid enlargement in cardiac rhabdomyoma during corticotropin therapy for infantile spasms. Can J Cardiol 13: 72–74.

Hunt A, Dennis J (1987). Psychiatric disorders among children with tuberous sclerosis. Dev Med Child Neurol 29: 190–198.

Hunt A, Shepherd C (1993). A prevalence study of autism in tuberous sclerosis. J Autism Dev Disord 23: 323–339.

Huttenlocher PR, Heydemann PT (1984). Fine structure of cortical tubers in tuberous sclerosis: a Golgi study. Ann Neurol 16: 595–602.

Jambaque I, Cusmai R, Curatolo P et al. (1991). Neuropsychological aspects of tuberous sclerosis in relation to epilepsy and MRI findings. Dev Med Child Neurol 33: 698–705.

Johnson MW, Emelin JK, Park SH et al. (1999). Colocalization of TSC1 and TSC2 gene products in tubers of patients with tuberous sclerosis. Brain Pathol 9: 45–54.

Johnson SR, Whale CI, Hubbard RB et al. (2004). Survival and disease progression in UK patients with lymphangioleiomyomatosis. Thorax 59: 800–803.

Jones AC, Shyamsundar MM, Thomas MW et al. (1999). Comprehensive mutation analysis of TSC1 and TSC2- and phenotypic correlations in 150 families with tuberous sclerosis. Am J Hum Genet 64: 1305–1315.

Kandt RS, Gebarski SS, Goetting MG (1985). Tuberous sclerosis with cardiogenic cerebral embolism: magnetic resonance imaging. Neurology 35: 1223–1225.

Kandt RS, Haines JL, Smith M et al. (1992). Linkage of an important gene locus for tuberous sclerosis to a chromosome 16 marker for polycystic kidney disease. Nat Genet 2: 37–41.

Kiribuchi K, Uchida Y, Fukuyama Y et al. (1986). High incidence of fundus hamartomas and clinical significance of a fundus score in tuberous sclerosis. Brain Dev 8: 509–517.

Konkol RJ, Walsh EP, Power T et al. (1986). Cerebral embolism resulting from an intracardiac tumor in tuberous sclerosis. Pediatr Neurol 2: 108–110.

Krueger DA, Northrup H (2013). Tuberous sclerosis complex surveillance and management: recommendations of the 2012 International Tuberous Sclerosis Complex Consensus Conference. Pediatr Neurol 49: 255–265.

Krueger DA, Care MM, Holland K et al. (2010). Everolimus for subependymal giant-cell astrocytomas in tuberous sclerosis. N Engl J Med 363: 1801–1811.

Krueger DA, Wilfong AA, Holland-Bouley K et al. (2013). Everolimus treatment of refractory epilepsy in tuberous sclerosis complex. Ann Neurol 74: 679–687.

Kwiatkowska J, Wigowska-Sowinska J, Napierala D et al. (1999). Mosaicism in tuberous sclerosis as a potential cause of the failure of molecular diagnosis. N Engl J Med 340: 703–707.

Leclezio L, Jansen A, Whittemore VH et al. (2015). Pilot validation of the tuberous sclerosis-associated neuropsychiatric disorders (TAND) checklist. Pediatr Neurol 52: 16–24.

Marcotte L, Aronica E, Baybis M et al. (2012). Cytoarchitectural alterations are widespread in cerebral cortex in tuberous sclerosis complex. Acta Neuropathol 123: 685–693.

Mirkin LD, Ey EH, Chaparro M (1999). Congenital subependymal giant-cell astrocytoma: case report with prenatal ultrasonogram. Pediatr Radiol 29: 776–780.

Mlynarczyk G (1991). Enamel pitting: a common symptom of tuberous sclerosis. Oral Surg Oral Med Oral Pathol 71: 63–67.

Monaghan HP, Krafchik BR, MacGregor DL et al. (1981). Tuberous sclerosis complex in children. Am J Dis Child 135: 912–917.

Nir A, Tajik AJ, Freeman WK et al. (1995). Tuberous sclerosis and cardiac rhabdomyoma. Am J Cardiol 76: 419–421.

Nixon JR, Miller GM, Okazaki H et al. (1989). Cerebral tuberous sclerosis: postmortem magnetic resonance imaging and pathologic anatomy. Mayo Clin Proc 64: 305–311.

Norio R, Oksanen T, Rantanen J (1996). Hypopigmented skin alterations resembling tuberous sclerosis in normal skin. J Med Genet 33: 184–186.

Northrup H, Krueger DA (2013). Tuberous sclerosis complex diagnostic criteria update: recommendations of the 2012 International Tuberous Sclerosis Complex Consensus Conference. Pediatr Neurol 49: 243–254.

Northrup H, Kwiatkowski DJ, Roach ES et al. (1992). Evidence for genetic heterogeneity in tuberous sclerosis: one locus on chromosome 9 and at least one locus elsewhere. Am J Hum Genet 51: 709–720.

Nyboer JH, Robertson DM, Gomez MR (1976). Retinal lesions in tuberous sclerosis. Arch Ophthalmol 94: 1277–1280.

O'Callaghan FJK, Shiell AW, Osborne JP et al. (1998). Prevalence of tuberous sclerosis estimated by capture-recapture analysis. Lancet 351: 1490.

Oesterling JE, Fishman EK, Goldman SM et al. (1986). The management of renal angiomyolipoma. J Urol 135: 1121–1124.

Osborne JP, Fryer A, Webb D (1991). Epidemiology of tuberous sclerosis. Ann N Y Acad Sci 615: 125–127.

Pack S, Turner ML, Zhuang Z et al. (1998). Cutaneous tumors in patients with multiple endocrine neoplasia type 1 show allelic deletion of the MEN1 gene. J Invest Dermatol 110: 438–440.

Pringle JJ (1890). A case of congenital adenoma sebaceum. Br J Dermatol 2: 1–14.

Reagan TJ (1988). Neuropathology. In: MR Gomez (Ed.), Tuberous Sclerosis, 2nd edn. Raven Press, New York, pp. 63–74.

Roach ES (1988). Diagnosis and management of neurocutaneous syndromes. Semin Neurol 8: 83–96.

Roach ES (1992). Neurocutaneous syndromes. Pediatr Clin North Am 39: 591–620.

Roach ES, Williams DP, Laster DW (1987). Magnetic resonance imaging in tuberous sclerosis. Arch Neurol 44: 301–303.

Roach ES, Kerr J, Mendelsohn D et al. (1991a). Detection of tuberous sclerosis in parents by magnetic resonance imaging. Neurology 41: 262–265.

Roach ES, Kerr J, Mendelsohn D et al. (1991b). Diagnosis of symptomatic and asymptomatic gene carriers of tuberous sclerosis by CCT and MRI. Ann N Y Acad Sci 615: 112–122.

Roach ES, Gomez MR, Northrup H (1998). Tuberous sclerosis complex consensus conference: revised clinical diagnostic criteria. J Child Neurol 13: 624–628.

Roberts PS, Dabora S, Thiele EA et al. (2004). Somatic mosaicism is rare in unaffected parents of patients with sporadic tuberous sclerosis. J Med Genet 41, e69.

Rose VM, Au KS, Pollom G et al. (1999). Germ-line mosaicism in tuberous sclerosis: how common? Am J Hum Genet 64: 986–992.

Roth J, Roach ES, Bartels U et al. (2013). Subependymal giant cell astrocytoma: diagnosis screening and treatment. Recommendations from the International Tuberous Sclerosis Complex Consensus Conference 2012. Pediatr Neurol 49: 439–444.

Sampson JR, Scahill SJ, Stephenson JB et al. (1989a). Genetic aspects of tuberous sclerosis in the west of Scotland. J Med Genet 26: 28–31.

Sampson JR, Yates JR, Pirrit LA et al. (1989b). Evidence for genetic heterogeneity in tuberous sclerosis. J Med Genet 26: 511–516.

Sampson JR, Attwood D, al Mughery AS et al. (1992). Pitted enamel hypoplasia in tuberous sclerosis. Clin Genet 42: 50–52.

Shepherd CW, Stephenson JB (1992). Seizures and intellectual disability associated with tuberous sclerosis complex in the west of Scotland. Dev Med Child Neurol 34: 766–774.

Shepherd CW, Gomez MR, Lie JT et al. (1991). Causes of death in patients with tuberous sclerosis. Mayo Clin Proc 66: 792–796.

Shepherd CW, Houser OW, Gomez MR (1995). MR findings in tuberous sclerosis complex and correlation with seizure development and mental impairment. AJNR Am J Neuroradiol 16: 149–155.

Smythe JF, Dyck JD, Smallhorn JF et al. (1990). Natural history of cardiac rhabdomyoma in infancy and childhood. Am J Cardiol 66: 1247–1249.

Sparagana SP, Delgado MR, Batchelor LL et al. (2003). Seizure remission and antiepileptic drug discontinuation in children with tuberous sclerosis complex. Arch Neurol 60: 1286–1289.

Steiner MS, Goldman SM, Fishman EK et al. (1993). The natural history of renal angiomyolipoma. J Urol 150: 1783–1786.

Tee AR, Fingar DC, Manning BD et al. (2002). Tuberous sclerosis complex-1 and -2 gene products function together to inhibit mammalian target of rapamycin (mTOR)-mediated downstream signaling. Proc Natl Acad Sci U S A 99: 13571–13576.

Thiele EA (2004). Managing epilepsy in tuberous sclerosis complex. J Child Neurol 19: 680–686.

Tien RD, Hesselink JR, Duberg A (1990). Rare subependymal giant-cell astrocytoma in a neonate with tuberous sclerosis. AJNR Am J Neuroradiol 11: 1251–1252.

Torres OA, Roach ES, Delgado MR et al. (1998). Early diagnosis of subependymal giant cell astrocytoma in patients with tuberous sclerosis. J Child Neurol 13: 173–177.

van Baal JG, Fleury P, Brummelkamp WH (1989). Tuberous sclerosis and the relation with renal angiomyolipoma. A genetic study on the clinical aspects. Clin Genet 35: 167–173.

van Slegtenhorst M, de Hoogt R, Hermans C et al. (1997). Identification of the tuberous sclerosis gene TSC1 on chromosome 9q34. Science 277: 805–808.

van Slegtenhorst M, Nellist M, Nagelkerken B et al. (1998). Interaction between hamartin and tuberin the TSC1 and TSC2 gene products. Hum Mol Genet 7: 1053–1057.

Vanderhooft SL, Francis JS, Pagon RA et al. (1996). Prevalence of hypopigmented macules in a healthy population. J Pediatr 129: 355–361.

Vaughn J, Hagiwara M, Katz J et al. (2013). MRI characterization and longitudinal study of focal cerebellar lesions in a young tuberous sclerosis cohort. AJNR Am J Neuroradiol 34: 655–659.

Vogt H (1908). Diagnostik der tuberösen Sklerose. Zeitschrift für die Erforschung und Behandlung des jugendlichen Schwachsinns auf Wissenschaftlicher Grundlage 2: 1–16.

Waga S, Yamamoto Y, Kojima T et al. (1977). Massive hemorrhage in tumor of tuberous sclerosis. Surg Neurol 8: 99–101.

Webb DW, Thomson JL, Osborne JP (1991). Cranial magnetic resonance imaging in patients with tuberous sclerosis and normal intellect. Arch Dis Child 66: 1375–1377.

Webb DW, Super M, Normand IC et al. (1993). Tuberous sclerosis and polycystic kidney disease. BMJ 306: 1258–1259.

Webb DW, Fryer AE, Osborne JP (1996). Morbidity associated with tuberous sclerosis: a population study. Dev Med Child Neurol 38: 146–155.

Weig SG, Pollack P (1993). Carbamazepine-induced heart block in a child with tuberous sclerosis and cardiac rhabdomyoma: implications for evaluation and follow-up. Ann Neurol 34: 617–619.

Weiner DM, Ewalt DH, Roach ES et al. (1998). The tuberous sclerosis complex: a comprehensive review. J Am Coll Surg 187: 548–561.

Wienecke R, Konig A, DeClue JE (1995). Identification of tuberin the tuberous sclerosis-2 product. J Biol Chem 270: 16409–16414.

Winterkorn EB, Pulsifer MB, Thiele EA (2007). Cognitive prognosis of patients with tuberous sclerosis complex. Neurology 68: 62–64.

Yates JRW, van Bakel I, Sepp T et al. (1997). Female germline mosaicism in tuberous sclerosis confirmed by molecular genetic analysis. Hum Mol Genet 6: 2265–2269.

Zhang Y, Gao X, Saucedo LJ et al. (2003). Rheb is a direct target of the tuberous sclerosis tumour suppressor proteins. Nat Cell Biol 5: 578–581.

Handbook of Clinical Neurology, Vol. 132 (3rd series)
Neurocutaneous Syndromes
M.P. Islam and E.S. Roach, Editors

Chapter 7

Neurocutaneous melanosis

MONICA P. ISLAM*

Department of Child Neurology, Nationwide Children's Hospital, Ohio State University College of Medicine, Columbus, OH, USA

INTRODUCTION

Among the individuals born with large congenital melanocytic nevi (LCMN), up to 12% will have neurocutaneous melanosis (NCM; OMIM #249400), a benign or malignant proliferation of melanocytes in the brain parenchyma and leptomeninges (Agero et al., 2005; Becher et al., 2009; Lovett et al., 2009; Ramaswamy et al., 2012). LCMN is estimated to occur in fewer than 1 in 20 000, and therefore the incidence of NCM is quite rare. As neuroimaging is performed in more individuals who are at high risk based on the distribution of the skin findings but asymptomatic, this percentage may be closer to 25% (Agero et al., 2005; Becher et al., 2009; Lovett et al., 2009). The condition appears to affect males and females equally (Agero et al., 2005; Hale et al., 2005). NCM has been reported more in individuals who are white, but it is unclear if this represents a true racial predilection (Lovett et al., 2009). Similar to other neurocutaneous syndromes, an abnormality in developing neuroectoderm manifests with brain as well as skin findings. The association for these individuals with NCM portends a high mortality, whether due to melanoma of the central nervous system (CNS) or skin, or to nonmalignant melanosis within the brain contributing to increased intracranial pressure. By some reports, 92% of individuals with neurocutaneous melanosis had died by the time they were reported (DeDavid et al., 1996). On the other hand, there are individuals with LCMN and abnormal imaging who may not develop neurologic symptoms or progression for decades – if ever. It is unclear if this represents a broad but nonspecific spectrum of disease or whether there are specific variables such as age or extent of CNS involvement that contribute to the prognosis or manifestations.

HISTORY

The first report of neurocutaneous melanosis appeared in 1861 with Baron Carl von Rokitansky's description of an adolescent with large cutaneous nevi, intellectual disability, hydrocephalus, and evidence of benign leptomeningeal melanocytic deposits at autopsy (Kadonaga and Frieden, 1991; Ramaswamy et al., 2012). Van Bogaert applied the current-day term in 1948 (Bett, 2006). The first formal definition appeared in 1972 and this was updated in 1991 to include quantifying parameters. Diagnostic criteria proposed by Kadonaga and Frieden consist of evidence of meningeal melanosis in the presence of large or multiple (> 3) congenital nevi. If cutaneous melanoma is present then meningeal lesions need to be benign, and conversely if meningeal melanoma is present then cutaneous lesions must be benign (DeDavid et al., 1991).

ALTERNATE TERMINOLOGY

To avoid confusing histologic implications of the terms "melanism" and "melanomatosis," Fox identified "melanosis" as a preferable denomination on the basis of descriptiveness (Fox et al., 1964). Neurocutaneous melanocytosis has been offered as an updated term. The term neurocutaneous melanosis alternatively and transiently was used to describe what now is known as heredofamilial melanosis, a benign and autosomal dominant condition. Other terms that may depict NCM are: giant melanocytic nevus, mélanose neurocutanée, dark hairy nevus syndrome, congenital melanocytic hairy nevi, giant hairy nevus, garment hairy pigmented nevus, nevus pigmentosus et pillosus, nevomelanocytic nevus, melanocytic nevus, and Becker hairy nevus (Miller, 2004).

*Correspondence to: Monica P. Islam, MD, Nationwide Children's Hospital, Division of Neurology, 700 Children's Drive, Columbus, OH 43205, USA. Tel: +1-614-722-4625, E-mail: Monica.Islam@nationwidechildrens.org

CLINICAL MANIFESTATIONS

Large nevi are present in the neonatal period, but the diagnosis historically was not made until neurologic symptoms manifested in the first few years of life. Autopsy would demonstrate leptomeningeal melanosis (Agero et al., 2005). More recently, magnetic resonance imaging (MRI) in the neonatal period is identifying the diagnosis early. A second peak of presentation can occur around puberty.

Skin

Large congenital melanocytic nevi measure at least 20 cm in adults, extrapolated to at least 9 cm on the scalp or head in infants and 6 cm on the body (Kadonaga and Frieden, 1991; Foster et al., 2001). Individuals with nevi over the upper back and neck regions or over the lumbar back are at highest risk to have NCM. These patterns of distribution are described as "cape" nevi and "bathing trunk" nevi, respectively (Fig. 7.1). Less commonly, individuals with LCMN on the head or scalp are at risk; lesions in these locations are not usually posterior (Bett, 2006). Rare reports have been made of findings in the nervous system in association with large nevi in the extremities alone (Becher et al., 2009). From a different perspective, those with posterior or midline lesions are at high risk, and especially those with lesions that are both posterior and midline (DeDavid et al., 1996; Marghoob et al., 2004; Agero et al., 2005). Multiple satellite nevi (more than 20 small or medium sized nevi surrounding a large congenital melanocytic nevus) also are a risk factor, and perhaps a more significant finding than posterior and midline location (DeDavid et al., 1996; Lovett et al., 2009). With multiple satellite nevi present, the odds ratio for developing NCM reaches 4.6 (Marghoob et al., 2004; Agero et al., 2005; Lovett et al., 2009). The absence of both posterior midline location and multiple satellite nevi demonstrates a strong negative correlation with NCM (Lovett et al., 2009). In the NYU-LCMN Registry, median size of LCMN in individuals who had NCM was 55 cm (Hale et al., 2005). Though not as high a risk, individuals with multiple congenital melanocytic nevi (MCMN) may also develop NCM. These individuals have at least three small or medium sized congenital melanocytic nevi, with medium sized defined as 1.5–19.9 cm (Lovett et al., 2009).

LCMN appear more prone to eczema, which can be a source of discomfort. Areas demonstrating change may also demonstrate decreased pigmentation and even vitiligo, which may be distant to the primary nevus. Vitiligo may represent an autoimmune response to repetitive trauma to the nevus (Lovett et al., 2009). Reports of evolution of LCMN into the appearance of vascular malformations have also been made (Lovett et al., 2009).

Nervous system

Neurologic symptoms, and too often death, result from symptomatic hydrocephalus that arises from the proliferation of leptomeningeal deposits. This can progress rapidly. Alternatively, early neurologic symptoms

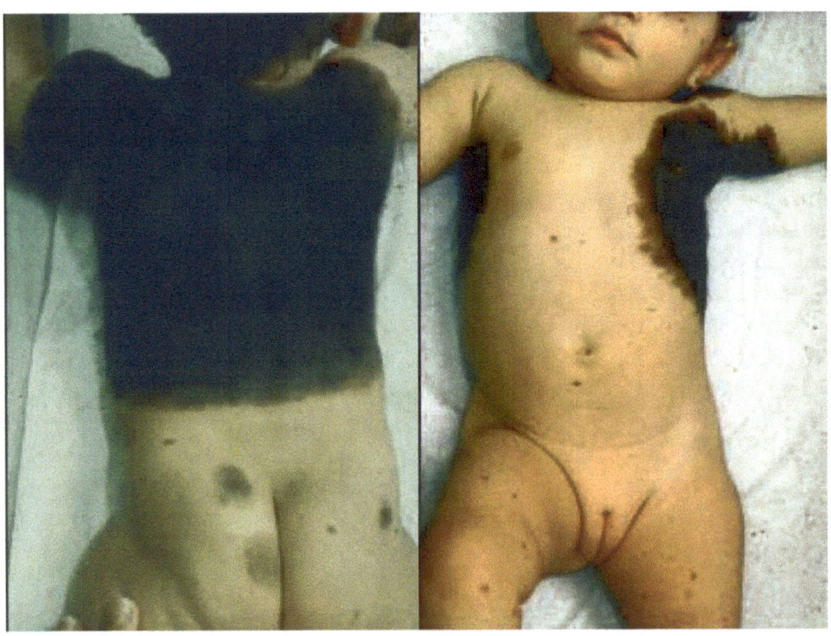

Fig. 7.1. Cape nevus in a 5-month-old infant. The posterior and midline location of the large nevus is a risk factor for neurocutaneous melanosis. Several satellite nevi are present.

typically consist of seizure and epilepsy (DeDavid et al., 1996). Epilepsy has about a 50–60% prevalence (Agero et al., 2005). Seizures are controlled on a single medication in some individuals, while others experience seizures that are resistant to medications (Ramaswamy et al., 2012). Seizures often present in infancy, and infantile spasms have been noted. Some individuals even have had spontaneous remission of seizures (Bett, 2006). The presence and severity of intellectual disability or delayed motor milestones is variable, and many are typically developing (Agero et al., 2005). Those with profound delay may be more likely to have diffuse leptomeningeal deposits or intractable seizures (Ramaswamy et al., 2012). Symptomatic NCM presents early in life, frequently by 3 years of age (DeDavid et al., 1996; Agero et al., 2005; Lovett et al., 2009).

Symptoms of increased intracranial pressure in an infant or young child include irritability, lethargy, emesis, accelerated increase in head circumference, bulging anterior fontanelle, and cranial nerve palsies. Melanocytic cells as they proliferate accumulate in the basal cisterns and prevent resorption of cerebrospinal fluid; this is the basis for communicating hydrocephalus. Less often, noncommunicating hydrocephalus occurs if melanocytic cells block the outflow foramina. Similarly, during proliferation along the leptomeninges of the spinal cord, symptoms of myelopathy or cauda equina syndrome may manifest (Miller, 2004).

Ophthalmologic

Fundoscopic examination sometimes demonstrates irregular retinal pigmentation. To date, ophthalmologic findings have not contributed to the diagnosis, prognosis or management (Lovett et al., 2009).

Other

There is an association with renal abnormalities such as abnormalities of the kidney or ureters and with renal cysts. Meckel's diverticulum has been documented (Kadonaga and Frieden, 1991).

TUMOR POTENTIAL

Although the proliferative potential of leptomeningeal melanosis itself carries significant morbidity and mortality, melanoma appears in many individuals with NCM. In the NYU-LCMN registry, 64% of 33 identified individuals had CNS melanomas – most had died by the time they were reported (DeDavid et al., 1996). Melanoma, whether intracranial or cutaneous in origin, may metastasize. Metastasis may arise in the abdominal cavity in relation to a ventriculoperitoneal or lumboperitoneal shunt. The incidence of transformation to melanoma is unknown but may approach 10–15% and occur in the first decade of life (Foster et al., 2001).

Evaluation for CNS melanoma consists of cerebrospinal fluid (CSF) analysis by lumbar puncture or leptomeningeal biopsy. This should be considered in individuals with rapidly proliferating nodules. Even in the absence of melanoma, lumbar puncture may – but certainly not reliably – reveal an elevated opening pressure, leukocytosis, and elevated protein. Increased red blood cells, hemosiderin or melanin may contribute to xanthochromia (Becher et al., 2009).

RADIOGRAPHIC FINDINGS

Magnetic resonance imaging reveals melanocytic deposits along the leptomeninges and within the parenchyma as hyperintensities on T1-weighted imaging and hypointensities on T2-weighted imaging (Agero et al., 2005). The most common locations (Fig. 7.2) have included the mesial/anterior temporal lobes, the cerebellum, and the leptomeninges (Agero et al., 2005; Ramaswamy et al., 2012). Mass effect is not present. Enhancement with the administration of contrast identifies early meningeal thickening (Agero et al., 2005). Enlarged subarachnoid spaces can also be evidence of meningeal thickening (Pavlidou et al., 2008). Skilled neuroradiology review may be necessary to distinguish normal myelin from melanocytic deposits or melanocytic deposits from artifact. Both false negative and false positive results may be reported by an inexperienced radiologist (Bett, 2006).

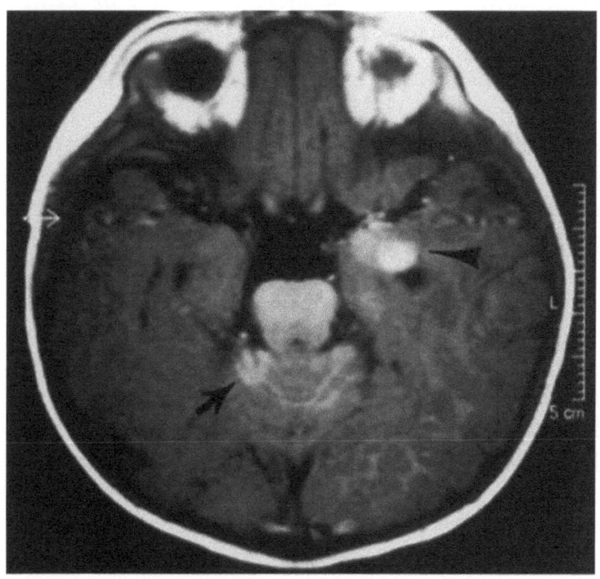

Fig. 7.2. T1-weighted noncontrast MRI demonstrating melanocytic lesions in common locations: cerebellum and mesial temporal lobe. (Reproduced from Miller and Roach, 2000, with permission.)

Spine imaging has demonstrated spinal arachnoid cysts, typically asymptomatic, in some (Ramaswamy et al., 2012). Serial imaging has suggested that these can remain unchanged for prolonged intervals, but the natural history is not fully understood (Ramaswamy et al., 2012). Imaging the spine when the LCMN occurs in the lumbosacral region may detect spinal cord tethering or other treatable conditions, though these are not common (Foster et al., 2001).

Reports of accompanying malformations have included Dandy–Walker malformation, lissencephaly, and Chiari I malformation (Ramaswamy et al., 2012). For Dandy–Walker malformation, the enlargement of the fourth ventricle and partial or complete absence of the cerebellar vermis may relate to impaired CSF flow through the foramina of Luschka and Magendie during development of the fetus.

DIFFERENTIAL DIAGNOSES

The primary differential diagnosis for NCM is metastatic melanoma arising either in the central nervous system or the skin as a primary site. Individuals with LCMN are at increased risk relative to the general population for melanoma, and the central nervous system is a common site for metastasis. Leptomeningeal melanoma metastatic to the skin is a less likely occurrence.

Neurofibromatosis type 1 is another differential diagnosis to consider as LCMN have been mistaken for large café-au-lait patches (Kadonaga and Frieden, 1991). Nevi also could be mistaken for a port-wine stain; the presence of hypertrichosis and presence of strong pigmentation at birth, however, would be more suggestive of LCMN (Lovett et al., 2009).

PATHOLOGY

At autopsy, patients with NCM demonstrate diffuse leptomeningeal deposits in the brain and spine (Fig. 7.3). The most common locations for melanosis have occurred along the interpeduncular fossa, ventral brainstem, upper cervical cord, and ventral lumbosacral cord. The distribution of either diffuse or nodular deposits follows a normal pattern but is pathologic on the basis of degree of infiltration. Melanocytic cells in NCM fill Virchow–Robin spaces, a finding not expected normally. Rarely, leptomeningeal melanosis has been evident at a histologic level in the absence of macroscopic presence. Leptomeningeal melanoma is not uncommon. Indicators of transformation to melanoma include the presence of hemorrhage, necrosis, invasion of blood vessels' basal lamina, cellular atypia, and high mitotic activity (Pavlidou et al., 2008).

Histology of cutaneous lesions in NCM is identical to the histology of melanocytic lesions occurring without CNS involvement. Melanocytic nevus cells are found in the reticular dermis and occasionally in the subcutis layers. Nevus cells appear in cords or nests with few mitoses and with large atypical cells that stain positive for S-100 and HMB-45 antibodies; they also demonstrate formaldehyde-induced green specific fluorescence (Miller, 2004).

MOLECULAR BASIS/PATHOPHYSIOLOGY

As a result of abnormal melanin-producing genes, abnormal migration of melanocyte precursors leads to deposits in the meninges (Lovett et al., 2009). One theory suggests pluripotential cells originating in the neural crest migrate via paraspinal ganglia cells and peripheral nerve sheaths to the skin and differentiate to melanocytes (Ramaswamy et al., 2012). These melanocytes subsequently have the potential to undergo transformation into melanoma (Agero et al., 2005). It has been suggested that the typical posterior and midline location of LCMN associated with NCM is an indicator that melanocyte development and migration from the neural crest are aberrant; LCMN that develop on the extremities may arise later (DeDavid et al., 1996). Similarly, the higher likelihood of NCM with satellite nevi may reflect the presence of multifocal melanocytic malformations (DeDavid et al., 1996).

GENETICS

Individuals with NCM may represent mosaicism of a yet to be determined gene. This may be autosomal dominant and otherwise lethal. Individuals with NCM do not have a family history of family members with NCM or LCNM. There are reports of individuals with unaffected dizygotic twins; no affected individual has been reported with an affected monozygotic twin. Congenital melanocytic nevi have been associated with BRAF and NRAS somatic mutations (Charbel et al., 2014; Salgado et al., 2015). NRAS is a GTPase protein impacting important RAS cell-signaling pathways: the RAF/MEK/ERK pathway and phosphatidylinositol 3-kinase-mediated Akt pathway (Kinsler et al., 2013). Identical NRAS codon 61 mutations have been found in the affected cutaneous and neurologic tissue from individuals with NCM but not in normal tissue, which supports a postzygotic process and somatic mosaicism. The mutation is heterozygous outside of melanoma, where it demonstrates homozygosity (Kinsler et al., 2013). Impacting growth from Ras to Mek1/2, the BRAF gene encodes a serine/threonine kinase; this mutation occurs less frequently and may be associated with a less severe phenotype (Salgado et al., 2015).

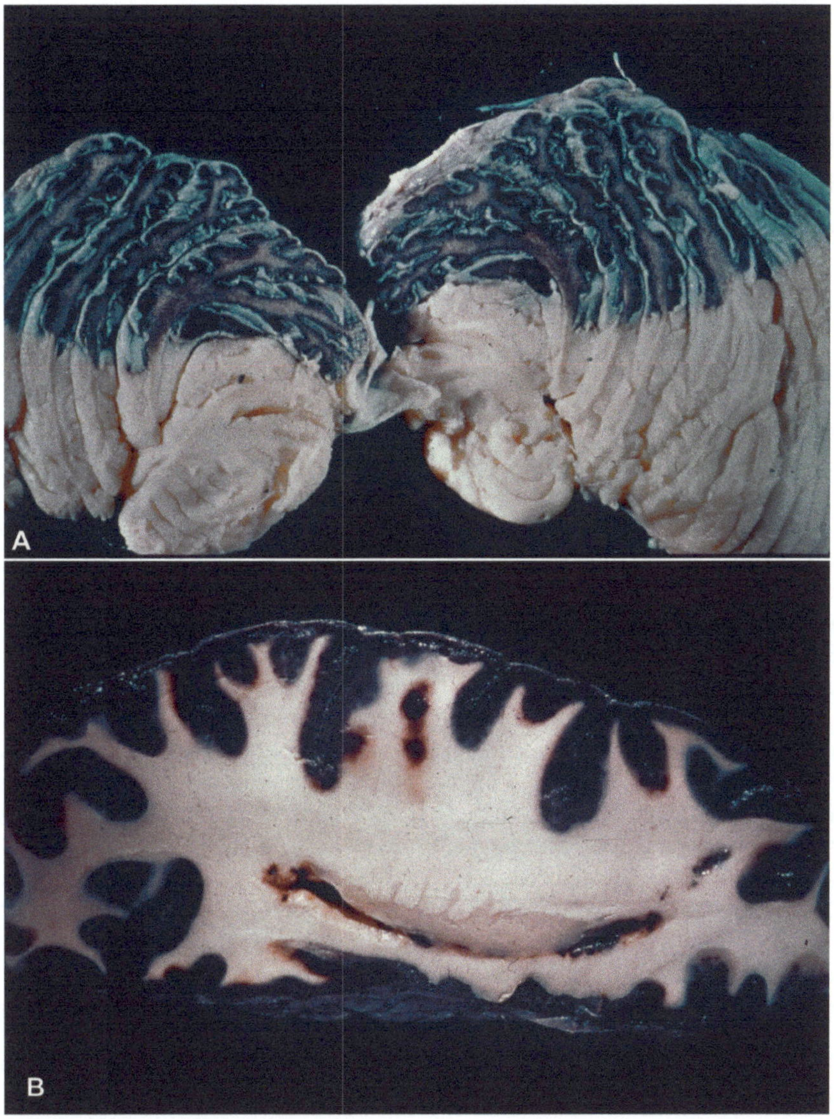

Fig. 7.3. Evidence at autopsy of melanin staining indicating disseminated leptomeningeal melanoma in the (**A**) cerebellum and (**B**) cerebral cortex. (Reproduced from Miller, 2004, with permission. Photographs courtesy of Drs John B. Bodensteiner and S.S. Schochet.)

TREATMENT AND MANAGEMENT

Nevus Outreach, Inc. (http:/www.nevus.org) is a nonprofit support group with an Internet-based registry for individuals with LCMN and their families (Marghoob et al., 2004). In use since 1998, questionnaires and invitations to participate anonymously have been the basis for several of the published reports of occurrence of radiographic findings and neurologic symptoms of NCM in individuals with LCMN (Bett, 2006). These data may be impacted by the awareness and motivation of individuals and their families to access the registry and their knowledge of self-reported information such as results of imaging (Agero et al., 2005).

Indications for surveillance imaging are not well established but baseline MR imaging of the brain and spine in the first year of life – even the first few months – may help identify the diagnosis and determine the timing and need for further imaging (Lovett et al., 2009). Melanocytosis may be more easily detected in the unmyelinated brain, younger than 4 months, given the similar characteristics of myelinated brain and melanocytic deposits that may be confused later (Ramaswamy et al., 2012). On this premise, serial imaging in some individuals has given the appearance of improving melanocytosis as they got older. Since the melanocytic deposits may proliferate rapidly and unpredictably, the role of surveillance imaging is limited as symptoms likely precede new findings on imaging. Some advocate for annual imaging for individuals whose screening imaging is abnormal (Agero et al., 2005). Rather, it may be advisable to have a low threshold for repeating imaging for clinical change. Neurodevelopmental testing may identify subtle neurologic involvement. Concerns on testing followed by change on imaging may prompt medical or surgical intervention. Imaging that has been repeated in asymptomatic individuals with abnormal prior imaging often has not demonstrated progression (Lovett et al., 2009). The demonstration of more than three foci of NCM on imaging carries a poor prognosis (Lovett et al., 2009).

The cost-effectiveness of MRI screening in infancy may be limited from the perspective of limitations of prognostic potential. Nevertheless, given the frequent association with fatal complications, a normal study can greatly reduce the family's anxiety and even provider concern. An abnormal study, on the other hand, may affect the need or timing for interventions with long-term versus short-term impact (Agero et al., 2005).

Evaluation for CNS melanoma consists of CSF analysis by lumbar puncture or leptomeningeal biopsy. This should be considered in individuals with rapidly proliferating nodules. Even in the absence of melanoma, lumbar puncture may reveal an elevated opening pressure, leukocytosis, and elevated protein. Round cells with numerous irregular finger-like projections are characteristic of NCM (Reyes-Mugica et al., 1993). Increased red blood cells, hemosiderin, or melanin may contribute to xanthochromia.

Frequent screening with neurologic examinations or baseline neurodevelopmental testing may allow for diagnosis of NCM early for symptomatic management. This can contribute to better quality of life though it may not slow the progression of the primary disease process (Agero et al., 2005).

Spine MRI at baseline may be helpful given the propensity for spinal arachnoid cysts; this should be considered especially in those with LCMN in the lumbosacral region (Ramaswamy et al., 2012). For spinal arachnoid cysts contributing to a myelopathy, surgical fenestration should improve symptoms.

Surgical excision of LCMN is controversial, although it may reduce the likelihood of transformation to melanoma and provides an obvious cosmetic benefit. However, surgery offers little protection from leptomeningeal melanosis. Complete excision of an LCMN can take multiple surgeries and reconstruction and result in extensive scarring. Excision rather than dermabrasion likely more effectively reduces the potential for malignant transformation. For the individual who is already symptomatic with intracranial melanosis, the risks of anesthesia and surgery may not be justified given the potential short life expectancy (Lovett et al., 2009). For individuals with abnormal imaging and no neurologic symptoms, waiting beyond the first few years of life for high-risk surgeries or elective surgeries is advisable as this might be the time that neurologic symptoms could develop and progress quickly (DeDavid et al., 1996; Agero et al., 2005).

Although some individuals with NCM have survived into adulthood, most die in childhood, between 7 and 10 years of age. The NYU-LCMN registry reported one-half died before age 5 years (DeDavid et al., 1996). Death seems to occur in up to 90% around 3 years following onset of neurologic symptoms (Foster et al., 2001; Agero et al., 2005; Hale et al., 2005). Rapid proliferation of leptomeningeal deposits leads to a communicating hydrocephalus and subsequently death. When melanoma has been identified, treatment with chemotherapy or radiation has not altered the mortality rate (Agero et al., 2005). Primary cutaneous melanoma is not likely to be the cause of death (Bett, 2006).

Radiotherapy, treatment with interferon-α and interleukin-2-based immunotherapy have been attempted. While these might impact cutaneous melanoma, they have not impacted intracranial melanosis or melanoma. One unconfirmed report documented clinical and radiographic improvement in an individual with proliferating NCM after undergoing treatment with interferon-α-2b and isotretinoin (Bett, 2006). For its ability to cross the blood–brain barrier, temozolomide has been considered. The iodine-131-labeled monoclonal antibody ^{131}I-8H9 has also been considered (Agero et al., 2005).

FUTURE DIRECTIONS

Deregulation of growth factor signaling occurring during a critical time in fetal development has produced a syndrome in animal models that resembles NCM (Pavlidou et al., 2008). As melanocyte development, proliferation, and migration are regulated by Kit and its ligand, hepatocyte growth factor, or the Met signaling pathway, therapies inhibiting these targets may be promising interventions (Agero et al., 2005). Therapies may also be developed to take advantage of the role that the proto-oncogene c-met may serve in the development of NCM (Agero et al., 2005). If these pathways lead to possibility of interventions, then MRI or other testing on a surveillance basis may become more effective in impacting prognosis.

References

Agero AL, Benvenuto-Andrade C, Dusza SW et al. (2005). Asymptomatic neurocutaneous melanocytosis in patients with large congenital melanocytic nevi: a study of cases from an Internet-based registry. J Am Acad Dermatol 53: 959–965.

Becher OJ, Souweidane M, Lavi E et al. (2009). Large congenital melanotic nevi in an extremity with neurocutaneous melanocytosis. Pediatr Dermatol 26: 79–82.

Bett BJ (2006). Large or multiple congenital melanocytic nevi: occurrence of neurocutaneous melanocytosis in 1008 persons. J Am Acad Dermatol 54: 767–777.

Charbel C, Fontaine RH, Malouf GG et al. (2014). NRAS mutation is the sole recurrent somatic mutation in large congenital melanocytic nevi. J Invest Dermatol 134: 1067–1074.

DeDavid M, Orlow SJ, Provost N et al. (1996). Neurocutaneous melanosis: clinical features of large congenital melanocytic nevi in patients with manifest central nervous system melanosis. J Am Acad Dermatol 35: 529–538.

Foster RD, Williams ML, Barkovich AJ et al. (2001). Giant congenital melanocytic nevi: the significance of neurocutaneous melanosis in neurologically asymptomatic children. Plast Reconstr Surg 107: 933–941.

Fox H, Emery JL, Goodbody RA et al. (1964). Neurocutaneous melanosis. Arch Dis Child 39: 508–516.

Hale EK, Stein J, Ben-Porat L et al. (2005). Association of melanoma and neurocutaneous melanocytosis with large congenital melanocytic naevi – results from the NYU-LCMN registry. Br J Dermatol 152: 512–517.

Kadonaga JN, Frieden IJ (1991). Neurocutaneous melanosis: definition and review of the literature. J Am Acad Dermatol 24: 747–755.

Kinsler VA, Thomas AC, Ishida M et al. (2013). Multiple congenital melanocytic nevi and neurocutaneous melanosis are caused by postzygotic mutations in codon 61 of NRAS. J Invest Dermatol 133: 2229–2236.

Lovett A, Maari C, Decarie JC et al. (2009). Large congenital melanocytic nevi and neurocutaneous melanocytosis: one pediatric center's experience. J Am Acad Dermatol 61: 766–774.

Marghoob AA, Dusza S, Oliveria S et al. (2004). Number of satellite nevi as a correlate for neurocutaneous melanocytosis in patients with large congenital melanocytic nevi. Arch Dermatol 140: 171–175.

Miller VS (2004). Neurocutaneous melanosis. In: ES Roach, VS Miller (Eds.), Neurocutaneous Disorders. Cambridge University Press, Cambridge, pp. 71–76.

Miller VS, Roach ES (2000). Chapter 69: Neurocutaneous syndromes. In: WG Bradley, RB Daroff, GM Fenichel, CD Marsden (Eds.), Bradley's Neurology in Clinical Practice. Butterworth-Heinemann, Boston, pp. 1665–1700.

Pavlidou E, Hagel C, Papavasilliou A et al. (2008). Neurocutaneous melanosis: report of three cases and up-to-date review. J Child Neurol 23: 1382–1391.

Ramaswamy V, Delaney H, Haque S et al. (2012). Spectrum of central nervous system abnormalities in neurocutaneous melanocytosis. Dev Med Child Neurol 54: 563–568.

Reyes-Mugica M, Chou P, Byrd S et al. (1993). Nevomelanocytic proliferations in the central nervous system of children. Cancer 72: 2277–2285.

Salgado CM, Basu D, Nikiforova M et al. (2015). BRAF mutations are also associated with neurocutaneous melanocytosis and large/giant congenital melanocytic nevi. Pediatr Dev Pathol 18: 1–9.

Handbook of Clinical Neurology, Vol. 132 (3rd series)
Neurocutaneous Syndromes
M.P. Islam and E.S. Roach, Editors

Chapter 8

Basal cell nevus syndrome or Gorlin syndrome

SRIKANTH THALAKOTI AND THOMAS GELLER*
Department of Child Neurology, St Louis University, St Louis, MO, USA

BACKGROUND

Basal cell nevus syndrome (BCNS), also called nevoid basal cell carcinoma (NBCC) or Gorlin syndrome is a rare hereditary neurocutaneous syndrome with well-defined diagnostic criteria. It is characterized by the presence of a range of developmental clinical features including multiple basal cell carcinomas, jaw cysts, palmar/plantar pits, radiologic abnormalities such as bifid ribs, hemi- or bifid vertebra, and calcification of the falx cerebri. The disorder may be associated with cognitive deficits and developmental delay in as many as 5% of patients (Gorlin, 2004). There is a predisposition to malignancies such as desmoplastic medulloblastoma (Kimonis et al., 1997).

This syndrome was first well-characterized by Gorlin and Goltz (1960) but first reported by Jarisch (1894) and White (1894). There was evidence that BCNS existed in Ancient Egyptian times from the assemblage of findings in archeological excavations of skeletons dating to 1000 BC (Satinoff and Wells, 1969).

Gorlin syndrome is defined clinically on the basis of major and minor criteria, first defined by Evans et al. in 1993 and later modified by Kimonis et al. in 1997. In 2011, the International BCNS colloquium stated in consensus that medulloblastoma should be added to the major criteria (Bree et al., 2011). Gorlin syndrome is an autosomal dominant condition with complete penetrance and variable expressivity. It is caused by a mutation of the *PTCH1* gene on chromosome 9p22.3 (Hahn et al., 1996).

EPIDEMIOLOGY

Based on the limited studies on prevalence of this condition in various countries, it is estimated to range from 1/55 600 to 1/256 000 (Evans et al., 1991; Lo Muzio, 2008). Estimated prevalence in a study in north-west England was 1/55 600, and a birth incidence of 1/18 000, was reported (Evans et al., 2010). The estimated prevalence of Gorlin syndrome was much lower, at 1/256 000, in Italy (Lo Muzio et al., 1999), and was 1/235 800 in Japan and 1/164 000 in Australia (Shanley et al., 1994). This syndrome occurs in all ethnic racial groups with no specific predilection to either gender. In all age groups with basal cell carcinoma, it is estimated that 1 in 200 (0.5%) has Gorlin syndrome, but when assessed in individuals under 19 years of age with basal cell carcinomas, the proportion is much higher, at 1 in 5 (Springate, 1986). The median age for BCNS diagnosis is 20–25 years but with the increasing awareness of the diagnosis in families, the presence of support groups, and defined diagnostic criteria leading to early detection in the pediatric population, the age at diagnosis is decreasing.

PATHOGENESIS

BCNS is an autosomal dominant condition with complete penetrance and variable expressivity caused by mutations in a PATCHED (*PTCH1*) gene, a tumor suppressor gene (Bale, 1997). Localization of the gene for BCNS chromosome 9q22.3-q31 was first shown by Farndon et al. in 1992 using linkage analysis. The gene was first isolated in Australia and in the USA in 1996 as the human analog of the drosophila PTCH1 gene (Farndon et al., 1992; Johnson et al., 1996).

Other terms used for this syndrome are:

Multiple hereditary cutaneomandibular polyoncosis
Multiple naevoid basal-cell carcinoma syndrome
Fifth phakomatosis
Epithelioma naevique syndrome
Nevoid basal cell carcinoma epithelioma- jaw cysts
Naevous epitheliomatodes multiplex
Multiple-basal-cell carcinoma syndrome
Multiple-basal-cell naevi syndrome

*Correspondence to: Thomas J. Geller, MD, Department of Child Neurology, 1465 S. Grand Ave, 6317 Southwood Ave, 2W, St Louis, MO 63104, USA. Tel: +1-314-268-4105, E-mail: gellertj@slu.edu

Multiple bifid rib syndrome
Ward syndrome II
Hereditary cutaneo-mandibular polyoncosis
Hermans-Grosfeld-Spaas-Valk syndrome

The *PTCH1* gene is composed of 23 exons, encoding a transmembrane receptor glycoprotein made of 1447 amino acids. This protein has 12 transmembrane domains and two large hydrophilic extracellular loops where the sonic hedgehog (SHH) ligand binds and activates the sonic hedgehog signal pathway. The sonic hedgehog signal pathway plays an important role in development of the axial skeleton, notochord, spinal cord, floor plate, gut endoderm, limbs and hair follicle growth and cycle. Interaction of sonic hedgehog ligand protein with PATCHED forms a complex which inhibits the activity of another transmembrane protein called Smoothened (SMO) and results in modulation of the balance between cell proliferation and differentiation. Mutations of *PTCH1* lead to disruption of second loop formation and SHH cannot bind (Fig. 8.1). In the absence of SHH, PATCHED inhibition on SMO is released resulting in unregulated cell proliferation and differentiation (Nikolaou et al., 2012).

In 60–80% of Gorlin syndrome patients, a single copy of the altered or mutated *PATCH1* gene is inherited by a germ line mutation causing developmental defects. In the individual's lifetime the other copy of *PATCH1* may be altered by various deletions, insertions, splice site alterations, nonsense and missense mutations, and cause other features of Gorlin syndrome such as BCCs and other tumors. Other *de novo* mutations of the *PTCH1* gene were identified in 20–40% of BCNS patients with no family history of this condition. A two-hit model, similar to Knudson's model for development of neoplasia from tumor suppressor genes, was proposed to explain both sporadic and familial anomalies in Gorlin syndrome but is lacking molecular evidence (Levanat, et al., 1996).

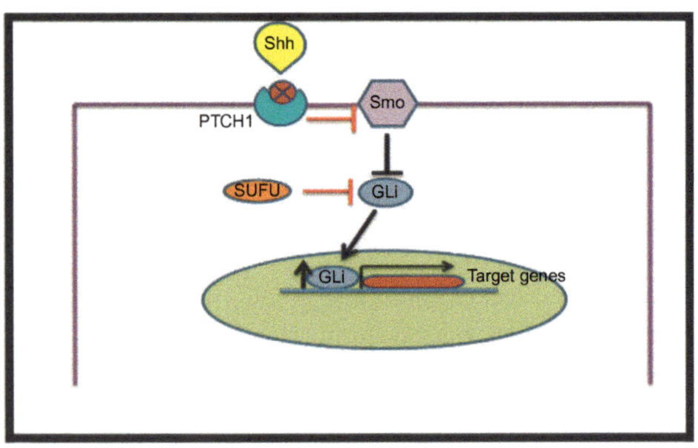

Fig. 8.1. Schematic representation of sonic hedgehog-PTCH1 and Smo complex interactions and effect on downstream pathway in modulation of cell proliferation and differentiation.

Recently, in a retrospective analysis of gene mutations in individuals with medulloblastoma, some were noted to have upregulation of the *PTCH2* gene with alterations in *PTCH1* gene. The *PTCH2* gene is highly homologous to *PTCH1* gene and has been mapped to chromosome lp32.1-p32.3. This indicates that PTCH2 modulates tumorogenesis associated with PTCH1 haploinsufficiency (Lee et al., 2007). An Italian family with PTCH1-negative genotyping and signs and symptoms of Gorlin syndrome (including desmoplastic medulloblastoma but without BCCs) was found to have mutations in SUFU, a negative regulator protein of the SHH pathway (Pastorino et al., 2009). These findings add to the evidence for genetic heterogeneity in BCNS presentations.

CLINICAL MANIFESTATIONS

Clinical manifestations of Gorlin syndrome are variable, some occurring very early and some developing later in life.

Dermatologic manifestations

Basal cell carcinoma

The most characteristic feature of Gorlin syndrome is the presence of multiple basal cell carcinomas (Fig. 8.2). These lesions can appear as early as 2–3 years of age, but the usual age of onset is between puberty and age 35. In Gorlin syndrome patients sometimes fail to develop BCCs in their lifetime despite having other clinical stigmata of the condition. BCCs were diagnosed in 90% of the Caucasian Gorlin patients age 40 or older, but in only 40% of the Gorlin African American population of all ages. Basal cell carcinomas vary in number from a few to thousands, range in size from 1 mm to 10 mm, and appear pearly to flesh colored to pale brown. Nodules or papules or ulcerating plaque forms have been documented. They often involve the face, back and chest. They are rarely seen below the waist or on the legs. There is a strong direct relationship to sun exposure and skin pigmentation. In Gorlin syndrome patients, BCCs rarely become locally aggressive. Radiation therapy causes proliferation of basal cell carcinomas with local invasion in Gorlin patients within a few years. Basal cell carcinomas occurring in BCNS are not histopathologically distinct from ordinary sporadic BCCs. Histologically these tumors are composed of nests and islands or sheets of large, darkly stained nuclei with indistinct cell membrane and well-polarized epithelial cells at the periphery of each lesion. In about 30% of Gorlin syndrome patients, two or more histologic patterns of BCCs

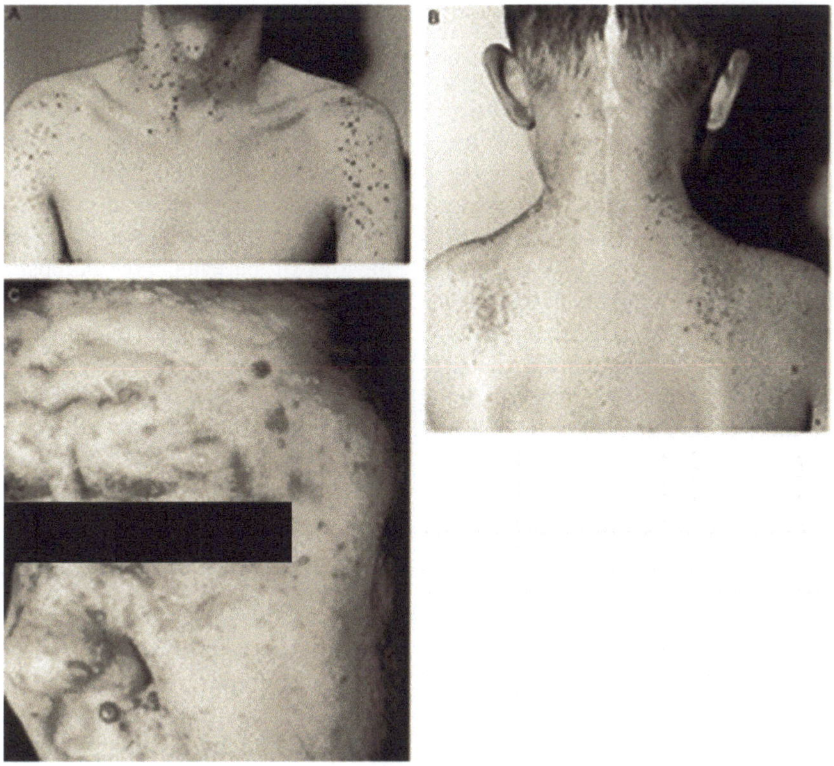

Fig. 8.2. Multiple basal cell carcinomas involving neck, shoulders, upper back and face.

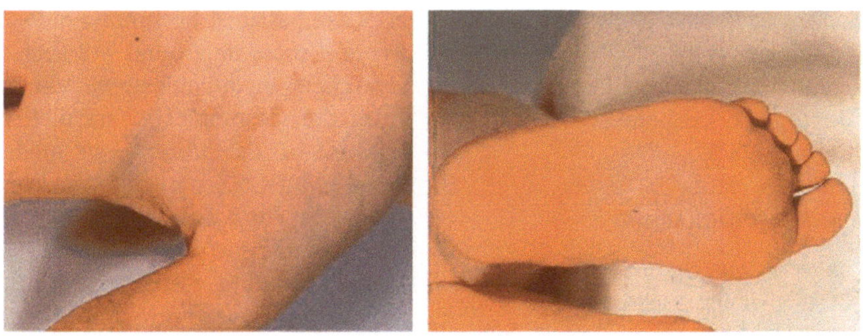

Fig. 8.3. Palmar and plantar pits. Reproduced from Kimonis et al., 2013.

occur (morphea like, solid, superficial, cystic, adenoid or fibroepithelial) (Gorlin, 2004).

Milia

Milia are small cysts filled with keratin, seen on the face, below the eyes or on the forehead. In about 30–50% of patients with Gorlin syndrome, BCCs are intermixed with milia and calcified foci, not only in the tumor but also in normal surrounding skin. This appears to reflect the "turning on" of the bone morphogenic protein (BMP) gene (Gorlin, 2004).

Palmar and plantar pits

Palmar and plantar pits (Fig. 8.3) are very strong diagnostic indicators when found in a child. They are small pits more commonly seen on the palmar surface (70%) than on the plantar surface (30%), generally presenting at 2–3 mm in diameter and 1–3 mm in depth. These pits may be easily seen if the hands are soaked in water for 10 minutes. Prevalence of pits ranges from 30% to 65% in children under age 10 and increases with age to 85% in individuals above 20 years of age (Gorlin, 2004).

Skeletal manifestations

Odontogenic keratocysts

These are one of the most important diagnostic features of Gorlin syndrome (Fig. 8.4). Multiple and recurrent jaw cysts (both in the upper and the lower jaw) occur in about 90% of patients above 40 years and in about 80% above 20 years of age. These cysts are seen with a mean age of occurrence around 15 years of age, but the youngest age of an individual with documented cysts was 5 years old. Keratocysts are most frequently seen in the mandibular angles (44%), followed by the canine–incisor region (18%). There is no racial predilection. Keratocysts are mostly asymptomatic until progression of size causes jaw swelling, displacement, or results in impacted teeth. Rupture of a keratocyst causes an unusual taste but has not resulted in jaw fracture. Radiologically, keratocysts are either unilocular or multilocular with smooth or scalloped cystic borders. Histologically, keratocysts are characterized by a thin external fibrous capsule and internal keratinized stratified squamous epithelial lining. These are believed to originate from dental lamina. There is a 60% risk of recurrence following surgery either due to incomplete removal or from proliferation of the basal layer of the epithelium (Dominguez and Keszler, 1988). Syndromic keratocysts appear at an earlier age than isolated nonsyndromic keratocysts. Recently, a newly described radiologic finding of bilateral hyperplasia of the mandibular coronoid processes has been identified as a useful early sign of BCNS in pediatric populations (Leonardi et al., 2002). A few individuals with ameloblastoma and squamous cell carcinomas arising from odontogenic keratocysts have been documented (Schultz et al., 1987).

Musculoskeletal and radiographic

Patients with BCNS have various characteristic skeletal manifestations. About 15% of patients are much taller than family members. Macrocephaly occurs in more than 50% of patients, with the head circumference above the 95th percentile, and frontal bossing is present in 25% of the patients. Widely separated eyes are described. Various rib anomalies like bifid, fused, wide or missing ribs (30–60%) and scoliosis (40%) are reported. Spina bifida occulta of cervical or thoracic vertebra occurs in 20–60% of cases, pectus deformities in 30–40%, Sprengel anomaly (elevation of scapula with rotation toward the spine with scoliosis) has been observed in 10–40% of patients. Pre- or postaxial polydactyly of hands or feet, hallux valgus, or syndactyly of finger 2 or 3 occurs in about 5% of patients. X-rays of the small hand bones show small flame shaped cysts, and the long bones, pelvis, and skull frequently exhibit bone cysts. These cysts usually do not cause any harm, but involvement of the skull can give a false impression that medulloblastoma has spread to the bone (Kimonis et al., 2004).

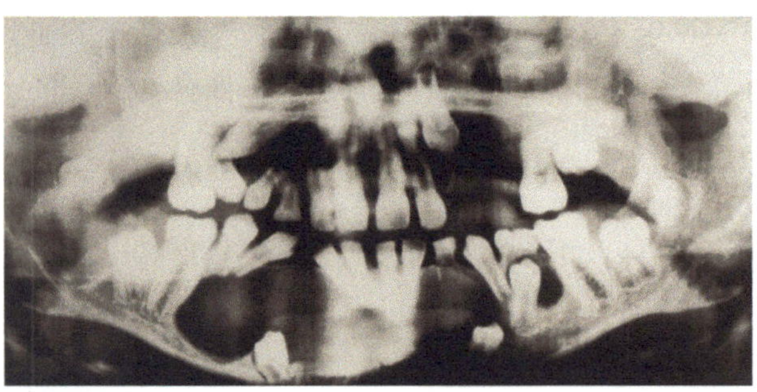

Fig. 8.4. Multiple keratocysts involving the jaw bone – Panorex view.

Central nervous system manifestations

Individuals with BCNS develop central nervous system tumors such as desmoplastic medulloblastoma, other tumors, and epileptic seizures. Medulloblastoma with BCNS was first described by Kimonis et al. in 1997. The incidence of medulloblastoma in BCNS patients is 3–5% in all age groups. In contrast, the incidence of BCNS in one cohort of 173 medulloblastoma patients was 1–2% (Evans et al., 1991). It is estimated to be more prevalent in females than males by 3:1. This tumor is characteristically presents in the first 2 years of life in BCNS patients, in contrast to the general population, where it presents around 7–8 years of age. Among the variants of medulloblastoma, almost all BNCS patients have the desmoplastic form (Evans et al., 1991). Because of the presence of medulloblastoma earlier in life in BCNS patients, children under 5 years of age with medulloblastoma should be carefully screened for other signs of the syndrome (Kimonis et al., 1997). Medulloblastoma associated with BCNS is more benign in nature than the nonsyndromic isolated medulloblastoma in the general population. Radiation therapy of medulloblastoma in individuals with BCNS may cause an increase in the number of invasive basal cell carcinomas in the radiation field. Thus care should be taken to avoid exposing these patients to radiation (Amlashi et al., 2003).

Other tumors such as meningioma, oligodendroglyoma, and craniopharyngioma are reported less frequently than medulloblastoma in Gorlin syndrome. Some patients with BCNS have exhibited colloid cysts of third ventricle, arachnoid cysts, intraparenchymal cysts, cysts of septum pellucidum, and vermian dysgenesis. Other asymptomatic brain signs that facilitate diagnosis are ectopic calcifications of the falx cerebri, tentorium cerebelli, the pectinoid ligament, diaphragm

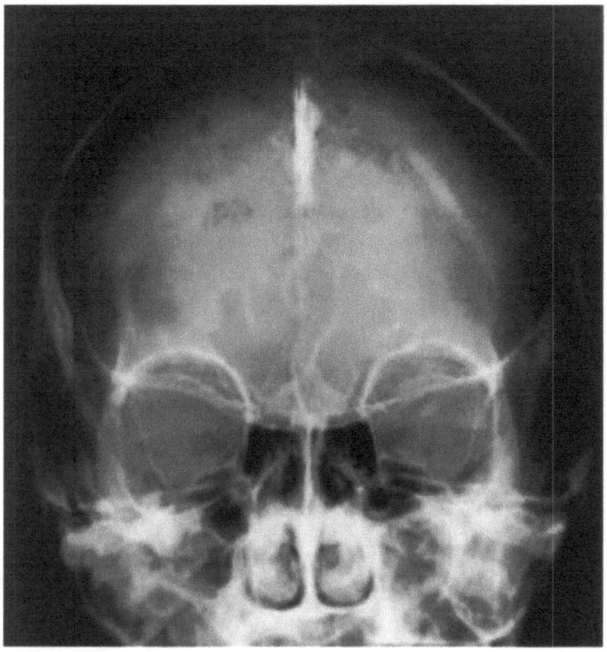

Fig. 8.5. Falx calcification.

sellae, and complete or partial bony bridging of the sella turcica (Fig. 8.5) (Kimonis et al., 1997).

Seizures are also reported but understudied in BCNS patients. Seizures in these individuals are mostly associated with tumors but may also result from focal neuronal heterotopia (Hogan et al., 1996). Intellectual disabilities occur in 5% of cases. Muller et al. (2012) have reported a microdeletion of 9q22.3 including the *PTCH1* gene in 10 patients who exhibited metopic craniosynostosis, severe obstructive hydrocephalus, macrosomia, and developmental delay or intellectual disability, none of which are typically observed in BCNS patients. These patients did not meet clinical criteria for BCNS but were found to have deletion of 9q22.3, which suggests there are genes other than *PTCH1* in the deleted gene region contributing to these nonsyndromic features (Muller et al., 2012).

Various ocular abnormalities have been linked to BCNS, including hypertelorism, exophthalmos, rotatory nystagmus, internal strabismus, congenital cataracts, orbital cysts, coloboma of the iris, choroid and optic nerve, microphthalmia, and chalazions (Salati et al., 1997; Gorlin, 2004).

Genitourinary system manifestations

Ovarian fibromas and cysts are present, often bilateral, in 25–50% of affected women from a 1993 population-based study (Evans et al., 1993). These do not present unless they become large, calcified, or twisted. Calcified ovarian cysts are usually misdiagnosed as uterine leiomyomas. They do not reduce fertility but are rarely virilizing or rennin secreting. Between 5% and 10% of males exhibit signs of hypogonadotropic hypogonadism such as anosmia, cryptorchidism, female pubic escutcheon, gynaecomastia, and scant facial or body hair (Shanley et al., 1994).

Cardiac manifestations

Between 3% and 5% of Gorlin patients have been reported to have cardiac fibromas (Evans et al., 1993). Fetal cardiac rhabdomyomas have been reported as inherited BCNS findings. The presence of mature striated muscle within the tumor is a characteristic feature of fetal rhabdomyomas. This tumor is often located in the left ventricular wall and can cause alteration in hemodynamics or cardiac conduction defects (Schweisguth et al., 1968; Bossert et al., 2006).

Renal manifestations

Various renal anomalies are reported in 5% of the individuals with BCNS. These include horseshoe kidney, L-shaped kidney, unilateral renal agenesis, renal cysts, and duplication of renal pelvis or ureters (Gorlin, 2004).

Gastrointestinal manifestations

Thin-walled mesenteric chylous or lymphatic cysts have been described (Fig. 8.6). Most of the affected individuals were asymptomatic (Gorlin, 2004).

Other tumors

Other isolated tumors have been documented in BCNS patients, but no specific association has been confirmed. Included are hamartomas, leiomyomas of bowel and menstery, leiomyosarcoma, Hodgkins lymphoma, non-Hodgkins lymphoma, nasal dermoid, seminoma, paratesticular pseudotumor, schwannoma, pleomorphic

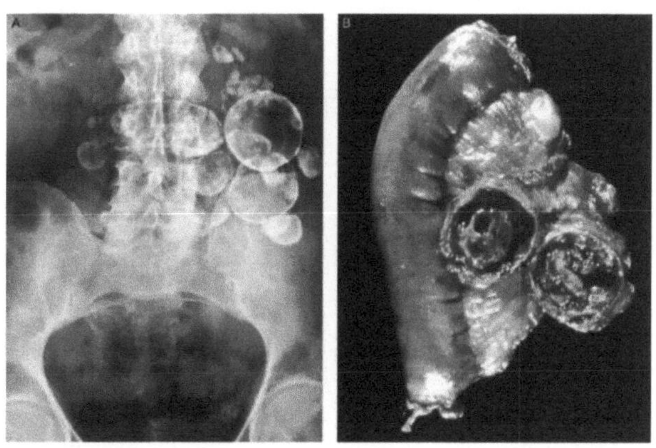

Fig. 8.6. Mesenteric cysts.

adenoma of parotid, adenoid cystic carcinoma of salivary gland, and adrenal tumors (Gorlin, 2004).

DIAGNOSIS

Diagnostic criteria

The initial diagnostic criteria were published by Evans et al. in 1993 and later modified by Kimonis et al. in 1997. In the first international colloquium on BCNS held in 2011 at St Louis University School of Medicine, previous diagnostic criteria and surveillance recommendations were reviewed and revised (Tables 8.1–8.3). The consensus statement included a small change in diagnostic criteria, moving medulloblastoma from minor to major criteria (Bree et al., 2011). BCNS diagnosis can be confirmed if an individual meets one of the following: (i) one major criterion and molecular confirmation; (ii) two major criteria; or (iii) one major and two minor criteria.

Differential diagnosis

Benign conditions such as seborrheic keratosis and benign naevi resemble basal cell carcinomas, but suspicion should be raised whenever other features of the syndrome are present. Multiple basal cell carcinomas occur in other rare dermatologic conditions, such as Brooke–Spiegler syndrome, Bazex syndrome, Rombo syndrome, melanocytic naevi, epithelioma adnoides cysticum, and Muir–Torre syndrome. Conditions such as Sotos, Beckwith–Wiedemann, isolated hydrocephaly and megalencephaly are considered in the differential diagnosis of macrocephaly. Isolated keratocysts of the jaw can occur outside of Gorlin syndrome.

Genetic counseling

Gorlin syndrome or BCNS is inherited as an autosomal dominant trait and manifests with multisystem involvement; a referral to geneticist is recommended. There is a 50% risk of inheriting the affected gene causing the syndrome for every child of a parent with Gorlin syndrome. In about 20–30% of isolated cases, *de novo* mutations are reported. Sometimes parents of the affected person may have the faulty gene causing the syndrome but due to variable expression may have only mild features and are missed clinically. Families of the affected person should be involved in regular screening. If a family member has not inherited the gene, then a surveillance program is not required. The definitive diagnostic test is detection of the *PTCH1* gene mutation on chromosome 9 using chromosomal analysis by array-based comparative genomic hybridization technology (CGH) (Klein et al., 2005). Recently, a new technique using high-resolution oligonucleotide microarray designed by Fujii et al. has been used to detect gene deletions in BCNS patients. The probe used in this technique helps to identify submicroscopic genomic deletions as small as 165 kb in the PTCH1 gene, which is sometimes missed by conventional chromosomal analysis (Fujii et al., 2007).

Antenatal diagnosis

In pregnant women with BCNS or in individuals with a family history of BCNS, ultrasound scanning may identify developmental malformations such as macrocephaly or fibromas of the heart. This can help in planning the delivery. Prenatal diagnosis for BCNS is available with DNA analysis on extracted fetal cells by amniocentesis (done usually between 15 and 18 weeks of gestational age) or by chorionic villus sampling (done at about 10–12 weeks of gestational age) but should be considered only when at least one of the affected family member has a proven mutation of the allele causing Gorlin syndrome (Petrikovsky et al., 1996).

Table 8.1

BCNS diagnostic criteria

Major criteria	Minor criteria
1. Multiple (>2) BCCs or one under 20 years of age	1. Rib anomalies (bifid, fused or markedly splayed ribs)
2. Odontogenic keratocysts of the jaws prior to 20 years of age	2. Cleft lip or palate
3. Palmar or plantar pits	3. Other specific skeletal malformations and radiologic changes (vertebral anomalies, kyphoscoliosis, short fourth metacarpals, postaxial polydactyly)
4. Calcification of the falx cerebri	4. Macrocephaly
5. Medulloblastoma, typically desmoplastic	5. Ovarian/cardiac fibroma
6. First-degree relatives with BCNS	6. Lymphomesenteric cysts
	7. Ocular abnormalities (i.e., strabismus, hypertelorism, congenital cataracts, glaucoma, coloboma)

Table 8.2

Education protocol for evaluation of patients with suspected BCNS

Medical history
- Birth history: macrocephaly, hydrocephalus, undescended testes, hernia
- Developmental history: developmental delay, school performance, height and weight as compared to siblings
- Medical/surgical history: history/treatment of brain tumor, history/treatment of strabismus, oral cleft repair, dental extractions or oral surgery, surgical treatment of skin lesions, cardiac problems, infertility, fractures
- Social history: environmental exposure, including radiation therapy, ultraviolet exposure

Examination by a medical/clinical geneticist
- Facial dysmorphology: macrocephaly, biparietal/frontal bossing, broad nasal root, mandibular prognathism, facial asymmetry, oral clefting, dental malocclusion, hypertelorism, synophrys, coloboma, epicanthal folds
- Skeletal: pectus anomalies, sprengel deformity, scoliosis
- Skin: pits of soles, palms, web spaces between fingers
- Inguinal hernia

Radiologic examination
- Panorex of jaw
- MRI of brain
- Additional studies if warranted:
  - PA and lateral skull for ectopic calcification
  - CXR for evaluation of bifid ribs
  - Full PA and lateral spine for scoliosis and vertebral anomalies
  - Long bones for bone cysts

Dermatologic examination
- Palmar/plantar pits
- Basal cell carcinomas
- Milia-like papules in periorbital and perinasal area
- Radiation port site for prior radiation
- Dermoid cysts especially at the web space of first and second fingers
- Dental examination
- Digital Panorex of jaw if not previously done
- Sinus films if symptomatic

Cardiology examination
- Cardiac ultrasound

Gynecology examination
- Pelvic ultrasound

TREATMENT AND MANAGEMENT

Management of Gorlin syndrome requires a multidiciplinary specialty approach. Early detection and management of the condition reduces morbidity and mortality.

Basal cell carcinoma

Basal cell carcinomas are more likely to occur in sun exposed areas. In BCNS patients basal cell carcinomas are very common. Prevention is better than cure, so BCNS patients should implement preventive measures such as avoidance of sun exposure. BCCs are rarely invasive and may be treated with either a surgical or a nonsurgical approach. Radiotherapy should be avoided if possible. Choosing the method of treatment depends on the location, number, recurrent or radiation induced lesions. The methods described in Table 8.4 have all been employed in BCCs treatment (Van der Geer et al., 2009).

Interferon injections into BCC lesions were previously done. Recently this has been revisited, combining with other nonsurgical methods, in the experimental stage; more studies are needed to confirm the effectiveness of this approach. It is associated with side-effects including fever, shivering, drop in leukocyte count, and pain at the site of injection (Chitwood et al., 2013). Sonic hedgehog antagonist in the form of cream (cyclopamine) together with oral medication vismodegib (GDC-0449) is a recently tested treatment modality. It is FDA approved for use in adults with advanced and inoperable BCCs. Phase 2 trials showed efficacy in advanced and metastatic BCCs as well as in Gorlin syndrome patients with recurrent BCCs after surgery. It is costly, and the side-effects include myalgia, alopecia, taste disturbance, weight loss, and fatigue (Tang et al., 2012; Poggi and Kolesar, 2013).

Keratocyst

Management of keratocysts depends on several factors including the age of the patient, the size, extent, number and location of the lesion, rupture, or infiltration of the adjacent soft tissue. Methods of treatment can be conservative, aggressive or radical. Conservative treatment consists of enucleation of the tumor from the bone. This method is associated with higher recurrence risk due to the presence of satellite microtumors in the surrounding bone (Mendes et al., 2010; Casaroto et al., 2011). To reduce the recurrence risk, enucleation is followed by cryotherapy using liquid nitrogen (−70°C) or chemical curettage using Carnoy's solution applied to the surrounding bone margins. However, cryotherapy is associated with 11% recurrence risk whereas Carnoy's solution causes only < 2% recurrence risk. Carnoy's solution is most often used in cysts in the mandible but not the maxilla because of the risk of mucosal necrosis. Radical treatment involves resection of the tumor along with a 5 mm margin of healthy bone tissue. This method has the lowest recurrence risk rate. In children with loose tooth eruptions or when the bone formation process has not finished, radical methods should be replaced by conservative methods (Stoelinga, 2005).

Table 8.3

Proposed management protocol for surveillance of patients with BCNS

Investigation/consultation	Follow-up
MRI brain with contrast and epilepsy protocol	Baseline, yearly thereafter until 8 years old, repeat sooner if symptomatic
Cardiac ECHO	Baseline, repeat if symptomatic
Dermatologic examination	Baseline, yearly until first BCC then every 6 months at least
Panorex	Baseline, yearly until first jaw cyst then every 6 months until no jaw cysts for 2 years or until 21; repeat more regularly if needed for symptoms or occurrence
Spine X-ray	Baseline at age of 1 or at time of diagnosis; repeat if symptomatic; if abnormal repeat per scoliosis protocol every 6 months
Pelvic ultrasound (girls)	At menarche or age 18, sooner if symptomatic, repeat if abnormal or symptomatic
Developmental screening	Routine with well child visits, further assessment and testing if delayed milestones
Vision, hearing and speech	Screen annually, continue in school age
Ophthalmologic evaluation	Baseline, repeat if symptomatic
Psychology evaluation	Initial evaluation for support and counseling, follow-up based on recommendations
Medical/clinical genetics evaluation	Baseline, repeat annually to ensure multidisciplinary care approach
Molecular diagnosis	If necessary to confirm diagnosis
Genetic counseling	Baseline, preconception/prenatal counseling for couples at risk
Ionizing radiation exposure	Minimize, radiographs warranted for evaluation of valid medical problems; utilize nonionizing/digital imaging modalities if possible

Medulloblastoma

Medulloblastoma is best treated by aggressive surgical resection of the tumor combined with chemo- and radiotherapy. The variant of medulloblastoma found in Gorlin syndrome patients is the desmoplastic type. Desmoplastic medulloblastomas found in Gorlin syndrome are more benign in nature and have a better prognosis compared to other types of medulloblastomas. Radiotherapy can induce invasive or multiple basal cell carcinomas in the irradiated skin area and also numerous neoplastic lesions in the surrounding tissues of patients with Gorlin syndrome. It can cause ototoxicity and radiation injury to the temporal lobes. Therefore, care must be taken to avoid radiation if possible, or to use very low doses of radiation (Marín-Gutzke et al., 2004; Lopes et al., 2010).

PROGNOSIS

Life expectancy in Gorlin syndrome patients is similar to that of the general population. Cosmetic disfigurement from treatment of BCCs is a common side-effect. There is no change in lifestyle if BCCs and neoplasms are detected early and treated. Early death was noted from invasive secondary metastasis from infiltrative BCCs only.

SUPPORT GROUPS

A number of support groups can be accessed online, including: Gorlin Syndrome Group (http:www.gorlingroup.org); National Basal Cell Nevus Support Network (http://www.bccns.org); BCCBS Life Support Network (http://www.gorlinsyndrome.org); BCCNS Patient Advocate website (http:/www.bccns.org).

CONCLUSIONS

Basal cell nevus syndrome or Gorlin syndrome is a rare dominantly inherited neurocutaneous syndrome characterized by the development of multiple basal cell carcinomas, jaw cysts, desmoplastic medulloblastoma, congenital skeletal abnormalities, and other neoplastic tumors. This condition results from *PTCH1* gene mutation and due to heterogeneity, variable expressivity of the syndrome features, and multisystem involvement a multidisciplinary multimodal approach by involvement of various specialties is recommended for early detection of the condition and treatment from early in life (Kiwilsza and Sporniak-Tutak, 2012). Life expectancy is not substantially altered compared to the general population if tumors and BCCs are detected and treated early. Gorlin syndrome patients may have cosmetic scarring effects from treatment of BCCs. Genetic

Table 8.4

BCCs treatment methods

Photodynamic therapy (PDT)	1. Newer method, very effective for superficial small or large lesions in low-risk areas 2. Photosensitizing agents applied locally or intravenously and activated by using corresponding absorbing spectrum wavelength radiation	1. Cosmetic side-effects 2. Not suitable for children
5% Imiquimod	1. Used in nodular BCCs 2. Used alone or in combination with curettage	1. Needs five applications in a week and at least 6 weeks of treatment
0.1% Tretinoin	Applied locally and daily	Needs follow-up with dermatology in 3 months
5-Fluorouracil	Applied locally twice daily for 6–12 weeks	
Oral retinoid	1. For chemoprevention or to delay development of BCCs 2. Used in very large doses for long periods	1. Causes damage to eyes, liver, bone, brain or muscle 2. Associated with very high recurrence rate
Excision	1. Most common approach 2. Good for less and small size lesions 3. For better outcome advised to excise BCC with 2–8 mm margin of normal surrounding skin	1. Cosmetic side-effects 2. High success rate
Excision	1. Most common approach 2. Good for fewer and small size lesions 3. For better outcomes advised to excise BCC with 2–8 mm margin of normal surrounding skin	1. Cosmetic side-effects 2. High success rate
Electrodessication and curettage	1. Good for greater number of lesions, lesions in low recurrence risk areas (neck, trunk or extremities) 2. Simple and fast method; needs 3–4 sittings for maximum benefit	1. Damage to nerves and numbness in operated area 2. Success rate of 92–93%
Cryosurgery	1. Quick, efficient and cost-effective procedure 2. Avoid in high recurrence areas 3. Use several cycles of freezing with liquid nitrogen to destroy neoplastic tissue	1. Cure rate of 92.5% 2. Scarring and numbness in the area of lesion
CO_2 Laser ablation	1. Used in combination with curettage or alone 2. Good for multiple superficial lesions only	1. Cosmetic side-effects
Mohs micrographic surgery	1. Technique involves excision of lesions with precise microscopic normal margin 2. Very costly – so use in BCCs with recurrence, in high risk sites, in infiltrating lesions and in area of radiation therapy	1. High success rate, long-lasting but costly

counseling and prenatal genetic testing is available to pregnant women with an inherited PATCH 1 gene or to affected family members.

References

Amlashi SF, Riffaud L, Brassier G et al. (2003). Nevoid basal cell carcinoma syndrome: relation with desmoplastic medulloblastoma in infancy: a population-based study and review of the literature. Cancer 98: 618–624.

Bale A (1997). The nevoid basal cell carcinoma syndrome: genetics and mechanism of carcinogenesis. Cancer Invest 15: 180–186.

Bossert T, Walther T, Vondrys D et al. (2006). Cardiac fibroma as an inherited manifestation of nevoid basal-cell carcinoma syndrome. Tex Heart Inst J 33: 88–90.

Bree AF, Shah MR, Group BC (2011). Consensus statement from the first international colloquium on basal cell nevus syndrome (BCNS). Am J Med Genet A 155A: 2091–2097.

Casaroto AR, Loures DC, Moreschi E et al. (2011). Early diagnosis of Gorlin–Goltz syndrome: case report. Head Face Med 7: 2.

Chitwood K, Etzkorn J, Cohen G (2013). Topical and intralesional treatment of nonmelanoma skin cancer: efficacy and cost comparisons. Dermatol Surg 39: 1306–1316.

Dominguez F, Keszler A (1988). Comparative study of keratocysts, associated and non-associated with nevoid basal cell carcinoma syndrome. J Oral Pathol 17: 39–42.

Evans D, Farndon PA, Burnell L et al. (1991). The incidence of Gorlin syndrome in 173 consecutive cases of medulloblastoma. Br J Cancer 64: 959–961.

Evans DG, Ladusans EJ, Rimmer S et al. (1993). Complications of the naevoid basal cell carcinoma syndrome: results of a population based study. J Med Genet 30: 460–464.

Evans DG, Howard E, Giblin C et al. (2010). Birth incidence and prevalence of tumor-prone syndromes: estimates from a UK family genetic register service. Am J Med Genet A 152A: 327–332.

Farndon PA, Del Mastro RG, Evans DG et al. (1992). Location of gene for Gorlin syndrome. Lancet 339: 581–582.

Fujii K, Ishikawa S, Uchikawa H et al. (2007). High-density oligonucleotide array with sub-kilobase resolution reveals breakpoint information of submicroscopic deletions in nevoid basal cell carcinoma syndrome. Hum Genet 122: 459–466.

Gorlin RJ (2004). Nevoid basal cell carcinoma (Gorlin) syndrome. Genet Med 6: 530–539.

Gorlin RJ, Goltz RW (1960). Multiple nevoid basal-cell epithelioma, jaw cysts and bifid rib. A syndrome. N Engl J Med 262: 908–912.

Hahn H, Wicking C, Zaphiropoulous PG et al. (1996). Mutations of the human homolog of Drosophila patched in the nevoid basal cell carcinoma syndrome. Cell 85: 841–851.

Hogan R, Tress B, Gonzales MF et al. (1996). Epilepsy in the nevoid basal-cell carcinoma syndrome (Gorlin syndrome): report of a case due to a focal neuronal heterotopia. Neurology 46: 574–576.

Jarisch W (1894). Zur Lehre von den Hautgeschwulsten. Arch Dermal Syphilol 162–222.

Johnson RL, Rothman A, Xie J et al. (1996). Human homolog of patched, a candidate gene for the basal cell nevus syndrome. Science 272: 1668–1671.

Kimonis VE, Goldstein AM, Pastakia B et al. (1997). Clinical manifestations in 105 persons with nevoid basal cell carcinoma syndrome. Am J Med Genet 69: 299–308.

Kimonis VE, Mehta SG, Digiovanna JJ et al. (2004). Radiological features in 82 patients with nevoid basal cell carcinoma (NBCC or Gorlin) syndrome. Genet Med 6: 495–502.

Kimonis VE, Singh KE, Zhong R et al. (2013). Clinical and radiological features in young individuals with basal cell carcinoma syndrome. Genet Med 15: 79–83.

Kiwilsza M, Sporniak-Tutak K (2012). Gorlin–Goltz syndrome – a medical condition requiring a multidisciplinary approach. Med Sci Monit 18: 145–153.

Klein RD, Dykas D, Bale A (2005). Clinical testing for the nevoid basal cell carcinoma syndrome in a DNA diagnostic laboratory. Genet Med 7: 611–619.

Lee YW, Roh BH, Ki CS et al. (2007). Identification of a novel mutation in the PTCH gene in a Korean family with naevoid basal cell carcinoma syndrome. Clin Exp Dermatol 32: 202–203.

Leonardi R, Caltabiano M, Lo Muzio L et al. (2002). Bilateral hyperplasia of the mandibular coronoid processes in patients with nevoid basal cell carcinoma syndrome: an undescribed sign. Am J Med Genet A 110: 400–403.

Levanat S, Gorlin RJ, Fallet S et al. (1996). A two-hit model for developmental defects in Gorlin syndrome. Nat Genet 12: 85–87.

Lo Muzio L (2008). Nevoid basal cell carcinoma syndrome (Gorlin syndrome). Orphanet J Rare Dis 3: 1–16.

Lo Muzio L, Nocini PF, Savoia A et al. (1999). Nevoid basal cell carcinoma syndrome. Clinical findings in 37 Italian affected individuals. Clin Genet 55: 34–40.

Lopes NN, Caran EM, Lee ML et al. (2010). Gorlin–Goltz syndrome and neoplasms: a case study. J Clin Pediatr Dent 35: 203–206.

Marín-Gutzke M, Sánchez-Olaso A, Berenguer B et al. (2004). Basal cell carcinoma in childhood after radiation therapy: case report and review. Ann Plast Surg 53: 593–595.

Mendes RA, Carvalho JF, van der Waal I (2010). Characterization and management of the keratocystic odontogenic tumor in relation to its histopathological and biological features. Oral Oncol 46: 219–225.

Muller EA, Aradhya S, Atkin JF et al. (2012). Microdeletion 9q22.3 syndrome includes metopic craniosynostosis, hydrocephalus, macrosomia, and developmental delay. Am J Med Genet A 158A: 391–399.

Nikolaou V, Stratigos A, Tsao H (2012). Hereditary nonmelanoma skin cancer. Semin Cutan Med Surg 31: 204–210.

Pastorino L, Ghiorzo P, Nasti S et al. (2009). Identification of a SUFU germline mutation in a family with Gorlin syndrome. Am J Med Genet A 149A: 1539–1543.

Petrikovsky BM, Bialer MG, McLaughlin JA (1996). Sonographic and DNA-based prenatal detection of Gorlin syndrome. J Ultrasound Med 15: 493–495.

Poggi L, Kolesar JM (2013). Vismodegib for the treatment of basal cell skin cancer. Am J Health Syst Pharm 70: 1033–1038.

Salati C, Virgili G, Menchini U et al. (1997). Gorlin's syndrome. Case report. Eur J Ophthalmol 7: 113–114.

Satinoff M, Wells C (1969). Multiple basal cell naevus syndrome in ancient Egypt. Med Hist 13: 294–297.

Schultz SM, Twickler DM, Wheeler DE et al. (1987). Ameloblastoma associated with basal cell nevus (Gorlin) syndrome: CT findings. J Comput Assist Tomogr 11: 901–904.

Schweisguth O, Gerard-Marchant R, Lemerle J (1968). Basal cell nevus syndrome. Association with congenital rhabdomyosarcoma. Arch Fr Pediatr 25: 1083–1093.

Shanley S, Ratcliffe J, Hockey A et al. (1994). Nevoid basal cell carcinoma syndrome: review of 118 affected individuals. Am J Med Genet 50: 282–290.

Springate JE (1986). The nevoid basal cell carcinoma syndrome. J Pediatr Surg 21: 908–910.

Stoelinga PJ (2005). The treatment of odontogenic keratocysts by excision of the overlying, attached mucosa, enucleation, and treatment of the bony defect with carnoy solution. J Oral Maxillofac Surg 63: 1662–1666.

Tang JY, Mackay-Wiggan JM, Aszterbaum M et al. (2012). Inhibiting the hedgehog pathway in patients with the basal-cell nevus syndrome. N Engl J Med 366: 2180–2188.

Van der Geer S, Ostertag JU, Krekels GA (2009). Treatment of basal cell carcinomas in patients with nevoid basal cell carcinoma syndrome. J Eur Acad Dermatol Venereol 23: 308–313.

White JC (1894). Multiple benign cystic epitheliomas. J Cutan Genitourin Dis 12: 447–484.

Handbook of Clinical Neurology, Vol. 132 (3rd series)
Neurocutaneous Syndromes
M.P. Islam and E.S. Roach, Editors

Chapter 9

PTEN hamartoma tumor syndrome

JESSICA MESTER AND CHARIS ENG*
Genomic Medicine Institute, Cleveland Clinic, Cleveland, OH, USA

INTRODUCTION

History

PTEN hamartoma tumor syndrome (PHTS) collectively includes patients with Cowden syndrome (CS, OMIM #158350), Bannayan–Riley–Ruvalcaba syndrome (BRRS, OMIM #153480), macrocephaly–autism (OMIM #605309), and any other clinical presentation with germline mutation of the *PTEN* tumor suppressor gene (Zbuk and Eng, 2007). While first thought to be disparate conditions, patients with both CS and BRRS were discovered to have germline *PTEN* mutations with overlapping mutational spectra (Marsh et al., 1998). The component phenotypes of PHTS have both age- and gender-related penetrance (Lachlan et al., 2007), leading many patients with PHTS to continue to be clinically referred to as having either CS or BRRS.

Origin of the name

PHTS is named for the gene involved (*PTEN*), nature of many of the component lesions (*hamartomatous*), and propensity for *tumor* formation. CS was named after Rachel Cowden, a young woman presented in the medical literature by Lloyd and Dennis in 1963 with a collection of features which came to be understood as components of this rare syndrome (Lloyd and Dennis, 1963). BRRS is sometimes referred to as Bannayan–Zonana, Riley–Smith, or Ruvalcaba–Myhre–Smith syndrome; all names come from the clinicians who reported on patients with unifying features early in the literature (Riley and Smith, 1960; Bannayan, 1971; Zonana et al., 1976).

Epidemiology

In the Dutch population, 22 persons were noted as having PHTS within a national database of 4.5 million people, leading the authors to estimate a disease frequency of 1 in 200 000–250 000 persons (Nelen et al., 1999). These patients were diagnosed with CS and then referred for *PTEN* mutation analysis based on presence of mucocutaneous features. Given our new understanding of the clinical variability of PHTS, the true disease incidence is likely to be higher than this early estimate.

CLINICAL MANIFESTATIONS

Typical age of presentation, progression, life expectancy

PHTS patients can present at any age. Many children with PHTS first come to medical attention due to the presence of macrocephaly and neurodevelopmental delays, benign skin lesions such as lipomas or penile freckling, all of which may also encourage the clinician to consider a PHTS diagnosis. Those diagnosed with PHTS in adulthood tend to be referred for evaluation based on a history of early-onset cancer or gastrointestinal polyposis. Whereas macrocephaly or neurodevelopmental delays may have been present earlier in life, these symptoms alone would not warrant genetics referral such that these patients are undiagnosed until other symptoms arise. Given their benign nature and subtle appearance, the mucocutaneous features of PHTS tend to go unrecognized before the PHTS diagnosis is made and, in our experience today, have not been major drivers for genetics referral and ultimately diagnosis.

Hallmark signs and symptoms

Characteristics which should raise strong clinical suspicion for PHTS, given their rarity in the general population or in other genetic syndromes, include:

*Correspondence to: Charis Eng, MD, PhD, Genomic Medicine Institute, Cleveland Clinic, 9500 Euclid Avenue, NE-50, Cleveland, OH 44195, USA. Tel: +1-216-444-3440, E-mail: engc@ccf.org

1. Lhermitte–Duclos disease (dysplastic cerebellar gangliocytoma)
2. extreme macrocephaly (head circumference: in children, ≥5 standard deviations above the mean; adult women ≥60 cm; adult men ≥63 cm)
3. oral mucosal papillomatosis
4. penile freckling
5. hamartomatous gastrointestinal polyps
6. pediatric nonmedullary thyroid carcinoma.

These findings were among the highest scoring components on the Cleveland Clinic PTEN Risk Calculation tool, which calculates a patient's *a priori PTEN* mutation risk (Tan et al., 2011). This tool awarded a weighted score for each characteristic after comparing age-related prevalence within mutation-positive and -negative research participants to expected community frequencies as derived from published literature and the Surveillance Epidemiology and End Results database.

Involvement by systems

Central nervous system

Macrocephaly. Approximately 94% of people with PHTS have head circumferences which measure > 2 standard deviations from the norm for age and gender. On average, head circumference measured +4.89 standard deviations above the mean for children. Adult men and women average head circumference measured 62.8 cm and 60.0 cm, respectively (Mester et al., 2011). Case reports and small imaging series have found the enlarged head circumference in PHTS is due to overgrowth of brain tissues as opposed to hydrocephalus (Padberg et al., 1991; Eng et al., 1994; Lok et al., 2005).

Autism/developmental delay. *PTEN* mutations have been identified in 10–17% of children with both macrocephaly and autism (Butler et al., 2005; Buxbaum et al., 2007; Varga et al., 2009; McBride et al., 2010). The range of developmental impact varies widely, with some patients having no developmental issues and normal to superior IQ whereas others are classified as severely impacted by autism with or without mental retardation (Lachlan et al., 2007; Busch et al., 2013).

Lhermitte±Duclos disease (LDD). In a case report describing two unrelated patients with both LDD and CS, the authors surmised that the co-occurrence of two seemingly rare conditions was owed to more than chance alone (Padberg et al., 1991). Several subsequent case reports and series have identified LDD in patients with PHTS (Eng et al., 1994; Zhou et al., 2003). Germline or somatic *PTEN* mutations have been identified in a majority of LDD tumors (Zhou et al., 2003). Fortunately, LDD seems to be an uncommon manifestation of PHTS, with one recent European series finding this diagnosis in 9% of cases (Nieuwenhuis et al., 2014).

Dermatologic/mucosal involvement

While the mucocutaneous lesions associated with PHTS are benign in nature, they can be important "red flags" that an underlying *PTEN* mutation is present. Prior to the advent of molecular testing, the diagnosis of CS relied on the presence of the following key features, resulting in a likely ascertainment bias reflected in data citing these features as present in nearly 100% of patients.

Facial papules. Early histologic examinations of facial papules from CS patients revealed most studied lesions to be varieties of trichilemmoma (Fig. 9.1) (Halevy et al., 1985; Starink et al., 1985).

Mucosal papillomas. Observed along the tongue, gums, or nasal epithelia. The tongues of many patients we have seen also appear thick and deeply furrowed (Figs. 9.2, 9.3).

Acral or palmoplantar keratoses. Most often identified on the hands and feet. On gross examination described as small verrucous hyperkeratotic papules similar to viral warts or small keratoses with a central depression (Halevy et al., 1985; Starink et al., 1985) (Figs. 9.4, 9.5).

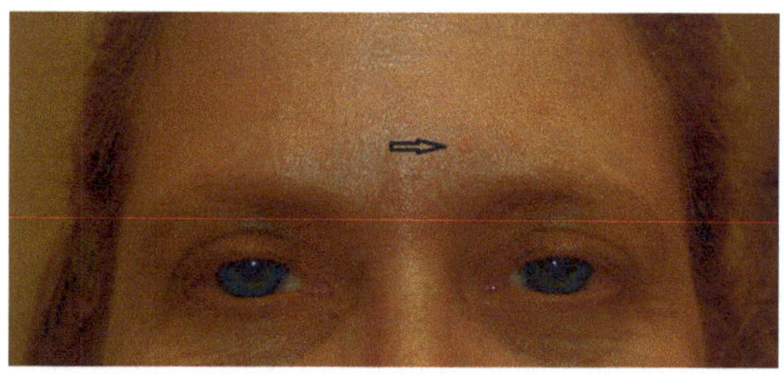

Fig. 9.1. Biopsy-proven trichilemmoma from 43-year-old female patient with PHTS.

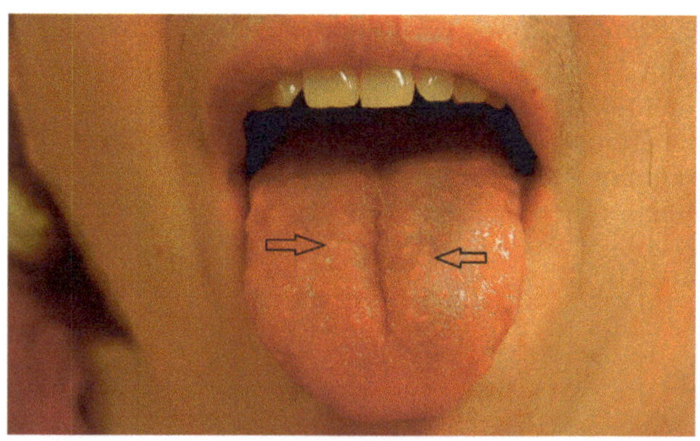

Fig. 9.2. Arrows pointing to papillomas of the tongue on a 57-year-old woman with PHTS; also notice deep tongue furrow.

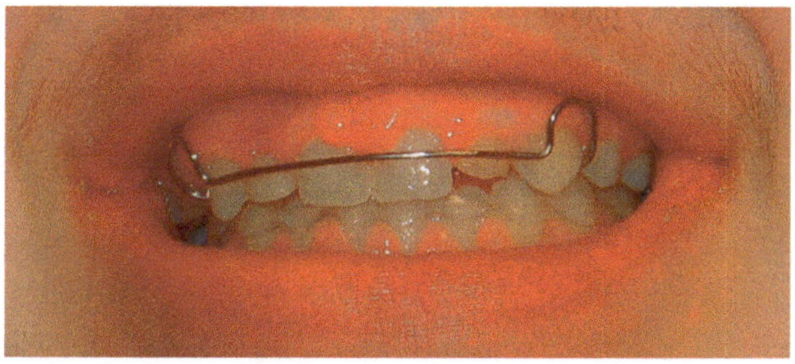

Fig. 9.3. Gingival papillomas in a young lady in her teens with PHTS; easily spotted when the overgrowth crosses the tooth border.

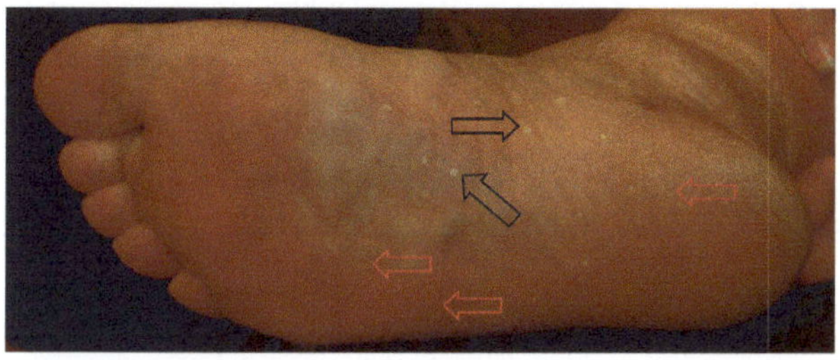

Fig. 9.4. Ventral surface of the right foot of a 54-year-old gentleman with PHTS. Notice the multiple keratoses (black arrows) and smaller, translucent lesions with a pitted appearance (red arrows).

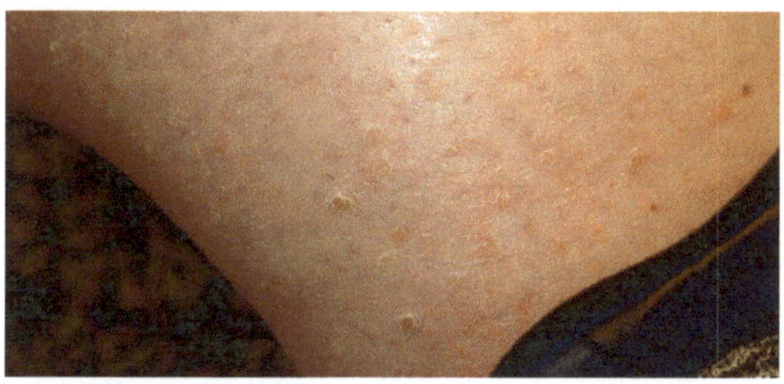

Fig. 9.5. Multiple hyperkeratotic papules present just above the right ankle on a 50-year-old lady with PHTS.

Other dermatologic lesions reported in many patients include lipomas and fibromas (sclerotic, oral, and dermato-). Men can develop freckling of the glans penis, a finding first introduced as associated with BRRS. Increased risk for melanoma has also been identified (Tan et al., 2012; Bubien et al., 2013).

Breast

Cancer. Risk for female breast cancer in PHTS is high, with recent studies predicting an up to 87% lifetime risk with diagnoses beginning around 30 years of age (Tan et al., 2012; Bubien et al., 2013; Nieuwenhuis et al., 2014).

Benign disease. Severe fibrocystic breast disease is what led Rachel Cowden to first present for medical attention; her breast disease was so severe that bilateral mastectomies became necessary at age 20, and on histologic examination some areas were noted as being "indistinguishable from carcinoma" (Lloyd and Dennis, 1963). Benign breast disease is a common finding in women with PHTS or CS with diverse types of benign lesions noted (Schrager et al., 1998). Among the first 100 CS cases described in the literature, 60% had fibrocystic changes or hypertrophy (Starink et al., 1986).

Thyroid/endocrine

Cancer. Lifetime risk for thyroid cancer in PHTS is estimated to be 34% with increased risk beginning in childhood (Tan et al., 2012; Bubien et al., 2013; Nieuwenhuis et al., 2014). Among a population of persons with thyroid cancer, follicular histology, male gender, and age at diagnosis of < 18 years were all highly predictive of underlying *PTEN* mutation (Ngeow et al., 2011).

Benign disease. In one small hospital-based series, 100% of those with PHTS by age 13 years had some kind of thyroid abnormality, most often multiple benign nodules or Hashimoto's thyroiditis (Milas et al., 2012).

Genitourinary

Cancer. People with PHTS have elevated risks for endometrial and renal cancer, with lifetime risks of 28% and 34%, respectively. Risk for endometrial cancer was found to begin to elevate at age 25 and risk for renal cancer started at 40 (Lachlan et al., 2007; Tan et al., 2012). The majority of the cancers were identified symptomatically, negative for metastasis, and were well-treated by conventional surgical methods. On review by a dedicated uropathologist, most of the renal tumors were either papillary (both type I and II) or chromophobe lesions as opposed to the more common clear cell histology (Mester et al., 2012).

Benign disease. Uterine fibroids and endometriosis have both been frequently reported in series of women with CS (Starink et al., 1986), but it is unknown whether the prevalence of these diagnoses would be impressive considering their common occurrence in the general population.

Vascular

Hemangiomas requiring surgical intervention have been described in series of patients with CS and PHTS. The unique nature of these lesions has recently been elucidated. As opposed to solely involving vascular tissue, lipomatous and muscular tissues in PHTS are also involved in a disorganized growth pattern, giving rise to the moniker PTEN-related hamartoma of soft tissue (PHOST) as an appropriate descriptor (Kurek et al.,

2012). Unfortunately some patients develop multiple troublesome PHOSTs which have not responded well to surgical management via resection or embolization and have tended to recur post-treatment (Greene and Orbach, 2011). PHOSTs affecting the limbs have caused some patients to have significantly reduced mobility and troubles with daily living activities. A couple of patients have reported to us lesions which sound like PHOSTs that resolved spontaneously, but unfortunately we cannot confirm these reports.

Cerebrovascular malformations including dural arteriovenous fistulas, small developmental venous anomalies, paraspinal arteriovenous malformations, and cavernous hemangiomas have also been reported in PHTS patients (Srinivasa and Burrows, 2006; Tan et al., 2007; Bae et al., 2011; Moon et al., 2013). As with PHOSTs, for lesions requiring treatment the course has been complicated. We are aware of patients for whom surgical intervention was not successful, and rupture of these lesions resulted in blindness, stroke, and death.

Gastrointestinal

Polyps were present in > 90% of a series of PHTS patients who had undergone endoscopy (Heald et al., 2010). The number of polyps varied from one to innumerable, and the histologic subtypes were quite diverse as well. Although hamartomas have traditionally been characterized as the polyp type most often seen in PHTS, only 18/62 (29%) of the patients in this study had a hamartomatous polyp. Hyperplastic polyps were the most common type identified, observed in about half of the patients studied. Among patients with ≥5 polyps including at least one hyperplastic or hamartomatous polyp, patients with either at least one ganglioneuroma or a mix of three or more polyp types were more likely to have mutations in *PTEN* as opposed to other nonadenomatous-polyposis genes (Ngeow et al., 2013). Colorectal cancer risk in PHTS patients has been found to be slightly elevated above the general population risk, with maximal lifetime risks cited as 18% (Nieuwenhuis et al., 2012; Tan et al., 2012; Bubien et al., 2013).

Tumor/neoplastic potential

As discussed above, elevated lifetime risks for malignancy in multiple organ systems for persons with PHTS – chiefly breast, thyroid, endometrial, and renal – have been identified.

Hamartomatous lesions have low potential to turn malignant. While it has been shown that trichilemmomas have the potential to transform into trichilemmal carcinoma (Kurokawa et al., 2006), we are unaware of such a phenomenon occurring in a patient with PHTS. We are also not aware of a lipoma or PHOST lesion in a PHTS patient which has turned into a malignancy, although we have in our clinical experience observed regeneration after surgery.

Imaging

A case series of 20 PHTS patients who underwent prospective brain MRI identified cerebral anomalies in seven (35%) including Lhermitte–Duclos, meningioma, vascular malformations, and cavernous angiomas (Lok et al., 2005). Within a series of 23 PHTS patients who were selected mainly due to presence of known white matter abnormalities, brain MRIs underwent retrospective review and either multifocal white matter lesions or enlarged perivascular spaces were found in every studied patient (Vanderver et al., 2014).

Differential diagnoses to consider

Differential diagnoses to consider vary greatly depending on the patient's age and clinical presentation.

Children with macrocephaly + neurodevelopmental delays. Per the American College of Medical Genetics, chromosome microarray is recommended as a first-line screening test for any child presenting with multiple anomalies not specific to a well-delineated syndrome, apparently nonsyndromic developmental delay, or an autism spectrum disorder (Manning and Hudgins, 2010). Neurofibromatosis 1 (NF1), tuberous sclerosis complex (TSC), and Sotos syndrome may all present with macrocephaly (although not to such a high degree as seen in PHTS) along with developmental delays and with NF1 and TSC, skin lesions, or pigmentary abnormalities. A few café-au-lait macules have been noted in some PHTS patients, but it is unknown whether the frequency is greater than could be observed in the general population. In our experience the number of café-au-lait macules does not approach the quantity seen with NF1.

Hamartomatous gastrointestinal polyps. Patients with mutations in *SMAD4* or *BMPR1A* present with multiple juvenile polyposis (juvenile referring to the type of polyp, not necessarily age at presentation) and have elevated cancer risks, but do not have the systemic complications observed in PHTS. Diffuse gastrointestinal ganglioneuromatosis may be observed in adults with NF1 or multiple endocrine neoplasia type 2B (MEN2B); in PHTS, the ganglioneuromas have a polypoid appearance. While both NF1 and MEN2B do have some clinical overlap with PHTS (NF1 – benign skin lesions, MEN2B – mucosal lesions), careful evaluation for presence of absence of features associated with each should enable a clinician to distinguish these syndromes from each other.

Early-onset cancer diagnoses. PHTS has overlapping cancer spectra with hereditary breast and ovarian cancer syndrome and Lynch syndrome. Given the greater populational frequencies of these two conditions in comparison with PHTS, many PHTS patients with early-onset cancer diagnoses are first tested for these syndromes. This approach is certainly reasonable when the patient's *a priori* risk to have either condition is equivocal or when the risk for another syndrome is higher than for PHTS. The advent of panel testing – where several genes associated with increased risk for cancer or another phenotype are tested simultaneously – may be especially advantageous in a circumstance where multiple diagnoses are equally likely.

PATHOLOGY

A hamartoma is defined as "a focal malformation that resembles a neoplasm, grossly and even microscopically, but results from faulty development, with a disproportion of abnormal mixture of tissue elements normally present at the site" (Stedman's, 4th edition). The PHOST lesions described previously are a prime example of a hamartoma.

Early studies of biopsies from facial and extrafacial skin lesions from CS patients found no evidence for a viral component. Most biopsies from facial lesions showed abnormalities which were entirely or predominantly follicular in nature, with cylindrical and lobulated types of trichilemmomas, keratinizing type trichilemmomas, tumors of the follicular infundibulum, specific follicular hyperplasia, and verruca vulgaris-like. Within the extrafacial lesions, again verruca vulgaris-like lesions were found along with hyperkeratotic papillomas, acanthomas, and fibromas (Starink and Hausman, 1984a, b; Starink et al., 1985).

Immunohistochemical (IHC) staining of various tumor types has been studied as a possible mechanism to identify patients with germline *PTEN* mutations, as is routinely done for Lynch syndrome. Complete loss of PTEN protein has been observed in papillary renal carcinomas from PHTS patients (Mester et al., 2012) as well as thyroid carcinomas (Ngeow et al., 2012). Larger prospective studies are needed to determine whether screening these tumor types via PTEN IHC would be a sensitive, specific, and cost-effective tool towards identifying patients at risk for PHTS.

MOLECULAR BASIS/PATHOLOGY

PTEN is a dual-specificity phosphatase with a core promoter region and nine exons. Its full-length protein comprises 403 amino acids. *PTEN* has a core promoter region followed by nine exons which encode phosphatase, C2, PDZ, and two PEST domains. The core promoter contains androgen receptor and p53 binding domains. Within the phosphatase domain lie two ATP-binding motifs, the second of which houses the catalytic core region. Mutations of all types – missense, nonsense, splice, large deletion, etc. – have been identified in each exon. Deletions have been identified which span one or a few exons as well as the entire gene plus the surrounding genomic region (Mester and Eng, 2013). Thus it is important that any laboratory seeking to provide *PTEN* mutation analysis also performs large rearrangement testing such as MLPA to detect such a large deletion or duplication which would not be detected by PCR-based Sanger sequencing alone.

As a lipid phosphatase, PTEN's major role is in the inhibition of Akt via the phosphoinositide 3-kinase (PI3K)/Akt pathway (Stambolic et al., 1998; Wu et al., 1998). PTEN dephosphorylates phosphatidylinositol 3,4,5-trisphosphate (PIP3) to phosphphatidylinositol 4,5-diphosphate (PIP2). When left unchecked through mutant PTEN, an overabundance of PIP3 leads to constitutive downstream activation of Akt (Weng et al., 1999), which in turn leads to increased cell growth and decreased apoptosis (Datta et al., 1999). PTEN also functions as a protein phosphatase and has been shown to act against both serine/threonine and tyrosine-phosphorylated proteins (Myers et al., 1997).

ANIMAL MODELS

Several different mouse models have been generated for *PTEN*. Homozygous knockout mice have not been found to survive past 7.5 days' gestation, confirming *PTEN*'s important role in embryonic development. Heterozygous knockouts survive but develop multiple tumors, commonly in the organs seen to be affected in patients with PHTS although the tumor spectrum is not entirely similar to human PHTS (Di Cristofano et al., 1998; Podsypanina et al., 1999). A knockin mouse model has also been generated for which both heterozygotes and homozygotes survive past birth, with brain weight and size increasing with loss of normal *Pten* allele and with age (Mester et al., 2011). Pten-shRNA mice have also been generated with resultant partial reduction of Pten dosage; these mice also developed multiorgan tumors, supporting the notion that even a partial loss (haploinsufficiency) of normal PTEN can lead to tumorigenesis (Shen-Li et al., 2010). The latter was recently shown to be likely in human PHTS (Tan et al., 2011).

GENETICS

PHTS is inherited in an autosomal dominant manner. Penetrance by early adulthood is believed to be 100% with variable expressivity. At minimum, 10% of patients' mutations occur *de novo*; the maximum *de novo* mutation rate could be as high as 44% (Mester and Eng, 2012). Somatic mosaicism has been proven and documented in three PHTS patients in the literature (Zhou et al., 2000; Gammon et al., 2013; Pritchard et al., 2013). Thus even when both parents test negative for a child's mutation, testing of the proband's at-risk siblings should still be offered, given the inherent cancer risks and changes in management that would result from a positive test result.

TREATMENT AND MANAGEMENT

Patients with features or symptoms concerning for PHTS should seek evaluation by a genetics provider. Those with a family history of PHTS should also seek genetics evaluation prior to testing. Any genetic testing should always include pre- and post-test counseling by a genetic counselor. Use of the Cleveland Clinic PTEN Risk Calculator can help clinicians judge when to refer for genetics evaluation or pursue genetic testing. This tool is freely available and web-based (www.lerner.ccf.org/gmi/ccscore) and calculates the *a priori* risk for any person to have a *PTEN* mutation based on personal history and physical findings (Tan et al., 2011).

After a PHTS diagnosis is made, a number of referrals to subspecialists for organ-specific screening will likely be necessary (see, e.g., http:/my.clevelandclinic.org/genomics-genetics/subspecialties/pten-clinic.aspx). Collaboration between the patient, their primary care provider, and the genetics team can be extremely helpful for patient care and referral management. In addition to genetics, patients frequently benefit from care from subspecialists in the following areas: endocrinology, gastroenterology, dermatology, high-risk breast management, gynecologic oncology, neurology, and developmental pediatrics.

A major benefit of making a PHTS diagnosis early in the patient's life is the ability to take advantage of high-risk screening to reduce the morbidity and mortality that could otherwise result from the patient's high cancer risks (Tan et al., 2012). By system, recommendations for care include:

- *Breast*. Women should seek consultation with a high-risk breast specialist to discuss their plans for high-risk screening starting at the latest by age 30 years. For women with a family history of early-onset breast cancer, screening should begin 10 years prior to the youngest affected relative's age at diagnosis (so at age 25 years if a relative was diagnosed at 35). The 87% lifetime breast cancer risk for PHTS exceeds the 25% threshold for offering surveillance via MRI as established by the American Cancer Society. Some patients at this risk level may also wish to consider prophylactic mastectomy.
- *Thyroid*. Screening via thyroid ultrasound should start at the age of diagnosis, with repeat imaging yearly if results are normal. Pediatric thyroid cancer diagnoses – the youngest at age 6 years – have been observed rarely (Smith et al., 2011) but given the lack of risk associated with thyroid ultrasound and great benefit to early diagnosis, we feel it is prudent to initiate this screening in childhood.
- *Renal*. Unless patients develop symptoms that lead to a need for earlier renal imaging, we recommend surveillance for renal tumors beginning at age 40 and if normal, repeated once every 2 years.
- *Endometrial*. Women with PHTS should be educated regarding the signs and symptoms of endometrial cancer and be encouraged to seek attention should abnormal bleeding occur. In conjunction with a gynecologic oncologist, screening with ultrasound or blind endometrial biopsies could be considered as options. After childbearing is complete, prophylactic hysterectomy may be an option for patients, especially those who may be taking tamoxifen for breast cancer treatment or prevention purposes.
- *Colorectal*. Unless patients develop symptoms for which earlier screening is indicated, we recommend that patients with PHTS have a baseline colonoscopy at age 35. Interval to decide upon repeat surveillance should be dictated based on the patient's polyp burden, both in type and number. This decision should be made by a gastrointestinal specialist with expertise in managing patients with polyposis syndromes. Whereas patients with several adenomas may be requested to have their next scope within a year, patients with a few hyperplastic polyps may go without need for repeat scope for 5 years or more, depending on their age. Likewise, decision regarding need for upper endoscopy should also be made by a gastroenterologist based on the patient's symptoms.
- *Dermatologic*. Patients with PHTS should consider seeing a dermatologist yearly to have any lesions present characterized and monitored over time. Unless a skin lesion is concerning for malignancy or highly bothersome to a patient, either physically or emotionally, we recommend against removal. In some patients regrowth of removed lesions has been reported as well as excess scarring during the healing process.
- *Central nervous system*. Routine brain imaging is not currently recommended for asymptomatic patients with PHTS. Patients should be educated about the signs and symptoms of Lhermitte–Duclos disease (seizure, tremor, headaches, balance difficulties) and cerebrovascular malformations (sharp headache, "whooshing" sound or sensation between the ears) and insist on quick medical attention should they appear.

- *Vascular*. Routine imaging for development of PHOSTs or other vascular lesions via whole-body MRI or other techniques is not currently recommended. Patients should be educated regarding the nature of these lesions (may be red/blue/purple in color, appear as a swelling, can appear suddenly or gradually, may or may not be painful) and referred to a tertiary care center with a vascular anomalies clinic, preferably one with demonstrated expertise in treating PHOSTs.

RECENT ADVANCES

One of PTEN's major roles is as an inhibitor of the AKT pathway, which is regulated upstream by phosphoinositide 3-kinase (PI3K) (Stambolic et al., 1998). Constitutive activation of AKT following *PTEN* mutation leads to a cascade of phosphorylation events involving multiple proteins, including mammalian target of rapamycin (mTOR) (Podsypanina et al., 2001). Drugs targeting the PI3K/AKT pathway in patients with germline *PTEN* mutations should theoretically lead to decreased tumor risks; such drugs are already used widely in clinical care and research trials (Kurtz and Ray-Coquard, 2012). The US National Institutes of Health conducted a clinical phase II open label trial utilizing the mTOR inhibitor sirolimus in a small group of patients; results from this study have yet to be published. Other mTOR inhibitors have been used with some success in patients with tuberous sclerosis complex and Peutz–Jeghers syndromes, conditions demonstrating overlapping clinical features with PHTS also caused by dysregulation of the AKT pathway (Crino et al., 2006; Wei et al., 2008; Krueger et al., 2013).

References

Bae BG, Kim HJ, Lee SG et al. (2011). A novel PTEN mutation in a Korean patient with Cowden syndrome and vascular anomalies. Acta Derm Venereol 91: 88–90.

Bannayan GA (1971). Lipomatosis, angiomatosis, and macrencephalia. A previously undescribed congenital syndrome. Arch Pathol 92: 1–5.

Bubien V, Bonnet F, Brouste V et al. (2013). High cumulative risks of cancer in patients with PTEN hamartoma tumour syndrome. J Med Genet 50: 255–263.

Busch RM, Chapin JS, Mester J et al. (2013). Cognitive characteristics of PTEN hamartoma tumor syndromes. Genet Med 15: 548–553.

Butler MG, Dasouki MJ, Zhou XP et al. (2005). Subset of individuals with autism spectrum disorders and extreme macrocephaly associated with germline PTEN tumour suppressor gene mutations. J Med Genet 42: 318–321.

Buxbaum JD, Cai G, Chaste P et al. (2007). Mutation screening of the PTEN gene in patients with autism spectrum disorders and macrocephaly. Am J Med Genet B Neuropsychiatr Genet 144B: 484–491.

Crino PB, Nathanson KL, Henske EP (2006). The tuberous sclerosis complex. N Engl J Med 355: 1345–1356.

Datta SR, Brunet A, Greenberg ME (1999). Cellular survival: a play in three Akts. Genes Dev 13: 2905–2927.

Di Cristofano A, Pesce B, Cordon-Cardo C et al. (1998). Pten is essential for embryonic development and tumour suppression. Nat Genet 19: 348–355.

Eng C, Murday V, Seal S et al. (1994). Cowden syndrome and Lhermitte–Duclos disease in a family: a single genetic syndrome with pleiotropy? J Med Genet 31: 458–461.

Gammon A, Jasperson K, Pilarski R et al. (2013). PTEN mosaicism with features of Cowden syndrome. Clin Genet 84: 593–595.

Greene AK, Orbach DB (2011). Management of arteriovenous malformations. Clin Plastic Surg 38: 95–106.

Halevy S, Sandbank M, Pick AI et al. (1985). Cowden's disease in three siblings: electron-microscope and immunological studies. Acta Derm Venereol 65: 126–131.

Heald B, Mester J, Rybicki L et al. (2010). Frequent gastrointestinal polyps and colorectal adenocarcinomas in a prospective series of PTEN mutation carriers. Gastroenterol 139: 1927–1933.

Krueger DA, Care MM, Agricola K et al. (2013). Everolimus long-term safety and efficacy in subependymal giant cell astrocytoma. Neurology 80: 574–580.

Kurek KC, Howard E, Tennant LB et al. (2012). PTEN hamartoma of soft tissue: a distinctive lesion in PTEN syndromes. Am J Surg Pathol 36: 671–687.

Kurokawa I, Senba Y, Nishimura K et al. (2006). Cytokeratin expression in trichilemmal carcinoma suggests differentiation towards follicular infundibulum. In Vivo 20: 583–585.

Kurtz JE, Ray-Coquard I (2012). PI3 kinase inhibitors in the clinic: an update. Anticancer Res 32: 2463–2470.

Lachlan KL, Lucassen AM, Bunyan D et al. (2007). Cowden syndrome and Bannayan Riley Ruvalcaba syndrome represent one condition with variable expression and age-related penetrance: results of a clinical study of PTEN mutation carriers. J Med Genet 44: 579–585.

Lloyd 2nd KM, Dennis M (1963). Cowden's disease. A possible new symptom complex with multiple system involvement. Ann Intern Med 58: 136–142.

Lok C, Viseux V, Avril MF et al. (2005). Brain magnetic resonance imaging in patients with Cowden syndrome. Medicine (Baltimore) 84: 129–136.

Manning M, Hudgins L (2010). Array-based technology and recommendations for utilization in medical genetics practice for detection of chromosomal abnormalities. Genet Med 12: 742–745.

Marsh DJ, Coulon V, Lunetta KL et al. (1998). Mutation spectrum and genotype-phenotype analyses in Cowden disease and Bannayan–Zonana syndrome, two hamartoma syndromes with germline PTEN mutation. Hum Mol Genet 7: 507–515.

McBride KL, Varga EA, Pastore MT et al. (2010). Confirmation study of PTEN mutations among individuals with autism or developmental delays/mental retardation and macrocephaly. Autism Res 3: 137–141.

Mester J, Eng C (2012). Estimate of de novo mutation frequency in probands with PTEN hamartoma tumor syndrome. Genet Med 14: 819–822.

Mester J, Eng C (2013). When overgrowth bumps into cancer: the PTEN-opathies. Am J Med Genet C Semin Med Genet 163C: 114–121.

Mester JL, Tilot AK, Rybicki LA et al. (2011). Analysis of prevalence and degree of macrocephaly in patients with germline PTEN mutations and of brain weight in Pten knock-in murine model. Eur J Hum Genet 19: 763–768.

Mester JL, Zhou M, Prescott N et al. (2012). Papillary renal cell carcinoma is associated with PTEN hamartoma tumor syndrome. Urology 79 (1187): e1181–e1187.

Milas M, Mester J, Metzger R et al. (2012). Should patients with Cowden syndrome undergo prophylactic thyroidectomy? Surgery 152: 1201–1210.

Moon K, Ducruet AF, Crowley RW et al. (2013). Complex dural arteriovenous fistula in Bannayan–Riley–Ruvalcaba syndrome. J Neurosurg Pediatrics 12: 87–92.

Myers MP, Stolarov JP, Eng C et al. (1997). P-TEN, the tumor suppressor from human chromosome 10q23, is a dual-specificity phosphatase. Proc Natl Acad Sci U S A 94: 9052–9057.

Nelen MR, Kremer H, Konings IB et al. (1999). Novel PTEN mutations in patients with Cowden disease: absence of clear genotype-phenotype correlations. EurJ Hum Genet 7: 267–273.

Ngeow J, Mester J, Rybicki LA et al. (2011). Incidence and clinical characteristics of thyroid cancer in prospective series of individuals with Cowden and Cowden-like syndrome characterized by germline PTEN, SDH, or KLLN alterations. J Clin Endocrinol Metab 96: E2063–E2071.

Ngeow J, He X, Mester JL et al. (2012). Utility of PTEN protein dosage in predicting for underlying germline PTEN mutations among patients presenting with thyroid cancer and Cowden-like phenotypes. J Clin Endocrinol Metab 97: E2320–E2327.

Ngeow J, Heald B, Rybicki LA et al. (2013). Prevalence of germline PTEN, BMPR1A, SMAD4, STK11, and ENG mutations in patients with moderate-load colorectal polyps. Gastroenterol 144: 1402–1409. 1409 e1401–1409 e1405.

Nieuwenhuis MH, Kets CM, Murphy-Ryan M et al. (2012). Is colorectal surveillance indicated in patients with PTEN mutations? Colorectal Dis 14: e562–e566.

Nieuwenhuis MH, Kets CM, Murphy-Ryan M et al. (2014). Cancer risk and genotype-phenotype correlations in PTEN hamartoma tumor syndrome. Fam Cancer 13: 57–63.

Padberg GW, Schot JD, Vielvoye GJ et al. (1991). Lhermitte–Duclos disease and Cowden disease: a single phakomatosis. Ann Neurol 29: 517–523.

Podsypanina K, Ellenson LH, Nemes A et al. (1999). Mutation of Pten/Mmac1 in mice causes neoplasia in multiple organ systems. Proc Natl Acad Sci U S A 96: 1563–1568.

Podsypanina K, Lee RT, Politis C et al. (2001). An inhibitor of mTOR reduces neoplasia and normalizes p70/S6 kinase activity in Pten+/− mice. Proc Natl Acad Sci U S A 98: 10320–10325.

Pritchard CC, Smith C, Marushchak T et al. (2013). A mosaic PTEN mutation causing Cowden syndrome identified by deep sequencing. Genet Med 15: 1004–1007.

Riley Jr HD, Smith WR (1960). Macrocephaly, pseudopapilledema and multiple hemangiomata: a previously undescribed heredofamilial syndrome. Pediatrics 26: 293–300.

Schrager CA, Schneider D, Gruener AC et al. (1998). Clinical and pathological features of breast disease in Cowden's syndrome: an underrecognized syndrome with an increased risk of breast cancer. Hum Pathol 29: 47–53.

Shen-Li H, Koujak S, Szablocs M et al. (2010). Reduction of Pten dose leads to neoplastic development in multiple organs of Pten (shRNA) mice. Cancer Biol Ther 10: 1194–1200.

Smith JR, Marqusee E, Webb S et al. (2011). Thyroid nodules and cancer in children with PTEN hamartoma tumor syndrome. J Clin Endocrinol Metab 96: 34–37.

Srinivasa RN, Burrows PE (2006). Dural arteriovenous malformation in a child with Bannayan–Riley–Ruvalcaba syndrome. Am J Neuroradiol 27: 1927–1929.

Stambolic V, Suzuki A, de la Pompa JL et al. (1998). Negative regulation of PKB/Akt-dependent cell survival by the tumor suppressor PTEN. Cell 95: 29–39.

Starink TM, Hausman R (1984a). The cutaneous pathology of extrafacial lesions in Cowden's disease. J Cutan Pathol 11: 338–344.

Starink TM, Hausman R (1984b). The cutaneous pathology of facial lesions in Cowden's disease. J Cutan Pathol 11: 331–337.

Starink TM, Meijer CJ, Brownstein MH (1985). The cutaneous pathology of Cowden's disease: new findings. J Cutan Pathol 12: 83–93.

Starink TM, van der Veen JP, Arwert F et al. (1986). The Cowden syndrome: a clinical and genetic study in 21 patients. Clin Genet 29: 222–233.

Tan WH, Baris HN, Burrows PE et al. (2007). The spectrum of vascular anomalies in patients with PTEN mutations: implications for diagnosis and management. J Med Genet 44: 594–602.

Tan MH, Mester J, Peterson C et al. (2011). A clinical scoring system for selection of patients for PTEN mutation testing is proposed on the basis of a prospective study of 3042 probands. Am J Hum Genet 88: 42–56.

Tan MH, Mester JL, Ngeow J et al. (2012). Lifetime cancer risks in individuals with germline PTEN mutations. Clin Cancer Res 18: 400–407.

Vanderver A, Tonduti D, Kahn I et al. (2014). Characteristic brain magnetic resonance imaging pattern in patients with macrocephaly and PTEN mutations. Am J Med Genet A 164A: 627–633.

Varga EA, Pastore M, Prior T et al. (2009). The prevalence of PTEN mutations in a clinical pediatric cohort with autism spectrum disorders, developmental delay, and macrocephaly. Genet Med 11: 111–117.

Wei C, Amos CI, Zhang N et al. (2008). Suppression of Peutz–Jeghers polyposis by targeting mammalian target of rapamycin signaling. Clin Cancer Res 14: 1167–1171.

Weng LP, Smith WM, Dahia PL et al. (1999). PTEN suppresses breast cancer cell growth by phosphatase activity-dependent G1 arrest followed by cell death. Cancer Res 59: 5808–5814.

Wu X, Senechal K, Neshat MS et al. (1998). The PTEN/MMAC1 tumor suppressor phosphatase functions as a negative regulator of the phosphoinositide 3-kinase/Akt pathway. Proc Natl Acad Sci U S A 95: 15587–15591.

Zbuk KM, Eng C (2007). Cancer phenomics: RET and PTEN as illustrative models. Nat Rev Cancer 7: 35–45.

Zhou XP, Marsh DJ, Hampel H et al. (2000). Germline and germline mosaic PTEN mutations associated with a Proteus-like syndrome of hemihypertrophy, lower limb asymmetry, arteriovenous malformations and lipomatosis. Hum Mol Genet 9: 765–768.

Zhou XP, Marsh DJ, Morrison CD et al. (2003). Germline inactivation of PTEN and dysregulation of the phosphoinositol-3-kinase/Akt pathway cause human Lhermitte–Duclos disease in adults. Am J Hum Genet 73: 1191–1198.

Zonana J, Rimoin DL, Davis DC (1976). Macrocephaly with multiple lipomas and hemangiomas. J Pediatr 89: 600–603.

Handbook of Clinical Neurology, Vol. 132 (3rd series)
Neurocutaneous Syndromes
M.P. Islam and E.S. Roach, Editors

Chapter 10

Von Hippel–Lindau disease

PRASHANT CHITTIBOINA[1]* AND RUSSELL R. LONSER[1,2]

[1]*Surgical Neurology Branch, National Institute of Neurological Disorders and Stroke, National Institutes of Health, Bethesda, MD, USA*

[2]*Department of Neurological Surgery, Ohio State University Wexner Medical Center, Columbus, OH, USA*

INTRODUCTION

Von Hippel–Lindau disease (VHL) is an inherited condition caused by genetic mutations of the *VHL* gene on chromosome 3 (Kaelin, 2002). Although the *de novo* manifestation of VHL has been documented in up to 20% of newly diagnosed cases, VHL typically presents as an inherited disease due to a germline mutation (Sgambati et al., 2000; Evans et al., 2010). VHL is transmitted through generations in an autosomal dominant fashion with near complete penetrance (Maher et al., 1991). VHL is characterized by formation of multiple benign and malignant tumors, as well as cysts in multiple organs (Fig. 10.1). Affected individuals frequently develop retinal and central nervous system hemangioblastomas, clear cell renal cell carcinomas (RCC), pheochromocytomas, pancreatic neuroendocrine tumors, and endolymphatic sac tumors (ELSTs) (Maher et al., 1990b; Lonser et al., 2003a).

HISTORY AND CLINICAL DESCRIPTION

History

Treacher Collins (1894) first described familial retinal hemangioblastomas in pathologic specimens from three enucleations from two siblings (Melmon and Rosen, 1964). Separately, Eugene von Hippel studied the clinical appearance and progression of retinal lesions (now known as hemangioblastomas) (Hippel et al., 1904). Later, with pathologic confirmation from an enucleated specimen, he cited Collins and Czemark, and named the lesions *angiomatosis retinae.* Brandt reported the autopsy performed on the same patient 4 years later (Brandt, 1921). Visceral features included renal cysts and tumors as well as epididymal cysts (Melmon and Rosen, 1964; Lamiell et al., 1989).

It was Arvid Lindau who first proposed the term *central nervous system angiomatosis*, based on a review of clinical and pathologic features of 16 of his patients and 24 patients from literature (Lindau, 1927). His patients had developed cysts in the cerebellum that were associated with mural nodules. He found that the cysts and tumors were found in medulla and spinal cord in association with retinal hemangioblastomas and visceral tumors. Since his publication, authors have shortened the descriptive term to the eponymous *von Hippel–Lindau disease* in acknowledgement of the two authors' descriptions: retinal hemangioblastomas (von Hippel tumors) and cerebellar hemangioblastomas (Lindau's tumors) (Melmon and Rosen, 1964).

Diagnostic criteria

Melmon and Rosen (1964) were the first to propose clinical criteria for the clinical diagnosis of VHL. They recognized cerebellar hemangioblastomas as the hallmark *Lindau's tumor*. Lindau's disease (now know as VHL) was defined as the association of a cerebellar hemangioblastoma with one of the following: retinal hemangioblastoma, pancreatic cyst, renal or epididymal abnormalities. They recognized the inherited nature of the disease, as well as the variable penetrance and delayed expression. They proposed expanding the diagnostic criteria to include patients with only one of the lesions of the complex in the presence of central nervous system hemangioblastomas in a family member. The definition for clinical diagnosis of VHL was later refined by Lamiell and others (Lamiell et al., 1989; Lonser et al., 2003a) to include an

*Correspondence to: Prashant Chittiboina, MD, MPH, Staff Clinician, Surgical Neurology Branch, National Institute of Neurological Disorders and Stroke, National Institutes of Health, Bethesda, MD 20892, USA. Tel: +1-301-496-5728, Fax: +1-301-402-0380, E-mail: prashant.chittiboina@nih.gov

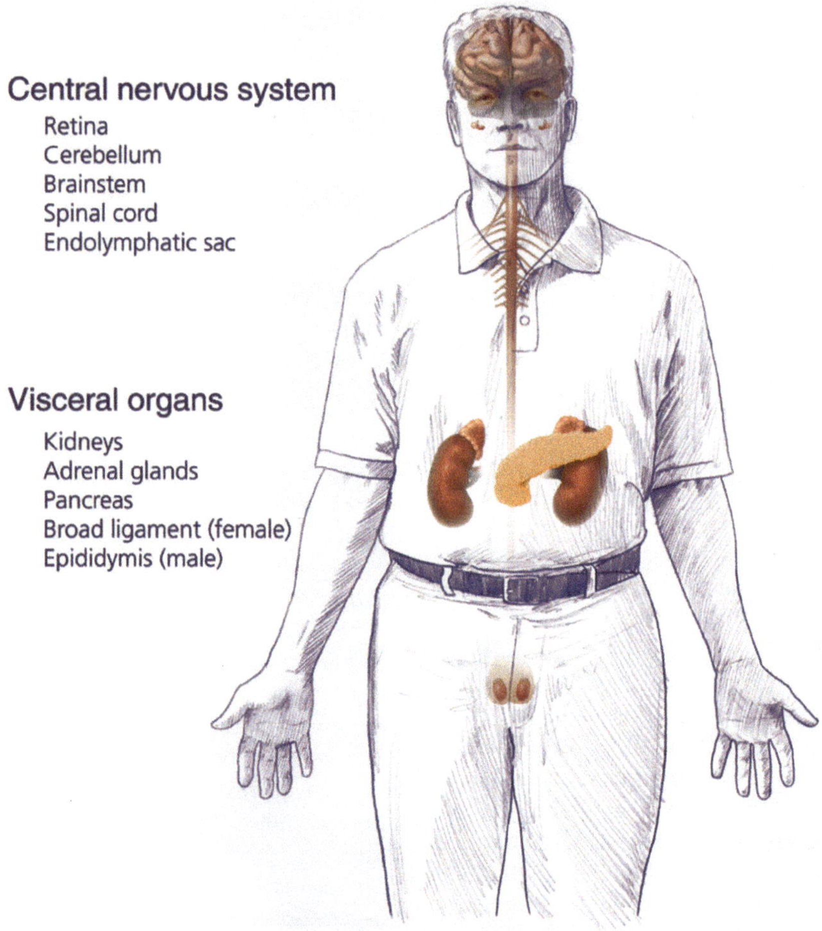

Fig. 10.1. Organs affected by von Hippel–Lindau disease.

expanded list of associated visceral tumors. Contemporary diagnostic criteria for VHL often include both genetic and clinical findings (Friedrich, 1999; Binderup et al., 2013; Lonser et al., 2014).

Currently, the following manifestations are included in clinical diagnostic critera (Lonser et al., 2003a; Butman et al., 2007; Maher et al., 2011; Binderup et al., 2013):

1. central nervous system hemangioblastoma (including retinal hemangioblastoma)
2. endolymphatic sac tumors (ELST) (Manski et al., 1997)
3. renal cell carcinoma (RCC)
4. pheochromocytoma, paraganglioma, and/or glomus tumor
5. neuroendocrine neoplasm and/or multiple cysts of the pancreas (Hammel et al., 2000).

The diagnosis of VHL is made if an individual has the following combination of manifestations of genetic/ family factors:

1. at least two central nervous system hemangioblastomas
2. at least one central nervous system hemangioblastoma and one other manifestation described above
3. at least one of the manifestations described above, and a pathogenic mutation in *VHL* gene or a first-degree relative with VHL.

Using advances in genetic testing, including DNA sequencing and semiquantitave Southern blotting, a *VHL* mutation can now be identified in nearly 100% patients (Stolle et al., 1998). Diagnosis of VHL is particularly challenging in apparent *de novo* appearance without a family history. In these cases, mosaicism in peripheral blood leukocytes that do not carry the VHL mutation could result in a false negative genetic test. Careful clinical testing and mutation analysis in other tissues or VHL manifestations may reveal the presence of VHL in these instances (Sgambati et al., 2000; Nordstrom-O'Brien et al., 2010).

EPIDEMIOLOGY

The incidence of VHL ranges from 1 in 36 000 to 1 in 45 000 live births (Maher et al., 1991; Maddock et al.,

1996; Evans et al., 2010). VHL has a point prevalence of 1 in 38000 to 1 in 91 000 persons (Maher et al., 1991; Neumann and Wiestler, 1991; Maddock et al., 1996; Evans et al., 2010). Typically, VHL first manifests in the second decade of life (Table 10.1). At the time of detection, 50% of patients are symptomatic (Poulsen et al., 2010). Cerebellar hemangioblastomas (35%) are the most common presenting manifestation of VHL (Maddock et al., 1996). Penetrance of VHL is almost complete by 70 years of age (Table 10.2) (Maher et al., 1991; Poulsen et al., 2010). With improved imaging and other diagnostic tools, first manifestations of VHL are diagnosed sooner. Analysis of 82 Dutch VHL patients suggests that over 95% of patients will have a first manifestation by 34 years of age (Kruizinga et al., 2014). Recently, researchers have reported evidence of progressively earlier age of onset and more severe presentation in successive generations. The phenomenon is known as genetic anticipation and may be due to shortening of telomere length (Ning et al., 2014).

Life expectancy for VHL patients historically has been between 40 and 52 years (Maher et al., 1990b; Akcaglar et al., 2008; Wilding et al., 2012). Currently, unlike the general population, male life expectancy (59.4 years) is significantly higher then female (48.4 years) (Wilding et al., 2012). Despite understanding and treatment advances, life expectancy in VHL remains the lowest among common inherited tumor syndromes (Wilding et al., 2012). Early reports established that the major cause of mortality could be either central nervous

Table 10.1

Reported mean age of onset and frequency of lesions

	Mean (range) of onset in years	Frequency in patients (%)
Central nervous system lesions		
Retinal hemangioblastoma (Ong et al., 2007; Webster et al., 1998; Wong et al., 2008a)	25–37 (9–84)	15–73%
Endolymphatic sac tumor (Butman et al., 2007; Choo et al., 2004; Manski et al., 1997)	22–40 (11–63)	3–16%
Cerebellar hemangioblastoma (Maher et al., 1990a, b; Ong et al., 2007; Poulsen et al., 2010)	29–30 (13–61)	35–79%
Brainstem hemangioblastoma (Filling-Katz et al., 1991; Poulsen et al., 2010; Wanebo et al., 2003)	25–38 (16–60)	4–22%
Spinal hemangioblastoma (Maher et al., 1990b; Ong et al., 2007; Poulsen et al., 2010)	33–34 (8–60)	7–53%
Supratentorial hemangioblastoma (Maddock et al., 1996; Peyre et al., 2010; Poulsen et al., 2010)	20–29 (14–48)	1–7%
Visceral lesions		
Renal cell carcinoma (Maher et al., 1990a, b; Ong et al., 2007)	40–45 (20–69)	30–70%
Renal cyst (Maddock et al., 1996; Poulsen et al., 2010)	34–39 (12–64)	60%
Pheochromocytoma (Binderup et al., 2013; Lonser et al., 2003a; Walther et al., 1999)	20–29 (5–62)	16%
Pancreatic neuroendocrine tumor (Charlesworth et al., 2012; Hough et al., 1994; Igarashi et al., 2014)	32–38 (16–68)	15–56%
Pancreatic cyst (Charlesworth et al., 2012; Hough et al., 1994; Igarashi et al., 2014)	29–37 (12–63)	21–72%
Epididymal cyst adenoma (Gruber et al., 1980; Lonser et al., 2003a)	24 (10–37)	25%

Table 10.2

Cumulative frequency of first VHL manifestations by age

Age	Maher et al., 1990b	Poulsen et al., 2010	Kruizinga et al., 2014
5	0.02	0	
15	0.19	0.31	
25	0.52	0.62	
35	0.78	0.70	0.95
45	0.91	0.95	
55	0.96	0.98	
65	0.99		
75+	1.00		

system (CNS) hemangioblastomas (Lamiell et al., 1989; Maddock et al., 1996) or RCC (Maher et al., 1990b). However, in VHL patients that manifest with CNS hemangioblastoma, these tumors remain the primary cause of death (Lonser et al., 2014). Death in VHL patients is more likely (73%) due to VHL-related causes than unrelated causes (Wilding et al., 2012).

MOLECULAR GENETICS

The *VHL* gene

Linkage studies indicate that the *VHL gene* is a tumor suppressor gene on chromosome 3p25-26. Genetic alterations leading to loss of function of this gene were observed in both VHL kindreds and in sporadic RCC patients (Seizinger et al., 1988). The *VHL* gene was isolated and sequenced in 1993 (Latif et al., 1993). It consists of three exons and encodes the VHL protein, a glycan-anchored membrane protein responsible for signal transduction (Latif et al., 1993). The gene is highly conserved across species (Woodward et al., 2000). VHL conforms to the Knudson "*two-hit*" model of hereditary cancer (Knudson and Strong, 1972; Knudson, 1986; Kaelin, 2002). A germline VHL mutation results in a defective allele in all cells of the body (first hit). However, a second, somatic event is required for tumor formation (second hit or loss of heterozygosity) (Knudson, 1986).

Hemangioblastomas and RCC in VHL patients follow a single-stage somatic mutation model (Maher et al., 1990a). Indeed, the second hit hypothesis is supported by studies showing somatic VHL gene mutations (two-stage) in about 20% of sporadic hemangioblastomas (Kanno et al., 1994; Tse et al., 1997; Lee et al., 1998) and up to 90% of sporadic clear cell RCC (Gnarra et al., 1994; Gallou et al., 1999; Kondo et al., 2002b). Nevertheless, although some VHL mutations can lead to increased incidence of pheochromocytomas, *VHL* gene mutations are rare (3%) in sporadic pheochromocytomas (Gnarra et al., 1994; Brauch et al., 1997).

The VHL protein

The *VHL* gene encodes the VHL protein (pVHL), which is composed of 213 amino acid residues and has a molecular weight of 24–30 kDa. A second, smaller isoform with a molecular weight of 19 kDa can be found in many tissues. This form of pVHL arises as a result of alternate translation initiation from codon 54 (Iliopoulos et al., 1998). VHL occurs when inactivating mutations occur between codon 54 and the carboxy terminal, affecting both isoforms of protein. pVHL is found in both nuclear and cytoplasmic compartments. Trafficking between these two compartments is essential for pVHL function (Lee et al., 1999).

The best characterized function of pVHL is the targeting of hypoxia inducible factor (HIF) for proteasomal degradation (Li and Kim, 2011). HIF is a heterodimeric transcription factor consisting of an unstable alpha (α) and a stable beta (β) subunit. There are three known human HIF-α genes. The isoforms are not completely redundant (Wang et al., 2005). In some conditions, HIF-1α and HIF-2α may be mutually antagonistic (Yuan et al., 2013). HIF-3α appears to be involved in a feedback mechanism controlling HIF-1α activity (Mandriota et al., 2002). pVHL binds elongin C, which in turns binds elongin B and CUL2 to form the VCB-CUL2 complex.

In normoxic conditions, VCB-CUL2 complex binds HIF-α and polyubiquitinates it, targeting it for proteasomal degradation (Fig. 10.2). HIF-α undergoes hydroxylation under normoxic conditions within the oxygen-dependent degradation domain (ODD). Hydroxylation allows binding of pVHL and VCB-CUL2 complex, leading to ubiquitination and proteasomal degradation. Under hypoxic conditions, HIF-α binds cyclic AMP response element binding protein and transcriptional coactivator p300 to activate transcription of HIF-regulated genes. When pVHL is inactivated, tumorigenesis may occur through HIF-mediated effects or through direct effects of pVHL (Kaelin, 2002; Lonser et al., 2003a; Maher et al., 2011).

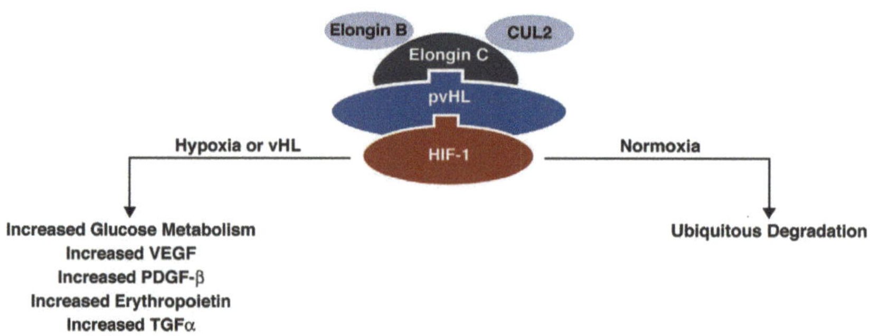

Fig. 10.2. VCB-CUL2 complex is formed with binding of VHL protein to elongin C, elongin B and CUL2. The complex then binds HIF and leads to degradation in normoxic conditions. (Adapted from Lonser et al., 2003a.)

Tumorigenesis in VHL

The pVHL/HIF pathway has been implicated in tumorigenesis for RCC (Krieg et al., 2000), retinal hemangioblastomas (Wang et al., 2014), CNS (Vortmeyer et al., 2006) hemangioblastomas, and neuroendocrine tumors of the pancreas (PNETs) (Périgny et al., 2009). pVHL inactivation leads to increased expression and stabilization of HIF proteins in VHL-related tumors (Krieg et al., 2000; Wang et al., 2014). Specifically, HIF-2α is necessary and sufficient for RCC (Kondo et al., 2002a, 2003; Zimmer et al., 2004). HIF-2α also may be required for CNS (Shively et al., 2008) and retinal (Wang et al., 2014) hemangioblastoma formation and growth. HIF can lead to tumorigenesis by differential upregulation of over 50 genes, including genes for growth factors (TGF, PDGF), angiogenesis (VEGF), and glucose metabolism (GLUT1, PFK1). Moreover, HIF can lead to activation of genes that affect tumor microenvironment (LOX, MMP1) (Lee et al., 1998; Wang et al., 2005). Loss of pVHL activity may lead to tumorigenesis via HIF-independent pathways as well. pVHL has been implicated in regulation of apoptosis (p53 inactivation and increased NF-κB activity), stabilization of microtubules and regulation of extracellular matrix (Li and Kim, 2011).

Mechanism of *VHL* gene inactivation

Most patients with VHL inherit a germline mutation from an affected parent and a wild type allele from the unaffected parent (Maddock et al., 1996). The most common germline mutations are missense mutations (27–38%) and nonsense mutations (13–27%). Large (9–20%) or microdeletions (10%), as well as rearrangements (25%), account for a majority of the remaining germline mutations. Insertions or splice-site mutations are rare (Stolle et al., 1998; Friedrich, 1999). Somatic mutation events leading to homozygous inactivation of the *VHL alleles* are tumorigenic. These "second hit" events are frequently allelic loss (49%) or hypermethylation (35%) of the wild type allele (Prowse et al., 1997). Somatic inactivation of the wild type allele may also occur due to point mutations, especially in instances of large deletion of 3p as a "first hit" (Wait et al., 2004).

Genotype–phenotype correlation

Subclassification of VHL into types 1, 2A, 2B or 2C was based initially on clinical phenotypic manifestations. Early kindred studies reported familial clustering of manifestations in VHL carriers. High frequency of retinal hemangioblastomas (51%), CNS hemangioblastomas (46%), and renal cell cancers (33%) were found in all kindreds. Alternatively, pheochromocytomas were seen in only 20% of kindreds. No members of other kindreds manifested pheochromocytomas, pointing to distinct subsets of patients with VHL. Specifically, type 1 kindreds are without pheochromocytomas and type 2 have pheochromocytomas (Table 10.3) (Lamiell et al., 1989; Neumann and Wiestler, 1991).

Type 1 disease is likely caused by mutations leading to severe disruption of protein activity, including deletions, nonsense mutations, and other microdeletions/insertions (Chen et al., 1995; Crossey et al., 1995). Missense mutations that lead to structural changes in pVHL also give rise to the type 1 VHL phenotype (Ong et al., 2007). Further, subclassifications have been suggested based on observations that large *VHL gene* (exon 1 only or exon

Table 10.3

Recognized subtypes of VHL disease

		Phenotype		Genotype	
		Frequency of pheochromocytoma	Frequency of RCC	Common features of mutations	Examples of commonly associated mutations
Type 1	A	Low	High	Germline deletions, truncating mutations, and missense mutations affecting structural integrity of pVHL	
	B	Low	Low		Large deletions involving C3orf10
Type 2	A	High	Low	Surface missense mutations	"Black Forest" founder mutation: p.Tyr169His
	B	High	High		Missense at codon 167: p.Arg167Trp and p. ArgGln
	C	High*			

*Type 2C manifests pheochromocytomas in absence of other tumors.

1,2,3) deletions that include the neighboring C3orf10 gene may lead to lower incidence of RCC. Concurrent VHL and C3orf10 inactivation in RCC is hypothesized to lead to a growth disadvantage for tumor cells. This in turn, may lead to a significantly lower lifetime risk of RCC (type 1B) (McNeill et al., 2009).

Type 2 disease nearly always (78–96%) is associated with missense mutations (Chen et al., 1995; Zbar et al., 1996). The missense mutations leading to type 2 disease result in a substitution of an amino acid on the surface of the protein and may predict a high lifetime risk of pheochromocytoma (Ong et al., 2007). VHL type 2 families from the Black Forest region of Germany all shared a low risk of RCC (hence designated type 2A). These 16 families were discovered to share a distinct missense mutation (p.Tyr169His) of the *VHL* gene with a founder effect confirmed on haplotype analysis (Brauch et al., 1995). Missense mutations of codon 167 (e.g., p. Arg167Gln or p.Arg167Trp) can lead to a VHL type 2B phenotype with a high risk of RCC (60% at 60 years) and pheochromocytomas (82% at 50 years) (Maher et al., 1996; Zbar et al., 1996). Rare kindreds with only pheochromocytomas were found to have specific missense mutations at VHL codons 238, 259 and other rare point mutations. Originally diagnosed with multiple endocrine neoplasia (MEN) type 2, the kindreds were reclassified as VHL type 2C (Crossey et al., 1995; Neumann et al., 1995; Woodward et al., 1997).

Genetic testing

Genetic testing is performed when VHL is suspected because of clinical presentation in a patient or a relative (Binderup et al., 2013). After obtaining a medical history and a family history, a clinical examination should be performed prior to mutation testing. Peripheral blood samples are obtained for mutation testing. To identify cases of mosaicism, peripheral blood from the affected parent is preferred. An analysis for duplication and deletions is performed by using Southern blotting to detect whole gene deletions, Southern blotting to detect gene rearrangements, and fluorescence *in situ* hybridization (FISH) to confirm deletions (Stolle et al., 1998). Alternatively, multiplex ligation-dependent probe amplification (MLPA) may be used to detect deletions and insertions (Binderup et al., 2013). Sequencing of the *VHL* gene is performed including the coding exons and intron–exon boundaries. Sequence variations are compared with the 377 known intragenic mutations associated with VHL disease (see Resources; Schmidt and Linehan, 2014). When mutation screening fails to identify a mutation in the peripheral blood, screening of DNA from other tissues and/or from VHL manifestations may be necessary to confirm diagnosis (Binderup et al., 2013).

CENTRAL NERVOUS SYSTEM LESIONS

Central nervous system hemangioblastomas

General features

CNS hemangioblastomas are among the most common manifestations in VHL disease. Up to 72% of VHL patients may present with a CNS hemangioblastoma in the cerebellum (16–69%), brainstem (5–22%), spinal cord (13–53%), cauda equina (11%), or supratentorial (1–7%) locations (Filling-Katz et al., 1991; Poulsen et al., 2010; Binderup et al., 2013; Huntoon and Lonser, 2014; Lonser et al., 2014). Patients present with these lesions in their second or third decade. CNS hemangioblastomas are the most common (30–35%) and often the first manifestation of VHL (Maddock et al., 1996; Poulsen et al., 2010). Although benign, hemangioblastomas are a significant source of morbidity and mortality in VHL patients due to mass effect on CNS structures (Wanebo et al., 2003; Lonser et al., 2014).

Clinical, imaging, and histologic findings

Mass effect occurs due to tumor growth, edema formation and/or cyst formation. When observed for longer than 5 years, over 50% of hemangioblastomas will grow. Tumor growth progresses commonly (72%) in a saltatory fashion with periods of quiescence, but tumors also can show linear (6%) or exponential (22%) growth patterns. A patient may harbor various tumors that demonstrate different rates and patterns of growth (Wanebo et al., 2003; Wind and Lonser, 2011; Lonser et al., 2014). Symptoms arise from mass effect on neighboring neural structures (Table 10.4) (Lonser et al., 2014). High tumor burden is associated with male sex and partial germline deletion. New tumors form more frequently in younger patients and those with a higher tumor burden at initial visit (Lonser et al., 2014).

Imaging. Contrast-enhanced magnetic resonance (MR) imaging is the gold standard for detecting and monitoring CNS hemangioblastomas (Filling-Katz et al., 1989; Choyke et al., 1995). Hemangioblastomas as small as 2 mm can be visualized as discrete enhancing lesions on contrast-enhanced T1-weighted images (Fig. 10.3). Fluid-attenuated inversion recovery (FLAIR) and T2-weighted images are used to detect hemangioblastoma-associated edema or cyst formation (Butman et al., 2008). Arteriography can be used to assess vascular supply and drainage for large tumors (Huson et al., 1986; Neumann et al., 1989; Peyre et al., 2010; Wind et al., 2011; Lonser et al., 2014).

Gross appearance and histology. During surgical resection, hemangioblastomas appear as distinct, red, vascular masses with a thin layer of capsule. Histologically, the tumors consist of polygonal stromal cells, and

Table 10.4

VHL surveillance guidelines*

Period	Surveillance test	Frequency	NIH	DCG	VHLFA
0–4 year	Ophthalmoscopy	Annual	+	+	+
0–4 years	Physical exam	Annual	+	+	+
2 years onwards	Plasma or urinary metanephrines	Annual	+	5 years onwards	8 years onwards or earlier if needed
5–15 years	Ophthalmoscopy	Annual	+	+	+
5–15 years	Physical exam	Annual	+	+	+
5–14 years	Hearing exam	Annual	When clinically indicated	+	Audiology assessment once. Annual assessments if any hearing loss, tinnitus or vertigo is found
8 years onwards	Ultrasound of abdomen	Annual	+	Once between 8 and 14 years of age	Annually from 8 years or earlier if clinically indicated
11 years onwards	MRI of craniospinal axis	Annual	+	MRI once between 8 and 14 years of age	
15 years onwards	Ophthalmoscopy	Annual	+	+	+
15 years onwards	Neurological exam	Annual		+	
15 years onwards	MRI of craniospinal axis including inner ears	Every 2 years	Annual	+	+
15 years onwards	Ultrasound or MRI of abdomen	Annual	Ultrasound annually, MRI as needed	+	Every 2 years
18 years onwards	CT abdomen	Annual	+		
15 years onwards	Plasma metanephrines, normetanephrines and chromogranin A	Annual	Plasma or urinary catecholamines annually	+	+
15 years onwards	Hearing exam	Annual	When clinically indicated	+	Every 2 years
Pregnancy	MRI craniospinal axis without contrast	Once at fourth month of pregnancy			+

*+, recommendation is substantially similar to the guideline by the organization.
NIH, US National Institutes of Health; DCG, Danish Coordination Group for VHL; VHLFA, VHL Family Alliance; MRI, magnetic resonance imaging.

numerous vascular structures (see Fig. 10.3), both of which carry the somatic second hit (Zhuang et al., 2014). Larger tumor-associated blood vessels derive from reactive migrating endothelial cells from the surrounding, normal tissues. Multipotent hemangioblasts are the cells of origin for the stromal cells, and *de novo* tumor blood vessels (by vasculogenesis) (Shively et al., 2008, 2011; Zhuang et al., 2014).

Treatment

Microsurgical resection is the treatment of choice for CNS hemangioblastomas. Safe and complete resection of hemangioblastomas can be performed in most cases (Lonser et al., 2003b, 2014; Weil et al., 2003; Peyre et al., 2010; Wind et al., 2011). Circumferential dissection using meticulous microsurgical technique allows resection of these tumors with minimal blood loss. Preoperative embolization to reduce intraoperative bleeding is not routinely used and can be associated with additional risk (Lonser et al., 2003a).

Stereotactic radiation or craniospinal radiation has been used in patients who are not surgical candidates. However, long term follow-up reveals that radiation-treated hemangioblastomas will continue to grow in a proportion similar to that expected by natural history

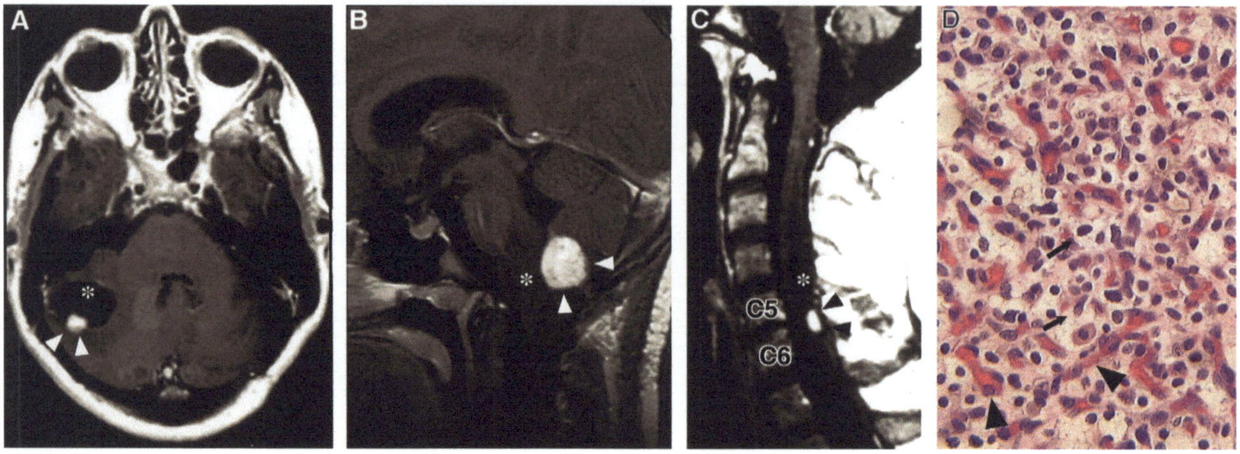

Fig. 10.3. (**A**) Axial T1-weighted postcontrast magnetic resonance image (MRI) of a cerebellar hemangioblastoma (arrowheads) with an associated cyst (asterisk) in a 40-year-old woman. (**B**) Sagittal T1-weighted postcontrast MRI of medullary hemangioblastoma (arrowheads) with associated brainstem edema (asterisk) in a 12-year-old girl. (**C**) Sagittal T1-weighted postcontrast MRI of the cervical spinal cord of a 50-year-old man. Hemangioblastoma (black arrowheads) is located in the dorsal spinal cord at C5 and C6, and is associated with a large syrinx (asterix). (**D**) Hematoxylin and eosin staining of a hemangioblastoma showing the lipid-laden stromal cells (arrows) distributed within a capillary network (arrowheads). (Adapted from Lonser et al., 2003a.)

studies (Chang et al., 1998; Asthagiri et al., 2010; Simone et al., 2011). Given the unpredictable nature of growth patterns in VHL-related hemangioblastomas, prophylactic radiation is not advised except for tumors that are not surgically accessible or that cannot be removed safely (Lonser et al., 2003a, 2014).

Effect of pregnancy

Multiple case reports have described an initial diagnosis of VHL or worsening of clinical manifestations of VHL with pregnancy (Ogasawara et al., 1995; Othmane et al., 1999; Capone et al., 2013). Nevertheless, a prospective natural history study of VHL failed to find any significant differences in either appearance of new hemangioblastomas ($p=0.6$), the rates of hemangioblastoma growth ($p=0.6$), development of new tumors/cysts (0.4 tumors/year; 0.1 ± -0.1 cysts/year) or the rate of growth of associated cysts ($p=0.4$). Since hemangioblastomas in VHL have an unpredictable growth pattern (Wanebo et al., 2003), management in pregnant patients can be guided by onset of symptoms rather than a change in the size of lesions on imaging (Ye et al., 2012).

Retinal hemangioblastomas

General features

Retinal hemangioblastomas occur very frequently in VHL patients. Between 49% and 62% of VHL patients will develop retinal hemangioblastomas (Poulsen et al., 2010; Toy et al., 2012; Wittström et al., 2014). Retinal hemangioblastomas are often the first manifestation of VHL, with the lowest median age of onset (21 years) among the various manifestations of VHL (Kreusel et al., 2006; Poulsen et al., 2010). Most hemangioblastomas (61%) are symptomatic at presentation and all patients with at least one retinal hemangioblastoma will develop bilateral disease by 56 years of age (Kreusel et al., 2006; Toy et al., 2012). The presence of retinal hemangioblastomas is highly suggestive of VHL (Singh et al., 2002; Binderup et al., 2013). Retinal hemangioblastomas can lead to blindness or severe visual deficits in 5–8% of VHL patients (Kreusel et al., 2006; Toy et al., 2012; Wittström et al., 2014).

Clinical findings

Retinal hemangioblastomas may occur at the periphery (50%) or at the optic nerve (50%) (Wittström et al., 2014). An increase in the number of lesions and complications from the retinal hemangioblastomas can lead to blindness in 15% of affected eyes (Wittström et al., 2014). Both peripheral and optic nerve tumors can lead to symptoms due to accumulation of subretinal fluid and formation of hard exudates at the macula (Fig. 10.4) (Lonser et al., 2003a). Lesions located in close proximity to the macula negatively impact visual function (Wong et al., 2008). Younger age at onset of ocular disease, involvement of the fellow eye and missense or protein truncating mutations can lead to worse visual outcomes (Wong et al., 2008a; Toy et al., 2012).

Fundoscopic evaluation

Dilated fundoscopy and slit lamp examinations are performed to analyze ocular manifestations of VHL. Fluorescein angiography is used to confirm the presence of

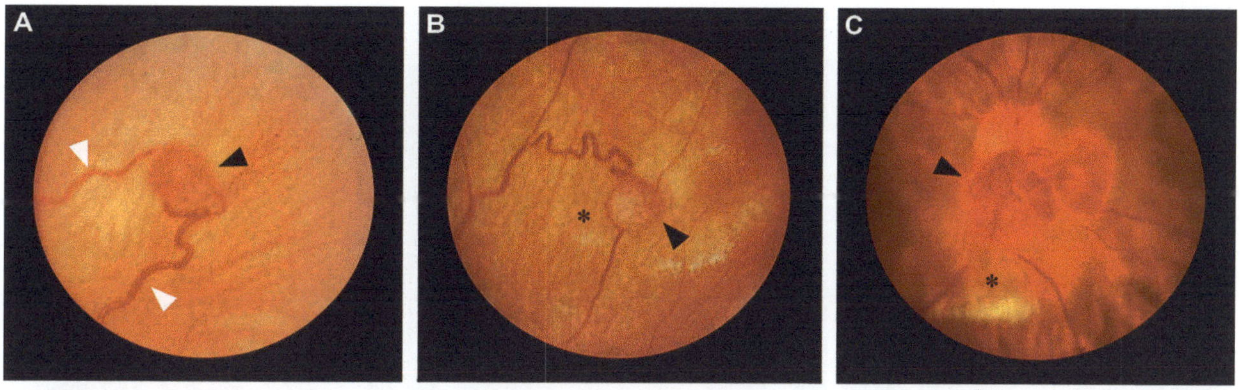

Fig. 10.4. (**A**) Peripheral retinal hemangioblastoma (black arrowhead) with enlarged vessels (white arrowheads) in a 22-year-old woman. (**B**) Peripheral retinal hemangioblastoma (arrowhead) with fibrous changes, and hard exudates and retinal edema (asterisk) in a 24-year-old man. (**C**) Retinal hemangioblastoma (arrowhead) on the optic nerve head with yellow retinal hard exudates (asterisk) in a 32-year-old man. (Adapted from Lonser et al., 2003a.)

newly diagnosed lesions. The patients are closely followed with wide-angle photography to document the lesions and changes due to disease progression (Kreusel et al., 2006; Wong et al., 2008a; Toy et al., 2012).

Histologic findings

Retinal hemangioblastomas are grossly and histologically similar to CNS hemangioblastomas. Retinal gliosis and hemorrhages are present in association with severe lesions. Differential upregulation of HIF-2α has been demonstrated in severe lesions that are refractory to anti-VEGF therapy (Lonser et al., 2003a; Wang et al., 2014).

Treatment

Laser photocoagulation and cryotherapy are the mainstays of surgical management of retinal hemangioblastomas. Either of these methods can be used to as a sole method for treatment of extrapapillary hemangioblastomas (Singh et al., 2002). Vitreoretinal surgery can be performed when preretinal and vitreal membranes, retinal detachment from traction, and exudation occur (Machemer and Williams, 1988; Johnson et al., 1992). For hemangioblastomas in close proximity to the optic nerve, intravitreal anti-VEGF therapy may arrest progression for small lesions and may help reverse exudates and edema in some cases (Wong et al., 2008b; Hrisomalos et al., 2010). Photodynamic therapy or plaque radiotherapy may have a limited role in the management of retinal hemangioblastomas (Kreusel et al., 1998; Lazzeri et al., 2011; Stattin et al., 2014). Salvage external beam radiation may be used for treatment-refractory retinal hemangioblastomas with reduction in tumor volume and improvement in visual acuity (Raja et al., 2004).

Endolymphatic sac tumors

General features

Endolymphatic sac tumors (ELSTs) arise from the endolymphatic epithelium within the vestibular aqueduct (Lonser et al., 2008). These tumors are rare in the general population but are found in up to 6–15% of VHL patients (Manski et al., 1997; Lonser et al., 2003a; Choo et al., 2004; Butman et al., 2007). Bilateral ELSTs are found only in VHL (Lonser et al., 2003a).

Clinical

ELSTs arise within the vestibular aqueduct, and can lead to hearing loss due to intralabrynthine hemorrhage, endolymphatic hydrops, and/or by otic capsule invasion (Butman et al., 2007). ELSTs can present with vestibular symptoms (62%), partial or complete hearing loss (95 to 100%), tinnitus (77%), and/or facial paresis (8%) (Manski et al., 1997).

Imaging

High resolution computed tomography (CT) scans of the temporal bone and pre-/postcontrast enhanced MR imaging of the internal auditory canals are necessary to detect and follow these tumors (Fig. 10.5). CT is used to detect the presence of otic capsule invasion and to evaluate the extent of bony erosion of the temporal bone (particularly, the vestibular aqueduct in very small tumors). Postcontrast T1-weighted MR imaging can reveal signs of small ELST tumors as subtle asymmetric enhancement of the endolymphatic sac. Larger tumors are heterogeneously enhancing on postcontrast T1W images. Intralabrynthine hemorrhage is detected by increased signal on precontrast T1W images (Lonser et al., 2003a; Kim et al., 2005; Butman et al., 2007).

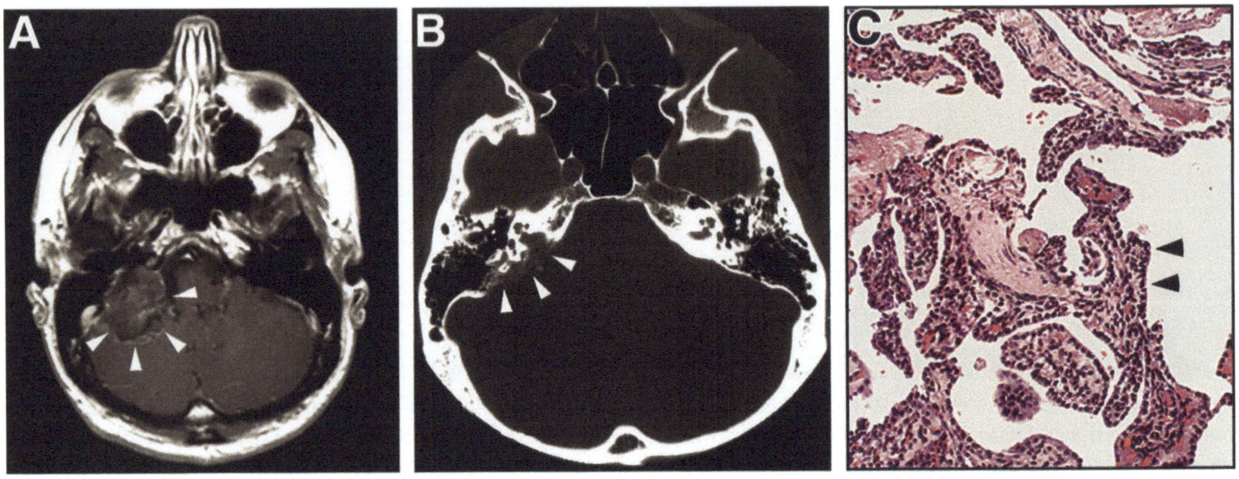

Fig. 10.5. (**A**) Axial T1-weighted postcontrast magnetic resonance image shows large heterogeneously enhancing tumor in the right mastoid region (arrowheads). (**B**) Axial computed tomography through the same region showing the bony erosion of posterior petrous region that often occurs in these tumors (arrowheads). (**C**) Hematoxylin and eosin stained section showing the typical histologic features of neoplasm, including cuboidal epithelium (arrowheads) in a papillary pattern. (Adapted from Lonser et al., 2003a.)

Histologic findings

ELSTs are papillary cystic glandular neoplasms with variegated patterns (see Fig. 10.5). Due to the presence of local bone invasion, these tumors were called low grade adenocarcinoma of probable endolymphatic origin (Heffner, 1989). ELSTs may be found entirely within the endolymphatic sac when small. When large, these inevitably invade and erode the surrounding temporal bone (Lonser et al., 2008).

Treatment

Surgery is the treatment of choice for ELSTs. Complete resection may be achieved in a majority of cases with minimal risk for recurrence (3%) after gross total resection (Kim et al., 2013). Surgical resection of the tumor can relieve audiovestibular symptoms and can preserve hearing in a majority of cases (Kim et al., 2005, 2013). Indications for treatment of ELSTs include progressive sensorineural hearing loss, vestibular symptoms, facial nerve compression, and local mass effect from tumor growth. Early resection of ELSTs is recommended to prevent sensorineural hearing loss and to alleviate vestibular symptoms. Facial nerve decompression (large tumors) performed during surgery can improve facial nerve function in some cases (Kim et al., 2013). The role of radiation in the management of ELST is unproven (Lonser et al., 2003a; Poletti et al., 2013).

VISCERAL LESIONS

Renal cell carcinomas and renal cysts

General features

RCC may affect up to 30% of VHL patients. Multiple renal cysts are often found along with RCC in up to 60% of VHL patients (Fig. 10.6). The patients typically develop RCC and renal cysts bilaterally starting in their third and fourth decades of life (Lonser et al., 2003a; Bausch et al., 2013; Binderup et al., 2013). The frequency of RCC increases with age and up to 70% of VHL patients develop RCC by 60 years of age (Maher et al., 1990b). VHL patients frequently develop lesions that range from simple cysts to entirely solid lesions. Solid components of cystic lesions invariably contain RCC (Choyke et al., 1995). Although simple cysts may form precursor lesions for later development of RCC, such transformation is rare (Neumann and Wiestler, 1991; Choyke et al., 1995). Each kidney may contain up to 1100 cysts and 600 RCC foci (Manski et al., 1997; Wind and Lonser, 2011). Solid masses and complex cysts may grow over time, but, simple cysts tend to remain indolent, or may even involute (Choyke et al., 1992).

Clinical

RCC or renal cysts are rarely (less than 7%) the first manifestation of VHL (Poulsen et al., 2010). When present, large RCCs may present with classic findings of a renal mass with flank pain, hematuria and/or a flank mass. Simple renal cysts tend to be asymptomatic, while complex cysts may progress to solid RCC masses (Choyke et al., 1992; Lonser et al., 2003a; Bausch et al., 2013). VHL patients may maintain renal function despite multiple renal cysts (Bausch et al., 2013).

Imaging

Surveillance imaging with abdominal CT is the most sensitive method to detect renal involvement (see Fig. 10.6). Multidetector CT scanning allows for discrimination of

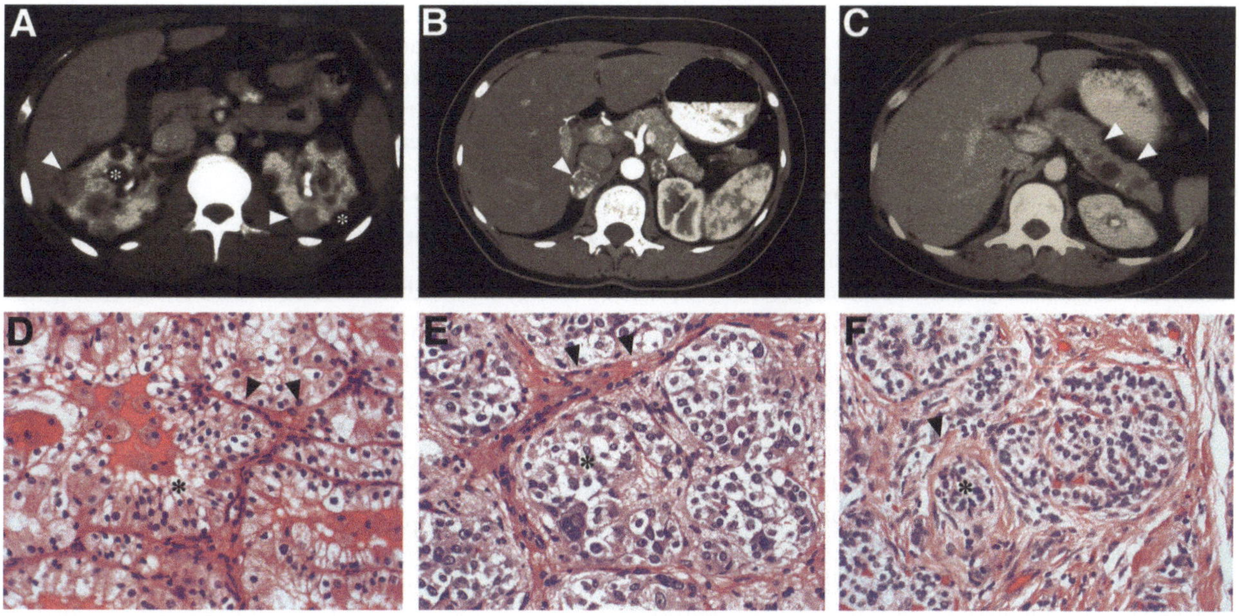

Fig. 10.6. (**A**) Bilateral multifocal renal cell carcinoma with both solid (arrowheads) and cystic (asterisk) disease in 22-year-old man. (**B**) Bilateral pheochromocytomas (arrowheads) with rim enhancement in the adrenal glands of a 29-year-old woman. (**C**) Pancreatic neuroendocrine tumor (arrowheads) in the body of the pancreas of a 26-year-old woman. (**D**) Renal cell carcinoma of the clear cell subtype (asterisk) with acinar and tubular architecture embedded in fibrovascular stroma (arrowheads). (**E**) Pheochromocytomas show a similar appearance and are composed of chromaffin cells. The tumor cells are arranged in rounded clusters (asterisk), separated by endothelial-lined spaces, and have vesicles containing norepinephrine and epinephrine. (**F**) Pancreatic neuroendocrine tumors show trabecular architecture, small nuclei, and abundant eosinophilic cytoplasm. Nests of tumor cells (asterisk) show focal nuclear atypia with surrounding stromal collagen bands (arrowhead). (Adapted from Lonser et al., 2003a.)

complex from simple cysts and identification of surgical lesions. Abdominal MR imaging is often accurate in differentiating between complex and simple cysts (Beer et al., 2006).

Histologic features

Both cysts and RCC in VHL are the clear cell type (see Fig. 10.6) (Walther et al., 1995). Although, histologically similar to sporadic clear cell carcinomas, VHL-related RCC is morphologically distinct in bilaterality, multifocality, and association with renal cysts. VHL-associated RRCs often have lower grade histology and are only known to metastasize when greater than 3 cm in diameter (Neumann et al., 1998).

Treatment

RCC lesions below 3 cm in diameter on CT images do not metastasize. Consequently, a 3 cm threshold is often recommended for surgical resection (Walther et al., 1999). Nephron sparing surgery performed to resect lesions larger than 3 cm allows long-term preservation of renal function and 10 year cancer specific survival rates of up to 81% (Steinbach et al., 1995; Duffey et al., 2004). First-line surgery for RCC has moved away from total nephrectomies to nephron sparing surgeries (partial nephrectomies). Recently, radiofrequency ablation has been used (Joly et al., 2011).

Pheochromocytomas

General features

Pheochromocytomas arise most commonly in the adrenal medulla and are found in up to 16% of VHL patients (Lonser et al., 2003a; Binderup et al., 2013). Pheochromocytomas in VHL can be bilateral and occasionally multifocal (see Fig. 10.6) (Lenders et al., 2005). Pheochromocytomas in extra-adrenal locations (paragangliomas), including carotid body, glomus jugulare, and periaortic tissues, are rare in VHL and are more likely to be due to germline succinate dehydrogenase (SDH) mutation or sporadic tumors (Neumann et al., 2002). Pheochromocytomas in VHL present in the second decade of life (mean age 18–20 years), and rarely (1–5%) transform to malignant tumors (Walther et al., 1999; Neumann et al., 2002; Lenders et al., 2005; Bausch et al., 2014).

Clinical

Pheochromocytomas in VHL produce norepinephrine in excess and may result in paroxysmal or sustained hypertension, palpitations, tachycardia, headaches, sweating, pallor, and nausea (Lonser et al., 2003a; Lenders et al., 2005). Occasionally, these tumors may present as hypertensive crises in children younger than 10 years of age.

Laboratory and imaging findings

Diagnosis of pheochromocytomas is made by biochemical testing demonstrating excess catecholamines and imaging. Traditional tests include urinary and plasma catecholamines, urinary metanephrines (normetaneprhine and metanephrine), and urinary vanillylmandelic acid (VMA). Recently, plasma levels of free fractions of metanephrines have been used to reliably detect pheochromocytomas with high sensitivity (97%) and specificity (93%) (Pamporaki et al., 2013).

Imaging is used to localize the tumor once there is biochemical evidence for pheochrmocytomas. While post-contrast CT scans of the abdomen are sensitive in detecting adrenal and abdominal extra-adrenal pheochromocytomas, adrenergic blockade is needed before injecting iohexol contrast to prevent a hypertensive crisis or cardiac arrhythmia from catecholamine release.

Contrasted MR imaging of the abdomen has similar sensitivity (90–100%) to CT for detecting abdominal pheochromocytomas. MR imaging is often the preferred imaging modality (versus CT), since adrenergic blockade is not needed prior to gadolinium contrast injection. Nevertheless, CT and MRI imaging have a specificity that ranges from 70% to 80% (Lenders et al., 2005; Sahdev et al., 2005). To improve specificity and to detect extra-adrenal/multifocal pheochromocytomas, ^{131}I-metaiodobenzylguanidine (MIBG) scanning may be used (specificity 100%) (Shapiro et al., 1985).

Perioperative management

In VHL patients, pheochromocytomas can lead to catecholamine-induced, potentially life-threatening complications, including hypertensive crises, cardiac arrhythmias, pulmonary edema, and myocardial ischemia/infarction. Elective surgeries in VHL patients with pheochromocytomas may need preoperative α adrenergic blockade. Blockade can be achieved with phenoxybenzamine (nonselective α adrenergic blocker), prazosin, doxazosin or urapidil for 10–14 days. In patients who have tachyarrhythmias, a β-blocker may be added after a few days of α blockade. Postoperatively, patients with adrenergic blockade are monitored for hypotension and hypoglycemia (Kinney et al., 2000; Chen et al., 2010; King and Pacak, 2014).

Treatment

Laparoscopic surgery is the treatment of choice of pheochromocytomas and paragangliogliomas (Hwang et al., 2004; Chen et al., 2010; Moraitis et al., 2014). In VHL patients, partial adrenalectomies are performed via laparoscopic approach. Patients need long-term follow-up to evaluate for recurrence of pheochromocytomas in the remnant adrenal cortex (Lonser et al., 2003a; Chen et al., 2010).

Pancreatic neuroendocrine tumors and cysts

General features

VHL patients can develop a variety of pancreatic lesions. Up to 72% of VHL patients will have pancreatic cysts on autopsy. In nearly half (47%) of these patients, the cysts may be the only form of pancreatic pathology (Hough et al., 1994; Charlesworth et al., 2012). Cysts are commonly multiple (84%) and asymptomatic (94%). In 12% of patients, pancreatic cysts may be seen as the only manifestation of VHL, before other manifestations develop (Charlesworth et al., 2012). Serous cystadenomas are multiloculated lesions of the pancreas that are usually benign and asymptomatic (Mohr et al., 2000). Neuroendocrine tumors of the pancreas (PNET) are found in 15–56% of VHL patients (Hough et al., 1994; Blansfield et al., 2007; Charlesworth et al., 2012). PNETs can become malignant and metastasize in 8% of VHL patients (Blansfield et al., 2007). Occasionally, pancreatic adenocarcinomas or metastatic tumors from RCC can be found in the pancreas (Charlesworth et al., 2012).

Clinical

Cysts and cystadenomas can cause exocrine or endocrine deficiency by replacing the pancreatic parenchyma. Compression of intestine or the bile duct can lead to symptomatic pancreatic cysts (Blansfield et al, 2007). PNETs are mostly nonfunctional and are rarely the cause of major morbidity or mortality in VHL patients. Nevertheless, malignant conversion or metastasis of PNETs leads to poor prognoses (average survival of 1–3 years) (Weisbrod et al., 2014). VHL patients with metastatic PNETs are likely to have exon 3 mutations compared to those that do not have metastasis (80% versus 46%). Other indicators of tumor aggression include tumor size greater than 3 cm and a tumor doubling time of less than 500 days (Libutti et al., 2000; Blansfield et al., 2007; Charlesworth et al., 2012).

Imaging

Pre- and postcontrast abdominal CT and MR imaging can be used to diagnose pancreatic lesions. Abdominal CT images are favored for surveillance of simple and complex pancreatic lesions at intervals determined by presence of symptoms, previous evidence of growth or solid lesions (see Fig. 10.6) (Weisbrod et al., 2014).

Recently, FDG positron emission tomography (PET) has been used to detect PNETs not visible by CT. Furthermore, FDG PET can detect metastatic lesions (Sadowski et al., 2014).

Histologic features

True cysts and cystadenomas contain prominent fibrous stroma, clear, glycogen-rich epithelial cells with numerous epithelial cells forming small vessels in the tumors (Mohr et al., 2000). PNETs historically were called islet cell tumors. PNETs arise from the pancreatic islets and the tumor cells stain for pancreatic and gastrointestinal hormones (Hough et al., 1994; Lonser et al., 2003a).

Treatment

Treatment is usually not required for asymptomatic cysts. Laparoscopic surgical decompression is performed for cysts that are causing obstructive symptoms. Enucleation or selective pancreatic resections are performed for PNETs that have a higher likelihood for metastasis (greater than 3 cm, exon 3 mutation, and doubling time more than 500 days) (Blansfield et al., 2007; Libutti et al., 2000; Charlesworth et al., 2012; Weisbrod et al., 2014).

Epididymal cystadenomas

General features

Cystadenomas develop in the epididymis unilaterally or bilaterally in 25–60% of males who have VHL (Gruber et al., 1980; Lonser et al., 2003a). These tumors are benign and rarely need surgical resection. The tumors are derived from the mesonephric ducts and share the morphologic features of other VHL manifestations (Mehta et al., 2008).

Clinical, imaging, and histologic findings

Epididymal cystadenomas are typically asymptomatic. Patients may become aware of firm masses above the testes. In about 17% of the patients, the lesions are too small to be palpated, but can be detected by ultrasonography (Choyke et al., 1997). These masses may also be a target for metastasis of RCC tumors (Mehta et al., 2007).

Epididymal cystadenomas are grossly solid but contain multiple cysts filled with colloid material. Microscopically, the cysts are lined by clear epithelial cells, and are surrounded by dense stroma.

Treatment

Epididymal cystadenomas are typically monitored by physical examination and ultrasonography. Surgery is needed rarely for symptomatic tumors. Surgical indications for removal of epididymal cystadenomas can include local discomfort (Gruber et al., 1980; Choyke et al., 1997; Lonser et al., 2003a).

Broad ligament cystadenomas

General features

Papillary cystadenomas of the broad ligament are rare, and appear to arise from the mesonephric remnants. Papillary cystadenomas are found in the mesosalpinx and the broad ligament (Gersell and King, 1988; Werness and Guccion, 1997). The tumors may be found unilaterally or bilaterally in women with VHL disease (Zanotelli et al., 2010; Nogales et al., 2012). The incidence and prevalence of this tumor in VHL remains unknown (Lonser et al., 2003a).

Clinical, imaging, and histologic findings

Papillary cystadenomas are usually asymptomatic. Occasionally, the tumor may present as an abdominopelvic mass with symptoms of abdominal discomfort or present as a painful adnexal mass (Lonser et al., 2003a; Nogales et al., 2012). The masses may be detected by abdominal MR imaging or pelvic ulstrasonography (Zanotelli et al., 2010).

Papillary cystadenomas have a prominent papillary architecture with fibrovascular and hyalinized stromal cores similar to cystadenomas or the epididymis (Zanotelli et al., 2010; Brady et al., 2012; Nogales et al., 2012).

Treatment

Papillary cystadenomas are resected typically when found incidentally during caesarean section surgery or rarely due to symptoms of pelvic mass (Zanotelli et al., 2010; Nogales et al., 2012). In most cases, papillary cystadenomas remain asymptomatic and are followed with serial imaging (Lonser et al., 2003a).

SCREENING AND SURVEILLANCE

Screening for VHL

Initial screening workup includes a thorough history and clinical examination, an ophthalmologic examination including fundoscopy, MR imaging of the craniospinal axis, kidneys, pancreas, and liver, an audiologic examination and laboratory tests (to detect pheochromocytomas) (Binderup et al., 2013). Genetic testing of the peripheral leukocytes and/or other body tissues is performed to confirm the presence of a *VHL* gene mutation (Nordstrom-O'Brien et al., 2010). Genetic testing may be useful in patients who present with known manifestations of VHL (Binderup et al., 2013). The incidence of VHL in patients presenting with hemangioblastomas is as high as 30–40% (Huson et al., 1986; Neumann et al., 1989; Sora et al., 2001). A higher percentage of patients with retinal hemangioblastomas (up to 81%) may be VHL carriers (Wanebo et al., 2003; Kreusel et al., 2006). The frequency of VHL among patients with pheochromocytomas (2–11%), ELST (5–15%) or PNET (1%) is lower but may warrant screening for VHL (Binderup et al., 2013). Genetic testing can allow for discontinuation of surveillance for family members that do not carry VHL mutations. Early diagnosis of VHL mutation carriers allows initiation of appropriate screening procedures, and to improve morbidity and mortality (Gläsker et al., 1999; Rasmussen et al., 2010).

Surveillance

Recommendations for surveillance of patients and carriers with VHL have been generated by the US National Institutes of Health (Choyke et al., 1995; Lonser et al., 2003a), the Danish VHL Coordination Group (Poulsen et al., 2010, 2011; Binderup et al., 2013), and the VHL Family Alliance (see Resources and Table 10.4).

PREVENTION AND GENETIC COUNSELING

Prenatal diagnosis and preimplantation genetic diagnosis for at-risk pregnancies (either parent with known VHL disease or mutation) should be provided (Simpson et al., 2005). To prevent misinterpretation of test results, referral to a genetic counselor is recommended. Secondary effects from VHL manifestations are mitigated by routine surveillance and early detection. Risk of renal cancer may be increased by exposure to tobacco products and other toxins. Contact sports should be avoided if adrenal or pancreatic lesions are present (Frantzen et al., 2012).

References

Akcaglar S, Yavascaoglu I, Vuruskan H et al. (2008). Genetic evaluation of von Hippel–Lindau disease for early diagnosis and improved prognosis. Int Urol Nephrol 40: 615–620.

Asthagiri AR, Mehta GU, Zach L et al. (2010). Prospective evaluation of radiosurgery for hemangioblastomas in von Hippel–Lindau disease. Neuro Oncol 12: 80–86.

Bausch B, Jilg C, Gläsker S et al. (2013). Renal cancer in von Hippel–Lindau disease and related syndromes. Nat Rev Nephrol 9: 529–538.

Bausch B, Wellner U, Bausch D et al. (2014). Long-term prognosis of patients with pediatric pheochromocytoma. Endocr Relat Cancer 21: 17–25.

Beer AJ, Dobritz M, Zantl N et al. (2006). Comparison of 16-MDCT and MRI for characterization of kidney lesions. Am J Roentgenol 186: 1639–1650.

Binderup MLM, Bisgaard ML, Harbud V et al. (2013). Von Hippel–Lindau disease (vHL). National clinical guideline for diagnosis and surveillance in Denmark. 3rd edition. Dan Med J 60: B4763.

Blansfield JA, Choyke L, Morita SY et al. (2007). Clinical, genetic and radiographic analysis of 108 patients with von Hippel–Lindau disease (VHL) manifested by pancreatic neuroendocrine neoplasms (PNETs). Surgery 142: 814–818, discussion 818.e811–812.

Brady A, Nayar A, Cross P et al. (2012). A detailed immunohistochemical analysis of 2 cases of papillary cystadenoma of the broad ligament: an extremely rare neoplasm characteristic of patients with von Hippel–Lindau disease. Int J Gynecol Pathol 31: 133–140.

Brandt R (1921). Zur Frage der Angiomatosis retinae. Graefes Archiv Clin Exp Ophthalmol 106: 127–165.

Brauch H, Kishida T, Glavac D et al. (1995). Von Hippel–Lindau (VHL) disease with pheochromocytoma in the Black Forest region of Germany: evidence for a founder effect. Hum Genet 95: 551–556.

Brauch H, Hoeppner W, Jähnig H et al. (1997). Sporadic pheochromocytomas are rarely associated with germline mutations in the VHL tumor suppressor gene or the ret protooncogene. J Clin Endocrinol Metab 82: 4101–4104.

Butman JA, Kim HJ, Baggenstos M et al. (2007). Mechanisms of morbid hearing loss associated with tumors of the endolymphatic sac in von Hippel–Lindau disease. JAMA 298: 41–48.

Butman JA, Linehan WM, Lonser RR (2008). Neurologic manifestations of von Hippel–Lindau disease. JAMA 300: 1334–1342.

Capone F, Profice P, Pilato F et al. (2013). Spinal hemangioblastoma presenting with low back pain in pregnancy. Spine J 13: e27–e29.

Chang SD, Meisel JA, Hancock SL et al. (1998). Treatment of hemangioblastomas in von Hippel–Lindau disease with linear accelerator-based radiosurgery. Neurosurgery 43: 28–34, discussion 34–25.

Charlesworth M, Verbeke CS, Falk GA et al. (2012). Pancreatic lesions in von Hippel–Lindau disease? A systematic review and meta-synthesis of the literature. J Gastrointest Surg 16: 1422–1428.

Chen F, Kishida T, Yao M et al. (1995). Germline mutations in the von Hippel–Lindau disease tumor suppressor gene: correlations with phenotype. Hum Mutat 5: 66–75.

Chen H, Sippel RS, O'Dorisio MS et al. (2010). The North American Neuroendocrine Tumor Society consensus guideline for the diagnosis and management of neuroendocrine tumors: pheochromocytoma, paraganglioma, and medullary thyroid cancer. Pancreas 39: 775–783.

Choo D, Shotland L, Mastroianni M et al. (2004). Endolymphatic sac tumors in von Hippel–Lindau disease. J Neurosurg 100: 480–487.

Choyke PL, Glenn GM, Walther MM et al. (1992). The natural history of renal lesions in von Hippel–Lindau disease: a serial CT study in 28 patients. Am J Roentgenol 159: 1229–1234.

Choyke PL, Glenn GM, Walther MM et al. (1995). von Hippel–Lindau disease: genetic, clinical, and imaging features. Radiology 194: 629–642.

Choyke PL, Glenn GM, Wagner JP et al. (1997). Epididymal cystadenomas in von Hippel–Lindau disease. Urology 49: 926–931.

Collins TE (1894). Two cases, brother and sister, with peculiar vascular new growth, probably primarily retinal, affecting both eyes. Trans Ophthalmol Soc U K 14 (141–149): 1894.

Crossey PA, Eng C, Ginalska-Malinowska M et al. (1995). Molecular genetic diagnosis of von Hippel– Lindau disease in familial phaeochromocytoma. J Med Genet 32: 885–886.

Duffey BG, Choyke PL, Glenn G et al. (2004). The relationship between renal tumor size and metastases in patients with von Hippel–Lindau disease. J Urol 172: 63–65.

Evans DG, Howard E, Giblin C et al. (2010). Birth incidence and prevalence of tumor-prone syndromes: estimates from a UK family genetic register service. Am J Med Genet A 152A: 327–332.

Filling-Katz MR, Choyke PL, Patronas NJ et al. (1989). Radiologic screening for von Hippel–Lindau disease: the role of Gd-DTPA enhanced MR imaging of the CNS. J Comput Assist Tomogr 13: 743–755.

Filling-Katz MR, Choyke PL, Oldfield E et al. (1991). Central nervous system involvement in Von Hippel–Lindau disease. Neurology 41: 41–46.

Frantzen C, Links TP, Giles RH (2012). Von Hippel–Lindau Disease, University of Washington, Seattle.

Friedrich CA (1999). Von Hippel–Lindau syndrome. A pleomorphic condition. Cancer 86: 2478–2482.

Gallou C, Joly D, Méjean A et al. (1999). Mutations of the VHL gene in sporadic renal cell carcinoma: definition of a risk factor for VHL patients to develop an RCC. Hum Mutat 13: 464–475.

Gersell DJ, King TC (1988). Papillary cystadenoma of the mesosalpinx in von Hippel–Lindau disease. Am J Surg Pathol 12: 145–149.

Gläsker S, Bender BU, Apel TW et al. (1999). The impact of molecular genetic analysis of the VHL gene in patients with haemangioblastomas of the central nervous system. J Neurol Neurosurg Psychiatry 67: 758–762.

Gnarra JR, Tory K, Weng Y et al. (1994). Mutations of the VHL tumour suppressor gene in renal carcinoma. Nat Genet 7: 85–90.

Gruber MB, Healey GB, Toguri AG et al. (1980). Papillary cystadenoma of epididymis: component of von Hippel–Lindau syndrome. Urology 16: 305–306.

Hammel PR, Vilgrain V, Terris B et al. (2000). Pancreatic involvement in von Hippel–Lindau disease. Gastroenterology 119: 1087–1095.

Heffner DK (1989). Low-grade adenocarcinoma of probable endolymphatic sac origin. A clinicopathologic study of 20 cases. Cancer 64: 2292–2302.

Hippel E, Netzhaut ED, Beobachtungen K (1904). Über eine sehr seltene Erkrankung der Netzhaut. Graefes Archiv Clin Exp Ophthalmol 59: 83–106.

Hough DM, Stephens DH, Johnson CD (1994). Pancreatic lesions in von Hippel–Lindau disease: prevalence, clinical significance, and CT findings. Am J Roentgenol 162: 1091–1094.

Hrisomalos FN, Maturi RK, Pata V (2010). Long-term use of intravitreal bevacizumab (Avastin) for the treatment of von Hippel–Lindau associated retinal hemangioblastomas. Open Ophthalmol J 4: 66–69.

Huntoon K, Lonser RR (2014). Findings from the natural history of central nervous system hemangioblastomas in von Hippel–Lindau disease. Neurosurgery 61 (Suppl 1): N159–N162.

Huson SM, Harper PS, Hourihan MD et al. (1986). Cerebellar haemangioblastoma and von Hippel–Lindau disease. Brain 109: 1297–1310.

Hwang J, Shoaf G, Uchio EM et al. (2004). Laparoscopic management of extra-adrenal pheochromocytoma. J Urol 171: 72–76.

Igarashi H, Ito T, Nishimori I et al. (2014). Pancreatic involvement in Japanese patients with von Hippel–Lindau disease: results of a nationwide survey. J Gastroenterol 49: 511–516.

Iliopoulos O, Ohh M, Kaelin WG (1998). pVHL19 is a biologically active product of the von Hippel–Lindau gene arising from internal translation initiation. Proc Natl Acad Sci U S A 95: 11661–11666.

Johnson MW, Flynn HW, Gass JD (1992). Pars plana vitrectomy and direct diathermy for complications of multiple retinal angiomas. Ophthalmic Surg 23: 47–50.

Joly D, Méjean A, Corréas J-M et al. (2011). Progress in nephron sparing therapy for renal cell carcinoma and von Hippel–Lindau disease. J Urol 185: 2056–2060.

Kaelin WG (2002). Molecular basis of the VHL hereditary cancer syndrome. Nat Rev Cancer 2: 673–682.

Kanno H, Kondo K, Ito S et al. (1994). Somatic mutations of the von Hippel–Lindau tumor suppressor gene in sporadic central nervous system hemangioblastomas. Cancer Res 54: 4845–4847.

Kim HJ, Butman JA, Brewer C et al. (2005). Tumors of the endolymphatic sac in patients with von Hippel–Lindau disease: implications for their natural history, diagnosis, and treatment. J Neurosurg 102: 503–512.

Kim HJ, Hagan M, Butman JA et al. (2013). Surgical resection of endolymphatic sac tumors in von Hippel–Lindau disease: findings, results, and indications. Laryngoscope 123: 477–483.

King KS, Pacak K (2014). Familial pheochromocytomas and paragangliomas. Mol Cell Endocrinol 386: 92–100.

Kinney MA, Warner ME, vanHeerden JA et al. (2000). Perianesthetic risks and outcomes of pheochromocytoma and paraganglioma resection. Anesth Analg 91: 1118–1123.

Knudson AG (1986). Genetics of human cancer. Ann Rev Genet 20: 231–251.

Knudson Jr AG, Strong LC (1972). Mutation and cancer: neuroblastoma and pheochromocytoma. Am J Hum Genet 24: 514–532.

Kondo K, Klco J, Nakamura E et al. (2002a). Inhibition of HIF is necessary for tumor suppression by the von Hippel–Lindau protein. Cancer Cell 1: 237–246.

Kondo K, Yao M, Yoshida M et al. (2002b). Comprehensive mutational analysis of the VHL gene in sporadic renal cell carcinoma: relationship to clinicopathological parameters. Genes Chromosomes Cancer 34: 58–68.

Kondo K, Kim WY, Lechpammer M et al. (2003). Inhibition of HIF2alpha is sufficient to suppress pVHL-defective tumor growth. PLoS Biol 1. E83.

Kreusel KM, Bornfeld N, Lommatzsch A et al. (1998). Ruthenium-106 brachytherapy for peripheral retinal capillary hemangioma. Ophthalmology 105: 1386–1392.

Kreusel K-M, Bechrakis NE, Krause L et al. (2006). Retinal angiomatosis in von Hippel–Lindau disease: a longitudinal ophthalmologic study. Ophthalmology 113: 1418–1424.

Krieg M, Haas R, Brauch H et al. (2000). Up-regulation of hypoxia-inducible factors HIF-1 a and HIF-2 a under normoxic conditions in renal carcinoma cells by von Hippel–Lindau tumor suppressor gene loss of function. Oncogene 19: 5435–5443.

Kruizinga RC, Sluiter WJ, de Vries EGE et al. (2014). Calculating optimal surveillance for detection of von Hippel–Lindau-related manifestations. Endocr Relat Cancer 21: 63–71.

Lamiell JM, Salazar FG, Hsia YE (1989). von Hippel–Lindau disease affecting 43 members of a single kindred. Medicine 68: 1–29.

Latif F, Tory K, Gnarra J et al. (1993). Identification of the von Hippel–Lindau disease tumor suppressor gene. Science 260: 1317–1320.

Lazzeri S, Figus M, Di Bartolo E et al. (2011). Verteporfin photodynamic therapy for retinal hemangioblastoma associated with von Hippel–Lindau disease in a 9-year-old child. Clin Experiment Ophthalmol 39: 179–181.

Lee JY, Dong SM, Park WS (1998). Loss of heterozygosity and somatic mutations of the VHL tumor suppressor gene in sporadic cerebellar hemangioblastomas. Cancer Res 58: 504–508.

Lee S, Neumann M, Stearman R et al. (1999). Transcription-dependent nuclear-cytoplasmic trafficking is required for the function of the von Hippel–Lindau tumor suppressor protein. Mol Cell Biol 19: 1486–1497. 1999.

Lenders JWM, Eisenhofer G, Mannelli M et al. (2005). Phaeochromocytoma. Lancet 366: 665–675.

Li M, Kim WY (2011). Two sides to every story: the HIF-dependent and HIF-independent functions of pVHL. J Cell Mol Med 15: 187–195.

Libutti SK, Choyke PL, Alexander HR et al. (2000). Clinical and genetic analysis of patients with pancreatic neuroendocrine tumors associated with von Hippel–Lindau disease. Surgery 128: 1022–1027. discussion 1027–1028.

Lindau A (1927). Zur Frage der Angiomatosis Retinae und Ihrer Hirnkomplikationen. Acta Ophthalmol 4: 193–226.

Lonser RR, Glenn GM, Walther M et al. (2003a). von Hippel–Lindau disease. Lancet 361: 2059–2067.

Lonser RR, Weil RJ, Wanebo JE et al. (2003b). Surgical management of spinal cord hemangioblastomas in patients with von Hippel–Lindau disease. J Neurosurg 98: 106–116.

Lonser RR, Baggenstos M, Kim HJ et al. (2008). The vestibular aqueduct: site of origin of endolymphatic sac tumors. J Neurosurg 108: 751–756.

Lonser RR, Butman JA, Huntoon K et al. (2014). Prospective natural history study of central nervous system hemangioblastomas in von Hippel–Lindau disease. J Neurosurg 120: 1055–1062.

Machemer R, Williams JM (1988). Pathogenesis and therapy of traction detachment in various retinal vascular diseases. Am J Ophthalmol 105: 170–181.

Maddock IR, Moran A, Maher ER et al. (1996). A genetic register for von Hippel–Lindau disease. J Med Genet 33: 120–127.

Maher ER, Yates JR, Ferguson-Smith MA (1990a). Statistical analysis of the two stage mutation model in von Hippel–Lindau disease, and in sporadic cerebellar haemangioblastoma and renal cell carcinoma. J Med Genet 27: 311–314.

Maher ER, Yates JR, Harries R et al. (1990b). Clinical features and natural history of von Hippel–Lindau disease. Q J Med 77: 1151–1163.

Maher ER, Iselius L, Yates JR et al. (1991). Von Hippel–Lindau disease: a genetic study. J Med Genet 28: 443–447.

Maher ER, Webster AR, Richards FM et al. (1996). Phenotypic expression in von Hippel–Lindau disease: correlations with germline VHL gene mutations. J Med Genet 33: 328–332.

Maher ER, Neumann HP, Richard S (2011). von Hippel–Lindau disease: a clinical and scientific review. Eur J Hum Genet 19: 617–623.

Mandriota SJ, Turner KJ, Davies DR et al. (2002). HIF activation identifies early lesions in VHL kidneys: evidence for site-specific tumor suppressor function in the nephron. Cancer Cell 1: 459–468.

Manski TJ, Heffner DK, Glenn GM et al. (1997). Endolymphatic sac tumors. A source of morbid hearing loss in von Hippel–Lindau disease. JAMA 277: 1461–1466.

McNeill A, Rattenberry E, Barber R et al. (2009). Genotype-phenotype correlations in VHL exon deletions. Am J Med Genet A 149A: 2147–2151.

Mehta GU, Shively SB, Gläsker S et al. (2007). von Hippel–Lindau disease: epididymal cystadenoma targeted by metastatic events. Urology 69. 1209.e1209–e1212.

Mehta GU, Shively SB, Duong H et al. (2008). Progression of epididymal maldevelopment into hamartoma-like neoplasia in VHL disease. Neoplasia 10: 1146–1153.

Melmon KL, Rosen SW (1964). Lindau's disease. Review of the literature and study of a large kindred. Am J Med 36: 595–617.

Mohr VH, Vortmeyer AO, Zhuang Z et al. (2000). Histopathology and molecular genetics of multiple cysts and microcystic (serous) adenomas of the pancreas in von Hippel–Lindau patients. Am J Pathol 157: 1615–1621.

Moraitis AG, Martucci VL, Pacak K (2014). Genetics, diagnosis, and management of medullary thyroid carcinoma and pheochromocytoma/paraganglioma. Endocr Pract 20: 176–187.

Neumann HP, Wiestler OD (1991). Clustering of features of von Hippel–Lindau syndrome: evidence for a complex genetic locus. Lancet 337: 1052–1054.

Neumann HP, Eggert HR, Weigel K et al. (1989). Hemangioblastomas of the central nervous system. A 10-year study with special reference to von Hippel–Lindau syndrome. J Neurosurg 70: 24–30.

Neumann HP, Eng C, Mulligan LM et al. (1995). Consequences of direct genetic testing for germline mutations in the clinical management of families with multiple endocrine neoplasia, type II. JAMA 274: 1149–1151.

Neumann HP, Bender BU, Berger DP et al. (1998). Prevalence, morphology and biology of renal cell carcinoma in von Hippel–Lindau disease compared to sporadic renal cell carcinoma. J Urol 160: 1248–1254.

Neumann HPH, Bausch B, McWhinney SR et al. (2002). Germ-line mutations in nonsyndromic pheochromocytoma. N Engl J Med 346: 1459–1466.

Ning XH, Zhang N, Li T et al. (2014). Telomere shortening is associated with genetic anticipation in Chinese Von Hippel–Lindau disease families. Cancer Res 74: 3802–3809.

Nogales FF, Goyenaga P, Preda O et al. (2012). An analysis of five clear cell papillary cystadenomas of mesosalpinx and broad ligament: four associated with von Hippel–Lindau disease and one aggressive sporadic type. Histopathology 60: 748–757.

Nordstrom-O'Brien M, van der Luijt RB, van Rooijen E et al. (2010). Genetic analysis of von Hippel–Lindau disease. Hum Mutat 31: 521–537.

Ogasawara KK, Ogasawara EM, Hirata G (1995). Pregnancy complicated by von Hippel–Lindau disease. Obstet Gynecol 85: 829–831.

Ong KR, Woodward ER, Killick P et al. (2007). Genotype-phenotype correlations in von Hippel–Lindau disease. Hum Mutat 28: 143–149.

Othmane IS, Shields C, Singh A et al. (1999). Postpartum cerebellar herniation in von Hippel–Lindau syndrome. Am J Ophthalmol 128: 387–389.

Pamporaki C, Därr R, Bursztyn M et al. (2013). Plasma-free vs deconjugated metanephrines for diagnosis of phaeochromocytoma. Clin Endocrinol (Oxf) 79: 476–483.

Périgny M, Hammel P, Corcos O et al. (2009). Pancreatic endocrine microadenomatosis in patients with von Hippel–Lindau disease: characterization by VHL/HIF pathway proteins expression. Am J Surg Pathol 33: 739–748.

Peyre M, David P, Van Effenterre R et al. (2010). Natural history of supratentorial hemangioblastomas in von Hippel–Lindau disease. Neurosurgery 67: 577–587, discussion 587.

Poletti AM, Dubey SP, Barbò R et al. (2013). Sporadic endolymphatic sac tumor: its clinical, radiological, and histological features, management, and follow-up. Head Neck 35: 1043–1047.

Poulsen ML, Budtz-Jørgensen E, Bisgaard ML (2010). Surveillance in von Hippel–Lindau disease (vHL). Clin Genet 77: 49–59.

Poulsen ML, Gimsing S, Kosteljanetz M et al. (2011). von Hippel–Lindau disease: surveillance strategy for endolymphatic sac tumors. Genet Med 13: 1032–1041.

Prowse AH, Webster AR, Richards FM et al. (1997). Somatic inactivation of the VHL gene in Von Hippel–Lindau disease tumors. Am J Hum Genet 60: 765–771.

Raja D, Benz MS, Murray TG et al. (2004). Salvage external beam radiotherapy of retinal capillary hemangiomas secondary to von Hippel–Lindau disease: visual and anatomic outcomes. Ophthalmology 111: 150–153.

Rasmussen A, Alonso E, Ochoa A et al. (2010). Uptake of genetic testing and long-term tumor surveillance in von Hippel–Lindau disease. BMC Med Genet 11: 4.

Sadowski SM, Weisbrod AB, Ellis R et al. (2014). Prospective evaluation of the clinical utility of 18-fluorodeoxyglucose PET CT scanning in patients with von Hippel–Lindau-associated pancreatic lesions. J Am Coll Surg 218: 997–1003.

Sahdev A, Sohaib A, Monson JP et al. (2005). CT and MR imaging of unusual locations of extra-adrenal paragangliomas (pheochromocytomas). Eur Radiol 15: 85–92.

Schmidt LS, Linehan WM (2014). Editorial comment. Urology 83. 675.e675; discussion 675.e675–676.

Seizinger BR, Rouleau GA, Ozelius LJ et al. (1988). Von Hippel–Lindau disease maps to the region of chromosome 3 associated with renal cell carcinoma. Nature 332: 268–269.

Sgambati MT, Stolle C, Choyke PL et al. (2000). Mosaicism in von Hippel–Lindau disease: lessons from kindreds with germline mutations identified in offspring with mosaic parents. Am J Hum Genet 66: 84–91.

Shapiro B, Copp JE, Sisson JC et al. (1985). Iodine-131 metaiodobenzylguanidine for the locating of suspected pheochromocytoma: experience in 400 cases. J Nucl Med 26 (576–585): 1985.

Shively SB, Beltaifa S, Gehrs B et al. (2008). Protracted haemangioblastic proliferation and differentiation in von Hippel–Lindau disease. J Pathol 216: 514–520.

Shively SB, Falke EA, Li J et al. (2011). Developmentally arrested structures preceding cerebellar tumors in von Hippel–Lindau disease. Mod Pathol 24: 1023–1030.

Simone CB, Lonser RR, Ondos J et al. (2011). Infratentorial craniospinal irradiation for von Hippel–Lindau: a retrospective study supporting a new treatment for patients with CNS hemangioblastomas. Neuro Oncol 13: 1030–1036.

Simpson JL, Carson SA, Cisneros P (2005). Preimplantation genetic diagnosis (PGD) for heritable neoplasia. J Natl Cancer Inst Monogr 34: 87–90.

Singh AD, Nouri M, Shields CL et al. (2002). Treatment of retinal capillary hemangioma. Ophthalmology 109: 1799–1806.

Sora S, Ueki K, Saito N et al. (2001). Incidence of von Hippel–Lindau disease in hemangioblastoma patients: the University of Tokyo Hospital experience from 1954–1998. Acta Neurochir (Wien) 143: 893–896.

Stattin M, Kralinger M, Haas G et al. (2014). Photodynamic therapy for retinal capillary hemangioblastoma. Can J Ophthalmol 49: e32–e35.

Steinbach F, Novick AC, Zincke H et al. (1995). Treatment of renal cell carcinoma in von Hippel–Lindau disease. J Urol 4: 1812–1816.

Stolle C, Glenn G, Zbar B et al. (1998). Improved detection of germline mutations in the von Hippel–Lindau disease tumor suppressor gene. Hum Mutat 12: 417–423.

Toy BC, Agrón E, Nigam D et al. (2012). Longitudinal analysis of retinal hemangioblastomatosis and visual function in ocular von Hippel–Lindau disease. Ophthalmology 119: 2622–2630.

Tse JY, Wong JH, Lo KW et al. (1997). Molecular genetic analysis of the von Hippel–Lindau disease tumor suppressor gene in familial and sporadic cerebellar hemangioblastomas. Am J Clin Pathol 107: 459–466.

Vortmeyer AO, Tran MGB, Zeng W et al. (2006). Evolution of VHL tumourigenesis in nerve root tissue. J Pathol 210: 374–382.

Wait SD, Vortmeyer AO, Lonser RR et al. (2004). Somatic mutations in VHL germline deletion kindred correlate with mild phenotype. Ann Neurol 55: 236–240.

Walther MM, Lubensky IA, Venzon D et al. (1995). Prevalence of microscopic lesions in grossly normal renal parenchyma from patients with von Hippel–Lindau disease, sporadic renal cell carcinoma and no renal disease: clinical implications. J Urol 154: 2010–2014, discussion 2014–2015.

Walther MM, Reiter R, Keiser HR et al. (1999). Clinical and genetic characterization of pheochromocytoma in von Hippel–Lindau families: comparison with sporadic pheochromocytoma gives insight into natural history of pheochromocytoma. J Urol 162: 659–664.

Wanebo JE, Lonser RR, Glenn GM et al. (2003). The natural history of hemangioblastomas of the central nervous system in patients with von Hippel–Lindau disease. J Neurosurg 98: 82–94.

Wang V, Davis DA, Haque M et al. (2005). Differential gene up-regulation by hypoxia-inducible factor-1alpha and hypoxia-inducible factor-2alpha in HEK293T cells. Cancer Res 65: 3299–3306.

Wang Y, Abu-Asab MS, Shen D et al. (2014). Upregulation of hypoxia-inducible factors and autophagy in von Hippel–Lindau-associated retinal hemangioblastoma. Graefes Archiv Clin Exp Ophthalmol 252: 1319–1327.

Webster AR, Richards FM, MacRonald FE et al. (1998). An analysis of phenotypic variation in the familial cancer syndrome von Hippel–Lindau disease: evidence for modifier effects. Am J Hum Genet 63: 1025–1035.

Weil RJ, Lonser RR, DeVroom HL et al. (2003). Surgical management of brainstem hemangioblastomas in patients with von Hippel–Lindau disease. J Neurosurg 98: 95–105.

Weisbrod AB, Kitano M, Thomas F et al. (2014). Assessment of tumor growth in pancreatic neuroendocrine tumors in von Hippel Lindau syndrome. J Am Coll Surg 218: 163–169.

Werness BA, Guccion JG (1997). Tumor of the broad ligament in von Hippel–Lindau disease of probable mullerian origin. Int J Gynecol Pathol 16: 282–285.

Wilding A, Ingham SL, Lalloo F et al. (2012). Life expectancy in hereditary cancer predisposing diseases: an observational study. J Med Genet 49: 264–269.

Wind JJ, Lonser RR (2011). Management of von Hippel–Lindau disease-associated CNS lesions. Expert Rev Neurother 11: 1433–1441.

Wind JJ, Bakhtian KD, Sweet JA et al. (2011). Long-term outcome after resection of brainstem hemangioblastomas in von Hippel–Lindau disease. J Neurosurg 114: 1312–1318.

Wittström E, Nordling M, Andréasson S (2014). Genotype-phenotype correlations, and retinal function and structure in von Hippel–Lindau disease. Ophthal Genet 35: 91–106.

Wong WT, Agrón E, Coleman HR et al. (2008a). Clinical characterization of retinal capillary hemangioblastomas in a large population of patients with von Hippel–Lindau disease. Ophthalmology 115: 181–188.

Wong WT, Liang KJ, Hammel K et al. (2008b). Intravitreal ranibizumab therapy for retinal capillary hemangioblastoma related to von Hippel–Lindau disease. Ophthalmology 115: 1957–1964.

Woodward ER, Eng C, McMahon R et al. (1997). Genetic predisposition to phaeochromocytoma: analysis of candidate genes GDNF, RET and VHL. Hum Mol Genet 6: 1051–1056.

Woodward ER, Buchberger A, Clifford SC et al. (2000). Comparative sequence analysis of the VHL tumor suppressor gene. Genomics 65: 253–265.

Ye DY, Bakhtian KD, Asthagiri AR et al. (2012). Effect of pregnancy on hemangioblastoma development and progression in von Hippel–Lindau disease. J Neurosurg 117: 818–824.

Yuan G, Peng YJ, Reddy VD et al. (2013). Mutual antagonism between hypoxia-inducible factors 1α and 2α regulates oxygen sensing and cardio-respiratory homeostasis. Proc Natl Acad Sci U S A 110: E1788–E1796.

Zanotelli DB, Bruder E, Wight E et al. (2010). Bilateral papillary cystadenoma of the mesosalpinx: a rare manifestation of Von Hippel–Lindau disease. Arch Gynecol Obstet 282: 343–346.

Zbar B, Kishida T, Chen F et al. (1996). Germline Mutations in the von Hippel–Lindau disease (VrwL) gene families from North America, Europe, and Japan. Hum Mutat 8: 348–357.

Zhuang Z, Frerich JM, Huntoon K et al. (2014). Tumor derived vasculogenesis in von Hippel–Lindau disease-associated tumors. Sci Rep 4: 4102.

Zimmer M, Doucette D, Siddiqui N et al. (2004). Inhibition of hypoxia-inducible factor is sufficient for growth suppression of VHL −/− tumors. Mol Cancer Res 2: 89–95.

Resources

Suggested Screening Guidelines – VHL Alliance, VHL homepage – Zhejiang University Center for Genetic and Genomic Medicine (ZJU-CGGM) – Leiden Open Variation Database.

Handbook of Clinical Neurology, Vol. 132 (3rd series)
Neurocutaneous Syndromes
M.P. Islam and E.S. Roach, Editors

Chapter 11

Sturge–Weber syndrome

ANNE M. COMI*
Hunter Nelson Sturge-Weber Center, Kennedy Krieger Institute, Baltimore, MD, USA

CLINICAL DEFINITION

Sturge–Weber syndrome is the association of the facial port-wine birthmark (capillary malformation) (Fig. 11.1A) with abnormal blood vessels in the brain (leptomeningeal "angioma" Fig. 11.1B–D) and the eye (Table 11.1). The syndrome exists as a spectrum ranging from the isolated brain involvement seen in about 10% of the cases, to the birthmark associated with brain and skin involvement; isolated eye involvement has also been observed (personal communication, Dr. Harry A. Quigley). That Sturge–Weber syndrome is a spectrum is now even more clearly the case since the causative somatic mosaic mutation has been identified and is the same for both Sturge–Weber syndrome and isolated port-wine birthmarks (Shirley et al., 2013). Since the isolated birthmark is not usually referred to as Sturge–Weber syndrome, the typical brain involvement is most frequently used to make the diagnosis. Since the eponyms Schirmer's syndrome and Lawford's syndrome are not currently widely recognized, the small subset of patients who have the birthmark and eye involvement are also often said to have Sturge–Weber syndrome, but if they are well past 1 year of age and their contrast-enhanced MRI is normal (performed when past 1 year of age) then they are very unlikely to develop Sturge–Weber syndrome brain involvement.

The port-wine birthmark is present at birth; this can be helpful for differentiating this birthmark from a hemangioma which is not present at birth. The port-wine birthmark grows proportionately with the child and it is a vascular malformation and not a tumor (Willenberg and Baumgartner, 2008; Aboutalebi et al., 2012). Sturge–Weber syndrome is the third most common neurocutaneous disorder after neurofibromatosis (NF) and tuberous sclerosis complex (TSC). However, unlike NF and TSC (Laycock-van Spyk et al., 2011; Mettin et al., 2014), Sturge–Weber syndrome is not associated with a greatly increased risk of tumors. Sturge–Weber syndrome can also be confused with macrocephaly-capillary malformation syndrome; however, the intracranial findings are different: while Sturge–Weber syndrome is diagnosed by visualizing leptomeningeal enhancement, intracranial abnormalities typical of macrocephaly-capillary malformation syndrome include ventriculomegaly, cerebral and/or cerebellar asymmetric overgrowth, high frequency of cavum septum pellucidum vergae, thickened corpus callosum, prominent optic nerve sheaths, and Chiari type 1 malformation (Moffitt et al., 1999). Patients with Sturge–Weber syndrome can also have extensive areas with increased skin melanosis; this is referred to as phakomatosis pigmentovascularis (PPV), a rare syndrome in which a vascular nevus is associated with an extensive pigmentary nevus (Al Robaee et al., 2004). Further research is needed to determine whether the hypermelanosis is also caused by a somatic mutation in *GNAQ*; however, given that the same mutation in adults can cause a rare form of ocular melanoma, and given the frequent association of Sturge–Weber syndrome with PPV, is it likely that for at least a subset of patients with PPV, their melanocytic birthmark is caused by a somatic mutation in *GNAQ*. It is likely that the challenging array of syndromes associated with capillary malformations (including Klippel–Trénaunay–Weber syndrome, which is also not infrequently seen concurrently in patients with Sturge–Weber syndrome) currently diagnosed primarily based on the clinical phenotype will one day be grouped, diagnosed, and treated based on genotype, as well as phenotype.

Now that the somatic mosaic mutation in *GNAQ* has been identified (Shirley et al., 2013), it is tempting to consider how this information may be used for diagnosis. At this point, however, the role for *GNAQ* gene mutation

*Correspondence to: Anne M. Comi, Kennedy Krieger Institute, 801 North Broadway, Baltimore, MD 21205, USA. Tel: +1-443-923-9127, Fax: +1-443-923-9540, E-mail: comi@kennedykrieger.org

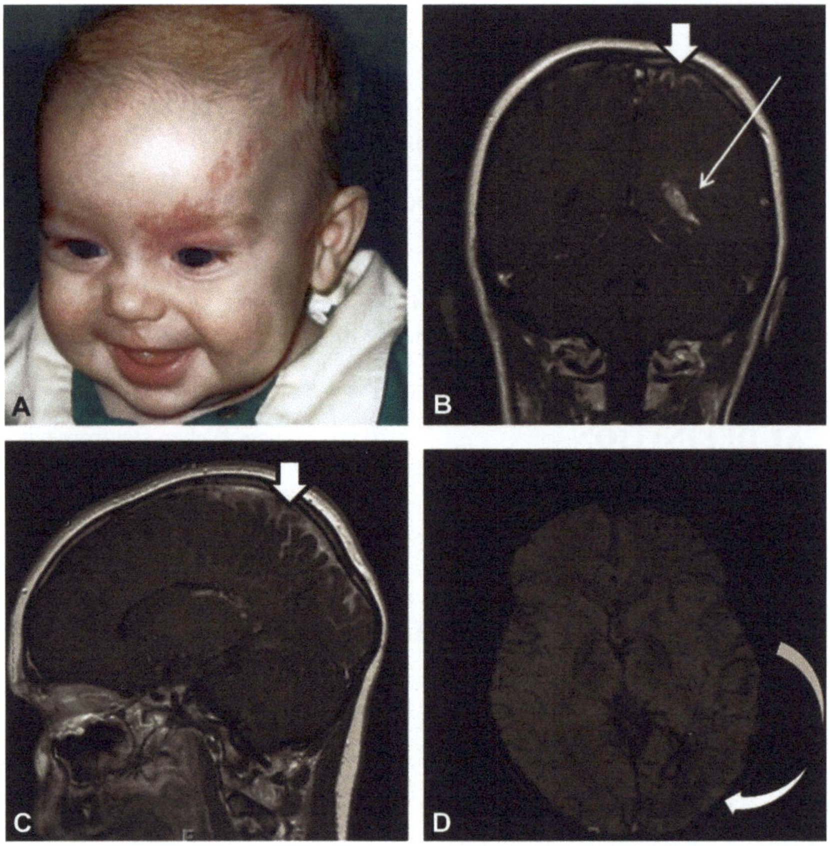

Fig. 11.1. Images of a port-wine birthmark and of neuroimaging from patients with Sturge–Weber syndrome. (**A**) Male child with Sturge–Weber syndrome who was born with a port-wine birthmark as seen in this picture on the forehead over the left eye; the birthmark is flat and reddish pink and grew commensurate with the boy's growth. (**B–D**) Representative images from Magnetic resonance imaging (MRI) of a 10-year-old girl with Sturge–Weber syndrome. (**B**) T1-weighted postcontrast coronal MRI showing leptomeningeal enhancement over the left cortex (thick arrow), and an enlarged and enhancing choroid plexus (thin arrow). (**C**) T1-weighted postcontrast sagittal MRI showing leptomeningeal enhancement over the left parietal-occipital cortex (thick arrow). (**D**) Susceptibility-weighted image demonstrating increased number of deep draining vessels in the left hemisphere, particularly posteriorly (curved arrow).

testing is limited. Although this needs additional study, it is likely that the gene mutation is rarely present in the blood of affected individuals, so skin or other abnormal tissue would be required for testing. Testing skin tissue from the port-wine birthmark is not helpful in determining whether an infant with the birthmark has brain and/or eye involvement; the same mutation is found both in patients with an isolated port wine birthmark and in those with the full syndrome. It may be that in the future genetic testing will be helpful in distinguishing patients with Sturge–Weber syndrome who have the *GNAQ* mutation from those with another capillary malformation-related syndrome with clinical overlap but a different genetic basis, but further study of the *GNAQ* mutation-negative samples, and of other capillary malformation-related syndromes is needed before this approach can be clinically useful. Certainly this is an area of active research interest.

HISTORICAL PERSPECTIVE

A large number of eponyms are associated with Sturge–Weber syndrome: these include Dimitri's hemangiomatosis syndrome, Jahnke syndrome, Kalischer syndrome, Krabbe syndrome II, Lawford's meningocutaneous syndrome, Miller syndrome, Parkes Weber–Dimitri syndrome, Schirmer syndrome, and Sturge–Kalischer–Weber syndrome, all designated for the various physicians who contributed to the description of the syndrome and the various manifestations of the spectrum with which it can present.

In 1879 the first description of the syndrome was made by William Allen Sturge; the patient was a 6½-year-old girl whose birthmark, referred to as a "mother's mark" and which Sturge called a port-wine stain, was demarcated at the midline and extended from the forehead on the right side down onto the chest and up

Table 11.1

Sturge–Weber syndrome synopsis

Clinical definition	Leptomeningeal "angioma" diagnosed by neuroimaging or neuropathology. Usually (90%) associated with a facial port-wine birthmark, and often (30–70%) with a choroidal "angioma." Vascular malformations NOT tumors
Clinical presentation	Birthmark present at birth Seizures: 90% with onset first 2 years of life; usually but not always focal Early handedness or visual gaze preference less common presentations Glaucoma (peak periods in infancy and young adulthood)
Diagnosis	Referral to vascular specialist/dermatologist to confirm port-wine birthmark. Referral of neonate to ophthalmologist for exclusion of early glaucoma. Neurologic history, exam, EEG and *consider* contrast enhanced MRI with postcontrast FLAIR and SWI, MRA, MRV although sensitivity in early infancy is not high. If early neuroimaging not obtained, or if negative, obtain when symptoms start or after 1 year of age
Cause	Somatic mosaic mutation in *GNAQ* causing hyperactivation of pathways downstream of coded protein Gαq
Treatment	Consider presymptomatic treatment with low-dose aspirin (3–5 mg/kg/day) and/or and anticonvulsant (i.e. levetiracetam), especially if bilateral brain involvement Aggressive treatment with anticonvulsants (see p. 165 for description of aggressive treatment of seizures) Low-dose aspirin (3–5 mg/mg day) somewhat controversial although evidence growing If two or more anticonvulsants (and aspirin) fail to control seizures adequately consider surgical options: hemispherectomy, focal resection, VNS Atkins or ketogenic diet for those who fail anticonvulsants and are not good surgical candidates Other: annual flu shot, maintain hydration, treat migraines with preventative medications, avoid triggers

EEG, electroencephalogram; FLAIR, fluid-attenuated inversion recovery; SWI, susceptibility-wighted imaging; MRA, magnetic resonance angiography; MRV, magnetic resonance venography; VNS, Vagal nerve stimulator.

into the scalp on that side to the neckline. The child had had "twitching of (the) left side of her body starting at six months old." Sturge hypothesized that the neurologic involvement was caused by a lesion on the surface of the brain on the same side as the birthmark (Sturge, 1879).

Parkes Weber added the feature of the "sclerosing" brain atrophy in 1922; he noted hyperdensity on the brain radiogram of a 22-year-old woman with a bilateral port-wine birthmark extending onto her torso and upper limbs, left side greater than right, with loss of vision in her left eye due to glaucoma and congenital right sided hemiparesis (Weber, 1922). Rudolf Schirmer (1831–1896) had earlier described a 36-year-old man with a left facial nevus and buphthalmos, but he made no reference to epilepsy or brain disease (Schirmer, 1860); Schirmer syndrome has therefore been used to indicate the presence of uculocutaneous angiomatosis with early glaucoma.

In 1987, Rudolf Happle applied the principle of somatic mosaic mutation to explain the pathogenesis of several syndromes with cutaneous manifestations, including Sturge–Weber syndrome (Happle, 1987). He hypothesized that given the localized, asymmetric area of involvement and the lack of a familial pattern of inheritance, Sturge–Weber syndrome was likely to be caused by a mutation occurring postconception, in the developing fetus, in a progenitor cell. This hypothesis was critical to the eventual discovery of the underlying cause, because this proposed mechanism dictated the basic approach needed to ascertain the putative mutation: DNA from an affected area of the body had to be compared to that from an unaffected area of the body (Comi et al., 2013). It was not until 25 years after the time this hypothesis was proposed that molecular techniques sufficiently evolved to test it and prove it correct.

MOLECULAR BASIS, HISTOLOGY AND PATHOGENESIS

GNAQ codes for Gαq, a guanine nucleotide protein which is part of a trimeric G protein complex that couples with certain G protein coupled receptors (GPCR), such as the endothelin 1 receptor, angiotensin 2 receptor type I, α-1 adrenergic receptors, vasopressin type 1A and 1B receptors, and others. The somatic mosaic mutation in *GNAQ* that causes Sturge–Weber syndrome and isolated port-wine birthmarks results in a c.548G→A p.Arg183Gln mutation that affects the autohydrolysis site of the protein, thereby impairing the ability of Gαq to bind to guanosine diphosphate (GDP) and return to the inactive state. As predicted by the mutational site, functional studies with mutant constructs transiently transfected into 293T cells demonstrated that the downstream pathways were constitutively hyperactivated (Shirley et al., 2013).

Interestingly the same R183Q mutation in *GNAQ* occurring in melanocytes rather than in a cell type related to vascular development and function, and in adults rather than the developing fetus, results in uveal melanoma (Harbour, 2012). This is the less common mutation; the more common mutation in *GNAQ* known to cause uveal melanoma is the Q209L mutation. It is likely that it is the different timing in development that results in the occurrence of a vascular malformation, rather than cancer, in the case of Sturge–Weber syndrome or port-wine birthmark as opposed to uveal melanoma. In the same vein, most likely it is the timing of the mutation, during early postconceptual fetal development, that determines the extent of involvement (Poduri et al., 2013) in Sturge–Weber syndrome and port-wine birthmark; a somatic mosaic mutation occurring in an earlier progenitor and therefore impacting more structures (brain, skin, and eye) would result in Sturge–Weber syndrome (Fig. 11.2), while a somatic mutation occurring later in fetal development after these structures have separated would have a more limited impact on, for example, only the skin for isolated, nonsyndromic port-wine birthmarks, or only in the brain in the case of those patients with the variant of typical Sturge–Weber syndrome brain involvement without skin or eye involvement (see Fig. 11.2).

An important question currently being addressed in the lab is which cell type(s) are affected by the somatic mutation in *GNAQ*; this has implications for understanding how the mutation results in the vascular malformation and for creating suitable animal and *in vitro* models. For now we can only speculate that abnormal signaling through GPCRs may be associated with GNAQ. These receptors are important to blood vessel development and function, and their abnormal signaling may result in abnormal fetal vascular development in the area affected by the somatic mutation (Kimple et al., 2011). Since the cortex develops in tandem with the local maturation of the vasculature, abnormal vascular development would be expected to negatively impact cortical development as well (Ward and Lamanna, 2004). This has been shown to be the case in Sturge–Weber surgical samples demonstrating cortical malformations, focal cortical dysgenesis, and evidence of white matter neurons (Murakami et al., 2012).

In Sturge–Weber syndrome, histologic studies have reported an increased number of enlarged, dilated leptomeningeal vessels where normally much fewer arterioles would be present (Norman and Schoene, 1977). Within the cortex itself is described a range of atrophy with cortical and subcortical deposition of calcium frequently around blood vessels (Maton et al., 2010). Numbers of cortical draining vessels are decreased; however, since the brain tissue does not start out atrophied, but rather becomes that way, it is unclear whether that observation is primary to the condition or rather is secondary to the loss of brain tissue. Gliosis and loss of cortical neurons is also described. On neuroimaging, calcification is often not visible initially and becomes apparent as the hemisphere atrophies (Wu et al., 2011). Therefore, calcification, neuronal loss, and gliosis are likely secondary to

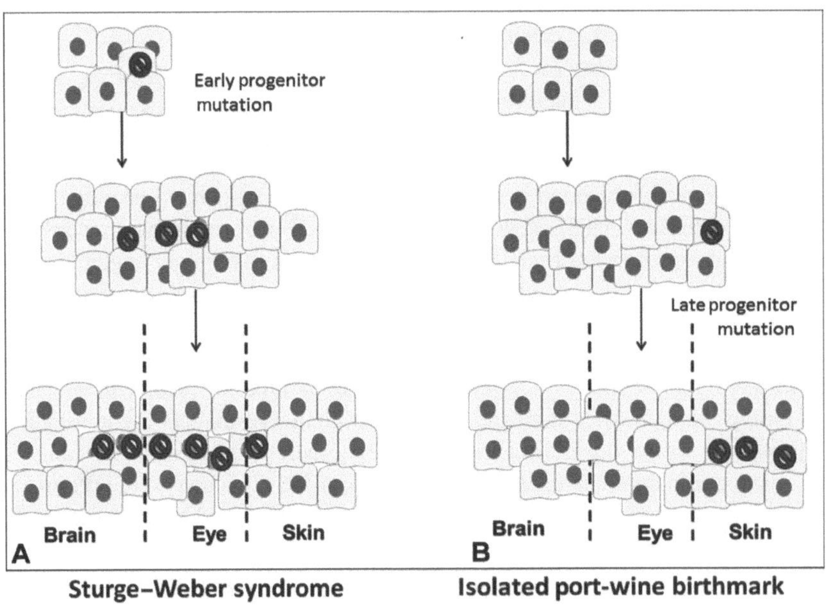

Fig. 11.2. Timing of somatic mutation resulting in Sturge–Weber syndrome versus isolated port-wine birthmark. (**A**) When the somatic mutation in *GNAQ* occurs in an early progenitor then the mutation will impact multiple structures including the brain, skin, and eye, resulting in the full spectrum of Sturge–Weber syndrome. (**B**) When the somatic mutation in *GNAQ* occurs in a later progenitor then the mutation has a more restricted impact and manifests just as an isolated port-wine birthmark, as shown in the figure; depending on the location of the later somatic mutation the structure impacted may alternatively be just the brain (as in isolated Sturge–Weber brain involvement) or just the eye.

brain injury resulting from venous stasis and impaired brain perfusion rather than being primarily the result of the somatic mutation.

Abnormal vascular innervation almost certainly has an important role in the pathophysiology of both Sturge–Weber syndrome and port-wine birthmarks. Aberrant vascular innervation has been consistently reported in studies of port-wine birthmarks and in the abnormal Sturge–Weber blood vessels of the brain (Smoller and Rosen 1986; Rosen and Smoller, 1987; Rydh et al., 1991; Cunha e Sá et al., 1997). Cunha e Sá et al. noted that the malformative cortical vessels in Sturge–Weber syndrome were only innervated by noradrenergic nerve fibers and lacked the normal cholinergic and sensory fibers (Cunha e Sá et al., 1997). Furthermore, the vasoactive responses of both the birthmark and the brain vascular malformation are abnormal: the birthmark can have abnormal responses to temperature or may significantly darken with seizures, and ictal single-photon emitted computed tomography (SPECT) blood flow studies in patients with Sturge–Weber syndrome have shown that, during a seizure, instead of blood flow increasing in response to a seizure, which is the normal response, blood flow can decrease to ischemic levels, a phenomenon which likely contributes to the strokes and stroke-like episodes these patients demonstrate (Griffiths et al., 1997; Namer et al., 2005). It is likely that the somatic mutation lies within the vascular innervation, the vascular supporting cells such as fibroblasts, smooth muscle cells, pericytes, and/or the endothelial cells; all of these cell types interact in the formation of blood vessels, their normal innervation, and in their normal function.

There is evidence of ongoing vascular remodeling as well. Modestly increased proliferative index was noted within the leptomeningeal blood vessels, suggesting ongoing vascular remodeling at a low level (Comati et al., 2007). This vascular remodeling was not suggestive of a tumor but rather suggested a response to a trigger such as chronic ischemia. Individuals with Sturge–Weber syndrome have been shown to have more vascular factors in their urine than normal controls (Sreenivasan et al., 2013). Furthermore, vascular endothelial growth factor (VEGF) expression is increased in the cortical neurons and glia underlying the abnormal leptomeningeal vessels, and VEGFR-1, VEGFR-2, HIF-1α, and HIF-2α expression is increased in the endothelial cells lining the abnormal leptomeningeal vessels, indicating that chronic tissue hypoxia and VEGF may be involved in driving ongoing vascular remodeling (Comati et al., 2007). Given that stimulation of the RAS-RAF-MEK-ERK pathway can increase both VEGF (Argraves and Drake, 2005) and hypoxia-inducible factor (HIF) activity (Jiang and Liu, 2008), the underlying somatic mutation also likely has a role in the vascular remodeling.

With the discovery of the somatic mosaic mutation underlying Sturge–Weber syndrome, the condition joins a growing group of capillary malformation-related disorders with known genetic, frequently somatic causes (Fig. 11.3). This insight is important for two reasons. First, increasingly these conditions will be grouped by their genetic cause in addition to, or instead of, their clinical phenotypes; this approach is already being taken for Cowden syndrome where genetic testing of *PTEN* is expanding the clinical phenotype and leading to discussions on how best to define, diagnose, and name this spectrum of conditions (Lachlan, 2013). Some patients with Sturge–Weber syndrome also have Klippel–Trénaunay–Weber syndrome or PPV (Kramer, 1963; Furukawa et al., 1970; Leung et al., 1988; Hagiwara et al., 1998; Kiley et al., 2002; Al Robaee et al., 2004). It is likely that many of these patients also have the same somatic mutation in *GNAQ*, although this remains to be determined. Secondly, understanding the interrelated pathways involved in these various capillary malformation syndromes means that as therapies are developed for the treatment of one or more of these conditions, these new treatments may be relevant to some of the other capillary malformation-related conditions with mutations in the same or related molecular pathways. For example, the mutation for familial capillary malformation involves *RASA1* (Eerola et al., 2003), a gene coding for a protein which regulates RAS (see Fig. 11.3); the mutation results in impaired regulation of the downstream pathways which are probably also impacted by the *GNAQ* mutation that causes Sturge–Weber syndrome. Therefore, a therapy useful in treating one of these conditions may be useful for the other.

CLINICAL PRESENTATIONS

Increasingly the diagnosis of Sturge–Weber syndrome is being considered when an infant is born with a port-wine birthmark in the upper portion of the face. From large dermatologic clinics (Enjolras et al., 1985; Tallman et al., 1991; Ray et al., 2010) it has been demonstrated that while a patient with a capillary malformation anywhere on the face has approximately an 8% risk of having Sturge–Weber syndrome brain and/or eye involvement, most of the risk is really associated with the presence of the birthmark in the distribution of the forehead and upper eyelid. Risk increases with the size of the birthmark: to about 25% in those with a capillary malformation covering most of one side of the face and to 35% when most of both sides of the face are involved with the birthmark. Usually the birthmark is on the same side

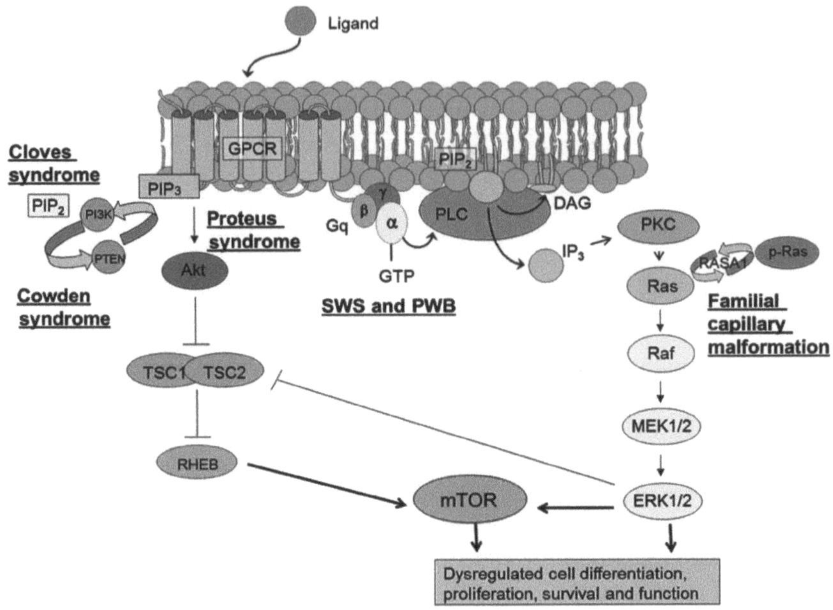

Fig. 11.3. Genetic causes of capillary malformation-related disorders. The underlying mutations for many syndromes associated with capillary malformations have now been identified, including Proteus syndrome (*AKT*), Cowden syndrome (*PTEN*), CLOVES syndrome (*PIK3CA*) and familial capillary malformation (*RASA1*) in addition to Sturge–Weber syndrome (SWS) and isolated port-wine birthmarks (PWB) (*GNAQ*). The molecular interrelationships between these syndromes can now be studied and novel targeted treatments considered based on an understanding of the pathogenesis.

as brain and eye involvement, but the brain or eye is rarely involved contralateral to the birthmark, and the brain involvement can be bilateral when the skin involvement is only unilateral, and vice versa.

Eye involvement in infancy can present with increased vascularity of the conjunctiva which gives the eye an injected appearance, with buphthalmos (eye enlargement), and/or with increased tearing of the involved eye (Sullivan et al., 1992). Risk of Sturge–Weber syndrome eye involvement increases greatly when the birthmark involves the upper and/or lower eyelids (ranges from 35% to 75%). There are two peak periods of onset of glaucoma: infancy and young adulthood. Therefore newborns should be promptly evaluated and these infants regularly seen by ophthalmology every few months over the first several years of life (Ray et al., 2010). Even if no evidence of glaucoma is found, annual examinations for life are recommended.

Neurologic presentation is typically with the onset of seizures in the first year or two of life (Sujansky and Conradi, 1995b), although increasingly patients are being diagnosed presymptomatically with neuroimaging. Seizures are typically of focal onset, and 75% of patients with brain involvement will present with seizures in the first year of life, 90% within the first 2 years of life. Nevertheless, individuals have been known to present with seizures later in childhood or even in adulthood; this is particularly common with the subset of patients with the intracranial variant of Sturge–Weber syndrome who do not have the birthmark (see below) (Ambrosetto et al., 1983; Comi et al., 2003). Occasional infants present with early handedness or with a visual gaze preference prior to seizure onset and these signs. Evidence of asymmetry in frequency or amplitude can also be noted on electroencephalography (EEG) prior to the onset of seizures. A history of abnormal spells suggestive of seizures, abnormal development or neurologic examination, and/ or abnormal EEG are all possible evidence of Sturge–Weber syndrome brain involvement and support the need for early neuroimaging.

Epilepsy in Sturge–Weber syndrome is frequently difficult to control, occurring in clusters of seizures and episodes of status epileptics (Kossoff et al., 2009). While generalized seizures can occur, most seizures are focal, with or without impaired consciousness, in semiology. A small subset of patients develop infantile spasms (Fukuyama and Tsuchiya, 1979) that respond to hormone therapy or vigabatrin. Another small percentage of patients taking anticonvulsants such as oxcarbazepine, carbamazepine, or lamotrigine develop a generalized spike and wave pattern on EEG associated with myoclonic seizures (Ewen et al., 2007). Approximately 90% of patients will obtain reasonable seizure control on one or two anticonvulsants plus low dose aspirin (seizure-free for 6 months or more, or sporadic but not monthly brief seizures in the past 6 months) (Lance et al., 2013). Some, however, are unable to obtain reasonable seizure control with a combination of multiple anticonvulsants and low dose aspirin or have intolerable side-effects from medications; many of these also have a significant hemiparesis and visual field deficits, and significant developmental and learning impairment. For these patients hemispherectomy or other surgical interventions become options.

Cognitive impairments in Sturge–Weber syndrome range from mild to moderate learning disabilities (Zabel et al., 2010), attention problems, social impairments, to much more significant intellectual impairments. One study noted that in young children with Sturge–Weber syndrome, hemiparesis was associated with cognitive impairments detected with neuropsychological testing (Reesman et al., 2009). Therefore hemiparesis is a sign that can be used to detect children with Sturge–Weber syndrome in need of more intensive testing.

A range of endocrinological abnormalities has been noted with increased frequency over the general population in patients with Sturge–Weber syndrome, including growth hormone deficiency and central hypothyroidism (Miller et al., 2006). Growth hormone deficiency is generally noted in childhood when the young child begins to fall off their growth curve. An IGF-1 level can be used to screen children over the age of 2 years for growth hormone deficiency, and if abnormally low then formal testing is required to establish the diagnosis. The diagnosis of growth hormone deficiency can also be made in short adults with Sturge–Weber syndrome and treatment with low dose growth hormone is helpful in preventing mood problems and the negative effects of the deficiency upon other organ systems. Central hypothyroidism or primary hypothyroidism are also occasionally seen (Comi et al., 2008; Siddique et al., 2013), and may be associated with anticonvulsant use. When suspected, because of symptoms such as excessive weight gain, dry skin, sleepiness, and constipation, then patients should be tested by sending blood for free T4 by equilibrium dialysis. Treatment with levothyroxine will lead to resolution of these symptoms.

DIAGNOSIS OF BRAIN INVOLVEMENT

Diagnosis of Sturge–Weber syndrome brain involvement is made on neuroimaging with contrast-enhanced magnetic resonance imaging (MRI) (see Fig. 11.1B–D); without contrast the abnormal blood vessels will not be seen. MRI sequences that can increase the sensitivity for diagnosis and the extent of brain involvement include postcontrast flair and susceptibility-weighted imaging (Hu et al., 2008). Contrast-enhanced MRI has lower sensitivity in newborns and young infants and therefore when neuroimaging is negative in young infants it must be repeated after the first year of age (Fischbein et al., 1998). It is helpful in any case to repeat the MRI after the first year because the extent of brain involvement can become more apparent after that time (Porto et al., 2006). Diagnosis of Sturge–Weber syndrome brain involvement is made on the basis of visualizing the enhancing dilated leptomeningeal vessels, often associated with enlarged deep draining vessels and a choroid plexus glomus (an enlarged and enhancing choroid plexus on the same side). Head CT is often used in the emergency room setting and the cortical calcifications noted can sometimes be confused for intracranial hemorrhage. Head CT is more sensitive for visualizing the extent of calcifications, which are only present after brain injury has occurred in the involved brain regions.

Because MRI imaging in the newborn has low sensitivity, there is controversy in the different approaches taken to making early diagnosis of brain involvement. Some clinicians image all newborns an effort to make the diagnosis. Another approach is to evaluate newborns, take a history, do an exam and an EEG and only do an early MRI if an abnormality is noted. In addition, imaging of newborns with a large facial port-wine birthmark, who are at greater risk of brain involvement, is a reasonable approach. While presymptomatic treatment is being attempted at some centers, in order to delay or decrease early brain injury, further work in this area is needed. When a truly effective treatment becomes available the need to make an early diagnosis will become truly imperative; currently it is still useful, if only to prepare the family by providing anticipatory guidance.

Because of concerns regarding low sensitivity of neuroimaging in young infants, efforts have been made to develop quantitative EEG as an approach to aid screening for brain involvement. In addition to history, examination, and routine EEG, decreased power on the affected side can be used to detect brain involvement in infants with Sturge–Weber syndrome (Ewen et al., 2009). Currently research is underway to determine the negative and positive predictive value of this approach.

Perfusion imaging has demonstrated that decreased venous drainage results in impaired blood flow to the affected brain regions (Lin et al., 2006). Ictal SPECT studies in patients with Sturge–Weber syndrome have revealed that during seizures blood flow does not increase greatly as expected during the seizure, but rather may increase only marginally or even decrease to ischemic levels (Griffiths et al., 1997; Aylett et al., 1999; Namer et al., 2005). It is likely that the abnormal automatic regulation of blood flow during seizures in patients with Sturge–Weber syndrome contributes to the stroke-like episodes which they suffer frequently in association with seizures. Infants who have prolonged and repeated seizures develop hemiparesis, visual gaze preference, and developmental delays.

NEUROLOGIC PROGRESSION AND NATURAL HISTORY

Neurologic progression is extremely variable (Sujansky and Conradi, 1995a). When seizures are infrequent and not prolonged, then neurologic and developmental progress over the first few years of life tends to be better. However, specific learning disabilities and attention problems are common by school age even among the higher functioning children performing at approximately age level or just below. Early age of seizure onset and extent of brain involvement on neuroimaging are both correlated with the severity of epilepsy and neurologic outcome with regards to hemiparesis and visual fields (Kramer et al., 2000). By school age seizures tend to plateau and not be as severe but headaches and migraines can become a bigger problem. Migraines can begin at an early age (Klapper, 1994; Dora and Balkan, 2001; Kossoff et al., 2005) and, if improperly treated, can serve to trigger seizures and stroke-like episodes. The newer anticonvulsants now available, combined with low-dose aspirin and migraine therapy, have probably improved the overall prognosis for these patients. However, a small subset of patients still require hemispherectomy or other surgery for their seizures (Hoffman et al., 1979; Arzimanoglou et al., 2000). Many patients, upon reaching their adolescent and young adult years, still face many challenges and some will fail to cross the threshold into successful adult, independent life. Lack of sleep is an important trigger for migraines and seizures in young adults (Kossoff et al., 2005) and working on this along with diet, exercise, and avoiding triggers can make a big difference to the quality of life of some patients. Not enough information is understood or published about the natural history in older adults with Sturge–Weber syndrome (Sujansky and Conradi, 1995a; Lamy et al., 2003); this is an area of need for future work and research.

TREATMENT OF STURGE–WEBER SYNDROME BRAIN INVOLVEMENT

Given that impaired blood flow and seizures contribute to the early brain injury and neurologic morbidity in infants with Sturge–Weber syndrome, those diagnosed with brain involvement presymptomatically may be offered either treatment with anticonvulsants or low-dose aspirin. However, data demonstrating the benefit of presymptomatic treatment is still being gathered. Phenobarbital was studied for the presymptomatic treatment of infants with Sturge–Weber syndrome and a trend was noted for improved cognitive outcome. However, the infants receiving presymptomatic phenobarbital had less extensive brain involvement compared to those who were not treated presymptomatically (Ville et al., 2002). Low-dose aspirin is another reasonable approach as it has been used safely in affected infants as young as 1 month. More experience is needed to established whether or not this approach is beneficial in all cases (Lance et al., 2013). In infants with extensive bilateral brain involvement it is reasonable to consider presymptomatic treatment with both an anticonvulsant and low-dose aspirin since these infants are at highest risk for poor outcome and can suffer ischemic injury to most of their brain. This group of patients has the least to lose and the most to gain from progress in the field and new potential treatment options should be offered to them as they become available.

It is important that families with infants at risk for Sturge–Weber syndrome brain involvement receive anticipatory guidance regarding what seizures in infants look like and how their caretaker should respond to them. This guidance is important because, without detailed information, some parents will think that seizures have to be generalized jerking. They may watch twitching of a foot for a long time before seeking help if they do not realize these movements are seizures (author's experience). After 3 months of age, parents may administer rectal diazepam. Because seizures that are prolonged and repeated in young children with Sturge–Weber syndrome can contribute to infarction, one should try to attain and maintain complete seizure control, or at least limit prolonged seizures. The decompensations and seizures are most likely to occur during illnesses. Reasonable efforts to keep the child well include the administration of annual influenza vaccines, regularly increasing anticonvulsant doses for weight gain particularly in infants and young children, screening for and treating iron deficiency anemia, and taking history to screen for excessive snoring and sleep apnea (Irving et al., 2010). Minor head injury is also a common cause of triggering seizures in toddlers and children with Sturge–Weber syndrome (Zolkipli et al., 2007). Therefore, parents need to be instructed on how to guide their children into suitable sports and activities.

As previously stated, one of the greatest challenges in the treatment of Sturge–Weber syndrome is the tremendous variability in the severity of neurologic involvement and outcome. It is very difficult early on to provide a detailed prognosis and therefore to titrate optimally aggressive treatment. The only factors which have been shown to provide prognostic value are the extent of brain involvement, especially unilateral versus bilateral brain involvement, and the early onset of difficult to control severe seizures during the first few months of life (before 9 months of age). The period of greatest risk for severe deterioration is during the first 4–5 years of life. FDG (fluorodeoxyglucose) positron emission tomography (PET) imaging has demonstrated that glucose hypometabolism tends to stabilize by around 4 years of age (Lee et al., 2001). Around this age, the patient's seizure status also tends to stabilize at whatever their severity has been, or even improve, along with stabilization of their neurologic status. Therefore, once seizure freedom is attained then anticonvulsants are usually continued until at least 4 years of age. Frequently, instead of weaning their child off the anticonvulsants completely, parents will decide to not continue to increase the dose of medication for weight gain if the child has been seizure-free for several years and is at least 3 or 4 years old. FDG PET imaging has also demonstrated that when a child with Sturge–Weber syndrome remains seizure-free for a year, then their glucose hypometabolism can improve. Seizure-freedom is also associated with improvements in hemiparesis and visual field deficits and in cognitive progress.

Use of low-dose aspirin in children with Sturge–Weber syndrome to reduce stroke-like episodes and seizures, and thereby improve neurologic and cognitive outcome, has been controversial. Initially proposed by Garcia and colleagues (1981), the use of low-dose aspirin has been studied by several groups. Maria et al. reported patients with Sturge–Weber syndrome having fewer stroke-like episodes when treated with low-dose aspirin (Maria et al., 1998). Udani et al. reported that children taking low-dose aspirin also had greatly reduced seizures (Udani et al., 2007). Bay et al. reported results of an online survey demonstrating that patients with Sturge–Weber syndrome reported significantly fewer stroke-like episodes and seizures in the year following initiation of low-dose aspirin compared to the year prior to starting low-dose aspirin (Bay et al., 2011). Lance et al. reported the clinical experience over the past decade at their Sturge–Weber syndrome clinic in almost 60 children with Sturge–Weber syndrome; and low-dose aspirin was safely given to infants as young as 1 month of age. The most common side-effects were increased bruising and nose bleeds and most patients with these issues did not have to stop the aspirin (Lance et al., 2013). Subdural and subgaleal hematomas have rarely been reported in patients with Sturge–Weber syndrome both on and off low-dose aspirin (Greco et al., 2008; Lopez et al., 2013). There is a need for continued study of this approach.

For those patients whose seizures fail to respond to medical management with anticonvulsants, additional treatment options include hemispherectomy, focal resection, the ketogenic or Atkins diet (Kossoff et al., 2010), and the vagus nerve stimulator. For unilaterally involved patients, surgery should be considered for those who have failed two or more anticonvulsants combined with low-dose aspirin (Arzimanoglou, et al., 2000; Kossoff et al., 2002). The decision to proceed with surgery is easier in those patients who also have hemiparesis and a significant visual field deficit in addition to medically refractory seizures. Surgery should also be seriously considered in patients whose cognitive development is falling further and further behind normal, even if their seizures and other neurologic symptoms are not so very severe. The most common scenario where surgery is pursued, in my experience, is the young child with frequent seizures or clusters of severe seizures, a significant hemiparesis, and visual field deficits. For these children the need for surgery is obvious and the risk:benefit ratio greatly supports their having the surgery. Experience supports the conclusion that seizure control results in the best chance of optimal cognitive and neurologic development whether medically managed or treated with the surgery; when medical management fails surgery should be considered, especially when hemiparesis and visual deficits are already present. Small published series of patients with Sturge–Weber syndrome suggest that hemispherectomy is very successful in eliminating seizures in most cases (about 90% effective), focal resections significantly less so (Arzimanoglou et al., 2000).

The child with Sturge–Weber syndrome and extensive bilateral brain involvement has the highest risk of poor neurologic and cognitive outcome. Essentially most of their cortex is at risk for ischemic brain injury, atrophy, and calcification. For these infants, very aggressive treatment with anticonvulsants and low-dose aspirin is warranted. These children are often not considered good surgical candidates; hemispherectomy has only been recommended in bilaterally affected children with very severe disabling seizures primarily coming from one hemisphere and the surgery is considered palliative rather than potentially curative (Tuxhorn and Pannek, 2002).

A significant percentage of school-age children with Sturge–Weber syndrome also have attention problems which require treatment in order for the child to perform optimally in school. In a small series of patients, stimulant medications improved impaired attention in most individuals (Lance et al., 2014). The side-effects noted were typical of those seen with the use of stimulants; appetite suppression, headaches, and impaired sleep. Eight out of the 13 patients benefited from the stimulant with improved ability to focus during class, ability to complete homework, and improved grades in school. No severe complications were noted. Neuropsychological testing is helpful for evaluating for evidence of attention deficit and need for medical treatment.

CONTROVERSIES AND GOALS FOR FUTURE RESEARCH

The tremendous variability in the extent of clinical involvement with the vascular malformation in the brain, skin, and eye, and the difficulty in controlling the seizures results in a great range in severity of symptoms, difficulty in providing prognosis in infants and young children, and problems with developing treatment strategies that are universally successful. For this reason, several significant controversies exist and require further and multicentered evaluation. The use of low-dose aspirin, while gaining traction, has continued to be controversial, even in centers recommending it for most patients. Questions regarding its use for older children and adults remain. Timing of, and patient selection for, surgery is also a complex question. Infants with bilateral involvement continue to have a poor prognosis, and side-effects from medications or morbidity from surgery are persistent issues as well.

The landscape of important future research questions has been clarified by the recent gene mutation discovery. Analysis of brain, eye, and skin tissue immunohistochemistry can now be done in the context of the underlying somatic mosaic mutation in *GNAQ*. Research is already underway to determine the cell type(s) hosting the mutation in Sturge–Weber syndrome; this discovery is essential to understanding pathogenesis and to developing further *in vitro* and animal models. The putative downstream pathways impacted by the mutation require further study and will suggest the most relevant clinical trials for the future. Identification of the underlying causative somatic mutation in *GNAQ* is likely to lead to small clinical trials on a compassionate use basis of agents likely to address the pathways impacted by the mutation. These efforts will lead to more controversies in the short term unless centers work together to study sufficient numbers of subjects to draw substantial conclusions on which to base treatment recommendations. Planning for multicentered studies is currently underway. Biomarker development, already underway for the last several years, gains further importance as novel clinical strategies are discussed and attempted. Urine vascular biomarkers have been shown to be present with abnormally high frequency and amounts in subjects with Sturge–Weber syndrome and to correlate with clinical status (Sreenivasan et al., 2013). Quantitative EEG (Hatfield et al., 2004; Ewen et al., 2009), transcranial Doppler, and various neuroimaging techniques also show promise as biomarkers to monitor treatment response or to select patients for treatment. It is likely that the next decade will bring significant advances in our understanding and treatment of Sturge–Weber syndrome.

ACKNOWLEDGMENTS

We acknowledge editorial assistance from Cathy Bachur. We are also grateful for funding from the National Institute of Neurological Disorders and Stroke (NINDS) (National Institutes of Health [NIH] U54NS065705) and from Hunter's Dream for a Cure Foundation. The Brain Vascular Malformation Consortium (U54NS065705) is a part of the NIH Rare Disease Clinical Research Network, supported through a collaboration between the NIH Office of Rare Diseases Research at the National Center for Advancing Translational Science and the NINDS.

References

Aboutalebi A, Jessup CJ, North PE et al. (2012). Histopathology of vascular anomalies. Facial Plast Surg 28: 545–553.

Al Robaee A, Banka N, Alfadley A (2004). Phakomatosis pigmentovascularis type IIb associated with Sturge–Weber syndrome. Pediatr Dermatol 21: 642–645.

Ambrosetto P, Ambrosetto G, Michelucci R et al. (1983). Sturge–Weber syndrome without port-wine facial nevus. Report of 2 cases studied by CT. Childs Brain 10: 387–392.

Argraves WS, Drake CJ (2005). Genes critical to vasculogenesis as defined by systematic analysis of vascular defects in knockout mice. Anat Rec A Discov Mol Cell Evol Biol 286: 875–884.

Arzimanoglou AA, Andermann F, Aicardi J et al. (2000). Sturge–Weber syndrome: indications and results of surgery in 20 patients. Neurology 55: 1472–1479.

Aylett SE, Neville BG, Cross JH et al. (1999). Sturge–Weber syndrome: cerebral haemodynamics during seizure activity. Dev Med Child Neurol 41: 480–485.

Bay MJ, Kossoff EH, Lehmann CU et al. (2011). Survey of aspirin use in Sturge–Weber syndrome. J Child Neurol 26: 692–702.

Comati A, Beck H, Halliday W et al. (2007). Upregulation of hypoxia-inducible factor (HIF)-1alpha and HIF-2alpha in leptomeningeal vascular malformations of Sturge–Weber syndrome. J Neuropathol Exp Neurol 66: 86–97.

Comi AM, Fischer R, Kossoff EH (2003). Encephalofacial angiomatosis sparing the occipital lobe and without facial nevus: on the spectrum of Sturge–Weber syndrome variants? J Child Neurol 18: 35–38.

Comi AM, Bellamkonda S, Ferenc LM et al. (2008). Central hypothyroidism and Sturge–Weber syndrome. Pediatr Neurol 39: 58–62.

Comi AM, Marchuk DA, Pevsner J (2013). A needle in a haystack: Sturge–Weber syndrome gene discovery. Pediatr Neurol 49: 391–392.

Cunha e Sá M, Barroso CP, Caldas MC et al. (1997). Innervation pattern of malformative cortical vessels in Sturge–Weber disease: an histochemical, immunohistochemical, and ultrastructural study. Neurosurgery 41: 872–876.

Dora B, Balkan S (2001). Sporadic hemiplegic migraine and Sturge–Weber syndrome. Headache 41: 209–210.

Eerola I, Boon LM, Mulliken JB et al. (2003). Capillary malformation-arteriovenous malformation, a new clinical and genetic disorder caused by RASA1 mutations. Am J Hum Genet 73: 1240–1249.

Enjolras O, Riche MC, Merland JJ (1985). Facial port-wine stains and Sturge–Weber syndrome. Pediatrics 76: 48–51.

Ewen JB, Comi AM, Kossoff EH (2007). Myoclonic-astatic epilepsy in a child with Sturge–Weber syndrome. Pediatr Neurol 36: 115–117.

Ewen JB, Kossoff EH, Crone NE et al. (2009). Use of quantitative EEG in infants with port-wine birthmark to assess for Sturge–Weber brain involvement. Clin Neurophysiol 120: 1433–1440.

Fischbein NJ, Barkovich AJ, Wu Y et al. (1998). Sturge–Weber syndrome with no leptomeningeal enhancement on MRI. Neuroradiology 40: 177–180.

Fukuyama Y, Tsuchiya S (1979). A study on Sturge–Weber syndrome. Report of a case associated with infantile spasms and electroencephalographic evolution in five cases. Eur Neurol 18: 194–204.

Furukawa T, Igata A, Toyokura Y (1970). Ikeda S: Sturge–Weber and Klippel–Trénaunay syndrome with nevus of ota and ito. Arch Dermatol 102: 640–645.

Garcia JC, Roach ES, McLean WT (1981). Recurrent thrombotic deterioration in the Sturge–Weber syndrome. Childs Brain 8: 427–433.

Greco F, Fiumara A, Sorge G et al. (2008). Subgaleal hematoma in a child with Sturge–Weber syndrome: to prevent stroke-like episodes, is treatment with aspirin advisable? Childs Nerv Syst 24: 1479–1481.

Griffiths PD, Boodram MB, Blaser S et al. (1997). 99mTechnetium HMPAO imaging in children with the Sturge–Weber syndrome: a study of nine cases with CT and MRI correlation. Neuroradiology 39: 219–224.

Hagiwara K, Uezato H, Nonaka S (1998). Phacomatosis pigmentovascularis type IIb associated with Sturge–Weber syndrome and pyogenic granuloma. J Dermatol 25: 721–729.

Happle R (1987). Lethal genes surviving by mosaicism: a possible explanation for sporadic birth defects involving the skin. J Am Acad Dermatol 16: 899–906.

Harbour JW (2012). The genetics of uveal melanoma: an emerging framework for targeted therapy. Pigment Cell Melanoma Res 25: 171–181.

Hatfield LA, Crone NE, Kossoff EH et al. (2004). Quantitative EEG analysis correlation with clinical severity in unilateral Sturge–Weber syndrome. Epilepsia 45 (s7): 67.

Hoffman HJ, Hendrick EB, Dennis M et al. (1979). Hemispherectomy for Sturge–Weber syndrome. Childs Brain 5: 233–248.

Hu J, Yu Y, Juhasz C et al. (2008). MR susceptibility weighted imaging (SWI) complements conventional contrast enhanced T1 weighted MRI in characterizing brain abnormalities of Sturge–Weber syndrome. J Magn Reson Imaging 28: 300–307.

Irving ND, Lim JH, Cohen B et al. (2010). Sturge–Weber syndrome: ear, nose, and throat issues and neurologic status. PediatrNeurol 43: 241–244.

Jiang BH, Liu LZ (2008). Role of mTOR in anticancer drug resistance: perspectives for improved drug treatment. Drug Resist Updat 11: 63–76.

Kiley MA, Oxbury JM, Coley SC (2002). Intracranial hypertension in Sturge–Weber/Klippel–Trénaunay–Weber overlap syndrome due to impairment of cerebral venous outflow. J Clin Neurosci 9: 330–333.

Kimple AJ, Bosch DE, Giguere PM et al. (2011). Regulators of G-protein signaling and their Galpha substrates: promises and challenges in their use as drug discovery targets. Pharmacol Rev 63: 728–749.

Klapper J (1994). Headache in Sturge–Weber syndrome. Headache 34: 521–522.

Kossoff EH, Buck C, Freeman JM (2002). Outcomes of 32 hemispherectomies for Sturge–Weber syndrome worldwide. Neurology 59: 1735–1738.

Kossoff EH, Hatfield LA, Ball KL et al. (2005). Comorbidity of epilepsy and headache in patients with Sturge–Weber syndrome. J Child Neurol 20: 678–682.

Kossoff EH, Ferenc L, Comi AM (2009). An infantile-onset, severe, yet sporadic seizure pattern is common in Sturge–Weber syndrome. Epilepsia 50: 2154–2157.

Kossoff EH, Borsage JL, Comi AM (2010). A pilot study of the modified Atkins diet for Sturge–Weber syndrome. Epilepsy Res 92: 240–243.

Kramer W (1963). Syndromes of Klippel–Trénaunay and Sturge–Weber in the same patient. Psychiatr Neurol Neurochir 66: 362–370.

Kramer U, Kahana E, Shorer Z et al. (2000). Outcome of infants with unilateral Sturge–Weber syndrome and early onset seizures. Dev Med Child Neurol 42: 756–759.

Lachlan KL (2013). Cowden syndrome and the PTEN hamartoma tumor syndrome: how to define rare genetic syndromes. J Natl Cancer Inst 105: 1595–1597.

Lamy C, Domigo V, Semah F et al. (2003). Early and late seizures after cryptogenic ischemic stroke in young adults. Neurology 60: 400–404.

Lance EI, Sreenivasan AK, Zabel TA et al. (2013). Aspirin use in Sturge–Weber syndrome: side effects and clinical outcomes. J Child Neurol 28: 213–218.

Lance EI, Lanier KE, Zabel TA, Comi AM (2014). Stimulant use in patients with sturge-weber syndrome: safety and efficacy. Pediatr Neurol 51 (5): 675–680.

Laycock-van Spyk S, Thomas N, Cooper DN et al. (2011). Neurofibromatosis type 1-associated tumours: their somatic mutational spectrum and pathogenesis. Hum Genomics 5: 623–690.

Lee JS, Asano E, Muzik O et al. (2001). Sturge–Weber syndrome: correlation between clinical course and FDG PET findings. Neurology 57: 189–195.

Leung AK, Lowry RB, Mitchell I et al. (1988). Klippel–Trénaunay and Sturge–Weber syndrome with extensive Mongolian spots, hypoplastic larynx and subglottic stenosis. Clin Exp Dermatol 13: 128–132.

Lin DD, Barker PB, Hatfield LA et al. (2006). Dynamic MR perfusion and proton MR spectroscopic imaging in Sturge–Weber syndrome: correlation with neurological symptoms. J Magn Reson Imaging 24: 274–281.

Lopez J, Yeom KW, Comi A et al. (2013). Case report of subdural hematoma in a patient with Sturge–Weber syndrome and literature review: questions and implications for therapy. J Child Neurol 28: 672–675.

Maria BL, Neufeld JA, Rosainz LC et al. (1998). High prevalence of bihemispheric structural and functional defects in Sturge–Weber syndrome. J Child Neurol 13: 595–605.

Maton B, Krsek P, Jayakar P et al. (2010). Medically intractable epilepsy in Sturge–Weber syndrome is associated with cortical malformation: implications for surgical therapy. Epilepsia 51: 257–267.

Mettin RR, Merkenschlager A, Bernhard MK et al. (2014). Wide spectrum of clinical manifestations in children with tuberous sclerosis complex – follow-up of 20 children. Brain Dev 36: 306–314.

Miller RS, Ball KL, Comi AM et al. (2006). Growth hormone deficiency in Sturge–Weber syndrome. Arch Dis Child 91: 340–341.

Moffitt DL, Kennedy CT, Newbury-Ecob R (1999). What syndrome is this? Macrocephaly with cutis marmorata, hemangioma, and syndactyly syndrome. Pediatr Dermatol 16: 235–237.

Murakami N, Morioka T, Suzuki SO et al. (2012). Focal cortical dysplasia type IIa underlying epileptogenesis in patients with epilepsy associated with Sturge–Weber syndrome. Epilepsia 53: e184–e188.

Namer IJ, Battaglia F, Hirsch E et al. (2005). Subtraction ictal SPECT co-registered to MRI (SISCOM) in Sturge–Weber syndrome. Clin Nucl Med 30: 39–40.

Norman MG, Schoene WC (1977). The ultrastructure of Sturge–Weber disease. Acta Neuropathol (Berl) 37: 199–205.

Poduri A, Evrony GD, Cai X et al. (2013). Somatic mutation, genomic variation, and neurological disease. Science 341: 1237758.

Porto L, Kieslich M, Yan B et al. (2006). Accelerated myelination associated with venous congestion. Eur Radiol 16: 922–926.

Ray D, Mandal AK, Chandrasekhar G et al. (2010). Port-wine vascular malformations and glaucoma risk in Sturge–Weber syndrome. J AAPOS 14: 105.

Reesman J, Gray R, Suskauer SJ et al. (2009). Hemiparesis is a clinical correlate of general adaptive dysfunction in children and adolescents with Sturge–Weber syndrome. J Child Neurol 24: 701–708.

Rosen S, Smoller BR (1987). Port-wine stains: a new hypothesis. J Am Acad Dermatol 17: 164–166.

Rydh M, Malm M, Jernbeck J et al. (1991). Ectatic blood vessels in port-wine stains lack innervation: possible role in pathogenesis. Plast Reconstr Surg 87: 419–422.

Schirmer R (1860). Ein Fall von Telangiektasia. Albrecht von Graefes Archiv für Ophtalmologie 7: 119–121.

Shirley MD, Tang H, Gallione CJ et al. (2013). Sturge–Weber syndrome and port-wine stains caused by somatic mutation in GNAQ. N Engl J Med 368: 1971–1979.

Siddique L, Sreenivasan A, Comi AM et al. (2013). Importance of utilizing a sensitive free thyroxine assay in Sturge–Weber syndrome. J Child Neurol 28: 269–274.

Smoller BR, Rosen S (1986). Port-wine stains. A disease of altered neural modulation of blood vessels? Arch Dermatol 122: 177–179.

Sreenivasan AK, Bachur CD, Lanier KE et al. (2013). Urine vascular biomarkers in Sturge–Weber syndrome. Vasc Med 18: 122–128.

Sturge WA (1879). A case of partial epilepsy, apparently due to a lesion of one of the vasomotor centres of the brain. Trans Clin Soc Lond 12: 162–167.

Sujansky E, Conradi S (1995a). Outcome of Sturge–Weber syndrome in 52 adults. Am J Med Genet 57: 35–45.

Sujansky E, Conradi S (1995b). Sturge–Weber syndrome: age of onset of seizures and glaucoma and the prognosis for affected children. J Child Neurol 10: 49–58.

Sullivan TJ, Clarke MP, Morin JD (1992). The ocular manifestations of the Sturge–Weber syndrome. J Pediatr Ophthalmol Strabismus 29: 349–356.

Tallman B, Tan OT, Morelli JG et al. (1991). Location of port-wine stains and the likelihood of ophthalmic and/or central nervous system complications. Pediatrics 87: 323–327.

Tuxhorn IE, Pannek HW (2002). Epilepsy surgery in bilateral Sturge–Weber syndrome. Pediatr Neurol 26: 394–397.

Udani V, Pujar S, Munot P et al. (2007). Natural history and magnetic resonance imaging follow-up in 9 Sturge–Weber syndrome patients and clinical correlation. J Child Neurol 22: 479–483.

Ville D, Enjolras O, Chiron C et al. (2002). Prophylactic antiepileptic treatment in Sturge–Weber disease. Seizure 11: 145–150.

Ward NL, Lamanna JC (2004). The neurovascular unit and its growth factors: coordinated response in the vascular and nervous systems. Neurol Res 26: 870–883.

Weber FP (1922). Right-sided hemi-hypertrophy resulting from right-sided congenital spastic hemiplegia, with a morbid condition of the left side of the brain, revealed by radiograms. J Neurol Psychopathol 3: 134–139.

Willenberg T, Baumgartner I (2008). Vascular birthmarks. Vasa 37: 5–17.

Wu J, Tarabishy B, Hu J et al. (2011). Cortical calcification in Sturge–Weber syndrome on MRI-SWI: relation to brain perfusion status and seizure severity. J Magn Reson Imaging 34: 791–798.

Zabel TA, Reesman J, Wodka EL et al. (2010). Neuropsychological features and risk factors in children with Sturge–Weber syndrome: four case reports. Clin Neuropsychol 24: 841–859.

Zolkipli Z, Aylett S, Rankin PM et al. (2007). Transient exacerbation of hemiplegia following minor head trauma in Sturge–Weber syndrome. Dev Med Child Neurol 49: 697–699.

Handbook of Clinical Neurology, Vol. 132 (3rd series)
Neurocutaneous Syndromes
M.P. Islam and E.S. Roach, Editors

Chapter 12

PHACE(S) syndrome

GEOFFREY L. HEYER*

Division of Pediatric Neurology, Nationwide Children's Hospital and Department of Neurology, Ohio State University, Columbus, OH, USA

INTRODUCTION

PHACE(S) is a neurocutaneous disorder of unknown etiology. The PHACE acronym refers to the association of *p*osterior fossa malformations, large facial *h*emangiomas, cerebral *a*rterial anomalies, *c*ardiovascular anomalies, and *e*ye anomalies. (Frieden et al., 1996) When ventral developmental defects such as sternal clefting or supraumbilical raphe occur, the PHACES acronym is used.

History

Several early reports demonstrated an association between cutaneous hemangiomas affecting the head and neck and a spectrum of extracutaneous anomalies. In 1978, Pascual-Castroviejo and colleagues published a series of seven female patients with infantile hemangiomas and malformations of the cerebral vasculature (Pascual-Castroviejo, 1978). The report emphasized three basic types of cerebral vascular anomaly: (1) anomalous origin or hypoplasia of the principal cerebral arteries; (2) persistence of embryonic arteries; and (3) "angiomatous" arterial malformations, referring to aneurysmal dilatations and anomalous vessels predominantly located at the carotid siphon and the hypothalamic region. Three of the seven patients in their series had abnormalities of the posterior fossa: Dandy–Walker malformation, partial agenesis of the cerebellar vermis, and a posterior fossa arachnoid cyst. In the same year, Stoeter and Marquardt described a 3-year-old girl with a large facial hemangioma, Dandy–Walker malformation, and a persistent trigeminal artery(Stoeter and Marquardt, 1978). Hirsch and colleagues reviewed the clinical features of 40 patients with Dandy–Walker malformations in 1984, noting that four patients in their series had associated facial hemangiomas (Hirsch et al., 1984).

Origin of the name

In 1996, Frieden and colleagues summarized the clinical features of two new and 41 published cases with large facial hemangiomas and associated brain, heart, eye, and arterial abnormalities. Their summary revealed that all patients had hemangiomas; 88% were female; 74% had posterior fossa malformations; 41% had cerebral vascular anomalies (although not all cases had adequate cerebral vascular imaging); 26% had cardiac or aortic malformations; 10% had eye abnormalities; and 7% had sternal clefting or supraumbilical raphe (Frieden et al., 1996). They proposed the PHACES acronym to emphasize the characteristic findings of this neurocutaneous disorder (Frieden et al., 1996). Various other names have been suggested including 3-C syndrome (*c*erebellar hypoplasia, *c*avernous hemangioma, and *c*oarctation of the aorta), cutaneous hemangioma–vascular complex syndrome, and Pascual-Castroviejo syndrome type II (Goh and Lo, 1993; Pascual-Castroviejo et al., 1996, 2005).

Epidemiology

Infantile hemangiomas are the most common benign tumors of infancy with an estimated incidence of 4–5% of all live births (Kilcline and Frieden, 2008). A subgroup of these patients has been found to have associated extracutaneous anomalies of PHACE, but the overall incidence of PHACE syndrome is not known. In a prospective, multi-institutional study of 1096 children 12 years of age and younger with infantile hemangiomas, 25 (2.3%) met diagnostic criteria for PHACE syndrome (Metry et al., 2006). These 25 patients represented 20% of all children with large, segmental facial hemangiomas. PHACE diagnoses outnumbered

*Correspondence to: Geoffrey L. Heyer, MD, Departments of Pediatrics and Neurology, Nationwide Children's Hospital, 700 Children's Drive, ED-5, Columbus, OH 43205, USA. Tel: +1-614-722-4625, Fax: +1-614-722-4633, E-mail: geoffrey.heyer@nationwidechildrens.org

Sturge–Weber diagnoses during the study period, suggesting that PHACE syndrome may be as common (or more common) than the better-characterized Sturge–Weber syndrome (Metry et al., 2006). Large facial hemangiomas appear to confer greater risk of associated extracutaneous anomalies. A subsequent multicenter, prospective study of 108 children with large, segmental facial hemangiomas showed that 33 (31%) met diagnostic criteria for PHACE syndrome (Haggstrom et al., 2010). Infants with PHACE syndrome had significantly larger hemangiomas than those without PHACE syndrome (mean 120 cm^2 versus 68 cm^2). The risk of extracutaneous manifestations of PHACE increased with increasing hemangioma size and when the hemangioma(s) affected multiple facial segments (see facial segment description below) (Haggstrom et al., 2010). While infantile hemangiomas generally affect girls more than boys, a striking female predominance is seen among PHACE patients (Gorlin et al., 1994; Haggstrom et al., 2007). The female-to-male ratio for patients with PHACE syndrome ranges from 5.6 to 9:1 (Haggstrom et al., 2010; Metry et al., 2006).

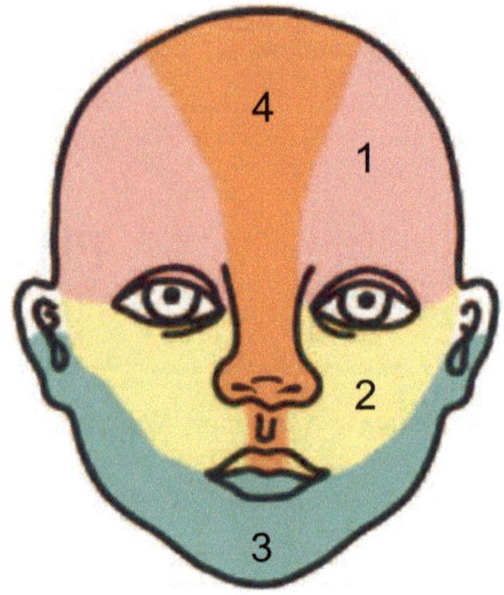

Fig. 12.1. Diagram illustrates the proposed hemangioma segmental patterns. Segment 1 follows a frontotemporal distribution; segment 2 has a maxillary distribution; segment 3 has a mandibular distribution; and segment 4 involves the frontonasal region and includes the philtrum. (Reproduced from Haggstrom et al., 2006, with permission from the American Academy of Pediatrics.)

CLINICAL MANIFESTATIONS

Typical age of presentation

In contrast to the capillary malformation, often referred to as a port wine stain, of Sturge–Weber syndrome, which is always present at birth, grows commensurately with the child, and never regresses, infantile hemangiomas may not be conspicuous at birth, exhibit a rapid growth phase during the neonatal period, and tend to regress slowly over a course of months or years. Cutaneous hemangiomas may begin as reticulated or confluent erythematous patches which resemble capillary malformations or as telangiectatic regions surrounded by pallor. Within weeks, erythematous papules and plaques develop confirming the hemangioma diagnosis. In some instances the hemangioma fails to proliferate and maintains a reticular or telangiectatic appearance (Corella et al., 2008). The infantile hemangioma also differs from the facial birthmarks associated with Wyburn-Mason syndrome, which represents a quiescent Schobinger type-1 arteriovenous malformation, and the capillary malformation–AVM syndrome (Theron et al., 1974; Garzon et al., 2007).

Facial hemangiomas do not follow dermatomal distributions or lines of Blaschko. They can be categorized as focal lesions with clear spatial containment or as plaque-like segmental lesions. Studies of hemangioma patterns have identified four primary facial segments (Fig. 12.1): the frontotemporal region (segment 1); the maxillary region respecting the nasomedial sulcus (segment 2); the mandible, chin, and lower lip (segment 3); and the medial frontal scalp, nasal bridge, and philtrum (segment 4) (Haggstrom et al., 2006). The segmental patterns help to standardize the nomenclature of facial segmental hemangiomas. PHACE patients can have hemangiomas that follow one or more of the segmental patterns and occur unilaterally or bilaterally. The presence of large, plaque-like hemangiomas occupying at least one of the facial segments warrants an evaluation for the associated extracutaneous manifestations of PHACE syndrome.

Hallmark signs and symptoms

Over 300 patients with PHACE syndrome have been described, mostly as single case reports and small case series (Metry et al., 2006). The hallmark of the disorder is the facial hemangioma which is typically large and segmental. The characteristic features of the infantile hemangioma distinguish PHACE syndrome from other neurocutaneous disorders that have associated red birthmarks. The extracutaneous features include abnormalities of brain, artery, heart, eye, and chest wall abnormalities. However, not all PHACE patients will have anomalies of every organ system. In 2008, an expert panel was convened to develop a consensus statement for the diagnostic criteria (Table 12.1) (Metry et al., 2009). The relatively inclusive criteria distinguish patients with cutaneous and extracutaneous disease manifestations as PHACE syndrome or *possible* PHACE syndrome. Major and minor diagnostic criteria are described. PHACE syndrome is characterized by a segmental hemangioma or a facial or scalp hemangioma (>5 cm in diameter) plus one major criterion

Table 12.1

Diagnostic criteria for PHACE syndrome*

PHACE syndrome				
Facial hemangioma > 5 cm in diameter Plus One major or two minor criteria				
Possible PHACE syndrome				
Facial hemangioma > 5 cm in diameter Plus One minor criterion	OR	Hemangioma of the neck or upper torso Plus One major or two minor criteria	OR	No hemangioma Plus Two major criteria

Organ system	Major criteria	Minor criteria
Cerebrovascular	Anomaly of major cerebral artery(ies)[1] Dysplasia[2] Arterial stenosis or occlusion with or without moyamoya collaterals Absence or hypoplasia Aberrant origin or course Persistent trigeminal artery Saccular aneurysm(s)	Persistent embryonic artery (other than trigeminal) Proatlantal, hypoglossal, or otic arteries
Structural brain	Posterior fossa anomaly Dandy–Walker complex or cerebellar hypo-/dysplasia	Enhancing extra-axial lesion(s) consistent with intracranial hemangioma Midline anomaly[3] Neuronal migration disorder[4]
Cardiovascular	Aortic arch anomaly Coarctation Aortic aneurysm Aberrant origin of the subclavian artery with or without a vascular ring	Ventricular septal defect Right aortic arch
Ocular	Posterior segment abnormality Persistent fetal vasculature Retinal vascular anomalies Morning Glory disc anomaly Optic nerve hypoplasia Peripapillary staphyloma	Anterior segment abnormality Sclerocornea Cataract Coloboma Microphthalmia
Ventral or midline	Sternal defect(s) Sternal cleft Supraumbilical raphe	Hypopituitarism Ectopic thyroid

*Reproduced from Metry et al., Consensus Statement on Diagnostic Criteria for PHACE Syndrome. Pediatrics, 124, (5) 1447–1456, Copyright 2009 American Academy of Pediatrics.

[1]Internal carotid, middle cerebral, anterior cerebral, or posterior cerebral artery, or vertebrobasilar system.

[2]Includes arterial kinking, looping, tortuosity, or dolichectasia.

[3]Callosal agenesis or dysgenesis, septum pellucidum agenesis, or pituitary malformation or ectopia.

[4]Polymicrogyria, cortical dysplasia, or gray matter heterotopia(s).

or two minor criteria (Metry et al., 2009). Possible PHACE syndrome can be diagnosed in three circumstances: (1) facial or scalp hemangioma (>5 cm in diameter) plus one minor criterion; (2) hemangioma of the neck or upper torso plus one major criterion or two minor criteria; or (3) no hemangioma but two major criteria.

Involvement by systems

Multiple associated anomalies have been described with PHACE syndrome. The anomalies that occur commonly are emphasized in the diagnostic criteria (Metry et al., 2009) and summarized below. While determining major and minor diagnostic criteria, members of the expert panel calculated the relative frequencies of each criterion as reported in the literature and from existing PHACE syndrome registries (Table 12.2) (Metry et al., 2009). Notably, adequate brain, cerebral vascular, and cardiac imaging was not performed in all published and registry patients, lowering the relative frequencies of those clinical features that require dedicated testing to detect.

Brain anomalies

The spectrum of posterior fossa abnormalities ranges from focal cerebellar lesions to the Dandy–Walker complex. In the initial series of seven patients with hemangiomas and associated cerebral vascular lesions reported by Pascual-Castroviejo and colleagues, three had posterior fossa anomalies: Dandy–Walker malformation, partial agenesis of the cerebellar vermis, and a posterior fossa arachnoid cyst (Pascual-Castroviejo, 1978). In 1984, Hirsch and colleagues documented facial hemangiomas in four (10%) of their 40 patients with Dandy–Walker

Table 12.2

Relative frequencies of major and minor PHACE diagnostic criteria*

Parameter	Published (n = 317), %	Texas Children's Registry (n = 46), %	Genetic Analysis of PHACE OHSU (n = 65), %
Major			
Segmental infantile hemangioma of the head	92.4	95.7	100
Infantile hemangioma of head > 5 cm	92.4	95.7	100
Dysplasia of large cerebral artery[1]	15.4	30.5	13.9
Stenosis, occlusion, absence, or hypoplasia of large cerebral artery[2]	18.3	52.2	60.4
Aberrant origin or course of large cerebral artery	4.4	10.9	2
Persistent trigeminal artery	4.7	13	2
Posterior fossa anomaly	32.2	30.4	34.5
Coarctation of the aorta	14.5	6.5	28.3
Anomalous subclavian	6.3	15.2	13.9
Posterior segment anomaly of the eye	5.7	8.7	7.7
Sternal defect	19.5	10.9	21.6
Umbilical raphe	9.1	4.3	9.2
Minor			
Segmental infantile hemangioma of the neck or chest only	0.3	4.3	0
Persistent embryonic artery other than trigeminal artery	1.6	8.7	6.9
Midline anomalies[3]	6.3	6.5	1.5
Disorders of migration[4]	2.8	4.3	6.2
Ventricular septal defect	7.6	8.7	13.2
Anterior segment defect of the eye	6.6	8.7	7.7
Sternal papule	1.6	2.2	1.5

*Reproduced from Metry et al., Consensus Statement on Diagnostic Criteria for PHACE Syndrome. Pediatrics, 124, (5) 1447–1456, Copyright 2009 American Academy of Pediatrics.

[1]Includes kinking, looping, fusiform aneurysm, and/or dolichoectasia.

[2]Internal carotid, middle cerebral, anterior cerebral, or posterior cerebral arteries, or vertebrobasilar system.

[3]Callosal agenesis or dysgenesis, septum pellucidum agenesis, pituitary malformations, or pituitary ectopia.

[4]Polymicrogyria, cortical dysplasia, or gray matter heterotopia(s).

malformations (Hirsch et al., 1984). They identified an additional seven patients from the extant literature with Dandy–Walker malformations and facial hemangiomas (including the case reported by Pascual-Castroveio et al. and described above) (Carmel et al., 1977; Pascual-Castroviejo, 1978; Stoeter and Marquardt, 1978; Sawaya and McLaurin, 1981). Nine (73%) of the 11 patients were female, highlighting a female predominance in PHACE syndrome that would be corroborated by every large case series to follow. In 1993, Reese and colleagues described nine additional patients with posterior fossa abnormalities, facial hemangiomas, and associated cerebrovascular, heart, and eye abnormalities (Reese et al., 1993). Their patients had a range of posterior fossa abnormalities that included Dandy–Walker malformation, cerebellar hypoplasia, and posterior fossa arachnoid cyst(s). The focal regions of cerebellar hypoplasia or dysplasia tend to occur on the same side as the cutaneous hemangioma. When neuroimaging is performed early after the appearance of a facial hemangioma, regions of contrast enhancement may overlie the hypoplastic cerebellar (or cerebral) region (Heyer et al., 2008). This enhancement likely represents slow flow through small, dysplastic blood vessels; however, these regions have not yet been pathologically evaluated. The overall prevalence of posterior fossa anomalies in PHACE patients varies by case series and registry from 30.4% to 81% (Frieden et al., 1996; Metry et al., 2001, 2009; Poetke et al., 2002). In a study of the neuroimaging features of PHACE syndrome, 24 (41%) of 59 patients had structural brain (infra- or supratentorial) abnormalities detected (Hess et al., 2010). All brain lesions that were unilateral had associated cerebral arteriopathy and cutaneous hemangioma on the ipsilateral side.

Cerebral lesions occur less commonly than lesions of the posterior fossa. These abnormalities include cerebral dysplasia, neuronal migration abnormalities, and midline defects. Brain magnetic resonance imaging (MRI) has shown a variety of cerebral abnormalities including pachygyria, polymicrogyria, thickening of the cerebral cortex, heterotopic gray matter, and cerebral volume loss (Pascual-Castroviejo et al., 1995; Aeby et al., 2003; Grosso et al., 2004; Poindexter et al., 2007; Heyer et al., 2008; Oza et al., 2008; Hess et al., 2010; Tangtiphaiboontana et al., 2013). Similar to focal regions of cerebellar dysplasia, when a focal cerebral lesion is present, it tends to occur on the same side as the facial hemangioma (Heyer et al., 2006). Midline defects including absence or hypoplasia of the corpus callosum, absence of the septum pellucidum, and structural (as well as functional) pituitary lesions have been reported (Geller et al., 1991; Burrows et al., 1998; Metry et al., 2001; Goddard et al., 2006; Tangtiphaiboontana et al., 2013). An endocrinology evaluation should be considered in all patients with large hemangiomas and PHACE syndrome, particularly those with radiographic findings of empty sella or an abnormal pituitary structure (Huang et al., 2000; Goddard et al., 2006).

Neurodevelopmental outcomes have not been well studied in this neurocutaneous disorder. Substantial cerebral malformation, ischemic stroke, the presence of seizures, or the finding of progressive cerebral vasculopathy can help to predict poor neurologic outcomes among PHACE patients, but motor and cognitive impairments can also be seen with less extensive brain or cerebral vascular involvement. In a study of 29 PHACE syndrome patients aged more than 1 year (median age: 4 years, 2 months) who underwent neurologic evaluation, 20 children had abnormal neurodevelopment (Tangtiphaiboontana et al., 2013). Brain structural abnormalities were present in 59% of those with abnormal development, compared to 33% of those with normal evaluations. Supratentorial brain lesions were predictive of abnormal development, while posterior fossa lesions were not predictive. Notably, the relatively small sample size prevented statistical detection of any modest effects (Tangtiphaiboontana et al., 2013).

Cerebral vascular anomalies

The cerebral vasculopathy of PHACE syndrome comprises a spectrum of congenital and progressive arterial abnormalities with rare reports of associated venous and cerebral sinus malformations (Pascual-Castroviejo et al., 1996, 2005; Metry et al., 2001; Lin et al., 2003; Bhattacharya et al., 2004; Smith et al., 2004; Drolet et al., 2006; Drosou et al., 2006; Heyer et al., 2008; Hess et al., 2010). When all studied cases undergo adequate vascular imaging, arterial anomalies are probably the most common extracutaneous feature of PHACE syndrome. The primary involvement of the arterial system is one of the features that distinguishes PHACE syndrome from other neurocutaneous disorders such as Sturge–Weber syndrome and cerebrofacial arteriovenous metameric syndrome which involve capillaries and veins (Metry et al., 2009). Based on case descriptions from published series, patient registries, and literature reviews (Heyer et al., 2008; Oza et al., 2008; Metry et al., 2009; Hess et al., 2010), the arterial anomalies can be classified according to their imaging appearance and their tendency to progress:

- *Dysplasia* includes arterial coiling, looping, kinking, elongation, ectasia, and focal or fusiform aneurysmal enlargement. Arterial dysplasia probably represents the most common type of cerebral vascular anomaly in PHACE syndrome.
- *Hypoplasia* is characterized by the abnormal narrowing of an arterial segment without signs of progressive narrowing.

- *Aberrant origin or course* describes substantial differences in the expected branching pattern or course of principal cervical or cerebral arteries.
- *Absence or agenesis* describes the lack of normal embryonic development or abnormal embryonic involution of a principal cervical or cerebral artery.
- *Persistent embryonic artery* constitutes the abnormal persistence of an embryonic anastomosis between anterior and posterior circulations. These anastomoses are referred to as presegmental arteries during the embryonic period. They include trigeminal, otic (acoustic), hypoglossal, and proatlantal arteries.
- *Stenosis or occlusion* is an abnormal arterial narrowing or closure that is not congenital. Arterial stenosis or occlusion can be difficult to distinguish from narrowing or agenesis, respectively, when serial neuroimaging studies have not been performed.

The ability to distinguish each type of arterial anomaly depends on the type of vascular imaging performed, the age of the patient, and the presence of serial studies. The relative frequencies of specific arterial anomalies can vary across published reports for a variety reasons: case inclusion criteria, the presence of adequate vascular imaging, and publication biases.

Arterial abnormalities in PHACE syndrome often correspond with the location of the cervicofacial hemangioma(s). In an analysis of reported PHACE cases, cardiac anomalies, aortic coarctation, and lesions of the supra-aortic vessels tended to coincide with cutaneous hemangiomas that involved the mandibular segment (S3), neck, shoulder, or chest (Heyer et al., 2008). Conversely, when hemangiomas were restricted to the temporal, periorbital, and maxillary locations, the cardiac, aortic, and supra-aortic anomalies occurred less frequently. Supra-aortic vessels refer to the subclavian and brachiocephalic arteries and the origins of the common carotid and vertebral arteries. A similar regional correlation could be demonstrated based on hemangioma laterality (Heyer et al., 2008). When the facial hemangioma was unilateral, the arteriopathy was present on the ipsilateral side; when bilateral hemangiomas were present, the arteriopathy tended to be bilateral.

Several of the common congenital arterial anomalies in PHACE syndrome can help to determine the embryonic period in which an insult might occur. For example, normal regression of the primitive (presegmental) vessels happens early in embryogenesis, prior to 36 days gestation (Padget, 1948; Lie, 1968). Disruption of normal vascular development during this period can result in persistence of one or more primitive arteries. The three Saltzman types of persistent trigeminal artery can be distinguished, in part, by hypoplasia or absence of the corresponding posterior communicating, basilar, or vertebral arteries (Saltzman, 1959; Lie, 1968). Persistent trigeminal arteries are detected in 2–17% of registries and published reports (Heyer et al., 2008; Metry et al., 2009). Similar to the persistence of embryonic arteries, congenital absence of the internal carotid artery relates to an early gestational insult. Primary agenesis of the internal carotid can be confirmed by the absence or aplasia of the corresponding carotid canal (Fig. 12.2). The basal mesenchyme consolidates to form the sphenoid cartilage at 35–36 days gestation (Padget, 1948; Streeter, 1949). If the internal carotid artery is not present when the skull base forms, the carotid canal will not develop.

Burrows and colleagues were the first to describe a progressive arteriopathy in PHACE syndrome (Burrows et al., 1998). Four of their eight reported cases had acquired arterial stenoses or occlusions on serial neuroimaging, three of whom had a moyamoya-like vasculopathy with large-vessel steno-occlusive disease and the development of small collateral vessels. Progressive vascular changes and moyamoya in PHACE syndrome have been further characterized since the original description in single case reports and small case series

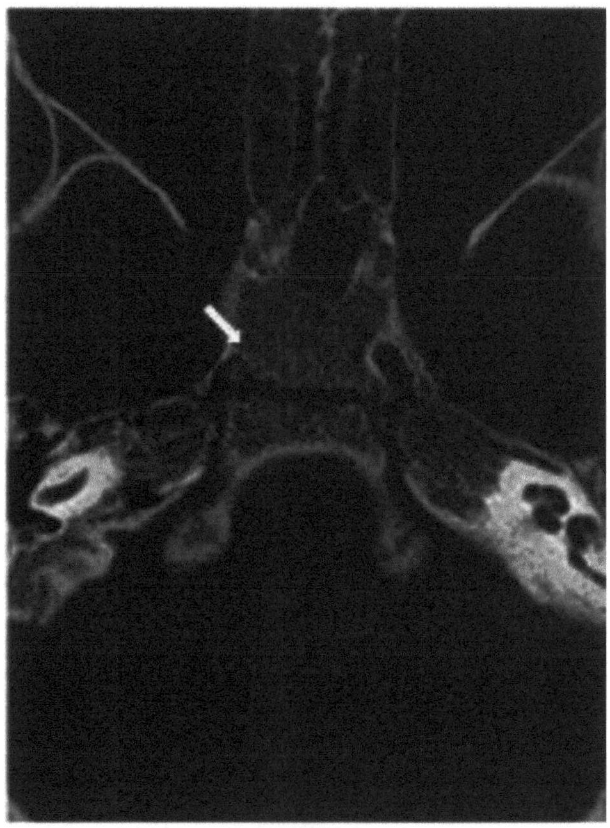

Fig. 12.2. Computed tomography (CT) image of the basal skull of a PHACE patient with agenesis of the right internal carotid artery. The arrow points to where a carotid canal should have formed. A normal carotid canal is present on the left side. (Modified from Heyer et al., 2008, with permission from Wolters Kluwer Health.)

(Luo et al., 2003; Bhattacharya et al., 2004; Drolet et al., 2006; Heyer et al., 2006). Unfortunately, when vascular imaging is performed late in the course of disease, often following the acute presentation of ischemic stroke, without baseline imaging, distinguishing between progressive stenosis or occlusion versus congenital arterial hypoplasia or agenesis can be difficult. The development of collateral vessels suggests progressive changes. Risk factors for progressive arteriopathy have not been determined. PHACE patients found to have moyamoya syndrome tend to have other congenital vascular anomalies also.

PHACE patients have several potential risk factors for arterial ischemic stroke. Stroke can occur from thromboembolic disease in patients with cardiac abnormalities, artery-to-artery thromboembolism when dysplastic large vessels provide a thrombus origin, and perfusion deficits related to the development of arterial occlusions or critical stenoses. Several cases of definite or probable ischemic stroke have been reported (Fisher, 1959; Isler 1984; Burrows et al., 1998; Luo et al., 2003; Bhattacharya et al., 2004; Lasky et al., 2004; Drolet et al., 2006; Heyer et al., 2006, 2008; Siegel et al., 2012). Steno-occlusive vasculopathy led to ischemic stroke in four of the patients reported by Burrows and colleagues; three of the four had developed moyamoya collaterals (Burrows et al., 1998). Pial synangiosis and other bypass procedures have successfully prevented repeat stroke in PHACE patients presenting with moyamoya and arterial ischemic stroke (Burrows et al., 1998; Heyer et al., 2008). A recent literature review and registry search led to the identification of 22 patients with PHACE syndrome and stroke (Siegel et al., 2012). Of the 22 cases, 20 (95%) had MR angiography and/or catheter-based angiography evidence of narrowing or nonvisualization of at least one principal vessel; 15 (75%) of 20 had evidence of narrowing or nonvisualization $\geq$2 principal vessels. Moreover, 16 (80%) of 20 had anterior and posterior circulation arteriopathy, and four (20%) of 20 had moyamoya collaterals (Siegel et al., 2012). Cardiovascular findings were reported in 15 (68%) of the 22 cases, most of which involved the aortic arch. Acute stroke was diagnosed in 15 (68%) of the 22 patients, with the remaining seven diagnosed after the acute period. Among those with acute stroke diagnoses the age ranged from 3 months to 5 years, median age 10.5 months (Siegel et al., 2012). It is presumed that stroke risk in PHACE syndrome continues into adulthood (Arora et al., 2011; Chou and Guo, 2012). As the hemangioma involutes with age, the hallmark cutaneous feature of PHACE syndrome may not be conspicuous. Clinicians may not suspect a neurocutaneous disorder in older children, adolescents, and adults. The subtlety of a regressed facial hemangioma is demonstrated in a 4-year-old boy with PHACE syndrome (Fig. 12.3A) and multiple congenital and progressive vascular anomalies (Fig. 12.3B). The presence of multiple cerebral vascular anomalies, with or without other extracutaneous PHACE features, in older patients should always prompt questioning about a childhood hemangioma. To our knowledge, hemorrhagic stroke has not been reported as a complication of PHACE syndrome. However, the presence of saccular aneurysms, the friable collateral vessels of moyamoya, and the rare finding of arteriovenous malformations and arteriovenous dural fistulae (Metry et al., 2008; Wang et al., 2009; Brandon et al., 2011) may increase the hemorrhage risk in some patients, particularly in adulthood. Severe headaches have been reported among PHACE patients, with some clinicians and investigators presuming that a relationship exists between headache pain and vascular anomalies. However, such a relationship has not yet been well characterized.

Cardiac and aortic malformations

Cardiac, aortic, and supra-aortic vessel anomalies are common extracutaneous features of PHACE syndrome, and they constitute a substantial source of morbidity and mortality in this neurocutaneous disorder. Bayer and colleagues reviewed the cardiovascular findings in 150 PHACE patients identified from the PHACE Syndrome International Clinical Registry (Bayer et al., 2013). Sixty-two (41%) of the 150 cases had cardiac, aortic arch, and/or brachiocephalic anomalies. These findings were further classified by specific lesion: aberrant subclavian artery in 31 individuals; aortic coarctation in 28; ventricular septal defect in 19; venous anomaly in 12; right aortic arch in 10; other structural heart defect in 9; and a vascular ring in 5 (Bayer et al., 2013). The aortic coarctation in PHACE syndrome is often complex with long-segment hypoplasia, interruption of the transverse arch, and varying degrees of aneurysmal dilatation of the adjacent arch segments (Quecedo et al., 1997; Yates et al., 2000; Metry et al., 2001; Wong et al., 2001; Gargiulo et al., 2002; Bronzetti et al., 2004; Wendelin et al., 2004; Bijulal et al., 2008; Rao et al., 2008). Unlike coarctation in non-PHACE patients, the aortic arch anomalies with PHACE syndrome rarely cause mitral or aortic valve incompetence or lead to abnormalities of left ventricular inflow or outflow. A spectrum of associated vascular anomalies can occur with aortic coarctation. Right-sided aortic arches and double aortic arches occur much more commonly among PHACE syndrome patients than in the general population (Yates et al., 2000; Bronzetti et al., 2004). The incidence of a right-sided and/or a double aortic arch among PHACE patients with aortic arch pathology has been estimated at 33%, while only 0.01% of normal individuals have these, suggesting that either finding should constitute a minor diagnostic criterion for PHACE syndrome (Metry et al., 2009).

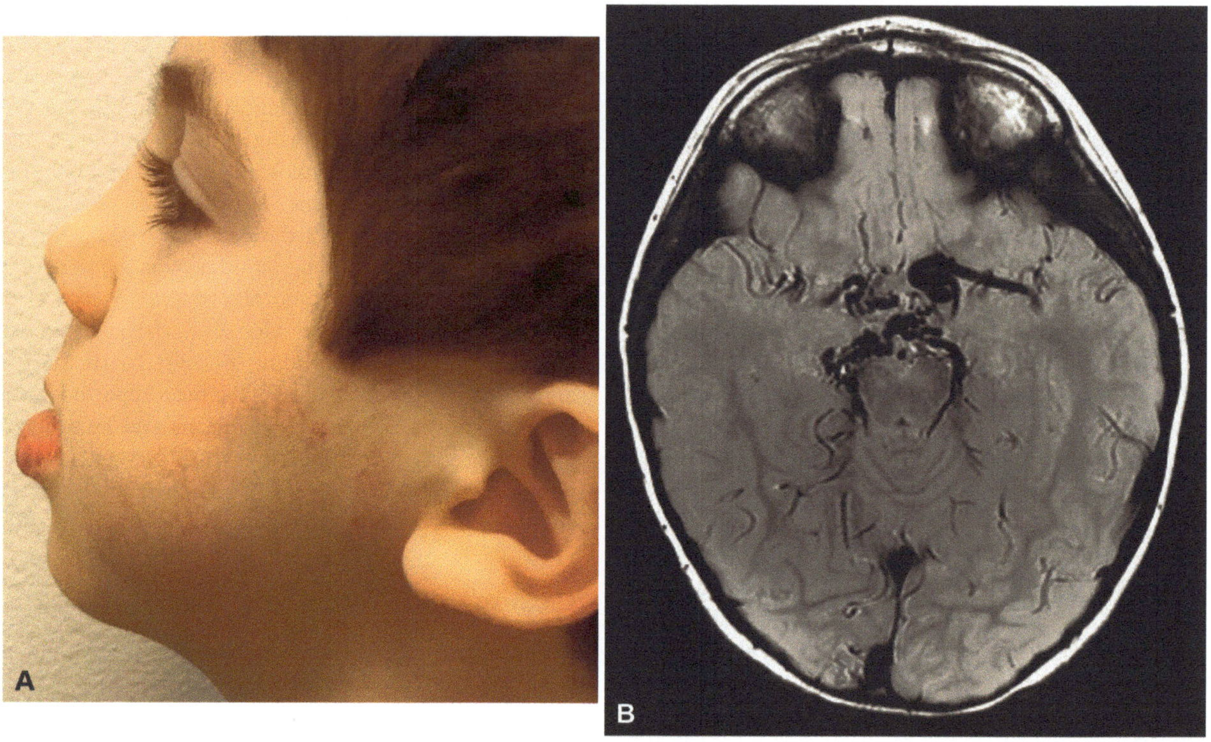

Fig. 12.3. (**A**) Remnants of a facial infantile hemangioma in 4-year-old boy with PHACE syndrome. Note the segment 3 (mandibular) distribution of small telangiectasias and the prominence of the lower lip. As the hemangioma continues to regress the cutaneous features will become less conspicuous. (Author's previously unpublished photo.) (**B**) Magnetic resonance imaging (MRI) findings in the 4-year-old boy with PHACE syndrome. There are multiple cerebral vascular anomalies. The left internal carotid artery is coiled and ectatic. Vascular anomalies fill the interpeduncular and ambient cisterns, and there is asymptomatic blunting of the right cerebral peduncle. Multiple collateral vessels have developed. (The MRI sequence is proton density-weighted fast-spin echo.) (Author's previously unpublished image.)

Complex congenital heart disease is relatively rare among PHACE patients. The PHACE Syndrome International Clinical Registry review identified only three cases of complex congenital heart disease: two had tetralogy of Fallot and one had tricuspid atresia (Bayer et al., 2013). The most common heart anomaly was isolated ventricular septal defect. Other detected cardiac anomalies included pulmonic stenosis, atrial septal defect, patent ductus arteriosis, and bicuspid aortic valve (without associated aortic arch pathology).

Eye abnormalities

A spectrum of ophthalmologic features has been described among PHACE patients (Geller et al., 1991; Fernandez et al., 1995; Pascual-Castroviejo et al., 1995; Burrows et al., 1998; Fard and Traboulsi, 1998; Holmstrom and Taylor, 1998; Coats et al., 1999; Lasky et al., 2004). These can be categorized based on the structure(s) affected in the anterior or the posterior segments of the eye. Anterior segment anomalies include sclerocornea, congenital cataract, lens coloboma, and microphthalmos. The posterior segment anomalies comprise optic nerve atrophy, retinal vascular anomalies, persistent primary vitreous, peripapillary staphyloma, retinal coloboma and the morning glory anomaly. Congenital third nerve palsy and Horner syndrome have also been reported (Burrows et al., 1998). In keeping with the recognized laterality of brain and cerebral vascular findings in PHACE syndrome, most eye abnormalities occur ipsilateral to the facial hemangioma (Metry et al., 2001). When the excessive growth of a hemangioma leads to eye findings such as proptosis or vision deficits due to eyelid closure, this should not be considered as a primary feature of PHACE syndrome because it can occur with isolated periorbital hemangiomas. Posterior segment eye anomalies constitute major diagnostic criteria for PHACE syndrome and have estimated relative frequencies of 5–8.7% among published PHACE cases and registry participants; anterior segment anomalies occur in 6.6–8.7% of patients and are deemed minor diagnostic criteria (Metry et al., 2009).

Sternal clefting and supraumbilical raphe

Sternal clefting and supraumbilical abdominal raphe represent midline defects of ventral development (Metry et al., 2001, 2009). Sternal clefting can range in degree from nonunion of the manubrium and abdominal extension to cutaneous midline pits, dimples, or papules without underlying soft tissue or bony abnormality.

Imaging

The neuroimaging features of PHACE syndrome vary according to the number and extent of associated brain and cerebrovascular anomalies and the imaging modalities used. It is reasonable to perform brain MRI and MR angiography on all patients with large, segmental hemangiomas of the face. The MRI should include sequences with gadolinium because hemangiomas demonstrate contrast enhancement and some dysplastic brain regions may have overlying pial enhancement. It is not clear why pial enhancement occurs, and it tends to resolve over time. In rare cases, we have observed dysplastic cerebral arteries with perivascular regions of contrast enhancement. At 2 months of age the child portrayed in Figure 12.4A began to have regression of her frontonasal (segment 4) hemangioma. Her MRI at 4 days of age showed multiple abnormalities including agenesis of the corpus callosum, diffusely dysplastic cerebral hemispheres with bilateral regions of polymicrogyria, and multiple arterial anomalies including nodular enhancement of the dysplastic left anterior cerebral artery (Fig. 12.4B). Note the enhancement of the scalp hemangioma on this image. The repeat MRI at 5 months showed substantially less enhancement in the path of the anterior cerebral artery and the scalp hemangioma is much less conspicuous (Fig. 12.4C). The reason for enhancement is not clear. It may represent intracranial hemangioma that regressed at the same pace as the scalp lesion or meningeal engorgement from abnormal arterial blood flow in the area.

PATHOLOGY

Little is known about the histopathologic changes in PHACE syndrome. A full, postmortem evaluation is available for the single reported case of a 5-year-old girl with a facial hemangioma and multiple extracutaneous anomalies associated with PHACE syndrome (Chad et al., 2012). The cause of death was rupture of a stented aorta in the setting of bacterial aortitis. Brain MRI performed during the initial diagnostic evaluation, well before death, showed signs of a chronic stroke of the left posterior limb of the internal capsule. The authors presumed that the stroke was perinatal in origin. MR angiography, followed by catheter-based angiography, showed basilar artery tortuosity and the absence of internal carotid artery flow bilaterally. Extensive collateral vessels were present. On post mortem, the stented region of the aorta showed chronic fibroinflammatory changes with degeneration of the vessel media, coupled with adventitial hemorrhage. The rupture site had superimposed suppurative inflammatory changes associated with fibronecrosis of the vessel wall, mural fibrosis, and hemorrhage. Neuropathologic examination demonstrated a cavitary lesion of the left internal capsule and putamen consistent with chronic stroke. Comprehensive histologic sampling of the principal cerebral vessels revealed widespread arteriopathic changes which included thickening and fibrosis of the media and intima, variable fragmentation of the elastica, distortion of the mural architecture, and multiple dystrophic collateral vessels (Chad et al., 2012). Only medium- and large-sized arteries were involved. Pathologic changes were not appreciated in veins, venules, arterioles, or capillaries. Interestingly, the right internal carotid artery was not occluded on pathologic examination, although it did not fill during the earlier catheter-based angiogram (Chad et al., 2012).

Bayer and colleagues found similar histologic abnormalities upon examination of surgically excised aortic coarctation tissues from seven PHACE syndrome patients (Bayer et al., 2013). In five of seven specimens there were large, well-delineated mural zones of scarring and necrosis with near complete loss of arterial smooth muscle cells and elastic fibers in the intima and media. The remaining two cases had decreased smooth muscle cells in the tunica media and evidence of increased adventitial collagen deposition. These findings among PHACE patients were notably different from the excised coarctation tissues from 52 control patients without PHACE syndrome (Bayer et al., 2013). The control specimens exhibited short, asymmetric, postductal narrowing, described as an intimal shelf. Additionally, the controls had intimal fibrosis with the tunica media showing multiple fascicles of muscular elastic tissue. Only five of the 52 non-PHACE patient specimens contained the zones of medial vessel necrosis found in PHACE patient tissues.

Interestingly, the pathologic features of the placenta differed between 13 women who delivered very low birthweight neonates who developed infantile hemangiomas compared to 13 women who delivered very low birthweight neonates without hemangioma development (Lopez Gutierrez et al., 2007). All placentas from children who developed hemangiomas showed gross lesions with disturbance of the uteroplacental circulation including massive retroplacental hemorrhage, extensive ischemic infarction, dilated vascular communications, vasculitis, chorioamnionitis, and funiculitis. Placentas from newborns without hemangioma showed only abnormal placentation in one specimen and isolated villous dysmaturity in two others. Although none of the newborns in the study developed PHACE syndrome, identifying the placental pathology that associates with hemangioma development may help to elucidate the syndrome origin. The authors infer from their findings that

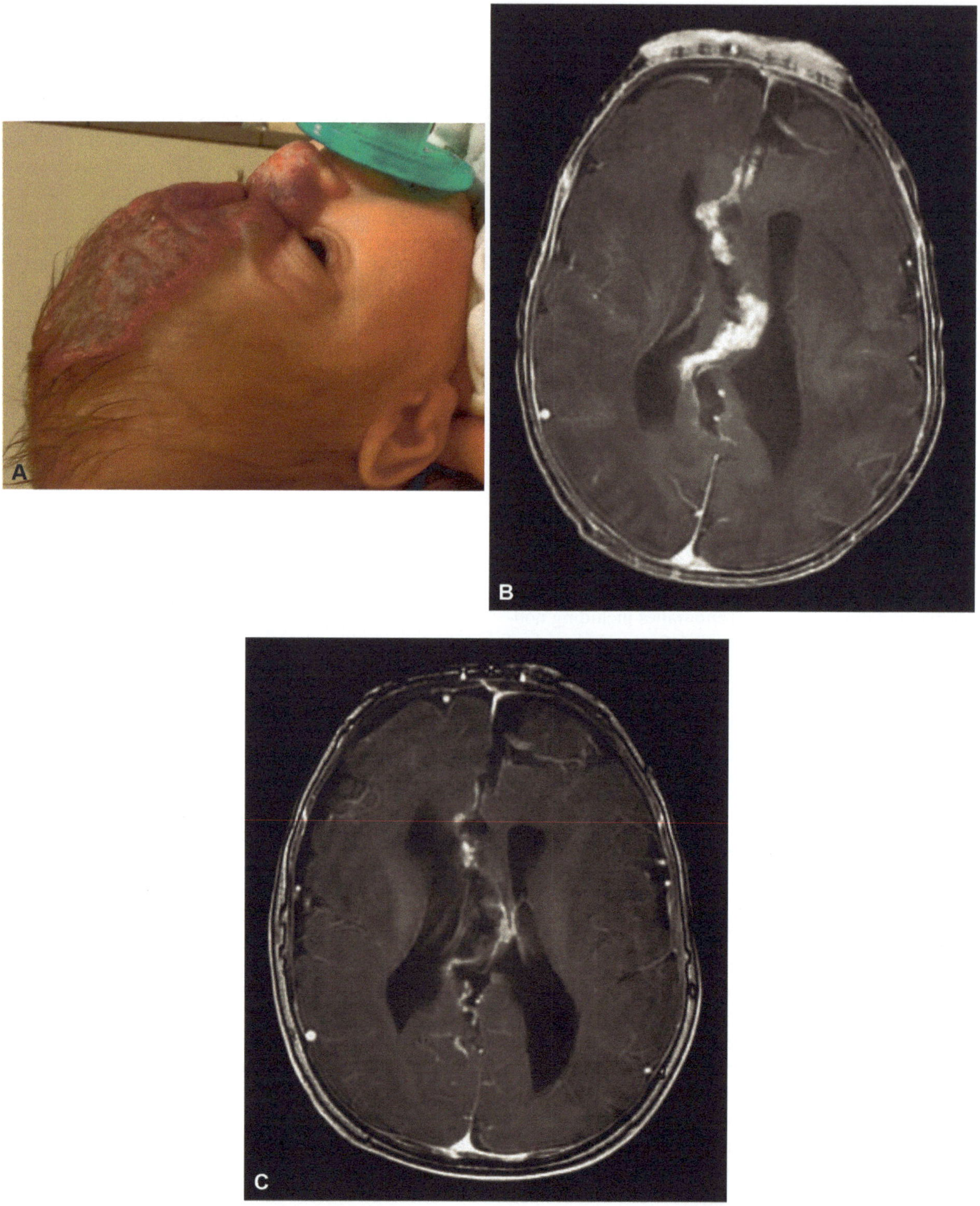

Fig. 12.4. (**A**) Large, frontonasal (segment 4) hemangioma in a child 2 months of age. (**B**) Magnetic resonance imaging (MRI) findings in child at 4 days of age. Multiple MRI abnormalities were present including agenesis of the corpus callosum, cerebral dysplasia, and multiple cerebral vascular anomalies. The ectatic and tortuous left anterior cerebral artery has multiple regions of nodular contrast enhancement. Note also the contrast enhancement of the scalp hemangioma. (**C**) MRI findings at 5 months of age. The scalp and the perivascular contrast enhancement have diminished.

fetal hypoxic stress and the related production of vascular and placental growth factors may determine development of infantile hemangiomas (Lopez Gutierrez et al., 2007).

GENETICS

The etiology of PHACE syndrome is not known. To date all published cases of PHACE have been sporadic. Genetic studies have not yet produced a causative gene. Comparative genomic hybridization microarray in a single PHACE patient identified a complete deletion of *SLC35B4* on chromosome 7q33 (Mitchell et al., 2012). Sequencing of this genetic region in a cohort of 33 PHACE patients identified a second patient with a nonoverlapping deletion 80 kb upstream (Mitchell et al., 2012). While these deletions are unlikely to represent a singular cause of PHACE syndrome, the authors suggest that this chromosomal region may provide some form of disease susceptibility. A study of genomic copy number variation performed in 98 PHACE patients demonstrated 10 rare variations, but none of these occurred in more than one patient (Siegel et al., 2013).

The striking female predominance in PHACE syndrome has caused some researchers to question if the disorder could be related to an X-linked dominant mutation with potential lethality in boys. A clinical comparison of disease severity between male and female PHACE patients did not show a convincing trend toward greater (or lesser) disease severity in boys (Metry et al., 2008). Levin and colleagues reported studies of X-chromosome inactivation in a female PHACE patient and her mother (Levin and Kaler, 2007). The female proband had a random pattern of X-chromosome inactivation, while the mother had a skewed pattern, suggesting that skewing in the mother could have been protective. Sullivan and colleagues conducted a follow-up study of X inactivation in 31 female PHACE patients and their mothers (Sullivan et al., 2013). While skewed X-chromosome inactivation was found in 19% of the mothers with informative (nonhomozygous) alleles and 10% of the female probands with informative (nonhomozygous) alleles, this was not statistically different from healthy populations.

TREATMENT AND MANAGEMENT

Diagnostic approach

The presence of one or more large facial hemangiomas should prompt a diagnostic evaluation for PHACE syndrome. As above, infantile hemangiomas differ clinically from the capillary malformations of Sturge–Weber syndrome and capillary malformation–AVM syndrome as well as the Schobinger stage-1 arteriovenous malformation of Wyburn-Mason syndrome. An adequate diagnostic evaluation of a patient with suspected PHACE syndrome should include brain MRI with MR angiography of the head and neck, transthoracic echocardiography, and evaluations by pediatric dermatology, neurology, and ophthalmology. The MRI should be done with gadolinium as intracranial hemangiomas demonstrate contrast enhancement and pial enhancement can be seen overlying dysplastic regions of the cerebrum or cerebellum. Moreover, the gadolinium study will allow evaluation of the intraorbital extent of periorbital hemangiomas. Computed tomography scanning should not be done as an initial screen for PHACE syndrome as the imaging modality is not adequate to detect smaller regions of cortical or cerebellar dysplasia and the radiation exposure is not necessary. Computed tomography angiography should also be avoided as the high dose of radiation in the neonatal period rarely outweighs the potential benefits of high-resolution vascular imaging. When cutaneous midline defects are present – sternal clefting, pitting, dimpling, or supraumbilical raphe – a radiographic study to assess chest wall development should be considered. An endocrinology evaluation may be reasonable, particularly for those with abnormal pituitary structure or empty sella on MRI or with extensive hemangiomatosis. Minimally, laboratory screening for thyroid function should be performed.

Some clinicians and investigators (myself included) recommend serial neuroimaging, specifically repeat brain MRI and MR angiography, when cerebral vascular anomalies are detected. The interval between imaging studies and the age when serial imaging should be stopped have not been well studied. Although the age at acute ischemic stroke varies, most well-described cases of progressive cerebral vasculopathy began prior to 1 year (Burrows et al., 1998; Drolet et al., 2006; Heyer et al., 2008; Siegel et al., 2012). A lack of radiographic progression by 1 year of age probably lowers the risk of later progression. However, certain arterial defects including saccular aneurysms and presumed stenoses should be followed long term. Repeat imaging should be done after any acute neurologic changes. It would be reasonable to reimage a child with PHACE syndrome and a new onset of severe headaches, a change in the character of prior headaches, or with any new neurologic sign(s) or symptom(s).

Treatment

While the medical treatment of infantile hemangiomas is beyond the scope of this chapter, the recent discovery that propranolol can effectively hasten hemangioma involution is pertinent to the management of patients with PHACE syndrome. Oral corticosteroids had been used as the first-line treatment of hemangiomas that proliferate rapidly, threaten the airway, or threaten vision (Bennett et al., 2001; Frieden et al., 2005). In 2008, Léauté-Labrèze and colleagues published their experience with propranolol in two index cases (Léauté-Labrèze et al., 2008). The first patient developed obstructive hypertrophic myocardiopathy while treated for a nasal hemangioma with corticosteroids; the second had intraorbital hemangioma involvement, also treated with corticosteroids, and was found to have abnormally increased cardiac output. Propranolol was initiated for cardiac disease in both patients. Within days of starting propranolol, both children had substantial hemangioma involution. Nine additional patients with hemangioma were studied, and each demonstrated rapid improvements with propranolol (Léauté-Labrèze et al., 2008). Since this discovery several groups have published their experiences with propranolol as a hemangioma treatment. It remains unclear how propranolol arrests hemangioma proliferation and speeds involution (Drolet et al., 2013). Potential drug mechanisms include vasoconstriction, decreased renin production, inhibition of angiogenesis, and stimulation of apoptosis (Sans et al., 2009; Schiestl et al., 2011; Chim et al., 2012).

The potential side-effects of propranolol include hypotension and hypoglycemia. In a patient with PHACE syndrome and arterial hypoplasia or stenosis, fluctuations in blood pressure could, theoretically, cause a critical change in blood flow across the narrowed vessel, resulting in ischemic stroke. There are two reports of ischemic stroke occurring in PHACE patients while on propranolol therapy (Siegel et al., 2012). However, the contribution of propranolol to the ischemic process is not clear. Metry and colleagues conducted a retrospective review of 32 patients with PHACE syndrome and associated arterial abnormalities treated with propranolol between 2008 and 2011 (Metry et al., 2013). One patient developed a change in neurologic status during treatment. Baseline MRI revealed hypoplasia of the left cerebral peduncle. At 6 months of age a left hand preference (suggesting right-side hemiparesis) was noted. Per the report, neuroimaging remained stable (Metry et al., 2013). The hemiparesis may have emerged at 6 months of age as a result of the congenital brainstem lesion. A consensus statement addressing the general use of propranolol in hemangioma treatment and the specific use in the setting of PHACE syndrome emphasizes the notion that clinical trials have not been conducted to assess overall safety (Drolet et al., 2013). The consensus group recommended that propranolol should be considered in PHACE patients when the potential benefits outweigh the risks. For example, when an infantile hemangioma distorts the airway or prevents an eyelid from opening as it proliferates, the impending risk of respiratory compromise or vision loss may outweigh the theoretic risk of lowering the critical threshold for ischemic stroke.

The clinical management of patients with PHACE syndrome requires a multidisciplinary approach. When the risk of thromboembolic disease is thought to be high, either from a primary cardiac malformation or a large vessel anomaly, aspirin or anticoagulation should be considered. It is probably not beneficial to give aspirin prophylaxis to all PHACE patients with minor cerebral vascular findings. There are no studies of primary medical prevention of stroke with aspirin or any other medicine. The detection of a clot in the cardiovascular system may warrant further hematologic investigation for an increased clotting susceptibility. In cases where progressive cerebral vasculopathy is suspected, neurosurgery should be consulted. Arterial bypass procedures, specifically pial synangiosis and enecephaloduralarteriosynangiosis (EDAS), have been performed successfully in PHACE patients with moyamoya vasculopathy with secondary prevention of recurrent strokes (Burrows et al., 1998; Heyer et al., 2006). Ideally, a progressive vasculopathy could be identified by serial neuroimaging prior to an ischemic event and primary stroke prevention attempted. A clinical trial of primary stroke prevention using a surgical bypass procedure has not been attempted in this rare disorder.

Migraine-like headaches and seizures can occur in PHACE syndrome. While not well studied, some patients with this neurocutaneous disorder develop severe headaches at a relatively young age. Headache treatment should be based on headache burden and disability. As headaches can be a presenting symptom of cerebral vasculopathy, repeat neuroimaging should be considered with new-onset headaches, a change in headache quality, or the emergence of concerning signs or symptoms. Localization-related epilepsy and infantile spasms have been reported in PHACE syndrome (Heyer et al., 2008). The approach to epilepsy in PHACE patients should not deviate from the approach to epilepsy in patients without PHACE syndrome. Patients with cortical dysplasia and neuronal migration defects should be monitored closely for signs of seizures.

References

Aeby A, Guerrini R, David P et al. (2003). Facial hemangioma and cerebral corticovascular dysplasia: a syndrome associated with epilepsy. Neurology 60: 1030–1032.

Arora SS, Plato BM, Sattenberg RJ et al. (2011). Adult presentation of PHACES syndrome. Interv Neuroradiol 17: 137–146.

Bayer ML, Frommelt PC, Blei F et al. (2013). Congenital cardiac, aortic arch, and vascular bed anomalies in PHACE syndrome (from the International PHACE Syndrome Registry). Am J Cardiol 112: 1948–1952.

Bennett ML, Fleischer Jr AB, Chamlin SL et al. (2001). Oral corticosteroid use is effective for cutaneous hemangiomas: an evidence-based evaluation. Arch Dermatol 137: 1208–1213.

Bhattacharya JJ, Luo CB, Alvarez H et al. (2004). PHACES syndrome: a review of eight previously unreported cases with late arterial occlusions. Neuroradiology 46: 227–233.

Bijulal S, Sivasankaran S, Krishnamoorthy KM et al. (2008). Unusual coarctation – the PHACE syndrome: report of three cases. Congenit Heart Dis 3: 205–208.

Brandon K, Burrows P, Hess C et al. (2011). Arteriovenous malformation: a rare manifestation of PHACE syndrome. Pediatr Dermatol 28: 180–184.

Bronzetti G, Giardini A, Patrizi A et al. (2004). Ipsilateral hemangioma and aortic arch anomalies in posterior fossa malformations, hemangiomas, arterial anomalies, coarctation of the aorta, and cardiac defects and eye abnormalities (PHACE) anomaly: report and review. Pediatrics 113: 412–415.

Burrows PE, Robertson RL, Mulliken JB et al. (1998). Cerebral vasculopathy and neurologic sequelae in infants with cervicofacial hemangioma: report of eight patients. Radiology 207: 601–607.

Carmel PW, Antunes JL, Hilal SK et al. (1977). Dandy–Walker syndrome: clinico-pathological features and re-evaluation of modes of treatment. Surg Neurol 8: 132–138.

Chad L, Dubinski W, Hawkins C et al. (2012). Postmortem vascular pathology in PHACES syndrome: a case report. Pediatr Dev Pathol 15: 507–510.

Chim H, Armijo BS, Miller E et al. (2012). Propranolol induces regression of hemangioma cells through HIF-1alpha-mediated inhibition of VEGF-A. Ann Surg 256: 146–156.

Chou PS, Guo YC (2012). Limb-shaking transient ischemic attacks in an adult PHACE syndrome: a case report and review of the literature. Neurol Sci 33: 305–307.

Coats DK, Paysse EA, Levy ML (1999). PHACE: a neurocutaneous syndrome with important ophthalmologic implications: case report and literature review. Ophthalmology 106: 1739–1741.

Corella F, Garcia-Navarro X, Ribe A et al. (2008). Abortive or minimal-growth hemangiomas: immunohistochemical evidence that they represent true infantile hemangiomas. J Am Acad Dermatol 58: 685–690.

Drolet BA, Dohil M, Golomb MR et al. (2006). Early stroke and cerebral vasculopathy in children with facial hemangiomas and PHACE association. Pediatrics 117: 959–964.

Drolet BA, Frommelt PC, Chamlin SL et al. (2013). Initiation and use of propranolol for infantile hemangioma: report of a consensus conference. Pediatrics 131: 128–140.

Drosou A, Benjamin L, Linfante I et al. (2006). Infantile midline facial hemangioma with agenesis of the corpus callosum and sinus pericranii: another face of the PHACE syndrome. J Am Acad Dermatol 54: 348–352.

Fard AK, Traboulsi EI (1998). Coloboma of the lens, optic nerve hypoplasia, and orbital hemangioma – a possible developmental field defect. Ophthalmic Genet 19: 209–212.

Fernandez GR, Munoz FJ, Padron C et al. (1995). Microphthalmos, facial capillary hemangioma and Dandy–Walker malformation. Acta Ophthalmol Scand 73: 173–175.

Fisher CM (1959). Early-life carotid-artery occlusion associated with late intracranial hemorrhage; observations on the ischemic pathogenesis of mantle sclerosis. Lab Invest 8: 680–700.

Frieden IJ, Reese V, Cohen D (1996). PHACE syndrome. The association of posterior fossa brain malformations, hemangiomas, arterial anomalies, coarctation of the aorta and cardiac defects, and eye abnormalities. Arch Dermatol 132: 307–311.

Frieden IJ, Haggstrom AN, Drolet BA et al. (2005). Infantile hemangiomas: current knowledge, future directions. Proceedings of a research workshop on infantile hemangiomas, April 7–9, 2005, Bethesda, Maryland, USA. Pediatr Dermatol 22: 383–406.

Gargiulo G, Pace NC, Giardini A et al. (2002). Repair of a complex aortic arch anomaly associated with cutaneous hemangioma. Ann Thorac Surg 74: 245–246.

Garzon MC, Huang JT, Enjolras O et al. (2007). Vascular malformations: Part I. J Am Acad Dermatol 56: 353–370.

Geller JD, Topper SF, Hashimoto K (1991). Diffuse neonatal hemangiomatosis: a new constellation of findings. J Am Acad Dermatol 24: 816–818.

Goddard DS, Liang MG, Chamlin SL et al. (2006). Hypopituitarism in PHACES association. Pediatr Dermatol 23: 476–480.

Goh WH, Lo R (1993). A new 3C syndrome: cerebellar hypoplasia, cavernous haemangioma and coarctation of the aorta. Dev Med Child Neurol 35: 637–641.

Gorlin RJ, Kantaputra P, Aughton DJ et al. (1994). Marked female predilection in some syndromes associated with facial hemangiomas. Am J Med Genet 52: 130–135.

Grosso S, De Cosmo L, Bonifazi E et al. (2004). Facial hemangioma and malformation of the cortical development: a broadening of the PHACE spectrum or a new entity? Am J Med Genet A 124A: 192–195.

Haggstrom AN, Lammer EJ, Schneider RA et al. (2006). Patterns of infantile hemangiomas: new clues to hemangioma pathogenesis and embryonic facial development. Pediatrics 117: 698–703.

Haggstrom AN, Drolet BA, Baselga E et al. (2007). Prospective study of infantile hemangiomas: demographic, prenatal, and perinatal characteristics. J Pediatr 150: 291–294.

Haggstrom AN, Garzon MC, Baselga E et al. (2010). Risk for PHACE syndrome in infants with large facial hemangiomas. Pediatrics 126: e418–e426.

Hess CP, Fullerton HJ, Metry DW et al. (2010). Cervical and intracranial arterial anomalies in 70 patients with PHACE syndrome. AJNR Am J Neuroradiol 31: 1980–1986.

Heyer GL, Millar WS, Ghatan S et al. (2006). The neurologic aspects of PHACE: case report and review of the literature. Pediatr Neurol 35: 419–424.

Heyer GL, Dowling MM, Licht DJ et al. (2008). The cerebral vasculopathy of PHACES syndrome. Stroke 39: 308–316.

Hirsch JF, Pierre-Kahn A, Renier D et al. (1984). The Dandy–Walker malformation. A review of 40 cases. J Neurosurg 61: 515–522.

Holmstrom G, Taylor D (1998). Capillary haemangiomas in association with morning glory disc anomaly. Acta Ophthalmol Scand 76: 613–616.

Huang SA, Tu HM, Harney JW et al. (2000). Severe hypothyroidism caused by type 3 iodothyronine deiodinase in infantile hemangiomas. N Engl J Med 343: 185–189.

Isler W (1984). Stroke in childhood and adolescence. Eur Neurol 23: 421–424.

Kilcline C, Frieden IJ (2008). Infantile hemangiomas: how common are they? A systematic review of the medical literature. Pediatr Dermatol 25: 168–173.

Lasky JB, Sandu M, Balashanmugan A (2004). PHACE syndrome: association with persistent fetal vasculature and coloboma-like iris defect. J AAPOS 8: 495–498.

Léauté-Labrèze C, Dumas DLR, Hubiche T et al. (2008). Propranolol for severe hemangiomas of infancy. N Engl J Med 358: 2649–2651.

Levin JH, Kaler SG (2007). Non-random maternal X-chromosome inactivation associated with PHACES. Clin Genet 72: 345–350.

Lie TA (1968). Congenital Anomalies of the Carotid Arteries. An Angiographic Study and Review of the Literature, Excerpta Medical Foundation, Amsterdam, pp. 52–94.

Lin MC, Chen CH, Chi CS (2003). PHACE syndrome: report of one case. Acta Paediatr Taiwan 44: 379–382.

Lopez Gutierrez JC, Avila LF, Sosa G et al. (2007). Placental anomalies in children with infantile hemangioma. Pediatr Dermatol 24: 353–355.

Luo CB, Lasjaunias P, Teng MM et al. (2003). Cervicocerebrovascular anomalies in children with PHACE syndrome. J Formos Med Assoc 102: 379–386.

Metry DW, Dowd CF, Barkovich et al. (2001). The many faces of PHACE syndrome. J Pediatr 139: 117–123.

Metry DW, Haggstrom AN, Drolet BA et al. (2006). A prospective study of PHACE syndrome in infantile hemangiomas: demographic features, clinical findings, and complications. Am J Med Genet A 140: 975–986.

Metry DW, Siegel DH, Cordisco MR et al. (2008). A comparison of disease severity among affected male versus female patients with PHACE syndrome. J Am Acad Dermatol 58: 81–87.

Metry D, Heyer G, Hess C et al. (2009). Consensus statement on diagnostic criteria for PHACE syndrome. Pediatrics 124: 1447–1456.

Metry D, Frieden IJ, Hess C et al. (2013). Propranolol use in PHACE syndrome with cervical and intracranial arterial anomalies: collective experience in 32 infants. Pediatr Dermatol 30: 71–89.

Mitchell S, Siegel DH, Shieh JT et al. (2012). Candidate locus analysis for PHACE syndrome. Am J Med Genet A 158A: 1363–1367.

Oza VS, Wang E, Berenstein A et al. (2008). PHACES association: a neuroradiologic review of 17 patients. AJNR Am J Neuroradiol 29: 807–813.

Padget DH (1948). The development of the cranial arteries in the human embryo. Contr Embryol 32: 205–262.

Pascual-Castroviejo I (1978). Vascular and nonvascular intracranial malformation associated with external capillary hemangiomas. Neuroradiology 16: 82–84.

Pascual-Castroviejo I, Viano J, Pascual-Pascual SI et al. (1995). Facial haemangioma, agenesis of the internal carotid artery and dysplasia of cerebral cortex: case report. Neuroradiology 37: 692–695.

Pascual-Castroviejo I, Viano J, Moreno F et al. (1996). Hemangiomas of the head, neck, and chest with associated vascular and brain anomalies: a complex neurocutaneous syndrome. AJNR Am J Neuroradiol 17: 461–471.

Pascual-Castroviejo I, Pascual-Pascual SI, Velazquez-Fragua R et al. (2005). Cutaneous hemangiomas and vascular malformations and associated pathology (Pascual-Castroviejo type II syndrome). Study of 41 patients. Rev Neurol 41: 223–236.

Poetke M, Frommeld T, Berlien HP (2002). PHACE syndrome: new views on diagnostic criteria. Eur J Pediatr Surg 12: 366–374.

Poindexter G, Metry DW, Barkovich AJ et al. (2007). PHACE syndrome with intracerebral hemangiomas, heterotopia, and endocrine dysfunction. Pediatr Neurol 36: 402–406.

Quecedo E, Gil-Mateo MP, Pont V et al. (1997). Giant hemifacial angioma and PHACE syndrome. Br J Dermatol 136: 649–650.

Rao RP, Drolet BA, Holland KE et al. (2008). PHACES association: a vasculocutaneous syndrome. Pediatr Cardiol 29: 793–799.

Reese V, Frieden IJ, Paller AS et al. (1993). Association of facial hemangiomas with Dandy–Walker and other posterior fossa malformations. J Pediatr 122: 379–384.

Saltzman GF (1959). Patent primitive trigeminal artery studied by cerebral angiography. Acta Radiol 51: 329–336.

Sans V, de la Roque ED, Berge J et al. (2009). Propranolol for severe infantile hemangiomas: follow-up report. Pediatrics 124: e423–e431.

Sawaya R, McLaurin RL (1981). Dandy–Walker syndrome. Clinical analysis of 23 cases. J Neurosurg 55: 89–98.

Schiestl C, Neuhaus K, Zoller S et al. (2011). Efficacy and safety of propranolol as first-line treatment for infantile hemangiomas. Eur J Pediatr 170: 493–501.

Siegel DH, Tefft KA, Kelly T et al. (2012). Stroke in children with posterior fossa brain malformations, hemangiomas, arterial anomalies, coarctation of the aorta and cardiac defects, and eye abnormalities (PHACE) syndrome: a systematic review of the literature. Stroke 43: 1672–1674.

Siegel DH, Shieh JT, Kwon EK et al. (2013). Copy number variation analysis in 98 individuals with PHACE syndrome. J Invest Dermatol 133: 677–684.

Smith DS, Lee KK, Milczuk HA (2004). Otolaryngologic manifestations of PHACE syndrome. Int J Pediatr Otorhinolaryngol 68: 1445–1450.

Stoeter P, Marquardt B (1978). Combined deformity of the brain and cerebral vessels (congenital hydrocephalus) in a 3-year-old child: Dandy–Walker syndrome, arteria primitiva trigemina and haemangioma. Rofo 128: 680–683.

Streeter GL (1949). Developmental horizons in human embryos; a review of the histogenesis of cartilage and bone. Contrib Embryol 33: 149–168.

Sullivan CT, Christian SL, Shieh JT et al. (2013). X chromosome-inactivation patterns in 31 Individuals with PHACE syndrome. Mol Syndromol 4: 114–118.

Tangtiphaiboontana J, Hess CP, Bayer M et al. (2013). Neurodevelopmental abnormalities in children with PHACE syndrome. J Child Neurol 28: 608–614.

Theron J, Newton TH, Hoyt WF (1974). Unilateral retinocephalic vascular malformations. Neuroradiology 7: 185–196.

Wang H, Oh AK, Orbach DB (2009). Arteriovenous shunting as a new feature of PHACES. Case report. J Neurosurg Pediatr 3: 53–56.

Wendelin G, Kitzmuller E, Salzer-Muhar U (2004). PHACES: a neurocutaneous syndrome with anomalies of the aorta and supraaortic vessels. Cardiol Young 14: 206–209.

Wong CH, Wright JG, Silove ED et al. (2001). A new syndrome of multiple hemangiomas, right dominant double aortic arch, and coarctation. J Thorac Cardiovasc Surg 121: 1207–1209.

Yates R, Syed S, Tsang V et al. (2000). Haemangioma of the head and neck with subglottic involvement and atypical coarctation. Br J Dermatol 143: 686–688.

Handbook of Clinical Neurology, Vol. 132 (3rd series)
Neurocutaneous Syndromes
M.P. Islam and E.S. Roach, Editors

Chapter 13

Hereditary hemorrhagic telangiectasia

MELISSA G. CHUNG*
Division of Pediatric Neurology and Critical Care Medicine, Department of Pediatrics, Nationwide Children's Hospital, Columbus, OH, USA

INTRODUCTION

Hereditary hemorrhagic telangiectasia (HHT), also known as Osler–Weber–Rendu syndrome, is an autosomal dominant disorder that is characterized by multiple arteriovenous malformations (AVMs) involving the skin, mucosal surfaces, and internal organs. HHT has an age-dependent penetrance and usually initially presents with recurrent epistaxis followed by the characteristic telangiectasias of the face, oropharynx, and hands over time. Patients often have vascular malformations involving the lungs, brain, spinal cord, and gastrointestinal tract as well, which are the main causes of morbidity in patients with HHT. The sequelae of visceral organ involvement include ischemic stroke, cerebral bacterial abscesses, intracranial hemorrhage, chronic hypoxia, dyspnea with exertion, pulmonary hypertension, high output heart failure, gastrointestinal bleeds, and liver failure.

HISTORY

This association between gastrointestinal bleeding, epistaxis, and mucosal and skin telangiectasias was first recognized in the 19th century (Begbie et al., 2003). The earliest known report was published in 1864 by Henry Sutton (Prasad and Chudley, 2002). In October 1896, Henri Rendu presented a 52-year-old man with recurrent nosebleeds to the Medical Society of Parisian Hospitals. Rendu highlighted that the patient had anemia due to his bleeding from multiple mucosal and skin telangiectasias rather than from hemophilia (Gibbs, 1986). He also postulated that the disease might be familial since the patient's father had a history of repeated melena and the patient's mother and brother had recurrent epistaxis. It is the renowned physician Sir William Osler, though, who is credited with bringing the disease to the attention of the wider medical community when he presented three patients with multiple vascular lesions and recurrent epistaxis in 1901 (Gibbs, 1986; Prasad and Chudley, 2002). He established that this syndrome was most likely familial, pointing out that two of the patients were siblings and that seven of their family members across three generations had similar symptoms. In 1907, Frederick Weber further solidified the inherited nature of the disease when he published a report of a woman from London and her children who had multiple cutaneous and mucosal angiomas (Gibbs, 1986). Despite the initial contributions to literature by Dr. Sutton, this syndrome became known by the eponym Osler–Weber–Rendu syndrome. The term "hereditary hemorrhagic telangiectasia" was coined in 1909 by a Johns Hopkins University medical student named Frederic Hanes (Gibbs, 1986).

EPIDEMIOLOGY

The reported prevalence of HHT varies significantly in the literature; many authors postulate that the disease is underdiagnosed due to the variability in its clinical presentation, even within a single family, and its age-dependent penetrance. The highest known point prevalence is in the Afro-Caribbean population in the Netherlands Antilles, where approximately 1 in 1331 people over the age of 12 years have HHT based upon clinical criteria (Westermann et al., 2003). The reported prevalence in other populations ranges from 1:5000–8000 in Japan; 1:5000 in Ontario, Canada; 1:6410 in Fyn County in Denmark; 1:2351 in Ain, France; 1:16 500 in Vermont; 1 in 39 216 in northern England to 1:9400 in the UK (Porteous et al., 1992; Guttmacher et al., 1994; Dakeishi

*Correspondence to: Melissa G. Chung, MD, Assistant Professor, Department of Pediatrics, Division of Pediatric Neurology and Critical Care Medicine, Nationwide Children's Hospital, 700 Children's Drive ED 566, Columbus OH 43205, USA. Tel: +1-614-722-4625, E-mail: Melissa.chung@nationwidechildrens.org

et al., 2002; Donaldson et al., 2014; Latino et al., 2014b; Fish and Wythe, 2015; Kjeldsen et al., 2000). Overall, the estimated prevalence worldwide is thought to be between 1:5000 and 1:8000.

CLINICAL MANIFESTATIONS

HHT originally was loosely defined as a syndrome of skin and cutaneous malformations, recurrent nosebleeds and resulting anemia. Then, in 2000, the Scientific Advisory Board of the HHT Foundation International Inc. created the "Curaçao Criteria", which are now widely recognized as the primary clinical criteria used to diagnosis HHT (Shovlin et al., 2000). The diagnosis of HHT is based upon four criteria: (1) epistaxis, (2) telangiectasias (multiple at characteristic sites such as lips, oral cavity, fingers and nose); (3) visceral lesions (such as pulmonary, hepatic, cerebral or spinal arteriovenous malformations or gastrointestinal telangiectasias); (4) family history of a first degree relative with HHT. A diagnosis is considered definite if three criteria are fulfilled, possible if two criteria are met, and unlikely if fewer than two criteria are fulfilled. The sensitivity of the criteria is age-dependent, however; very few patients will meet the criteria during the first decade of life due to the gradual emergence of clinical symptoms over time (Faughnan et al., 2011).

Early symptoms

In Henri Rendu's initial paper, he noted that his patient's symptoms began with recurrent nosebleeds at the age of 12 years (Gibbs, 1986). Subsequent studies have shown that epistaxis is usually the earliest manifestation of HHT. A third of patients report that they develop recurrent nosebleeds before the age of 10 years (Berg et al., 2003). About 50–95% of patients develop nosebleeds before the age of 20 years (Plauchu et al., 1989; Berg et al., 2003). The severity of the nosebleeds often increases over time and may lead to iron deficiency anemia (AAssar et al., 1991; Folz et al., 2005; Dheyauldeen et al., 2011). The characteristic cutaneous telangiectasias that involve the face, oral mucosa, and hands usually develop at a later age. In one study, 60% of patients reported the telangiectasias did not develop until 10–30 years after their initial nosebleeds (Plauchu et al., 1989). Visceral organ involvement, or at least symptomatic involvement, also increases with age, peaking around the age of 70 years (Plauchu et al., 1989; Al-Saleh et al., 2009; Faughnan et al., 2011). However, early screening of children with HHT for pulmonary and cerebral AVMs suggests the time of diagnosis may reflect the onset of clinical symptoms rather than actual onset of the visceral AVMs (Giordano et al., 2015). One notable exception is that AVMs, particularly pulmonary AVMs, do seem to increase during pregnancy, particularly during the second and third trimester (Ference et al., 1994; Shovlin et al., 2008b; de Gussem et al., 2014).

Pulmonary

Approximately 15–50% of patients with HHT have pulmonary AVMs (Fig. 13.1) (Gossage and Kanj, 1998; Kjeldsen et al., 2000; Cottin et al., 2007). Conversely, approximately 80–90% of patients with pulmonary AVMS have HHT (Dines et al., 1983; Gossage and Kanj, 1998; Shovlin et al., 2008a). The pulmonary AVMs in HHT are often multiple and preferentially affect the lower lobes whereas non-HHT-related AVMs tend to be singular (Dines et al., 1983; White et al., 1988).

Most patients with pulmonary AVMs are asymptomatic (White et al., 1988; Shovlin et al., 2008a). However, these vascular malformations are the cause of significant morbidity and mortality in a number of patients with HHT. Pulmonary AVMs can cause right-to-left shunting with associated hypoxia, dyspnea, and ultimately pulmonary hypertension (Shovlin and Letarte, 1999; Begbie et al., 2003; Bhatia et al., 2003; Bayrak-Toydemir et al., 2006b). The most common symptom of the pulmonary AVMs is dyspnea upon exertion.

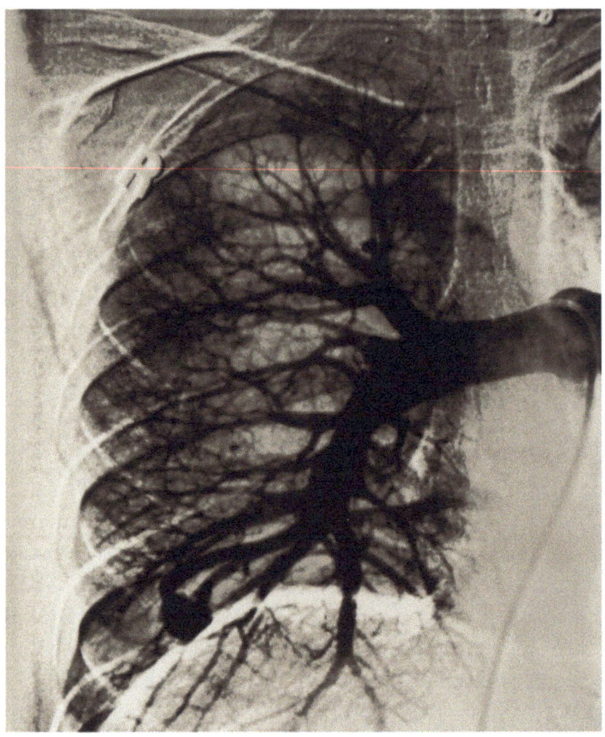

Fig. 13.1. This pulmonary angiogram in a patient with hereditary hemorrhagic telangiectasia shows a typical pulmonary arteriovenous fistula. (Reproduced from Dowling, 2004, with permission.)

However, the most feared complication of right-to-left shunting is neurologic injury due to paradoxic emboli (further described below). The fragile vasculature of the malformations may also rupture and cause hemoptysis or in rare cases (~8%) significant and life-threatening pulmonary hemorrhage (Ference et al., 1994; Sharma and Forkert, 2004; Berg et al., 2010).

Neurologic

The neurologic complications of HHT can be divided into two main categories: those secondary to pulmonary AVMs and those due to central nervous system vascular malformations.

Right-to-left shunting from a pulmonary AVM can allow passage of paradoxic bacterial or clot emboli to the brain with resulting brain abscesses, transient ischemic attacks, or ischemic strokes (Roman et al., 1978; Maher et al., 2001; Espinosa et al., 2008; Shovlin et al., 2008a). One study estimated that 61% of patients with HHT who have neurologic involvement were secondary to pulmonary AVMs instead of cerebral vascular malformations (Roman et al., 1978). A more recent study of 321 patients with HHT found that almost 30% of the patients with pulmonary AVMs had a history of transient ischemic attacks; the mean age in this cohort was only 50.3 years (Maher et al., 2001). In another study of patients with HHT and pulmonary AVMs at a London hospital, over a 6 year period, approximately one-third of patients had either a brain abscess or an ischemic stroke before or during the study period; the authors also noted that development of brain abscesses often was preceded by a history of procedures (e.g. dental work) that create a risk for bacteremia (Shovlin et al., 2008a). Case reports of cerebral air emboli have also been reported with HHT and pulmonary AVMs (Petrillo et al., 2001; Hsu et al., 2004). Overall, ischemic strokes and brain abscesses occur frequently in patients with HHT and may be the initial event that leads to the diagnosis of pulmonary AVMs and ultimately HHT, including neonates and children (White et al., 1988; Maher et al., 2001; Shovlin et al., 2008a). Some authors have also postulated that pulmonary right-to-left shunts are a key feature in the development of migraines in patients with HHT (Steele et al., 1993; Post et al., 2005; Thenganatt et al., 2006; Marziniak et al., 2009). This theory is supported by the fact that embolization of pulmonary AVMs decreases the incidence of migraines in patients with HHT (Post et al., 2006).

A smaller group of patients with HHT develop neurologic complications from cerebral vascular malformations. These malformations are most commonly found in the brain and less frequently in the spinal cord (Roman et al., 1978; Krings et al., 2005; Espinosa et al., 2008). The true incidence of cerebral vascular malformations in HHT likely is underestimated in the literature because, historically, screening imaging of the central nervous system was not done routinely. One study that reviewed the medical records of patients with HHT concluded that only 3.7% of them had cerebral vascular malformations but this center did not do routine brain imaging of their patients unless they were symptomatic (Maher et al., 2001). In contrast, a study that screened 196 patients with intravenous digital subtraction angiography of the brain reported that 12.2% of patients had a cerebral vascular malformation; many of these patients had multiple brain lesions (Willemse et al., 2000). Use of magnetic resonance imaging (MRI) with dedicated susceptibility-weighted imaging further increases the detection of cerebral vascular malformations. When brain MRIs were done on 184 patients with HHT, 23% of the patients were found to have some type of cerebral vascular malformation (Fulbright et al., 1998). AVMS are the most common type of cerebral vascular malformation seen in patients with HHT (Fig. 13.2). Less common vascular anomalies include cerebral telangiectasias, cavernous malformations, vein of Galen

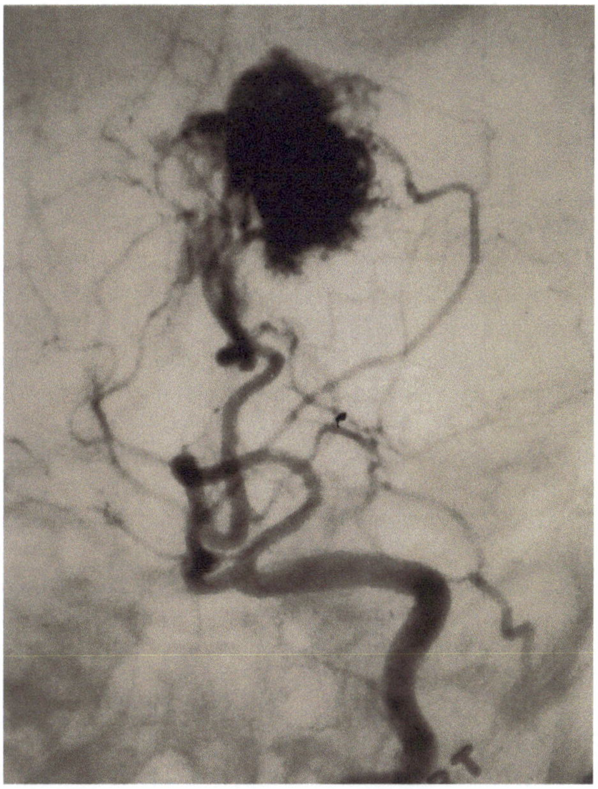

Fig. 13.2. An arteriovenous malformation revealed by cerebral angiogram in a patient with hereditary hemorrhagic telangiectasia. (Reproduced from Dowling, 2004, with permission.)

malformations, arteriovenous fistuale, venous angiomas, and aneurysms. These vascular malformations, especially the high pressure lesions, can bleed and cause symptomatic intracranial hemorrhage and/or seizures. The average age for patients with intracranial hemorrhage (ICH) with HHT is 25 years but catastrophic intracranial hemorrhage from AVMs has been described as early as the neonatal period (Morgan et al., 2002; Maher et al., 2001; Saleh et al., 2013). Overall, though, one retrospective study estimated that the bleeding risk from the HHT-related cerebral vascular malformations is as low as 0.27–0.46% per year compared to the usual bleeding risk of 2–4% per year for most AVMs; the authors postulated that there is a lower risk of bleeding in patients with HHT because their malformations tend to be low-flow, nidus-type AVMs (Willemse et al., 2000; Faughnan et al., 2011).

Symptomatic spinal AVMs are thought to be rare but may cause sensory loss and paralysis depending upon their level and degree of involvement (Espinosa et al., 2008).

Gastrointestinal

Approximately 15–45% of patients with HHT develop gastrointestinal (GI) hemorrhage at some point in their life due to fragile telangiectasias that develop along the GI tract (Plauchu et al., 1989; Korzenik, 1996; Sharma and Howden, 1998; Kjeldsen and Kjeldsen, 2000; Longacre et al., 2003; Grève et al., 2010). The stomach and small intestine are the parts of the GI system that are most frequently involved (McDonald and Pyeritz, 1993; Korzenik, 1996; Kjeldsen and Kjeldsen, 2000; Longacre et al., 2003). Using small capsule endoscopy, one study found that almost 80% of patients had small telangiectasias in the stomach or small intestine. The discrepancy between the incidence of telangiectasias and the percentage of patients that complain of GI bleeds suggests that most of the telangiectasias are asymptomatic or never cause clinically significant bleeding (Ingrosso et al., 2004). The patients who do develop GI bleeds usually present later in life, around 50–60 years of age (Plauchu et al., 1989; Sharma and Howden, 1998; Kjeldsen and Kjeldsen, 2000). No correlation between lifestyle, medications, or exposures has been found to explain why some patients develop GI bleeds whereas others never do.

Liver involvement is less well understood in HHT. The vascular malformations within the liver can affect different vascular beds causing shunting between the hepatic artery to the portal venous system, portal vein to hepatic vein, or from hepatic arteries to the hepatic veins. Screening studies of small groups of patients suggest that up to three-fourths of patients may have liver disease (Buscarini et al., 1997; Buscarini et al., 2004, 2005, 2008, 2011; Ravard et al., 2004). Only 8% of patients, however, develop symptomatic liver disease (Buscarini et al., 2006). Depending upon which type of malformation is present, the patients may present with biliary dysfunction and nodular liver disease, portal hypertension including esophageal varices, or congestive heart failure. Some of the patients may proceed to end stage liver disease and require liver transplantation.

Heart

The cardiac structures and large vessels are usually normal. However, patients can develop high output heart failure from high output AVMs, most commonly from a liver AVM. Excessive shunting of blood into the venous system leads to congestive right heart failure (Livneh et al., 1988; Garcia-Tsao et al., 2000; Garcia-Tsao, 2007; Goussous et al., 2009; Al-Saleh et al., 2011; Cho et al., 2012; Ginon et al., 2013). Patients are particularly vulnerable during pregnancy; the normal physiologic changes in pregnancy, including increased intravascular volume and increased cardiac output, are thought to worsen or to enlarge existing shunting lesions (Livneh et al., 1988; Goussous et al., 2009). Women with peripartum heart failure may take weeks to months after delivery before their cardiac function improves. If liver disease is severe, though, then congestive heart failure may become intractable unless the patient receives a liver transplantation.

Ears, nose, and throat

The most common feature of HHT is recurrent nosebleeds secondary to telangiectasias in the nasal mucosa. Patients can develop nosebleeds spontaneously or in response to minor irritants, such as dry mucosa or trauma. Many of these patients experience worsening in their nosebleeds over time, which can lead to significant blood loss. In the 19th and early 20th century, it was not uncommon for patients with HHT to die of blood loss from severe epistaxis. Development of blood transfusions has significantly reduced mortality but epistaxis continues to be a leading cause of morbidity and decreased quality of life amongst patients (Geirdal et al., 2012). Patients with recurrent epistaxis often develop iron deficiency anemia with associated pallor and fatigue (AAssar et al., 1991; Dheyauldeen et al., 2011). A survey of patients with HHT found that approximately 70% of them were on iron supplements and almost 20% had received at least 10 blood transfusions (Elphick and Shovlin, 2014).

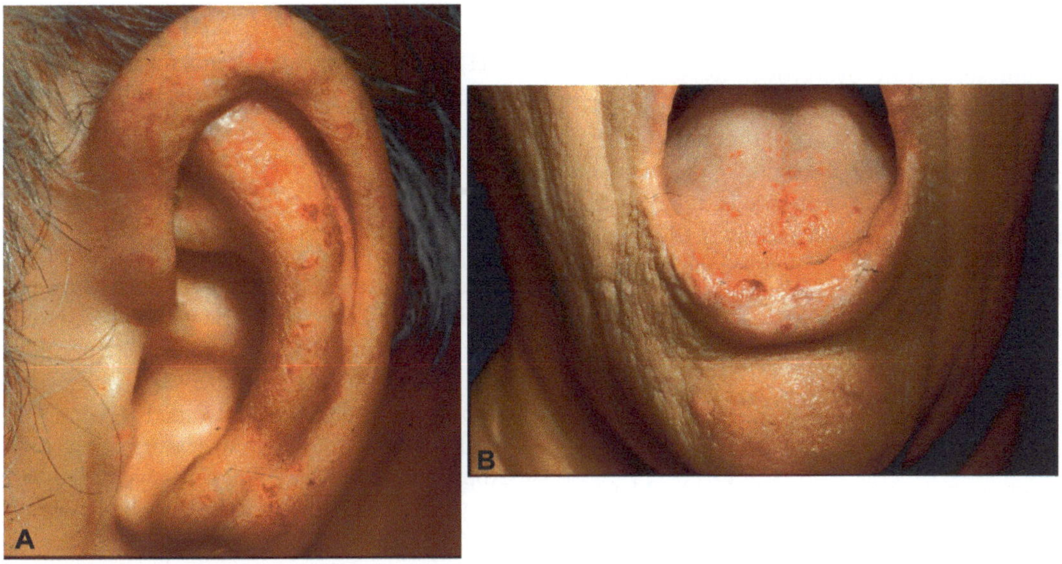

Fig. 13.3. (**A, B**) Classic telangiectasias on the ear and in the oropharynx in a patient with HHT. Reproduced by courtesy of Jack B. Cohen, D.O., from Dowling, MM. Hereditary Hemorrhagic Telangiectasia in Neurocutaneous Disorders. E. S. Roach & V. S. Miller, eds., 2004, with permission from Cambridge University Press.

Skin

Classically, patients develop multiple telangiectasias of the face, mouth, and fingers, including under the nail-beds (Fig. 13.3A, B). On the face, the telangiectasias tend to favor sun-exposed area of skin. These vascular malformations can be differentiated from petechiae by the fact that they blanch when pressure is applied to the lesion. As previously noted, these telangiectasias usually appear and increase in number as the patients get older.

Life expectancy

Limited epidemiologic data is available about the overall life expectancy of patients with HHT. A small study of 70 subjects with HHT found that their average lifespan was 63.2 years versus 70.0 years in subjects without HHT. However, overall mortality also was not different between the groups (Sabbà et al., 2006). The difference in lifespan appears to be due to a peak in mortality in younger patients with HHT under the age of 50 and then in patients over the age of 70. Early mortality in patients under 50 years of age has been linked to complications from visceral organ involvement (Kjeldsen et al., 1999; Sabbà et al., 2006). Without complications from pulmonary and cerebral vascular malformations, life expectancy may be normal for a given patient with HHT.

HHT and juvenile polyposis variant

In 1980, physicians first noted that a group of patients seemed to have an overlap syndrome between HHT and juvenile polyposis syndrome (Cox et al., 1980). The patients had the typical vascular malformations seen with HHT but were also found to have multiple hamartomatous polyps in their GI tract, particularly the colorectal region. These patients also demonstrated an increased risk of colorectal cancers. Researchers later discovered that this overlap syndrome occurs when patients have a mutation in *SMAD4,* which can cause juvenile polyposis syndrome as well as HHT independently or rarely, both diseases together (Gallione et al., 2004; Iyer et al., 2010; Williams et al., 2012).

Differential diagnosis

HHT can be confused with other syndromes that cause multiple skin telangiectasias; however, most of these other syndromes involve very different associated features. Hereditary benign telangiectasia is also inherited in an autosomal dominant fashion but patients present only with skin telangiectasias; patients have neither mucosal involvement nor any visceral organ involvement (Brancati et al., 2003). Also, patients with HHT usually present with epistaxis prior to appearance of their skin telangiectasias so a history of recurrent nosebleeds easily helps differentiate between these two diagnoses. Another group of autosomal dominant disorders, *RASA*-1 related disorders, including Parkes Weber syndrome, can be more difficult to differentiate from HHT since both diagnoses present with multiple telangiectasias and AVMs (Revencu et al., 2013). Patients with *RASA*-1 tend to have head and neck AVMs, which can lead to congestive heart failure; however, they do not tend to have the pulmonary, gastrointestinal, and nasal mucosal involvement seen in HHT. These patients may also have limb/soft tissue overgrowth associated with their capillary malformations that is not seen in HHT. Two other vastly different diseases that are characterized by telangiectasias are ataxia telangiectasia and CREST syndrome. Patients with ataxia telangiectasia have prominent cerebellar ataxia and many of them also have recurrent sinopulmonary infections; the inheritance of the disease is also autosomal recessive rather than autosomal dominant (Hoche et al., 2012). Patients with CREST syndrome may have telangiectasias, including pulmonary vascular malformations, but the other defining characteristics of this syndrome (calcinosis, Raynaud phenomenon, esophageal dysmotility, and sclerodactaly) are not seen in patients with HHT (McGuire et al., 2010).

PATHOLOGY/PATHOPHYSIOLOGY

The fundamental problem in HHT is the formulation of abnormal connections between postcapillary venules and arterioles. Instead of being normally separated by a capillary bed, the venules are in direct connection with the arterioles (Shovlin, 2010). A study by Braverman et al. examined sections of telangiectasias in patients with HHT under light and electron microscopy (Braverman et al., 1990). The authors found that the lesions in HHT begin with venules that are normally connected to the arterioles via a capillary bed. However, the venules then became dilated and develop lymphocytic perivascular infiltrates. With further dilation of the venules, the capillary beds are lost and the venules and arterioles form a direct connection. These venule walls lack the internal structure needed to tolerate the high pressure blood flow and thus are prone to hemorrhage.

Current research theorizes that patients with HHT are unable to adequately recruit smooth muscle cells during sprouting of new blood vessels triggered by angiogenesis or wound healing which results in fragile, abnormal vascular structures (Mancini et al., 2009; Shovlin, 2010). This theory stems from the fact that all of the known gene mutations that cause HHT affect TGFβ signaling pathways, particularly those involved with angiogenesis. The gene in HHT1, *endoglin*, also normally plays a role in regulation of endothelial NO synthase; loss of function mutations lead to increased production of oxidative species and impaired myogenic response, which may explain the venule dilation and the perivascular lymphocytic infiltration seen under the microscope in patients with HHT (Toporsian et al., 2005; Mancini et al., 2009).

GENETICS

HHT is an autosomal dominant disease with variable penetrance. Homozygosity of HHT is considered to be lethal (Karabegovic et al., 2004; El-Harith et al., 2006). Thus far, a total of five mutations have been described that cause HHT. Ninety percent of patients that meet clinical criteria for HHT based on the Curacao criteria are positive for one of these genes. All of these mutations are involved in TGFβ signaling pathway. Subtypes of HHT are based upon the patient's gene mutation.

Patients with HHT1 carry the first gene mutation that was discovered: *endoglin (ENG)*, which is located on chromosome 9q.34.1 (McDonald et al., 1994; Shovlin et al., 1994). *ENG* is an accessory receptor for TGFβ. Approximately 53% of patients with HHT are positive for the ENG mutation (Bayrak-Toydemir et al., 2006b). Genotype-phenotype studies have found that patients with HHT1 have a higher incidence of pulmonary malformations than other subtypes of HHT (Letteboer et al., 2006, 2008; Lesca et al., 2007).

The second mutation described is a mutation in the gene that encodes for activin-receptor-like-kinase-1 (*ALK-1*, now known as ACVRL1) on 12q11-14 (Heutink et al., 1994; Letteboer et al., 2006; Sabbà et al., 2006). Approximately 47% of patients with HHT have ALK-1 mutations. Like ENG, ALK-1 is a receptor for TGFβ. Onset of symptoms seems to occur slightly later in patients with HHT2 than HHT1 and more of them have evidence of liver disease (Bayrak-Toydemir et al., 2006b; Letteboer et al., 2008).

Genome and family linkage studies have identified target gene mutations on chromosome 5q31.3-32 (HHT3) and also 7p14 (HHT4) that cause classic HHT but researchers have yet to fully elucidate the details of these proteins (Cole et al., 2005; Bayrak-Toydemir et al., 2006a).

Recently a mutation in growth/differentiation factor-2 (*GDF2*, also known as bone morphogenetic protein 9) on 10q11.22 was discovered in a group of patients with HHT. This gene mutation accounts for mutations in 1–2% of patients with HHT (Wooderchak-Donahue et al., 2013). *GDF2* or *BMP9* are also members of the TGFβ family. Patients with GDF-2 mutations are considered to have the HHT5 subtype of HHT.

Mutations in the SMA- and MAD-related protein 4 (SMAD4) on 18q21.2 cause a unique form of HHT. SMAD4 is a protein that regulations transcription responses to TGFβ and BMP. Patients with SMAD4 mutations have the typical features of HHT but may also have features of juvenile polyposis (Cox et al., 1980; Gallione et al., 2004; Cole et al., 2005; Lesca et al., 2006; Iyer et al., 2010; Williams et al., 2012).

TREATMENT AND MANAGEMENT

Treatment and management of HHT focus on prevention of complications from its manifestations; there is no curative therapy available at this time for HHT. With this in mind, many physicians have begun to call for early screening for visceral AVMS, particularly pulmonary or cerebral, in patients with HHT or a family history of HHT. The first step in treatment is early identification of patients with HHT. As previously discussed, the Curacao criteria have very poor sensitivity for the disease in young children. Consequently, recently published international guidelines on the treatment and management of HHT recommend that children with a family history of HHT be considered to have the disease unless they are known to be negative for their family's gene mutation (Faughnan et al., 2011).

Pulmonary

Since untreated pulmonary AVMs cause significant morbidity and mortality in patients with HHT, the international guidelines on management on HHT recommend that all patients with possible or confirmed HHT be screened for pulmonary AVMs. The writers also recommend repeat screening after puberty, pregnancy, within 5 years preceding pregnancy and otherwise every 5–10 years (Faughnan et al., 2011). There was nearly complete agreement amongst the writers on this recommendation. A number of different diagnostic modalities have been used to screen for pulmonary AVMs including simple chest radiographs, pulse oximetry, contrast tomography (CT), angiography, and transthoracic contrast echocardiography with bubble studies. Simple chest radiographs and evaluation for hypoxia on pulse oximetry lack adequate sensitivity and specificity to serve as screening tests (Haitjema et al., 1995). CT, digital subtraction angiography, formal angiography, and nuclear medicine scans all put the patient at risk of repeated exposure to radiation. Though formal angiography logically may seem like the most sensitive test, it actually misses a number of smaller AVMs; approximately 40% of pulmonary AVMS detected on CT were missed on angiography in one study (Remy et al., 1992). Transthoracic contrast echocardiography (TTCE) with bubble studies allows for detection of right-to-left shunting and is considered a highly sensitive screening test (sensitivity 93%), though it lacks specificity (53%) since intracardiac shunts will also yield a positive bubble study (Shub et al., 1976; Nemec et al., 1991). The recommended screening test by the international guidelines at this time is TTCE with the caveat that no data exist on the sensitivity and specificity of these tests in children (Faughnan et al., 2011).

Once a significant pulmonary AVM is discovered (usually an AVM greater than 3 mm), embolization of the AVM is recommended, even in the asymptomatic adult patient (Faughnan et al., 2011). The main risks involved with embolization are pleurisy, angina, pulmonary infarction, transient ischemic attack (TIA), and stroke, all of which are reported to occur in < 1–2% of patients, except for pleurisy. Surgical management is no longer recommended for pulmonary AVMs. Embolization in children can be considered on an individual case basis. Symptomatic patients may have symptomatic relief of dyspnea and hypoxia after embolization (White et al., 1988; Gupta et al., 2002). Perhaps more importantly, embolization of these malformations significantly decreases the risk of transient ischemic attacks, ischemic stroke, and brain abscesses from right-to-left shunting (Shovlin et al., 2008a). Embolization also appears to decrease the incidence of migraines in patients with HHT (Post et al., 2006). Despite initial embolization, pulmonary AVMs may reperfuse or patients may have other AVMs that increase in size over time. Thus a repeat CT scan of the chest is recommended within 6–12 months of initial embolization and every 3 years thereafter (Faughnan et al., 2011). In addition, patients with treated and untreated pulmonary AVMs are advised to avoid situations that place them at risk for air emboli (scuba diving or intravenous air in an intravenous line) and to use antibiotic prophylaxis for procedures that place them at risk for bacteremia.

Neurologic

The most effective preventative measure for neurologic complications in patients with HHT is embolization of pulmonary AVMs. The evaluation and management of cerebral vascular malformations due to HHT is more controversial. The international guidelines officially recommend that all children with possible or definite HHT be screened for cerebral vascular malformations with a noncontrast magnetic resonance imaging (MRI) of the brain within the first 6 months of life but only 64% of the writers agreed upon this recommendation (Faughnan et al., 2011). Catheter angiography is the most sensitive method of detection of cerebral vascular malformations but also carries higher risks than other imaging modalities, including risk of stroke, which is why it is not recommended as a screening tool (Willinsky et al., 2003). In the body of the guidelines, the authors propose that the risk of sedation for a MRI in an infant is balanced by the potential benefit of diagnosing a high pressure cerebral malformation that might have catastrophic effect if left untreated (Morgan et al., 2002; Al-Saleh et al., 2009; Saleh et al., 2013; Latino et al., 2014a).

A retrospective study of children at the Hospital for Sick Children in Toronto, Canada, demonstrates this point: three of the 11 children in their study with HHT and brain AVMs were not diagnosed with the AVMs until their ICH despite having a family history of the disease (Latino et al., 2014a). For adults with possible or definite HHT, screening with a MR brain with and without contrast with dedicated sequences to look for blood, such as susceptibility-weighted imaging, is recommended (Faughnan et al., 2011). There are no data to suggest that clinically significant vascular lesions develop in the brain over time. Thus, repeated screening is not recommended after a normal study except for perhaps a repeat study done as an adult if the initial scan was done during infancy.

The management of these cerebral lesions depends upon the location and size. Unlike with embolization of pulmonary AVMs, embolization, surgical excision, and radiosurgery all carry significant risks even though the intervention may decrease the risk of ICH. A recent study of 223 adult patients with unruptured brain arteriovenous malformations (not limited to patients with HHT) was recently stopped early when an interim analysis found that medical management alone was superior to medical management with interventional therapy. The overall risk of ICH in patients with HHT is thought to be even lower than that in the general population of patients with non-HHT-related cerebral AVMs, which creates further clinical equipoise regarding optimal management of these lesions (Willemse et al., 2000; Maher et al., 2001). Thus, each cerebral vascular malformation should be discussed on a case-by-case basis with the patient with full disclosure of the possible risks and benefits of any interventional procedures before proceeding.

Gastrointestinal

Since most GI telangiectasias never become symptomatic, the international guidelines on management of HHT recommend checking a yearly hemoglobin and hematocrit after the age of 35 years and then to consider upper endoscopy only if anemia is out of proportion to epistaxis. For those patients found to have GI bleeds, iron supplementation is recommended. Some small studies have suggested that hormonal therapy, such as danocrine or estrogen/progesterone preparations, or antifibrinolytic therapy such as tranexamic acid may be helpful in reduction of GI bleeds but the data are inadequate to support their widespread use at this time (van Cutsem et al., 1990; Faughnan et al., 2011). Laser photocoagulation or bipolar electrocoagulation has been used for direct therapy to stop bleeding but the benefits are transient (Rutgeerts et al., 1985).

Similarly, while a large number of patients may have liver involvement in screening imaging, only about 8% of them are symptomatic (Buscarini et al., 2004; Ianora et al., 2004; Al-Saleh et al., 2011). Thus, liver imaging is only recommended in patients with symptoms or abnormal liver function testing. The gold standard for imaging is mesenteric catheter angiography but this is invasive and involves radiation. Ultrasound, including Doppler studies, provides reasonable sensitivity for a screening test and is noninvasive. In patients whose body habitus prohibits adequate sonographic windows, a CT scan with contrast can be done for evaluation instead of ultrasound. The international guidelines for management of HHT recommend that hepatic artery embolization not be pursued in patients with intractable symptomatic liver AVMs because embolization may cause liver necrosis and death in upwards of 30% of cases. Instead the authors recommend that patients proceed to liver transplantation (Chavan et al., 1998, 2004; Odorico et al., 1998; Le Corre et al., 2000; Lerut et al., 2006; Faughnan et al., 2011). Unfortunately, at this time, no effective medical therapy exists for intractable liver disease in patients with HHT. A small study of three patients and a single center phase 2 trial of 25 patients both reported that treatment with six cycles of bevacizumab showed improvement in their congestive heart failure due to liver involvement; however, larger randomized control trials are needed before this drug can be widely recommended for treatment (Dupuis-Girod et al., 2012; Chavan et al., 2013).

Ears, nose, and throat

While it may appear to be one of the more minor symptoms of HHT, patients report that epistaxis significantly impacts their quality of life (Pasculli et al., 2004; Sabbà et al., 2006). Conservative measures, such as humidification and lubricants, may help protect nasal mucosa and prevent bleeding. For those with ongoing bleeding, care should be taken with nasal packing because it might cause rebleeding during its removal. A number of local procedural therapies are available for treatment of nasal telangiectasias including sodium tetradecyl sulfate injections, septodermoplasty, laser coagulation, cryotherapy, and chemical cautery, but each procedure bears its own set of possible side-effects (McCaffrey et al., 1977; Shah et al., 2002; Fiorella et al., 2005; Mahoney and Shapshay, 2005; Lesnik et al., 2007; Boyer et al., 2011; Hanks et al., 2014; Kim et al., 2014). Similar to studies of control of GI bleeding, small studies have explored the possible use of local treatment with estrogen/progesterone products and antifibrinolytics without evidence of significant improvement in patients (Klepfish et al., 2001; Bergler et al., 2002). The two studies mentioned above that evaluated the use of systemic bevacizumab in patients with HHT did report improvement in epistaxis so there is hope that further trials may prove the drug's efficacy (Dupuis-Girod et al., 2012; Chavan et al., 2013). Since epistaxis is frequently a source of iron deficiency anemia in patients, adequate iron supplementation is also indicated in patients.

Targeted drug therapy

With the discovery of the genetic basis for HHT, attention has turned towards development of drug therapies that target the angiogenesis. Bevacizumab is one such drug that has been mentioned above. It is an antivascular endothelial growth factor (VEGF) drug that is currently approved as a chemotherapy agent (Dupuis-Girod et al., 2012). GlaxoSmithKline are in the process of recruiting patients with HHT for a clinical trial that will evaluate the safety and efficacy of another drug that blocks VEGF, pazopanib (clinicaltrials.gov study identifier NCT02204371). Other researchers are approaching the problem from a different angle; they postulate that the problem in HHT is not unchecked angiogenesis but rather the failure of the new blood vessels to mature and recruit mural cells. Thalidomide enhances pericyte recruitment so has been proposed as a possible therapy for HHT (Lebrin et al., 2010; Alam et al., 2011; Thalgott et al., 2015). A small study of seven patients showed improved hemoglobin levels and decreased hemorrhage after treatment with thalidomide compared to control patients not treated with the drug. The authors also reported that mice models treated with the drug and nasal biopsies of the treated patients also showed increased platelet-derived growth factor expression and increased mural cell activation (Lebrin et al., 2010).

While we await new drug therapies for HHT, perhaps the greatest stride that we might make in the management and treatment of the disease remains early recognition and screening of patients. Despite its initial description so many years ago, this neurocutaneous disorder remains underdiagnosed even within families with known disease. Only with more timely diagnosis can physicians strive to apply the recommended screening and hopefully prevent the complications of undiagnosed and unrecognized visceral vascular malformations.

References

AAssar OS, Friedman CM, White RI (1991). The natural history of epistaxis in hereditary hemorrhagic telangiectasia. Laryngoscope 101: 977–980.

Alam MA, Sami S, Babu S (2011). Successful treatment of bleeding gastro-intestinal angiodysplasia in hereditary haemorrhagic telangiectasia with thalidomide. BMJ Case Rep. Nov 8. pii: bcr0820114585.

Al-Saleh S, Mei-Zahav M, Faughnan ME et al. (2009). Screening for pulmonary and cerebral arteriovenous malformations in children with hereditary haemorrhagic telangiectasia. Eur Respir J 34: 875–881.

Al-Saleh S, John PR, Letarte M et al. (2011). Symptomatic liver involvement in neonatal hereditary hemorrhagic telangiectasia. Pediatrics 127: e1615–e1620.

Bayrak-Toydemir P, McDonald J, Akarsu N et al. (2006a). A fourth locus for hereditary hemorrhagic telangiectasia maps to chromosome 7. Am J Med Genet A 140: 2155–2162.

Bayrak-Toydemir P, McDonald J, Markewitz B et al. (2006b). Genotype-phenotype correlation in hereditary hemorrhagic telangiectasia: mutations and manifestations. Am J Med Genet A 140: 463–470.

Begbie ME, Wallace GM, Shovlin CL (2003). Hereditary haemorrhagic telangiectasia (Osler–Weber–Rendu syndrome): a view from the 21st century. Postgrad Med J 79: 18–24.

Berg J, Porteous M, Reinhardt D et al. (2003). Hereditary haemorrhagic telangiectasia: a questionnaire based study to delineate the different phenotypes caused by endoglin and ALK1 mutations. J Med Genet 40: 585–590.

Berg AM, Amirbekian S, Mojibian H et al. (2010). Hemothorax due to rupture of pulmonary arteriovenous malformation: an interventional emergency. Chest 137: 705–707.

Bergler W, Sadick H, Gotte K et al. (2002). Topical estrogens combined with argon plasma coagulation in the management of epistaxis in hereditary hemorrhagic telangiectasia. Ann Otol Rhinol Laryngol 111: 222–228.

Bhatia S, Morrison JF, Bower TC et al. (2003). Pulmonary hypertension in the setting of acquired systemic arteriovenous fistulas. Mayo Clin Proc 78: 908–912.

Boyer H, Fernandes P, Duran O et al. (2011). Office-based sclerotherapy for recurrent epistaxis due to hereditary hemorrhagic telangiectasia: a pilot study. Int Forum Allergy Rhinol 1: 319–323.

Brancati F, Valente EM, Tadini G et al. (2003). Autosomal dominant hereditary benign telangiectasia maps to the CMC1 locus for capillary malformation on chromosome 5q14. J Med Genet 40: 849–853.

Braverman IM, Keh A, Jacobson BS (1990). Ultrastructure and three-dimensional organization of the telangiectases of hereditary hemorrhagic telangiectasia. J Invest Dermatol 95: 422–427.

Buscarini E, Buscarini L, Danesino C et al. (1997). Hepatic vascular malformations in hereditary hemorrhagic telangiectasia: Doppler sonographic screening in a large family. J Hepatol 26: 111–118.

Buscarini E, Danesino C, Plauchu H et al. (2004). High prevalence of hepatic focal nodular hyperplasia in subjects with hereditary hemorrhagic telangiectasia. Ultrasound Med Biol 30: 1089–1097.

Buscarini E, Danesino C, Olivieri C et al. (2005). Liver involvement in hereditary haemorrhagic telangiectasia or Rendu–Osler–Weber disease. Dig Liver Dis 37: 635–645.

Buscarini E, Plauchu H, Garcia Tsao G et al. (2006). Liver involvement in hereditary hemorrhagic telangiectasia: consensus recommendations. Liver Int 26: 1040–1046.

Buscarini E, Gebel M, Ocran K et al. (2008). Interobserver agreement in diagnosing liver involvement in hereditary hemorrhagic telangiectasia by Doppler ultrasound. Ultrasound Med Biol 34: 718–725.

Buscarini E, Leandro G, Conte D et al. (2011). Natural history and outcome of hepatic vascular malformations in a large cohort of patients with hereditary hemorrhagic teleangiectasia. Dig Dis Sci 56: 2166–2178.

Chavan A, Galanski M, Wagner S et al. (1998). Hereditary hemorrhagic telangiectasia: effective protocol for embolization of hepatic vascular malformations – experience in five patients. Radiology 209: 735–739.

Chavan A, Caselitz M, Gratz KF et al. (2004). Hepatic artery embolization for treatment of patients with hereditary hemorrhagic telangiectasia and symptomatic hepatic vascular malformations. Eur Radiol 14: 2079–2085.

Chavan A, Schumann-Binarsch S, Luthe L et al. (2013). Systemic therapy with bevacizumab in patients with hereditary hemorrhagic telangiectasia (HHT). Vasa 42: 106–110.

Cho D, Kim S, Kim M et al. (2012). Two cases of high output heart failure caused by hereditary hemorrhagic telangiectasia. Korean Circ J 42: 861–865.

Cole SG, Begbie ME, Wallace GM et al. (2005). A new locus for hereditary haemorrhagic telangiectasia (HHT3) maps to chromosome 5. J Med Genet 42: 577–582.

Cottin V, Chinet T, Lavolé A et al. (2007). Pulmonary arteriovenous malformations in hereditary hemorrhagic telangiectasia: a series of 126 patients. Medicine (Baltimore) 86: 1–17.

Cox KL, Frates RC, Wong A et al. (1980). Hereditary generalized juvenile polyposis associated with pulmonary arteriovenous malformation. Gastroenterology 78: 1566–1570.

Dakeishi M, Shioya T, Wada Y et al. (2002). Genetic epidemiology of hereditary hemorrhagic telangiectasia in a local community in the northern part of Japan. Hum Mutat 19: 140–148.

de Gussem EM, Lausman AY, Beder AJ et al. (2014). Outcomes of pregnancy in women with hereditary hemorrhagic telangiectasia. Obstet Gynecol 123: 514–520.

Dheyauldeen S, Abdelnoor M, Bachmann-Harildstad G (2011). The natural history of epistaxis in patients with hereditary hemorrhagic telangiectasia in the Norwegian population: a cross-sectional study. Am J Rhinol Allergy 25: 214–218.

Dines DE, Seward JB, Bernatz PE (1983). Pulmonary arteriovenous fistulas. Mayo Clin Proc 58: 176–181.

Donaldson JW, McKeever TM, Hall IP et al. (2014). The UK prevalence of hereditary haemorrhagic telangiectasia and its association with sex socioeconomic status and region of residence: a population-based study. Thorax 69: 161–167.

Dowling MM (2004). Hereditary hemorrhagic telangiectasia. In: ES Roach, VS Miller (Eds.), Neurocutaneous Disorders. Cambridge University Press, Cambridge, pp. 159–165 (Chapter 20).

Dupuis-Girod S, Ginon I, Saurin JC et al. (2012). Bevacizumab in patients with hereditary hemorrhagic telangiectasia and severe hepatic vascular malformations and high cardiac output. JAMA 307: 948–955.

El-Harith e-HA, Kühnau W, Schmidtke J et al. (2006). Hereditary hemorrhagic telangiectasia is caused by the Q490X mutation of the ACVRL1 gene in a large Arab family: support of homozygous lethality. Eur J Med Genet 49: 323–330.

Elphick A, Shovlin CL (2014). Relationships between epistaxis migraines and triggers in hereditary hemorrhagic telangiectasia. Laryngoscope 124: 1521–1528.

Espinosa PS, Pettigrew LC, Berger JR (2008). Hereditary hemorrhagic telangectasia and spinal cord infarct: case report with a review of the neurological complications of HHT. Clin Neurol Neurosurg 110: 484–491.

Faughnan ME, Palda VA, Garcia-Tsao G et al. (2011). International guidelines for the diagnosis and management of hereditary haemorrhagic telangiectasia. J Med Genet 48: 73–87.

Ference BA, Shannon TM, White RI et al. (1994). Life-threatening pulmonary hemorrhage with pulmonary arteriovenous malformations and hereditary hemorrhagic telangiectasia. Chest 106: 1387–1390.

Fiorella ML, Ross D, Henderson KJ et al. (2005). Outcome of septal dermoplasty in patients with hereditary hemorrhagic telangiectasia. Laryngoscope 115: 301–305.

Fish JE, Wythe JD (2015). The molecular regulation of arteriovenous specification and maintenance. Dev Dyn 244: 391–409.

Folz BJ, Tennie J, Lippert BM et al. (2005). Natural history and control of epistaxis in a group of German patients with Rendu–Osler–Weber disease. Rhinology 43: 40–46.

Fulbright RK, Chaloupka JC, Putman CM et al. (1998). MR of hereditary hemorrhagic telangiectasia: prevalence and spectrum of cerebrovascular malformations. AJNR Am J Neuroradiol 19: 477–484.

Gallione CJ, Repetto GM, Legius E et al. (2004). A combined syndrome of juvenile polyposis and hereditary haemorrhagic telangiectasia associated with mutations in MADH4 (SMAD4). Lancet 363: 852–859.

Garcia-Tsao G (2007). Liver involvement in hereditary hemorrhagic telangiectasia (HHT). J Hepatol 46: 499–507.

Garcia-Tsao G, Korzenik JR, Young L et al. (2000). Liver disease in patients with hereditary hemorrhagic telangiectasia. N Engl J Med 343: 931–936.

Geirdal A, Dheyauldeen S, Bachmann-Harildstad G et al. (2012). Quality of life in patients with hereditary hemorrhagic telangiectasia in Norway: a population based study. Am J Med Genet A 158A: 1269–1278.

Gibbs DD (1986). Rendu–Osler–Weber disease: a triple eponymous title lives on. J R Soc Med 79: 742–743.

Ginon I, Decullier E, Finet G et al. (2013). Hereditary hemorrhagic telangiectasia liver vascular malformations and cardiac consequences. Eur J Intern Med 24: e35–e39.

Giordano P, Sangerardi M, Suppressa P et al. (2015). Pulmonary arteriovenous malformations and cerebral abscess recurrence in a child with hereditary hemorrhagic telangiectasia. J Pediatr Hematol Oncol 37(3): e200–e203.

Gossage JR, Kanj G (1998). Pulmonary arteriovenous malformations. A state of the art review. Am J Respir Crit Care Med 158: 643–661.

Goussous T, Haynes A, Najarian K et al. (2009). Hereditary hemorrhagic telangiectasia presenting as high output cardiac failure during pregnancy. Cardiol Res Pract 2009: 437237.

Grève E, Moussata D, Gaudin JL et al. (2010). High diagnostic and clinical impact of small-bowel capsule endoscopy in patients with hereditary hemorrhagic telangiectasia with overt digestive bleeding and/or severe anemia. Gastrointest Endosc 71: 760–767.

Gupta P, Mordin C, Curtis J et al. (2002). Pulmonary arteriovenous malformations: effect of embolization on right-to-left shunt hypoxemia and exercise tolerance in 66 patients. AJR Am J Roentgenol 179: 347–355.

Guttmacher AE, McKinnon WC, Upton MD (1994). Hereditary hemorrhagic telangiectasia: a disorder in search of the genetics community. Am J Med Genet 52: 252–253.

Haitjema T, Disch F, Overtoom TT et al. (1995). Screening family members of patients with hereditary hemorrhagic telangiectasia. Am J Med 99: 519–524.

Hanks JE, Hunter D, Goding GS et al. (2014). Complications from office sclerotherapy for epistaxis due to hereditary hemorrhagic telangiectasia (HHT or Osler–Weber–Rendu). Int Forum Allergy Rhinol 4: 422–427.

Heutink P, Haitjema T, Breedveld GJ et al. (1994). Linkage of hereditary haemorrhagic telangiectasia to chromosome 9q34 and evidence for locus heterogeneity. J Med Genet 31: 933–936.

Hoche F, Seidel K, Theis M et al. (2012). Neurodegeneration in ataxia telangiectasia: what is new? What is evident? Neuropediatrics 43: 119–129.

Hsu YL, Wang HC, Yang PC (2004). Desbaric air embolism during diving: an unusual complication of Osler–Weber–Rendu disease. Br J Sports Med 38: E6.

Ianora AA, Memeo M, Sabba C et al. (2004). Hereditary hemorrhagic telangiectasia: multi-detector row helical CT assessment of hepatic involvement. Radiology 230: 250–259.

Ingrosso M, Sabbà C, Pisani A et al. (2004). Evidence of small-bowel involvement in hereditary hemorrhagic telangiectasia: a capsule-endoscopic study. Endoscopy 36: 1074–1079.

Iyer NK, Burke CA, Leach BH et al. (2010). SMAD4 mutation and the combined syndrome of juvenile polyposis syndrome and hereditary haemorrhagic telangiectasia. Thorax 65: 745–746.

Karabegovic A, Shinawi M, Cymerman U et al. (2004). No live individual homozygous for a novel endoglin mutation was found in a consanguineous Arab family with hereditary haemorrhagic telangiectasia. J Med Genet 41: e119.

Kim JY, Oh JH, Kim GT et al. (2014). Endoscopic cryotherapy for the treatment of epistaxis due to hereditary hemorrhagic telangiectasia. J Craniofac Surg 25: e120–e122.

Kjeldsen AD, Kjeldsen J (2000). Gastrointestinal bleeding in patients with hereditary hemorrhagic telangiectasia. Am J Gastroenterol 95: 415–418.

Kjeldsen AD, Vase P, Green A (1999). Hereditary haemorrhagic telangiectasia: a population-based study of prevalence and mortality in Danish patients. J Intern Med 245: 31–39.

Kjeldsen AD, Oxhøj H, Andersen PE et al. (2000). Prevalence of pulmonary arteriovenous malformations (PAVMs) and occurrence of neurological symptoms in patients with hereditary haemorrhagic telangiectasia (HHT). J Intern Med 248: 255–262.

Klepfish A, Berrebi A, Schattner A (2001). Intranasal tranexamic acid treatment for severe epistaxis in hereditary hemorrhagic telangiectasia. Arch Intern Med 161: 767.

Korzenik JR (1996). Hereditary hemorrhagic telangiectasia and other intestinal vascular anomalies. Gastroenterologist 4: 203–210.

Krings T, Ozanne A, Chng SM et al. (2005). Neurovascular phenotypes in hereditary haemorrhagic telangiectasia patients according to age. Review of 50 consecutive patients aged 1 day – 60 years. Neuroradiology 47: 711–720.

Latino GA, Al-Saleh S, Carpenter S et al. (2014a). The diagnostic yield of rescreening for arteriovenous malformations in children with hereditary hemorrhagic telangiectasia. J Pediatr 165: 197–199.

Latino GA, Brown D, Glazier RH et al. (2014b). Targeting under-diagnosis in hereditary hemorrhagic telangiectasia: a model approach for rare diseases? Orphanet J Rare Dis 9: 115.

Le Corre F, Golkar B, Tessier C et al. (2000). Liver transplantation for hepatic arteriovenous malformation with high-output cardiac failure in hereditary hemorrhagic telangiectasia: hemodynamic study. J Clin Anesth 12: 339–342.

Lebrin F, Srun S, Raymond K et al. (2010). Thalidomide stimulates vessel maturation and reduces epistaxis in individuals with hereditary hemorrhagic telangiectasia. Nat Med 16: 420–428.

Lerut J, Orlando G, Adam R et al. (2006). Liver transplantation for hereditary hemorrhagic telangiectasia: report of the European liver transplant registry. Ann Surg 244: 854–862, discussion 862–864.

Lesca G, Burnichon N, Raux G et al. (2006). Distribution of ENG and ACVRL1 (ALK1). mutations in French HHT patients. Hum Mutat 27: 598.

Lesca G, Olivieri C, Burnichon N et al. (2007). Genotype–phenotype correlations in hereditary hemorrhagic telangiectasia: data from the French-Italian HHT network. Genet Med 9: 14–22.

Lesnik GT, Ross DA, Henderson KJ et al. (2007). Septectomy and septal dermoplasty for the treatment of severe transfusion-dependent epistaxis in patients with hereditary hemorrhagic telangiectasia and septal perforation. Am J Rhinol 21: 312–315.

Letteboer TG, Mager JJ, Snijder RJ et al. (2006). Genotype-phenotype relationship in hereditary haemorrhagic telangiectasia. J Med Genet 43: 371–377.

Letteboer TG, Mager HJ, Snijder RJ et al. (2008). Genotype-phenotype relationship for localization and age distribution of telangiectases in hereditary hemorrhagic telangiectasia. Am J Med Genet A 146A: 2733–2739.

Livneh A, Langevitz P, Morag B et al. (1988). Functionally reversible hepatic arteriovenous fistulas during pregnancy in patients with hereditary hemorrhagic telangiectasia. South Med J 81: 1047–1049.

Longacre AV, Gross CP, Gallitelli M et al. (2003). Diagnosis and management of gastrointestinal bleeding in patients with hereditary hemorrhagic telangiectasia. Am J Gastroenterol 98: 59–65.

Maher CO, Piepgras DG, Brown RD et al. (2001). Cerebrovascular manifestations in 321 cases of hereditary hemorrhagic telangiectasia. Stroke 32: 877–882.

Mahoney EJ, Shapshay SM (2005). Nd-YAG laser photocoagulation for epistaxis associated with hereditary hemorrhagic telangiectasia. Laryngoscope 115: 373–375.

Mancini ML, Terzic A, Conley BA et al. (2009). Endoglin plays distinct roles in vascular smooth muscle cell recruitment and regulation of arteriovenous identity during angiogenesis. Dev Dyn 238: 2479–2493.

Marziniak M, Jung A, Guralnik V et al. (2009). An association of migraine with hereditary haemorrhagic telangiectasia independently of pulmonary right-to-left shunts. Cephalalgia 29: 76–81.

McCaffrey TV, Kern EB, Lake CF (1977). Management of epistaxis in hereditary hemorrhagic telangiectasia. Review of 80 cases. Arch Otolaryngol 103: 627–630.

McDonald J, Pyeritz RE (1993–2015). In: RA Pagon, MP Adam, HH Ardinger, SE Wallace, A Amemiya, LJH Bean, TD Bird, CR Dolan, CT Fong, RJH Smith, K Stephens (Eds.), GeneReviews® [Internet]. University of Washington, Seattle, Seattle (WA).

McDonald MT, Papenberg KA, Ghosh S et al. (1994). A disease locus for hereditary haemorrhagic telangiectasia maps to chromosome 9q33-34. Nat Genet 6: 197–204.

McGuire F, Kennelly T, Tillack T et al. (2010). Pulmonary capillary hemangiomatosis associated with CREST syndrome: a case report and review of the literature. Respiration 80: 435–438.

Morgan T, McDonald J, Anderson C et al. (2002). Intracranial hemorrhage in infants and children with hereditary hemorrhagic telangiectasia (Osler–Weber–Rendu syndrome). Pediatrics 109: E12.

Nemec JJ, Davison MB, Marwick TH et al. (1991). Detection and evaluation of intrapulmonary vascular shunt with "contrast Doppler" transesophageal echocardiography. J Am Soc Echocardiogr 4: 79–83.

Odorico JS, Hakim MN, Becker YT et al. (1998). Liver transplantation as definitive therapy for complications after arterial embolization for hepatic manifestations of hereditary hemorrhagic telangiectasia. Liver Transpl Surg 4: 483–490.

Pasculli G, Resta F, Guastamacchia E et al. (2004). Health-related quality of life in a rare disease: hereditary hemorrhagic telangiectasia (HHT) or Rendu–Osler–Weber disease. Qual Life Res 13: 1715–1723.

Petrillo TM, Fortenberry JD, Chambliss CR et al. (2001). Radiological case of the month. Cerebral air embolism in hereditary hemorrhagic telangiectasia (Osler–Weber–Rendu disease). Arch Pediatr Adolesc Med 155: 847–848.

Plauchu H, de Chadarévian JP, Bideau A et al. (1989). Age-related clinical profile of hereditary hemorrhagic telangiectasia in an epidemiologically recruited population. Am J Med Genet 32: 291–297.

Porteous ME, Burn J, Proctor SJ (1992). Hereditary haemorrhagic telangiectasia: a clinical analysis. J Med Genet 29: 527–530.

Post MC, Letteboer TG, Mager JJ et al. (2005). A pulmonary right-to-left shunt in patients with hereditary hemorrhagic telangiectasia is associated with an increased prevalence of migraine. Chest 128: 2485–2489.

Post MC, Thijs V, Schonewille WJ et al. (2006). Embolization of pulmonary arteriovenous malformations and decrease in prevalence of migraine. Neurology 66: 202–205.

Prasad C, Chudley AE (2002). History of genetics through philately: Sir William Osler (1849–1919) and the Osler–Weber–Rendu syndrome. Clin Genet 61: 404–407.

Ravard G, Soyer P, Boudiaf M et al. (2004). Hepatic involvement in hereditary hemorrhagic telangiectasia: helical computed tomography features in 24 consecutive patients. J Comput Assist Tomogr 28: 488–495.

Remy J, Remy-Jardin M, Wattinne L et al. (1992). Pulmonary arteriovenous malformations: evaluation with CT of the chest before and after treatment. Radiology 182: 809–816.

Revencu N, Boon LM, Mendola A et al. (2013). RASA1 mutations and associated phenotypes in 68 families with capillary malformation-arteriovenous malformation. Hum Mutat 34: 1632–1641.

Roman G, Fisher M, Perl DP et al. (1978). Neurological manifestations of hereditary hemorrhagic telangiectasia (Rendu–Osler–Weber disease): report of 2 cases and review of the literature. Ann Neurol 4: 130–144.

Rutgeerts P, Van Gompel F, Geboes K et al. (1985). Long term results of treatment of vascular malformations of the gastrointestinal tract by neodymium Yag laser photocoagulation. Gut 26: 586–593.

Sabbà C, Pasculli G, Suppressa P et al. (2006). Life expectancy in patients with hereditary haemorrhagic telangiectasia. QJM 99: 327–334.

Saleh M, Carter MT, Latino GA et al. (2013). Brain arteriovenous malformations in patients with hereditary hemorrhagic telangiectasia: clinical presentation and anatomical distribution. PediatrNeurol 49: 445–450.

Shah RK, Dhingra JK, Shapshay SM (2002). Hereditary hemorrhagic telangiectasia: a review of 76 cases. Laryngoscope 112: 767–773.

Sharma KB, Forkert L (2004). Massive pulmonary hemorrhage from dual circulation pulmonary arteriovenous malformations in hereditary hemorrhagic telangiectasia. Can Respir J 11: 597–598.

Sharma VK, Howden CW (1998). Gastrointestinal and hepatic manifestations of hereditary hemorrhagic telangiectasia. Dig Dis 16: 169–174.

Shovlin CL (2010). Hereditary haemorrhagic telangiectasia: pathophysiology diagnosis and treatment. Blood Rev 24: 203–219.

Shovlin CL, Letarte M (1999). Hereditary haemorrhagic telangiectasia and pulmonary arteriovenous malformations: issues in clinical management and review of pathogenic mechanisms. Thorax 54: 714–729.

Shovlin CL, Hughes JM, Tuddenham EG et al. (1994). A gene for hereditary haemorrhagic telangiectasia maps to chromosome 9q3. Nat Genet 6: 205–209.

Shovlin CL, Guttmacher AE, Buscarini E et al. (2000). Diagnostic criteria for hereditary hemorrhagic telangiectasia (Rendu–Osler–Weber syndrome). Am J Med Genet 91: 66–67.

Shovlin CL, Jackson JE, Bamford KB et al. (2008a). Primary determinants of ischaemic stroke/brain abscess risks are independent of severity of pulmonary arteriovenous malformations in hereditary haemorrhagic telangiectasia. Thorax 63: 259–266.

Shovlin CL, Sodhi V, McCarthy A et al. (2008b). Estimates of maternal risks of pregnancy for women with hereditary haemorrhagic telangiectasia (Osler–Weber–Rendu syndrome): suggested approach for obstetric services. BJOG 115: 1108–1115.

Shub C, Tajik AJ, Seward JB et al. (1976). Detecting intrapulmonary right-to-left shunt with contrast echocardiography. Observations in a patient with diffuse pulmonary arteriovenous fistulas. Mayo Clin Proc 51: 81–84.

Steele JG, Nath PU, Burn J et al. (1993). An association between migrainous aura and hereditary haemorrhagic telangiectasia. Headache 33: 145–148.

Thalgott J, Dos-Santos-Luis D, Lebrin F (2015). Pericytes as targets in hereditary hemorrhagic telangiectasia. Front Genet 6: 37.

Thenganatt J, Schneiderman J, Hyland RH et al. (2006). Migraines linked to intrapulmonary right-to-left shunt. Headache 46: 439–443.

Toporsian M, Gros R, Kabir MG et al. (2005). A role for endoglin in coupling eNOS activity and regulating vascular tone revealed in hereditary hemorrhagic telangiectasia. Circ Res 96: 684–692.

van Cutsem E, Rutgeerts P, Vantrappen G (1990). Treatment of bleeding gastrointestinal vascular malformations with oestrogen-progesterone. Lancet 335: 953–955.

Westermann CJ, Rosina AF, De Vries V et al. (2003). The prevalence and manifestations of hereditary hemorrhagic telangiectasia in the Afro-Caribbean population of the Netherlands Antilles: a family screening. Am J Med Genet A 116A: 324–328.

White RI, Lynch-Nyhan A, Terry P et al. (1988). Pulmonary arteriovenous malformations: techniques and long-term outcome of embolotherapy. Radiology 169: 663–669.

Willemse RB, Mager JJ, Westermann CJ et al. (2000). Bleeding risk of cerebrovascular malformations in hereditary hemorrhagic telangiectasia. J Neurosurg 92: 779–784.

Williams JC, Hamilton JK, Shiller M et al. (2012). Combined juvenile polyposis and hereditary hemorrhagic telangiectasia. Proc (Bayl Univ Med Cent) 25: 360–364.

Willinsky RA, Taylor SM, TerBrugge K et al. (2003). Neurologic complications of cerebral angiography: prospective analysis of 2899 procedures and review of the literature. Radiology 227: 522–528.

Wooderchak-Donahue WL, McDonald J, O'Fallon B et al. (2013). BMP9 mutations cause a vascular-anomaly syndrome with phenotypic overlap with hereditary hemorrhagic telangiectasia. Am J Hum Genet 93: 530–537.

Handbook of Clinical Neurology, Vol. 132 (3rd series)
Neurocutaneous Syndromes
M.P. Islam and E.S. Roach, Editors

Chapter 14

Ataxia telangiectasia

ANDREEA NISSENKORN[1,2,3*] AND BRURIA BEN-ZEEV[1,3]

[1]*Pediatric Neurology Unit, Edmond and Lilly Safra Children Hospital, Chaim Sheba Medical Center, Ramat Gan, Israel*

[2]*National AT Center, Edmond and Lilly Safra Children Hospital, Chaim Sheba Medical Center, Ramat Gan, Israel*

[3]*Sackler School of Medicine, Tel Aviv University, Tel Aviv, Israel*

INTRODUCTION AND HISTORY

Ataxia telangiectasia is an autosomal recessive neurodegenerative disorder that is characterized by progressive cerebellar atrophy and ataxia, radiosensitivity, immune deficiency, and a tendency to develop lymphoreticular malignancies (Boder and Sedgwick, 1958; Crawford, 1998; Perlman et al., 2003; Chun and Gatti, 2004). The disorder is relatively rare, with an incidence of 1:40 000 to 1:100 000 (Crawford, 1998), but in several populations the incidence is considerably higher due to founder mutations (Gilad et al., 1996a). It is the second most frequent cause of recessive ataxia in childhood after Freidreich ataxia, but the most frequent in children with onset of ataxia within the first years of life.

The course of disease is ominous, leading to wheelchair dependency towards the second decade of life (Crawford et al., 2000; Nissenkorn et al., 2011), and early death (Crawford et al., 2006; Micol et al., 2011). The 20 year survival rate is about 50%, and this has remained largely unchanged despite supportive treatment (Crawford et al., 2006; Micol et al., 2011).

The earliest reports mentioning patients who could fit the description of the disease date as early as 1926 (Syllaba and Henner) and 1941 (Louis-Bar), but it was Elena Boder who established it as a distinctive syndrome (Boder and Sedgwick, 1958). The syndrome was named ataxia telangiectasia because of its two striking features, cerebellar ataxia and conjunctival telangiectasia. The acronyms A-T or AT, as well as the term Louis–Bar syndrome, are equally popular.

Boder and Segwick in 1958 meticulously described the main clinical features of AT (Boder and Sedgwick, 1958), and during the next several years further delineated the clinical as well as the main pathologic features of the disorder (Boder and Sedgwick, 1970; Boder, 1975; Baloh et al., 1978).

The gene responsible for the disease was identified and sequenced in 1995 in the Shiloh laboratories and named *ATM* (ataxia telangiectasia mutated) (Savitsky et al., 1995). Since the ATM protein encoded by this gene has a central role in DNA repair, most of the efforts in the following years were devoted to understanding the pathophysiologic mechanism of ATM (Gatti, 1993; Savitsky et al., 1995; Taylor and Byrd, 2005; McKinnon, 2012; Shiloh and Ziv, 2013; Cremona and Behrens, 2014).

PATHOPHYSIOLOGY

ATM is a member of the phosphatidylinositol 3-kinase like protein kinase (PIKK) family, which includes other proteins involved in DNA repair (Gatti, 1993; Savitsky et al., 1995; Taylor and Byrd, 2005; McKinnon, 2012; Shiloh and Ziv, 2013; Cremona and Behrens, 2014). It is a large protein (350 kDa) with three functional domains (Fig. 14.1); the kinase domain is crucial for phosphorylation of dozens of substrates involved in repair of DNA double strand breaks (Gatti, 1993; Savitsky et al., 1995; Taylor and Byrd, 2005; McKinnon, 2012; Shiloh and Ziv, 2013; Cremona and Behrens, 2014).

Breaks in double stranded DNA occur spontaneously each day as a result of normal cell metabolism, as well as from ionizing radiation and alkylating agents (Savitsky et al., 1995; Taylor and Byrd, 2005; McKinnon, 2012; Ambrose and Gatti, 2013). Replication of the damaged DNA of tumor suppressor genes would promote carcinogenesis, and resulting unstable and nonfunctional proteins could lead to cell death.

*Correspondence to: Andreea Nissenkorn, Pediatric Neurology Unit, Edmond and Lilly Safra Children Hospital, Chaim Sheba Medical Center, Ramat Gan, Israel. E-mail: Andreea.Nissenkorn@sheba.health.gov.il

Fig. 14.1. ATM protein domains and their function. The ATM protein has an N terminal site and three functional domains located at the C terminal site. The PIK3 is the ATM kinase domain which has a role in phosphorylating dozens of substrates of the ATM protein. The FAT domain contains the autophosphorylation site essential for activation of the ATM dimers. Cysteine residues within the FATC domain create disulfitic bounds which confer to the ATM an intrinsic antioxidant activity.

In response to DNA damage, ATM levels increase immediately and activate via autophosphorylation (Bakkenist and Kastan, 2003). Activated ATM binds to the MRN complex (composed of MRE11, RAD50, and NBS1) and subsequently phosphorylates dozens of substrates involved in repair of DNA breaks (Gatti, 1993; Savitsky et al., 1995; Taylor and Byrd, 2005; McKinnon, 2012; Shiloh and Ziv, 2013; Cremona and Behrens, 2014). Amongst the ATM targets are tumor suppressor genes as p53, cell cycle point kinases as ChK2, and DNA repair kinases. ATM arrests cells from entering the S and M1 phase of the cell cycle in order to prevent replication of damaged DNA; it also activates the kinases that will correct DNA (Gatti, 1993; Savitsky et al., 1995; Taylor and Byrd, 2005; McKinnon, 2012; Shiloh and Ziv, 2013; Cremona and Behrens, 2014). In case DNA cannot be repaired, ATM pushes cells towards apoptosis (Fig. 14.2).

In the absence of ATM, cells will enter the replication cycle with damaged DNA, and cells will be driven towards carcinogenesis. In cells of the immune system, ATM is essential for the V(D)J recombinations and diversity of the humoral and cellular arm (Bredemeyer et al., 2008; Gapud and Sleckman, 2011; Driessen et al., 2013). Therefore, features of AT such as radiosensitivity, tendency to malignancy, and immune deficiency are readily explained by this mechanism.

ATM response to DNA damage within nuclei of neuronal cells has been shown to follow the same pathway as in dividing cells outside the nervous system (Biton et al., 2006, 2008; Dar et al., 2006). However, since the neuronal cells do not divide, several additional mechanisms may explain neurodegeneration of Purkinje cells of the cerebellum.

A central factor contributing to neurodegeneration is oxidative stress. ATM acts as a sensor for reactive oxygen species (Ditch and Paull, 2012), and high levels of oxidative damage were found in $ATM^{-/-}$ mice, as well as in AT patients (Kamsler et al., 2001). This damage may be explained by at least two antioxidant pathways which are negatively affected by ATM deficiency. Cysteine residues on the ATM protein create intermolecular disulfite bounds which act as an antioxidant; absence of ATM will augment the oxidative stress by losing its intrinsic antioxidant activity (Guo et al., 2010). Furthermore, ATM plays a role in regulating glutathione synthesis, which is an antioxidant, via pentosophosphate pathway (Cosentino et al., 2011).

In the presence of oxidative stress ATM deficient neurons are prone to glutamatergic toxicity and might defend themselves by inducing the autophagy process that removes dysfunctional organelles (Kim et al., 2010). While autophagy is increased and protective in eliminating damaged neurons, elimination of defective mitochondria by mitophagy is impaired in ATM deficient cells (Valentin-Vega and Kastan, 2012; Valentin-Vega et al., 2012).

Recently, Li et al. (Li et al., 2012; Herrup, 2013; Herrup et al., 2013) described a new role for ATM in maintaining HDAC4 in cytoplasm by phosphorylation. In the absence of ATM, HDAC4 accumulates in the nucleus, binds to chromatin, and impairs protein translation by an epigenetic mechanism.

While most studies concentrate on the pathology of neurons, lack of ATM has been shown to affect astrocyte morphology and functioning, possibly adding to toxicity on the neurons (Raz-Prag et al., 2011; Meshulam et al., 2012).

There is growing evidence that ATM might also be important during intrauterine stages of neurogenesis and migration. Ectopic Purkinje cells have been found in brains of $Atm^{-/-}$ mice (Vinters et al., 1985; Bottini et al., 2012), suggesting abnormal development of the cerebellum. Mild microcephaly is a common early feature in children with AT, again indicating abnormal neurodevelopment (Nissenkorn et al., 2011). Therefore, ataxia telangiectasia might also be viewed as a neurodevelopmental disorder.

In addition, ATM exerts role in synaptic vesicle transmission (Li et al., 2009), mTOR pathway regulation (Ditch and Paull, 2012; Shen and Houghton, 2013), insulin resistance, and the metabolic syndrome (Schneider et al., 2006; Ching et al., 2013).

PATHOLOGY

Most of the data on the neuropathology in AT are based on autopsies dating to the 1970s and 1980s, before the discovery of the *ATM* gene (Boder, 1975; Amromin et al., 1979; Paula-Barbosa et al., 1983; Gatti and Vinters, 1985; Vinters et al., 1985), but more recent papers reconfirm the earlier findings (Bottini et al., 2012; Verhagen et al., 2012b).

The hallmark of the disease is neurodegeneration of the cerebellar cortex, with relative preservation of the

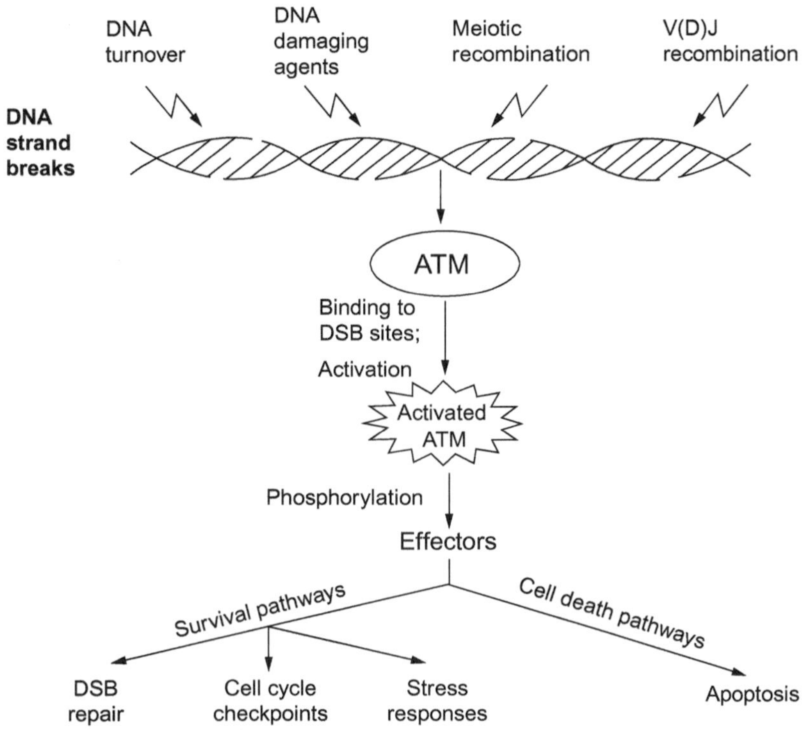

Fig. 14.2. ATM response to DNA double strand breaks. Following DSB induction, activated ATM binds to the damaged sites through mediators. Activated ATM in turn phosphorylates substrates involved in DNA repair, cell cycle checkpoint arrest and cell survival. In face of irreparable damage, cells will be driven to apoptosis. DSB, double strand breaks. (Reproduced from Barzilai et al., 2002.)

deep cerebellar nuclei, the dentate, and the inferior olive. Within the cerebellar cortex, there is especially progressive loss of Purkinje cells and granular cells (Gatti and Vinters, 1985; Vinters et al., 1985). The presence of empty basket cells suggests that Purkinje cells were present at some point in development but disappeared due to a neurodegenerative process. The remaining Purkinje cells have reduced dendritic ramifications (Boder, 1975; Amromin et al., 1979; Paula-Barbosa et al., 1983; Gatti and Vinters, 1985; Vinters et al., 1985). Additional data support the presence of dysgenetic Purkinje cells misplaced within the molecular layer, which suggest an earlier neurodevelopmental process during their migration (Gatti and Vinters, 1985).

With increasing age, degeneration of the posterior columns with gliosis, loss of myelination in the spinocerebellar and corticospinal tracts, and progressive degeneration of the anterior horn motor neuron appear (Boder, 1975; Amromin et al., 1979; Paula-Barbosa et al., 1983; Gatti and Vinters, 1985; Vinters et al., 1985; Verhagen et al., 2012b). Degeneration of the brainstem with involvement of cranial nerves, inferior olives, locus ceruleum, and substantia nigra appear in late stages (Boder, 1975; Amromin et al., 1979; Paula-Barbosa et al., 1983; Gatti and Vinters, 1985; Vinters et al., 1985).

NEUROLOGIC PRESENTATION

The main feature of patients with AT is neurodegeneration of the cerebellum, with ataxia, dysarthria, dysmetria, oculomotor apraxia, and extrapyramidal movements. All symptoms are progressive, but their prevalence differs by age.

Cerebellar

Ataxia is the most frequent symptom (87–100%), being present in virtually all individuals with the classic phenotype (Woods and Taylor, 1992; Crawford, 1998; Nissenkorn et al., 2011). Ataxia is usually the presenting symptom of the disease, typically starting at about 1 year of age when children start to walk, but it may appear up to 5 years of age (Crawford, 1998; Nissenkorn et al., 2011). In children with later onset of ataxia, chorea or myoclonus may be the presenting symptom (Woods and Taylor, 1992; Crawford, 1998).

The majority of patients begin to walk on time, but walking is unstable from the beginning (Crawford, 1998; Nissenkorn et al., 2011). In about 20% of children, walking age is delayed up to 4 years of age, but all children acquire this milestone before starting to deteriorate (Nissenkorn et al., 2011). At a very young age (before age 2 years) they tend to be toe walkers.

Children with AT do not widen the gait base and tend to walk on a narrow base, but wobble and have deviation of their gait path and frequent falls (Crawford, 1998; Nissenkorn et al., 2011). Interestingly, they are more stable while running, therefore they prefer to run rather than walk as if they are running after their gait center. This pattern of ataxia without a wide base differs from other etiologies of ataxia and is due probably to a component of gait apraxia or parkinsonism. Clinicians might overlook these symptoms and misdiagnose ataxia, and, since telangiectasias are not present at a young age, the diagnosis might be unnecessarily delayed.

At about 3 or 4 years of age the gait seems to improve and they have fewer falls, but after a few years of "honeymoon" gait deterioration resumes (Crawford, 1998; Nissenkorn et al., 2011). At about 6–7 years of age they require increasing degrees of support from an accompanying person, and between 7 and 10 years of age they experience a steep decline and become wheelchair bound at a mean age of 8 years (Woods and Taylor, 1992; Crawford et al., 2006; Nissenkorn et al., 2011). During the following years, they can still walk for short distances with an accompanying person or with walking aids, and they can make short transfers unaided. However, severe fatigue prevents them from walking longer distances. Deterioration continues into the second decade of life with loss of postural reflexes, which prevents them from walking even short distances with support, although transfers from the wheelchair can still be made with massive vertical support.

Truncal ataxia is another prominent feature of the cerebellar ataxia (Woods and Taylor, 1992; Nissenkorn et al., 2011). In many patients (especially younger affected siblings of already diagnosed children), parents describe ataxia already at 6 months of age when the child starts to sit (Crawford, 1998; Nissenkorn et al., 2011). Those babies have a tendency to sway while sitting and have head titubation. Later, patients sit while slowly swaying from side to side, and after age 10 years they might require some lateral support, but never lose the ability to sit. In contrast, ataxia of stance deteriorates more: during the first 4–5 years of life children stand on a narrow basis while slowly swaying from side to side, from 5–10 years they stand on a widened base with staggering steps, and later they need lateral support (Nissenkorn et al., 2011). Ultimately they fail to stand without total vertical support due to loss of postural reflexes.

Dysarthria is the next most frequent cerebellar symptom, present in 80% of the children with AT (Woods and Taylor, 1992; Crawford, 1998; Nissenkorn et al., 2011). Dysarthria appears later than ataxia, but preschool children already have unclear speech, with words which are not understandable and need to be repeated and progressing to unclear whole phrases. In addition to the cerebellar features of slurred staccato speech, intelligibility of speech is impaired by a weak dysphonic voice, drooling, and tendency to speak slowly in short sentences, and refraining from engaging in conversation. Phrasing and receptive language are unimpaired, but overall verbal communication skills are compromised and may be severely disabling.

Dysmetria is the next most frequent symptom, occurring in 75% of children; surprisingly, overshooting while trying to reach a target is not severe (Woods and Taylor, 1992; Crawford, 1998; Nissenkorn et al., 2011). The upper limb dysfunction which appears towards the second decade of life is more influenced by dyspraxia, severe bradykinesia, and movement disorder (chorea, myoclonus, dystonia, or tremor), which may be responsible for severe disability (Nissenkorn et al., 2011; Shaikh et al., 2013b).

Eye movements

Disorders of eye movement, and especially oculomotor apraxia, are one of the hallmarks of AT, while nystagmus, which is a main feature of most cerebellar ataxias of other etiologies, is a less prominent feature. Oculomotor apraxia is probably caused by disinhibition of the fastigial nucleus due to degeneration of the Purkinje layer (Shaikh et al., 2009). There is no direct correlation between the severity of eye movements and the severity of other neurologic symptoms (Crawford et al., 2000).

Eighty percent of patients have variousabnormalities of saccades and gaze control. These usually appear after 5 years of age and tend to progress from saccadic breakdown of smooth pursuit to hypometric saccades with increased latency and thereafter to horizontal oculomotor apraxia (Stell et al., 1989; Lewis et al., 1999; Farr et al., 2002; Nissenkorn et al., 2011). In younger children, oculomotor apraxia may be intermittent, while in older patients it is constant. In the most severe cases it leads to total ophthalmoplegia. A characteristic feature is intermittent strabismus or gaze deviation (especially of upper gaze) during motor activity or thinking (i.e. solving math problems). Upward gaze limitation is another characteristic finding for AT (Nissenkorn et al., 2011).

Impairment of gaze fixation and nystagmus are usually later findings in AT, resulting from loss of inhibition by the cerebellar cortex on the vestibular nuclei (Farr et al., 2002; Riise et al., 2007; Shaikh et al., 2009). In the most severely affected patients, attempts to fixate cause periodic alternating nystagmus. Another characteristic finding in AT due to disinhibition of the vestibular nuclei is an excessive vestibulo-ocular reflex (VOR) (Shaikh et al., 2011).

Extrapyramidal

Basal ganglia look normal by pathologic examination and imaging, but an extrapyramidal movement disorder is a major clinical feature and may be more disabling than the ataxia itself. This is not surprising since recent studies link the cerebellum as a causative factor of basal ganglia dysfunction (Bostan and Strick, 2010; Calderon et al., 2011; Wilson and Hess, 2013; Song et al., 2014). Aberrant cerebellothalamocortical circuits via abnormal superior cerebellar peduncle, as well as ectopic Purkinje cells, have been found in animal models of idiopathic dystonia (Bostan and Strick, 2010; Calderon et al., 2011; Wilson and Hess, 2013; Song et al., 2014).

Hyperkinetic or adventitious movements such as chorea or myoclonus are extremely frequent (60% of patients) and usually appear at a young age (Woods and Taylor, 1992; Crawford, 1998). In some cases these may be the presenting symptoms, before ataxia or telangiectasia appear, and therefore mislead the diagnosis (Nissenkorn et al., 2011). Patients with certain mutations might be more prone to present with chorea rather than ataxia (Nissenkorn et al., 2011). Chorea and myoclonus tend to appear upon motor activation or during cognitive tasks (i.e. solving math problems), involve especially distal extremities and face, and are rarely disabling (Nissenkorn et al., 2011). Sometimes myoclonic jerks appear in proximal limbs and may interfere with standing.

Focal dystonia of the limbs, neck, or trunk appears usually in older patients, during the second to third decade of life, and may cause severe disability (Nissenkorn et al., 2011). Dystonia is elicited upon motor activation in earlier stages, but may become constant and accompanied by severe dystonic tremor later.

A unique feature is tilting of the head, usually forward (Nissenkorn et al., 2011). This posturing is not related to upward gaze deviation or neck muscle weakness and might reflect neck dystonia.

In older patients forceful back-thrusts of the neck are characteristic, usually whenever the patient attempts to use upper limbs; it is not known whether this represents a myoclonic jerk or dystonia, but it causes severe impairment of upper limb function.

Postural, kinetic, or resting tremors appear, especially in older patients, and may be severe (Shaikh et al., 2013b).

A distinctive feature in AT children is bradykinesia, which affects 70% of patients (Nissenkorn et al., 2011). Bradykinesia appears in a mild form during the early years, and becomes more prominent during the second to third decade of life. It is manifested as slowness in performing motor tasks, cognitive tasks, and communication. Other parkinsonian features are present: paucity of spontaneous movements, stooped posture with small paces, and postural instability with retropulsion when pulled on shoulders (Nissenkorn et al., 2011). The face is hypomimic, with partially parted lips and a dystonic smile (Nissenkorn et al., 2011).

Neuropathy and motor neuron involvement

Starting the second decade of life, patients develop an axonal neuropathy which causes loss of deep tendon reflexes, but rarely produces disabling weakness (Woods and Taylor, 1992; Crawford, 1998; Perlman et al., 2003). Deformations of lower limbs such as pes cavus or tightened tendons are common. Older patients may develop a component of spinal motor neuron involvement (Verhagen et al., 2012b), which may be difficult clinically to differentiate from neuropathy.

Despite the lower motor neuron involvement, Babinski reflexes are not common (Woods and Taylor, 1992).

Cognitive and scholastic

The general role of the cerebellum in cognition, especially in attention and executive function, language, social recognition, and memory, is more recognized lately (Buckner, 2013; Koziol et al., 2014; Van Overwalle et al., 2014). Patients with AT do show a cognitive decline, but compared to the motor deterioration it is mild and plateaus at about the age of 10 years. The spectrum of cognitive involvement in AT has not been studied extensively; however, a few studies revealed below average intelligence in almost half of the patients (McFarlin et al., 1972), as well as language and attention deficits (Mostofsky et al., 2000; Vinck et al., 2011). In addition, they have speech impairment, problems in solving mathematics, and slowness of cognitive processing.

Although cognitive decline is mild and nonprogressive, scholastic performance is steeply impeded by motor deterioration. Progressive involvement of upper limbs due to dysmetria, bradykinesia, dyspraxia, and adventitious movements severely affects graphomotor abilities, and older children lose the ability to write. In a similar manner, oculomotor apraxia hampers the ability to read. Children in lower grades can read phrases and short stories in an age-matched manner, but older children can read only enlarged fonts due to eye saccadic failure.

Most patients in our experience need special schooling due to the multitude of their learning and motor disabilities, but occasional patients successfully attend regular schools and graduate from college.

Head circumference

A surprising feature in children with AT is small head circumference, with microcephaly in 17% of the patients (versus 3% of the general population) (Nissenkorn et al., 2011). Microcephaly is proportionate to height and weight and is acquired postnatally.

Many genes related to DNA repair are known to cause microcephaly; intense activation of DNA repair processes is taking place during neurogenesis, especially during the intrauterine process of division of periventricular progenitor cells into neurons (O'Driscoll and Jeggo, 2008). Examples of such disorders presenting with microcephaly at birth are defects in the *ATR* and *NBS1* genes, which are closely related to ATM pathway (Antoccia et al., 2006; O'Driscoll and Jeggo, 2008; Chrzanowska et al., 2012).

However, in AT microcephaly is acquired after the second year of life, therefore it is more plausible that it is related to processes that affect neurite growth and synaptogenesis (Nissenkorn et al., 2011). It is possible that a cytoplasmic ATM defect has a role in synaptic transmission through mTOR, a key regulator of neural cell size, dendritic arborization, and brain size (Shen and Houghton, 2013).

Cutaneous

Ocular telangiectasias are present in 97% of individuals with AT, and these were described as the hallmark of the disease in the original description of the disease (Boder and Sedgwick, 1958; Greenberger et al., 2013) (Fig. 14.3A). When present, telangiectasias facilitate the clinical diagnosis since they are virtually pathognomonic for AT, but usually they appear after 6 years of age. Conjunctival telangiectasias are bothersome to parents and patients, but beyond the cosmetic significance they are largely harmless; they do not interfere with vision or eye movement and do not bleed. Interestingly, ocular telangiectasias may be present in some of the carrier parents (Greenberger et al., 2013). Skin telangiectasias on the face, earlobes, and chest are less frequent, and they may appear with advancing age (Fig. 14.3B).

In some individuals, visceral telangiectasias may be present in brain, liver, or bladder. While brain telangiectasias seem to be innocent, bladder telangiectasia may cause life-threatening bleeding and require treatment (Cohen et al., 2008).

Pigmentary anomalies are frequent and include café-au-lait macules, hypopigmented macules, and melanocytic nevi (37%) (Ortonne et al., 1980; Greenberger et al., 2013). Cutaneous granulomas have been reported in several patients (Paller et al., 1991; Mitra et al., 2011). Additional findings include premature aging with bird-like facies, disappearance of facial subcutaneous tissue, and premature graying.

Malignancy

ATM has a central role in signaling to the cell cycle checkpoints, and many of its substrates are tumor suppressor genes as p53 or cancer susceptibility genes such as BRCA1 (Gatti, 1993; Taylor and Byrd, 2005; Iyama and Wilson, 2013; Shiloh and Ziv, 2013; Cremona and Behrens, 2014). Dysfunction of ATM is one of the major pathophysiologic mechanisms in cancer (Gatti, 1993; Taylor and Byrd, 2005; Iyama and Wilson, 2013; Shiloh and Ziv, 2013; Cremona and Behrens, 2014).

In the classic phenotype of AT about 30% of patients develop malignancy, which represents a major cause of mortality (Micol et al., 2011). Most of these malignancies are lymphoid hematopoietic, T cell origin acute lymhocytic leukemias and non-Hodgkin lymphomas of B cell types, but Hodgkin lymphoma and acute myeloid leukemia have

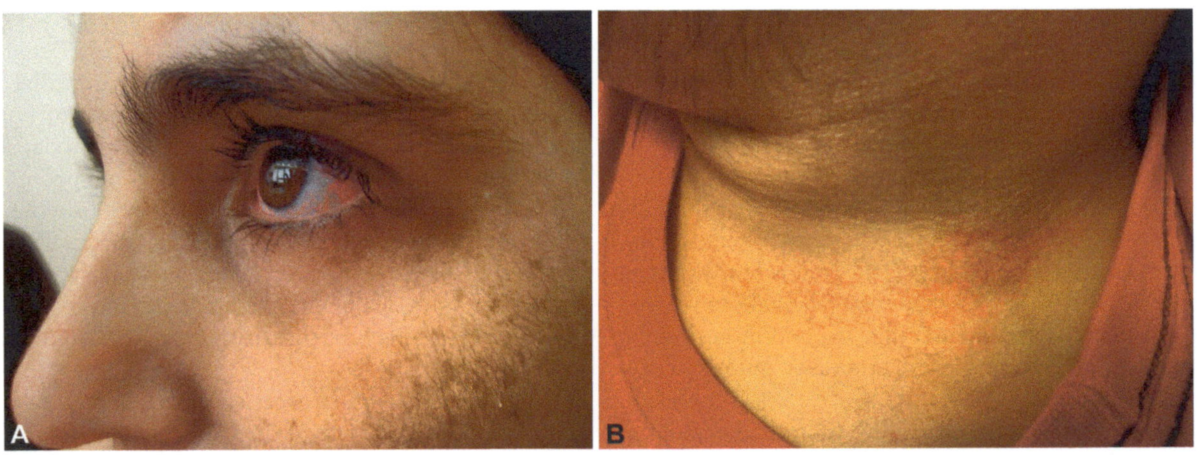

Fig. 14.3. Distribution of telangiectasia in affected patients: (**A**) conjunctiva; (**B**) neck. (Reproduced from Greenberger et al., 2013.)

also been reported (Gatti, 1993; Murphy et al., 1999; Loeb et al., 2000; Olsen et al., 2001). Treatment of these malignancies is complicated by the extreme radiosensitivity; the dosage of radiotherapy should be decreased and special chemotherapeutic protocols should be used in order to reduce toxicity and still maintain effectiveness (Sandlund et al., 2014). In older children and adults, solid tumors are more likely to appear – breast cancer, gastric adenocarcinoma, ovarian cancer, and melanoma (Murphy et al., 1999, Olsen et al., 2001). However, tumors of the nervous system are not a feature of AT.

Some double missense mutations may cause mild neurologic disease but have a high propensity towards developing malignancies (Mitui et al., 2009; Bielorai et al., 2013). A controversial issue is the tendency of heterozygous mothers or sisters to develop breast cancer. *ATM* is one of the well-known susceptibility genes for breast cancer, but the most common mutation known to increase the susceptibility to cancer is not a frequent mutations that causes AT (Micol et al., 2011; Gatti, 1993).

Immune deficiency, recurrent infections, and pulmonary disease

Immune deficiency and recurrent sinopulmonary infections are one of the core symptoms of AT; these result from ATM's role in nonhomologous end joining and V(D)J recombinations, which are vital for the diversity of the immune response (Bredemeyer et al., 2008; Gapud and Sleckman, 2011; Driessen et al., 2013). The severity of the immune deficiency is not directly related to the severity of neurologic phenotype and, as opposed to other features of the disease, is not progressive (Nowak-Wegrzyn et al., 2004). Patients commonly have low levels of immune globulins, especially immunoglobulin A, E and G2, as well as poor antibody response to pneumococcal polysaccharide vaccines (Nowak-Wegrzyn et al., 2004; Schubert et al., 2005; Guerra-Maranhao et al., 2006). About 30% of patients have deficits in the cellular arm of the immune system, with moderate lymphopenia and reduction in the number of absolute and naïve CD4 cells (Lantelme et al., 2003). Surprisingly for a primary immune deficiency, patients have mostly sinopulmonary infections with the common pathogens (Nowak-Wegrzyn et al., 2004). Replacement therapy with immune globulins improves the tendency to have recurrent infections; patients with a low lymphocyte count may require prophylactic antibiotics (Gatti, 1993; Nowak-Wegrzyn et al., 2004).

Despite treatment, pulmonary disease is still a common morbidity and a major cause of mortality (McGrath-Morrow et al., 2010, 2014). Recurrent infections due to immune deficiency, aspirations, and weak cough lead to bronchiectasis (McGrath-Morrow et al., 2010, 2014). In addition, patients with AT also have pulmonary interstitial disease (McGrath-Morrow et al., 2010, 2014; Schroeder and Zielen, 2014).

Other

Gastrointestinal symptoms are especially frequent after the age of 10 years. Patients have dysphagia and may choke on liquids and solid food (Lefton-Greif et al., 2000). Poor weight gain and failure to thrive are common (Lefton-Greif et al., 2000). Gastrostomy feeding tubes may facilitate weight gain and general well-being, but their placement might be useless when done after 15 years of age (Lefton-Greif et al., 2011). Patients over 15 years of age can have dyslipidemia and fatty liver, with elevation of liver enzymes, partially responsive to bile acid salts. Physicians should be alerted not to search for malignancy and liver biopsies are not warranted. Rather, hypercholesterolemia and fatty liver, along with insulin resistance, are a manifestation of a metabolic syndrome which develops due to the involvement of ATM in insulin and mTOR signaling (Woods and Taylor, 1992; Schneider et al., 2006; Ditch and Paull, 2012; Ching et al., 2013).

In addition to insulin resistance and diabetes, endocrine problems include gonadal failure and infertility (Woods and Taylor, 1992; Schneider et al., 2006; Ditch and Paull, 2012; Ching et al., 2013). Growth hormone deficiency and low stature are additional problems, but recombinant replacement therapy might be problematic because of increased risk of malignancy (Schubert et al., 2005).

Urological problems with incontinence are frequent, though often unreported (A. Nissenkorn, personal communication), probably due the involvement of the spinal cord in the degenerative process and unrelated to urinary tract infections. Fatigue is a special concern; it appears early in the course of the disease and limits participation in every domain (A. Nissenkorn, personal communication).

AT VARIANTS

After the discovery of the *ATM* gene, along with the era of molecular diagnosis, it became evident that the spectrum of AT is wider than the classic phenotype and encompasses milder cases, called AT variants.

In milder variants, cerebellar signs including ataxia and oculomotor apraxia appear later, during the second decade of life or even adulthood, progress more slowly, and do not lead to severe disability (Gilad et al., 1996b; McConville et al., 1996; Chun and Gatti, 2004; Taylor and Byrd, 2005; Alterman et al., 2007; Verhagen et al., 2012a; Worth et al., 2013). Adults with mutations in *ATM* may also present with cervical dystonia, myoclonic jerks, or motor neuron disease, mimicking other

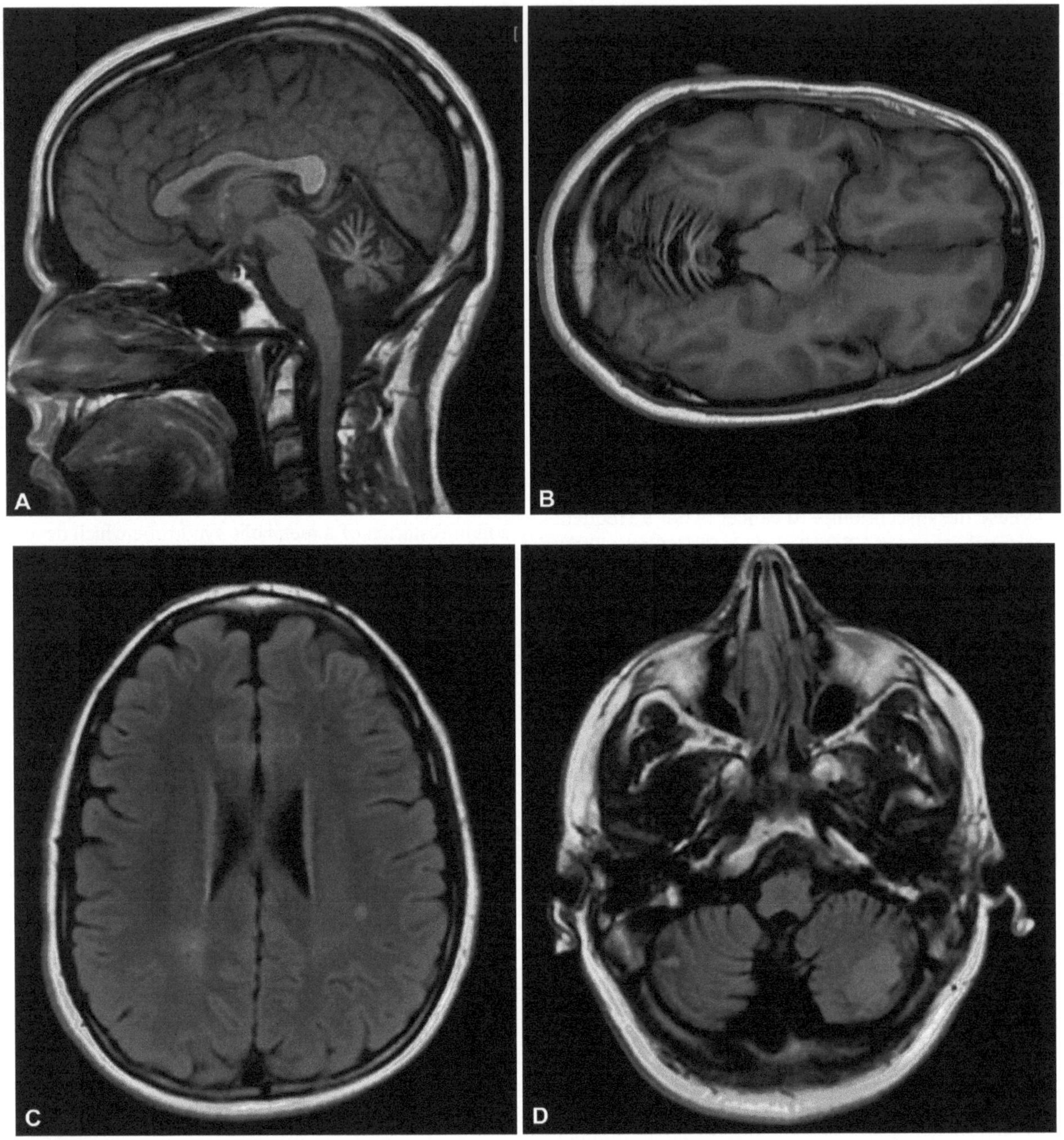

Fig. 14.4. Magnetic resonance imaging of patients with ataxia telangiectasia. (**A**) Midline sagittal T1 imaging reveals severe atrophy of the superior vermis in a 14-year-old patient; (**B**) axial T1 cut through midbrain shows severe atrophy of vermis and cerebellar hemispheres in the same patient. (**C**) Supratentorial white matter changes scattered on axial T2 FLAIR. (**D**) Axial T2 FLAIR reveals extensive cerebellar white matter changes mimicking leukoencephalopathy in a 20-year-old patient; atrophy of the cerebellar hemispheres and enlargement of the fourth ventricle are evident.

neurologic disorders as such idiopathic dystonia (Cummins et al., 2013) or adult-onset spinal muscular atrophy (Hiel et al., 2006). These patients still have the other characteristics of AT such as elevated α-fetoprotein, radiosensitivity, and the tendency to malignancy. Occasionally, mild cases without high α-fetoprotein are documented.

The mild AT variants are caused either by missense or splicing ("leaky site") mutations in which residual protein may be detected (Gilad et al., 1996b; McConville et al., 1996; Chun and Gatti, 2004; Taylor and Byrd, 2005; Verhagen et al., 2012a).

In the older literature, the term AT variant or AT-like disease (ATLD) has been attributed to two entities caused by mutant NBS1 (nibrin) or MRE11. Both proteins are components of the NRM complex that binds to *ATM*. However, the clinical picture of the disorders is entirely different. The NBS1 (nibrin) defect causes Nijmaegen breakage syndrome, presenting with primary microcephaly, immune deficiency, and chromosomal breakage (Antoccia et al., 2006; Chrzanowska et al., 2012). MRE11 deficiency causes a syndrome very much resembling mild AT variant progressive ataxia with cerebellar atrophy and oculomotor apraxia, but no immune deficiency, malignancy, or elevated α-fetoprotein (Stewart et al., 1999; Taylor et al., 2004).

IMAGING

Characteristically, brain imaging is normal during the first 2 years of life, even though children are already symptomatic. Between 2 and 8 years of age, virtually all children with the classic phenotype will develop cerebellar atrophy, which is progressive and parallels the progression of the pathology (Farina et al., 1994; Sardanelli et al., 1995) (Fig. 14.4A, B). Atrophy is more evident in the vermis than cerebellar hemispheres; it starts in the superior vermis and progresses to all parts of the vermis and lateral hemispheres (Tavani et al., 2003). Occasionally, on MRI imaging, white matter lesions may be present; these may mimic leukoencephalopathy but probably have no clinical consequences (Ciemins and Horowitz, 2000; Lin et al., 2014) (Fig. 14.4C, D). Punctate hemosiderin deposits provide evidence of telangiectasias but have no clinical consequences (Lin et al., 2014).

Magnetic resonance spectroscopy (MRS) studies confirm the presence of cortical cerebellar atrophy by a low *N*-acetylaspartate/choline ratio but offer no explanation for extrapyramidal involvement (Lin et al., 2006; Wallis et al., 2007). Newer neuroimaging techniques, such as diffusion tensor imaging, could improve our understanding of the pathology of the circuits between the cerebellum and the basal ganglia.

DIAGNOSIS

AT diagnosis should be suspected in all children with progressive ataxia starting in the first 5 years of life, accompanied by cerebellar atrophy, immune deficiency and elevated α-fetoprotein. One should be aware of the many caveats of the clinical diagnosis: cerebellar atrophy appears later than the clinical symptoms, and immune deficiency may not be present. Telangiectasias, the cutaneous hallmark of the disease, appear after the age of 6 years. The milder clinical cases usually present at a later age, may present with a movement disorder, may progress slowly, not exhibit elevated α-fetoprotein, and have no telangiectasias.

Elevated α-fetoprotein is a sensitive biochemical marker of the disease, but the levels increase with age, therefore it may be normal in younger children and in AT variants. Additionally, α-fetoprotein should not be used as a screening test in neonates with affected family members as it is always high during the first month of life in healthy infants. Elevation of α-fetoprotein is not specific for AT; it may be encountered in other DNA repair-related ataxias (such as ataxia oculomotor apraxia 1 and ataxia oculomotor apraxia 2) as well as in liver malignancies.

The clinical suspicion of AT should be supported by molecular diagnosis. Firm diagnosis of AT should be made as soon as possible, since delay in diagnosis may expose a patient to unnecessary ionizing radiation and prevent prenatal diagnosis.

AT is a monogenic disorder caused by mutations in the *ATM* gene. Point mutations in the *ATM* gene are responsible for over 95% of individuals with AT; rare cases are caused by genomic or intragenic deletions (Gatti, 1993; Gilad et al., 1996b). Most patients are double heterozygotes for different mutations, but in populations with founder mutations or a high rate of intermarriage, homozygotes are prevalent (Gilad et al., 1996a; McConville et al., 1996). In about 10% of patients only one mutation can be identified.

Most mutations of *ATM* are severe truncating mutations (primary stop codon or frameshift) (Gatti, 1993; Gilad et al., 1996a, b), but in certain populations missense or splicing site mutations are more prevalent (McConville et al., 1996; Verhagen et al., 2012a). In 90% of the cases with truncating and part of the missense mutations, the protein is unstable and its level is undetectable, but some splicing site mutations and missense mutations have detectable levels of protein (Gatti, 1993; Gilad et al., 1996a, b; McConville et al., 1996; Taylor and Byrd, 2005; Verhagen et al., 2012a). The phenotype of severe classic AT is caused by mutations with no residual protein, while mutations allowing residual protein cause milder clinical variants (Gilad et al., 1996b; McConville et al., 1996; Chun and Gatti, 2004; Taylor and Byrd, 2005; Verhagen et al., 2012a). There are several exceptions to this rule, with mild clinical phenotypes being described in the presence of severe truncating mutations (Alterman et al., 2007; Worth et al., 2013). Another exception to this rule is missense mutation in the kinase domain, where the protein level is normal but is nonfunctional (kinase dead) (Stewart et al., 2001). It has been suggested that residual kinase activity may influence not only milder neurologic function but may offer protection from other symptoms such as malignancy, immune deficiency, and diabetes (Verhagen et al., 2012a). In contrast, individuals with mild neurologic features with double missense mutations but a high tendency to cancer have been reported (Mitui et al., 2009; Bielorai et al., 2013).

ATM is a large gene spanning 150 kb of genomic DNA and encompassing 66 exons; until recently, sequencing was expensive and time-consuming, but different laboratory methods measure the level or function of the ATM protein (Gatti, 1993; Savitsky et al., 1995; McKinnon, 2012).

Since intracellular ATM is usually severely depleted, the most sensitive and specific detection method is immunoblotting for the ATM protein (Gatti, 1993). The gold standard is considered to be Western blot analysis, which may accurately predict absence of the ATM protein in 90% of the cases and trace levels in about 10%, but experienced personnel are needed to interpret this test (Gatti, 1993). Since the level of ATM in blood lymphocytes is low, lymphoblastoid cell lines or fibroblasts from a skin biopsy should be used, a technique that requires a prolonged preparation process (30–60 days). More accurate rapid immunoassays to measure ATM levels have been developed, but, as with the Western blot, they are not readily available in diagnostic laboratories (Nahas et al., 2009).

Several other tests measure the function rather then the level of protein, either directly by assay of the ATM-related phosphorylation or indirectly by measuring chromosomal breakage or radiosensitivity. These tests are mainly used in research laboratories:

1. Methods to assess phosphorylation of the ATM substrates were developed by rapid flow cytometry (Gatti, 1993). Alternatively, the ATM kinase activity can be measured using immunoblotting or antibodies to phosphorylated targets of ATM (Gatti, 1993).
2. Regular karyotyping is useless in the diagnosis, but chromosomal analysis under special conditions may reveal breaks at 14 q11 (T cell receptor) and 14q32 (B cell receptor) as well as 7;14 translocation in 15% of the cases (Gatti, 1993).
3. Radiosensitivity assays reveal reduced colony survival of patient-derived lymphoblastoid cell lines when exposed to standard radiation (Gatti, 1993).

In recent years, tremendous development occurred in the field of molecular genetic analysis by introduction of high throughput sequencing techniques. Whole exome sequencing as well as Sanger sequencing for a single gene are now easy to perform in a multitude of diagnostic laboratories, at a reasonable price and turnover time, and do not require specialization in AT. Whenever the clinical diagnosis of AT is highly suspected, Sanger sequencing for *ATM* should be done first. However, in clinical practice we are usually faced with patients with ataxia in whom a wide differential diagnosis is possible. In those patients, whole exome sequencing should be done first (Nemeth et al., 2013).

Interpretation of the results of sequencing should be done with caution, in order to differentiate between benign polymorphisms and pathogenic mutations. The problem is less meaningful whenever nonsense mutations are present, but the significance of missense mutations is more difficult to interpret, and occasionally functional tests such as phosphorylation or radioimmunassay may be needed. Fortunately, several bioinformatics tools have been developed to assess the pathogenicity of a mutation; these tools are easy to use with minimal training and are free of charge on the internet (e.g. Polyphen2: http://genetics.bwh.harvard.edu/pph2/, Mutation Taster: http://www.mutationtaster.org/ and Condel: http://bg.upf.edu/fannsdb/).

Differential diagnosis

The etiology of progressive ataxia with cerebellar atrophy is diverse; with the advent of whole exome sequencing in the last years dozens of genetic defects have been identified. The disorders most closely resembling AT are caused by genes either involved in DNA repair, or in mitochondrial homeostasis.

1. AT-like disorder (ATLD) is caused by a genetic defect that impairs production of the MRE11 protein which binds to *ATM* through its pathway in DNA double strand break repair. The phenotype is milder but similar to AT (Stewart et al., 1999; Taylor et al., 2004). Usually α-fetoprotein is not elevated, and there is no immune deficiency or tendency to malignancy.
2. Ataxia oculomotor apraxia 1 caused by mutation in the *AOA1* gene, which encodes for aprataxin, a protein involved in DNA repair (Takahashi et al., 2007b). Ataxia with cerebellar atrophy, oculomotor apraxia, and frequently elevated α-fetoprotein and hypoalbuminemia, appear during childhood, but later than AT. The main neurologic disability is caused by severe neuropathy leading to quadriparesis (Coutinho and Barbot, 1993).
3. Ataxia oculomotor apraxia 2 results from mutations in the *AOA2* gene encoding for senataxin. Affected individuals usually present during the second decade of life or early adulthood with progressive cerebellar atrophy and ataxia, neuropathy, oculomotor apraxia, and sometimes high α-fetoprotein (Moreira et al., 2004).
4. Ataxias can be caused by mitochondrial dysfunction such as mutations of mitochondrial DNA-dependent DNA polymerase γ (POLG), mutations in mitochondrial helicase Twinkle or deficiency of coenzyme Q10 (CoQ10). These present with early

onset ataxia, cerebellar atrophy, white matter pathology, ophthalmoplegia, neuropathy, and multisystem involvement (Zeviani et al., 2012). α-Fetoprotein levels are normal, but lactate levels may be elevated, immune deficiency and malignancies are absent, as are telangiectasias.

5. Friedreich ataxia is caused by an intronic expansion in the gene encoding for frataxin, a protein involved in mitochondrial iron homeostasis (Marmolino, 2011). Ataxia, along with spinal cord involvement and neuropathy, appears during the second to third decade of life, but cerebellar atrophy is a late finding. Nystagmus and optic atrophy are the characteristic eye findings.

ANIMAL MODELS

Shortly after the discovery of the *ATM* gene in 1995, the opportunity arose to create animal models for AT. Human ATM bears an 85% homology with the mouse protein, therefore this rodent seems to be an ideal choice for an animal model. The first mouse model was created in 1996 by Barlow et al. (Barlow et al., 1996), with a few other knockout or knockin models in the following years (Lavin, 2013). Most models were created by exon deletions, leading to an unstable truncated protein, but a model with low levels of protein was created by in-frame deletion (Barlow et al., 1996; Lavin, 2013). These models recapitulate many of the systemic manifestations of AT: immune deficiency, tendency to malignancy, radiosensitivity, high oxidative stress, and chromosomal breakage, proving themselves to be a reliable model in the investigation of certain aspects of the disease such as cancer (Barlow et al., 1996; Lavin, 2013).

Nevertheless, none of these models succeeded in creating significant neurologic involvement. On histologic examination of the cerebellum, there are large vacuoles and reduced dendritic spine densities in the Purkinje layer, with ectopic Purkinje cells, but severe neurodegenerative changes with cerebellar atrophy are lacking (Barlow et al., 1996; Lavin, 2013). Additionally, on neurologic examination, while having some minor aberrations, $Atm^{-/-}$ mice fail to develop ataxia (Barlow et al., 1996; Lavin, 2013).

Obviously, the existent mouse models of AT are not well suited to explore neurodegeneration or to assess neurologic disease-modifying treatments. Interestingly, other mouse models, related in their pathophysiologic pathway to ATM, such as $NBS^{-/-}$ (Demuth et al., 2004) and $HDAC^{-/-}$ (Majdzadeh et al., 2008) mice, reveal cerebellar atrophy and degeneration. Current efforts focus on producing models with different types and locations of mutations in order to create mouse or rat models compatible with the neurologic morbidity in humans.

Li et al. (2011) found that in, mouse mutant *ATM*, brain-specific alternative splicing is present, leading to near fully length and catalytically active protein. Understanding the mechanism that provides neuroprotection to the $Atm^{-/-}$ mouse might offer a unique opportunity to develop disease modifying treatments for patients with AT.

Developing animal models in lower species, such as Drosophila (Petersen et al., 2012, 2013), may broaden the understanding of disease biology but also may be suitable for cost-effective high throughput drug screening.

In addition to animal models, *in vitro* studies on human cells might be equally important for finding biomarkers for disease severity, establishing nonexpensive systems for drug screening, or elucidating the pathophysiology of disease in humans. Until recently, these efforts were hampered by the unavailability of human nervous system tissue for research purposes, as opposed to lymphocytes or fibroblasts.

The entorhinal tissue is rich in olfactory neurons, and a growing literature supports the use of nasal biopsies from patients for investigating neurodegenerative disorders, including AT (Ronnett et al., 2003; Arnold et al., 2010). Exciting new horizons in neurology have been opened by the discovery that human somatic cells can be reprogrammed into induced pluripotent stem cells (iPSCs), which can further differentiate into neuronal precursor cells or adult differentiated GABA-ergic or glutaminergic neurons (Takahashi et al., 2007a; Takahashi and Yamanaka, 2013). Recently, reprogrammed neuronal stem cells were derived from patients with AT, and these become a leading *in vitro* model (Carlessi et al., 2013a, b).

TREATMENT

Currently, the main treatment in AT is supportive, since no therapy seems to change the disease course. There are well developed guidelines for treating immune deficiency, cancer, and pulmonary disease. Treatment of neurologic symptoms is mostly anecdotal, and no drug is suitable to treat all aspects of the disease.

Patients should undergo annual functional evaluations and treatment should be adjusted according to their needs. Physical therapy has no value in improving ataxia, but it may help to prevent tendon shortening and contractures. Exercise, hydrotherapy, and hypotherapy improve general well-being and fatigue. Nonambulatory patients should be positioned erect a few times a week to prevent scoliosis and muscle contractures. Assistive technology for walking should be adapted, taking into consideration the special demands for AT: heavy walkers should be used to overcome ataxia, and powered wheelchairs may not be suitable for all patients with oculomotor apraxia. Computers should be adapted as early as possible to enable writing and communication. Fonts should be enlarged on computers as well as on printed pages in order to overcome reading difficulties imposed by oculomotor apraxia.

Recently, Nissenkorn et al. (2013) showed efficacy of treatment with amantadine on ataxia, bradykinesia, and hyperkinetic movement in a small pilot study. Shaikh et al. (2013a) demonstrated improvement of nystagmus in four adults with 4-aminopyridine. Several studies, including a double-blinded placebo-controlled trial, supported improvement in ataxia with steroid treatment (betamethasone) (Zannolli et al., 2012). Long-term administration might be hampered by side-effects, but newer technologies that ensure administration of steroids with fewer side-effects are under investigation (Chessa et al., 2014). Preliminary work found that dexamethasone induces alternative noncanonical splicing of ATM, creating a mini protein with retained kinase activity that may explain a positive clinical effect (Menotta et al., 2012).

A future endeavor would be correcting the ATM deficiency either by giving ATM via viral vectors or by correcting the molecular defect. Several exciting developments have been reported lately, by *in vitro* correction of primary stop codons with read-through compounds (Du et al., 2013) and splicing site mutations with antisense oligonucleotides (Nakamura et al., 2012). Better understanding of the pathways of neurodegeneration would enable additional disease-modifying intervention by the alteration of oxidative stress, mitophagy, glutamatergic toxicity, or histone deacetylation. Up to the present, antioxidant treatment has proven beneficial in animal models but failed in humans (Reliene and Schiestl, 2008). Further research on various cell lines as well as proper animal models will ultimately lead to a new era of combined gene therapy and neurodegeneration pathway intervention that may be applied in clinical trials.

References

Alterman N, Fattal-Valevski A, Moyal L et al. (2007). Ataxia-telangiectasia: mild neurological presentation despite null ATM mutation and severe cellular phenotype. Am J Med Genet A 143A: 1827–1834.

Ambrose M, Gatti RA (2013). Pathogenesis of ataxia-telangiectasia: the next generation of ATM functions. Blood 121: 4036–4045.

Amromin GD, Boder E, Teplitz R (1979). Ataxia-telangiectasia with a 32 year survival. A clinicopathological report. J Neuropathol Exp Neurol 38: 621–643.

Antoccia A, Kobayashi J, Tauchi H et al. (2006). Nijmegen breakage syndrome and functions of the responsible protein, NBS1. Genome Dyn 1: 191–205.

Arnold SE, Lee EB, Moberg PJ et al. (2010). Olfactory epithelium amyloid-beta and paired helical filament-tau pathology in Alzheimer disease. Ann Neurol 67: 462–469.

Bakkenist CJ, Kastan MB (2003). DNA damage activates ATM through intermolecular autophosphorylation and dimer dissociation. Nature 421: 499–506.

Baloh RW, Yee RD, Boder E (1978). Eye movements in ataxia-telangiectasia. Neurology 28: 1099–1104.

Barlow C, Hirotsune S, Paylor R et al. (1996). ATM-deficient mice: a paradigm of ataxia telangiectasia. Cell 86: 159–171.

Barzilai A, Rotman G, Shiloh Y (2002). ATM deficiency and oxidative stress: a new dimension of defective response to oxidative stress. DNA Repair (Amst) 1: 3–25.

Bielorai B, Fisher T, Waldman D et al. (2013). Acute lymphoblastic leukemia in early childhood as the presenting sign of ataxia-telangiectasia variant. Pediatr Hematol Oncol 30: 574–582.

Biton S, Dar I, Mittelman L et al. (2006). Nuclear ataxia-telangiectasia mutated (ATM) mediates the cellular response to DNA double strand breaks in human neuron-like cells. J Biol Chem 281: 17482–17491.

Biton S, Barzilai A, Shiloh Y (2008). The neurological phenotype of ataxia-telangiectasia: solving a persistent puzzle. DNA Repair (Amst) 7: 1028–1038.

Boder E (1975). Ataxia-telangiectasia: some historic, clinical and pathologic observations. Birth Defects Orig Artic Ser 11: 255–270.

Boder E, Sedgwick RP (1958). Ataxia-telangiectasia; a familial syndrome of progressive cerebellar ataxia, oculocutaneous telangiectasia and frequent pulmonary infection. Pediatrics 21: 526–554.

Boder E, Sedgwick RP (1970). Ataxia-telangiectasia (clinical and immunological aspects). Psychiatr Neurol Med Psychol Beih 13–14: 8–16.

Bostan AC, Strick PL (2010). The cerebellum and basal ganglia are interconnected. Neuropsychol Rev 20: 261–270.

Bottini AR, Gatti RA, Wirenfeldt M et al. (2012). Heterotopic Purkinje cells in ataxia-telangiectasia. Neuropathology 32: 23–29.

Bredemeyer AL, Huang CY, Walker LM et al. (2008). Aberrant V(D)J recombination in ataxia telangiectasia mutated-deficient lymphocytes is dependent on nonhomologous DNA end joining. J Immunol 181: 2620–2625.

Buckner RL (2013). The cerebellum and cognitive function: 25 years of insight from anatomy and neuroimaging. Neuron 80: 807–815.

Calderon DP, Fremont R, Kraenzlin F et al. (2011). The neural substrates of rapid-onset dystonia-parkinsonism. Nat Neurosci 14: 357–365.

Carlessi L, Fusar Poli E, De Filippis L et al. (2013a). ATM-deficient human neural stem cells as an in vitro model system to study neurodegeneration. DNA Repair (Amst) 12: 605–611.

Carlessi L, Fusar Poli E, Delia D (2013b). Brain and induced pluripotent stem cell-derived neural stem cells as an in vitro model of neurodegeneration in ataxia-telangiectasia. Exp Biol Med (Maywood) 238: 301–307.

Chessa L, Leuzzi V, Plebani A et al. (2014). Intra-erythrocyte infusion of dexamethasone reduces neurological symptoms in ataxia teleangiectasia patients: results of a phase 2 trial. Orphanet J Rare Dis 9: 5. http://dx.doi.org/10.1186/1750-1172-9-5. PubMed PMID: 24405665; PubMed Central PMCID: PMC3904207.

Ching JK, Luebbert SH, Collins 4th RL et al. (2013). Ataxia telangiectasia mutated impacts insulin-like growth factor 1 signalling in skeletal muscle. Exp Physiol 98: 526–535.

Chrzanowska KH, Gregorek H, Dembowska-Baginska B et al. (2012). Nijmegen breakage syndrome (NBS). Orphanet J Rare Dis 7: 13.

Chun HH, Gatti RA (2004). Ataxia-telangiectasia, an evolving phenotype. DNA Repair (Amst) 3: 1187–1196.

Ciemins JJ, Horowitz AL (2000). Abnormal white matter signal in ataxia telangiectasia. AJNR Am J Neuroradiol 21: 1483–1485.

Cohen JM, Cuckow P, Davies EG (2008). Bladder wall telangiectasis causing life-threatening haematuria in ataxia-telangiectasia: a new observation. Acta Paediatr 97: 667–669.

Cosentino C, Grieco D, Costanzo V (2011). ATM activates the pentose phosphate pathway promoting anti-oxidant defence and DNA repair. EMBO J 30: 546–555.

Coutinho P, Barbot C (1993). Ataxia with oculomotor apraxia type 1. In: RA Pagon, MP Adam, TD Bird et al. (Eds.), GeneReviews, University of Washington, Seattle, Seattle WA.

Crawford TO (1998). Ataxia telangiectasia. Semin Pediatr Neurol 5: 287–294.

Crawford TO, Mandir AS, Lefton-Greif MA et al. (2000). Quantitative neurologic assessment of ataxia-telangiectasia. Neurology 54: 1505–1509.

Crawford TO, Skolasky RL, Fernandez R et al. (2006). Survival probability in ataxia telangiectasia. Arch Dis Child 91: 610–611.

Cremona CA, Behrens A (2014). ATM signalling and cancer. Oncogene 33: 3351–3360.

Cummins G, Jawad T, Taylor M et al. (2013). Myoclonic head jerks and extensor axial dystonia in the variant form of ataxia telangiectasia. Parkinsonism Relat Disord 19: 1173–1174.

Dar I, Biton S, Shiloh Y et al. (2006). Analysis of the ataxia telangiectasia mutated-mediated DNA damage response in murine cerebellar neurons. J Neurosci 26: 7767–7774.

Demuth I, Frappart PO, Hildebrand G et al. (2004). An inducible null mutant murine model of Nijmegen breakage syndrome proves the essential function of NBS1 in chromosomal stability and cell viability. Hum Mol Genet 13: 2385–2397.

Ditch S, Paull TT (2012). The ATM protein kinase and cellular redox signaling: beyond the DNA damage response. Trends Biochem Sci 37: 15–22.

Driessen GJ, Ijspeert H, Weemaes CM et al. (2013). Antibody deficiency in patients with ataxia telangiectasia is caused by disturbed B- and T-cell homeostasis and reduced immune repertoire diversity. J Allergy Clin Immunol 131: 1367–1375 e9.

Du L, Jung ME, Damoiseaux R et al. (2013). A new series of small molecular weight compounds induce read through of all three types of nonsense mutations in the ATM gene. Mol Ther 21: 1653–1660.

Farina L, Uggetti C, Ottolini A et al. (1994). Ataxia-telangiectasia: MR and CT findings. J Comput Assist Tomogr 18: 724–727.

Farr AK, Shalev B, Crawford TO et al. (2002). Ocular manifestations of ataxia-telangiectasia. Am J Ophthalmol 134: 891–896.

Gapud EJ, Sleckman BP (2011). Unique and redundant functions of ATM and DNA-PKcs during V(D)J recombination. Cell Cycle 10: 1928–1935.

Gatti R (1993). Ataxia-t6elangiectasia. In: RA Pagon, MP Adam, TD Bird et al. (Eds.), GeneReviews, University of Washington, Seattle, Seattle WA.

Gatti RA, Vinters HV (1985). Cerebellar pathology in ataxia-telangiectasia: the significance of basket cells. Kroc Found Ser 19: 225–232.

Gilad S, Bar-Shira A, Harnik R et al. (1996a). Ataxia-telangiectasia: founder effect among North African Jews. Hum Mol Genet 5: 2033–2037.

Gilad S, Khosravi R, Shkedy D et al. (1996b). Predominance of null mutations in ataxia-telangiectasia. Hum Mol Genet 5: 433–439.

Greenberger S, Berkun Y, Ben-Zeev B et al. (2013). Dermatologic manifestations of ataxia-telangiectasia syndrome. J Am Acad Dermatol 68: 932–936.

Guerra-Maranhao MC, Costa-Carvalho BT, Nudelman V et al. (2006). Response to polysaccharide antigens in patients with ataxia-telangiectasia. J Pediatr (Rio J) 82: 132–136.

Guo Z, Kozlov S, Lavin MF et al. (2010). ATM activation by oxidative stress. Science 330: 517–521.

Herrup K (2013). ATM and the epigenetics of the neuronal genome. Mech Ageing Dev 134: 434–439.

Herrup K, Li J, Chen J (2013). The role of ATM and DNA damage in neurons: upstream and downstream connections. DNA Repair (Amst) 12: 600–604.

Hiel JA, Van Engelen BG, Weemaes CM et al. (2006). Distal spinal muscular atrophy as a major feature in adult-onset ataxia telangiectasia. Neurology 67: 346–349.

Iyama T, Wilson 3rd DM (2013). DNA repair mechanisms in dividing and non-dividing cells. DNA Repair (Amst) 12: 620–636.

Kamsler A, Daily D, Hochman A et al. (2001). Increased oxidative stress in ataxia telangiectasia evidenced by alterations in redox state of brains from Atm-deficient mice. Cancer Res 61: 1849–1854.

Kim TS, Kawaguchi M, Suzuki M et al. (2010). The ZFHX3 (ATBF1) transcription factor induces PDGFRB, which activates ATM in the cytoplasm to protect cerebellar neurons from oxidative stress. Dis Model Mech 3: 752–762.

Koziol LF, Budding D, Andreasen N et al. (2014). Consensus paper: the cerebellum's role in movement and cognition. Cerebellum 13: 151–177.

Lantelme E, Turinetto V, Mantovani S et al. (2003). Analysis of secondary V(D)J rearrangements in mature, peripheral

T cells of ataxia-telangiectasia heterozygotes. Lab Invest 83: 1467–1475.

Lavin MF (2013). The appropriateness of the mouse model for ataxia-telangiectasia: neurological defects but no neurodegeneration. DNA Repair (Amst) 12: 612–619.

Lefton-Greif MA, Crawford TO, Winkelstein JA et al. (2000). Oropharyngeal dysphagia and aspiration in patients with ataxia-telangiectasia. J Pediatr 136: 225–231.

Lefton-Greif MA, Crawford TO, McGrath-Morrow S et al. (2011). Safety and caregiver satisfaction with gastrostomy in patients with ataxia telangiectasia. Orphanet J Rare Dis 6: 23.

Lewis RF, Lederman HM, Crawford TO (1999). Ocular motor abnormalities in ataxia telangiectasia. Ann Neurol 46: 287–295.

Li J, Han YR, Plummer MR et al. (2009). Cytoplasmic ATM in neurons modulates synaptic function. Curr Biol 19: 2091–2096.

Li J, Chen J, Vinters HV et al. (2011). Stable brain ATM message and residual kinase-active ATM protein in ataxia-telangiectasia. J Neurosci 31: 7568–7577.

Li J, Chen J, Ricupero CL et al. (2012). Nuclear accumulation of HDAC4 in ATM deficiency promotes neurodegeneration in ataxia telangiectasia. Nat Med 18: 783–790.

Lin DD, Crawford TO, Lederman HM et al. (2006). Proton MR spectroscopic imaging in ataxia-telangiectasia. Neuropediatrics 37: 241–246.

Lin DD, Barker PB, Lederman HM et al. (2014). Cerebral abnormalities in adults with ataxia-telangiectasia. AJNR Am J Neuroradiol 35: 119–123.

Loeb DM, Lederman HM, Winkelstein JA (2000). Lymphoid malignancy as a presenting sign of ataxia-telangiectasia. J Pediatr Hematol Oncol 22: 464–467.

Louis-Bar D (1941). Sur un syndrome progressif cormprenant des télangiectasies capillaires cutanées et conjonctivales symétriques, à disposition naevoïde et des troubles cérébelleux. Confin Neurol 4: 32–42.

Majdzadeh N, Wang L, Morrison BE et al. (2008). HDAC4 inhibits cell-cycle progression and protects neurons from cell death. Dev Neurobiol 68: 1076–1092.

Marmolino D (2011). Friedreich's ataxia: past, present and future. Brain Res Rev 67: 311–330.

McConville CM, Stankovic T, Byrd PJ et al. (1996). Mutations associated with variant phenotypes in ataxia-telangiectasia. Am J Hum Genet 59: 320–330.

McFarlin DE, Strober W, Waldmann TA (1972). Ataxia-telangiectasia. Medicine (Baltimore) 51: 281–314.

McGrath-Morrow SA, Gower WA, Rothblum-Oviatt C et al. (2010). Evaluation and management of pulmonary disease in ataxia-telangiectasia. Pediatr Pulmonol 45: 847–859.

McGrath-Morrow SA, Lederman HM, Aherrera AD et al. (2014). Pulmonary function in children and young adults with ataxia telangiectasia. Pediatr Pulmonol 49: 84–90.

McKinnon PJ (2012). ATM and the molecular pathogenesis of ataxia telangiectasia. Annu Rev Pathol 7: 303–321.

Menotta M, Biagiotti S, Bianchi M et al. (2012). Dexamethasone partially rescues ataxia telangiectasia-mutated (ATM) deficiency in ataxia telangiectasia by promoting a shortened protein variant retaining kinase activity. J Biol Chem 287: 41352–41363.

Meshulam L, Galron R, Kanner S et al. (2012). The role of the neuro-astro-vascular unit in the etiology of ataxia telangiectasia. Front Pharmacol 3: 157.

Micol R, Ben Slama L, Suarez F et al. (2011). Morbidity and mortality from ataxia-telangiectasia are associated with ATM genotype. J Allergy Clin Immunol 128: 382–389 e1.

Mitra A, Gooi J, Darling J, Newton-Bishop JA (2011). Infliximab in the treatment of a child with cutaneous granulomas associated with ataxia telangiectasia. J Am Acad Dermatol 65: 676–677.

Mitui M, Nahas SA, Du LT et al. (2009). Functional and computational assessment of missense variants in the ataxia-telangiectasia mutated (ATM) gene: mutations with increased cancer risk. Hum Mutat 30: 12–21.

Moreira MC, Klur S, Watanabe M et al. (2004). Senataxin, the ortholog of a yeast RNA helicase, is mutant in ataxia-ocular apraxia 2. Nat Genet 36: 225–227.

Mostofsky SH, Kunze JC, Cutting LE et al. (2000). Judgment of duration in individuals with ataxia-telangiectasia. Dev Neuropsychol 17: 63–74.

Murphy RC, Berdon WE, Ruzal-Shapiro C et al. (1999). Malignancies in pediatric patients with ataxia telangiectasia. Pediatr Radiol 29: 225–230.

Nahas SA, Butch AW, Du L et al. (2009). Rapid flow cytometry-based structural maintenance of chromosomes 1 (SMC1) phosphorylation assay for identification of ataxia-telangiectasia homozygotes and heterozygotes. Clin Chem 55: 463–472.

Nakamura K, Du L, Tunuguntla R et al. (2012). Functional characterization and targeted correction of ATM mutations identified in Japanese patients with ataxia-telangiectasia. Hum Mutat 33: 198–208.

Nemeth AH, Kwasniewska AC, Lise S et al. (2013). Next generation sequencing for molecular diagnosis of neurological disorders using ataxias as a model. Brain 136: 3106–3118.

Nissenkorn A, Levi YB, Vilozni D et al. (2011). Neurologic presentation in children with ataxia-telangiectasia: is small head circumference a hallmark of the disease? J Pediatr 159: 466–471 e1.

Nissenkorn A, Hassin-Baer S, Lerman SF et al. (2013). Movement disorder in ataxia-telangiectasia: treatment with amantadine sulfate. J Child Neurol 28: 155–160.

Nowak-Wegrzyn A, Crawford TO, Winkelstein JA et al. (2004). Immunodeficiency and infections in ataxia-telangiectasia. J Pediatr 144: 505–511.

O'Driscoll M, Jeggo PA (2008). The role of the DNA damage response pathways in brain development and microcephaly: insight from human disorders. DNA Repair (Amst) 7: 1039–1050.

Olsen JH, Hahnemann JM, Borresen-Dale AL et al. (2001). Cancer in patients with ataxia-telangiectasia and in their relatives in the Nordic countries. J Natl Cancer Inst 93: 121–127.

Ortonne JP, Claudy AL, Freycon F (1980). Cafe au lait spots in ataxia-telangiectasia (AT). Histochemical and ultrastructural study in one case. Arch Dermatol Res 268: 91–99.

Paller AS, Massey RB, Curtis MA et al. (1991). Cutaneous granulomatous lesions in patients with ataxia-telangiectasia. J Pediatr 119: 917–922.

Paula-Barbosa MM, Ruela C, Tavares MA et al. (1983). Cerebellar cortex ultrastructure in ataxia-telangiectasia. Ann Neurol 13: 297–302.

Perlman S, Becker-Catania S, Gatti RA (2003). Ataxia-telangiectasia: diagnosis and treatment. Semin Pediatr Neurol 10: 173–182.

Petersen AJ, Rimkus SA, Wassarman DA (2012). ATM kinase inhibition in glial cells activates the innate immune response and causes neurodegeneration in Drosophila. Proc Natl Acad Sci U S A 109: E656–E664.

Petersen AJ, Katzenberger RJ, Wassarman DA (2013). The innate immune response transcription factor relish is necessary for neurodegeneration in a Drosophila model of ataxia-telangiectasia. Genetics 194: 133–142.

Raz-Prag D, Galron R, Segev-Amzaleg N et al. (2011). A role for vascular deficiency in retinal pathology in a mouse model of ataxia-telangiectasia. Am J Pathol 179: 1533–1541.

Reliene R, Schiestl RH (2008). Experimental antioxidant therapy in ataxia telangiectasia. Clin Med Oncol 2: 431–436.

Riise R, Ygge J, Lindman C et al. (2007). Ocular findings in Norwegian patients with ataxia-telangiectasia: a 5 year prospective cohort study. Acta Ophthalmol Scand 85: 557–562.

Ronnett GV, Leopold D, Cai X et al. (2003). Olfactory biopsies demonstrate a defect in neuronal development in Rett's syndrome. Ann Neurol 54: 206–218.

Sandlund JT, Hudson MM, Kennedy W et al. (2014). Pilot study of modified LMB-based therapy for children with ataxia-telangiectasia and advanced stage high grade mature B-cell malignancies. Pediatr Blood Cancer 61: 360–362.

Sardanelli F, Parodi RC, Ottonello C et al. (1995). Cranial MRI in ataxia-telangiectasia. Neuroradiology 37: 77–82.

Savitsky K, Bar-Shira A, Gilad S et al. (1995). A single ataxia telangiectasia gene with a product similar to PI-3 kinase. Science 268: 1749–1753.

Schneider JG, Finck BN, Ren J et al. (2006). ATM-dependent suppression of stress signaling reduces vascular disease in metabolic syndrome. Cell Metab 4: 377–389.

Schroeder SA, Zielen S (2014). Infections of the respiratory system in patients with ataxia-telangiectasia. Pediatr Pulmonol 49: 389–399.

Schubert R, Reichenbach J, Zielen S (2005). Growth factor deficiency in patients with ataxia telangiectasia. Clin Exp Immunol 140: 517–519.

Shaikh AG, Marti S, Tarnutzer AA et al. (2009). Gaze fixation deficits and their implication in ataxia-telangiectasia. J Neurol Neurosurg Psychiatry 80: 858–864.

Shaikh AG, Marti S, Tarnutzer AA et al. (2011). Ataxia telangiectasia: a "disease model" to understand the cerebellar control of vestibular reflexes. J Neurophysiol 105: 3034–3041.

Shaikh AG, Marti S, Tarnutzer AA et al. (2013a). Effects of 4-aminopyridine on nystagmus and vestibulo-ocular reflex in ataxia-telangiectasia. J Neurol 260: 2728–2735.

Shaikh AG, Zee DS, Mandir AS et al. (2013b). Disorders of upper limb movements in ataxia-telangiectasia. PLoS One 8: e67042.

Shen C, Houghton PJ (2013). The mTOR pathway negatively controls ATM by up-regulating miRNAs. Proc Natl Acad Sci U S A 110: 11869–11874.

Shiloh Y, Ziv Y (2013). The ATM protein kinase: regulating the cellular response to genotoxic stress, and more. Nat Rev Mol Cell Biol 14: 197–210.

Song CH, Bernhard D, Hess EJ et al. (2014). Subtle microstructural defects of the cerebellum in a knock-in mouse model of DYT1 dystonia. Neurobiol Dis 62: 372–380.

Stell R, Bronstein AM, Plant GT et al. (1989). Ataxia telangiectasia: a reappraisal of the ocular motor features and their value in the diagnosis of atypical cases. Mov Disord 4: 320–329.

Stewart GS, Maser RS, Stankovic T et al. (1999). The DNA double-strand break repair gene hMRE11 is mutated in individuals with an ataxia-telangiectasia-like disorder. Cell 99: 577–587.

Stewart GS, Last JI, Stankovic T et al. (2001). Residual ataxia telangiectasia mutated protein function in cells from ataxia telangiectasia patients, with 5762ins137 and 7271 T->G mutations, showing a less severe phenotype. J Biol Chem 276: 30133–30141.

Syllaba L, Henner K (1926). Contribution à l'étude de l'indépendance de l'athétose double idiopathique et congénitale. Atteinte familiale, syndrome dystrophique, signe du résau vasculaire conjonctival, intégrité psychique. Rev Neurol (Paris) 1: 541–560.

Takahashi K, Yamanaka S (2013). Induced pluripotent stem cells in medicine and biology. Development 140: 2457–2461.

Takahashi K, Okita K, Nakagawa M et al. (2007a). Induction of pluripotent stem cells from fibroblast cultures. Nat Protoc 2: 3081–3089.

Takahashi T, Tada M, Igarashi S et al. (2007b). Aprataxin, causative gene product for EAOH/AOA1, repairs DNA single-strand breaks with damaged 3′-phosphate and 3′-phosphoglycolate ends. Nucleic Acids Res 35: 3797–3809.

Tavani F, Zimmerman RA, Berry GT et al. (2003). Ataxia-telangiectasia: the pattern of cerebellar atrophy on MRI. Neuroradiology 45: 315–319.

Taylor AM, Byrd PJ (2005). Molecular pathology of ataxia telangiectasia. J Clin Pathol 58: 1009–1015.

Taylor AM, Groom A, Byrd PJ (2004). Ataxia-telangiectasia-like disorder (ATLD)-its clinical presentation and molecular basis. DNA Repair (Amst) 3: 1219–1225.

Valentin-Vega YA, Kastan MB (2012). A new role for ATM: regulating mitochondrial function and mitophagy. Autophagy 8: 840–841.

Valentin-Vega YA, Maclean KH, Tait-Mulder J et al. (2012). Mitochondrial dysfunction in ataxia-telangiectasia. Blood 119: 1490–1500.

Van Overwalle F, Baetens K, Marien P et al. (2014). Social cognition and the cerebellum: a meta-analysis of over 350 fMRI studies. Neuroimage 86: 554–572.

Verhagen MM, Last JI, Hogervorst FB et al. (2012a). Presence of ATM protein and residual kinase activity correlates with the phenotype in ataxia-telangiectasia: a genotype-phenotype study. Hum Mutat 33: 561–571.

Verhagen MM, Martin JJ, Van Deuren M et al. (2012b). Neuropathology in classical and variant ataxia-telangiectasia. Neuropathology 32: 234–244.

Vinck A, Verhagen MM, Gerven M et al. (2011). Cognitive and speech-language performance in children with ataxia telangiectasia. Dev Neurorehabil 14: 315–322.

Vinters HV, Gatti RA, Rakic P (1985). Sequence of cellular events in cerebellar ontogeny relevant to expression of neuronal abnormalities in ataxia-telangiectasia. Kroc Found Ser 19: 233–255.

Wallis LI, Griffiths PD, Ritchie SJ et al. (2007). Proton spectroscopy and imaging at 3 T in ataxia-telangiectasia. AJNR Am J Neuroradiol 28: 79–83.

Wilson BK, Hess EJ (2013). Animal models for dystonia. Mov Disord 28: 982–989.

Woods CG, Taylor AM (1992). Ataxia telangiectasia in the British Isles: the clinical and laboratory features of 70 affected individuals. Q J Med 82: 169–179.

Worth PF, Srinivasan V, Smith A et al. (2013). Very mild presentation in adult with classical cellular phenotype of ataxia telangiectasia. Mov Disord 28: 524–528.

Zannolli R, Buoni S, Betti G et al. (2012). A randomized trial of oral betamethasone to reduce ataxia symptoms in ataxia telangiectasia. Mov Disord 27: 1312–1316.

Zeviani M, Simonati A, Bindoff LA (2012). Ataxia in mitochondrial disorders. Handb Clin Neurol 103: 359–372.

Handbook of Clinical Neurology, Vol. 132 (3rd series)
Neurocutaneous Syndromes
M.P. Islam and E.S. Roach, Editors

Chapter 15

Pseudoxanthoma elasticum

E. STEVE ROACH* AND MONICA P. ISLAM

Department of Child Neurology, Nationwide Children's Hospital, Ohio State University College of Medicine, Columbus, OH, USA

INTRODUCTION

Pseudoxanthoma elasticum (PXE; OMIM #264800) is an autosomal recessive connective tissue disease that is characterized by fragmentation ("elastorrhexia") and calcification of elastic fibers. The deterioration of the elastic fibers leads to dermatologic, ophthalmologic, and vascular dysfunction. The clinical presentation, manifestations, and rate of progression vary considerably, even among affected members of the same family, and heterozygous individuals can also exhibit some features of the disorder.

PXE results from a mutation of the *ABCC6* gene on chromosome 16p13.1 that codes for a transmembrane transporter protein of the ATP binding ABC family (Chassaing et al., 2005). The prevalence of PXE is around 1 in 25 000 people, but some individuals with milder phenotypes probably remain undiagnosed (Chassaing et al., 2005).

HISTORY

Balzer provided the first clinical description of pseudoxanthoma elasticum in 1884, but he mistakenly suggested that the lesions represent a form of xanthoma and failed to appreciate the distinctive elastic fiber mineralization (Balzer, 1884). Darier (1896) soon discovered the mineralization of the skin elastic fibers in these lesions. To emphasize the absence of a xanthomatous lesion, Darier devised the term "pseudoxanthoma" and added "elasticum" to identify the specific anatomic site of the pathology (Darier 1896).

Around this same time, angioid streaks of the retina were first described by Robert Doyne (1889). Another patient was described 2 years later by Otto Plange (1891), and Knapp (1892) coined the term "angioid streaks" (Knapp, 1892). However, it was not until 1929 that Ester Grönblad and James Strandberg first recognized the retinal complications as a manifestation of PXE (Grönblad, 1929; Strandberg, 1929). The term Grönblad–Strandberg syndrome is still occasionally used in Europe, but pseudoxanthoma elasticum is the preferred term. Several years later, Carlborg described the cardiovascular aspects of PXE (Carlborg and Grönblad, 1944).

CLINICAL MANIFESTATIONS

Dermatologic features

Although the diagnosis of PXE is most often made late in the second or third decade, characteristic yellowish papules or plaque-like skin lesions often appear even in children (Naouri et al., 2009). The lesions often appear first on the lateral neck (Fig. 15.1). Other common lesion sites are the antecubital fossae, axillae, popliteal spaces, periumbilical area, and groin (Neldner, 1988). Similar plaque-like lesions have been documented in the mucous membranes. The skin lesions advance most rapidly during adolescence and young adult life, and then more or less stabilize with advancing age. These lesions are typically asymptomatic but they may itch. Older patients share a facial resemblance due to the lax redundant cutaneous changes of the face and neck (Goodman et al., 1963).

Ophthalmologic features

Angioid streaks of the retina (Fig. 15.2) eventually develop in over 90% of individuals with PXE (Orssaud et al., 2015), although the retinal lesions of PXE may appear several years after the appearance of the typical skin lesions. Angioid streaks appear as dark irregular retinal lines emanating from the optic disc (Neldner, 1988). These lesions are usually bilateral, although occasional individuals with unilateral angioid streaks have been

*Correspondence to: E. Steve Roach, MD, Department of Neurology, 700 Children's Drive, Columbus, OH 43205, USA. Tel: +1-614-722-4689, E-mail: roache@nationwidechildrens.org

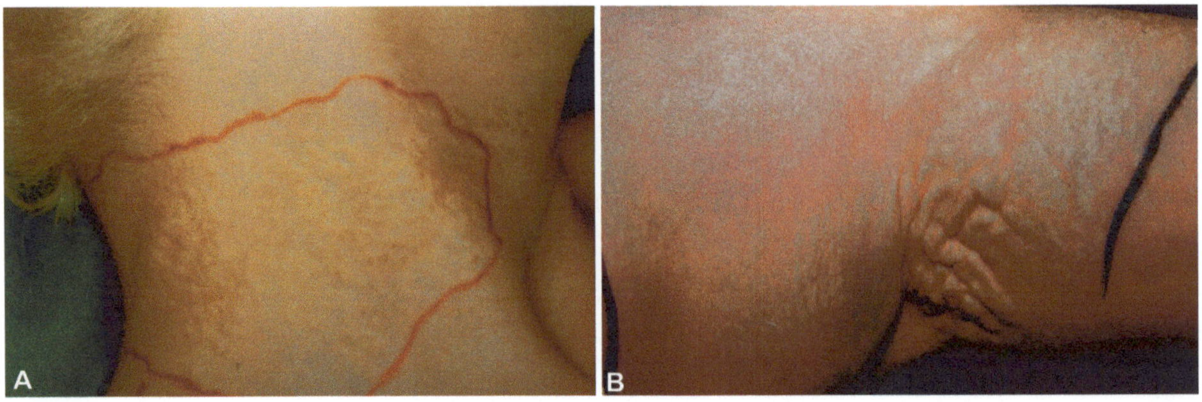

Fig. 15.1. (**A**) Characteristic cutaneous appearance of pseudoxanthoma elasticum (PXE) on the neck of an individual with PXE (outlined in red). (**B**) Redundant skin folds in the axilla in an individual with PXE. (Reproduced from Neldner and Roach, 2004, with permission from Cambridge University Press.)

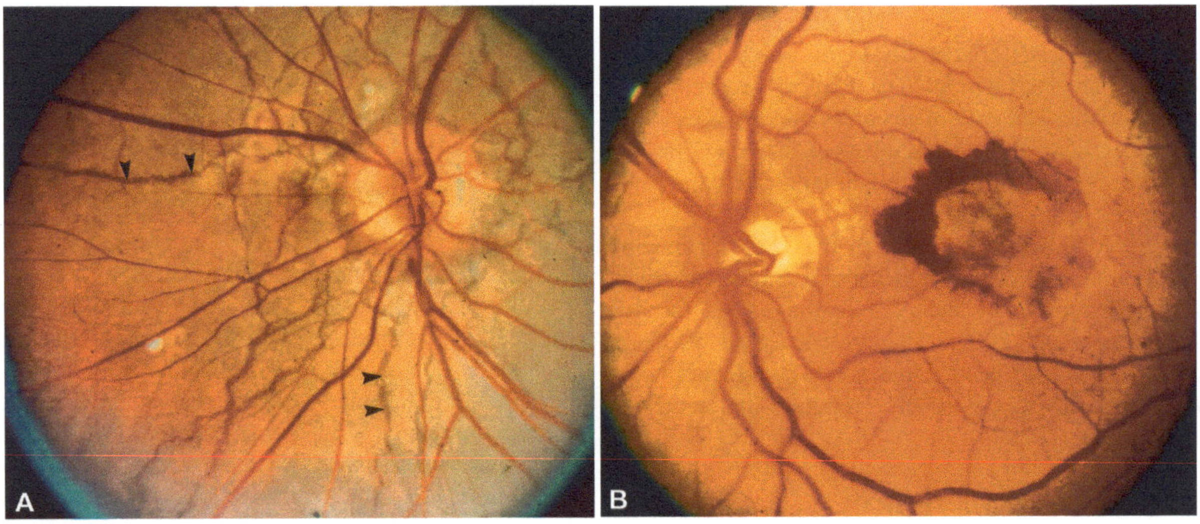

Fig. 15.2. (**A**) Angioid retinal streaks (arrowheads) in a patient with pseudoxanthoma elasticum (PXE). (**B**) Acute macular hemorrhage in another individual PXE. (Reproduced from Neldner and Roach, 2004, with permission from Cambridge University Press.)

documented (Orssaud et al., 2015). Angioid streaks are asymptomatic initially. The streaks represent disruption of Bruch's membrane due to the deterioration of the membrane's elastic fibers (Ebran et al., 2014; Spaide, 2015).

A dappled appearance of the macular region resulting from small dark areas within a slightly lighter portion of the retina, said to resemble the peel of an orange ("*peau d'orange*"), often precedes the angioid streaks (Naouri et al., 2009). Other ocular manifestations of PXE include macular atrophy with central visual loss, choroidal neovascularization, and subretinal hemorrhage and fibrosis (Spaide, 2015).

Retinal hemorrhage is common after age 40 years but has been documented as early as the second decade of life. Hemorrhage is usually unilateral but often occurs later in the other eye. Retinal hemorrhage may occur spontaneously or complicate accidental head trauma, weight lifting, or contact sports. Minor trauma sometimes initiates retinal hemorrhage. The hemorrhage usually affects the macula (see Fig. 15.2), leading to acute loss of central vision. Scarring after the hemorrhage may result in more chronic loss of central vision. Peripheral vision is usually spared. Some individuals experience gradual visual loss resulting from secondary macular degeneration (Gliem et al., 2013).

Cerebrovascular dysfunction

Pseudoxanthoma elasticum does not directly affect the nervous system, but secondary neurologic impairment due to cerebrovascular compromise is well documented.

Older individuals with PXE are more likely to develop cerebrovascular complications, typically resulting from progressive narrowing and occlusion of one or more cerebral arteries (Rios-Montenegro et al., 1972). The angiographic pattern of PXE resembles that of severe atherosclerosis, with arterial stenosis, tortuosity, and occlusion, especially in the internal carotid or vertebral arteries. Carotid occlusion with extensive collateral circulation is angiographically similar to moyamoya syndrome (Yasuhara et al., 2004; Del Zotto et al., 2012). Fibromuscular dysplasia has also been documented (Prick and Thijssen, 1977).

Intracranial arterial aneurysms occur with a slightly increased frequency in individuals with PXE, probably due to their abnormal elastic lamina (Iqbal et al., 1978). Aneurysms often develop in the intracranial portion of the internal carotid artery (Dixon, 1951; Scheie and Hogan, 1957; Munyer and Margulis, 1981). Aneurysms in other intracranial arteries are less common. One woman with subarachnoid bleeding and spinal cord dysfunction had an aneurysm of the anterior spinal artery (Kito et al., 1983). Both intracerebral and subarachnoid hemorrhage have been documented in individuals with PXE in the absence of an aneurysm (Sharma et al., 1974; Bock and Schwegler, 2008).

Other concerns

Gastrointestinal hemorrhage, primarily involving the stomach, occurs in about 10–15% of individuals with PXE (Neldner, 1988; Spinzi et al., 1996). Gastrointestinal hemorrhage can occur without warning and may be severe enough to require hospitalization and blood transfusion. Bleeding can occur at any age, but it is most frequently documented in young adults. Even slight vomiting of blood or the appearance of black tarry stools should be considered a potential emergency.

Individuals with PXE often experience peripheral arterial vascular disease, typically presenting as intermittent leg claudication (Neldner, 1988). Peripheral arteries may be palpably rigid, and pulses are often difficult to detect. Symptomatic peripheral vascular disease in individuals with PXE is most common after the fifth or sixth decades, but symptoms can begin even in the third or fourth decades. Radiographs of the extremities sometimes reveal arterial calcification (Wahlqvist et al., 1977).

Although there are reports documenting cardiac dysfunction in individuals with PXE, direct involvement of the heart by PXE is not common. Biere and colleagues did not demonstrate echocardiographic dysfunction in a cohort of 75 PXE patients without cardiac symptoms (Biere et al., 2014). Mitral valve prolapse occurs frequently in individuals with PXE, but it usually remains asymptomatic (Lebwohl et al., 1982). Fatal cardiomyopathy is rarely reported in infants and children (Nolte, 2000).

Generalized arterial calcification of infancy is characterized by extensive calcification of the medium-sized and large arteries and associated with various signs and symptoms of cardiovascular dysfunction (Nitschke and Rutsch, 2012). The clinical manifestations of generalized arterial calcification overlap those of PXE. Although most infants with generalized arterial calcification have a mutation of the *ENPP1* gene, some of these patients have an *ABCC6* mutation (Ferreira et al., 1993; Nitschke and Rutsch, 2012).

Pseudoxanthoma elasticum does not dramatically increase the complication rate of pregnancy and delivery, although pregnancy may increase the likelihood of gastrointestinal hemorrhage (Berde et al., 1983). Epistaxis, hematuria, and hemoptysis have also been documented (Tay, 1970).

Differential diagnosis

While angioid streaks are highly suggestive of PXE, they are not pathognomonic of the disorder. Angioid streaks have also been documented in individuals with Paget disease of bone, sickle-cell disease, β-thalassemia, and Marfan syndrome (Neldner, 1988; Lebwohl et al., 1993; Baccarani-Contri et al., 2001; Yu et al., 2009).

Several acquired skin lesions can resemble hereditary PXE. Skin lesions similar to those of PXE have been documented in farmers carrying bags of calcium fertilizer with their bare and moist arms (Christensen, 1964). Their lesions are confined to the sites where the calcium compounds were rubbed into the skin, which is histologically similar to hereditary PXE. Long-term use of penicillamine can produce similar findings (Coatesworth et al., 1998). Another unusual condition is localized acquired cutaneous or periumbilical PXE (Hicks et al., 1979; Neldner and Martinez-Hernandez, 1979). This condition mainly affects the abdominal skin of obese, hypertensive, multiparous females; the skin is histologically identical to hereditary PXE but is not inherited and does not share other clinical features of PXE. Bercovitch and colleagues described three patients who developed a PXE phenotype following liver transplantation (Bercovitch et al., 2011).

A PXE-like disorder with multiple coagulation deficiencies features yellowish papules or leathery plaque-like skin lesions, retinal angioid streaks or *peau d'orange* lesions, and fragmentation and calcification of elastic fibers in the dermis. However, these individuals experienced multiple vitamin K-dependent clotting abnormalities and exhibited much more obvious skin laxity with skin lesions primarily on the trunk and limbs. Their retinal lesions were not as pronounced and vision was preserved. Although light microscopy showed fragmented and mineralized elastic fibers, ultrastructural analysis revealed that the elastic fibers had a more fragmented and mottled appearance than is typical of PXE. Vanakker and colleagues (2007) identified mutations in the *GGCX* gene at 2p11.2. This gene encodes γ-glutamyl carboxylase, which is important for γ-carboxylation of gla-proteins.

There is genetic and phenotypic overlap between PXE and generalized arterial calcification of infancy, although the classic clinical profile of the two disorders is quite distinct. Individuals with generalized arterial calcification of infancy typically survive only a few months. Some *ABCC6* mutations are associated with premature atherosclerosis and arterial calcification, but affected individuals become symptomatic much later. Most individuals with generalized arterial calcification of infancy have a mutation of *ENPP1*, but about half of the remaining individuals have a mutation of *ABCC6* (Nitschke et al., 2012).

PATHOLOGY OF PSEUDOXANTHOMA ELASTICUM

Apart from the highly characteristic clinical appearance and location of the skin lesions, a skin biopsy showing calcified elastic fibers is diagnostic. The initial calcification occurs within the central core of the elastic fiber (Fig. 15.3), gradually becoming more pronounced and eventually leading to rupture of the fiber. Calcified elastic fibers are demonstrable even in early, barely visible, skin lesions. Increased concentrations of mucopolysaccharide (glycosaminoglycans) in the affected dermis have been demonstrated, but this measurement is difficult and thus not routinely performed.

Blood vessels exhibit elastic tissue degeneration, and marked subintimal and medial fibrosis may severely narrow the vessel's lumen (Goodman et al., 1963; Wahlqvist et al., 1977). Calcium deposition in the internal elastic lamina may be apparent. Fibrous thickening of the endocardium and calcification of the endocardial elastic fibers have been demonstrated (Akhtar and Brody, 1975).

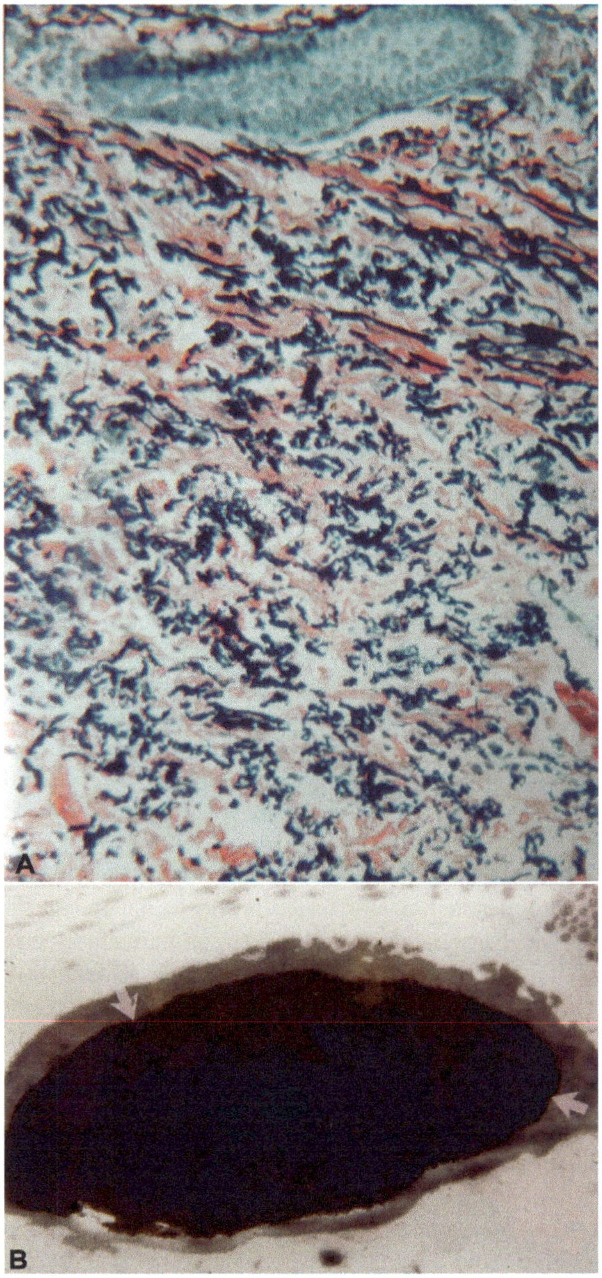

Fig. 15.3. (**A**) Skin biopsy specimen stained with elastic tissue stain shows irregular, degenerating elastic fibers in the mid and lower dermis (×225). (**B**) Electron micrograph of an elastic fiber from a pseudoxanthoma elasticum (PXE) patient shows central calcification (arrows) of the elastic fiber (×30000). (Reproduced from Neldner and Roach, 2004, with permission from Cambridge University Press.)

GENETICS OF PSEUDOXANTHOMA ELASTICUM

Pseudoxanthoma elasticum is best characterized as an autosomal recessive trait, although heterozygous individuals occasionally exhibit some signs of the disease. The chromosomal localization of the PXE gene to 16p 13.1 was demonstrated in 1997 (Struk et al., 1997), and the 31 exon gene (*ABCC6*) was identified simultaneously in 2000 by four separate groups (Bergen et al., 2000; Le Saux et al., 2000; Ringpfeil et al., 2000; Struk et al., 2000). A mutation in both *ABCC6* alleles can be demonstrated in most affected individuals. Over 300 *ABCC6* mutations have been documented (Hendig et al., 2013), and there is no consistent genotype-phenotype correlation (Pfendner et al., 2007). Mutational analysis is commercially available.

The *ABCC6* gene contains 31 exons and encodes a 1503 amino acid transmembrane transporter protein (known as the multidrug resistance-associated protein 6, or MRP6). *ABCC6* is a member of the ATP-binding cassette (ABC) family, subclass C (Uitto et al., 2010). The protein functions as a transmembrane transporter and is strongly expressed in the liver and kidney but weakly expressed in the retina, skin, or blood vessels – the tissues that bear the brunt of PXE. This pattern suggests the existence of one or more humoral factors that prevent the accumulation of calcium in elastic tissue rather than an abnormality of elastin *per se* (Uitto et al., 2010). Genes known to modulate elastin metabolism are not responsible for PXE (Uitto et al., 2010).

Generalized arterial calcification of infancy is an autosomal recessive disorder characterized by severe calcification of the internal elastic lamina of the medium- and large-sized arteries and intimal proliferation. Affected individuals typically survive only a few months due to the resulting arterial stenosis and cardiac failure. A mutation in the *ENPP1* gene can be demonstrated in about 75% of the individuals with generalized arterial calcification of infancy, and a mutation of *ABCC6* can be detected in about half of the remaining patients (Nitschke et al., 2012). Some specific *ABCC6* mutations are associated with premature atherosclerosis, even when present in a heterozygous state (Trip et al., 2002).

MANAGEMENT OF PSEUDOXANTHOMA ELASTICUM

There is no fully effective treatment for PXE. Many of the serious manifestations of PXE represent vascular dysfunction, so it is prudent to develop habits that facilitate lifelong general health and cardiovascular fitness. Affected individuals should avoid the use of tobacco, optimize the diet, and develop a lifelong fitness regimen. They should also maintain a normal lipid profile by whatever method is required.

It has been suggested that a high calcium diet early in life may facilitate vascular dysfunction due to PXE, and restriction of dietary calcium has long been recommended for these individuals (Renie et al., 1984; Neldner, 1988). A recent randomized trial of an oral phosphate binder demonstrated a reduction in both the calcification level and the clinical severity score, although the study cohort was relatively small and neither change reached statistical significance (Yoo et al., 2011). The current recommendation is to keep calcium intake at the basic adult level of 800 mg/day even during adolescence, when the recommended daily calcium allowance is 1200 mg/day.

Given the high risk of retinal hemorrhage, affected individuals should avoid contact and racket sports and other activities that might result in normally minor eye trauma. Boxing, football and rugby should be avoided, and even soccer and heavy weight lifting could lead to retinal hemorrhage in these individuals. Noncontact sports such as swimming and bicycling, however, should be encouraged.

Periodic intravitreal administration of agents derived from humanized monoclonal antibody to vascular endothelial growth factor (VGEF) has been shown to benefit choroidal neovascularization due to PXE (Finger et al., 2011a,b).

Aspirin and the nonsteroidal anti-inflammatory agents should be avoided because they could increase the likelihood of retinal and gastric hemorrhages.

Some *ABCC6* mutations are associated with premature atherosclerosis, even when present in a heterozygous state (Trip et al., 2002). Statin agents are thus often prescribed for individuals with PXE. In a survey of 1747 individuals with PXE, about a third of the patients reported current or prior use of cholesterol-lowering medications (Guo et al., 2013). Guo and colleagues administered two doses atorvastatin to *Abcc6* knockout mice either at 4 weeks of age or at 12 weeks of age (Guo et al., 2013). All of the animals exhibited lower cholesterol and triglyceride levels following medication administration, but only those who were exposed to the drug early in life experienced a reduction in vascular mineralization (Guo et al., 2013). These results suggest that atorvastatin is capable of preventing, but not reversing, the abnormal mineralization. If these similar results are realized in human trials, it seems likely that administration of the drug at an early age will be more useful than later treatment.

Genetic counseling may help the patient and family understand the likelihood of additional affected offspring and assist with family planning. Each full sibling of an affected child has a one in four chance of having PXE. Prenatal testing and gene carrier risk assessment can be offered if the affected family member's mutation has been identified.

References

Akhtar M, Brody H (1975). Elastic tissue in pseudoxanthoma elasticum. Arch Pathol 99: 667–671.

Baccarani-Contri M, Bacchelli B, Boraldi F et al. (2001). Characterization of pseudoxanthoma elasticum-like lesions in the skin of patients with beta-thalassemia. J Am Acad Dermatol 44: 33–39.

Balzer F (1884). Recherches sur les caracteres du xanthelasma. Archives of Physiology 4: 65–80.

Bercovitch L, Martin L, Chassaing N et al. (2011). Acquired pseudoxanthoma elasticum presenting after liver transplantation. J Am Acad Dermatol 64: 873–878.

Berde C, Willis DC, Sandberg EC (1983). Pregnancy in women with pseudoxanthoma elasticum. Obstet Gynecol Surv 38: 339–344.

Bergen AA, Plomp AS, Schuurman EJ et al. (2000). Mutations in ABCC6 cause pseudoxanthoma elasticum. Nat Genet 25: 228–231.

Biere L, Donal E, Terrien G et al. (2014). Left ventricular function in a large cohort of pseudoxanthoma elasticum patients. PLoS One 9: e90364.

Bock A, Schwegler G (2008). Intracerebral haemorrhage as first manifestation of pseudoxanthoma elasticum. Clin Neurol Neurosurg 110: 262–264.

Carlborg U, Grönblad E (1944). The incidence of arteriosclerosis in pseudoxanthoma elasticum. Acta Med Scand 308: 37–38.

Chassaing N, Martin L, Calvas P et al. (2005). Pseudoxanthoma elasticum: a clinical, pathophysiological and genetic update including 11 novel ABCC6 mutations. J Med Genet 42: 881–892.

Christensen OB (1964). An exogenous variety of pseudoxanthoma elasticum in old farmers. Acta Diabetol 89: 360–363.

Coatesworth AP, Darnton SJ, Green RM et al. (1998). A case of systemic pseudo-pseudoxanthoma elasticum with diverse symptomatology caused by long term penicillamine use. J Clin Pathol 51: 169–171.

Darier J (1896). Pseudo-xanthome élastique. III International Congress of Dermatology, London.

Del Zotto E, Ritelli M, Pezzini A et al. (2012). Clinical, neuroradiological and molecular features of a patient affected by pseudoxhantoma elasticum associated to carotid rete mirabile: case report. Clin Neurol Neurosurg 114: 758–761.

Dixon JM (1951). Angioid streaks and pseudoxanthoma elasticum with aneurysm of the internal carotid artery. Am J Ophthalmol 34: 1322–1323.

Doyne RW (1889). Choroidal and retinal changes – the result of blows to the eyes. Trans Ophthalmol Soc U K 9: 128–129.

Ebran JM, Milea D, Trelohan A et al. (2014). New insights into the visual prognosis of pseudoxanthoma elasticum. Br J Ophthalmol 98: 142–143.

Ferreira C, Ziegler S, Gahl W (1993). Generalized arterial calcification of infancy. GeneReviews® [Internet], University of Washington, Seattle, Seattle (WA). 1993–2015.

Finger RP, Charbel IP, Hendig D et al. (2011a). Monthly ranibizumab for choroidal neovascularizations secondary to angioid streaks in pseudoxanthoma elasticum: a one-year prospective study. Am J Ophthalmol 152: 695–703.

Finger RP, Charbel IP, Schmitz-Valckenberg S et al. (2011b). Long-term effectiveness of intravitreal bevacizumab for choroidal neovascularization secondary to angioid streaks in pseudoxanthoma elasticum. Retina 31: 1268–1278.

Gliem M, Zaeytijd JD, Finger RP et al. (2013). An update on the ocular phenotype in patients with pseudoxanthoma elasticum. Front Genet 4: 14.

Goodman RM, Smith EW, Paton D et al. (1963). Pseudoxanthoma elasticum: a clinical and histopathological study. Medicine 42: 297–334.

Grönblad E (1929). Angioid streaks – pseudoxanthoma elasticum. Acta Ophthalmol 7: 329.

Guo H, Li Q, Chou DW et al. (2013). Atorvastatin counteracts aberrant soft tissue mineralization in a mouse model of pseudoxanthoma elasticum (Abcc6(−)/(−)). J Mol Med 91: 1177–1184.

Hendig D, Knabbe C, Gotting C (2013). New insights into the pathogenesis of pseudoxanthoma elasticum and related soft tissue calcification disorders by identifying genetic interactions and modifiers. Front Genet 4: 114.

Hicks J, Carpenter Jr CL, Reed RJ (1979). Periumbilical perforating pseudoxanthoma elasticum. Arch Dermatol 115: 300–303.

Iqbal A, Alter M, Lee SH (1978). Pseudoxanthoma elasticum: a review of neurological complications. Ann Neurol 4: 18–20.

Kito K, Kobayashi N, Mori N et al. (1983). Ruptured aneurysm of the anterior spinal artery associated with pseudoxanthoma elasticum. J Neurosurg 58: 126–128.

Knapp H (1892). On the formation of dark angioid streaks as an unusual metamorphosis of retinal hemorrhage. Arch Ophthalmol 21: 289–292.

Le Saux O, Urban Z, Tschuch C et al. (2000). Mutations in a gene encoding an ABC transporter cause pseudoxanthoma elasticum. Nat Genet 25: 223–227.

Lebwohl MG, Distefano D, Prioleau PG et al. (1982). Pseudoxanthoma elasticum and mitral-valve prolapse. N Engl J Med 307: 228–231.

Lebwohl M, Halperin J, Phelps RG (1993). Occult pseudoxanthoma elasticum in patients with premature cardiovascular disease. N Engl J Med 329: 1237–1239.

Munyer TP, Margulis AR (1981). Pseudoxanthoma elasticum with internal carotid artery aneurysm. Am J Roentgenol 136: 1023–1029.

Naouri M, Boisseau C, Bonicel P et al. (2009). Manifestations of pseudoxanthoma elasticum in childhood. Br J Dermatol 161: 635–639.

Neldner KH (1988). Pseudoxanthoma elasticum. Clin Dermatol 6: 1–159.

Neldner KH, Martinez-Hernandez A (1979). Localized acquired cutaneous pseudoxanthoma elasticum. J Am Acad Dermatol 1: 523–530.

Neldner KH, Roach ES (2004). Pseudoxanthoma elasticum. In: ES Roach, VS Miller (Eds.), Neurocutaneous Disorders, 1st edn. Cambridge University Press, Cambridge, pp. 138–143.

Nitschke Y, Rutsch F (2012). Generalized arterial calcification of infancy and pseudoxanthoma elasticum: two sides of the same coin. Front Genet 3: 302.

Nitschke Y, Baujat G, Botschen U et al. (2012). Generalized arterial calcification of infancy and pseudoxanthoma elasticum can be caused by mutations in either ENPP1 or ABCC6. Am J Hum Genet 90: 25–39.

Nolte KB (2000). Sudden cardiac death owing to pseudoxanthoma elasticum: a case report. Hum Pathol 31: 1002–1004.

Orssaud C, Roche O, Dufier JL et al. (2015). Visual impairment in pseudoxanthoma elasticum: a survey of 40 patients. Ophthalmic Genet 21: 1–6.

Pfendner EG, Vanakker OM, Terry SF et al. (2007). Mutation detection in the ABCC6 gene and genotype-phenotype analysis in a large international case series affected by pseudoxanthoma elasticum. J Med Genet 44: 621–628.

Plange O (1891). Überstreifen förmige pigmentbildung mit sekundären veränderungen der netzhaut infolge von hämorrhagien. ArchAugenheilkd 23: 78–90.

Prick JJG, Thijssen HOM (1977). Radiodiagnostic signs in pseudoxanthoma elasticum generalisatum (dysgenesis elastofibrillaris mineralisans). Clin Radiol 28: 549–554.

Renie WA, Pyeritz RE, Combs J et al. (1984). Pseudoxanthoma elasticum: high calcium intake in early life correlates with severity. Am J Med Genet 19: 235–244.

Ringpfeil F, Lebwohl MG, Christiano AM et al. (2000). Pseudoxanthoma elasticum: mutations in the MRP6 gene encoding a transmembrane ATP-binding cassette (ABC) transporter. Proc Natl Acad Sci U S A 97: 6001–6006.

Rios-Montenegro EN, Behrens MM, Hoyt WF (1972). Pseudoxanthoma elasticum. Arch Neurol 26: 151–155.

Scheie HG, Hogan TF (1957). Angioid streaks and generalized arterial disease. Arch Ophthalmol 57: 855–868.

Sharma NGK, Ghosh SK, Beohar PC et al. (1974). Subarachnoid haemorrhage in pseudoxanthoma elasticum. Postgrad Med J 50: 774–776.

Spaide RF (2015). Peau d'orange and angioid streaks: manifestations of Bruch membrane pathology. Retina 35: 392–397.

Spinzi G, Strocchi E, Imperiali G et al. (1996). Pseudoxanthoma elasticum: a rare cause of gastrointestinal bleeding. Am J Gastroenterol 91: 1631–1634.

Strandberg J (1929). Pseudoxanthoma elasticum. Z Haut Geschlechtskr 31: 689–694.

Struk B, Neldner KH, Rao VS et al. (1997). Mapping of both autosomal recessive and dominant variants of pseudoxanthoma elasticum to chromosome 16p131. Hum Mol Genet 6: 1823–1828.

Struk B, Cai L, Zach S et al. (2000). Mutations of the gene encoding the transmembrane transporter protein ABC-C6 cause pseudoxanthoma elasticum. J Mol Med 78: 282–286.

Tay CH (1970). Pseudoxanthoma elasticum. Postgrad Med J 46: 97–108.

Trip MD, Smulders YM, Wegman JJ et al. (2002). Frequent mutation in the ABCC6 gene (R1141X) is associated with a strong increase in the prevalence of coronary artery disease. Circulation 106: 773–775.

Uitto J, Li Q, Jiang Q (2010). Pseudoxanthoma elasticum: molecular genetics and putative pathomechanisms. J Invest Dermatol 130: 661–670.

Vanakker OM, Martin L, Gheduzzi D et al. (2007). Pseudoxanthoma elasticum-like phenotype with cutis laxa and multiple coagulation factor deficiency represents a separate genetic entity. J Invest Dermatol 127: 581–587.

Wahlqvist ML, Fox RM, Beech AM et al. (1977). Peripheral vascular disease as a mode of presentation of pseudoxanthoma elasticum. Aust N Z J Med 7: 523–525.

Yasuhara T, Sugiu K, Kakishita M et al. (2004). Pseudoxanthoma elasticum with carotid rete mirabile. Clin Neurol Neurosurg 106: 114–117.

Yoo JY, Blum RR, Singer GK et al. (2011). A randomized controlled trial of oral phosphate binders in the treatment of pseudoxanthoma elasticum. J Am Acad Dermatol 65: 341–348.

Yu S, Ming A, Wegman A (2009). Pseudoxanthoma elasticum-like lesions in association with thalassaemia major. Australas J Dermatol 50: 186–189.

Handbook of Clinical Neurology, Vol. 132 (3rd series)
Neurocutaneous Syndromes
M.P. Islam and E.S. Roach, Editors

Chapter 16

Blue bleb rubber nevus syndrome

FANNY BALLIEUX[1], LAURENCE M. BOON[1*], AND MIIKKA VIKKULA[2]

[1]*Center for Vascular Anomalies, Division of Plastic and Reconstructive Surgery, Cliniques Universitaires Saint Luc, Université catholique de Louvain, Brussels, Belgium*

[2]*Laboratory of Human Molecular Genetics, de Duve Institute, Université catholique de Louvain, Brussels, Belgium*

INTRODUCTION

Vascular malformations are localized defects of vascular development and may affect arteries, capillaries, veins, or lymphatics. Multiple terms, such as "angioma" or "vascular birthmark," have been misused to define a heterogeneous group of vascular malformations generating diagnostic confusion. Notably, hemangioma was used to describe both vascular malformations and vascular tumors.

In 1982, Mulliken and Glowacki proposed a classification based on clinical, radiologic, histologic, and hemodynamic features to improve communication between specialists and to standardize management (Mulliken and Glowacki, 1982). The vascular anomalies were classified in two categories: vascular tumors and vascular malformations. This classification has since become even more precise and in 1996, it was accepted as the classification system by the International Society for the Study of Vascular Anomalies (ISSVA). In April 2014, during the 20th ISSVA workshop, the classification was updated and extended (http:/www.issva.org/classification). Most vascular anomalies (hemangioma, venous malformations, capillary malformations, lymphatic malformations, and arteriovenous malformations) can be correctly diagnosed and managed with this classification.

Vascular tumors, infantile hemangiomas, comprise the most common benign tumors of childhood and affect 10% of Caucasian neonates during the first year of life. There is female predominance. A hemangioma appears during the first month of life and develops following a three-phase cycle. The proliferating phase extends from age 6 months to 12 months. It is followed by an involuting phase with a spontaneous regression of the lesion, which becomes pale. Finally, by the age of 5–7 years, the hemangioma disappears: the involuted phase. Most hemangiomas involute without complication and do not require treatment. Hemangioma never appears in adulthood (Greene, 2011).

Vascular malformations are characterized by a focal increase in the number of vessels, which are abnormally tortuous and enlarged. They are usually present at birth but are not necessarily visible. They do not regress spontaneously. They may either be stable or expand. Vascular malformations are subdivided into fast-flow and slow-flow malformations, and may develop from arterial, venous, capillary, and lymphatic vessels. The lesions can be localized or diffuse, and can occur in any body part, including the viscera, the brain, or the rest of the neurologic system. Some cause cosmetic and others functional disabilities. Cutaneous lesions are the most frequently diagnosed due to visibility. Radiologic imaging is always required to characterize and delineate the lesion.

Venous malformations as part of a syndrome

Venous malformations (VM) are slow-flow malformations. Most are isolated, but some are part of a syndrome: *Klippel-Trénauney syndrome* (capillarolymphaticovenous malformation with overgrowth), *blue rubber bleb nevus syndrome* and *Maffucci syndrome* (association of spindle-cell hemangiomas with enchondromas) (Fig. 16.1).

About 5% of patients with VM have familial lesions. There are two known entities: *cutaneomucosal VM* (VMCM) (<1%) and *glomuvenous malformation* (GVM) (<5%) (see Fig. 16.1). The rest occur sporadically (Greene and Alomari, 2011).

*Correspondence to: Laurence M. Boon, MD, PhD, Division of Plastic Surgery, Center for Vascular Anomalies, Cliniques universitaires Saint Luc, Avenue Hippocrate 10, B-1200 Brussels, Belgium. Tel: +32-2-764-14-03, Fax: +32-2-762-62-84, E-mail: laurence.boon@uclouvain.be

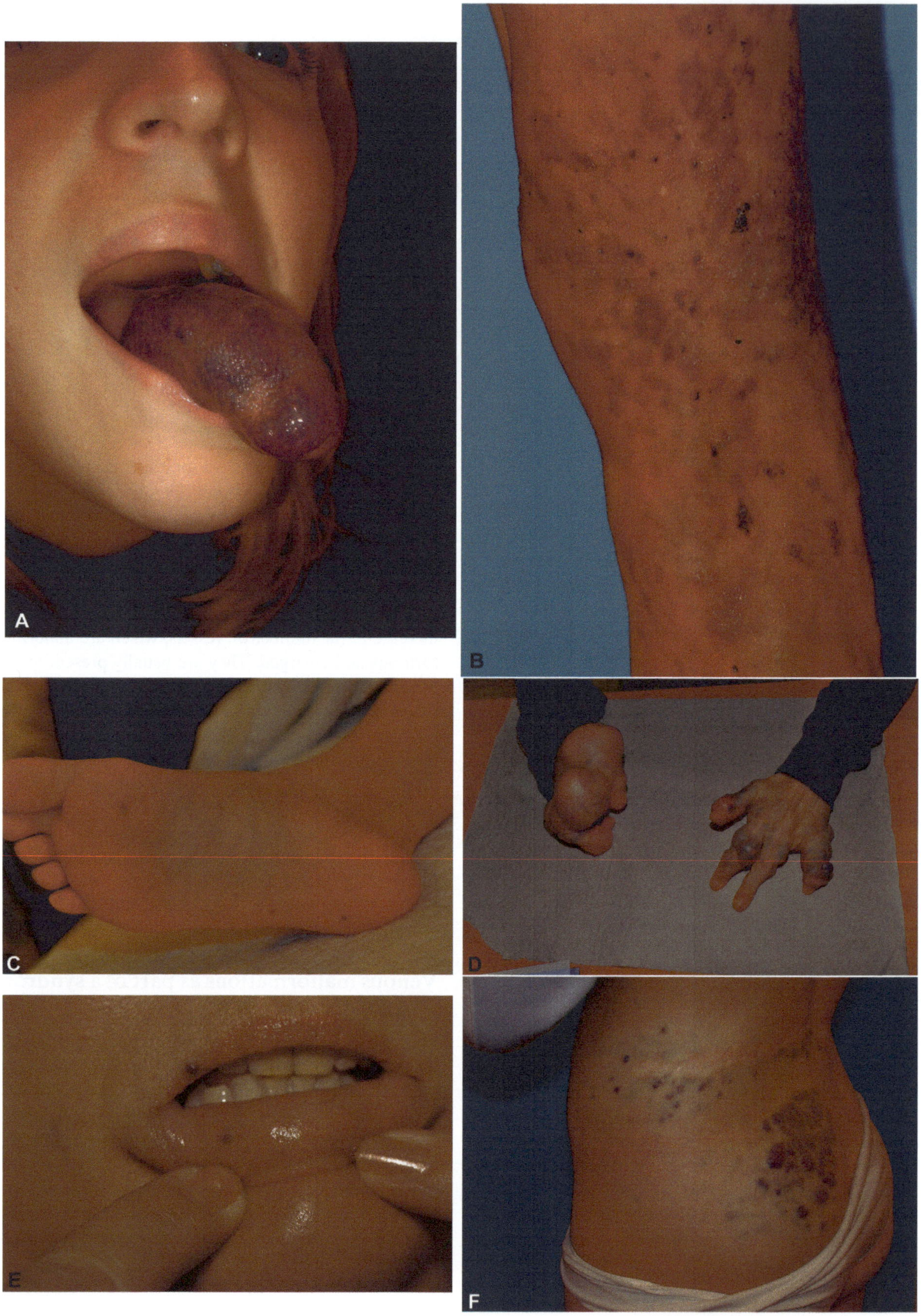

Fig. 16.1. (**A**) Common sporadic venous malformation of the tongue causing sleep apnea. (**B**) Combined capillarolymphaticovenous malformation with hypertrophy known as Klippel–Trénaunay syndrome. (**C**) Multifocal lesion on sole characteristic of blue rubber bleb nevus syndrome. (**D**) Maffucci syndrome. Note major deformation of right fingers. (**E**) Small mucosal venous malformations of the lip, typical of cutaneomucosal venous malformation (VMCM) phenotype. (**F**) Plaque-like glomuvenous malformation on back.

BLUE RUBBER BLEB NEVUS SYNDROME

Blue rubber bleb nevus syndrome (BRBN) is a rare vascular anomaly consisting of multifocal cutaneous venous malformations, especially located on the palms and soles, and associated with multiple gastrointestinal lesions. First reported by Gascoyen in 1860, this association was further characterized in 1958, by William Bennet Bean, who named the lesions "blue rubber bleb nevus" (Gascoyen, 1860; Bean, 1958). BRBN is sometimes referred to as Bean syndrome. To date, 269 cases of BRBN have been reported. The clinical literature is limited, especially regarding neurologic involvement.

Clinical presentation

Most patients are diagnosed at birth or within the first year of life. Females and males are affected equally. The lesions typically are small, circumscribed, and multifocal (see Fig. 16.1C). They are bluish to purple in color and rubbery on palpation, especially when located on palms and soles. Other VM lesions are compressible and refill slowly on release of pressure. The patient may present with a few to hundreds of cutaneous lesions.

Most lesions are present at birth, vary from 1 mm to 10 cm in diameter, and often increase in size and number with age. Most small VMs are asymptomatic. Patients with BRBN also often have one large VM that, depending on the size and location, can cause pain, anatomical dysfunction, and/or functional limitation. Hyperhidrosis may be noted. They rarely bleed spontaneously. They never undergo malignant transformation. The lesions principally affect the upper limbs, trunk, and perineum, although they may appear anywhere. Lesions of the lower limbs are unusual, except for the feet, and frequently associate with orthopedic involvement (Vanaman et al., 2010).

Orthopedic involvement is uncommon, but occurs especially in patients with acral cutaneous lesions. VM of the soft tissue may lead to bone deformity or hypertrophy and the intra-articular lesions may decrease the function of the joint. In some case, amputation of the limb may be required (McCarthy et al., 1982).

Gastrointestinal venous malformations occur commonly in the small intestines, but lesions may be found anywhere from the mouth to the anus (Fig. 16.2). They usually become evident in early adulthood. They typically cause chronic iron deficiency anemia and fatigue. These patients often need repetitive blood transfusions and are therefore at risk for hepatitis and allosensitization. Abdominal pain due to bowel intussusception or volvulus is also common as well as infarction or internal hemorrhage (Fishman et al., 2005).

VMs have also been described in the adrenal glands, kidneys, heart, lungs, bladder, penis, thyroid gland, spleen, ocular orbits, and muscles (Dobru et al., 2004). These lesions may result in a wide range of clinical features, such as menorrhagia, hematuria, epistaxis, hemoptysis, hemothorax, hemopericardium, thrombotic complications, spasticity, ataxia, and weakness (Fishman et al., 2005).

Central nervous system involvement is rare and leads to variable neurologic manifestations according to location and severity of the lesions. General physical examination including neurologic examination and a careful review of the patient's history is important. A Medline search reports 21 cases of BRBN with central nervous system involvement for 269 case reports.

Neurologic presentation can include vertigo, ataxia, aphasia, headache, ophthalmoplegia, lack of motor coordination, subarachnoid hemorrhage, or other lateralizing neurologic signs. Patient may present with a single seizure or epilepsy, and the seizure type may be variable including focal or myoclonic. Epilepsy is rare but can compromise cognitive and psychomotor status (Eiris-Punal et al., 2002). Kim and coworkers argue for an aggressive antiepileptic treatment to manage focal seizures in order to preserve normal psychomotor development (Kim, 2000).

Waybright and coworkers were the first to report a case of BRBN with central nervous system involvement. The patient was a 19-year-old man with bifrontal headaches, clonic movements, and dermatologic features of BRBN. The patient died in his sleep at home and postmortem findings delineated multiple vascular malformations in the skin, esophagus, right atrium, pericardium, pleura, and brain. Ectatic vessels were observed in the posterior regions of the brain. Multiple softened yellowish areas were seen along the surface of the posterior parietal and occipital lobes. In the region of the vein of Galen, an aneurysm of 2 cm in diameter, and containing a large thrombus, was found. The internal cerebral veins fed this bluish mass. Multiple capillary telangiectasias and "cavernous hemangiomas" were found in the cerebellum (Waybright et al., 1978).

Due to the slow flow of blood, thrombosis in the cutaneous lesions and the malformations of the cerebral surface and vein of Galen have been described (Waybright et al., 1978). In contrast, spontaneous thrombosis is rare for an arteriovenous malformation. Scattering of intravascular thrombi may induce strokes and dementia in BRBN (Waybright et al., 1978).

Therefore, although cerebral VM associated with BRBN are rare, the clinician has to keep in mind that given a patient with epilepsy or focal seizures associated with typical cutaneous or gastrointestinal (GI) lesions, brain lesions should be investigated. They present the

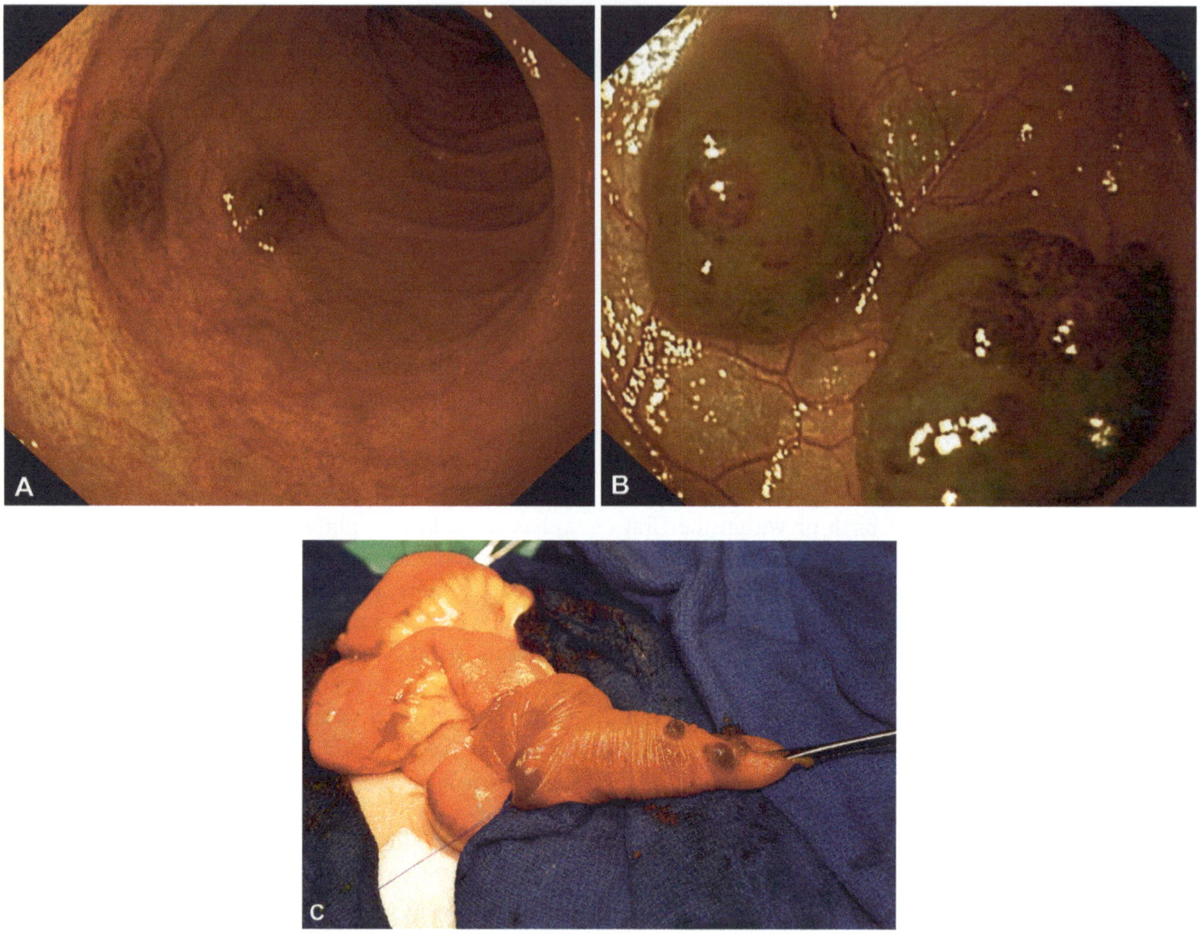

Fig. 16.2. A 10-year-old girl with blue rubber bleb nevus. (**A**) Gastroscopy identified several venous malformations on esophagus. (**B**) Gastroscopy identified several venous malformations on stomach. (**C**) Perioperative view showing several venous malformations on small intestine.

same tendency to thrombose as the cutaneous lesions. These cerebral VM should not be confused with developmental venous anomalies (DVA) that are an incidental finding, nonpathologic, and asymptomatic (Ruíz et al., 2009).

Histopathology and genetics

BRBN lesions histologically are composed of thin-walled ectatic veins. Thrombosis can be observed. There is no specific immunohistochemical marker to diagnose BRBN histologically.

BRBN is a sporadic disorder. The literature nevertheless contains reports of autosomal dominant transmission. Walshe and coworkers described two families with typical cutaneomucosal VMs without bleeding from the GI tract (Walshe et al., 1966). However, these most likely represent families with VMCM, an autosomal dominant cutaneomucosal venous malformation, and not BRBN.

Management

Investigation of patients with BRBN includes a systemic evaluation using blood count and coagulation tests, detection of fecal occult blood as a sign of gastrointestinal bleeding, and urine test to detect hematuria (Tzoufi et al., 2008). Positive occult blood test, anemia, and hematuria require investigation of the gastrointestinal tract to identify bleeding sites.

Blood tests can identify anemia, iron deficiency, thrombocytopenia, and disseminated intravascular coagulation (high D-dimer level with or without low fibrinogen level) which have been reported in BRBN. Dompmartin and coworkers reported two cases of BRBN with high D-dimer level and localized intravascular coagulopathy (Dompmartin et al., 2009, 2010).

Gastrointestinal investigations include abdominal radiography, which may detect phleboliths in intestinal lesions after thrombosis. Endoscopy is the best way to confirm diagnosis. Intestinal nodules appear bluish with normal mucosal walls (see Fig. 16.2). Upper GI endoscopy and colonoscopy are more useful than abdominal and barium radiography. Capsule endoscopy has recently been proposed as a new noninvasive endoscopic procedure for evaluating the small bowel, with a diagnostic rate of 50% in patients with occult bleeding. Magnetic resonance imaging (MRI) has been reported as a useful tool to screen asymptomatic family members (De Bona et al, 2005).

Diagnosis of BRBN involving the brain is based on electroencephalogram (EEG), computed tomography (CT), and MRI. An electroencephalogram may show spike discharges in the affected area. CT scan and angiogram of the brain vessels may reveal well-defined small densities that may or may not extend with contrast injection, characteristically with no mass effect or perilesional edema. These lesions have a tendency to calcify. Dilated vessels in the concerned area have also been described. Other reported findings include thrombosis or aneurysm of the vein of Galen (Walshe et al., 1966; Waybright et al., 1978; Gallmann and Boltshauser, 1987; Kim, 2000; Eiris-Punal et al., 2002; Tzoufi et al., 2008).

Therapeutic strategy

Treatment of cutaneous lesions is required when lesions are cosmetically unacceptable or functionally damaging. Laser treatments, surgical excision, and sclerotherapy are all potential treatment modalities. Pulsed-dye laser has been found to be the easiest way to remove hundreds of small superficial lesions without recurrence.

Symptomatic gastrointestinal lesions should be first treated conservatively with oral iron supplementation and blood transfusions, especially when there is occult blood loss. Management of serious chronic bleeding should be treated by endoscopic approaches with sclerotherapies and/or resection of VMs (Fishman et al., 2005). Systemic steroids, octreotide and interferon α-2a are currently not used as it has shown only limited success in managing the growth of the intestinal lesions (Boente et al., 1999; Gonzalez et al., 2001; Apak et al., 2004).

Extensive perineal VMs are best treated with several sessions of sclerotherapies using either foam aethoxysclerol 3% and/or bleomycin as sclerosing agent (Burrows, 2008).

Surgical resection of the GI lesions is recommended when conservative management fails, regardless of the number and location of lesions (Fig. 16.2) (Fishman et al., 2005). Complete eradication may avoid the need for repeated operations, but it is not always possible due to the extension of the lesion. Orthopedic lesions are treated with physical therapy, cast correction, surgical resection, and amputation if necessary (McCarthy et al., 1982).

Treatment of neurologic lesions depends on symptoms and localization of lesions. Focal seizures are treated with antiepileptic drugs, and intracranial hypertension or hemorrhage may require ventriculoperitoneal shunting. In most cases, no treatment is required, but follow-up with EEG and neurologic clinical examination is necessary.

Sirolimus (rapamycin) has been used successfully as an antiangiogenic agent to reduce the size of the VM lesion and the GI bleeding. One BRBN patient was successfully treated with low systemic dose of sirolimus (0.1 mg/kg). Chronic GI bleeding was controlled by the treatment and the patient's hemoglobin level rose significantly. There were no adverse effects (bone marrow alteration, renal dysfunction, increase of the cholesterol metabolism). This is consistent with the results obtained on sporadic venous malformations (Whiting et al., 1991; Yuksekkaya et al., 2012).

SPORADIC VENOUS MALFORMATION

Sporadic venous malformation (VM) is the most common cause for consultation at centers specializing in vascular anomalies. This is due to their propensity to cause disfigurement, bleeding, pain, and functional compromise. Within venous anomalies, sporadic VMs account for the large majority of patients (over 90%).

Clinical features

VMs are present at birth, but may become evident later in life. In contrast to vascular tumors, venous malformations generally grow with the child; they do not undergo involution. There is no sex-related difference in prevalence. VMs may be localized or extensive, single or sometimes multiple, and they are commonly located on the head and neck. They are inclined to enlarge with pregnancy and during puberty, and may be diagnosed only at this time. Patients can experience pain after activity or with hormonal and/or temperature variation. VMs have a characteristic bluish color (Fig. 16.1A) (Dompmartin et al., 2010).

Fifty percent of VM patients have associated localized intravascular coagulation (LIC) characterized by elevated D-dimer level with normal or low fibrinogen level. LIC often leads to acute pain due to the formation of thrombi which subsequently calcify into pathognomonic phleboliths. The latter are easily palpable on clinical examination and/or can be seen on plain X-ray. This localized coagulopathy can decompensate into disseminated intravascular coagulopathy (DIC) during surgery, causing severe hemorrhage due to consumption of coagulation factors. Therefore, patients with venous malformation and high D-dimer levels should be treated before any surgical procedure with low molecular weight heparin to avoid perioperative or postoperative complications (Dompmartin et al., 2008, 2009).

Histopathology and genetics

Histologically, VMs are composed of ectatic venous channels within the dermis. There is an uneven distribution of vascular smooth muscle cells around the lumen-forming endothelial cells. Somatic activating mutations in the TIE2 endothelial specific receptor account for 50% of sporadic venous malformations (Limaye et al., 2009; Soblet et al., 2013). The most common mutation is L914F, which accounts for 85% of the mutation-bearing lesions (Brouillard et al., 2013a). The aberrant TIE2 phosphorylation causes activation of PI3K-AKT signaling and inhibition of the transcription factor FOXO1 (Forkhead box O1). This results in diminished platelet-derived growth factor β (PDGF-β) production, likely playing a major role in the diminished recruitment of vascular smooth muscle cells (Uebelhoer et al., 2013).

Therapeutic strategy

Deformity and functional deficits may be treated conservatively by elastic compression. Low molecular weight heparin can be used to alleviate pain in case of associated high D-dimer levels, which may be an indicator of significant LIC or risk for disseminated intravascular coagulation. Sclerotherapy, sometimes followed by surgical resection, is the treatment of choice for most VMs. Some small lesions can benefit from complete surgical resection (Green and Alomari, 2011). Rapamycin seems to help patients with a VM refractory to standard care in a clinical trial (Boscolo et al., 2015). Bleeding is stopped, pain is reduced, functionality improves, and quality of life increases.

INHERITED VENOUS MALFORMATIONS

Venous lesions are seen in two inherited disorders: cutaneomucosal venous malformation (VMCM) and glomuvenous malformation (GVM) (Fig. 16.1E–F). Although the inheritance is autosomal dominant, there is large variability in expression. Thus, the parents may have such small lesions that they do not necessarily come to attention of the person. Therefore, familial history may be overlooked.

Cutaneomucosal venous malformation

Cutaneomucosal venous malformation (VMCM) accounts for 1% of venous malformations. Patients with VMCM classically have small (less than 5 cm) multifocal bluish lesions. Skin and oral mucosa are both involved, but in contrast to sporadic VM, joints, bones, and muscles and more rarely affected. Lesions can also be found in the gastrointestinal tract, lungs, and brain. Fifty percent of VMCM lesions are localized on the cervicofacial area and 37% on the extremities. As the lesions are small, they are often asymptomatic, and not painful on palpation. VMCM patients do not have the rubbery blue blebs on palms and soles, pathognomonic of BRBN (Dompmartin et al., 2010).

Histopathology and genetics

The lesions are inherited as an autosomal dominant trait with high penetrance. The mutated gene, *TIE2*, is located on the short arm of chromosome 9. VMCM is caused by an activating *TIE2* mutation found in a heterozygous state in all cells of the patient (Boon et al., 1994; Vikkula et al., 1996; Wouters et al., 2010). The most common mutation is an Arg849 to trypsin substitution. The multifocal localized lesions in VMCM likely result from somatic second hits on the normal allele. The single reported second hit was shown to result in loss of function, presumably abolishing the protective effect of the wild type allele (Limaye et al., 2009).

Therapeutic strategy

Treatment of VMCM is the same as for VM. Elastic compression can be used for large lesions. Low molecular weight heparin can be used to alleviate pain in case of associated high D-dimer levels. Small lesions can easily benefit from complete surgical resection (Greene and Alomari, 2011). Clinical trials are evaluating rapamycin as the first molecular therapy (Boscolo et al., 2015).

Glomuvenous malformation

GVMs, earlier called "glomangiomas", account for 5% of venous malformations. These lesions are bluish and multiple but, in contrast to VMCM, GVMs are often more purple and hyperkeratotic with a cobblestone appearance. They vary in size from a tiny point to a large "cobblestoned" area. They look like VMs and may be confused with them. In contrast to VMs, they are not compressible, and more likely involve the skin, rarely the mucosae or the muscle. D-dimer level is always normal in GVM, and thus a useful biomarker to differentiate GVM from VM (Boon et al., 2004).

Histopathology and genetics

More than 70% of GVMs are inherited as an autosomal dominant trait (Brouillard et al., 2005). This may be higher, as it was not always possible to evaluate family history. GVM is linked to 1p21-22 and is caused by loss of function of glomulin (Boon et al., 1999; Brouillard et al., 2002). In many lesions, the inherited mutation is duplicated and the wild type chromosomal copy is deleted, due to acquired uniparental isodisomy (Amyere et al., 2013). This results in complete localized loss of glomulin function (Brouillard et al., 2013b). Histologically, the lesions consist of large veins surrounded by abnormally differentiated vascular smooth muscle cells known as "glomus cells."

Therapeutic strategy

Treatment of GVM differs slightly from that of VMs. Pain is not alleviated but aggravated by elastic compression, and heparin injection is not useful either as there is no associated coagulopathy. The treatment of choice is surgical resection of lesions.

CONCLUSIONS

Central nervous system involvement in blue bleb rubber nevus syndrome is only rarely reported. The neurologic signs depend on the size and localization of the cerebral lesions. Diagnosis is based on clinical examination with characteristic blue, rubbery lesions on the skin associated with neurologic features and pathologic neurologic imaging. The involvement of multiple organ systems, such as the neurologic system, in BRBN underscores the need for careful examination for the various signs and symptoms, so that proper follow-up and timely management can be organized.

References

Amyere M, Aerts V, Brouillard P et al. (2013). Somatic uniparental isodisomy explains multifocality of glomuvenous malformations. Am J Hum Genet 92: 188–196.

Apak H, Celkan T, Ozkan A et al. (2004). Blue rubber bleb nevus syndrome associated with consumption coagulopathy: treatment with interferon. Dermatology (Basel, Switzerland) 208: 345–348.

Bean WB (1958). Blue bleb rubber nevi of the skin and gastrointestinal tract. In: CC Thomas (Ed.), Vascular Spiders and Related Lesions of the Skin. Oxford, Springfield, IL, pp. 17–185.

Boente MD, Cordisco MR, Frontini MD et al. (1999). Blue rubber bleb nevus (Bean syndrome): evolution of four cases and clinical response to pharmacologic agents. Pediatr Dermatol 16: 222–227.

Boon LM, Mulliken JB, Vikkula M et al. (1994). Assignment of a locus for dominantly inherited venous malformations to chromosome 9p. Hum Mol Genet 3: 1583–1587.

Boon LM, Brouillard P, Irrthum A et al. (1999). A gene for inherited cutaneous venous anomalies ("glomangiomas") localizes to chromosome 1p21-22. Am J Hum Genet 65: 125–133.

Boon LM, Mulliken JB, Enjolras O et al. (2004). Glomuvenous malformation (glomangioma) and venous malformation: distinct clinicopathologic and genetic entities. Arch Dermatol 140: 971–976.

Boscolo E, Limaye N, Huang L et al. (2015). Rapamycin improves TIE2-mutated venous malformation in murine model and human subjects. J Clin Invest 125 (9): 3491–3504.

Brouillard P, Boon LM, Mulliken JB et al. (2002). Mutations in a novel factor, glomulin, are responsible for glomuvenous malformations ("glomangiomas"). Am J Hum Genet 70: 866–874.

Brouillard P, Ghassibe M, Penington A et al. (2005). Four common glomulin mutations cause two thirds of glomuvenous malformations ("familial glomangiomas"): evidence for a founder effect. J Med Genet 42: e13.

Brouillard P, Boon LM, Revencu N et al. (2013a). Genotypes and phenotypes of 162 families with a glomulin mutation. Mol Syndromol 4: 157–164.

Brouillard P, Limaye N, Boon L et al. (2013b). Disorders of the venous system. In: DL Rimoin, RE Pyeritz, BR Korf (Eds.), Emery and Rimoin's Principles and Practice of Medical Genetics, 6th edn. 56, pp. 1–9.

Burrows PE (2008). Vascular malformations involving the female pelvis. Semin Intervent Radiol 25: 347–360.

De Bona M, Bellumat A, De Boni M (2005). Capsule endoscopy for the diagnosis and follow-up of blue rubber bleb nevus syndrome. Dig Liver Dis 37: 451–453.

Dobru D, Seuchea N, Dorin M et al. (2004). Blue rubber bleb nevus syndrome: case report and literature review. Rom J Gastroenterol 13: 237–240.

Dompmartin A, Acher A, Thibon P et al. (2008). Association of localized intravascular coagulopathy with venous malformations. Arch Dermatol 144: 873–877.

Dompmartin A, Ballieux F, Thibon P et al. (2009). Elevated D-dimer level in the differential diagnosis of venous malformations. Arch Dermatol 145: 1239–1244.

Dompmartin A, Vikkula M, Boon LM (2010). Venous malformation: update on aetiopathogenesis, diagnosis and management. Phlebology 25: 224–235.

Eiris-Punal J, Picon-Cotos M, Viso-Lorenzo A et al. (2002). Epileptic disorder as the first neurologic manifestation of blue rubber bleb nevus syndrome. J Child Neurol 17: 219–222.

Fishman SJ, Smithers CJ, Folkman J et al. (2005). Blue rubber bleb nevus syndrome: surgical eradication of gastrointestinal bleeding. Ann Surg 241: 523–528.

Gallmann T, Boltshauser E (1987). Blue rubber bleb nevus syndrome with CNS involvement. Klin Padiatr 199: 382–384.

Gascoyen M (1860). Case of naevus involving the parotid gland, and causing death from suffocation: naevi of the viscera. Trans Pathol Soc Lond 11: 267.

Gonzalez D, Elizondo BJ, Haslag S et al. (2001). Chronic subcutaneous octreotide decreases gastrointestinal blood loss in blue rubber-bleb nevus syndrome. J Pediatr Gastroenterol Nutr 33: 183–188.

Greene AK (2011). Management of hemangiomas and other vascular tumors. Clin Plast Surg 38: 45–63.

Greene AK, Alomari AI (2011). Management of venous malformations. Clin Plast Surg 38: 83–93.

Kim SJ (2000). Blue rubber bleb nevus syndrome with central nervous system involvement. Pediatr Neurol 22: 410–412.

Limaye N, Wouters V, Uebelhoer M et al. (2009). Somatic mutations in angiopoietin receptor gene TEK cause solitary and multiple sporadic venous malformations. Nat Genet 41: 118–124.

McCarthy JC, Goldberg MJ, Zimbler S (1982). Orthopaedic dysfunction in the blue rubber-bleb nevus syndrome. J Bone Joint Surg Am 64: 280–283.

Mulliken JB, Glowacki J (1982). Hemangiomas and vascular malformations in infants and children: a classification based on endothelial characteristics. Plast Reconstr Surg 69: 412–422.

Ruíz DS, Yilmaz H, Gailloud P (2009). Cerebral developmental venous anomalies: current concepts. Ann Neurol 66: 271–283.

Soblet J, Limaye N, Uebelhoer M et al. (2013). Variable somatic mutations in half of sporadic venous malformations. Mol Syndromol 4: 179–183.

Tzoufi MS, Sixlimiri P, Nakou I et al. (2008). Blue rubber bleb nevus syndrome with simultaneous neurological and skeletal involvement. Eur J Pediatr 167: 897–901.

Uebelhoer M, Natynki M, Kangas J et al. (2013). Venous malformation-causative TIE2 mutations mediate an AKT-dependent decrease in PDGFB. Hum Mol Genet 22: 3438–3448.

Vanaman MJ, Hervey-Jumper SL, Maher CO (2010). Pediatric and inherited neurovascular diseases. Neurosurg Clin N Am 21: 427–441.

Vikkula M, Boon LM, Carraway 3rd KL et al. (1996). Vascular dysmorphogenesis caused by an activating mutation in the receptor tyrosine kinase TIE2. Cell 87: 1181–1190.

Walshe MM, Evans CD, Warin RP (1966). Blue rubber bleb naevus. Br Med J 2: 931–932.

Waybright EA, Selhorst JB, Rosenblum WI, Suter CG (1978). Blue rubber bleb nevus syndrome with CNS involvement and thrombosis of a vein of Galen malformation. Ann Neurol 3: 464–467.

Whiting PH, Adam BJ, Woo J et al. (1991). The effect of rapamycin on renal function in the rat: a comparative study with cyclosporine. Toxicol Lett 58: 169–179.

Wouters V, Limaye N, Uebelhoer M et al. (2010). Hereditary cutaneomucosal venous malformations are caused by TIE2 mutations with widely variable hyper-phosphorylating effects. Eur J Hum Genet 18: 414–420.

Yuksekkaya H, Ozbek O, Keser M et al. (2012). Blue rubber bleb nevus syndrome: successful treatment with sirolimus. Pediatrics 129: e1080–e1084.

Handbook of Clinical Neurology, Vol. 132 (3rd series)
Neurocutaneous Syndromes
M.P. Islam and E.S. Roach, Editors

Chapter 17

Fabry disease

RAPHAEL SCHIFFMANN*
Institute of Metabolic Disease, Baylor Research Institute, Dallas, TX, USA

INTRODUCTION

History

Dermatologists Johannes Fabry and William Anderson first described "angiokeratoma corporis diffusum" in 1898 (Anderson, 1898; Fabry, 1898). It was recognized early as a systemic vascular disease and later as a storage disorder (Pompen et al., 1947) of lipids (Hornbostel and Scriba, 1953). The accumulation of the glycolipids ceramidetrihexoside (now called globotriaosylceramide (Gb_3 or GL-3)) and galabiosylceramide in a variety of different cell types was identified in 1963 (Sweeley and Klionsky, 1963). In addition, blood group antigens B, B1, and P1 also increase in certain individuals. The defect was established several years later as insufficient activity of the enzyme ceramidetrihexosidase which catalyzes the hydrolytic cleavage of the terminal molecule of galactose from Gb_3 (Brady et al., 1967). The anomeric specificity of ceramidetrihexosidase (α-galactosidase A) was determined in 1970 (Kint, 1970). The X-linked nature of the disease was first recognized in 1965 (Opitz et al., 1965).

Epidemiology

The disease incidence is about 1 in 117 000 live births for males (Meikle et al., 1999), although recent newborn screening surveys suggest that the incidence may be much higher, up to 1:3100 (Spada et al., 2006; Inoue et al., 2013; van der Tol et al., 2014). In Taiwan, the incidence of the relatively mild IVS4+919G>A mutation is 1:875 male live births and 1:399 female live births (Chien et al., 2012). Because of this recently discovered high incidence at birth, the nonspecific nature of the complications of Fabry disease and the common occurrence of single complications, it is likely that many undiagnosed patients exist. The incidence of Fabry disease is higher than in the general population in various at-risk populations, such as stroke in the young (Wozniak et al., 2010; De Brabander et al., 2013), hypertrophic cardiomyopathy of unknown cause (Sachdev et al., 2002; Chimenti et al., 2004; Elliott et al., 2011; Terryn et al., 2013), patients with chronic kidney disease (Nishino et al., 2012), and possibly even in patients with common heart disease (Schiffmann et al., 2014). Furthermore, the presence of equal numbers of females and males in large series suggests that up to 50% of the females with Fabry disease may be asymptomatic or are not identified (Mehta et al., 2004; Eng et al., 2007).

CLINICAL MANIFESTATIONS

Typical age of presentation, progression, life expectancy

Fabry disease may present at any age, in children and adults (Ries et al., 2005; Schiffmann et al., 2009). It is a progressive disease with a decreased life expectancy. Median survival is 50–55 years for males and 70 years for females (MacDermot et al., 2001; Branton et al., 2002; Vedder et al., 2007).

Hallmark signs and symptoms

Classic abnormalities

Patients with the classic form of the disease (with no residual α-galactosidase A activity). have typical dysmorphic abnormalities, particularly in the face. These dysmorphisms have been described quantitatively and in detail (Ries et al., 2006b). and include periorbital fullness, prominent lobules of the ears, bushy eyebrows, recessed forehead, pronounced nasal angle, generous nose/bulbous nasal tip, prominent supraorbital ridges, shallow midface, full lips, prominent nasal bridge, broad alar base, coarse features, posteriorly rotated ears, and

*Correspondence to: Raphael Schiffmann, MD, Institute of Metabolic Disease, Baylor Research Institute, 3812 Elm Street, Dallas, TX 75226, USA. Tel: +1-214-820-4533, Fax: +1-214-820-4853, E-mail: Raphael.Schiffmann@Baylorhealth.edu

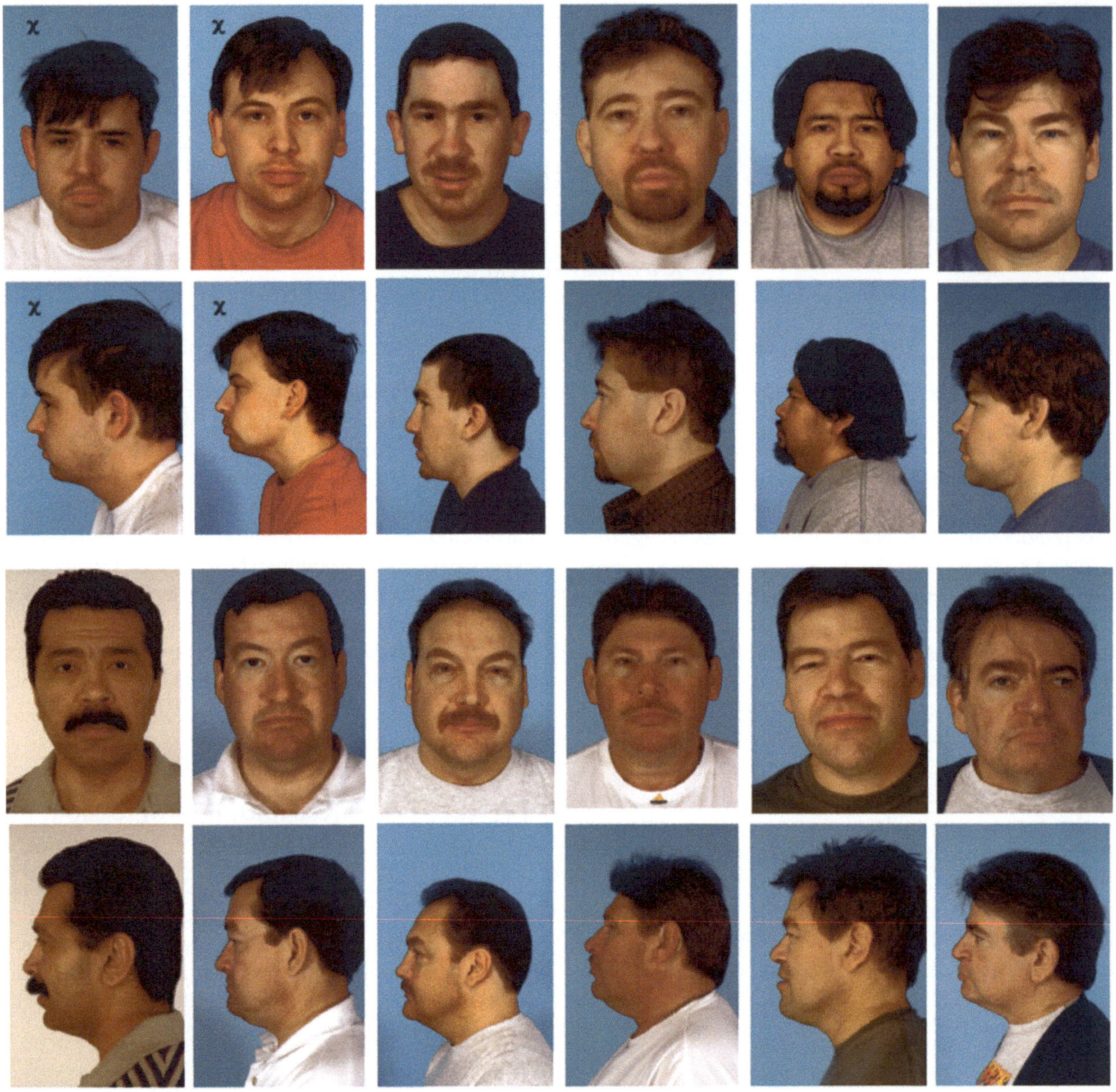

Fig. 17.1. Patients with Fabry disease, with frontal view of face directly above profile view of the same patient, arranged from youngest to oldest. X denotes cousins. (Reproduced from Ries et al., 2006b.)

prognathism (Fig. 17.1). Other abnormalities, including those of the extremities, have been identified (Ries et al., 2006b). The disease manifestations often start in childhood with episodes of extremity pain, fever of unknown origin, and hypohidrosis that often lead to decreased exercise tolerance (Cable et al., 1982a; Ries et al., 2005). Episodic diarrhea and abdominal pain are common, often associated with fatty foods (Rowe et al., 1974; Hoffmann and Keshav, 2007). These symptoms often reduce quality of life and school attendance in children but because of their relatively nonspecific nature usually do not lead to correct diagnosis in the absence of family history (Mehta et al., 2004; Ries et al., 2005). More specific disease manifestations that are usually present in late adolescence are typical vascular skin lesions termed angiokeratoma (Fig. 17.2A). and asymptomatic corneal opacities – cornea verticillata (Fig. 17.2B) (Weicksel, 1925; Hashimoto et al., 1965; van Mullem and Ruiter, 1970; Macrae et al., 1985; Yokota et al., 1995). These may lead to diagnosis by alert dermatologists or ophthalmologists. However, the importance of Fabry disease lies in the increased risk of developing a progressive renal insufficiency, cardiac abnormalities, and cerebrovascular stroke. These

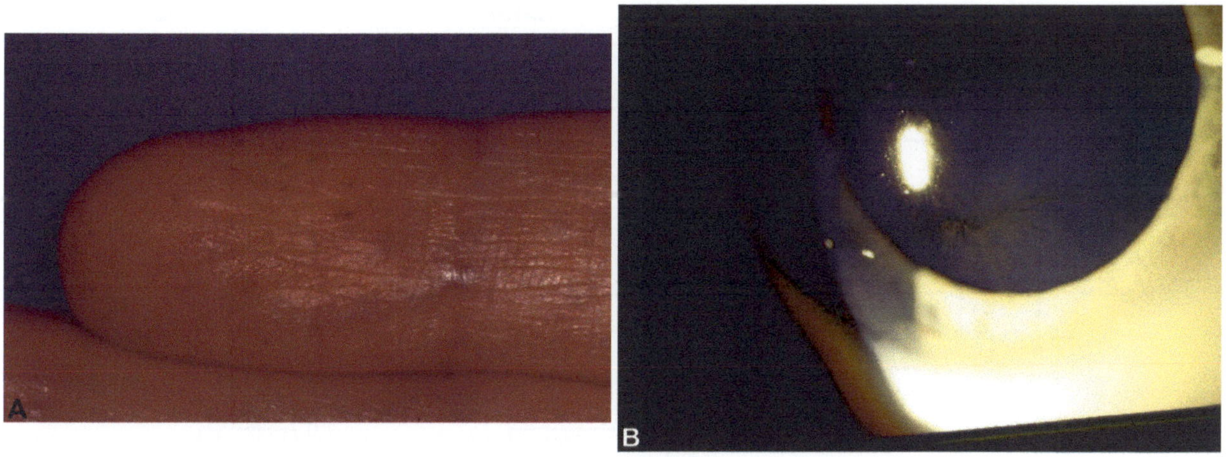

Fig. 17.2. (**A**) Angiokeratoma in fingers. (**B**) Cornea verticillata in a heterozygote female with Fabry disease. (Reproduced from Schiffmann, 2009, courtesy Drs Lorena Baccaglini and Janine Smith.)

complications may initially present themselves in the second decade of life, but more commonly in the third to fifth decade, resulting in decreased life expectancy (Branton et al., 2002; Mehta et al., 2004; Rolfs et al., 2005; Hopkin et al., 2008; Kampmann et al., 2008a).

The neuropathic pain (often referred to as acroparesthesia) is thought to be due to a length-dependent small fiber neuropathy (Attal and Bouhassira, 1999; Scott et al., 1999; Zimmermann, 2001). It often begins in childhood and is present in the vast majority of patients (MacDermot and MacDermot, 2001). It reaches its highest severity in the third and fourth decades of life and tends to diminish thereafter as the sensory function deteriorates. The pain may be continuous as well as episodic and often brought about by very low or very high environmental temperature, strenuous exercise, or stress. It usually begins in the feet, followed by the hands, and can become more generalized (Zarate and Hopkin, 2008). Joint pain is not uncommon in Fabry patients. The small-fiber neuropathy of Fabry disease is associated with an increased threshold of perception of cold and warm stimuli as well as heat pain (Dutsch et al., 2002; Luciano et al., 2002). Large fiber functions, such as vibration and position sensation, are usually intact and therefore peripheral nerve conduction velocity is typically normal unless compression neuropathy develops, particularly carpal tunnel syndrome (Fig. 17.3) (Scott et al., 1999; Luciano et al., 2002). Abnormalities suggesting a peripheral nerve dysfunction were also found in the Fabry mouse model (Rodrigues et al., 2009; Marshall et al., 2010). Progressive sensorineural hearing loss of early onset is very common in male and female patients with Fabry disease, and its severity is correlated with both the cerebrovascular and the peripheral nerve manifestations of the disease (Ries et al., 2007). This abnormality is often associated with disturbing tinnitus or episodes of vertigo that can begin in childhood (Conti and Sergi, 2003; Keilmann, 2003; Ramaswami et al., 2006; Vibert et al., 2006).

The kidney disease that complicates Fabry disease is usually associated with progressive proteinuria frequently associated with a decline in glomerular filtration rate, leading over a number of years to end-stage renal disease requiring dialysis and kidney transplantation (Donati et al., 1987; Branton et al., 2002; Ortiz et al., 2008; Schiffmann et al., 2009). There is evidence that hyperfiltration often precedes the decline in renal function, particularly in childhood (Ries et al., 2005; Vedder et al., 2007). Tubular dysfunction consists of hyposthenuria and polyuria that are nevertheless benign (Bichet, 2008).

A patient with Fabry disease may have virtually any cardiac complications (Linhart et al., 2000; Senechal and Germain, 2003). These include progressive hypertrophic cardiomyopathy (Fig. 17.4) with diastolic dysfunction, a variety of conduction defects and arrhythmias such as short P-R interval supraventricular and ventricular tachycardia. Other complications are atrial fibrillation as well as valvular disease (insufficiency or stenosis) and coronary vascular insufficiency of large or, more commonly, of small vessels (Weidemann et al., 2005; Kampmann et al., 2008b; Takenaka et al., 2008). Electrocardiogram is abnormal in most adults with Fabry disease and echography combined with a Doppler heart study is very helpful in characterizing the cardiac abnormalities of Fabry disease. Progressive bradycardia and decreased exercise capacity are very common, requiring the placement of a pacemaker in many patients (Nakayama et al., 1999; Lobo et al., 2008). Ejection fraction and cardiac output are often preserved but deteriorate in advanced cases (Kawano et al., 2007).

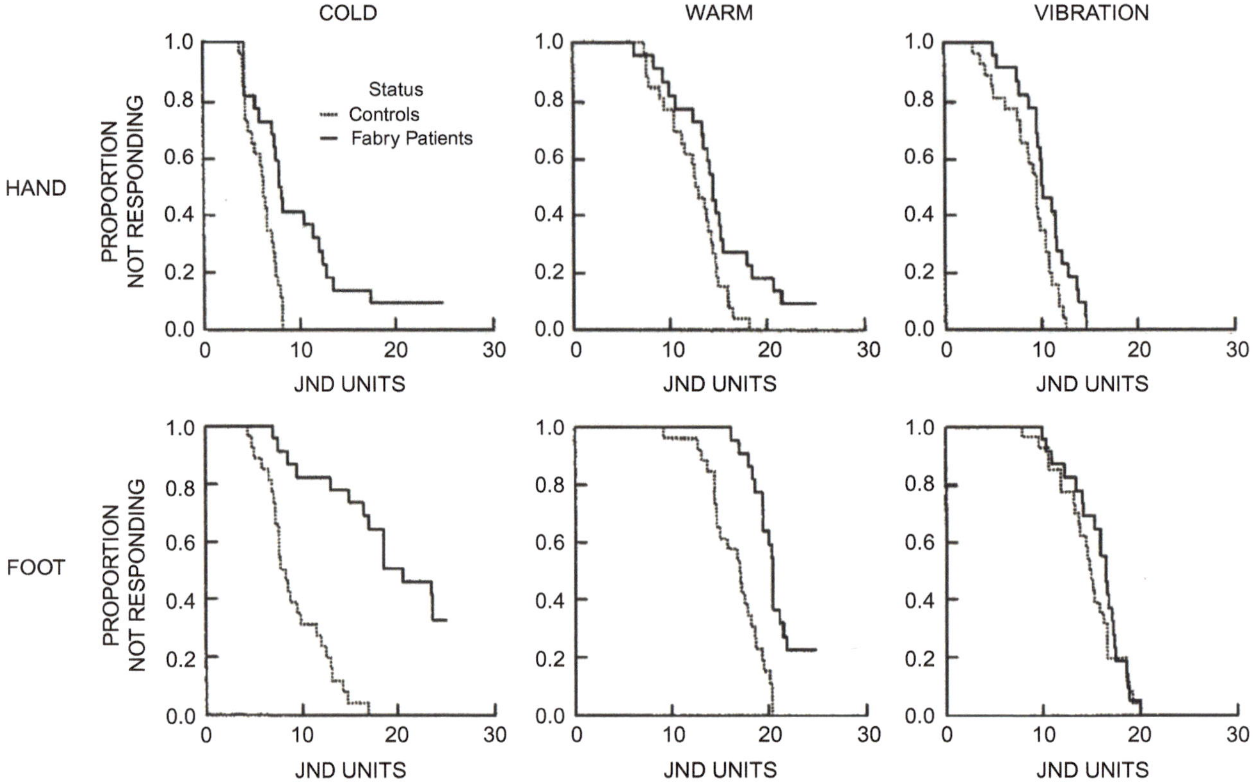

Fig. 17.3. Kaplan–Meier curves for patients and controls for the three sensory modalities in the hand and foot. Compared with controls, there are greater differences for cold than for warm sensation, and in the foot more than in the hand. Vibration was significantly different only for the hand. (Reproduced with permission from Luciano et al., 2002.)

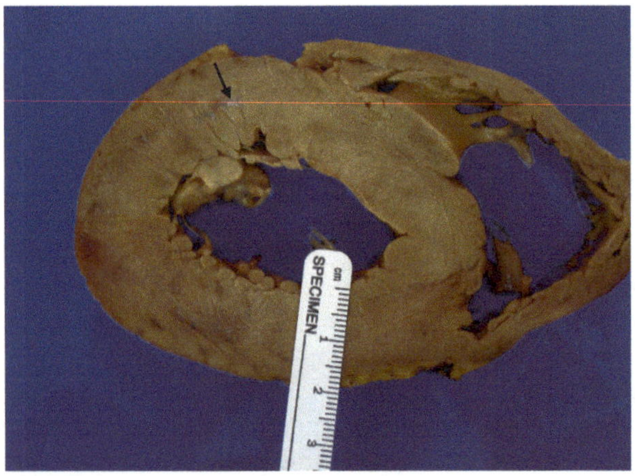

Fig. 17.4. Hypertrophic cardiomyopathy of the left ventricle in a 47-year-old male with Fabry disease who died after 2.5 years of enzyme replacement. Old scars from previous myocardial infarctions can be seen as well (arrow). The right ventricular wall is of normal thickness.

It is estimated that, depending on the age range cohort, patients with Fabry have a 5.5–12.2-fold increased risk of stroke compared to the general population (Sims et al., 2009). Both small and large vessel stroke occur, with brain regions perfused by the posterior circulation being affected more commonly than anterior circulation (Moore et al., 2001a, b, 2003b). The neurologic deficits in patients reflect the localization and extent of ischemic strokes and have no distinctive element (Fig. 17.5). Transient ischemic attacks are relatively common (Mitsias and Levine, 1996; Whybra et al., 2001; Mehta and Ginsberg, 2005; Moore et al., 2007a). Asymptomatic lesions on brain magnetic resonance imaging (MRI) are often present, usually in the white matter and may be a risk factor for vasculopathic complications in Fabry disease patients (Moore et al., 2003a; Kaneski et al., 2006). Large vessel embolic strokes of possible cardiac origin may occur (Mitsias and Levine, 1996).

Other abnormalities

A number of endocrine abnormalities have been described. The most common is antibody-negative hypothyroidism (Hauser et al., 2005; Faggiano et al., 2006). Asthenozoospermia, oligozoospermia and abnormal response to ACTH have also been described (Faggiano et al., 2006).

Other circulatory abnormalities can be found in Fabry disease. These include anemia, elevated sedimentation

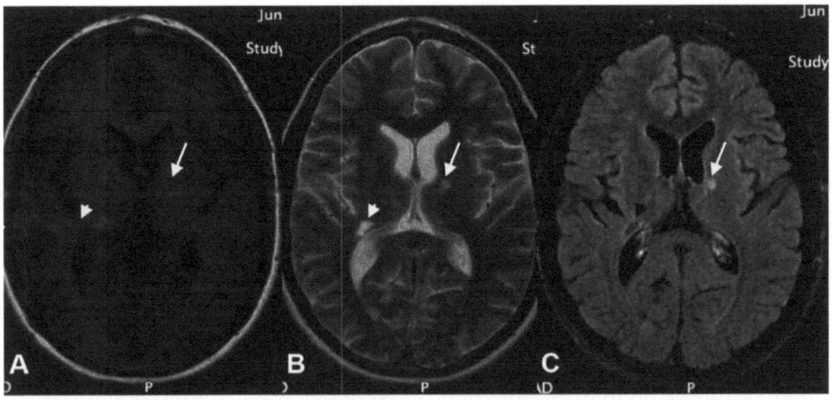

Fig. 17.5. Magnetic resonance image (MRI) of the brain of a 20-year-old man with a history of repeated strokes since age of 14 years, despite 3 years of enzyme replacement. (**A**) T1-weighted image. (**B**) T2-weighted image. (**C**) Fluid-attenuated inversion recovery image (FLAIR). Old lacunar infarcts are seen, including internal capsule (arrow), right thalamus, and a more recent lacunar infarct in the proximal right optic radiations (arrowhead). On the FLAIR image, a small amount of gliosis surrounds most of these lesions. (Reproduced from Schiffmann, 2009.)

rate, C-reactive protein, serum myeloperxidase (see also below)., and decreased α2-antiplasmin and plasminogen (Kleinert et al., 2005; Lacomis et al., 2005; Kaneski et al., 2006; Moore et al., 2007b). These findings define Fabry disease also as an inflammatory disorder. Any of these aberrations may serve as a biomarker in verifying the effect of therapy, alone or in combination.

Lower extremity edema commonly occurs in hemizygous males with Fabry disease. Abnormal lymphatic vessels have been described in one patient but it is likely that other vascular insufficiencies and cardiac disease contribute to this abnormality (Lozano et al., 1988; Chabas et al., 1994; Kampmann et al., 2005). Cardiac insufficiency is rarely associated with pedal edema but renal insufficiency is likely a contributing factor in some cases.

Effect of sex on clinical disease

Fabry disease can be considered an X-linked disorder with a high degree of intermediate penetrance in females. That term refers to the fact that only about 70% of females with *GLA* mutations have manifestations of Fabry disease while close to 100% of males have complications of the disease (Dobyns, 2006). Clinical symptoms are in part related to random X chromosome inactivation (Maier et al., 2006). Females are affected because of absent cross-correction between cells with normal α-galactosidase A activity (mutated X chromosome is inactivated) and enzyme-deficient cells (nonmutated X chromosome is inactivated) (Romeo and Migeon, 1970). Because of random X chromosome inactivation heterozygous females are essentially a "mosaic" of normal and mutant cells in varying proportions in various organ systems (Germain, 2010). The expression of the disease in female patients depends on the particular *GLA* mutation and especially on the pattern of X chromosome inactivation in each organ (Wang et al., 2007; Dobrovolny et al., 2005). Therefore, females can develop any of the complications that are seen in males, including strokes, cardiac disease and progressive renal insufficiency (Wang et al., 2007; Gupta et al., 2005; Ortiz et al., 2008; Laaksonen et al., 2008; Martin et al., 2007; Eng et al., 2007; Wilcox et al., 2008; Whybra et al., 2001). However, in general, the clinical abnormalities are more variable, less severe and of later onset compared to males with similar *GLA* mutations. Since large series of patients with Fabry disease usually have a roughly equal number of male and female patients, it is likely that a substantial number of female patients are not diagnosed or identified, but also that many females remain largely asymptomatic.

Disease modifiers

The clinical signs and symptoms described above can occur in any sequence and combination in any given patient with Fabry disease. A patient may have a number of complications in different organ systems, or just a few or even a single abnormality limited to one organ system. The clinical expression of Fabry disease is primarily modified by the residual α-galactosidase A activity (Desnick et al., 2001; Branton et al., 2002). For example, patients with some residual enzyme activity are often described as having a milder variant of Fabry disease with predominantly cardiac abnormalities, while having little or no kidney dysfunction and no painful acroparesthesia (Desnick et al., 2001). The onset of chronic renal insufficiency is significantly delayed in patients with more than 1% residual enzyme activity which results from having a *GLA* mutation consisting of a single amino acid change that is chemically conservative (Branton et al., 2002).

In addition to α-galactosidase A residual enzyme activity, it is very likely that there are many other genetic and nongenetic modifiers. Genotypes of polymorphisms G-174C of interleukin-6, G894T of endothelial nitric oxide synthase, factor V G1691A mutation (factor V Leiden), and the A-13G and G79A of protein Z all were associated significantly with the presence of presumably ischemic cerebral lesions on brain MRI (Altarescu et al., 2005). When the mouse model for Fabry disease was crossed with the mouse model for factor V Leiden mutation, the resulting double mutant had significantly increased vascular thrombi compared to either mouse model alone (Shen et al., 2006). Similarly, α-galactosidase A deficiency accelerated atherosclerosis in mice with apolipoprotein E deficiency (Bodary et al., 2005).

Fabry trait as a genetic risk factor

Patients with milder mutations and higher residual α-galactosidase A activity may have cardiac or renal manifestations only. The clinical expression in these patients of Fabry-related complications probably depends on the presence of other coexisting genetic, epigenetic and environmental modifiers (see also above). Furthermore, the clinical abnormalities described above are nonspecific in that they are clinically indistinguishable from similar complications of other etiologies that occur in the general population. For example, strokes in patients with Fabry disease may be of the small vessel or large vessel type and have similar neurologic and neuroimaging characteristics to hypertensive or embolic strokes (see Fig. 17.5) (Moore et al., 2007a). These features make *GLA* mutations a modifiable genetic risk factor, particularly for heart and kidney disease (Schiffmann et al., 2014).

Imaging

Patients with the classic form of the disease often have presumably ischemic white matter lesions on brain MRI that can be focal, multifocal and even confluent (Moore et al., 2003a). Rather specific are symmetric lesions in the posterior thalamus and pulvinar that have increased signal intensity on T1-weighted MRI image (Fig. 17.6) (Moore et al., 2003b; Takanashi et al., 2003; Burlina et al., 2008). These likely represent subtle dystrophic calcifications and end-organ damage associated with regional hyperperfusion (Moore et al., 2003b). Dystrophic calcifications can be observed in the gray–white matter junction, basal ganglia and cerebellum using MRI or computed tomography (CT) scan (Moore et al., 2003b). Fabry patients with strokes have typical findings on brain MRI that are no different from those found in ischemic strokes in patients who do not have Fabry disease (Mitsias and Levine, 1996). Elongation and distension of the vertebrobasilar artery (dolichoectasia) is commonly found in patients with Fabry disease (Garzuly et al., 2005; Fellgiebel et al., 2011), particularly in males (Uceyler et al., 2014).

Renal parapelvic cysts are rather typical in Fabry patients and can be diagnosed using ultrasound, CT or MRI (Glass et al., 2004; Ries et al., 2004). They do not seem to be associated with renal dysfunction and the mechanism of their development is not known. Cardiac hypertrophy with low T1 values (Sado et al., 2013; Thompson et al., 2013) or with late contrast enhancement on MRI are often seen in Fabry disease (Niemann et al., 2011).

Variants

(see also below)

Besides the classic form of the disease, patients with significant residual α-galactosidase A activity, up to 25% of normal values, often have a later onset, less severe form of the disease (von Scheidt et al., 1991; Branton et al., 2002; van der Tol et al., 2014). Often only one organ, most often the heart, is affected, which makes the diagnosis of Fabry disease particularly difficult.

Differential diagnoses to consider

The use of chloroquine or amiodarone can cause a corneal abnormality identical to the cornea verticillata of Fabry disease (D'Amico and Kenyon, 1981; Whitley et al., 1983; Inagaki et al., 1993). Exposure to silicon dust leads to a clinical and pathologic nephropathy that is very similar to the one seen in Fabry disease (Banks et al., 1983).

Cutaneous lesions identical to angiokeratomas occur in mannosidosis, fucosidosis, sialidosis, β-galactosidase deficiency, Schindler disease, and other disorders (Beratis et al., 1989; Gasparini et al., 1992; George and Graham-Brown, 1994; Calzavara-Pinton et al., 1995; Kawachi et al., 1998; Kodama et al., 2001; Suzuki et al., 2004). Isolated angiokeratomas without storage material have been described as well (Fimiani et al., 1997). One patient with Hodgkin lymphoma has developed angiokeratoma corporis diffusum that was thought to constitute a paraneoplastic manifestation (Han et al., 2013).

PATHOLOGY

The pathologic abnormalities can be divided into disease-specific and secondary changes that are not disease-specific but reflect organ abnormalities and

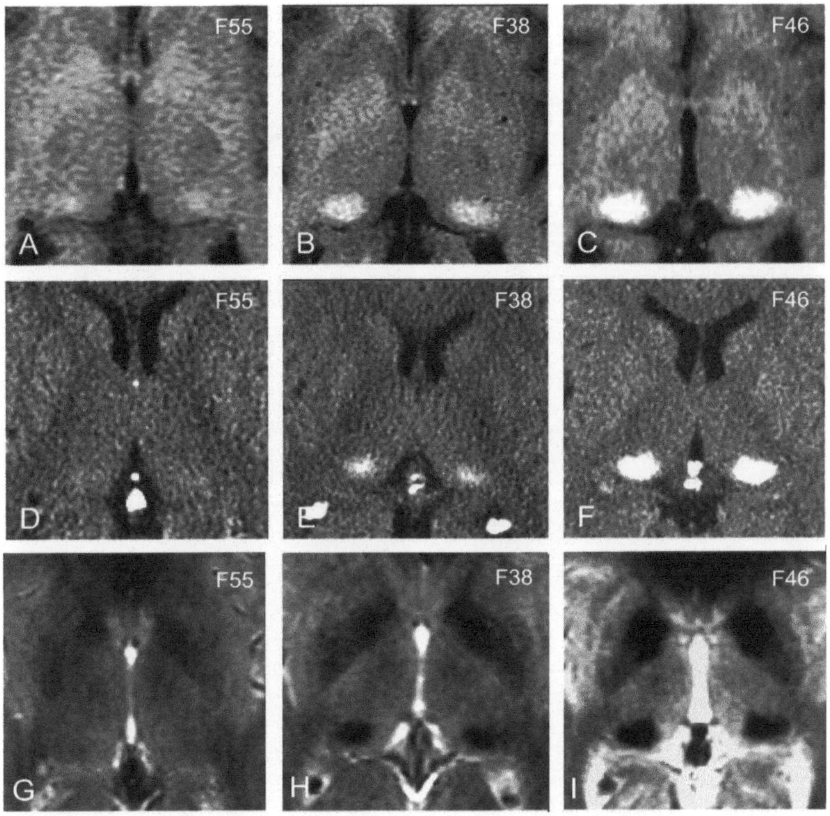

Fig. 17.6. Comparison of computed tomography (CT) and magnetic resonance imaging (MRI) findings in the posterior thalamus. (**A–C**) T1-weighted images through the thalamus in three patients with mild (**A**), moderate (**B**), and marked (**C**) hyperintensity, respectively. (**D–F**) Corresponding CT scans demonstrate increased attenuation, indicating calcification is present only in the moderate and marked cases. (**G–I**) Corresponding gradient-echo T2*-weighted images demonstrate that susceptibility-induced signal intensity loss is seen in the pulvinar in only the moderate and marked cases. (Numbers in upper right are patient identifiers.) (Reproduced with permission from Moore et al., 2003b, © American Society of Neuroradiology.)

dysfunction. Angiokeratoma consists of marked dilatation of the capillaries of the dermal papillae overlying keratotic epidermis (Massi et al., 2000). The most visually striking and historically important are lysosomal inclusions or lipid deposits that are seen in almost all cell types (Fig. 17.7). They are prominent in vascular cells, both endothelial and smooth muscle cells, cardiac cells including endocardial cells, cardiomyocytes and cardiac valves, kidney epithelial cells (tubular and glomerular cells and podocytes), brain (Fig. 17.8), and nerve cells, including dorsal root ganglia and some central nervous system neurons (Duncan, 1970; Gadoth and Sandbank, 1983; Nistal et al., 1983; Elleder, 2003).

The secondary pathologic changes are organ-specific but not necessarily disease-specific. Blood vessels may be thickened with a rather characteristic arteriosclerotic change that is different from typical atherosclerosis plaque (Case 2, 1984). True cardiac hypertrophy is often present with secondary fibrosis, and valves are often thickened (Schiffmann et al., 2005). The renal glomeruli undergo progressive change that starts with mesangial widening, followed by focal fibrosis ending with a completely fibrotic and obsolescent glomerulus (Chatterjee et al., 1984; Alroy et al., 2002). Tubular and interstitial fibrosis occurs as well. The brain may have rarefied and gliotic lesions secondary to ischemia, but spontaneous neuronal death and cerebral cortical atrophy have not been described (Tagliavini et al., 1982; Kaye et al., 1988; Schiffmann et al., 2005; Okeda and Nisihara, 2008).

MOLECULAR BASIS/PATHOPHYSIOLOGY

The α-galactosidase A gene (*GLA* – MIM no. 300644) is located on Xq22.1 (Bishop et al., 1988). It is 12 kb long with seven exons. *GLA* encodes for a 429 amino acid precursor protein that is processed to a 370 amino acid glycoprotein functioning as a homodimer (Garman and Garboczi, 2004). Based on the Human Gene Mutation Database at the Institute of Medical Genetics in Cardiff (http://www.hgmd.cf.ac.uk/ac/index.php), there are currently 596 mutations described. Of those, 416 are

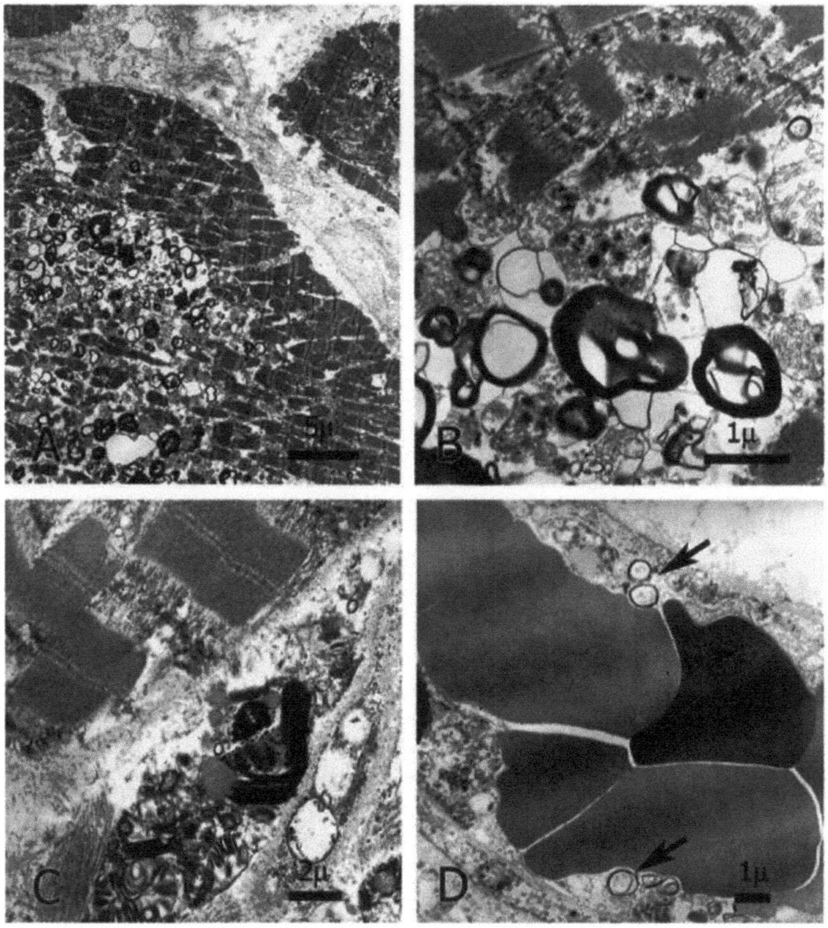

Fig. 17.7. Ultrastructure of the heart. (**A**) Large aggregates of concentric lamellated cytoplasmic inclusions displace cardiac myofibrils to the periphery of the cell. Interstitial fibrosis and increased numbers of mitochondria are secondary features of the cardiomyopathy. (**B**) Higher magnification of the lamellar inclusions and the accumulated mitochondria. The latter exhibit tubulovesicular cristae and cell size variation. (**C**) The myelinoid inclusions sometimes consisted of parallel stacked lamellae with a zebra-like pattern. (**D**) Inclusions were only rarely present in endothelial cells of adjacent interstitial capillaries (arrows). (Reproduced from Schiffmann, 2009.)

missense/nonsense type mutations, 83 small deletions, 19 large deletions, 32 splice defects, three complex rearrangements, and one large insertion. The cause of this large number of different mutations in the *GLA* gene is not known. One might speculate that having the Fabry trait presents a selective advantage such as resistance to certain types of bacterial infections, in particular those that express the *Escherichia coli* shiga-like toxin verotoxin (Cilmi et al., 2006). Patients with the classic, most severe form of Fabry disease almost always have a mutation that causes a total absence of α-galactosidase A activity, while patients with missense mutations often have some residual enzyme activity ranging from 2% to 25% (Desnick et al., 2001).

The mechanism by which α-galactosidase A deficiency and glycolipid accumulation cause such a wide variety of complications is not well understood. Based on the pathology of Fabry disease, the chronic accumulation of α-D-galactosyl moieties, particularly of Gb_3, appears to be a chronic toxicity state. There is no evidence of massive cell death although it is likely that there is increased turnover of some cell types such as vascular endothelial cells (Alroy et al., 2002). However, these findings do not describe the molecular cascades that lead to widespread cellular dysfunction in Fabry disease.

Over 150 years ago, Virchow described a triad of abnormalities (abnormal blood flow, abnormal vessel wall, and abnormal blood constituents) associated with thromboembolism formation (thrombogenesis) (Chung and Lip, 2003; Lowe, 2003; Bennett et al., 2009). The first two abnormalities are the most studied in Fabry disease. We previously demonstrated that patients with Fabry disease have abnormal functional vessel reactivity

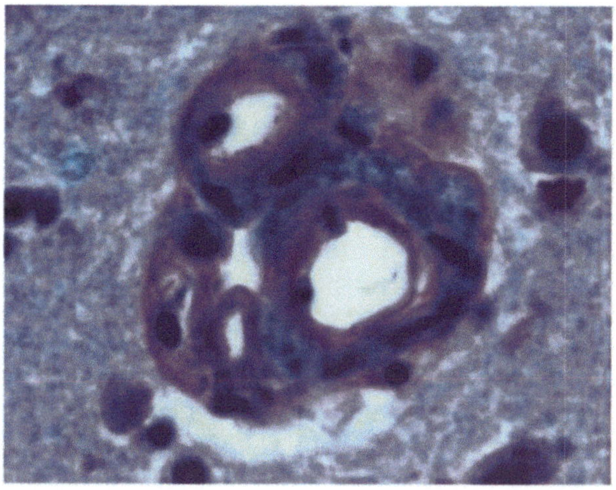

Fig. 17.8. Intracerebral blood vessels with Luxol fast blue (LFB)-stained deposits within vacuolated smooth muscle/adventitial cells. LFB-period acid Schiff (PAS), ×100 oil. (Reproduced from Schiffmann et al., 2006b, with permission from Springer Science and Business Media.)

secondary to endothelial dysfunction, together with cerebral hyperperfusion demonstrated using both the "gold standard" positron emission tomography (PET) and MR arterial spin labeling techniques (Altarescu et al., 2001; Moore et al., 2001b, 2004). Fabry patients also demonstrated delayed cerebral vasoreactivity, an elevated central nervous system (CNS) average diffusion constant and low serum ascorbate levels (Moore et al., 2002a, c, 2004). Furthermore, patients with Fabry disease have a dilated vasculopathy (arteriopathy and dolichoectasia) that is reflected in enlarged lumina of major intracranial vessels, particularly in the posterior circulation and the basilar artery (Fellgiebel et al., 2009). Vessel wall intimal medial thickness is also increased in association with smooth cell hypertrophy and presumed disturbed vessel mechanics (Rombach et al., 2012). This dilated vasculopathy may allow eddy formation and flow stagnation, resulting in an increased risk of artery-to-artery thromboembolism with development of clinical stroke. Small fiber neuropathy may also lead to changes in vascular reactivity and increased risk of ischemia (Stemper and Hilz, 2003).

Less attention has been paid to the role of blood constituents in the pathogenesis of the vasculopathy of Fabry disease. A likely prothrombotic state and an increased production of reactive oxygen species were found (DeGraba et al., 2000; Moore et al., 2001a; Shen et al., 2008). In the initial NIH randomized controlled trial, we found elevated 3-nitrotyrosine staining in dermal and cerebral blood vessel; this reduced significantly with enzyme replacement therapy (ERT), suggesting reduction in the potentially toxic effect of ROS secondary to peroxynitrite formation (Moore et al., 2001b). Evidence of activation of vascular endothelial cells and leukocytes, as well as a suggestion of abnormalities of fibrinolysis and angiogenesis factors, was observed in patients with Fabry disease (DeGraba et al., 2000; Kaneski et al., 2006; Gelderman et al., 2007; Moore et al., 2007b). However, direct functional measurement of the blood clotting system has not been examined systematically and, in particular, platelet function has not been studied. Chronic kidney insufficiency, which is common in patients with Fabry disease, is associated with high residual platelet functional reactivity despite dual antiplatelet therapy with both aspirin and clopidogrel (Muller et al., 2012; Aksu et al., 2013; Morel et al., 2013). Therefore, it is likely that patients with Fabry disease have dysfunctional platelets contributing to a hypercoagulable state (thrombophilia), increasing the risk of acute ischemic stroke. The findings by the Dutch Fabry group of a significant elevation of platelet activation markers – platelet factor 4 and β-thromboglobulin – in male and female patients also suggests the existence of platelet function abnormality in this disease (Vedder et al., 2009).

In addition to Gb_3, the deacylated form of this glycosphingolipid or lyso-Gb_3 (globotriaosylsphingosine) was recently found to be elevated in patients with Fabry disease, particularly in blood circulation, where it is over 50-fold higher than normal (Aerts et al., 2008). The plasma concentration of lyso-Gb_3 was in the nM range, while it was in the μM for Gb_3. Lyso-Gb_3 was found to be a potent inhibitor of α-galactosidase A and α-galactosidase B (N-acetylgalactosaminidase) and promoted smooth muscle cell proliferation *in vitro* at concentrations similar to the ones present in the plasma of Fabry patients (Aerts et al., 2008). This suggests a potential role of lyso-Gb_3 in the increased intima–media thickness seen in patients with Fabry disease. However, no correlation was noted between plasma lyso-Gb_3 with regard to age, the clinical severity score, or the cardiac mass in hemizygous male patients. Curiously, there was a correlation between plasma lyso-Gb_3 levels and cardiac mass in female heterozygotes. Lyso-Gb_3 levels were normal in patients with mutations leading to some residual activity and mainly cardiac manifestations, and they were also markedly elevated in the Fabry knockout mouse model. This animal model develops few Fabry-related phenotypic abnormalities such as small-fiber neuropathy and benign hypertrophic cardiomyopathy. Therefore, it remains to be seen whether lyso-Gb_3 plays an important role in the pathogenesis of Fabry disease. Thus far it seems that lyso-Gb_3 is helpful mainly in identifying patients with *GLA* mutations who have the classic form of Fabry disease (Niemann et al., 2014). Based on an extensive phenotypical characterization of the Fabry mouse model in comparison to wild-type control of identical genetic background, increased activity of androgen receptor (AR) signaling in Fabry disease was found (Shen et al., 2015). Blockade of AR signaling, either through castration or AR-antagonist, prevented and reversed cardiac and kidney hypertrophic phenotype (Shen et al., 2015).

ANIMAL MODELS

A knockout mouse model was developed a number of years ago (Ohshima et al., 1997, 1999). Although the mouse model only develops small-fiber peripheral neuropathy and hypertrophic cardiomyopathy, it does store Gb_3 and has been useful alone or when crossed with other models to study the pathogenesis of the disease and its therapy (see also above) (Ishii et al., 2004; Bodary et al., 2005; Shen et al., 2006; Park et al., 2008; Nguyen Dinh Cat et al., 2012; Kang et al., 2014). These included crossing the Fabry mouse with an apolipoprotein E deficient mouse showing that having both genetic defects accelerates atherosclerosis (Bodary et al., 2005), or introducing the factor V Leiden mutation into the Fabry mouse to show a synergistic interaction between α-galactosidase A and the clotting system to increase tissue fibrin deposition (Shen et al., 2006). More recently, a transgenic model with higher level of Gb_3 synthesis and deposition was created with apparently a useful renal phenotype (Taguchi et al., 2013).

GENETICS AND DIAGNOSIS

A presumed diagnosis of Fabry disease must be confirmed by the finding of low α-galactosidase A activity on peripheral blood white cells or cultured skin fibroblasts (Desnick et al., 2001). The latter usually have higher enzyme activity than peripheral white blood cells (Romeo et al., 1975). Generally, levels below 25% of normal should be considered diagnostic, and activity below 35% should lead to suspicion of Fabry disease (Desnick et al., 2001; Kitagawa et al., 2008). The enzyme assay is useful in males, but females often have mildly reduced or normal enzyme activity because of random X chromosome inactivation. Therefore, the finding of a mutated *GLA* is critical for confirmation of the diagnosis of Fabry disease in females (Wilcox et al., 2008). Increased urinary sediment Gb_3 measured in 24 h collection was thought to be useful in diagnosing Fabry disease, particularly in females who almost always have elevated levels (Cable et al., 1982b; Gupta et al., 2005). However, when whole urine is used, up to 40% of affected subjects may be missed (Fuller et al., 2005; Auray-Blais et al., 2008; Kitagawa et al., 2008). Furthermore, urinary Gb_3 may be elevated in patients with non-Fabry-related common heart disease (Schiffmann et al., 2014).

TREATMENT AND MANAGEMENT

Specific therapy

ERT is the first specific therapy for Fabry disease. It has been available since 2001, so it is a little early to reach any definitive conclusions as to whether this therapy can modify the natural history of Fabry disease. Two forms of α-galactosidase A for ERT are approved currently by regulatory authorities. These are agalsidase alfa (Replagal, Shire Human Genetic Therapies, Cambridge, MA, 0.2 mg/kg per infusion) and agalsidase beta (Fabrazyme, Genzyme Corporation, Cambridge, MA, 1 mg/kg per infusion). Both are approved in Europe and many other countries (Eng et al., 2001; Schiffmann et al., 2001), but in the US the Food and Drug Adminstration (FDA) approved only agalsidase beta (Eng et al., 2001). Both forms of the enzyme are usually administered every 2 weeks. No clear difference in clinical effect was demonstrated between the two enzyme preparations in a randomized controlled prospective study using either an identical dose or the approved dose (Vedder et al., 2008). The latter study was not randomized and has a number of flaws (Mehta et al., 2008).

One can describe the effect of ERT by looking at specific organ systems. Accumulating evidence suggests that ERT slows the decline of renal glomerular function (Schiffmann et al., 2006c, 2007; Banikazemi et al., 2007; Germain et al., 2007; Tahir et al., 2007). It may be particularly effective if initiated before kidney reserve is exhausted and glomerular filtration rate becomes significantly decreased (Banikazemi et al., 2007). In patients who have significant decline in kidney function despite ERT, weekly enzyme infusions may further slow the decline in the glomerular filtration rate (Schiffmann et al., 2007). The extent of proteinuria is an important factor in predicting the rate of decline in renal function and the response to ERT (Banikazemi et al., 2007).

We attempted to evaluate the effect of ERT on neuropathic pain. In the initial NIH sponsored 6 month randomized controlled study we found a significant reduction in pain scores, using the Brief Pain Inventory, in patients on ERT compared with placebo (Schiffmann et al., 2001). When the patients on placebo crossed over to receive agalsidase alfa, they exhibited a similar benefit (Schiffmann et al., 2003). When followed over a longer period of time, however, there was no further reduction in pain scores. For these studies, patients were selected for severe neuropathic pain, and neuropathic pain medication was stopped 1 week prior to pain scoring (Case 2, 1984). The clinical impression since then confirms these initial findings in adults (Guffon and Fouilhoux, 2004; Eto et al., 2005; Hoffmann et al., 2005) and children (Ries et al., 2006a). However, neuropathic pain is often not completely eliminated and patients commonly need to continue with their pain medication, albeit at a lower dose.

ERT with agalsidase alfa had no significant effect on warm and cold sensation over the 6 month randomized controlled trial (Schiffmann et al., 2001). Over the 3 years of open label treatment, however, there was a significant but modest reduction in the cold and warmth detection thresholds in the foot in both groups, and for warmth perception in the thigh (Schiffmann et al., 2003). There was also a trend toward reduction of cold detection thresholds in the hand. This effect took about 18 months to develop, and sensory function seemed to stabilize thereafter. Similar results, particularly regarding heat pain thresholds, were obtained by a group treating patients with agalsidase beta (Hilz et al., 2004). These authors also described an improvement in vibration detection thresholds. The functional improvement in cold perception of about 10% was not associated with an increase in epidermal innervation density (Schiffmann et al., 2006a).

We studied sweat function using the quantitative sudomotor sweat test (QSART) (Low and Opfer-Gehrking, 1999). As we did not have this technique at our disposal at the start of our initial randomized controlled study, the study of sweat function was started at the 3 year time point for this patient cohort. Sweat function was found to improve 24–72 hours after enzyme infusion compared with preinfusion values, while the QSART response normalized in four anhidrotic patients (Schiffmann et al., 2003). To date, however, many patients have remained anhidrotic despite years of ERT.

ERT reduced left ventricular mass in one small randomized controlled trial and in some noncontrolled studies, but there is so far no evidence that it can fundamentally change Fabry cardiac disease (Weidemann et al., 2003; Spinelli et al., 2004; Hughes et al., 2008). It is possible that once irreversible changes such as fibrosis occur, ERT is no longer effective. The clinical impression is that the cardiac abnormalities, including progressive bradycardia and conduction abnormalities, as well as valvular and coronary artery disease, continue to progress in many patients (Beer et al., 2006; Kalliokoski et al., 2006; Mougenot et al., 2008).

There has been no evidence so far for reduction of stroke risk with ERT in Fabry disease, although we found an improvement in the pattern of blood flow using $H_2{}^{15}O$ and PET (Moore et al., 2001b). The decrease in resting hyperperfusion was confirmed using transcranial Doppler and spin tagging MRI (Moore et al., 2002b, 2004). Improved acetazolamide response suggested a role for vascular endothelial cells in the CNS aspects of Fabry disease (Moore et al., 2002a). In pediatric patients we found increased circulating vascular endothelial microparticles that decreased to the normal range after 6 weeks of ERT, but the level of endothelial microparticles was below normal at the 12 months time point (see Fig. 17.7) (Gelderman et al., 2007). The significance of this finding is unclear at present (see also below).

Despite significant improvement in the function of the cerebral vasculature, four of 25 patients in our original study, followed for 4.5 years on ERT, developed nondebilitating strokes, and one patient had a transient ischemic attack (Schiffmann et al., 2006c). One stroke involved a large vessel (vertebral artery occlusion). and the others were small-vessel strokes. Based on our experience and that of others, strokes also continue to occur in patients on agalsidase beta (Wilcox et al., 2004; Buechner et al., 2008), and there is no indication of a marked reduction in stroke risk in the first few years following initiation of ERT in adulthood. A pediatric patient of ours continued to have strokes on ERT for a number of years (see Fig. 17.3) (Ries et al., 2006a). More recently, a number of meta-analytic studies showed little or no effect of ERT on Fabry disease (Alegra et al., 2012; El Dib et al., 2013; Rombach et al., 2014). The Canadian Fabry disease initiative confirmed that strokes continue on ERT. They found no significant difference in clinical outcome between the two approved enzyme preparations. Part of the difficulty in assessing the net effect of ERT is that it is virtually impossible to separate its effect from the impact of other standard of care medical therapies (Sirrs et al., 2014). Recent publications describing the long-term outcome of patients with Fabry disease on ERT confirmed its modest effect when initiated in adults (Alegra et al., 2012; El Dib et al., 2013; Anderson et al., 2014; Rombach et al., 2014; Sirrs et al., 2014).

The lack of an observed clinical effect of ERT, particularly on stroke, can be explained in a number of ways: First, there may be a small therapeutic effect that cannot be detected in the relatively small number of patients studied. Second, it is also possible that the infused α-galactosidase A does not have access to the entire thickness of cerebral vessels and therefore leaves untreated a significant component of the vessel wall (Murray et al., 2007). Third, the intermittent nature of ERT may be insufficient to prevent the chronic toxic process that is described in this review. In support of this hypothesis is the fact that low levels of enzyme activity in patients with mild mutations are sufficient to delay the onset of chronic insufficiency by at least one decade (Branton et al., 2002). The additional effect of weekly enzyme infusions further supports the need for continuous activity of α-galctosidase A in the cell (Schiffmann et al., 2007). Finally, the possibility of pre-existing irreversible structural changes in the vasculature may limit the effect of ERT when initiated in adulthood. It is unfortunate that we do not know whether earlier treatment will be more efficacious than later. A randomized controlled trial of early ERT versus delayed ERT in children with Fabry disease is probably the only way to answer that question with reliability.

ERT has been studied in children, but exclusively thus far in open-label clinical trials. It has been determined as safe and associated with reduced skin and urinary Gb_3, reduced neuropathic pain, fewer gastrointestinal symptoms, increased heart rate variability, stable renal function and cardiac structure and function (Ries et al., 2006a; Ramaswami, 2008; Wraith et al., 2008). However, one patient in our study continued to have strokes for a number of years on ERT (Ries et al., 2006a). As stated above, these studies cannot tell us when to initiate ERT in children in order to obtain maximal preventive effect.

Nonspecific therapy

Since Fabry disease causes medical complications that are nonspecific in nature and indistinguishable from similar complications occurring in the general population, it is expected that they will be responsive to standard medical and surgical care. This is indeed the case. Effective antiplatelet agents such as clopidogrel (or other ADP receptor inhibitors) or aspirin/long acting dipyridamole should be used to prevent strokes in all patients with Fabry disease, and in particular those with a family history of stroke. This is important since, as noted above, ERT on its own does not seem to appreciably reduce the incidence of stroke.

The kidney dysfunction in Fabry disease responds well to angiotensin converting enzyme inhibitors (ACEI) and angiotensin receptor blockers (ARB). They seem especially useful when used in conjunction with ERT with the goal of reducing proteinuria to less than 500 mg/24 h (Warnock et al., 2012). Normalization of blood pressure is very important to preserve kidney function and prevent other vascular events. Renal transplantation is as effective in Fabry disease patients who reached end-stage renal disease as in any other disorder (Ojo et al., 2000). Patients are increasingly transplanted early with kidneys from both live donors and cadavers (Ojo et al., 2000).

Neuropathic pain often is treated with relatively low doses of antiepileptic medications such as carbamazepine, neurontin, and lamotrigine (Tremont-Lukats et al., 2000; Ries et al., 2003; Horowitz, 2007). Nonsteroidal anti-inflammatory drugs are less effective and narcotics are effective but usually avoided (Gordon et al., 1995). Pancreatic enzymes are useful in the prevention of postprandial diarrhea in patients (personal observations).

UPCOMING TREATMENT OPTIONS

Other specific therapeutic approaches are being developed. Of particular interest are pharmacologic chaperones (Fan and Ishii, 2007). These are small molecules that are active-site inhibitors which, at low doses, can promote the normal folding and trafficking of misfolded wild-type and mutated proteins. Preliminary results of a phase 2 multicenter study showed that the drug AT1001 (migalastat hydrochloride) is safe and well tolerated (Schiffmann et al., 2008; Giugliani et al., 2013; Young-Gqamana et al., 2013). Significant increase in enzyme levels in peripheral white blood cells and kidney tissue was found which paralleled a decrease in urinary Gb_3. Two phase 3 studies have been completed and results will be reported in the near future. Substrate reduction (Marshall et al., 2010) and gene therapy (Pacienza et al., 2012) studies are planned for Fabry disease.

References

Aerts JM, Groener JE, Kuiper S et al. (2008). Elevated globotriaosylsphingosine is a hallmark of Fabry disease. Proc Natl Acad Sci U S A 105: 2812–2817.

Aksu HU, Oner E, Erturk M et al. (2013). Aspirin resistance in patients with impaired renal functions. Kardiol Pol 72: 331–338.

Alegra T, Vairo F, de Souza MV et al. (2012). Enzyme replacement therapy for Fabry disease: a systematic review and meta-analysis. Genet Mol Biol 35: 947–954.

Alroy J, Sabnis S, Kopp JB (2002). Renal pathology in fabry disease. J Am Soc Nephrol 13 (Suppl 2): S134–S138.

Altarescu G, Moore DF, Pursley R et al. (2001). Enhanced endothelium-dependent vasodilation in Fabry disease. Stroke 32: 1559–1562.

Altarescu G, Moore DF, Schiffmann R (2005). Effect of genetic modifiers on cerebral lesions in Fabry disease. Neurology 64: 2148–2150.

Anderson WA (1898). A case of angiokeratoma. Br J Dermatol 18: 113–117.

Anderson LJ, Wyatt KM, Henley W et al. (2014). Long-term effectiveness of enzyme replacement therapy in Fabry disease: results from the NCS-LSD cohort study. J Inherit Metab Dis 37: 969–978.

Attal N, Bouhassira D (1999). Mechanisms of pain in peripheral neuropathy. Acta Neurol Scand Suppl 173: 12–24.

Auray-Blais C, Cyr D, Ntwari A et al. (2008). Urinary globotriaosylceramide excretion correlates with the genotype in children and adults with Fabry disease. Mol Genet Metab 93: 331–340.

Banikazemi M, Bultas J, Waldek S et al. (2007). Agalsidase-beta therapy for advanced Fabry disease: a randomized trial. Ann Intern Med 146: 77–86.

Banks DE, Milutinovic J, Desnick RJ et al. (1983). Silicon nephropathy mimicking Fabry's disease. Am J Nephrol 3: 279–284.

Beer M, Weidemann F, Breunig F et al. (2006). Impact of enzyme replacement therapy on cardiac morphology and function and late enhancement in Fabry's cardiomyopathy. Am J Cardiol 97: 1515–1518.

Bennett PC, Silverman SH, Gill PS et al. (2009). Peripheral arterial disease and Virchow's triad. Thromb Haemost 101: 1032–1040.

Beratis NG, Varvarigou-Frimas A, Beratis S et al. (1989). Angiokeratoma corporis diffusum in GM1 gangliosidosis, type 1. Clin Genet 36: 59–64.

Bichet DG (2008). Vasopressin receptor mutations in nephrogenic diabetes insipidus. Semin Nephrol 28: 245–251.

Bishop DF, Kornreich R, Desnick RJ (1988). Structural organization of the human alpha-galactosidase A gene: further evidence for the absence of a 3′ untranslated region. Proc Natl Acad Sci U S A 85: 3903–3907.

Bodary PF, Shen Y, Vargas FB et al. (2005). Alpha-galactosidase A deficiency accelerates atherosclerosis in mice with apolipoprotein E deficiency. Circulation 111: 629–632.

Brady R, Gal AE, Bradley RM et al. (1967). Enzymatic defect in Fabry disease: ceramide trihexosidase deficiency. N Engl J Med 276: 1163–1167.

Branton MH, Schiffmann R, Sabnis SG et al. (2002). Natural history of Fabry renal disease: influence of alpha-galactosidase A activity and genetic mutations on clinical course. Medicine (Baltimore) 81: 122–138.

Buechner S, Moretti M, Burlina AP et al. (2008). Central nervous system involvement in Anderson–Fabry disease: a clinical and MRI retrospective study. J Neurol Neurosurg Psychiatry 79: 1249–1254.

Burlina AP, Manara R, Caillaud C et al. (2008). The pulvinar sign: frequency and clinical correlations in Fabry disease. J Neurol 255: 738–744.

Cable WJ, Kolodny EH, Adams RD (1982a). Fabry disease: impaired autonomic function. Neurology 32: 498–502.

Cable WJ, McCluer RH, Kolodny EH et al. (1982b). Fabry disease: detection of heterozygotes by examination of glycolipids in urinary sediment. Neurology 32: 1139–1145.

Calzavara-Pinton PG, Colombi M, Carlino A et al. (1995). Angiokeratoma corporis diffusum and arteriovenous fistulas with dominant transmission in the absence of metabolic disorders. Arch Dermatol 131: 57–62.

Case 2-1984 (1984). Fabry's disease. N Engl J Med 310: 1607.

Chabas A, Coll MJ, Aparicio M et al. (1994). Mild phenotypic expression of alpha-N-acetylgalactosaminidase deficiency in two adult siblings. J Inherit Metab Dis 17: 724–731.

Chatterjee S, Gupta P, Pyeritz RE et al. (1984). Immunohistochemical localization of glycosphingolipid in urinary renal tubular cells in Fabry's disease. Am J Clin Pathol 82: 24–28.

Chien YH, Lee NC, Chiang SC et al. (2012). Fabry disease: incidence of the common later-onset alpha-galactosidase A IVS4+919G–>A mutation in Taiwanese newborns – superiority of DNA-based to enzyme-based newborn screening for common mutations. Mol Med 18: 780–784.

Chimenti C, Pieroni M, Morgante E et al. (2004). Prevalence of Fabry disease in female patients with late-onset hypertrophic cardiomyopathy. Circulation 110: 1047–1053.

Chung I, Lip GY (2003). Virchow's triad revisited: blood constituents. Pathophysiol Haemost Thromb 33: 449–454.

Cilmi SA, Karalius BJ, Choy W et al. (2006). Fabry disease in mice protects against lethal disease caused by Shiga toxin-expressing enterohemorrhagic *Escherichia coli*. J Infect Dis 194: 1135–1140.

Conti G, Sergi B (2003). Auditory and vestibular findings in Fabry disease: a study of hemizygous males and heterozygous females. Acta Paediatr Suppl 92: 33–37, discussion 27.

D'Amico DJ, Kenyon KR (1981). Drug-induced lipidoses of the cornea and conjunctiva. Int Ophthalmol 4: 67–76.

De Brabander I, Yperzeele L, Ceuterick-De Groote C et al. (2013). Phenotypical characterization of alpha-galactosidase A gene mutations identified in a large Fabry disease screening program in stroke in the young. Clin Neurol Neurosurg 115: 1088–1093.

DeGraba T, Azhar S, Dignat-George F et al. (2000). Profile of endothelial and leukocyte activation in Fabry patients. Ann Neurol 47: 229–233.

Desnick RJ, Ioannou YA, Eng CM (2001). α-Galactosidase A deficiency: Fabry disease. In: CR Scriver, AL Beaudet, WS Sly et al. (Eds.), The Metabolic and Molecular Bases of Inherited Disease, 8th edn. McGraw–Hill, New York.

Dobrovolny R, Dvorakova L, Ledvinova J et al. (2005). Relationship between X-inactivation and clinical involvement in Fabry heterozygotes. Eleven novel mutations in the alpha-galactosidase A gene in the Czech and Slovak population. J Mol Med (Berl) 83: 647–654.

Dobyns WB (2006). The pattern of inheritance of X-linked traits is not dominant or recessive, just X-linked. Acta Paediatr Suppl 95: 11–15.

Donati D, Novario R, Gastaldi L (1987). Natural history and treatment of uremia secondary to Fabry's disease: an European experience. Nephron 46: 353–359.

Duncan C (1970). The renal lesion of angiokeratoma corporis diffusum (Fabry's disease): report of three cases including some electron microscopic findings. Pathology 2: 9–14.

Dutsch M, Marthol H, Stemper B et al. (2002). Small fiber dysfunction predominates in Fabry neuropathy. J Clin Neurophysiol 19: 575–586.

El Dib RP, Nascimento P, Pastores GM (2013). Enzyme replacement therapy for Anderson–Fabry disease. Cochrane Database Syst Rev 2. CD006663.

Elleder M (2003). Sequelae of storage in Fabry disease – pathology and comparison with other lysosomal storage diseases. Acta Paediatr Suppl 92: 46–53, discussion 45.

Elliott P, Baker R, Pasquale F et al. (2011). Prevalence of Anderson–Fabry disease in patients with hypertrophic cardiomyopathy: the European Anderson–Fabry Disease survey. Heart 97: 1957–1960.

Eng CM, Guffon N, Wilcox WR et al. (2001). Safety and efficacy of recombinant human alpha-galactosidase A–replacement therapy in Fabry's disease. N Engl J Med 345: 9–16.

Eng CM, Fletcher J, Wilcox WR et al. (2007). Fabry disease: baseline medical characteristics of a cohort of 1765 males and females in the Fabry Registry. J Inherit Metab Dis 30: 184–192.

Eto Y, Ohashi T, Utsunomiya Y et al. (2005). Enzyme replacement therapy in Japanese Fabry disease patients: the results of a phase 2 bridging study. J Inherit Metab Dis 28: 575–583.

Fabry J (1898). Ein Betrag zur Kenntnis der Ppurpura hammarrhagica nodularis. Arch Dermatol Syphilol 43: 187–200.

Faggiano A, Pisani A, Milone F et al. (2006). Endocrine dysfunction in patients with Fabry disease. J Clin Endocrinol Metab 91: 4319–4325.

Fan JQ, Ishii S (2007). Active-site-specific chaperone therapy for Fabry disease. Yin and Yang of enzyme inhibitors. Febs J 274: 4962–4971.

Fellgiebel A, Keller I, Marin D et al. (2009). Diagnostic utility of different MRI and MR angiography measures in Fabry disease. Neurology 72: 63–68.

Fellgiebel A, Keller I, Martus P et al. (2011). Basilar artery diameter is a potential screening tool for Fabry disease in young stroke patients. Cerebrovasc Dis 31: 294–299.

Fimiani M, Mazzatenta C, Rubegni P et al. (1997). Idiopathic angiokeratoma corporis diffusum. Clin Exp Dermatol 22: 205–206.

Fuller M, Sharp PC, Rozaklis T et al. (2005). Urinary lipid profiling for the identification of Fabry hemizygotes and heterozygotes. Clin Chem 51: 688–694.

Gadoth N, Sandbank U (1983). Involvement of dorsal root ganglia in Fabry's disease. J Med Genet 20: 309–312.

Garman SC, Garboczi DN (2004). The molecular defect leading to Fabry disease: structure of human alpha-galactosidase. J Mol Biol 337: 319–335.

Garzuly F, Marodi L, Erdos M et al. (2005). Megadolichobasilar anomaly with thrombosis in a family with Fabry's disease and a novel mutation in the alpha-galactosidase A gene. Brain 128: 2078–2083.

Gasparini G, Sarchi G, Cavicchini S et al. (1992). Angiokeratoma corporis diffusum in a patient with normal enzyme activities and Turner's syndrome. Clin Exp Dermatol 17: 56–59.

Gelderman MP, Schiffmann R, Simak J (2007). Elevated endothelial microparticles in Fabry children decreased after enzyme replacement therapy. Arterioscler Thromb Vasc Biol 27: e138–e139.

George S, Graham-Brown RA (1994). Angiokeratoma corporis diffusum in fucosidosis. J R Soc Med 87: 707.

Germain DP (2010). Fabry disease. Orphanet J Rare Dis 5: 30.

Germain DP, Waldek S, Banikazemi M et al. (2007). Sustained, long-term renal stabilization after 54 months of agalsidase beta therapy in patients with Fabry disease. J Am Soc Nephrol 18: 1547–1557.

Giugliani R, Waldek S, Germain DP et al. (2013). A phase 2 study of migalastat hydrochloride in females with Fabry disease: selection of population, safety and pharmacodynamic effects. Mol Genet Metab 109: 86–92.

Glass RB, Astrin KH, Norton KI et al. (2004). Fabry disease: renal sonographic and magnetic resonance imaging findings in affected males and carrier females with the classic and cardiac variant phenotypes. J Comput Assist Tomogr 28: 158–168.

Gordon KE, Ludman MD, Finley GA (1995). Successful treatment of painful crises of Fabry disease with low dose morphine. Pediatr Neurol 12: 250–251.

Guffon N, Fouilhoux A (2004). Clinical benefit in Fabry patients given enzyme replacement therapy – a case series. J Inherit Metab Dis 27: 221–227.

Gupta S, Ries M, Kotsopoulos S et al. (2005). The relationship of vascular glycolipid storage to clinical manifestations of Fabry disease: a cross-sectional study of a large cohort of clinically affected heterozygous women. Medicine 84: 261–268.

Han F, Wang P, Li Z et al. (2013). Angiokeratoma corporis diffusum in a patient with Hodgkin lymphoma: a new paraneoplastic skin manifestation? Br J Dermatol 168: 1351–1353.

Hashimoto K, Gross BG, Lever WF (1965). Angiokeratoma corporis diffusum (Fabry). Histochemical and electron microscopic studies of the skin. J Invest Dermatol 44: 119.

Hauser AC, Gessl A, Lorenz M et al. (2005). High prevalence of subclinical hypothyroidism in patients with Anderson–Fabry disease. J Inherit Metab Dis 28: 715–722.

Hilz MJ, Brys M, Marthol H et al. (2004). Enzyme replacement therapy improves function of C-, Adelta-, and Abeta-nerve fibers in Fabry neuropathy. Neurology 62: 1066–1072.

Hoffmann B, Keshav S (2007). Gastrointestinal symptoms in Fabry disease: everything is possible, including treatment. Acta Paediatr Suppl 96: 84–86.

Hoffmann B, Garcia de Lorenzo A, Mehta A et al. (2005). Effects of enzyme replacement therapy on pain and health related quality of life in patients with Fabry disease: data from FOS (Fabry Outcome Survey). J Med Genet 42: 247–252.

Hopkin RJ, Bissler J, Banikazemi M et al. (2008). Characterization of Fabry disease in 352 pediatric patients in the Fabry Registry. Pediatr Res 64: 550–555.

Hornbostel H, Scriba K (1953). Excision of skin in diagnosis of Fabry's angiokeratoma with cardio-vasorenal syndrome as phosphatide storage disease. Klin Wochenschr 31: 68–69.

Horowitz SH (2007). The diagnostic workup of patients with neuropathic pain. Med Clin North Am 91: 21–30.

Hughes DA, Elliott PM, Shah J et al. (2008). Effects of enzyme replacement therapy on the cardiomyopathy of Anderson–Fabry disease: a randomised, double-blind, placebo-controlled clinical trial of agalsidase alfa. Heart 94: 153–158.

Inagaki M, Katsumoto T, Nanba E et al. (1993). Lysosomal glycosphingolipid storage in chloroquine-induced alpha-galactosidase-deficient human endothelial cells with transformation by simian virus 40: in vitro model of Fabry disease. Acta Neuropathol 85: 272–279.

Inoue T, Hattori K, Ihara K et al. (2013). Newborn screening for Fabry disease in Japan: prevalence and genotypes of Fabry disease in a pilot study. J Hum Genet 58: 548–552.

Ishii S, Yoshioka H, Mannen K et al. (2004). Transgenic mouse expressing human mutant alpha-galactosidase A in an endogenous enzyme deficient background: a biochemical animal model for studying active-site specific chaperone therapy for Fabry disease. Biochim Biophys Acta 1690: 250–257.

Kalliokoski RJ, Kantola I, Kalliokoski KK et al. (2006). The effect of 12-month enzyme replacement therapy on myocardial perfusion in patients with Fabry disease. J Inherit Metab Dis 29: 112–118.

Kampmann C, Baehner FA, Whybra C et al. (2005). The right ventricle in Fabry disease. Acta Paediatr Suppl 94: 15–18, discussion 19–10.

Kampmann C, Linhart A, Baehner F et al. (2008a). Onset and progression of the Anderson–Fabry disease related cardiomyopathy. Int J Cardiol 130: 367–373.

Kampmann C, Wiethoff CM, Whybra C et al. (2008b). Cardiac manifestations of Anderson–Fabry disease in children and adolescents. Acta Paediatr 97: 463–469.

Kaneski CR, Moore DF, Ries M et al. (2006). Myeloperoxidase predicts risk of vasculopathic events in hemizgygous males with Fabry disease. Neurology 67: 2045–2047.

Kang JJ, Shu L, Park JL et al. (2014). Endothelial nitric oxide synthase uncoupling and microvascular dysfunction in the mesentery of mice deficient in alpha-galactosidase A. Am J Physiol Gastrointest Liver Physiol 306: G140–G146.

Kawachi Y, Matsu-ura K, Sakuraba H et al. (1998). Angiokeratoma corporis diffusum associated with galactosialidosis. Dermatology 197: 52–54.

Kawano M, Takenaka T, Otsuji Y et al. (2007). Significance of asymmetric basal posterior wall thinning in patients with cardiac Fabry's disease. Am J Cardiol 99: 261–263.

Kaye EM, Kolodny EH, Logigian EL et al. (1988). Nervous system involvement in Fabry's disease: clinicopathological and biochemical correlation. Ann Neurol 23: 505–509.

Keilmann A (2003). Inner ear function in children with Fabry disease. Acta Paediatr Suppl 92: 31–32, discussion 27.

Kint JA (1970). Fabry's disease: alpha-galactosidase deficiency. Science 167: 1268–1269.

Kitagawa T, Suzuki K, Ishige N et al. (2008). Non-invasive high-risk screening for Fabry disease hemizygotes and heterozygotes. Pediatr Nephrol 23: 1461–1471.

Kleinert J, Dehout F, Schwarting A et al. (2005). Anemia is a new complication in Fabry disease: data from the Fabry Outcome Survey. Kidney Int 67: 1955–1960.

Kodama K, Kobayashi H, Abe R et al. (2001). A new case of alpha-N-acetylgalactosaminidase deficiency with angiokeratoma corporis diffusum, with Meniere's syndrome and without mental retardation. Br J Dermatol 144: 363–368.

Laaksonen SM, Roytta M, Jaaskelainen SK et al. (2008). Neuropathic symptoms and findings in women with Fabry disease. Clin Neurophysiol 119: 1365–1372.

Lacomis D, Roeske-Anderson L, Mathie L (2005). Neuropathy and Fabry's disease. Muscle Nerve 31: 102–107.

Linhart A, Palecek T, Bultas J et al. (2000). New insights in cardiac structural changes in patients with Fabry's disease. Am Heart J 139: 1101–1108.

Lobo T, Morgan J, Bjorksten A et al. (2008). Cardiovascular testing in Fabry disease: exercise capacity reduction, chronotropic incompetence and improved anaerobic threshold after enzyme replacement. Intern Med J 38: 407–414.

Low PA, Opfer-Gehrking TL (1999). The autonomic laboratory. Am J Electroneurodiagn Technol 39: 65–76.

Lowe GD (2003). Virchow's triad revisited: abnormal flow. Pathophysiol Haemost Thromb 33: 455–457.

Lozano F, Garcia-Talavera R, Gomez-Alonso A (1988). An unusual cause of lymphoedema – confirmed by isotopic lymphangiography. Eur J Vasc Surg 2: 129–131.

Luciano CA, Russell JW, Banerjee T et al. (2002). Physiologic characterization of neuropathy in Fabry disease. Muscle Nerve 26: 622–629.

MacDermot J, MacDermot KD (2001). Neuropathic pain in Anderson–Fabry disease: pathology and therapeutic options. Eur J Pharmacol 429: 121–125.

MacDermot KD, Holmes A, Miners AH (2001). Anderson–Fabry disease: clinical manifestations and impact of disease in a cohort of 98 hemizygous males. J Med Genet 38: 750–760.

Macrae WG, Ghosh M, McCulloch C (1985). Corneal changes in Fabry's disease: a clinico-pathologic case report of a heterozygote. Ophthalmic Paediatr Genet 5: 185–190.

Maier EM, Osterrieder S, Whybra C et al. (2006). Disease manifestations and X inactivation in heterozygous females with Fabry disease. Acta Paediatr 95: 30–38.

Marshall J, Ashe KM, Bangari D et al. (2010). Substrate reduction augments the efficacy of enzyme therapy in a mouse model of Fabry disease. PLoS One 5: e15033.

Martin L, Sturgess A, Sillence D et al. (2007). Fabry disease in a heterozygote presenting as hand ischaemia and painful acroparaesthesia. Australas J Dermatol 48: 40–42.

Massi D, Martinelli F, Battini ML et al. (2000). Angiokeratoma corporis diffusum (Anderson–Fabry's disease): a case report. J Eur Acad Dermatol Venereol 14: 127–130.

Mehta A, Ginsberg L (2005). Natural history of the cerebrovascular complications of Fabry disease. Acta Paediatr Suppl 94: 24–27, discussion 29–10.

Mehta A, Ricci R, Widmer U et al. (2004). Fabry disease defined: baseline clinical manifestations of 366 patients in the Fabry Outcome Survey. Eur J Clin Invest 34: 236–242.

Mehta A, Beck M, Kampmann C et al. (2008). Enzyme replacement therapy in Fabry disease: comparison of agalsidase alfa and agalsidase beta. Mol Genet Metab 95: 114–115.

Meikle PJ, Hopwood JJ, Clague AE et al. (1999). Prevalence of lysosomal storage disorders. JAMA 281: 249–254.

Mitsias P, Levine SR (1996). Cerebrovascular complications of Fabry's disease. Ann Neurol 40: 8–17.

Moore DF, Herscovitch P, Schiffmann R (2001a). Selective arterial distribution of cerebral hyperperfusion in Fabry disease. J Neuroimaging 11: 303–307.

Moore DF, Scott LT, Gladwin MT et al. (2001b). Regional cerebral hyperperfusion and nitric oxide pathway dysregulation in Fabry disease: reversal by enzyme replacement therapy. Circulation 104: 1506–1512.

Moore DF, Altarescu G, Herscovitch P et al. (2002a). Enzyme replacement reverses abnormal cerebrovascular responses in Fabry disease. BMC Neurol 2: 4.

Moore DF, Altarescu G, Ling GS et al. (2002b). Elevated cerebral blood flow velocities in Fabry disease with reversal after enzyme replacement. Stroke 33: 525–531.

Moore DF, Schiffmann R, Ulug AM (2002c). Elevated CNS average diffusion constant in Fabry disease. Acta Paediatr Suppl 91: 67–68.

Moore DF, Altarescu G, Barker WC et al. (2003a). White matter lesions in Fabry disease occur in "prior" selectively hypometabolic and hyperperfused brain regions. Brain Res Bull 62: 231–240.

Moore DF, Ye F, Schiffmann R et al. (2003b). Increased signal intensity in the pulvinar on T1-weighted images: a pathognomonic MR imaging sign of Fabry disease. AJNR Am J Neuroradiol 24: 1096–1101.

Moore DF, Ye F, Brennan ML et al. (2004). Ascorbate decreases Fabry cerebral hyperperfusion suggesting a reactive oxygen species abnormality: an arterial spin tagging study. J Magn Reson Imaging 20: 674–683.

Moore DF, Kaneski CR, Askari H et al. (2007a). The cerebral vasculopathy of Fabry disease. J Neurol Sci 257: 258–263.

Moore DF, Krokhin OV, Beavis RC et al. (2007b). Proteomics of specific treatment-related alterations in Fabry disease: a strategy to identify biological abnormalities. Proc Natl Acad Sci U S A 104: 2873–2878.

Morel O, Muller C, Jesel L et al. (2013). Impaired platelet P2Y12 inhibition by thienopyridines in chronic kidney disease: mechanisms, clinical relevance and pharmacological options. Nephrol Dial Transplant 28: 1994–2002.

Mougenot P, Lidove O, Caillaud C et al. (2008). Fabry disease and treatment with agalsidase alpha: unsuspected cardiac arrhythmia in two heterozygous women. In reference to pharmacovigilance. Eur J Clin Pharmacol 64: 635–639.

Muller C, Caillard S, Jesel L et al. (2012). Association of estimated GFR with platelet inhibition in patients treated with clopidogrel. Am J Kidney Dis 59: 777–785.

Murray GJ, Anver MR, Kennedy MA et al. (2007). Cellular and tissue distribution of intravenously administered agalsidase alfa. Mol Genet Metab 90: 307–312.

Nakayama Y, Tsumura K, Yamashita N et al. (1999). Dynamic left ventricular arterial pressure gradient and sick sinus syndrome with heterozygous Fabry's disease improved following implantation of a dual chamber pacemaker. Pacing Clin Electrophysiol 22: 1114–1115.

Nguyen Dinh Cat A, Escoubet B, Agrapart V et al. (2012). Cardiomyopathy and response to enzyme replacement therapy in a male mouse model for Fabry disease. PLoS One 7: e33743.

Niemann M, Herrmann S, Hu K et al. (2011). Differences in Fabry cardiomyopathy between female and male patients: consequences for diagnostic assessment. JACC Cardiovasc Imaging 4: 592–601.

Niemann M, Rolfs A, Stork S et al. (2014). Gene mutations versus clinically relevant phenotypes: lyso-gb3 defines Fabry disease. Circ Cardiovasc Genet 7: 8–16.

Nishino T, Obata Y, Furusu A et al. (2012). Identification of a novel mutation and prevalence study for Fabry disease in Japanese dialysis patients. Ren Fail 34: 566–570.

Nistal M, Paniagua R, Picazo ML (1983). Testicular and epididymal involvement in Fabry's disease. J Pathol 141: 113–124.

Ohshima T, Murray GJ, Swaim WD et al. (1997). alpha-Galactosidase A deficient mice: a model of Fabry disease. Proc Natl Acad Sci U S A 94: 2540–2544.

Ohshima T, Schiffmann R, Murray GJ et al. (1999). Aging accentuates and bone marrow transplantation ameliorates metabolic defects in Fabry disease mice. Proc Natl Acad Sci U S A 96: 6423–6427.

Ojo A, Meier-Kriesche HU, Friedman G et al. (2000). Excellent outcome of renal transplantation in patients with Fabry's disease. Transplantation 69: 2337–2339.

Okeda R, Nisihara M (2008). An autopsy case of Fabry disease with neuropathological investigation of the pathogenesis of associated dementia. Neuropathology 28: 532–540.

Opitz JM, Stiles FC, Wise D et al. (1965). The genetics of angiokeratoma corporis diffusum (Fabry's disease), and its linkage with Xg (a) locus. Am J Hum Genet 17: 325–342.

Ortiz A, Oliveira JP, Waldek S et al. (2008). Nephropathy in males and females with Fabry disease: cross-sectional description of patients before treatment with enzyme replacement therapy. Nephrol Dial Transplant 23: 1600–1607.

Pacienza N, Yoshimitsu M, Mizue N et al. (2012). Lentivector transduction improves outcomes over transplantation of human HSCs alone in NOD/SCID/Fabry mice. Mol Ther 20: 1454–1461.

Park JL, Whitesall SE, D'Alecy LG et al. (2008). Vascular dysfunction in the alpha-galactosidase A-knockout mouse is an endothelial cell-, plasma membrane-based defect. Clin Exp Pharmacol Physiol 35: 1156–1163.

Pompen AW, Ruiter M, Wyers HJ (1947). Angiokeratoma corporis diffusum (universale) Fabry, as a sign of an unknown internal disease; two autopsy reports. Acta Med Scand 128: 234–255.

Ramaswami U (2008). Fabry disease during childhood: clinical manifestations and treatment with agalsidase alfa. Acta Paediatr Suppl 97: 38–40.

Ramaswami U, Whybra C, Parini R et al. (2006). Clinical manifestations of Fabry disease in children: data from the Fabry Outcome Survey. Acta Paediatr 95: 86–92.

Ries M, Mengel E, Kutschke G et al. (2003). Use of gabapentin to reduce chronic neuropathic pain in Fabry disease. J Inherit Metab Dis 26: 413–414.

Ries M, Bettis KE, Choyke P et al. (2004). Parapelvic kidney cysts: a distinguishing feature with high prevalence in Fabry disease. Kidney Int 66: 978–982.

Ries M, Gupta S, Moore DF et al. (2005). Pediatric Fabry disease. Pediatrics 115: e344–e355.

Ries M, Clarke JT, Whybra C et al. (2006a). Enzyme-replacement therapy with agalsidase alfa in children with Fabry disease. Pediatrics 118: 924–932.

Ries M, Moore DF, Robinson CJ et al. (2006b). Quantitative dysmorphology assessment in Fabry disease. Genet Med 8: 96–101.

Ries M, Kim HJ, Zalewski CK et al. (2007). Neuropathic and cerebrovascular correlates of hearing loss in Fabry disease. Brain 130: 143–150.

Rodrigues LG, Ferraz MJ, Rodrigues D et al. (2009). Neurophysiological, behavioral and morphological abnormalities in the Fabry knockout mice. Neurobiol Dis 33: 48–56.

Rolfs A, Bottcher T, Zschiesche M et al. (2005). Prevalence of Fabry disease in patients with cryptogenic stroke: a prospective study. Lancet 366: 1794–1796.

Rombach SM, van den Bogaard B, de Groot E et al. (2012). Vascular aspects of Fabry disease in relation to clinical manifestations and elevations in plasma globotriaosylsphingosine. Hypertension 60: 998–1005.

Rombach SM, Smid BE, Linthorst GE et al. (2014). Natural course of Fabry disease and the effectiveness of enzyme replacement therapy: a systematic review and meta-analysis: effectiveness of ERT in different disease stages. J Inherit Metab Dis 37: 341–352.

Romeo G, Migeon BR (1970). Genetic inactivation of the alpha-galactosidase locus in carriers of Fabry's disease. Science 170: 180–181.

Romeo G, D'Urso M, Pisacane A et al. (1975). Residual activity of alpha-galactosidase A in Fabry's disease. Biochem Genet 13: 615–628.

Rowe JW, Gilliam JI, Warthin TA (1974). Intestinal manifestations of Fabry's disease. Ann Intern Med 81: 628–631.

Sachdev B, Takenaka T, Teraguchi H et al. (2002). Prevalence of Anderson–Fabry disease in male patients with late onset hypertrophic cardiomyopathy. Circulation 105: 1407–1411.

Sado DM, White SK, Piechnik SK et al. (2013). Identification and assessment of Anderson–Fabry disease by cardiovascular magnetic resonance noncontrast myocardial T1 mapping. Circ Cardiovasc Imaging 6: 392–398.

Schiffmann R (2009). Fabry disease. Pharmacol Therapeut 122: 65–77.

Schiffmann R, Kopp JB, Austin 3rd HA et al. (2001). Enzyme replacement therapy in fabry disease: a randomized controlled trial. JAMA 285: 2743–2749.

Schiffmann R, Floeter MK, Dambrosia JM et al. (2003). Enzyme replacement therapy improves peripheral nerve and sweat function in Fabry disease. Muscle Nerve 28: 703–710.

Schiffmann R, Rapkiewicz A, Abu-Asab M et al. (2005). Pathological findings in a patient with Fabry disease who died after 2.5 years of enzyme replacement. Virchows Arch Nov 29: 1–7.

Schiffmann R, Hauer P, Freeman B et al. (2006a). Enzyme replacement therapy and intraepidermal innervation density in Fabry disease. Muscle Nerve 34: 53–56.

Schiffmann R, Rapkiewicz A, Abu-Asab M et al. (2006b). Pathological findings in a patient with Fabry disease who died after 2.5 years of enzyme replacement. Virchows Arch 448: 337–343.

Schiffmann R, Ries M, Timmons M et al. (2006c). Long-term therapy with agalsidase alfa for Fabry disease: safety and effects on renal function in a home infusion setting. Nephrol Dial Transplant 21: 345–354.

Schiffmann R, Askari H, Timmons M et al. (2007). Weekly enzyme replacement therapy may slow decline of renal function in patients with Fabry disease who are on long-term biweekly dosing. J Am Soc Nephrol 18: 1576–1583.

Schiffmann R, Germain DP, Castelli J et al. (2008). Phase 2 clinical trials of the pharmacological chaperone AT1001 for the treatment of Fabry disease. Paper presented at the Annual Meeting of the America Society of Human Genetics.

Schiffmann R, Warnock DG, Banikazemi M et al. (2009). Fabry disease: progression of nephropathy, and prevalence of cardiac and cerebrovascular events before enzyme replacement therapy. Nephrol Dial Transplant 24: 2102–2111.

Schiffmann R, Forni S, Swift C et al. (2014). Risk of death in heart disease is associated with elevated urinary globotriaosylceramide. J Am Heart Assoc 3: e000394.

Scott LJ, Griffin JW, Luciano C et al. (1999). Quantitative analysis of epidermal innervation in Fabry disease. Neurology 52: 1249–1254.

Senechal M, Germain DP (2003). Fabry disease: a functional and anatomical study of cardiac manifestations in 20 hemizygous male patients. Clin Genet 63: 46–52.

Shen Y, Bodary PF, Vargas FB et al. (2006). Alpha-galactosidase A deficiency leads to increased tissue fibrin deposition and thrombosis in mice homozygous for the factor V Leiden mutation. Stroke 37: 1106–1108.

Shen JS, Meng XL, Moore DF et al. (2008). Globotriaosylceramide induces oxidative stress and up-regulates cell adhesion molecule expression in Fabry disease endothelial cells. Mol Genet Metab 95: 163–168.

Shen JS, Meng XL, Wight-Carter M et al. (2015). Blocking hyperactive androgen receptor signaling ameliorates cardiac and renal hypertrophy in Fabry mice. Hum Mol Genet 24: 3181–3191.

Sims K, Politei J, Banikazemi M et al. (2009). Stroke in Fabry disease frequently occurs before diagnosis and in the absence of other clinical events: natural history data from the Fabry Registry. Stroke 40: 788–794.

Sirrs SM, Bichet DG, Casey R et al. (2014). Outcomes of patients treated through the Canadian Fabry disease initiative. Mol Genet Metab 111: 499–506.

Spada M, Pagliardini S, Yasuda M et al. (2006). High incidence of later-onset Fabry disease revealed by newborn screening. Am J Hum Genet 79: 31–40.

Spinelli L, Pisani A, Sabbatini M et al. (2004). Enzyme replacement therapy with agalsidase beta improves cardiac involvement in Fabry's disease. Clin Genet 66: 158–165.

Stemper B, Hilz MJ (2003). Postischemic cutaneous hyperperfusion in the presence of forearm hypoperfusion suggests sympathetic vasomotor dysfunction in Fabry disease. J Neurol 250: 970–976.

Suzuki N, Konohana I, Fukushige T et al. (2004). Beta-mannosidosis with angiokeratoma corporis diffusum. J Dermatol 31: 931–935.

Sweeley CC, Klionsky BL (1963). Fabry's disease: classification as a sphingolipidosis and partial characterization of a novel glycolipid. J Biol Chem 238 PC3148.

Tagliavini F, Pietrini V, Gemignani F et al. (1982). Anderson–Fabry's disease: neuropathological and neurochemical investigation. Acta Neuropathol 56: 93–98.

Taguchi A, Maruyama H, Nameta M et al. (2013). A symptomatic Fabry disease mouse model generated by inducing globotriaosylceramide synthesis. Biochem J 456: 373–383.

Tahir H, Jackson LL, Warnock DG (2007). Antiproteinuric therapy and Fabry nephropathy: sustained reduction of proteinuria in patients receiving enzyme replacement therapy with agalsidase-beta. J Am Soc Nephrol 18: 2609–2617.

Takanashi J, Barkovich AJ, Dillon WP et al. (2003). T1 hyperintensity in the pulvinar: key imaging feature for diagnosis of Fabry disease. AJNR Am J Neuroradiol 24: 916–921.

Takenaka T, Teraguchi H, Yoshida A et al. (2008). Terminal stage cardiac findings in patients with cardiac Fabry disease: an electrocardiographic, echocardiographic, and autopsy study. J Cardiol 51: 50–59.

Terryn W, Deschoenmakere G, De Keyser J et al. (2013). Prevalence of Fabry disease in a predominantly hypertensive population with left ventricular hypertrophy. Int J Cardiol 167: 2555–2560.

Thompson RB, Chow K, Khan A et al. (2013). T1 mapping with cardiovascular MRI is highly sensitive for Fabry disease independent of hypertrophy and sex. Circ Cardiovasc Imaging 6: 637–645.

Tremont-Lukats IW, Megeff C, Backonja MM (2000). Anticonvulsants for neuropathic pain syndromes: mechanisms of action and place in therapy. Drugs 60: 1029–1052.

Uceyler N, Homola GA, Guerrero Gonzalez H et al. (2014). Increased arterial diameters in the posterior cerebral circulation in men with Fabry disease. PLoS One 9: e87054.

van der Tol L, Smid BE, Poorthuis BJ et al. (2014). A systematic review on screening for Fabry disease: prevalence of individuals with genetic variants of unknown significance. J Med Genet 51: 1–9.

van Mullem PJ, Ruiter M (1970). Fine structure of the skin in angiokeratoma corporis diffusum (Fabry's disease). J Pathol 101: 221–226.

Vedder AC, Linthorst GE, van Breemen MJ et al. (2007). The Dutch Fabry cohort: diversity of clinical manifestations and Gb3 levels. J Inherit Metab Dis 30: 68–78.

Vedder AC, Breunig F, Donker-Koopman WE et al. (2008). Treatment of Fabry disease with different dosing regimens of agalsidase: effects on antibody formation and GL-3. Mol Genet Metab 94: 319–325.

Vedder AC, Biro E, Aerts JM et al. (2009). Plasma markers of coagulation and endothelial activation in Fabry disease: impact of renal impairment. Nephrol Dialysis, Transplant 24: 3074–3081.

Vibert D, Blaser B, Ozdoba C et al. (2006). Fabry's disease: otoneurologic findings in twelve members of one family. Ann Otol Rhinol Laryngol 115: 412–418.

von Scheidt W, Eng CM, Fitzmaurice TF et al. (1991). An atypical variant of Fabry's disease with manifestations confined to the myocardium. N Engl J Med 324: 395–399.

Wang RY, Lelis A, Mirocha J et al. (2007). Heterozygous Fabry women are not just carriers, but have a significant burden of disease and impaired quality of life. Genet Med 9: 34–45.

Warnock DG, Ortiz A, Mauer M et al. (2012). Renal outcomes of agalsidase beta treatment for Fabry disease: role of proteinuria and timing of treatment initiation. Nephrol Dial Transplant 27: 1042–1049.

Weicksel J (1925). Angiomatosis, bzw. Angiokeratosis universalis (eine sehr seltene Haut-und Gafasskrankheit). Dtsch Med Wochenschr 51: 898.

Weidemann F, Breunig F, Beer M et al. (2003). Improvement of cardiac function during enzyme replacement therapy in patients with Fabry disease: a prospective strain rate imaging study. Circulation 108: 1299–1301.

Weidemann F, Breunig F, Beer M et al. (2005). The variation of morphological and functional cardiac manifestation in Fabry disease: potential implications for the time course of the disease. Eur Heart J 26: 1221–1227.

Whitley CB, Tsai MY, Heger JJ et al. (1983). Amiodarone phenocopy of Fabry's keratopathy. JAMA 249: 2177–2178.

Whybra C, Kampmann C, Willers I et al. (2001). Anderson–Fabry disease: clinical manifestations of disease in female heterozygotes. J Inherit Metab Dis 24: 715–724.

Wilcox WR, Banikazemi M, Guffon N et al. (2004). Long-term safety and efficacy of enzyme replacement therapy for Fabry disease. Am J Hum Genet 75: 65–74.

Wilcox WR, Oliveira JP, Hopkin RJ et al. (2008). Females with Fabry disease frequently have major organ involvement: lessons from the Fabry Registry. Mol Genet Metab 93: 112–128.

Wozniak MA, Kittner SJ, Tuhrim S et al. (2010). Frequency of unrecognized Fabry disease among young European-American and African-American men with first ischemic stroke. Stroke 41: 78–81.

Wraith JE, Tylki-Szymanska A, Guffon N et al. (2008). Safety and efficacy of enzyme replacement therapy with agalsidase beta: an international, open-label study in pediatric patients with Fabry disease. J Pediatr 152: 563–570. 570 e561.

Yokota M, Koji M, Yotsumoto S (1995). Histopathologic and ultrastructural studies of angiokeratoma corporis diffusum in Kanzaki disease. J Dermatol 22: 10–18.

Young-Gqamana B, Brignol N, Chang HH et al. (2013). Migalastat HCl reduces globotriaosylsphingosine (lyso-Gb3) in Fabry transgenic mice and in the plasma of Fabry patients. PLoS One 8 e57631.

Zarate YA, Hopkin RJ (2008). Fabry's disease. Lancet 372: 1427–1435.

Zimmermann M (2001). Pathobiology of neuropathic pain. Eur J Pharmacol 429: 23–37.

Handbook of Clinical Neurology, Vol. 132 (3rd series)
Neurocutaneous Syndromes
M.P. Islam and E.S. Roach, Editors

Chapter 18

Hutchinson–Gilford progeria syndrome

NICOLE J. ULLRICH[1] AND LESLIE B. GORDON[2,3,4]*

[1]*Department of Neurology, Boston Children's Hospital and Harvard Medical School, Boston, MA, USA*

[2]*Department of Anesthesia, Boston Children's Hospital and Harvard Medical School, Boston, MA, USA*

[3]*Department of Pediatrics, Hasbro Children's Hospital and Warren Alpert Medical School of Brown University, Providence, RI, USA*

[4]*Progeria Research Foundation, Peabody, MA, USA*

INTRODUCTION

History

Hutchinson–Gilford progeria syndrome (HGPS) is an autosomal dominant, rare, fatal pediatric segmental premature aging disease (Gordon et al., 2003). The first description of patients with HGPS was in 1886, by Jonathan Hutchinson (Hutchinson, 1886), and again later by his colleague Hastings Gilford (Hutchinson and Gilford, 1897), who named the condition progeria (prematurely aged) in 1904 (Gilford, 1904). The word *progeria* comes from the Greek words *pro*, meaning "before" or "premature," and *gēras*, meaning "old age." These original case reports gave an excellent clinical picture as a syndrome resembling aging in many ways, with onset sometime in the first year after birth, including growth failure, loss of fat and hair, skin changes, bone and joint disease, and atherosclerosis. HGPS is also referred to as the Hutchinson–Gilford syndrome, progeria, or progeria of childhood.

EPIDEMIOLOGY

HGPS is an ultra-rare, sporadic, autosomal dominant orphan disease (Eriksson et al., 2003; De Sandre-Giovannoli and Levy, 2006), without gender or ethnic propensity (Gordon et al., 2014). This progressive, degenerative disorder has onset in early childhood with an estimated overall prevalence of 1 in 18 million (Gordon, 2013), and therefore 300–350 children were estimated to be living with progeria worldwide in 2014. The Progeria Research Foundation International Registry as of September, 2015 (http://www.progeriaresearch.org/patient_registry.html) has identified 125 children living in 43 countries. The estimated incidence is one in 4 million births (Hennekam, 2006), and average lifespan is 14.6 years (Gordon et al., 2014) (range 1.6–27.5 years).

CLINICAL FEATURES

Overview of disease course

Although some signs and symptoms vary in age of onset and degree of severity, clinical features of HGPS are remarkably consistent and individuals appear very similar in phenotype (Fig. 18.1) (Merideth et al., 2008). Birth weight and early postnatal development are usually normal. Typically within the first year of life, children develop profound failure to thrive with sclerodermatous skin changes (Fig. 18.2). Characteristic craniofacial features such as receding mandible, narrow nasal bridge, and pointed nasal tip develop, followed by partial then complete alopecia, loss of subcutaneous fat, progressive contractures of the joints, bony dysplasia, dystrophic nails, and delayed primary tooth eruption. Later features may include conductive hearing loss, dental crowding, excessive tearing, exposure keratitis and Raynaud's phenomenon, and insulin resistance without overt signs or development of diabetes mellitus. Affected individuals do not become sexually mature and, therefore, do not reproduce.

Cardiovascular disease and cerebrovascular disease constitute the major cause of morbidity and mortality. Early cardiac changes can manifest at ages 5–8 years, but usually begin after age 8 years. The underlying

*Correspondence to: Leslie B. Gordon, MD, PhD, Department of Pediatrics, Hasbro Children's Hospital, 593 Eddy Street, Providence, RI, 02903, USA. Tel: +1-978-535-2594, E-mail: Leslie_gordon@brown.edu

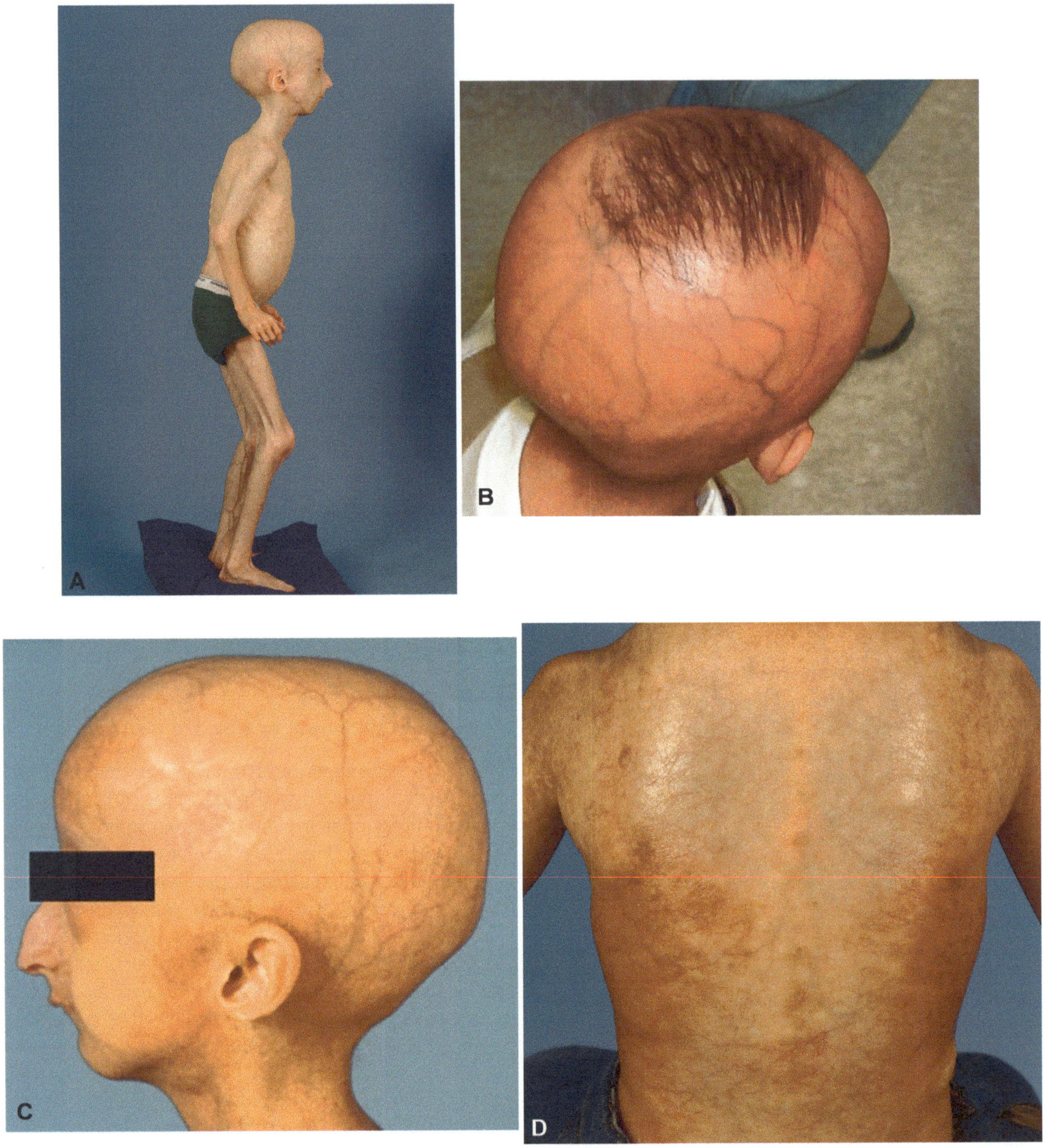

Fig. 18.1. Some distinguishing clinical features of HGPS. (**A**) Whole body side view showing generalized lipoatrophy leaves muscular prominence, knee contracture; (**B**) hair loss begins in the temporal area in the process of becoming total alopecia; prominent scalp veins; (**C**) lateral facial view with retrognathia, craniofacial disproportion; (**D**) skin tightening and mottling; *(Continued)*

disease is primarily from vascular changes, which are characterized by early and accelerated vascular stiffening, followed by small and large vessel occlusive disease due to atherosclerotic plaque formation, and, in later years, valvular and cardiac insufficiency (Merideth et al., 2008; Gerhard-Herman et al., 2012). Typical cardiac manifestations include increased afterload, angina, and late findings such as dyspnea on exertion. Hypertension is considered a later sign of vascular disease. Gerhard-Herman and colleagues found evidence of vascular dysfunction in 100% of a cohort of 26 children with HGPS as young as age 3 years (Gerhard-Herman et al., 2012), with dramatically elevated carotid–femoral pulse wave velocity and changes to the carotid artery wall. Death

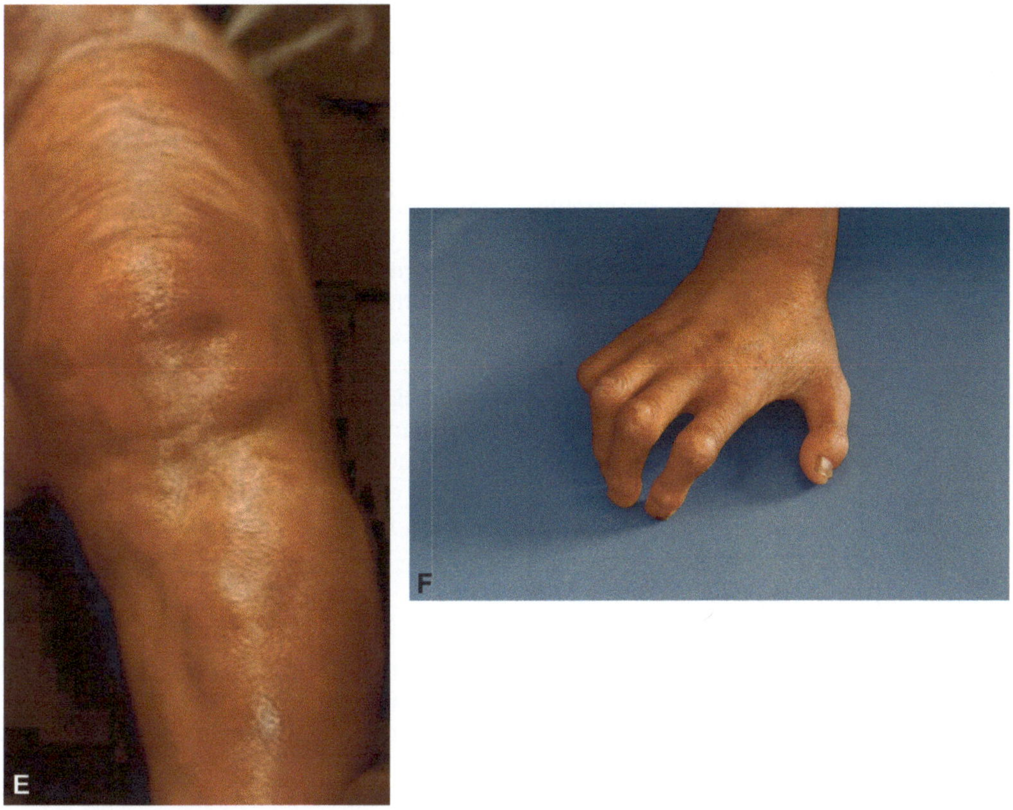

Fig. 18.1—Cont'd (**E**) small, soft outpouchings with a pseudo-cellulite appearance; (**F**) digital joint contractures. (Photographs courtesy of Progeria Research Foundation Medical and Research Database and Boston Children's Hospital.)

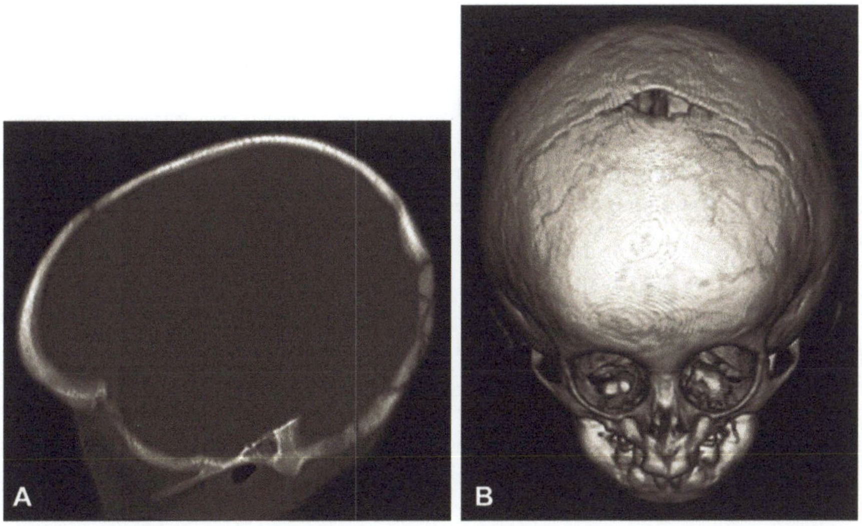

Fig. 18.2. Computed tomography (CT) scan (**A**) and anterior projection of a 3D CT shaded surface display of the head (**B**) demonstrates prominent craniofacial features findings of HGPS in a 10-year-old boy. (**A**) Thinning of the calvarium, a paucity of scalp fat, and mottling of the skull; (**B**) open fontanelle, midface hypoplasia, increased cranial vascular markings, craniofacial disproportion, narrowly spaced eyes and dental crowding. (Scans courtesy of Progeria Research Foundation Medical and Research Database.)

occurs by myocardial infarction or stroke, generally between ages 6 and 20 years, with an average lifespan of 14.6 years (Gordon et al., 2014).

HGPS is a syndrome of premature aging with many, but not all, clinical features of aging/senescence. Features that resemble the normal aging process include alopecia, joint contractures, low bone density, lipoatrophy with limb wasting, and global atherosclerosis. Motor and intellectual development is typically normal, unless there has been history of many strokes. Dementia, osteoarthritis, and cancer are absent. Immune systems of children with HGPS are considered normal, as is wound healing. Generally, liver, kidney and gastrointestinal systems remain intact. Other disease manifestations that differ from aging include growth failure, skeletal dysplasia, and lack of secondary sex characteristics or pubertal development.

Dermatologic changes

Skin findings are often apparent as initial signs of HGPS, with findings usually apparent before age 12 months (Rork et al., 2014). These are variable in severity and include areas of discoloration, stippled pigmentation, tightened areas that can restrict movement, and small (1–2 cm) areas of the trunk or legs where soft bulging skin is present (see Fig. 18.1). Though born with normal hair presence, cranial hair is slowly lost within the first few years, leaving soft, downy, sparse, immature hair on the scalp, no eyebrows, and scant eyelashes. Scalp alopecia progresses in a distinct pattern, with preservation of the hair over the midscalp and vertex areas for the longest period of time. Prominent superficial vasculature is manifested as circumoral cyanosis and pronounced veins on the scalp and body. Hypo- and hyperpigmentation is often observed over areas of sclerodermoid change. Tightened skin along with decreased fat around the eyes results in nocturnal lagophthalmos, with increased risk of corneal dryness and resultant exposure keratitis with opacity.

Failure to thrive

Children with HGPS experience apparently normal fetal and early postnatal development. The absence of overt symptoms at birth probably correlates with the absence of progerin during early fetal development, when cells are not fully differentiated (Liu et al., 2011a). Within the first year of life, abnormalities in growth and body composition become readily apparent. Severe failure to thrive ensues, heralding generalized lipoatrophy (Badame, 1989; Pesce and Rothe, 1996), with apparent wasting of limbs, circumoral cyanosis, and prominent veins around the scalp, neck, and trunk. Clinical problems caused by the lack of subcutaneous fat include insulin resistance, sensitivity to cold temperatures, and foot discomfort due to lack of fat cushioning. Though overt diabetes is very unusual in HGPS, about 30–40% of children suffer from insulin resistance (Merideth et al., 2008; Gerhard-Herman et al., 2012). Children reach a final height of approximately 1 meter and weight of approximately 14 kg. The mean weight, which is normal at birth, decreases to below the third percentile of normal children despite adequate caloric intake for normal growth and normal resting energy expenditure (Gordon et al., 2007; Merideth et al., 2008). A retrospective data set of 35 children showed an average weight increase of only 0.44 kg/year, beginning at 24 months of age and persisting through life, with remarkable intra-patient linearity (Gordon et al., 2007), which contrasts sharply with the parabolic growth pattern for normal age- and gender-matched children.

Neurologic features

The neurologic manifestations can be divided into (1) craniofacial/structural features (Ullrich et al., 2012) and (2) cerebrovascular features (Silvera et al., 2013).

Craniofacial features

Craniofacial disease-related abnormalities of HGPS can be divided into abnormalities of the scalp, calvarium and skull base, facial bones, soft tissues and orbit. Typical craniofacial features include craniofacial disproportion, thinned skull, prominent eyes, narrow nasal bridge with broad nasal tip, disproportionally small mid and lower face, hypognathia and retrognathia, lack of facial fat, and prominent scalp veins.

Thinning of the calvarium is accompanied by a paucity of scalp fat. In addition, mottling of the skull is common, most frequently observed in the frontal, parietal and sphenoid regions (see Fig. 18.2). This is thought to represent bony osteolysis. Prominent vascular markings of the bony calvarium are also common, as is delayed closure of the anterior and posterior fontanelle, which remained patent at a mean age of 7.4 years on a recent retrospective study of craniofacial findings (Ullrich et al., 2012). Widened sagittal, squamous and lambdoid sutures were also observed. The appearance of macrocrania and craniofacial disproportion is frequently noted in children with HGPS, with relatively large heads in relation to the small, narrow facies with retrognathia. Although a large cranium relative to size of the face is typical, mean head circumference normalized to height–age is over the 50th percentile, suggesting that the progressive remodeling of the facial bones gives rise to the appearance of craniofacial disproportion rather than the presence of macrocephaly.

More recently described features of the jaw include short mandibular rami, shallow glenoid fossae, and flattening of the mandibular condyles as well as the known V-shaped palate and disorganized dentition. Most children with HGPS have limited oral opening on clinical examination and these findings suggest possible causes for the deficit due to limited range of motion at the abnormally configured temporomandibular joint.

Brain and cerebrovascular manifestations of HGPS

Stroke or cerebrovascular disease is a leading cause of morbidity and cause of mortality in about 10% of children with HGPS (Gordon et al., 2014). Stroke or recurrent transient ischemic attacks (TIA) are frequently accompanied by headaches, muscular weakness or paralysis affecting one side of the body, or seizures (Silvera et al., 2013). By the time children with HGPS present with neurologic symptoms from stroke, there is often evidence of prior "silent" infarcts that have occurred in the past. These "silent" infarcts have occurred presumably without dramatic clinical symptoms but, over time, may accumulate and lead to more fixed neurologic symptoms.

A retrospective study to identify the neurovascular features, infarct type, and natural history of stroke in a cohort of 25 children ages birth to 14.1 years with a confirmed diagnosis of classic HGPS identified features suggestive of a vasculopathy unique to HGPS (Silvera et al., 2013). Radiographic evidence of infarction was identified in 60% of the patients, in whom half were thought to be clinically silent. The mean age of infarct based on imaging was 6.8 years and the earliest age at infarction was 0.4 years.

The topographic distribution of infarcts included arterial territory, white matter, lacunar and watershed infarcts; however, no brainstem, cerebellar, or hemorrhagic infarcts were observed (Silvera et al., 2013) (Fig. 18.3). Stroke symptoms were similar to those typically found in children with stroke secondary to other etiologies such as embolism, dissection, and focal cerebral arteriopathy of childhood (Tsze and Valente, 2011).

Both large and small vessel disease was observed. The most frequent intracranial stenoses involved the precavernous and supraclinoid internal carotid arteries (see Fig. 18.3). Narrowing of the anterior carotid and middle carotid arteries was also common. A particularly robust finding was narrowing of the distal V2/proximal V3 segment of the vertebral artery centered at C2. Notably, collateral vessel formation was extensive and involved the suprasellar, subfrontal, and perisplenial regions (see Fig. 18.3). Unlike the aging population, the severity of intracranial arterial disease was distinctly more severe than disease within the neck, as opposed to adult atherosclerosis, in which cervical occlusive disease predominates. Also, the distribution observed in children with HGPS is distinct from cerebrovascular disease of aging. No aneurysm, arterial dissection, arteriovenous malformation (AVM), arteriovenous fistula, sinus thrombosis or subarachnoid hemorrhage was observed. Additionally, the collateral vessel formation suggests a unique pattern of intracranial arteriopathy in HGPS distinct from atherosclerosis of aging and other vasculopathies of childhood, such as moyamoya.

Despite the presence of multisystem premature aging, progeria patients do not appear to have cognitive deterioration, suggesting that some aspects of brain function may be protected from the expression of the progeria mutation. Based on limited autopsy material, there is no pathologic evidence of dementia or

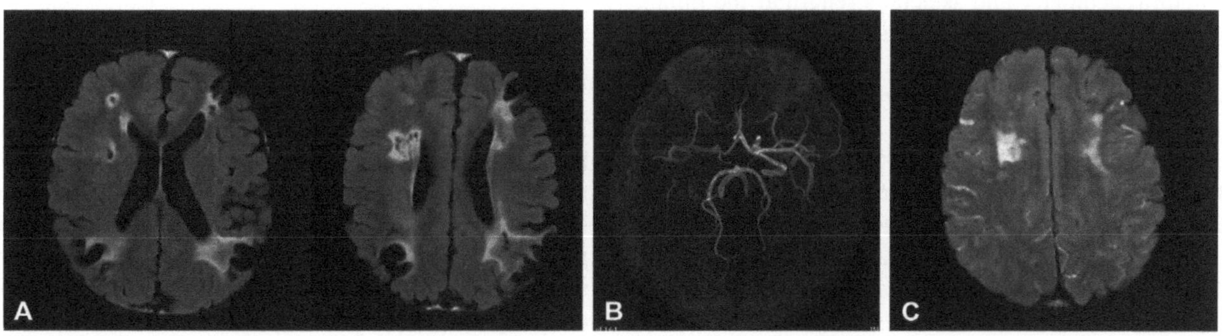

Fig. 18.3. Magnetic resonance imaging (MRI) demonstrates vascular and stroke findings in HGPS. (**A**) Axial FLAIR MRI images demonstrating divergent patterns of infarct topography in an 11-year-old, which include large vessel, white matter and watershed. (**B**) Collapsed view of the MR angiography of the circle of Willis demonstrates bilateral short segment high grade stenosis of the cavernous internal carotid arteries. (**C**) Bright signal in the sulci on FLAIR imaging indicates slow cortical collateral flow with an adjacent subacute infarction. (Scans courtesy of Progeria Research Foundation Medical and Research Database.)

Alzheimer-type changes. In a transgenic mouse model with expression of progerin in brain, skin, bone, and heart, there are distortions of neuronal nuclei at the ultrastructural level with irregular shape and severe invaginations, but no evidence of inclusions or aberrant tau in brain sections. Importantly, the nuclear distortions did not result in significant changes in gene expression in hippocampal neurons (Baek et al., 2015). In another progerin-expressing mouse model, the levels of progerin in brain tissue were extremely low due to downregulated expression by a brain-specific microRNA, miR-9 (Jung et al., 2012). If true in humans, expression of miR-9 in brain may protect patients from extravascular central nervous system pathology.

Bone and cartilaginous abnormalities

Disease in HGPS affects the extracellular matrix, and manifests as significant disease of the bones and joints. Development of bone structure and bone density represents a unique skeletal dysplasia which is not based in malnutrition. Abnormalities on X-ray include clavicular resorption, coxa valga, distal phalangeal resorption, facial disproportion (a pointed nose and receding mandible), and short stature (Cleveland et al., 2012). Importantly, all children with HGPS suffer from progressive, bilateral coxa valga. Hip dysplasia is often progressive and may result in avascular necrosis, hip dislocation, and inability to bear weight. Pyriform chest structure and a small clavicle can lead to reducible shoulder dislocations. Low tone conductive hearing loss is pervasive in HGPS, and is indicative of a stiff tympanic membrane, and/or deficits in the middle ear bony and ligamentous structures.

Bone age is generally normal, with variable readings within a single X-ray. Bone structure assessed by peripheral quantitative computed tomography of the radius demonstrates distinct and severe abnormalities in bone structural geometry and skeletal strength at the radius compared with healthy age-matched controls. Areal bone mineral density (aBMD) Z scores by dual-energy X-ray absorptiometry are abnormally low, but improve after adjustment for height-age. This knowledge is crucial since the two-dimensional DXA scan will exaggerate BMD deficits due to inability to adjust for small size (Gordon et al., 2011). Thus it is not surprising that fracture rates in HGPS are normal, unlike more severely affected diseases of BMD such as osteogenesis imperfecta.

Joint contractures, due to ligamentous and skin changes, limit range of motion. Contractures in multiple joints including fingers, elbows, hips, knees, and ankles may be present at birth and/or in later years.

Normally functioning systems

Generally, liver, kidney, immune, gastrointestinal, and neurologic systems (other than stroke-related) remain intact. Intellectually, children are normal for age. This may be because the cell types involved with function of these systems, such as neurons (Jung et al., 2012) or immune cells, have lower progerin expression, or because there is adequate reserve organ function to remain in the normal clinical range.

Genetic testing

HGPS is diagnosed by phenotype recognition plus progerin-producing mutation in the *LMNA* gene either within exon 11 (classic form) or at the exon 11 intronic border (atypical form) (Table 18.1). *LMNA* is the only gene known to cause HGPS, which is an autosomal dominant disorder with complete penetrance. HGPS typically is caused by a *de novo* mutation, though one case of somatic and gonadal mosaicism is documented (Wuyts et al., 2005). The risk to a sibling of a proband is estimated at one in 500. There are two known sets of identical twins with HGPS. Targeted mutation analysis can be used to identify mutations in *LMNA* associated with HGPS (Cao and Hegele, 2003; Eriksson et al., 2003). Sequence analysis of the entire coding region and associated splice junctions may be used to identify the common mutation of classic HGPS as well as the mutations that define atypical HGPS and other sequence variants that may be associated with other progeroid syndromes (Moulson et al., 2007). It is likely that the severity of disease is determined in part by progerin levels, which are regulated by the specific mutation, the tissue type, and other factors that influence the internal splice site within exon 11. Disease in HGPS is produced by a dominant negative mechanism, wherein it is the action of progerin, not the diminution of lamin A, that leads to disease phenotype.

Variant presentations/genetically related allelic disorders

There are more than 10 other diseases and conditions that have been identified that are characterized by mutations or variations in the *LMNA* gene. Progeroid laminopathy is used to describe phenotypes that resemble HGPS but are clinically distinguishable, in which an *LMNA* mutation other than those of classic and atypical HGPS has been identified. To date there are 18 referenced mutations identified that cause progeroid laminopathies (see www.umd.be/LMNA). Children with atypical mutations involving *LMNA* have demonstrated more or less severe phenotypes.

Table 18.1

HGPS causative mutations and comparative clinical phenotypes

Phenotype	Causative mutations in *LMNA*	Protein amino acid change	Clinical features compared to HGPS	Number of affected individuals identified	Reference
Classic HGPS	c.1824C>T	p.Gly608Gly	NA	NA	(De Sandre-Giovannoli et al., 2003; Eriksson et al., 2003)
Atypical HGPS	c.1822G>A	p.Gly608Ser	Slightly more severe	4	PRF
	c.1968 +1G>A	Splice donor site mutation	Slightly to moderately more severe	4	(Moulson et al., 2007); PRF
	c.1968 +2 T>A	Splice donor site mutation	Similar	2	PRF
	c.1968 +2 T>C	Splice donor site mutation	Similar	1	PRF
	c.1968 +5G>C	Splice donor site mutation	Similar	3	PRF
	c.1968 +5G>A	Splice donor site mutation	Very mild	1	(Hisama et al., 2011)

HGPS, Hutchinson–Gilford progeria; NA, not applicable; PRF, Progeria Research Foundation Genetic Testing Program.

HGPS is a member of the family of genetic diseases known as the laminopathies, with mutations along the *LMNA* gene (Cau et al., 2014). There is overlap between these diseases, which have common features that include involvement of the striated muscle, lipodystrophy, peripheral neuropathy, and accelerated aging. The family of laminopathies include autosomal dominant and autosomal recessive Emery–Dreifuss muscular dystrophy, autosomal dominant familial dilated cardiomyopathy and conduction system defects (CMD1A), autosomal dominant Dunnigan-type familial partial lipodystrophy (FPLD), autosomal dominant limb-girdle muscular dystrophy 1B (LGMD1B), autosomal recessive axonal neuropathy Charcot–Marie–Tooth disease 2B1 (CMT2B1), and autosomal recessive mandibuloacral dysplasia (MAD).

DIFFERENTIAL DIAGNOSES TO CONSIDER

A variety of other syndromes include some features of premature aging (Coppedè, 2013) (Table 18.2) and should be considered in the differential diagnosis when HGPS is not phenotypically obvious.

Table 18.2

Some syndromes with features of premature aging

Acrogeria
Berardinelli–Seip congenital lipodystrophy (congenital generalized lipodystrophy)
Cockayne syndrome
Ehlers–Danlos syndrome, progeroid form
Gerodermia osteodysplastica
Hallermann–Streif syndrome
Mandibuloacral dysplasia
Neonatal progeroid syndrome (Weidemann–Rautenstrauch syndrome)
Nestor–Guillermo progeria syndrome
Mulvihill–Smith syndrome
Petty–Laxova–Weidemann progeroid syndrome
Rothmund–Thomson syndrome
Werner syndrome
Mutations in *ZMPSTE24*

CARDIOVASCULAR AND STROKE PATHOLOGY

Death in HGPS occurs primarily from myocardial infarction (estimated at 90%), and less often from strokes (estimated at 10%), though neurovascular morbidity is high, with up to 50% of children with HGPS experiencing silent strokes (Silvera et al., 2013).

There are several prior reports that did not examine genetically confirmed HGPS and therefore are more challenging to interpret (Reichel and Garcia-Bunuel, 1970; Stehbens et al., 2001).

Isolated from risk factors such as hypercholesterolemia and increased high sensitivity C-reactive protein (Gordon et al., 2005), early stage hypertension, and smoking, the study of HGPS provides an opportunity to discover new elements that influence the vascular disease of aging, such as progerin. Pathologic evaluation of HGPS autopsies reveals that concordance among many, but not all, aspects of the cardiovascular pathology are found in HGPS and elderly patients (Reichel and Garcia-Bunuel, 1970; Stehbens et al., 2001; Olive et al., 2010). Late stage plaques in HGPS reveal large and small vessel classic-looking plaque formation, prominent adventitial fibrosis, and intense progerin staining (Olive et al., 2010). Importantly, progerin is present in non-HPGS individuals, and increases at a rate of 3.34% per year, implying that progerin has a role in cardiovascular aging of the general population. Smooth muscle cell dropout from the vascular media is featured both in the human (Stehbens et al., 2001) and mouse model vasculature (Varga et al., 2006). In contrast, the pattern of cranial vasculopathy is unlike that of aging, where extensive collateral vessel formation supplies essential blood supply to the brain in response to diminished supply by stiffened and obstructed primary vasculature in HGPS (Silvera et al., 2013).

MOLECULAR PATHOGENESIS

An understanding of *LMNA* transcription, translation, and post-translational processing has been absolutely crucial to the progress made towards understanding and treating HGPS. This section will first describe the normal processing and functions of the *LMNA* gene and resulting protein, followed by an overview of how alterations in those normal pathways result in disease.

LMNA

The *LMNA* gene, also called the *LMNA/C* gene, is located at 1q21.2 (Haithcock et al., 2005; Capell and Collins, 2006; Rankin and Ellard, 2006). The *LMNA* gene includes 12 exons and codes for at least four isoforms: two major (lamin A and lamin C) and two minor (lamin AΔ10 and lamin C2) (Furukawa et al., 1994; Machiels et al., 1996). These isoforms diverge in structure, function, expression pattern, and binding partners. Only the lamin A isoform is associated with mammalian disease.

The lamin proteins are the principal proteins of the nuclear lamina, a complex molecular interface located between the inner membrane of the nuclear envelope and chromatin (Goldman et al., 2004). The integrity of the lamina is central to many cellular functions, including mitosis, creating and maintaining structural integrity of the nuclear scaffold, DNA replication, RNA transcription, organization of the nucleus, nuclear pore assembly, chromatin function, cell cycling, and apoptosis. Lamin A molecules are intermediate filament proteins which assemble as head-to-tail dimers that associate laterally along the inner nuclear membrane (Shumaker et al., 2003). All 12 exons are transcribed to produce the primary RNA transcript prelamin A. This precursor terminates in a CaaX box (cysteine, 2 aliphatic amino acids, any amino acid) and four post-translational processing stages: farnesylation, cleavage of the last three amino acids residues at its carboxy terminal, methyl esterification, and finally proteolysis of its C-terminal 18 amino acids, which includes the farnesyl group, to become mature lamin A (Weber et al., 1989; Sinensky et al., 1994; Kilic et al., 1997) (Fig. 18.4). The farnesyl group is a lipophilic moiety which facilitates intercalation of prelamin A into the inner nuclear membrane where most of its functions are performed. Subsequent loss of the farnesyl anchor releases prelamin from the nuclear membrane, rendering it soluble for autophagic degradation (Sinensky et al., 1994).

Subsequent to all HGPS mutations, translation followed by post-translational processing of the altered mRNA produces progerin, a shortened abnormal protein with a 50 amino acid deletion near its C-terminal end. It is likely that severity of disease is determined by progerin levels, which are regulated by the particular mutation, tissue type, or other factors influencing use of the internal splice site. The 50 amino acid deletion does not affect the ability of progerin to localize to the nucleus or to complex with normal lamin A protein, because the necessary components for these functions are not deleted (Sinensky et al., 1994). Importantly, however, it does remove the recognition site that leads to proteolytic cleavage of the terminal 18 amino acids and the farnesyl side group of prelamin A, along with the phosphorylation site(s) involved in the dissociation and reassociation of the nuclear membrane at each cell division (Sinensky et al., 1994; Kilic et al., 1997). Thus progerin retains the farnesyl group that was added during post-translational processing (see Fig. 18.4) (Capell and Collins, 2006). Farnesyltransferase inhibitors (FTIs) that prevent the initial attachment of the farnesyl group to newly synthesized preprogerin molecules were the first drugs suggested for treatment of HGPS (see Management and therapy section).

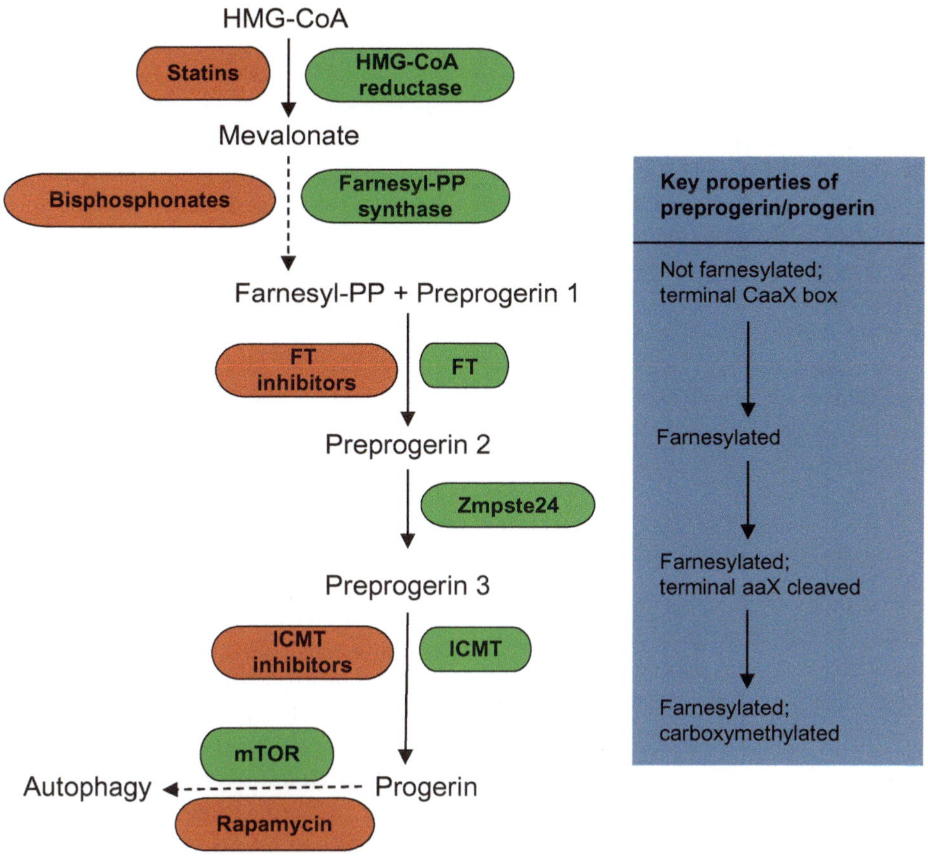

Fig. 18.4. Progerin post-translational processing pathway and potential target points for treatment. Enzymes shown in green; enzyme inhibitors shown in red; solid arrow, single step; dashed arrow, multiple steps; FT, farnesyltransferase; ICMT, S-isoprenylcysteineO-methyltransferase.

Lamin A expression is developmentally regulated and displays cell and tissue specificity, primarily in differentiated cells including fibroblasts, vascular smooth muscle cells, and vascular endothelial cells (Machiels et al., 1996; Tilli et al., 2003; McClintock et al., 2006). In HGPS, one unmutated *LMNA* gene on chromosome 1 yields the normal amount of transcript that encodes lamin A, while expression of the mutated *LMNA* gene results in both mutant (85%) and normal (15%) transcript (Reddel and Weiss, 2004). Though the alternate splicing in HGPS leads to decreased levels of lamin A, lower lamin A levels do not seem to be the disease-causing factor. In fact, a mouse model entirely lacking lamin A shows no signs of disease (Fong et al., 2006b).

PRECLINICAL DISEASE MODELS

A variety of *in vitro* and *in vivo* murine model systems exist for use in recapitulating aspects of human disease in order to investigate the biology of disease, natural history, and provide preclinical treatment testing systems for HGPS.

In vitro assays

There are a number of *in vitro* assessments that have overlap with cellular aging and are used as disease markers for HGPS.

The most overt cellular defect in HGPS is dramatic nuclear blebbing (Eriksson et al., 2003) by nonautomated or automated (Driscoll et al., 2012) image analysis. While the lamin proteins usually move dynamically between the nuclear lamina polymer at the periphery of the nucleus and the nucleoplasm, they become immobilized in HGPS patient cells leading to thickening of the lamina (Goldman et al., 2004). Mechanical properties of HGPS nuclei are subsequently altered and patient cells exhibit increased stiffness compared with cells from healthy individuals (Dahl et al., 2010). These alterations may affect the response of cells in tissues that are particularly exposed to mechanical stress such as the vasculature, bone, and joints.

Additional trackable abnormalities include mislocalized protein import into the cell nucleus (Snow et al., 2013); widespread alterations of chromatin structure, such as the loss of heterochromatin domains, and changes in epigenetic markers (Pegoraro et al., 2009);

DNA repair defects due to reduced recruitment of DNA damage response (DDR) factors such as 53BP1 to site of damage and slowed repair kinetics (Musich and Zou, 2009; Benson et al., 2010); altered regulatory and stress-response pathways (Varela et al., 2005; Liu et al., 2006; Kudlow et al., 2008); attrition of adult stem cells (Espada et al., 2008; Scaffidi and Misteli, 2008; Rosengardten et al., 2011), dysregulation of the somatotrophic axis and several miRNA-controlled circuits (Marino et al., 2010; Ugalde et al., 2011). Progerin also appears to modulate telomere function and to induce DDR signaling from telomeres via activation of p53 and Rb pathways (Benson et al., 2010). Interestingly, telomere dysfunction during senescence appears to feed back and promote progerin expression (Cao et al., 2011a).

Animal models

Though there are several mouse models that can be considered progeroid but that do not produce progerin (fully reviewed in Zhang et al., 2013), this section will focus on progerin-producing mouse models. Genetically engineered progerin-expressing HGPS mouse models have been generated using either mouse or human *LMNA* mutations, with variable recapitulation of human disease, making them useful for comparative studies. An embryonic stem cell gene-targeted model expresses bone fractures and neurologic deficits without cerebrovascular abnormalities (Fong et al., 2006a); a human bacterial artificial chromosome-transgenic model exhibits large vessel disease and blood pressure deficits without other obvious disease manifestations (Varga et al., 2006); two conditional transgenic lines carry minigenes specific to keratin 5 (Sagelius et al., 2008) or to osteoblasts (Schmidt et al., 2012) and exhibit skin, hair adipocyte and bone abnormalities, respectively; and a knock-in model exhibits loss of vascular smooth muscle cells in the aortic arch, bradycardia, prolonged QRS waves, marked curvature of the spine (cervicothoracic lordokyphosis), reduction in bone density and cortical thickness, cardiovascular disease, and spinal curvature (Osorio et al., 2011). Each of these mouse models expresses progerin with a different phenotype as noted above; however, as a group, homozygotes demonstrate progressive weight loss and decreased overall lifespan of unknown cause, with milder phenotypes in the heterozygotes. Most importantly, several models have cardiovascular disease (CVD) that overlaps with human disease phenotype, and CVD is generally difficult to generate in murine models.

Stem cells

Aside from fibroblast lines, tissue-specific cell lines from HGPS patients are not easily obtainable. Patient-derived induced pluripotent stem cells (iPSC) provide a unique tool to study differentiation pathways of tissues affected by the disease such as vascular smooth muscle cells (VSMCs), adipose tissue, and bone, as well as to evaluate the contribution of stem cells to tissue regeneration. HGPS iPS cells have successfully been differentiated along multiple lineages, including the particularly disease-relevant VSMCs and mesenchymal stem cells (Zhang et al., 2011). Induced pluripotent stem cells (iPSC) have been generated from HGPS fibroblasts and used as an *in vitro* model of HGPS (Liu et al., 2012 and www.progeriaresearch.org). At initial induction, the iPSC do not produce progerin or display hallmark features such as nuclear blebbing or other markers for HGPS or aging; however, as they differentiate into smooth muscle cells, the expression of progerin and other markers of HGPS are recapitulated, including accelerated senescence, reduced telomere length and other hallmarks of aging *in vitro* (Liu et al., 2011a). Patient-derived iPS cells have also been used to develop strategies that can successfully correct the HGPS mutation (Liu et al., 2011b), and, as with many other diseases, they will also be invaluable in screening of potential new, possibly tissue-specific, drugs. Finally, normal and gene-corrected HGPS iPS-derived tissue-specific cells could provide a foundation for cell replacement therapy approaches to HGPS.

MANAGEMENT AND THERAPY

A complete, system-based management guide is available from the Progeria Research Foundation or at http:/www.progeriaresearch.org/patient_care.html.

General guidelines and management

In order to establish the extent of disease in an individual diagnosed with HGPS, the following evaluations are recommended at time of diagnosis: (1) weight and height plotted on standardized growth charts to evaluate growth over time; (2) baseline electrocardiogram and echocardiogram; (3) baseline carotid artery duplex scans, in order both to establish baseline cardiovascular status and to establish the size of the lumen as well as intimal thickness; (4) baseline magnetic resonance imaging/angiography (MRI/MRA) of the head and neck to assess for vascular changes and silent strokes; (5) skeletal X-ray to evaluate for known findings such as acro-osteolysis, clavicular resorption, and coxa valga; (6) assessment of bone structure and bone mineral density using dual energy X-ray densitometry (DEXA) and peripheral quantitative computed tomography (CT); (7) assessments by physical therapy and occupational therapy to assess joint mobility; (8) nutritional assessment; (9) audiologic evaluation, with attention to possible low frequency conductive hearing loss;

(10) ophthalmologic evaluation for possible exposure keratopathy; (11) dental examination and cleaning; (12) insulin and glucose sensitivity, lipid profiles.

There is no current evidence that a specific diet influences the clinical course of HGPS. Therefore, a regular diet is indicated, unless there is evidence for elevated lipids/cholesterol, at which time diet modification and/or medications may be indicated. Maximizing caloric intake is advised and multiple small meals is often effective.

As a general recommendation, children should be encouraged to be as physically active as possible, taking into account limitations resulting from joint mobility and/or hip dislocation. Shoe padding, particularly around the heel, may be helpful, as lipoatrophy in that area can lead to foot discomfort. Age appropriate schooling is usually indicated, as children have normal intellectual function and normal social maturity. Modifications for seating and mobility in the classroom and school setting may be indicated.

Both physical and occupational therapy may be helpful to maintain range of motion. Some combination of active stretching/strengthening and aquatherapy may be of further benefit. Passive exercises are not recommended due to increased risk of shoulder and hip dislocations.

Many young children will require sedation for procedures or for imaging studies. For any individual patient, an evaluation by a qualified provider, such as an anesthesiologist or intensivist, is recommended prior to any planned sedation to discuss airway issues and fluid and blood pressure management plans. General anesthesia and intubation should be used with extreme caution, as narrow airway and retrognathia make intubation complicated.

Medication doses are based on body weight or body surface area, rather than age.

Surveillance

Surveillance for detection and prevention of complications is imperative. This includes, but is not limited to, serial cardiac assessments, including electrocardiogram, blood pressure measurement, echocardiogram and carotid duplex ultrasound annually (or more) to monitor for cardiovascular disease. These are important evaluations as children may demonstrate severe carotid artery blockage from atherosclerosis prior to any significant changes on electrocardiogram. In addition, neurologic assessment with MRI/MRA of the head and neck to assess for vascular changes and silent infarction is recommended. Ongoing monitoring of lipid profiles, dental examinations, audiologic assessments, and ophthalmologic evaluations are recommended. Hip X-rays to evaluate for avascular necrosis and progressive coxa valga are also important.

Management of neurologic issues/ cerebrovascular disease

Strokes remain a frequent cause of concern and morbidity in children with HGPS. If a stroke or acute neurologic symptom occurs that raises concern for TIA or stroke, then management of blood pressure – specifically, maintenance of adequate blood pressure and cerebral perfusion – is imperative. In the case of a larger stroke, monitoring in an intensive care setting is often indicated until the child's condition is stabilized. Pharmacologic treatments are often considered at that time.

Antiplatelet agents (such as aspirin) are often given to prevent clot formation and, ideally, to prevent future strokes from occurring. Based on evidence from adult studies, a dose of 3–5 mg/kg body weight per day is recommended. Some physicians believe that all patients with known narrowing of the great vessels should be on an antiplatelet agent permanently as a preventative step. The decision to start aspirin and/or to add another medication, such as clopidogrel or acetazolamide to the regimen, should always be made by consultation with a neurologist/stroke specialist. Safety of many of these medications and guidelines for use are not well-established in pediatric patients; therefore, careful consideration is required.

Acute neurologic symptoms are often brought on by activities that involve hyperventilation, reduction in blood pressure, or by dehydration. For these reasons, it is important that children remain well hydrated at all times. This is particularly important during times of illness and/or travel. We recommend that children with HGPS who plan to travel increase their hydration and fluid intake in the 24–48 hours prior to planned start of the trip. As a rough estimate, minimum fluid requirements are about 1 liter daily, with a goal closer to 1.5 liters.

Headaches

Headaches are frequently observed in children with progeria. This is, at least in part, likely related to the vasculopathy. Headaches can be single or recurrent in nature, and localized to one or more areas of the head and face. The exact etiology of headaches is not completely understood. It is thought that many are the result of tight muscles and dilated, or expanded, blood vessels in the head.

The ultimate goal is headache control. Supportive therapy may include rest in a quiet, dark environment, avoidance of known triggers such as certain foods and beverages, lack of sleep or fasting. Symptomatic treatment with nonsteroidal anti-inflammatory medications such as ibuprofen or with Tylenol (paracetamol) may be necessary to treat acute headaches if they occur frequently.

DISEASE MODIFYING THERAPY AND POTENTIAL FUTURE TREATMENT STRATEGIES

Treatment strategies for HGPS are derived from an ever-growing understanding of the genetic and biological basis for disease in HGPS, adapted from preclinical studies of aging or derived from chemical screening assays. To date, there is one treatment in humans with favorable published outcomes (Gordon et al., 2012, 2014; Ullrich et al., 2013) and two treatment trials currently in progress.

Farnesyltransferase inhibitors

The disease-causing gene product causing HGPS is progerin, a protein that lacks 50 amino acids and is permanently farnesylated, unlike its normal lamin A counterpart. This property tethers progerin to a cell's inner nuclear membrane, likely permanently (see Fig. 18.4). Protein farnesyltransferase inhibitors (FTIs) are small molecules that reversibly bind to the farnesyltransferase binding site, thereby inhibiting progerin farnesylation and interposition into the nuclear membrane (Cox and Der, 2002). Preclinical studies inhibiting post-translational protein farnesylation using farnesyltransferase inhibitors (FTIs) yielded improvements in both cellular and murine HGPS models. *In vitro*, nuclear deformations characteristic of HGPS fibroblasts were normalized, and a variety of signaling abnormalities were corrected (Capell et al., 2005; Glynn and Glover, 2005; Yang et al., 2005). *In vivo*, FTI treatment in several HGPS mouse models yielded dramatically improved disease phenotypes including reduced bone fractures, improved weight gain, increased lifespan (Fong et al., 2006a), and improved vascular pathology (Capell et al., 2008).

Most importantly, a 2 year prospective single-arm clinical trial treating 25 HGPS patients aged between 3 and 16 years with the oral FTI lonafarnib yielded positive results with treatment for a minimum of 2 years (Gordon et al., 2012). Subgroups of patients experienced increased rate of weight gain, decreased vascular stiffness measured via decreased carotid-femoral pulse wave velocity (PWV_{cf}) and carotid artery echodensity, increased radial bone structural rigidity, or improved sensorineural hearing (Gordon et al., 2012). Before treatment, more than half of the study participants had a history of headaches, four had a history of TIA or strokes, and four had a history of prior seizure. During treatment, stroke frequency was decreased and prevalence and frequency of headaches were reduced (Ullrich et al., 2013). Though neurologic outcomes were not identified as primary or secondary outcomes in the parent study, largely because baseline prevalence rates were unknown, these data suggest that disease modifying treatment may also alter the progression of underlying cerebrovascular disease. Finally, estimated lifespan was improved (Gordon et al., 2014), likely as a result of lonafarnib but also possibly due in part to additional farnesylation inhibition using pravastatin and aminobisphosphonate (see below). Dermatologic and dental symptoms, joint contractures, insulin resistance, lipodystrophy, and bone mineral density were unaffected by drug treatment. The improved cardiovascular status of children with HGPS is a potentially important finding because failure of this organ system is the ultimate cause of mortality. Arterial stiffness normally increases with age, and although the HGPS cohort initially exhibited PWV_{cf} equivalent to the age range of 60–69 years, median end-of-therapy PWV_{cf} was in the range of a typical 40–49-year-old. Based on studies of non-HGPS patients with diabetes, where aortic PWV increments as small as 1 meters per second have a significant independent effect on reduced mortality (Cruickshank et al., 2002), the median 4.5 meters per second decrease detected in this study implies a potential for benefit to cardiovascular mortality in children with HGPS.

Statins and aminobisphosphonates

Two additional drug classes that work further upstream of lonafarnib have promise for generating meaningful farnesylation inhibition in HGPS. Statins and aminobisphosphonates inhibit generation of the farnesyl moiety through the pathway of farnesyl pyrophosphate synthesis (see Fig. 18.4). Additionally, these drugs inhibit both farnesylation and geranylgeranylation, and may thwart an alternative progerin prenylation pathway called geranylgeranylation. *In vitro* studies as well as a therapeutic trial in a nonprogerin producing model mouse with progeroid features demonstrated were carried out using a combination of the farnesylation inhibitors pravastatin (a statin) and zoledronic acid (an aminobisphosphonate) (Varela et al., 2008). Treatment led to marked improvement of the nuclear abnormalities *in vitro*, as well as improved mouse model bone deficits, weight, and lifespan. Combination drug trials administering pravastatin and zoledronic acid with or without lonafarnib (an FTI) are ongoing (www.clinicaltrials.gov).

Methylation inhibitors

Similar to the farnesylation inhibitor treatment strategy, another potential treatment avenue is based on the understanding of lamin A post-translational modification. After farnesylation, the last three amino acids of prelamin A and progerin are cleaved, and the farnesylcysteine is methylated. Recently, inhibition of the post-translational methylation step improved HGPS cellular senescence *in vitro*, and ameliorated disease and early death in a progeroid mouse model (Ibrahim et al., 2013).

Aging treatment repurposing: rapamycin and resveratrol

Treatments with potential in the aging field may also benefit children with HGPS. The macrolide antibiotic rapamycin, which extends lifespan and improves stroke volume in several model organisms including aging mice, has been shown to decrease nuclear blebbing, increase degradation of progerin, reduce formation of insoluble progerin aggregates, and postpone senescence in HGPS fibroblasts by enhancing progerin clearance through autophagy (Cao et al., 2011b; Cenni et al., 2011; Blondel et al., 2014).

Another compound of interest is resveratrol, a natural plant compound derived from the skins of red grapes, which is thought to increase lifespan in yeast, worms, flies, and rodents. Resveratrol acts as a protein deacetylase SIRT1 activator that interacts with lamin A. In the presence of progerin, SIRT1 exhibits reduced association with the nuclear matrix and decreased deacetylase activity, leading to rapid depletion of adult stem cells in an HGPS model mouse. Resveratrol enhances the binding between SIRT1 and A-type lamins to increase its deacetylase activity. Treatment with resveratrol improved cellular phenotypes and extended lifespan in this mouse model (Liu et al., 2012).

Other studies using stem cell replenishment for HGPS are also being investigated in mouse models (Rosengardten et al., 2011).

RNA and gene therapy

Because the HGPS mutation results from the activation of an alternative pre-mRNA splice site, HGPS is a prime candidate for treatment via inhibition or elimination of this site. Both *in vitro* (Scaffidi and Misteli, 2005) and knockin HGPS mouse model (Osorio et al., 2011) disease improvement have been accomplished using morpholino antisense treatment, the latter resulting in improved body weight, extended lifespan, and correction of several mutant phenotypes (Osorio et al., 2011). An alternative approach is to eliminate the progerin mRNA using siRNA-based methods (Kole et al., 2012).

Correction of the genetic defect in HGPS patient-derived iPS cells has been achieved using adenovirus-based approach (Liu et al., 2011b). In future, zinc finger, TALEN or CRISPR-based approaches, in which the *LMNA* is repaired *ex vivo* and corrected cells are reintroduced into patients is a possibility, albeit a technically challenging one.

The microRNA miR-9 downregulates lamin and progerin synthesis in the neural cells (Jung et al., 2012; Nissan et al., 2012), which may explain the lack of nonvascular CNS disease in HGPS. Since miR-9 is also expressed peripherally, this could be a systemic treatment target in future.

Stem cell therapy

There is a mounting body of evidence for depletion of stem cells as a significant disease factor in HGPS (Halaschek-Wiener and Brooks-Wilson, 2007; Wenzel et al., 2012). *In vitro*, progerin affects the multipotency and differentiation of human mesenchymal stem cells (Scaffidi and Misteli, 2008) and HGPS patient-derived iPS cells exhibit differentiation defects (Zhang et al., 2011). In a murine HGPS model, the epidermal population of adult stem cells is depleted (Rosengardten et al., 2011). Once technical challenges for human treatment are overcome, stem cell therapy for HGPS may be a viable option.

CONCLUSION

As we continue to learn more about the pathobiology and natural history of HGPS, we move closer to positively influencing the clinical care, morbidity and mortality of children with HGPS. Importantly, HGPS serves as a model for research in rare diseases, with the rapid evolution from the discovery of the causative gene to bedside treatment. The trajectory is promising as we strive for a cure. As progerin is expressed at low levels in normal human cells and levels of progerin increase with advancing age (McClintock et al., 2007; Olive et al., 2010), elucidation of the role of progerin in the normal aging process may improve our understanding of a variety of age-dependent disorders.

References

Badame AJ (1989). Progeria. Arch Dermatol 125: 540–544.

Baek JH, Schmidt E, Viceconte N et al. (2015). Expression of progerin in aging mouse brains reveals structural nuclear abnormalities without detectible significant alterations in gene expression, hippocampal stem cells or behavior. Hum Mol Genet 24: 1305–1321.

Benson EK, Lee SW, Aaronson SA (2010). Role of progerin-induced telomere dysfunction in HGPS premature cellular senescence. J Cell Sci 123: 2605–2612.

Blondel S, Jaskowiak AL, Egesipe AL et al. (2014). Induced pluripotent stem cells reveal functional differences between drugs currently investigated in patients with

Hutchinson–Gilford progeria syndrome. Stem Cells Transl Med 3: 510–519.

Cao H, Hegele RA (2003). Lmna is mutated in hutchinson-gilford progeria (mim 176670) but not in wiedemann-rautenstrauch progeroid syndrome (mim 264090). J Hum Genet 48: 271–274.

Cao K, Blair CD, Faddah DA et al. (2011a). Progerin and telomere dysfunction collaborate to trigger cellular senescence in normal human fibroblasts. J Clin Invest 121: 2833–2844.

Cao K, Graziotto JJ, Blair CD et al. (2011b). Rapamycin reverses cellular phenotypes and enhances mutant protein clearance in Hutchinson–Gilford progeria syndrome cells. Sci Transl Med 3: 89ra58.

Capell BC, Collins FS (2006). Human laminopathies: nuclei gone genetically awry. Nat Rev Genet 7: 940–952.

Capell BC, Erdos MR, Madigan JP et al. (2005). Inhibiting farnesylation of progerin prevents the characteristic nuclear blebbing of Hutchinson–Gilford progeria syndrome. Proc Natl Acad Sci U S A 102: 12879–12884.

Capell BC, Olive M, Erdos MR et al. (2008). A farnesyltransferase inhibitor prevents both the onset and late progression of cardiovascular disease in a progeria mouse model. Proc Natl Acad Sci U S A 105: 15902–15907.

Cau P, Navarro C, Harhouri K et al. (2014). Nuclear matrix, nuclear envelope and premature aging syndromes in a translational research perspective. Semin Cell Dev Biol 29: 125–147.

Cenni V, Capanni C, Columbaro M et al. (2011). Autophagic degradation of farnesylated prelamin A as a therapeutic approach to lamin-linked progeria. Eur J Histochem 55. E36.

Cleveland RH, Gordon LB, Kleinman ME et al. (2012). A prospective study of radiographic manifestations in Hutchinson–Gilford progeria syndrome. Pediatr Radiol 42: 1089–1098.

Coppedè F (2013). The epidemiology of premature aging and associated comorbidities. Clin Interv Aging 8: 1023–1032.

Cox AD, Der CJ (2002). Farnesyltransferase inhibitors: promises and realities. Curr Opin Pharmacol 2: 388–393.

Cruickshank K, Riste L, Anderson SG et al. (2002). Aortic pulse-wave velocity and its relationship to mortality in diabetes and glucose intolerance: an integrated index of vascular function? Circulation 106: 2085–2090.

Dahl KN, Kalinowski A, Pekkan K (2010). Mechanobiology and the microcirculation: cellular nuclear and fluid mechanics. Microcirculation 17: 179–191.

De Sandre-Giovannoli A, Levy N (2006). Altered splicing in prelamin A-associated premature aging phenotypes. Prog Mol Subcell Biol 44: 199–232.

De Sandre-Giovannoli A, Bernard R, Cau P et al. (2003). Lamin A truncation in Hutchinson–Gilford progeria. Science 300: 2055.

Driscoll M, Albanese J, Xiong Z et al. (2012). Automated image analysis of nuclear shape: what can we learn from a prematurely aged cell? Aging (Albany NY). 4: 119–132.

Eriksson M, Brown WT, Gordon L et al. (2003). Recurrent de novo point mutations in lamin A cause Hutchinson–Gilford progeria syndrome. Nature 423: 293–298.

Espada J, Varela I, Flores I et al. (2008). Nuclear envelope defects cause stem cell dysfunction in premature-aging mice. J Cell Biol 181: 27–35.

Fong LG, Frost D, Meta M et al. (2006a). A protein farnesyltransferase inhibitor ameliorates disease in a mouse model of progeria. Science 311: 1621–1623.

Fong LG, Ng JK, Lammerding J et al. (2006b). Prelamin A and lamin A appear to be dispensable in the nuclear lamina. J Clin Invest 116: 743–752.

Furukawa K, Inagaki H, Hotta Y (1994). Identification and cloning of an MRNA coding for a germ cell-specific A-type lamin in mice. Exp Cell Res 212: 426–430.

Gerhard-Herman M, Smoot LB, Wake N et al. (2012). Mechanisms of premature vascular aging in children with Hutchinson–Gilford progeria syndrome. Hypertension 59: 92–97.

Gilford H (1904). Progeria: a form of senilism. Practitioner 73: 188–217.

Glynn MW, Glover TW (2005). Incomplete processing of mutant lamin A in Hutchinson–Gilford progeria leads to nuclear abnormalities which are reversed by farnesyltransferase inhibition. Hum Mol Genet 14: 2959–2969.

Goldman RD, Shumaker DK, Erdos MR et al. (2004). Accumulation of mutant lamin A causes progressive changes in nuclear architecture in Hutchinson–Gilford progeria syndrome. Proc Natl Acad Sci U S A 101: 8963–8968.

Gordon LB (2013). PRF by the numbers. Progeria Research Foundation, Available: http://www.progeriaresearch.org/prf-by-the-numbers.html (accessed November 27, 2014).

Gordon L, Brown WT, Collins FS (2003). Hutchinson–Gilford progeria syndrome. GeneReviews®: 1–15.

Gordon LB, Harten IA, Patti ME et al. (2005). Reduced adiponectin and HDL cholesterol without elevated C-reactive protein: clues to the biology of premature atherosclerosis in Hutchinson–Gilford progeria syndrome. J Pediatr 146: 336–341.

Gordon LB, Mccarten KM, Giobbie-Hurder A et al. (2007). Disease progression in Hutchinson–Gilford progeria syndrome: impact on growth and development. Pediatrics 120: 824–833.

Gordon CM, Gordon LB, Snyder BD et al. (2011). Hutchinson–Gilford progeria is a skeletal dysplasia. J Bone Miner Res 26: 1670–1679.

Gordon LB, Kleinman ME, Miller DT et al. (2012). Clinical trial of a farnesyltransferase inhibitor in children with Hutchinson–Gilford progeria syndrome. Proc Natl Acad Sci U S A 109: 16666–16671.

Gordon LB, Massaro J, D'Agostino Sr RB et al. (2014). Impact of farnesylation inhibitors on survival in Hutchinson–Gilford progeria syndrome. Circulation 130: 27–34.

Haithcock E, Dayani Y, Neufeld E et al. (2005). Age-related changes of nuclear architecture in *Caenorhabditis elegans*. Proc Natl Acad Sci U S A 102: 16690–16695.

Halaschek-Wiener J, Brooks-Wilson A (2007). Progeria of stem cells: stem cell exhaustion in Hutchinson–Gilford progeria syndrome. J Gerontol A Biol Sci Med Sci 62: 3–8.

Hennekam RC (2006). Hutchinson–Gilford progeria syndrome: review of the phenotype. Am J Med Genet A 140: 2603–2624.

Hisama FM, Lessel D, Leistritz D et al. (2011). Coronary artery disease in a Werner syndrome-like form of progeria characterized by low levels of progerin A splice variant of lamin A. Am J Med Genet A 155A: 3002–3006.

Hutchinson J (1886). Congenital absence of hair and mammary glands with atrophic condition of the skin and its appendages in a boy whose mother had been almost wholly bald from alopecia areata from the age of six. Med Chir Trans 69: 473–477.

Hutchinson J, Gilford H (1897). On a condition of mixed premature and immature development. Med Chir Trans 80: 2–42.

Ibrahim M, Sayin V, Akula M et al. (2013). Targeting isoprenylcysteine methylation ameliorates disease in a mouse model of progeria. Science 340: 1330–1333.

Jung HJ, Coffinier C, Choe Y et al. (2012). Regulation of prelamin A but not lamin C by miR-9, a brain-specific microRNA. Proc Natl Acad Sci U S A 109: E423–E431.

Kilic F, Dalton MB, Burrell SK et al. (1997). In vitro assay and characterization of the farnesylation-dependent prelamin A endoprotease. J Biol Chem 272: 5298–5304.

Kole R, Krainer AR, Altman S (2012). RNA therapeutics: beyond RNA interference and antisense oligonucleotides. Nat Rev Drug Discov 11: 125–140.

Kudlow BA, Stanfel MN, Burtner CR et al. (2008). Suppression of proliferative defects associated with processing-defective lamin A mutants by hTERT or inactivation of p53. Mol Biol Cell 19: 5238–5248.

Liu Y, Rusinol A, Sinensky M et al. (2006). DNA damage responses in progeroid syndromes arise from defective maturation of prelamin A. J Cell Sci 119: 4644–4649.

Liu GH, Barkho BZ, Ruiz S et al. (2011a). Recapitulation of premature ageing with iPSCs from Hutchinson–Gilford progeria syndrome. Nature 472: 221–225.

Liu GH, Suzuki K, Qu J et al. (2011b). Targeted gene correction of laminopathy-associated LMNA mutations in patient-specific iPSCs. Cell Stem Cell 8: 688–694.

Liu B, Ghosh S, Yang X et al. (2012). Resveratrol rescues SIRT1-dependent adult stem cell decline and alleviates progeroid features in laminopathy-based progeria. Cell Metab 16: 738–750.

Liu G-H, Ding Z, Izpisua Belmonte JC (2012). Ipsc technology to study human aging and aging-related disorders. Curr Opin Cell Biol 24: 765–774.

Machiels BM, Zorenc AH, Endert JM et al. (1996). An alternative splicing product of the lamin A/C gene lacks exon 10. J Biol Chem 271: 9249–9253.

Marino G, Ugalde AP, Fernandez AF et al. (2010). Insulin-like growth factor 1 treatment extends longevity in a mouse model of human premature aging by restoring somatotroph axis function. Proc Natl Acad Sci U S A 107: 16268–16273.

McClintock D, Gordon LB, Djabali K (2006). Hutchinson–Gilford progeria mutant lamin A primarily targets human vascular cells as detected by an anti-lamin A G608G antibody. Proc Natl Acad Sci U S A 103: 2154–2159.

McClintock D, Ratner D, Lokuge M et al. (2007). The mutant form of lamin A that causes Hutchinson–Gilford progeria is a biomarker of cellular aging in human skin. Plos One 2: 1–9.

Merideth MA, Gordon L, Clauss S et al. (2008). Phenotype and course of Hutchinson–Gilford progeria syndrome. New Engl J Med 358: 592–604.

Moulson CL, Fong LG, Gardner JM et al. (2007). Increased progerin expression associated with unusual LMNA mutations causes severe progeroid syndromes. Hum Mutat 28: 882–889.

Musich PR, Zou Y (2009). Genomic instability and DNA damage responses in progeria arising from defective maturation of prelamin A. Aging (Albany NY) 1: 28–37.

Nissan X, Blondel S, Navarro C et al. (2012). Unique preservation of neural cells in Hutchinson–Gilford progeria syndrome is due to the expression of the neural-specific miR–9 microRNA. Cell Rep 2: 1–9.

Olive M, Harten I, Mitchell R et al. (2010). Cardiovascular pathology in Hutchinson–Gilford progeria: correlation with the vascular pathology of aging. Arterioscler Thromb Vasc Biol 30: 2301–2309.

Osorio FG, Navarro CL, Cadinanos J et al. (2011). Splicing-directed therapy in a new mouse model of human accelerated aging. Sci Transl Med 3: 106–107.

Pegoraro G, Kubben N, Wickert U et al. (2009). Ageing-related chromatin defects through loss of the nurd complex. Nat Cell Biol 11: 1261–1267.

Pesce K, Rothe MJ (1996). The premature aging syndromes. Clin Dermatol 14: 161–170.

Rankin J, Ellard S (2006). The laminopathies: a clinical review. Clin Genet 70: 261–274.

Reddel CJ, Weiss AS (2004). Lamin a expression levels are unperturbed at the normal and mutant alleles but display partial splice site selection in hutchinson-gilford progeria syndrome. J Med Genet 41: 715–717.

Reichel W, Garcia-Bunuel R (1970). Pathologic findings in progeria: myocardial fibrosis and lipofuscin pigment. Am J Clin Path 53: 243–253.

Rork JF, Huang JT, Gordon LB et al. (2014). Initial cutaneous manifestations of Hutchinson–Gilford progeria syndrome. Pediatr Dermatol 31: 196–202.

Rosengardten Y, Mckenna T, Grochova D et al. (2011). Stem cell depletion in Hutchinson–Gilford progeria syndrome. Aging Cell 10: 1011–1020.

Sagelius H, Rosengardten Y, Schmidt ES et al. (2008). Reversible phenotype in a mouse model of Hutchinson–Gilford progeria syndrome. J Med Genet 45: 794–801.

Scaffidi P, Misteli T (2005). Reversal of the cellular phenotype in the premature aging disease Hutchinson–Gilford progeria syndrome. Nat Med 11: 440–445.

Scaffidi P, Misteli T (2008). Lamin A-dependent misregulation of adult stem cells associated with accelerated ageing. Nat Cell Biol 10: 452–459.

Schmidt E, Nilsson O, Koskela A et al. (2012). Expression of the Hutchinson–Gilford progeria mutation during osteoblast development results in loss of osteocytes irregular mineralization and poor biomechanical properties. J Biol Chem 287: 33512–33522.

Shumaker D, Kuczmarski E, Goldman R (2003). The nucleoskeleton: lamins and actin are major players in essential nuclear functions. Curr Opin Cell Biol 15: 358–366.

Silvera VM, Gordon LB, Orbach DB et al. (2013). Imaging characteristics of cerebrovascular arteriopathy and stroke in Hutchinson–Gilford progeria syndrome. Am J Neuroradiol 34: 1091–1097.

Sinensky M, Fantle K, Trujillo M et al. (1994). The processing pathway of prelamin A. J Cell Sci 107: 61–67.

Snow CJ, Dar A, Dutta A et al. (2013). Defective nuclear import of Tpr in progeria reflects the Ran sensitivity of large cargo transport. J Cell Biol 201: 541–557.

Stehbens W, Delahunt B, Shozawa T et al. (2001). Smooth muscle cell depletion and collagen types in progeric arteries. Cardiovasc Pathol 10: 133–136.

Tilli CM, Ramaekers FC, Broers JL et al. (2003). Lamin expression in normal human skin actinic keratosis squamous cell carcinoma and basal cell carcinoma. Br J Dermatol 148: 102–109.

Tsze DS, Valente JH (2011). Pediatric stroke: a review. Emerg Med Int 2011: 734506.

Ugalde AP, Ramsay AJ, De La Rosa J et al. (2011). Aging and chronic DNA damage response activate a regulatory pathway involving miR-29 and p53. EMBO J 30: 2219–2232.

Ullrich NJ, Silvera VM, Campbell SE et al. (2012). Craniofacial abnormalities in Hutchinson–Gilford progeria syndrome. AJNR Am J Neuroradiol 33: 1512–1518.

Ullrich NJ, Kieran MW, Miller DT et al. (2013). Neurologic features of Hutchinson–Gilford progeria syndrome after lonafarnib treatment. Neurology 81: 427–430.

Varela I, Cadinanos J, Pendas AM et al. (2005). Accelerated ageing in mice deficient in Zmpste24 protease is linked to p53 signalling activation. Nature 437: 564–568.

Varela I, Pereira S, Ugalde AP et al. (2008). Combined treatment with statins and aminobisphosphonates extends longevity in a mouse model of human premature aging. Nat Med 14: 767–772.

Varga R, Eriksson M, Erdos MR et al. (2006). Progressive vascular smooth muscle cell defects in a mouse model of Hutchinson–Gilford progeria syndrome. Proc Natl Acad Sci U S A 103: 3250–3255.

Weber K, Plessmann U, Traub P (1989). Maturation of nuclear lamin A involves a specific carboxy–terminal trimming which removes the polyisoprenylation site from the precursor; implications for the structure of the nuclear lamina. Febs Lett 257: 411–414.

Wenzel V, Roedl D, Gabriel D et al. (2012). Naive adult stem cells from patients with Hutchinson–Gilford progeria syndrome express low levels of progerin in vivo. Biol Open 1: 516–526.

Wuyts W, Biervliet M, Reyniers E et al. (2005). Somatic and gonadal mosaicism in Hutchinson–Gilford progeria. Am J Med Genet A 135: 66–68.

Yang SH, Bergo MO, Toth JI et al. (2005). Blocking protein farnesyltransferase improves nuclear blebbing in mouse fibroblasts with a targeted Hutchinson–Gilford progeria syndrome mutation. Proc Natl Acad Sci U S A 102: 10291–10296.

Zhang J, Lian Q, Zhu G et al. (2011). A human Ipsc model of Hutchinson Gilford progeria reveals vascular smooth muscle and mesenchymal stem cell defects. Cell Stem Cell 8: 31–45.

Zhang H, Kieckhaefer JE, Cao K (2013). Mouse models of laminopathies. Aging Cell 12: 2–10.

Handbook of Clinical Neurology, Vol. 132 (3rd series)
Neurocutaneous Syndromes
M.P. Islam and E.S. Roach, Editors

Chapter 19

Encephalocraniocutaneous lipomatosis

MOISE L. LEVY[1,2,3]* AND CATHERINE MASSEY[4]

[1]*Dell Children's Medical Center, Austin, TX, USA*

[2]*Dell Medical School, University of Texas, Austin, TX, USA*

[3]*Baylor College of Medicine, Houston, TX, USA*

[4]*New Braunfels Pediatric Associates, New Braunfels, TX, USA*

INTRODUCTION

Encephalocraniocutaneous lipomatosis (ECCL) is a sporadic, congenital neurocutaneous syndrome first described by Haberland and Perou (1970). Although it was briefly referred to as Fishman's syndrome after a publication by Fishman and colleagues (Fishman et al., 1978), it is most commonly referred to by its more descriptive name of ECCL (or otherwise referred to as Haberland syndrome). Since its initial characterization following the autopsy of a 5-year-old boy with unilateral lipomas of the head and neck, lipodermoids of the eye, and ipsilateral brain anomalies, less than 60 cases have been described along with the development of diagnostic criteria (Haberland and Perou, 1970; Fishman, 1987; Hunter, 2006; Ayer and Zouros, 2011; Delfino et al., 2011). Although extremely rare, a few preliminary conclusions about the etiology of ECCL can be drawn and will be discussed later. Males and females appear to be equally affected, and no geographic or ethnic background associations have been established. Thus far, no recurrence in siblings or offspring has been documented. A clear pattern of inheritance is therefore not known (Rubegni et al., 2003; Moog, 2009).

CASE HISTORY

A 1-month-old girl was seen for evaluation of skin lesions and abnormal hair growth. She was the product of a term uncomplicated pregnancy. Examination showed a series of left facial soft papules and a single (clinically) subcutaneous soft papule, also on the left face. A single patch of darker hair was noted on the left frontal scalp, and an apparent lipodermoid was seen to involve the left eye (Figs. 19.1, 19.2).

A diagnosis of ECCL was considered, and neuroimaging was ordered. Consultations with genetics, ophthalmology, and neurosurgery were requested. MRI of the head showed a left cerebellopontine angle lipoma and a left cranial arachnoid cyst. Ophthalmology confirmed the diagnosis of an oculodermoid and neurosurgery has been following the lipoma and arachnoid cyst. The lipoma has shown slight enlargement by imaging. Chromosomal microarray analysis revealed no known microdeletion or microduplication syndromes, and no deletions of the mitochondrial genome were noted. No seizures or cardiovascular dysfunction have occurred. A spine survey was normal. Developmentally, the child is progressing normally.

CLINICAL MANIFESTATIONS

The clinical manifestations of ECCL include dermatologic, ocular, and central nervous system (CNS) anomalies. Although findings are present at the time of birth, the diagnosis of ECCL is frequently delayed until about 4–5 years of life or later and may have significant variation from patient to patient. By definition, it is a nonprogressive disorder felt to be related to a specific window in embryologic development involving neural crest cells and the closure of the anterior neural tube. Skin lesions are predominantly located in the craniofacial region with the findings of nonscarring alopecia (with or without underlying fatty tissue known as nevus psiloliparus) and subcutaneous fatty masses among the most important skin anomalies seen with or in ECCL. It is the smooth, hairless, fatty tissue nevus of the scalp

*Correspondence to: Moise L. Levy, MD, Dell Children's Medical Center, 4900 Mueller Blvd, Administration, Austin, Texas 78723, USA. Tel: +1-512-324-0089, E-mail: mllevy@seton.org

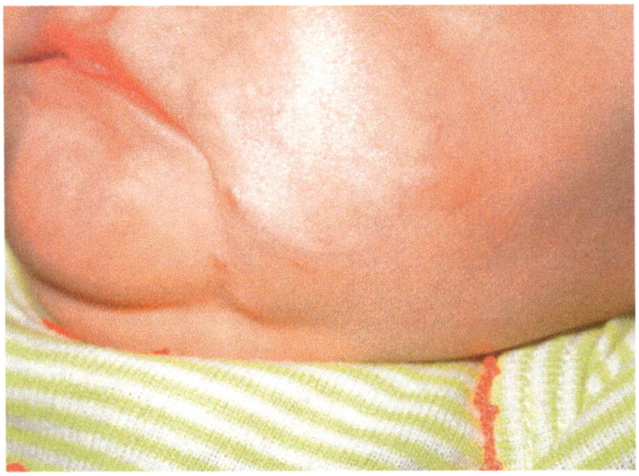

Fig. 19.1. Photograph of child showing left-sided soft papules and left-sided facial lipoma.

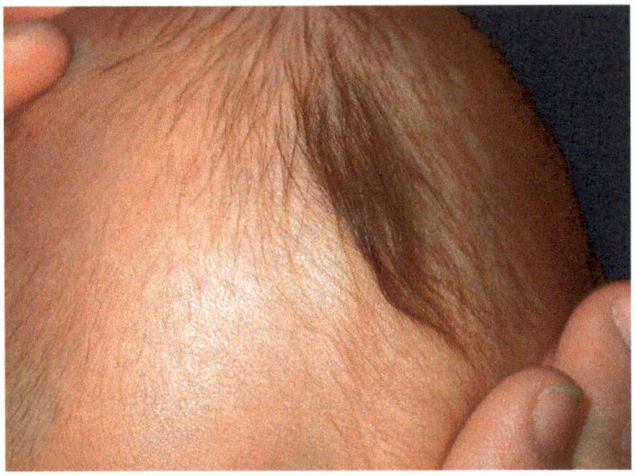

Fig. 19.2. Photograph of left frontal scalp illustrating single patch of darker and longer hair.

known as the nevus psiloliparus (NP) that is deemed the dermatologic hallmark of this syndrome, and Moog (2009) noted this in 44 of 54 patients described. Although previously considered pathognomonic for ECCL, two healthy girls with isolated NP were identified by Happle and Kuster (Happle and Kuster, 1998; Happle and Horster, 2004). Our illustrative patient likely has ECCL without NP but with abnormal hair growth and typical skin and neurologic features.

Alopecia that is not associated with a fatty nevus and is linear or patchy may follow the lines of Blaschko, further supporting the hypothesis of neural crest cell origin. Although some patients have scarring alopecia from focal skin aplasia or hypoplastic skin defects, such defects other than those on the scalp are very rare. In general, subcutaneous fatty tumors occur in the frontotemporal or zygomatic regions and may be bilateral or unilateral. Subcutaneous lipomas outside the craniofacial region are rare. Lastly, 74% of the patients in one report had small nodular skin tags (histologically fibroma, lipoma, fibrolipoma, or choristomas), most often on the eyelids or in the lateral periocular region. Some patients have several café-aulait spots (Moog, 2009).

The ocular findings of ECCL are typically benign tumors that include epibulbar or limbal dermoids (derived from epidermis and dermis) and lipodermoids (consisting mainly of fat) which rarely result in decreased vision. Choristomas with or without other eye anomalies are seen in most patients (Hunter, 2006). In contrast, ocular dermolipomas have been associated with significant corneal and scleral anomalies, anterior chamber abnormalities, ocular and palpebral colobomas, aniridia, microphthalmia, and calcification of the globe. The ocular findings of ECCL are the most easily recognizable characteristics, and 40% of patients have bilateral abnormalities of the eyes (and/or skin) (Moog, 2009).

As with other aspects of ECCL, lipomas are the primary CNS anomaly and occurred in 33 of the 54 cases reviewed by Moog (2009). Intracranial lipomas are frequently located in the cerebellopontine angle, a somewhat unique feature as it is more typical to find lipomas between cerebral hemispheres (Truwit and Barkovich, 1990). These lipomas may also be spinal and can extend the length of the cord, prompting some to suggest imaging of the spine as well (Dhouib et al., 2013). Subcutaneous fatty masses may overlie or occur adjacent to these spinal lipomas. Additional CNS lesions include congenital abnormalities of the meninges, asymmetric cerebral atrophy, dilated ventricles/hydrocephalus, porencephalic cysts, cortical dysplasia, parenchymal calcifications, and intracranial vessel anomalies. Frequently these intracranial findings are ipsilateral to unilateral skin findings (Fishman, 1987; Delfino et al., 2011), which is consistent with a mosaic condition. Given the variability of CNS findings, the neurologic presentation also varies greatly from patient to patient. It is important to note that the severity of the intracranial anomalies with ECCL does *not* correspond precisely with the extent of the associated skin and eye findings. There may be accompanying epilepsy with a neurodevelopmental spectrum that ranges from normal development to severe learning impairment. However, mild or severe cognitive impairment occurs in roughly two-thirds of ECCL patients, but there is insufficient long-term follow-up to detect progressively abnormal neurodevelopment.

Additional associated findings include skeletal anomalies including macrocephaly, jaw tumors (such as osteomas or odontoma), or ossifying fibromas which may be progressive. Three patients with a severe form of ECCL notably had multiple lytic bone lesions, only one of which was found to be progressive (Moog et al., 2007b). Congenital heart disease, particularly aortic coarctation, has been documented (Moog, 2009).

Evaluation, treatment, and management should ideally involve specialists from neurosurgery, pediatrics, neurology, plastic surgery, ophthalmology, and dermatology, depending on the clinical presentation and syndrome manifestations.

DIAGNOSTIC CRITERIA

Hunter (2006) compared findings in 54 cases to determine the initial diagnostic criteria for ECCL. Notably, the nevus psiloliparus and choristoma were outlined as major criteria at that time. However, Moog (2009) introduced modified and more stringent criteria to reflect the significance of intracranial and intraspinal lipomas as well as the rare, but highly specific, features of aortic coarctation, jaw tumors, and bone cysts. Furthermore, these modified criteria reflect the finding of a nevus psiloliparus and one minor criterion as insufficient to establish a definite diagnosis. These criteria may again change as the molecular understanding of ECCL is better appreciated (Moog, 2009). The reader is referred to the complete diagnostic criteria (Table 19.1), which include ocular, dermatologic, and CNS features. Ocular choristomas are considered a major criterion, as are proven nevus psiloliparis and either intracranial or spinal lipomas. Our illustrative patient fulfills the clinical criteria for the diagnosis of ECCL.

GENETICS

The genetic basis and pathogenesis of ECCL remains unknown. However, given its sporadic and varied presentation, it has been postulated that ECCL results from a mosaic autosomal gene mutation. The often unilateral and "patterned" expression of the skin, eye, and internal findings suggests such. This mutation might occur as the result of a heterozygous mutation or through a somatic second hit type mechanism. An example of the same phenomenon is, perhaps, illustrated in the finding of RAS mutations in patients with extensive epidermal nevi, skeletal dysplasia, and rickets (Lim et al., 2014). While the precise genetic defect remains unknown, the affected tissues arise mostly from neural crest cells. One gene, architectural transcription factor HMG2A (involved in vasculogenesis and mesenchymal tumors), has been implicated in other overgrowth syndromes such as familial multiple lipomatosis and has been identified as a likely contributor in ECCL, but further study is necessary (Moog, 2009; Prontera et al., 2009).

DIFFERENTIAL DIAGNOSIS/VARIANTS

ECCL has been compared to various syndromes that appear to have overlapping phenotypes, including Proteus syndrome, oculoectodermal syndrome, oculocerebrocutaneous syndrome, and epidermal nevus syndromes and familial multiple lipomatosis. However, when the more strict diagnostic criteria of Moog 2009 are applied,

Table 19.1

Revised diagnostic criteria for encephalocraniocutaneous lipomatosis*

Eye	*Major criteria*
	1. Choristoma, with or without associated anomalies
	Minor criteria
	1. Corneal and other anterior chamber anomalies
	2. Ocular or eyelid coloboma
	3. Calcification of globe
Skin	*Major criteria*
	1. Proven naevus psiloliparis (NP)
	2. Possible NP and ≥ 1 of minor criteria 2–5
	3. ≥ 2 of minor criteria 2–5
	Minor criteria
	1. Possible NP
	2. Patchy or streaky nonscarring alopecia (without fatty naevus)
	3. Subcutaneous lipoma(s) in frontotemporal region
	4. Focal skin aplasia/hypoplasia on scalp
	5. Small nodular skin tags on eyelids or between outer canthus and tragus
Central nervous system	*Major criteria*
	1. Intracranial lipoma
	2. Intraspinal lipoma
	3. ≥ 2 of minor criteria
	Minor criteria
	1. Abnormal intracranial vessels, e.g. angioma, excessive vessels
	2. Arachnoid cyst or other abnormality of meninges
	3. Complete or partial atrophy of a hemisphere
	4. Porencephalic cyst(s)
	5. Asymmetrically dilated ventricles or hydrocephalus
	6. Calcification (not basal ganglia)
Other	*Major criteria*
	1. Jaw tumor (osteoma, odontoma or ossifying fibroma)
	2. Multiple bone cysts
	3. Aortic coarctation
Application of criteria to the diagnosis of ECCL	*Definite case*
	1. Three systems involved, major criteria in ≥ 2, or
	2. Three systems involved, proven NP or possible NP+≥ 1 of minor skin criteria 2–5
	3. Two systems involved with major criteria, one of which proven NP or possible NP +≥ 1 of minor skin criteria 2–5
	Probable case
	1. Two systems involved, major criteria in both
	2. Two systems involved, proven or possible NP

*Reproduced from Moog, 2009.

ECCL, encephalocraniocutaneous lipomatosis; NP, naevus psiloliparis.

it becomes apparent that ECCL remains a distinct diagnostic entity.

It has been suggested that ECCL could represent as a form of localized Proteus syndrome. And while hyperostosis of the skull, dysregulation of adipose tissue, and vascular malformations are present in both, the specifics of these findings separate these two diagnoses. In addition, Proteus syndrome is characterized by progressive, disproportionate, and distorting overgrowth while ECCL (with the exception of skeletal cysts and jaw tumors) is characterized by congenital, nonprogressive findings that are typically associated with overlying alopecia or NP. Genetic diagnosis, as well, is available for PS (Haberland and Perou, 1970; Weidemann and Burgio, 1986; McCall et al., 1992; Rizzo et al., 1993; Cohen, 2005; Moog et al., 2007a; Moog, 2009).

Oculocerebrocutaneous syndrome (OCCS) is also frequently on the differential given the characteristic triad of congenital brain, eye, and skin anomalies predominantly affecting the craniofacies; however, it is again the specific type and location of these anomalies which set these two syndromes apart. In OCCS, a postauricular presentation of aplasia cutis and cystic micropthalmia are contrasted with the findings of NP and epibulbar dermoids in ECCL. Additionally, OCCS presents with a different characteristic pattern of primary brain malformations consisting of frontal predominant polymicrocgyria and agenesis of the corpus callosum, findings not typical of ECCL brain malformations. Lastly, from a genetic standpoint, OCCS is presently felt to be nonmosaic and possibly X-linked in origin (Moog, 2009).

Oculoectodermal syndrome, with its characteristic findings of epibulbar dermoids, focal aplastic skin defects, and nonscarring alopecia of the scalp, has been proposed as a mild form of ECCL. However, a lack of intracranial lipomas in this syndrome separate it from ECCL. It is notable that insufficient numbers of patients with oculoectodermal syndrome have undergone brain/spine imaging to permit any final conclusions on the matter (Ardinger et al., 2007; Moog, 2009).

Epidermal naevus syndromes are sometimes compared to ECCL as both are felt to be mosaic neurocutaneous conditions. And while epidermal naevus syndromes are also characterized by skin, congenital eye, CNS, and skeletal findings, the epidermal nevi characteristic of these syndromes are significantly more complex than the NP found in ECCL and change over time (Moog, 2009). As noted above, a recent report has noted mosaic mutations in RAS to explain overlapping features seen in cases of the epidermal naevus syndromes (Lim et al., 2014). These syndromes are felt to be mosaic conditions resulting from postzygotic mutations involving multiple tissues. This could explain how differing tissues could be affected with different morphologic findings.

IMAGING

When considering the diagnosis of ECCL, a careful eye examination and neuroimaging should be performed. Porencephalic cysts, atrophy, and other brain malformations are seen in a variety of patients. MRI of the head and spine should be done looking for ipsilateral CNS lipomas. Other brain malformations have also been noted in some patients, such as vascular defects and arachnoid cysts (Moog, 2009; Dhouib et al., 2013). Importantly, the CNS findings do not correlate with clinical features such as seizures or learning differences.

PATHOLOGY

The pathologic features of most of the skin, eye, and CNS lesions are those of lipomas. Other benign skin tumors such as angiofibromas, "papillomas," or other hamartomas have been noted. Bone tumors such as osteomas have been found in some patients, and odontomas of the jaw have been noted as well. Either total lack of, or decrease in the amount of hair in the nevus psiloliparus of the scalp is seen when this lesion is present. Lastly, some patients have been diagnosed with astrocytomas (Valera et al., 2012).

MANAGEMENT

Individuals without symptoms due to the lipomatous masses or other tumors associated with ECCL can be followed clinically. Those with spinal cord compression from such lesions, as was reported in a patient with a radiculopathy (Chiang et al., 2014), might require surgical decompression. Management of epilepsy is to be accomplished as indicated. Epilepsy occurs in approximately 50% of patients with ECCL (Chiang et al., 2014). Other neurologic findings that may require intervention include hydrocephalus and tethered cord. A multidisciplinary approach to management can include dermatology, ophthalmology, neurology, neurosurgery, and others, as indicated clinically. The primary care physician should become well informed of the patient's needs. Lastly, psychosocial support should be a part of the management of the families and the patients as they mature.

References

Ardinger HH, Horii KA, Begleiter ML (2007). Expanding the phenotype of oculoectodermal syndrome: possible relationship to encephalocraniocutaneous lipomatosis. Am J Med Genet A 143A: 2959–2962.

Ayer RE, Zouros A (2011). Case report encephalocraniocutaneous lipomatosis: a review of its clinical pathology and neurosurgical indications. J Neurosurg Pediatr 8: 316–320.

Chiang CC, Lin SC, Wu HM et al. (2014). Clinical manifestation and neurosurgical intervention of encephalocraniocutaneous lipomatosis – a case report and review of the literature. Childs Nerv Syst 30: 13–17.

Cohen Jr MM (2005). Proteus syndrome: an update. Am J Med Genet 137C: 38–52.

Delfino LN, Fareillo G, Quattrocchi CC et al. (2011). Encephalocraniocutaneous lipomatosis (ECCL): neuroradiological findings in three patients and a new association with fibrous dysplasia. Am J Med Genet 155A: 1690–1696.

Dhouib A, Hanquinet S, La Scala GC (2013). Encephalocraniocutaneous lipomatosis: magnetic resonance imaging findings in a child. J Pediatr 163: 297.

Fishman MA (1987). Encephalo-cranio-cutaneous lipomatosis. J Child Neurol 2: 186–193.

Fishman MA, Chang CS, Miller JE (1978). Encephalocraniocutaneous lipomatosis. Pediatrics 61: 580–583.

Haberland C, Perou M (1970). Encephalocraniocutaneous lipomatosis. A new example of ectomesodermal dysgenesis. Arch Neurol 22: 144–155.

Happle R, Horster S (2004). Nevus psiloliparus: report of two nonsyndromic cases. Eur J Dermatol 14: 314–316.

Happle R, Kuster W (1998). Nevus psiloliparus? A distinct fatty tissue nevus. Dermatology 197: 6–10.

Hunter A (2006). Oculocerebrocutaneous and encephalocraniocutaneous lipomatosis syndromes: blind men and an elephant or separate syndromes? Am J Med Genet A 140: 709–726.

Lim YH, Ovejero D, Sugarman JS et al. (2014). Multilineage somatic activating mutations in HRAS and NRAS cause mosaic cutaneous and skeletal lesions, elevated FGF23 and hypophosphatemia. Hum Mol Genet 23: 397–407.

McCall S, Ramzy MI, Cure JK et al. (1992). Encephalocraniocutaneous lipomatosis and the Proteus syndrome: distinct entities with overlapping manifestations. Am J Med Genet 43: 662–668.

Moog U (2009). Encephalocraniocutaneous lipomatosis. J Med Genet 46: 721–729.

Moog U, Jones MC, Viskochil DH et al. (2007a). Brain anomalies in encephalocraniocutaneous lipomatosis. Am J Med Genet A 143: 2963–2972.

Moog U, Roelens F, Mortier GR et al. (2007b). Encephalocraniocutaneous lipomatosis accompanied by the formation of bone cysts: harboring clues to pathogenesis? Am J Med Genet A 143A: 2973–2980.

Prontera P, Stangoni G, Manes I et al. (2009). Encephalocraniocutaneous lipomatosis (ECCL) in a patient with history of familial multipla lipmatosis (FML). Am J Med Genet A Part A: 543–545.

Rizzo R, Stangoni G, Manes I et al. (1993). Encephalocraniocutaneous lipomatosis, Proteus syndrome and somatic mosaicism. Am J Med Genet 47: 653–655.

Rubegni P, Risulo M, Sbano P et al. (2003). Encephalocraniocutaneous lipomatosis (Haberland syndrome) with bilateral cutaneous and visceral involvement. Clin Exp Dermatol 28: 387–390.

Truwit CL, Barkovich AJ (1990). Pathogenesis of intracranial lipoma: an MR study in 42 patients. AJR Am J Roentgenol 155: 855–864.

Valera ET, Brassesco MS, Scrideli CA et al. (2012). Are patients with encephalocraniocutaneous lipomatosis at increased risk of developing low-grade gliomas? Childs Nerv Syst 28: 19–22.

Weidemann HR, Burgio GR (1986). Encephalocraniocutaneous lipomatosis and Proteus syndrome. Am J Med Genet 25: 403–404.

Handbook of Clinical Neurology, Vol. 132 (3rd series)
Neurocutaneous Syndromes
M.P. Islam and E.S. Roach, Editors

Chapter 20

Incontinentia pigmenti (Bloch–Sulzberger syndrome)

MOHAN J. NARAYANAN[1], SAMPATHKUMAR RANGASAMY[2], AND VINODH NARAYANAN[2,3,4]*
[1]*LSU Health Sciences Center, New Orleans School of Medicine, New Orleans, LA, USA*
[2]*Neurology Research, Barrow Neurological Institute, Phoenix, AZ, USA*
[3]*Center for Rare Childhood Disorders, Translational Genomics Research Institute, Phoenix, AZ, USA*
[4]*Arizona Pediatric Neurology and Neurogenetics Associates, Phoenix, AZ, USA*

HISTORY AND EPIDEMIOLOGY

Incontinentia pigmenti (IP), also known as Bloch–Sulzberger syndrome (OMIM #308300), is named for Bruno Bloch (1878–1933), who was Chair of Dermatology at the University of Zurich, and Marion B. Sulzberger (1895–1983), who worked with Bloch (Sulzberger, 1980). Bruno Bloch described the first patient at a meeting of the Swiss Dermatologic Society in 1925 and coined the term incontinentia pigmenti; Sulzberger later published the first detailed clinicopathologic study of this patient and reviewed four others (Sulzberger, 1928). Sulzberger reported yet another patient in 1938, pointing out the involvement of multiple female members of the family and summarizing the features in previously described individuals (Sulzberger, 1938). Common features included pigmented macules of "peculiar size, color, and form, resembling the veins seen in certain types of marble" (Sulzberger, 1938). Histologic studies of these areas of pigmentation showed "pigment within phagocytic chromatophores in the cutis, as though the epidermis had become incontinent of its own pigment and allowed this to descend into, and thus tattoo the cutis." Carney, in 1951, reported that the pigmentary abnormalities were often preceded by inflammatory vesicles and bullae in lines and patches, appearing soon after birth (Carney, 1951). These were sometimes followed by linear verrucous lesions (Mendelsohn et al., 1959). The association with a variety of other developmental defects (including mental deficiency) and the female predominance were emphasized. In 1970, Carney and Carney Jr. reviewed the literature and reported 21 additional patients (Carney and Carney, 1970). In this series, only three of 26 patients had mental deficiency. Also described in these early patients were lesions of the retina, sharply demarcated areas of alopecia, involvement of teeth and nails, and an association with mental subnormality (Bambridge, 1953; Oldfelt, 1959). Differentiation from epidermolysis bullosa and dermatitis herpetiformis was emphasized.

Early patients were reported from Europe and North America, but IP has been recognized throughout the world. IP is rare, estimated to occur in approximately 1 of 50 000 newborns. In a review of the world's literature through 1976 Carney identified 682 reported cases of IP and excluded 65 as being too atypical (Carney, 1976). In the remaining 653 cases, 593 were female and 16 were male. Most were caucasian, but the disorder was also reported in various other racial and ethnic groups. Over 1300 cases have been reported to date.

CLINICAL STAGES AND EVOLUTION

The diagnosis of IP is typically based on the unusual pattern and form of pigmented macules on the skin usually found in older children and sometimes based on pathologic study of skin biopsy specimens. The skin manifestations evolve from birth to adulthood, going through specific stages of development (Fig. 20.1) (Bodemer, 2013), although not all phases are documented in all individuals.

The earliest signs of IP are often present at or shortly after birth, with an erythematous, vesicular or bullous eruption. These lesions typically appear on the extremities or trunk, may be arranged in a linear fashion, and may persist for a few months (Carney and Carney, 1970). The vesicular or bullous lesions occur in about

*Correspondence to: Vinodh Narayanan, MD, 3330 North 2nd Street, Suite 402, Phoenix, AZ 85012, USA. Tel: +1-602-687-8555, E-mail: Vinodh@azpediatricneurology.com

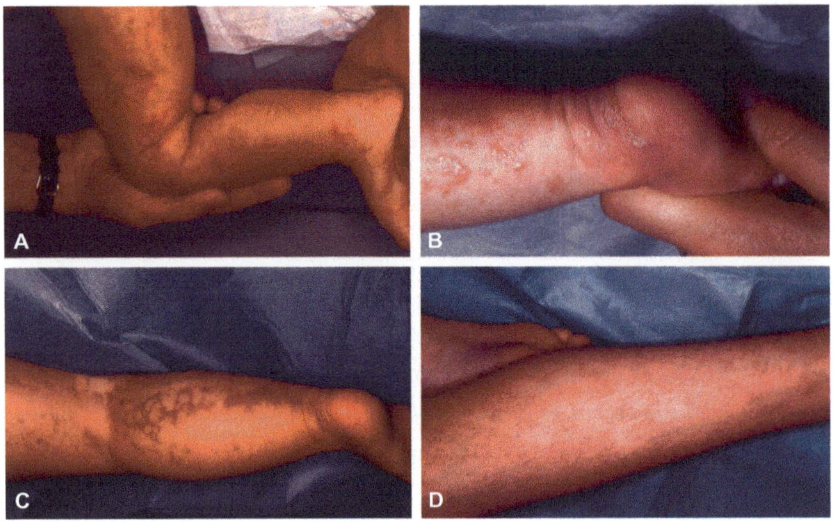

Fig. 20.1. Stages in the evolution of skin features in incontinentia pigmenti. (**A**) Erythematous, vesicular, bullous stage; (**B**) scarring, verrucous stage; (**C**) hyperpigmentation of the skin, following Blaschko's lines; (**D**) atrophic, hairless, linear depigmented lesions. (Reproduced from Bodemer, 2013, with permission.)

85% of patients (Carney, 1976). Included in the differential diagnosis of vesicular lesions in neonates are epidermolysis bullosa, herpes simplex, and chickenpox.

The vesicular phase is followed by scarring and a verrucous phase, which may also last weeks to months. The dyskeratotic, verrucous lesions are seen in up to 70% of individuals with IP. As the verrucous lesions regress, they may be replaced by areas of atrophy or depigmentation, but dyskeratotic papules may persist in bands on the feet or toes.

The pigmentary phase typically begins between several months to a few years of age; the hyperpigmented areas are not necessarily in the areas of inflammation and vesicular or verrucous eruption. This suggests that the hyperpigmentation is itself not a direct result of inflammation. Occasional newborns exhibit pigmented lesions in whorls and streaks over the trunk and extremities, without going through a vesicular or verrucous phase (Lerer et al., 1973). The pigmented macules are described as tan, brown, or chocolate in color, with sharply outlined jagged margins, in streaks or whorls, not conforming to any segmentation, lines of cleavage, or nerve or blood vessel distribution. Hyperpigmented streaks may follow a pattern known as Blaschko's lines, which may represent the end result of mosaicism or chimerism, combined with patterns of cellular (neural crest) migration during development (Happle, 1985). These pigmented lesions may become less prominent with time and eventually disappear. In affected adults, the only skin features may be areas of alopecia or atrophic, hairless, hypopigmented skin lesions (Hadj-Rabia et al. 2011).

Scalp, hair, and nail abnormalities, when present, persist into adulthood. The vesicular, verrucous, and pigmented stages may affect the scalp, but the most frequent finding is alopecia and scarring, usually of the vertex (Phan et al., 2005). This is found in 40–65% of cases. Nail dystrophy has been reported in 7–10% of cases of IP.

Dental features

The frequency of dental abnormalities in IP is probably underestimated; in a review of patients over 1 year of age, about 65% had major dental abnormalities (Carney, 1976). In other studies, dental abnormalities occur in up to 80% of cases of IP (Phan et al., 2005). A systematic review of patients published between 1993 and 2010 identified 513 (out of 1286 IP cases) with adequate information about dental or oral abnormalities (Minic et al., 2013a). In this cohort, 55% of patients had dental or oral abnormalities. Dental examination is important, as it may provide important diagnostic clues in older patients, where skin findings may be subtle or hidden. Dental findings in IP are similar to those in the anhidrotic form of ectodermal dysplasia and include abnormalities of dental shape (cone or peg shape, accessory cusps, tulip-shaped teeth, microdontia), hypodontia, and delayed teeth eruption (Gorlin and Anderson, 1960; Minic et al., 2013b).

Ophthalmologic features

Ocular disorders are among the most serious manifestation of IP, causing lifelong morbidity. Retinal vascular abnormalities may involve either the peripheral retina or the macula. Retinal abnormalities in infants may be missed without a dilated fundus exam and retinal angiography.

Table 20.1

Clinical diagnostic criteria for incontinentia pigmenti*

Major criteria	Minor criteria
Typical neonatal rash	Dental involvement (hypodontia, anodontia, peg shaped teeth)
Erythema	Alopecia (especially in vertex)
Vesicles	Woolly hair
Eosinophilia	Abnormal nails (ridging or pitting)
Typical hyperpigmentation	Retinal disease (neovascularization)
Mainly trunk	
Blaschko's lines	
Fading in adolescence	
Linear, atrophic, hairless lesions	

*Modified from Landy and Donnai, 1993.
Diagnosis of sporadic incontinentia pigmenti (IP) (absence of a first-degree female relative with IP): at least one major criterion is required. Minor criteria support the diagnosis. Diagnosis is questionable in absence of even a single minor criterion.
Diagnosis of familial IP in a female: first-degree female relative with IP, plus at least one major or minor criterion, or a history of multiple male miscarriages.

Angiography may demonstrate abnormal vessels traversing the center of the macula or neovascularization around the disc (Goldberg, 1994). In a detailed study which included serial retinal angiography in 12 girls with IP, findings included enlarged foveal avascular zones, progressive macular ischemia, and preretinal neovascularization (Goldberg, 1998). Retinal disease may lead to ischemic infarction, retinal detachment, and blindness (Goldberg, 2004). Based on these varied clinical features of patients with IP, clinical diagnostic criteria have been proposed (Table 20.1) (Landy and Donnai, 1993).

Neurologic involvement

Neurologic problems occur in about 30% of patients with IP, resulting in mental deficiency, motor abnormalities (hemiplegia, diplegia, or tetraplegia), and convulsive disorders (Meuwissen and Mancini, 2012). White matter abnormality (Fig. 20.2), cerebral infarction, thin corpus callosum, and cerebral atrophy are reported in older patients with IP, and there is a correlation between neurologic involvement and retinal disease (Lee et al., 1995; Bryant and Rutledge, 2007). Neonatal seizures with typical skin lesions (vesicular eruption) are probably the most common findings that lead to a diagnosis of IP, and occur in about 30% of cases (McPherson and Auth, 1963; Phan et al., 2005). Neonatal seizures are often focal, and are associated with imaging abnormalities demonstrated by magnetic resonance imaging (MRI) or angiography (MRA). MRI findings include areas of diffusion restriction in the subcortical or deep white matter, and areas of petechial hemorrhage (Hsieh and Chang, 2011; Venugopalan and Pang, 2012). Stroke, including recurrent acute strokes, can occur in newborns with IP, and is a manifestation of vaso-occlusive disease affecting both small and large vessels (Lee et al., 1995; Maingay-deGroof et al., 2008; Cartwright et al., 2009). In one well documented case of a neonate with IP (Hennel et al., 2003), serial imaging studies showed evolving areas of hemorrhagic infarction in the periventricular white matter associated with vascular disease (Fig. 20.3). These various studies demonstrate a significant cerebral macro- and microangiopathy in neonates with IP.

GENETICS

The familial nature and female preponderance of IP were recognized by Sulzberger who suggested that this was part of a hereditary familial syndrome (Sulzberger, 1938). Roughly half are isolated cases and the rest are familial.

In a study of familial cases, the distribution of normal males:normal females:affected females was close to 1:1:1; this distribution was consistent with dominant transmission (Carney and Carney, 1970). Possible explanations included autosomal dominant inheritance with prenatal lethality in males or X-linked dominant inheritance with prenatal lethality in males. In families with at least two affected children, the frequency of miscarriages approaches 25% (Carney, 1976). Female-to-female transmission, and an increase in the rate of miscarriages suggested X-linked transmission. IP has been reported in males, including several with Klinefelter syndrome (47, XXY) (Scheuerle, 1998). Rare examples of mother-to-son transmission escaping male lethality and two instances of father-to-daughter transmission have been reported (Hecht et al., 1982; Emery et al., 1993). Various mechanisms have been proposed to account for males with IP and a normal karyotype,

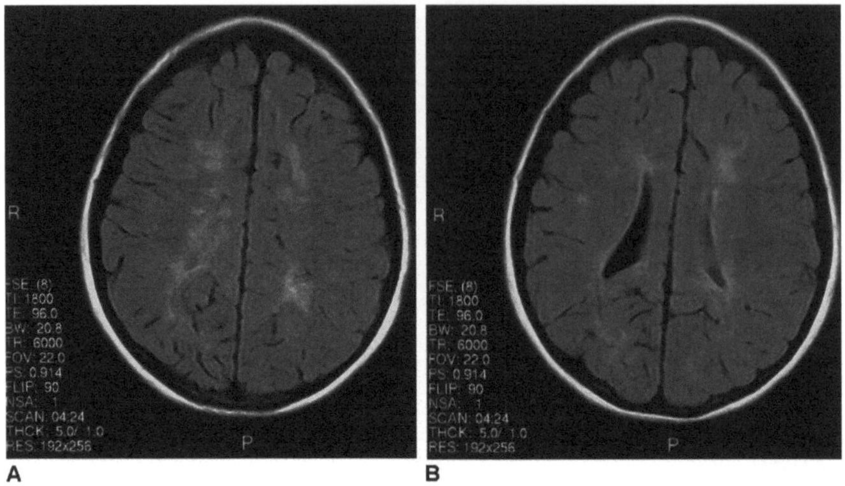

Fig. 20.2. Axial FLAIR (fluid-attenuated inversion recovery) MR images (**A**) and (**B**) in an 8-year-old girl with incontinentia pigmenti. This shows typical hyperintensities in the subcortical and periventricular white matter. (Reproduced from Bryant and Rutledge, 2007, with permission.)

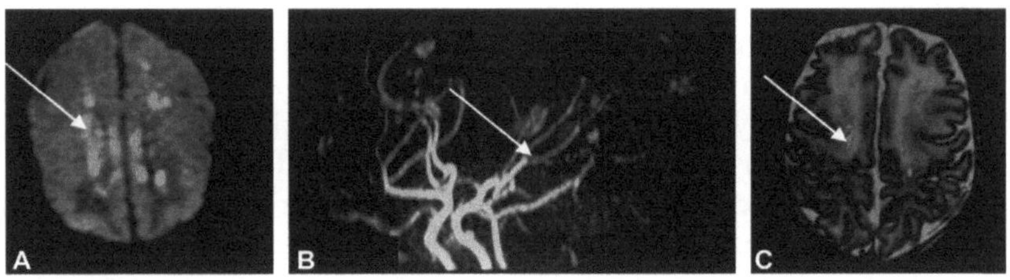

Fig. 20.3. Serial neuroimaging findings in a neonate with incontinentia pigmenti and focal seizures. (**A**) Day 8 of life: areas of diffusion restriction on diffusion weighted imaging; (**B**) day 13 of life: magnetic resonance angiography showed decreased branching and filling defects in the middle cerebral artery; (**C**) day 21 of life: magnetic resonance imaging shows cyst formation in an area of T1 and T2 hyperintensity in white matter. (Reproduced from Hennel et al., 2003, with permission.)

including postzygotic somatic mutation and mosaicism (Pacheco et al., 2006; Fusco et al., 2007).

Gene mapping

Based on a series of reports of females with IP with X-autosome translocation, it was proposed that the IP gene was located at Chr Xp11. However, linkage analysis and exclusion mapping studies in familial cases excluded the Xp11 region (Sefiani et al., 1988) and assigned the IP gene to Xq28 (Sefiani et al., 1989). This was confirmed in two large linkage studies which narrowed the critical region to a single segment of the distal Xq28 (Sefiani et al., 1991; Smahi et al.,1994). The idea that there are two genetic forms of incontinentia pigmenti, IP1 or sporadic IP (OMIM #308300), mapping to Xp11, and IP2 or familial IP (OMIM #308310), has been disputed (Happle, 1998). It has been suggested that the pigmentation changes in the Xp11 autosome translocation cases were a result of somatic mosaicism as might happen in hypomelanosis of Ito.

Gene identification

The studies that mapped the IP locus to a distal segment of Xq28 also demonstrated linkage to the factor VIII locus (Smahi et al., 1994). The gene encoding NEMO (NF-κB essential modulator)/IKKγ (IκB kinase-γ)/IKBKG had also been mapped proximal to the factor VIII locus, making it a candidate gene for IP (Jin and Jeang, 1999). The *NEMO* gene is a 23 kbp gene containing 10 exons. By studying cDNA from male and female fetuses, and examining genomic DNA in female patients, the International Incontinentia Pigmenti Consortium was able to demonstrate that mutations in *NEMO* were the cause of IP (International IP Consortium, 2000). The most common mutation causing IP (about 80% of cases) was a deletion of exons 4-10 (NEMOΔ4-10), a result of nonallelic recombination between repeated sequences within intron 3 and 3′ of exon 10. In fibroblasts from a male and female fetus with this deletion, there was a complete defect in NF-κB activation in response to treatment with tumor necrosis factor (TNF) (International IP Consortium, 2000).

Mutation spectrum, genotype–phenotype correlation

In a study of 357 unrelated IP patients, mutation of *NEMO* was found in about 80% of cases. The common NEMOΔ4-10 deletion was documented in about 90% of the cases with a *NEMO* mutation (Aradhya et al., 2001a). This was the most common mutation in both familial and sporadic (*de novo* mutation) cases. In a subset of patients with the NEMOΔ4-10 deletion where the parental origin of the mutation could be determined, it was on the paternal chromosome in 70% and maternal chromosome in 30% of cases. Other mutations documented in IP cases included primarily nonsense, frameshift (small insertions or deletions), and splice site mutations resulting in protein truncation (Aradhya et al., 2001a; Fusco et al., 2004). These mutations are scattered throughout the gene. Rare missense mutations have been reported, and can cause a mild IP phenotype or typical IP phenotype. Mutations affecting the C-terminus of NEMO (duplication or deletion within a cytosine tract in exon 10) cause atypical forms of IP (Aradhya et al., 2001b).

Allelic disorders

X-linked and autosomal dominant forms of hypohydrotic ectodermal dysplasias are known to be caused by mutations in genes linked to the TNF–NFκB signaling pathway (Smahi et al., 2002; Courtney et al., 2005). The X-linked recessive disorder, hypohydrotic ectodermal dysplasia with immune deficiency (HED-ID; OMIM 300291) is caused by hypomorphic mutations in exon 10 of the *NEMO* gene (Zonana et al., 2000). These mutations (missense, nonsense, frameshift) affect a putative zinc finger motif at the C-terminus of the NEMO protein, implying at least partially functional NEMO protein. In a recent report, an association between IP and systemic lupus erythematosus has been noted (Piccoli et al., 2012).

Skewed X-inactivation and selection against mutant alleles

Female carriers of X-linked disorders are mosaic, with populations of cells expressing the mutant or normal allele, as a result of X-chromosome inactivation (Franco and Ballabio, 2006). The deleterious effect of the mutation might be further ameliorated if mutation-expressing cells were eliminated over time, resulting in extreme skewing of X-inactivation. In one early study, skewing of X-inactivation (moderate or extreme) was found in 19/31 IP patients (60%) and extreme skewing in 11/31 (35%) of cases (Harris et al., 1992). A study of skin fibroblasts from affected females from three families showed that the normal allele was active in majority of cells (Migeon et al., 1989). The pattern of X-inactivation did not vary by site of the skin biopsy (normal or hyperpigmented skin). Supporting the idea of selection against cells expressing the mutant allele, serial studies of X-inactivation in blood cells from a female patient demonstrated progression from a random inactivation pattern at 24 months to completely skewed X-inactivation at 48 months (Martinez-Pomar et al., 2005). In another study, differential methylation at the androgen receptor locus was used to quantify skewed X-inactivation. In 21 multigenerational families, 98% of affected females showed complete skewing of X-inactivation, and in each case the inactive X was inherited form the mother with the IP phenotype (Parrish et al., 1996). Among the sporadic cases, 85% showed skewed X-inactivation. In a subset of these isolated cases, the inactivated chromosome was paternal, although there are cases described where the mutation arose on the paternal or the maternal chromosome.

Molecular mechanisms – IKKγ/NEMO, IKKs, IκB, NF-κB

The nuclear factor kappa B (NF-κB) family of inducible transcription factors plays a central role in the regulation of immune response to proinflammatory stimuli such as cytokines (TNF-α, interleukin-1 (IL-1)). These proinflammatory factors activate NF-κB mediated gene expression of chemokines, cytokines, and adhesion molecules that are critical for innate and adaptive immune responses (Baeuerle, 1998; Ghosh and Karin, 2002; Hayden and Ghosh, 2008). NF-κB also coordinates key developmental processes involving cell growth, proliferation, survival and apoptosis (Karin and Lin, 2002). Impairment in NF-κB function is likely to alter cellular homeostasis, because of its role in regulating diverse molecular pathways. Dysregulation of the NF-κB pathway is found in genetic disorders such as incontinentia pigmenti (IP), ectodermal dysplasias (EDs), cancer, and multifactorial biochemical disorders such as diabetes and metabolic syndrome (Courtois and Israel, 2000; Aradhya et al., 2001b; Courtois and Gilmore, 2006; Courtois and Smahi, 2006; Karin, 2006; Nelson, 2006).

The NF-κB pathway is a three-step cascade involving (1) the IκB kinase complex (IKK), (2) IκB proteins, and (3) NF-κB dimers (Table 20.2). NF-κB dimers (homo- or heterodimers) exist in an inactive state in the cytoplasm, bound to IκB proteins (inhibitors of NF-κB), which mask the nuclear localization signal and prevent nuclear translocation of NF-κB (Meffert and Baltimore, 2005). Multiple stimuli activate specific cell surface receptors, which transduce the signal to trigger IKK protein activation. The IKK complex (kinase) phosphorylates specific serine residues on IκB proteins, resulting in

Table 20.2

Members of IKK, IκB and NF-κB protein families

IKK family	IκB family	NF-κB family
IKKα	IκBα, IκBβ	P65
IKKβ	IκBϵ, IκBγ	RelB
NEMO (IKKγ)	IκB, Bcl-3	c_Rel
	IκBς	p100/p52(NF-κB2)
		p105/p50 (NF-κB1)

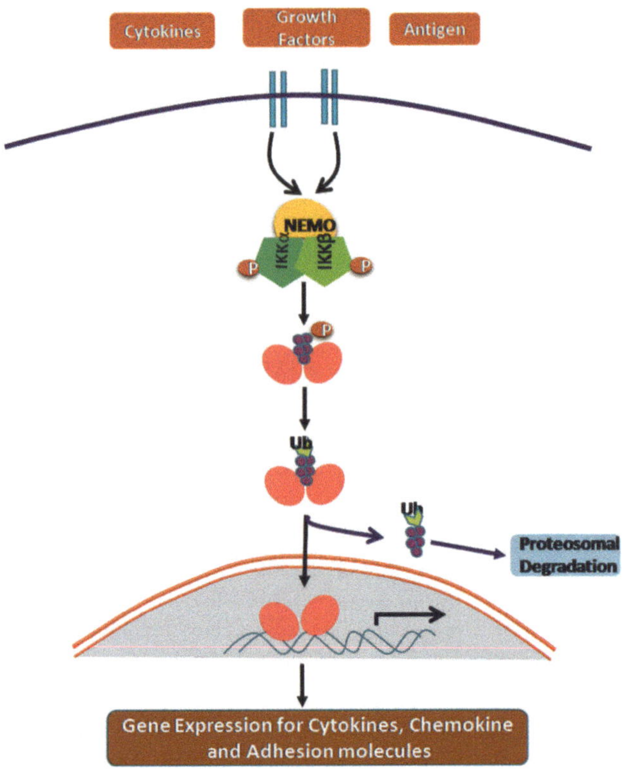

Fig. 20.4. Simplified schematic view of NF-κB canonical/classic pathway. In the absence of any cellular stimuli, NF-κB remains in the cytoplasm in an inactive state, bound to IκB proteins (inhibitors of NF-κB). Extracellular stimuli (antigens, growth factors, cytokines), through cell surface receptors, activate IKK complex consisting of NEMO, IKKα and IKKβ. IKK activation results in IκB phosphorylation and ubiquitin mediated proteosomal degradation. NF-κB dimers are steadily translocated to the nucleus, where they bind to specific DNA sequences and promote transcription of many target genes.

ubiquitination and degradation of IκB proteins. Once dissociated from IκB proteins, the NF-κB dimers are steadily translocated to the nucleus, where they bind to specific DNA sequences and promote transcription of many target genes (Fig. 20.4) (Karin, 2006).

The IκB kinase complex (IKK) contains two catalytic subunits, IKKα and IKKβ, which are serine/threonine kinases, and a noncatalytic, regulatory subunit, NEMO (IKKγ) (Chen et al., 1996; Didonato et al., 1997; Mercurio et al., 1999; May et al., 2000). The IKK complex can activate NF-κB through two distinct signaling pathways: (1) a canonical or classic pathway, activated by transient inflammatory stimuli, and (2) a noncanonical pathway that is activated by developmental signals, resulting in cell proliferation, differentiation, and maturation. The canonical pathway is regulated exclusively by NEMO-dependent IKK. Loss of NEMO in mice leads to complete inhibition of NF-κB activation through canonical pathways and to embryonic hepatocyte apoptosis (Rudolph et al., 2000; Schmidt-Supprian et al., 2000). The noncanonical pathway is mediated through a NEMO-independent mechanism, in which extracellular stimuli activate only IKKα via NF-κB-inducing kinase (NIK) (Häcker and Karin, 2006). IKKα-deficient mice do not survive past the immediate postnatal period and have extensive morphologic defects, without impairment of NF-κB activation in response to TNF-α or IL-1 (Hu et al., 1999; Takeda et al., 1999). Homozygous IKKβ-deficient mice also die during the embryonic period due to liver degeneration, but TNF and IL-1 mediated activation of NF-κB are severely impaired (Tanaka et al., 1999).

Studies done in human patients and in mouse models have contributed to our understanding of the pathogenesis of specific lesions in IP. In IP patients, expression of IKKγ/NEMO is significantly reduced in the affected areas without any changes in the level of IKKα and IKKβ (International IP Consortium, 2000). Mutant male offspring of female heterozygous IKKγ+/− mice die at day E12 from hepatocyte apoptosis, while heterozygous female carriers survive and develop a dermatosis with features typical of IP (Makris et al., 2000). Conditional knockout of NEMO or IKKβ in epidermal keratinocytes reproduced the dermatologic features of IP (Pasparakis et al., 2002; Nenci et al., 2006).

Female IKKγ+/− mice are mosaics with both *Ikkγ*− and *Ikkγ*+ keratinocytes. Loss of NF-κB activity in IKKγ+/− mice results in hyperproliferation and necrosis of *Ikkγ*− keratinocytes, releasing heat shock and other cellular proteins (Barnes and Karin, 1997). The released proteins activate Toll-like receptors in *Ikkγ*+ cells and induce cytokine production, creating a proinflammatory milieu (Asea et al., 2000; Ohashi et al., 2000). This increased generation of chemokines results in amplified infiltration of granulocytes into the epidermis and bullous lesions (Machado-Pinto et al., 1996). This process may account for the hyperproliferation, coupled with normal keratinocyte differentiation, resulting in hyperkeratotic lesions observed in both Ikkγ+/− mice and IP patients.

The mouse model is also characterized by increased sensitivity of keratinocytes towards apoptosis. It is proposed that the increased apoptosis is determined by TNF family of cytokines produced by adjacent IKKγ+ cells (Makris et al., 2000). NF-κB has been shown to protect cells from TNF induced apoptosis (Beg and Baltimore, 1996; Van Antwerp et al., 1996). Cells lacking NEMO undergo TNF-α, interleukin-1β (IL-1β) or lipopolysaccharide (LPS) induced apoptosis. Apoptosis of NEMO-deficient cells accounts for the disappearance of verrucous lesions and scarring in the second stage of disease, as well as for the progressive skewing of X-chromosome inactivation (Makris et al., 2000). Apoptosis and necrosis of Ikkγ− keratinocytes results in release of melanin, which may contribute to hyperpigmentation. This interaction between Ikkγ+ cells (capable of producing inflammatory factors) and Ikkγ− cells, causing a cyclic process of proliferation, necrosis, and apoptosis, continues as long as residual Ikkγ− cells remain (Fig. 20.5).

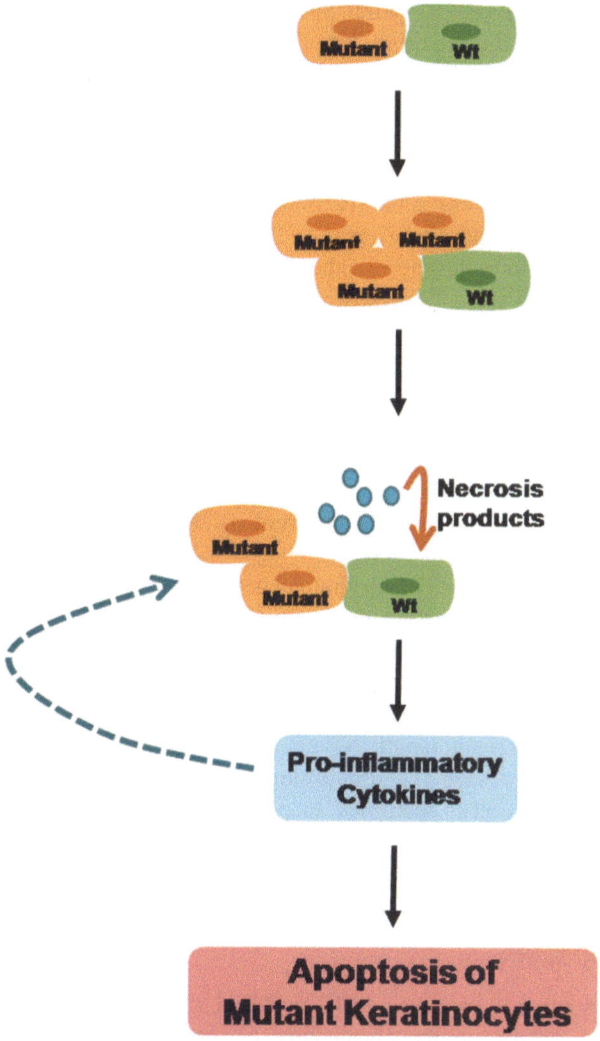

Fig. 20.5. Schematic diagram modeling effect of *NEMO* mutation in keratinocytes. Because of X-inactivation, epidermal keratinocytes express either wild-type (wt) or mutant X chromosome. Loss of NF-κB activity in the mutant cells leads to hyperproliferation and then necrosis of some of the mutant keratinocytes, which release heat shock proteins (HSP). The released proteins can activate Toll-like receptors in the adjacent wild-type cells, leading to secretion of proinflammatory cytokines. These cytokines in turn act on mutant cells lacking NF-κB activity and induce apoptosis (Makris et al., 2000).

TREATMENT, PROSPECTS FOR THE FUTURE

Treatment of specific organ system problems in IP generally follows standard practices. In most cases, the skin lesions do not require specific treatment as they tend to scar and involute with age, although steroids, tacrolimus, and other anti-inflammatory drugs have been used during the acute vesicular stage (Jessup et al., 2009; Kaya et al., 2009). Retinal and neurologic disease, both a result of vasculopathy, are probably the most serious, causing lifelong morbidity. Early ophthalmologic evaluation of infants diagnosed with IP is critical for detection of retinal ischemia and vascular proliferation. Treatment follows standard ophthalmologic practices, with photocoagulation and vitrectomy, among other surgical procedures. Progress in the treatment of diabetic retinopathy (vascular endothelial growth factor (VEGF) inhibitors) may lead to new approaches for treatment of IP related retinal disease (Rangasamy et al., 2012).

Neonatal seizures are often the initial symptom of IP; infants may be treated with anticonvulsants and for neonatal herpes encephalitis until CSF studies return normal. By this time, cerebral imaging studies may already show signs of microangiopathy with hemorrhagic infarction. In one reported case, treatment with aspirin did not prevent recurrent ischemia in that infant (Cartwright et al., 2009).

IP is the first defined human genetic disease involving the NF-κB signaling pathway. Over the past 27 years, much has been learned about the molecules and pathways that modulate NF-κB. The hope is that biological treatments discovered for cancer or autoimmune disorders (currently mostly NF-κB inhibitors) may be applied to prevent retinal and neurologic disease in IP. Instituting biological therapies to prevent cerebral vascular pathology in neonates with IP presents unique problems, unless we develop rapid clinical tests that predict the development of vasculopathy.

References

Aradhya S, Woffendin H, Jakins T et al. (2001a). A recurrent deletion in the ubiquitously expressed NEMO (IKK-gamma) gene accounts for the vast majority of incontinentia pigmenti mutations. Hum Mol Genet 10: 2171–2179.

Aradhya S, Courtois G, Rajkovic A et al. (2001b). Atypical forms of incontinentia pigmenti in male individuals result from mutations of a cytosine tract in exon 10 of NEMO (IKK-gamma). Am J Hum Genet 68: 765–771.

Asea A, Kraeft ST, Kurt-Jones EA et al. (2000). Hsp70 stimulates cytokine production through a CD14-dependant pathway, demonstrating its dual role as a chaperone and cytokine. Nat Med 6: 435–442.

Baeuerle PA (1998). Pro-inflammatory signaling: last pieces in the NF-κB puzzle? Curr Biol 8: R19–R22.

Bambridge H (1953). Incontinentia pigmenti: (Bloch–Sulzberger syndrome). Pediatrics 12: 622.

Barnes PJ, Karin M (1997). Nuclear factor-κB – a pivotal transcription factor in chronic inflammatory diseases. N Engl J Med 336: 1066–1071.

Beg AA, Baltimore D (1996). An essential role for NF-κB in preventing TNFα induced cell death. Science 274: 782–784.

Bodemer C (2013). Incontinentia pigmenti and hypomelanosis of Ito. In: O Dulac, M Lassonde, HB Sarnat (Eds.), Handbook of Clinical Neurology 111 (Third Series), Pediatric Neurology Part I, Elsevier, Edinburgh, pp. 341–347.

Bryant SA, Rutledge SL (2007). Abnormal white matter in a neurologically intact child with incontinentia pigmenti. Pediatr Neurol 36: 199–201.

Carney RG (1951). Incontinentia pigmenti. Arch Derm Syphilol 64: 126–135.

Carney Jr RG (1976). Incontinentia pigmenti: a world statistical analysis. Arch Dermatol 112: 535–542.

Carney RG, Carney Jr RG (1970). Incontinentia pigmenti. Arch Dermatol 102: 157–162.

Cartwright MS, White DL, Miller III LM et al. (2009). Recurrent stroke in a child with incontinentia pigmenti. J Child Neurol 24: 603–605.

Chen ZJ, Parent L, Maniatis T (1996). Site-specific phosphorylation of IκBα by a novel ubiquitination-dependent protein kinase activity. Cell 84: 853–862.

Courtney JM, Blackburn J, Sharpe PT (2005). The ectodysplasin and NF-κB pathways in odontogenesis. Arch Oral Biol 50: 159–163.

Courtois G, Gilmore TD (2006). Mutations in the NF-κB signaling pathway: implications for human disease. Oncogene 25: 6831–6843.

Courtois G, Israël A (2000). NF-κB defects in humans: the NEMO/incontinentia pigmenti connection. Sci STKE 58: pe1.

Courtois G, Smahi A (2006). NF-κB-related genetic diseases. Cell Death Differ 13: 843–851.

Didonato JA, Hayakawa M, Rothwarf DM et al. (1997). A cytokine-responsive IκB kinase that activates the transcription factor NF-κB. Nature 388: 548–554.

Emery MM, Siegfried EC, Stone MS et al. (1993). Incontinentia pigmenti: transmission from father to daughter. J Am Acad Dermatol 29: 368–372.

Franco B, Ballabio A (2006). X-inactivation and human disease: X-linked dominant male-lethal disorders. Curr Opin Genet Dev 16: 254–259.

Fusco F, Bardaro T, Fimiani G et al. (2004). Molecular analysis of the genetic defect in a large cohort of IP patients and identification of novel NEMO mutations interfering with NF-κB activation. Hum Mol Genet 13: 1763–1773.

Fusco F, Fimiani G, Tadini G et al. (2007). Clinical diagnosis of incontinentia pigmenti in a cohort of male patients. J Am Acad Dermatol 56: 264–267.

Ghosh S, Karin M (2002). Missing pieces in the NF-κB puzzle. Cell 109: S81–S96.

Goldberg MF (1994). The blinding mechanisms of incontinentia pigmenti. Trans Am Ophthalmol Soc 62: 167–179.

Goldberg MF (1998). Macular vasculopathy and its evolution in incontinentia pigmenti. Ophthalmic Genet 19: 141–148.

Goldberg MF (2004). The skin is not the predominant problem in incontinentia pigmenti. Arch Dermatol 140: 748–750.

Gorlin RJ, Anderson JA (1960). The characteristic dentition of incontinentia pigmenti. J Pediatr 57: 78–85.

Häcker H, Karin M (2006). Regulation and function of IKK and IKK-related kinases. Sci STKE 357: re13.

Hadj-Rabia S, Rimella A, Smahi A et al. (2011). Clinical and histologic features of incontinentia pigmenti in adults with nuclear factor-κB essential modulator gene mutations. J Am Acad Dermatol 64: 508–515.

Happle R (1985). Lyonization and the lines of Blaschko. Hum Genet 70: 200–206.

Happle R (1998). Incontinentia pigmenti versus hypomelanosis of Ito: the whys and wherefores of a confusing issue. Am J Med Genet 79: 64–65.

Harris A, Collins J, Vetrie D et al. (1992). X inactivation as a mechanism of selection against lethal alleles: further investigation of incontinentia pigmenti and X linked lymphoproliferative disease. J Med Genet 29: 608–614.

Hayden MS, Ghosh S (2008). Shared principles in NF-κB signaling. Cell 132: 344–362.

Hecht F, Hecht BK, Austin WJ (1982). Incontinentia pigmenti in Arizona Indians including transmission from mother to son inconsistent with the half chromatid mutation model. Clin Genet 21: 293–296.

Hennel SJ, Ekert PG, Volpe JJ et al. (2003). Insights into the pathogenesis of cerebral lesions in incontinentia pigmenti. Pediatr Neurol 29: 148–150.

Hsieh DT, Chang T (2011). Incontinentia pigmenti: skin and magnetic resonance imaging findings. Arch Neurol 68: 1080–1081.

Hu Y, Baud V, Delhase M et al. (1999). Abnormal morphogenesis but intact IKK activation in mice lacking the IKKα subunit of IκB kinase. Science 284: 316–320.

International Incontinentia Pigmenti (IP) Consortium (2000). Genomic rearrangement in NEMO impairs NF-κB activation and is a cause of incontinentia pigmenti. Nature 405: 466–472.

Jessup CJ, Morgan SC, Cohen LM et al. (2009). Incontinentia pigmenti: treatment of IP with topical tacrolimus. J Drugs Dermatol 8: 944–946.

Jin DY, Jeang KT (1999). Isolation of full-length cDNA and chromosomal localization of human NF-κB modulator NEMO to Xq28. J Biomed Sci 6: 115–120.

Karin M (2006). Nuclear factor-κB in cancer development and progression. Nature 441: 431–436.

Karin M, Lin A (2002). NF-κB at the crossroads of life and death. Nat Immunol 3: 221–227.

Kaya TI, Tursen U, Ikizoglu G (2009). Therapeutic use of corticosteroids in the vesiculobullous lesions of incontinentia pigmenti. Clin Exp Dermatol 34: e611–e613.

Landy SJ, Donnai D (1993). Incontinentia pigmenti (Bloch–Sulzberger syndrome). J Med Genet 30: 53–59.

Lee AG, Goldberg MF, Gillard JH et al. (1995). Intracranial assessment of incontinentia pigmenti using magnetic resonance imaging, angiography, and spectroscopic imaging. Arch Pediatr Adolesc Med 149: 573–580.

Lerer RJ, Ehrenkranz RA, Campbell AGM (1973). Pigmented lesions of incontinentia pigmenti in a neonate. J Pediatr 83: 503–505.

Machado-Pinto J, McCalmont TH, Golitz LE (1996). Eosinophilic and neutrophilic spongiosis: clues to the diagnosis of immunobullous diseases and other inflammatory disorders. Semin Cutan Med Surg 15: 308–316.

Maingay-de Groof F, Lequin MH, Roofthooft DW et al. (2008). Extensive cerebral infarction in the newborn due to incontinentia pigmenti. Eur J Paediatr Neurol 12: 284–289.

Makris C, Godfrey VL, Krähn-Senftleben G et al. (2000). Female mice heterozygous for IKK gamma/NEMO deficiencies develop a dermatopathy similar to the human X-linked disorder incontinentia pigmenti. Mol Cell 5: 969–979.

Martinez-Pomar N, Munoz-Saa I, Heine-Suner D et al. (2005). A new mutation in exon 7 of NEMO gene: late skewed X-chromosome inactivation in an incontinentia pigmenti female patient with immunodeficiency. Hum Genet 188: 458–565.

May MJ, D'Acquisto F, Madge LA et al. (2000). Selective inhibition of NF-κB activation by a peptide that blocks the interaction of NEMO with the IκB kinase complex. Science 289: 1550–1554.

McPherson A, Auth T (1963). Bloch–Sulzberger syndrome (incontinentia pigmenti): report of a case with prominent neurological features. Arch Neurol 8: 332–339.

Meffert MK, Baltimore D (2005). Physiological functions for brain NF-κB. Trends Neurosci 28: 37–43.

Mendelsohn RS, Medansky RS, Lazar MP (1959). Incontinentia pigmenti (Bloch–Sulzberger syndrome): a generalized disease. AMA J Dis Child 97: 348–352.

Mercurio F, Murray BW, Shevchenko A et al. (1999). IκB kinase (IKK)-associated protein 1, a common component of the heterogeneous IKK complex. Mol Cell Biol 19: 1526–1538.

Meuwissen MEC, Mancini GMS (2012). Neurological findings in incontinentia pigmenti: a review. Eur J Med Genet 55: 323–331.

Migeon BR, Axelman J, Jan de Beur S et al. (1989). Selection against lethal alleles in females heterozygous for incontinentia pigmenti. Am J Hum Genet 44: 100–106.

Minic S, Trpinac D, Gabriel H et al. (2013a). Dental and oral anomalies in incontinentia pigmenti: a systematic review. Clin Oral Investig 17: 1–8.

Minic S, Trpinac D, Obradovic M (2013b). Systematic review of central nervous system anomalies in incontinentia pigmenti. Orphanet J Rare Dis 8: 25.

Nelson DL (2006). NEMO, NF-κB signaling and incontinentia pigmenti. Curr Opin Genet Dev 16: 282–288.

Nenci A, Huth M, Funteh A et al. (2006). Skin lesion development in a mouse model of incontinentia pigmenti is triggered by NEMO deficiency in epidermal keratinocytes and requires TNF signaling. Hum Mol Genet 15: 531–542.

Ohashi K, Burkart V, Flohé S et al. (2000). Cutting edge heat shock protein 60 is a putative endogenous ligand of the Toll-like receptor-4 complex. J Immunol 164: 558–561.

Oldfelt V (1959). Incontinentia pigmenti. J Pediatr 54: 446–458.

Pacheco TR, Levy M, Collyer JC et al. (2006). Incontinentia pigmenti in male patients. J Am Acad Dermatol 55: 251–255.

Parrish JE, Scheuerle AE, Lewis RA et al. (1996). Selection against mutant alleles in blood leukocytes is a consistent feature in incontinentia pigmenti type 2. Hum Mol Genet 5: 1777–1783.

Pasparakis M, Courtois G, Hafner M et al. (2002). TNF-mediated inflammatory skin disease in mice with epidermis-specific deletion of IKK2. Nature 417: 861–866.

Phan TA, Wargon O, Turner AM (2005). Incontinentia pigmenti case series: clinical spectrum of incontinentia pigmenti in 53 female patients and their relatives. Clin Exp Dermatol 30: 474–480.

Piccoli GB, Attini R, Vigotti FN et al. (2012). NEMO syndrome (incontinentia pigmenti) and systemic lupus erythematosus: a new disease association. Lupus 21: 675–681.

Rangasamy S, McGuire PG, Das A (2012). Diabetic retinopathy and inflammation: novel therapeutic targets. Middle East Afr J Ophthalmol 19: 52–59.

Rudolph D, Yeh WC, Wakeham A et al. (2000). Severe liver degeneration and lack of NF-κB activation in NEMO/IKKγ-deficient mice. Genes Dev 14: 854–862.

Scheuerle AE (1998). Male cases of incontinentia pigmenti: case report and review. Am J Med Genet 77: 201–218.

Schmidt-Supprian M, Bloch W, Courtois G et al. (2000). NEMO/IKK γ-deficient mice model incontinentia pigmenti. Mol Cell 5: 981–992.

Sefiani A, Sinnett D, Abel L et al. (1988). Linkage studies do not confirm the cytogenetic location of incontinentia pigmenti on Xp11. Hum Genet 80: 282–286.

Sefiani A, Abel L, Heuertz S et al. (1989). The gene for incontinentia pigmenti is assigned to Xq28. Genomics 4: 427–429.

Sefiani A, M'rad R, Simard L et al. (1991). Linkage relationship between incontinentia pigmenti (IP2) and nine terminal X long arm markers. Hum Genet 86: 297–299.

Smahi A, Hyden-Granskog C, Peterlin B et al. (1994). The gene for the familial form of incontinentia pigmenti (IP2) maps to the distal part of Xq28. Hum Mol Genet 3: 273–278.

Smahi A, Courtois G, Rabia SH et al. (2002). The NF-κB signaling pathway in human diseases: from incontinentia pigmenti to ectodermal dysplasias and immune-deficiency syndromes. Hum Mol Genet 11: 2371–2375.

Sulzberger MB (1928). Uber eine bisher nicht beschriebene congenital Pigmentanomalie (incontinentia pigmenti). Archiv fur Dermatologie und Syphilologie 154: 19–32.

Sulzberger MB (1938). Incontinentia pigmenti (Bloch–Sulzberger): report of an additional case, with comment on possible relation to a new syndrome of familial and congenital anomalies. Arch Dermatol Syphilol 38: 57–69.

Sulzberger MB (1980). Three lessons learned in Bloch's clinic. Am J Dermatopathol 2: 321–325.

Takeda K, Takeuchi O, Tsujimura T et al. (1999). Limb and skin abnormalities in mice lacking IKKα. Science 284: 313–316.

Tanaka M, Fuentes ME, Yamaguchi K et al. (1999). Embryonic lethality, liver degeneration and impaired NF-κB activation in IKK-β-deficient mice. Immunity 10: 421–429.

Van Antwerp DJ, Martin SJ, Kafri T et al. (1996). Suppression of TNF-alpha-induced apoptosis by NF-κB. Science 274: 787–789.

Venugopalan P, Pang K (2012). Incontinentia pigmenti: presenting with neonatal seizures and diffuse MRI brain changes. Clin Dysmorphol 21: 231–233.

Zonana J, Elder ME, Schneidere LC et al. (2000). A novel X-linked disorder of immune deficiency and hypohidrotic ectodermal dysplasia is allelic to incontinentia pigmenti and due to mutations in IKK-gamma (NEMO). Am J Hum Genet 67: 1555–1562.

Handbook of Clinical Neurology, Vol. 132 (3rd series)
Neurocutaneous Syndromes
M.P. Islam and E.S. Roach, Editors

Chapter 21

Hypomelanosis of Ito

MARGIE REAM*
Department of Neurology, Nationwide Children's Hospital, Columbus, OH, USA

INTRODUCTION

Hypomelanosis of Ito (HI) is a neurocutaneous disorder with hypopigmented lesions following the lines of Blaschko (Fig. 21.1). Variable systemic findings can be associated with the skin lesions affecting the central nervous system (CNS), eye, musculoskeletal system, and kidney. HI was formerly known as incontinentia pigmenti achromians, but today is referred to by its eponym to avoid confusion with incontinentia pigmenti (IP).

HISTORY

Origin of the name

Dr. Minor Ito published a series of "Studies on Melanin" and ultimately a textbook in the 1950s. Therein he described a 22-year-old woman with hypopigmentation in linear and irregularly shaped patterns on her trunk and arms. She also had asymmetry of breast size but no other abnormalities were reported. Ito wrote that the findings were due to a condition that he called "incontinentia pigmenti achromians" (Ito, 1952). He coined this term because the areas of depigmentation were in the same pattern as areas of inflammation and hyperpigmentation seen in IP (Bloch–Sulzberger type) (Bloch, 1926). Interestingly, it has been proposed that his initial patient may actually have had incontinentia pigmenti, but the condition he thought he was describing remains. Today the term hypomelanosis of Ito is preferred to avoid confusion with IP (Jelinek et al., 1973). There continues to be some debate in the literature over whether HI is a distinct entity or a symptom associated with a variety of diseases (Sybert, 1994). Many studies reporting conditions associated with HI are affected by referral bias (e.g. neurologists tend to see patients with neurologic comorbidities). However, when diagnosed according to published criteria that include extracutaneous findings, HI stands as a neurocutaneous disorder deserving clinical attention (Ruiz-Maldonado et al., 1992).

Alfred Blaschko, a German dermatologist, presented his observation of curvilinear pigmentary patterns in 1901 (Fig. 21.2). It has since been proposed that the lines of Blaschko represent migratory pathways for fetal epidermal cell migration. The patterns of skin discoloration seen in hypomelanosis of Ito follow the lines of Blaschko (Blaschko, 1901), as they do in several other neurocutaneous disorders. Today there are a variety of conditions known with skin findings present along the lines of Blaschko that can be congenital or acquired, localized to the skin, or systemic (Bolognia et al., 1994). HI is one of the conditions that are usually present from birth.

Epidemiology

Studies of the epidemiology of HI were conducted in Spain and Italy. The estimated prevalence is 1 case per 7540 births and 1 in 82 000 people in the total population of Catania, Italy (Ruggieri and Pavone, 2000). Of patients admitted to a pediatric hospital in Spain, there was 1 case per 8000–10 000 new patients (Pascual-Castroviejo et al., 1998). Reports of the female to male ratio have varied between 0.7 and 1.7. HI is the third most common neurocutaneous syndrome, behind neurofibromatosis type 1 and tuberous sclerosis; it has been reported to be more common than Sturge–Weber syndrome (Pascual-Castroviejo et al., 1998).

CLINICAL MANIFESTATIONS

Typical age of presentation, diagnostic criteria

Skin lesions are visible at birth in 64–80% of patients with HI (Pascual-Castroviejo et al., 1998; Ruggieri and

*Correspondence to: Margie Ream, MD, PhD, Nationwide Children's Hospital, 700 Children's Dr., Columbus, OH 43205, USA. Tel: +1-614-722-4625, E-mail: Margie.ream@nationwidechildrens.org

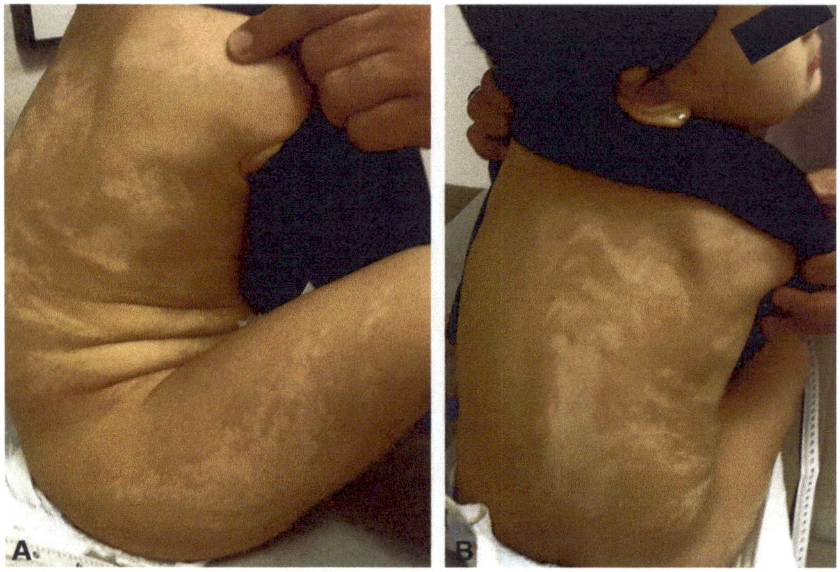

Fig. 21.1. Hypopigmentation following the lines of Blaschko in a patient with hypomelanosis of Ito. (Reproduced from Khaku et al., 2013, courtesy of Wolters Kluwer Health.)

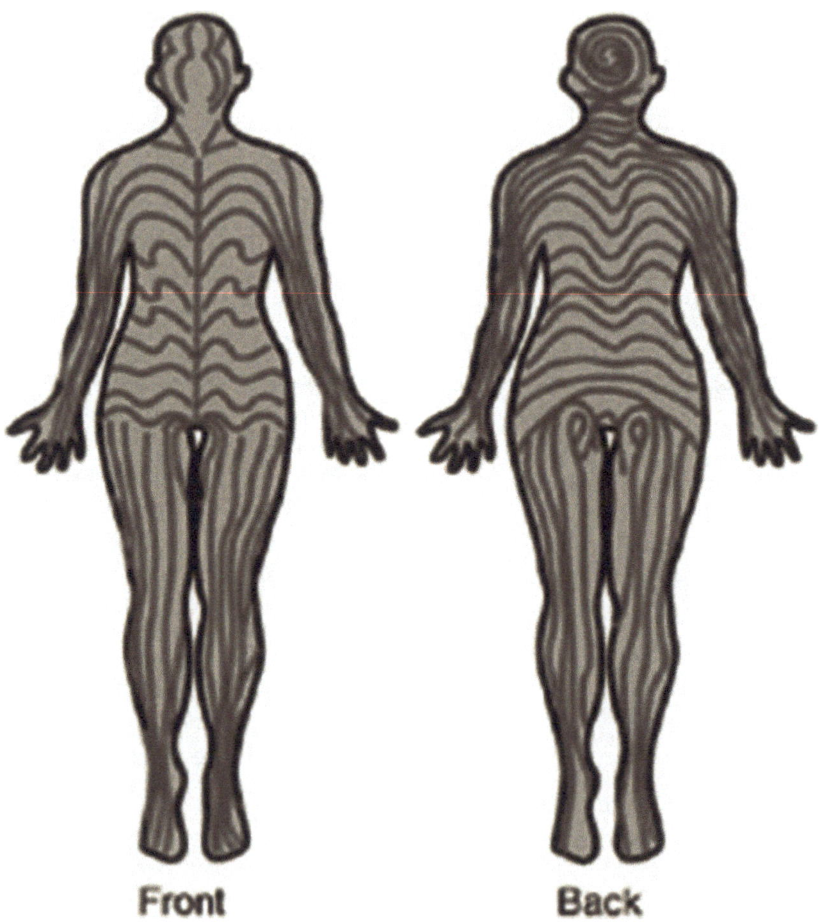

Fig. 21.2. Lines of Blaschko. (Adapted from Sun and Tsao, 2008, courtesy of Macmillan Publishers Ltd.)

Pavone, 2000). Prognosis and life expectancy are determined by the extent of multisystem involvement, particularly brain and eye, but generally HI is not thought to be a life-limiting condition. Some authors argue that published series describing patients with HI are biased by ascertainment of individuals with multisystem disease, those requiring hospitalization, or those requiring subspecialty referral. The retrospective assessment of patients through a dermatology clinic has shown that a majority of patients with HI and other conditions along the lines of Blaschko do not have extracutaneous involvement (Nehal et al., 1996). However, most series describing comorbidities of HI with patients diagnosed through specialty clinics and hospital admissions suggest a more serious condition.

To aid in the discussion of HI, diagnostic criteria have been proposed (Ruiz-Maldonado et al., 1992). The primary requirement is the early appearance of nonhereditary hypopigmented lesions in streaks or patches involving at least two body segments. Major criteria are at least one nervous system anomaly or at least one musculoskeletal anomaly. Minor criteria include at least two congenital malformations of other systems or chromosomal anomalies. A definite diagnosis consists of skin lesions plus one major or two minor criteria. A presumptive diagnosis consists of skin lesions with or without one minor criterion.

Hallmark signs and symptoms

Skin findings are required for the diagnosis of HI. Hypopigmented markings follow a variety of patterns in thin or thick stripes, usually stopping at the midline, but approximately one fifth of patients have patchy lesions without a specific pattern (Ruiz-Maldonado et al., 1992). They are most often seen on the trunk (anterior and posterior) and limbs. Involvement of face, scalp, mucous membranes, and volar surfaces is not typical. Hypopigmentation related to HI usually involves more than three body areas and there is an approximately even split between unilateral and bilateral lesions (Ruiz-Maldonado et al., 1992). In addition to lacking pigment, these regions are anhidrotic.

Lesions tend to become more obvious during the first year of life, particularly once exposed to sun when tanning occurs more in normal skin than in hypomelanotic skin. A Wood's lamp may aid in the detection of involved areas (Ardinger and Bell, 1986). Lesions stay stable during childhood and may decrease after adolescence (Ruggieri and Pavone, 2000). There may be a correlation between extent of cutaneous and extracutaneous involvement (Ruggieri and Pavone, 2000), although not all authors have found this association (Ruiz-Maldonado et al., 1992).

Multisystem involvement

Brain

Nervous system involvement is the most common extracutaneous manifestation of HI, present in 90% of patients in one 20 year prospective study (Ruiz-Maldonado et al., 1992). Intellectual disability is common, with 60–70% of patients having IQ $<$ 70 (Pascual-Castroviejo et al., 1988, 1998; Ruiz-Maldonado et al., 1992). Autism occurs in 10% of HI patients (Akefeldt and Gillberg, 1991; Zappella, 1992, 1993; Pascual-Castroviejo et al., 1998). Other developmental disorders can occur including hyperactivity, speech delay, motor delay, and hypotonia.

Epilepsy occurs in approximately 50% of patients with HI. A variety of semiologies can occur including infantile spasms (8% of 76 cases identified in infancy), myoclonic (4%), generalized tonic clonic (25%), and complex partial (12%) (Pascual-Castroviejo et al., 1998; Assogba et al., 2010). Seizures are easily controlled in some patients, but often require multiple medications or surgery to alleviate them (Placantonakis et al., 2005; Assogba et al., 2010; Manjila et al., 2014). Most patients with epilepsy have intellectual disability, with only 7% of HI patients having both seizures and normal intellect (Ruiz-Maldonado et al., 1992; Pascual-Castroviejo et al., 1998); however, there is one report of a girl of normal intelligence with HI skin lesions and juvenile myoclonic epilepsy (JME) (Bhattacharya, 2014).

Most patients with HI and epilepsy have normal neuroimaging (71%) (Pascual-Castroviejo et al., 1998). When seen, abnormalities on magnetic resonance imaging (MRI) are often migrational in nature. A review of computed tomography (CT) and MRI findings in 12 patients with HI found three patients with normal imaging, one each with lissencephaly, polymicrogyria and hemimegalencephaly, one with medulloblastoma, and five with minor abnormalities (atrophy, ventricular dilation, and enlargement of Virchow–Robin spaces) (Steiner et al., 1996). A review of MRI in 13 patients found four were normal, one had vermian atrophy, one had cerebral asymmetry, and seven had white matter abnormalities in the parietal periventricular and subcortical white matter in both hemispheres. White matter involvement correlated with developmental delay (Ruggieri et al., 1996). Cerebellar atrophy also has been found (Pini and Faulkner, 1995). Other associated abnormalities include hemimegalencephaly (Fig. 21.3), schizencephaly, anterior pachygyria with posterior lissencephaly (Iype et al., 2007) and small bilateral periventricular cysts (Williams and Elster, 1990).

Other neurologic involvement includes sensorineural hearing loss, hypotonia, hypertonia, macrocephaly, microcephaly, and hydrocephalus (Golden and Kaplan,

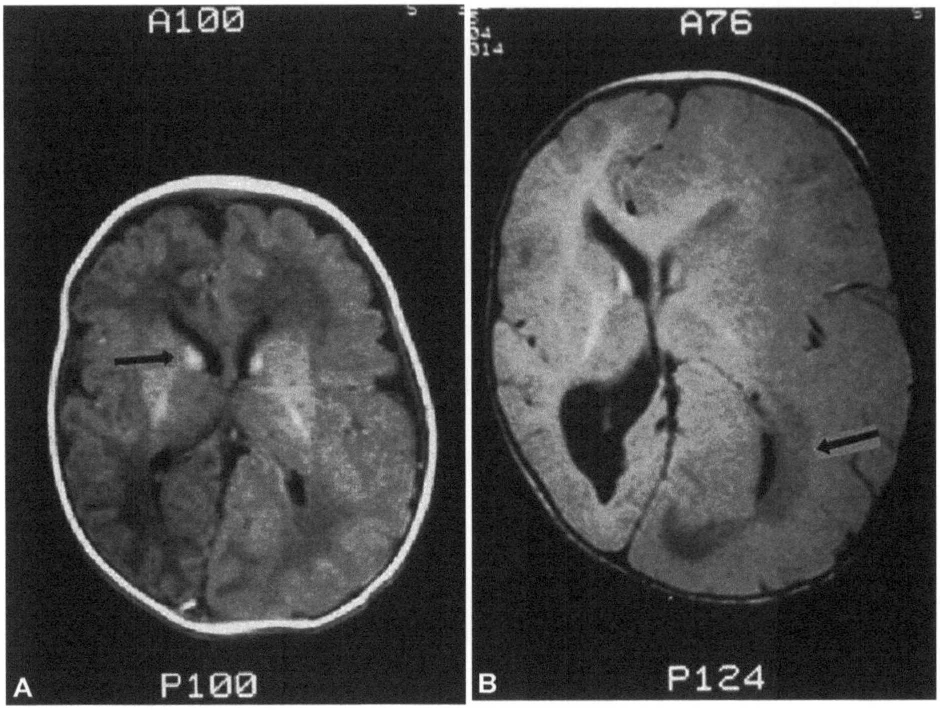

Fig. 21.3. Progressive left hemimegalencephaly. (**A**) T1 image at 22 days. (**B**) T1 image at 7 months. There is left-sided pachygyria, white matter hypointensity on the left, bilateral basal ganglia hyperintensity (arrow in (**A**)), and abnormal myelination with gliosis (arrow in (**B**)). (Reproduced from Auriemma et al., 2000.)

1986; Glover et al., 1989; Ruiz-Maldonado et al., 1992; Pascual-Castroviejo et al., 1998).

Musculoskeletal

Musculoskeletal abnormalities are second to neurologic alterations in the frequency of extracutaneous manifestations of HI, with 70% of patients affected. Most patients with musculoskeletal abnormalities also have nervous system involvement, but the reverse is not true. Scoliosis, chest wall deformity, and finger abnormalities (syndactyly, polydactyly, clinodactyly, brachydactyly) are the most common (Ruiz-Maldonado et al., 1992). Face or trunk asymmetry occurs in 16%, often expressed as hemihypertrophy (Pascual-Castroviejo et al., 1998).

Skin/scalp/hair

Skin abnormalities other than the pathognomonic hypopigmentation occur in 20% of patients. Alopecia, patchy depigmentation, and texture changes following lines of Blaschko can be seen (Ruiz-Maldonado et al., 1992). There are reports of café-au-lait spots, ichthyosis, morphea (reddish patches), cutis marmorata (reticulated red/blue pattern), nevus of Ota (facial patch of bluish hyperpigmentation), alopecia, hypertrichosis, increased facial hair, depigmentation of scalp hair, abnormal sweating, and nail dystrophy (Ruiz-Maldonado et al., 1992; Ballmer-Weber et al., 1996). Hair may also be more fragile as a result of abnormal grooves and twisting along the hair shaft (Selvaag et al., 2000).

Eye

Approximately a quarter of HI patients have some form of ocular involvement. Retinal hypopigmentation is the most commonly reported and can also involve the iris. The patchy hypopigmentation of the fundus becomes striated in the periphery, suggesting mosaicism of the retinal pigment epithelium – a pattern also seen in female carriers of X-linked albinism (Rott et al., 1990). Multiple ocular anomalies have been reported including strabismus, hypertelorism, amaurosis, microphthalmia, myopia, optic nerve hypoplasia, retinal hypopigmentation, irregular pupils, keritits, atrophy of the iris, cataract, and retinal detachment (Weaver et al., 1991). The optic changes are not progressive and often do not interfere with vision, despite obvious changes to the retinal pigment cells (Ikeda et al., 1999).

Oral

High arched palate, changes in enamel, and abnormalities in tooth development have been reported. Teeth may be misshapen or missing. There can be molar agenesis, tooth malformations, dental dysplasia, hypoplasia, and anodontia (Browne and Byrne, 1976; Happle and Vakilzadeh, 1982; Bartholomew et al., 1987; Tsutsumi and Oguchi, 1991).

Cardiac

Congenital heart disease has been reported. Three of 76 patients had cardiovascular malformations (one with tetralogy of Fallot and two with atrial septal defect (ASD)) (Pascual-Castroviejo et al., 1998). Another series of 41 patients found four with congenital heart disease including atrial and vetricular septal defect (VSD), tetralogy of Fallot, congenital heart block, and an unspecified lesion (Ruiz-Maldonado et al., 1992).

Genitourinary

Renal agenesis, ureteral duplications, cystic kidney disease, and focal segmental glomerulosclerosis have been reported in HI (Chevalier et al., 1997; Coward et al., 2001; Gatter et al., 2007; Vergine et al., 2008). Evaluation of kidney structure and function in all patients with HI has been recommended (Coward et al., 2001). Genital abnormalities were found in 5% of 76 cases and included microphallus, and hypospadias has also been seen (Ruiz-Maldonado et al., 1992; Pascual-Castroviejo et al., 1998).

Endocrine

Precocious puberty has been reported in five females with HI (Daubeney et al., 1993; Park et al., 2011). Most patients reported had peripheral or "pseudoprecocity" that is considered gonadotropin-independent and occurs as a result of excessive peripheral production of sex steroids (Wierman et al., 1985). Central precocity also has been observed and is due to early onset of otherwise normal hypothalamic release of gonadotropin-releasing hormone. Onset of puberty was reported as early as 4½ years old. Detailed endocrine evaluation is needed to determine central from peripheral precocious puberty, which in turn dictates treatment.

Vascular

There are rare reports of vascular abnormalities in patients with HI. Moyamoya syndrome, which has a known association with neurofibromatosis type 1 (NF1), was reported in two female patients with HI (Echenne et al., 1995; Rafay et al., 2005). Abdominal aortic hypoplasia and brain arteriovenous malformations (AVM) have each been reported in separate cases (Urgelles et al., 1996; Vivas et al., 2009).

TUMOR/NEOPLASTIC POTENTIAL

There are a few case reports of various tumors associated with HI; however, a general predisposition to the development of neoplasm has not been reported. Associated tumors include retinoblastoma (El-Sawy et al., 2011), a single patient with both cystic teratoma in the posterior mediastinum and an epidermoid cyst of the parietal bone (Ishikawa et al., 1985), a patient with primary meningeal rhabdomyosarcoma (Xu et al., 2000), dermoid cyst, teratoma, dysembryoma, and ganglioneuroblastoma (Ruggieri et al., 2001). Most tumor types associated with HI can be postulated to be due to abnormalities in neural crest development and as therefore having a common pathology with the skin lesions seen in HI (Xu et al., 2000; Ruggieri et al., 2001). Neuroblastoma, also of neural crest origin, was reported in one patient with HI (Oguma et al., 1996). Routine screening for tumors or neoplasia is not recommended.

DIFFERENTIAL DIAGNOSES TO CONSIDER

A practical approach to hypomelanotic pediatric skin disorders is available (Tey, 2010). In general, when evaluating skin lesions, it is important to understand the timing of the appearance of the lesions – were they congenital or acquired? It is also important to note if they have changed over time, how well demarcated they are, the pattern of distribution and associated symptoms, to suggest an isolated cutaneous disorder or a multisystem disorder. Dermatology referral and/or biopsy is sometimes needed to clarify the skin findings.

The most important diagnostic distinction for HI is IP. In IP, lesions occur in stages, starting with inflammation causing blistering, followed by hyperpigmentation and finally hypopigmentation in a whorled pattern along lines of Blaschko. HI never has an inflammatory skin stage.

There are several other causes of hypopigmentation that can occur in unusual patterns (Bonifazi, 2013). Nevus depigmentosus (achromic or hypochromic nevus) consists of localized, nonlinear lesions present at birth with hypopigmentation and no associated extracutaneous findings (Ruiz-Maldonado et al., 1992). Ash leaf macules in tuberous sclerosis are ovoid, generally well demarcated areas that may be present at birth, or noticed after tanning occurs in light-skinned individuals. Nevus anemicus causes hypopigmented patches due to vasoconstriction of blood vessels in the dermis. Vitiligo results in acquired areas of depigmentation due to autoimmunity, with half of cases starting in childhood (Silverberg, 2010). Piebaldism consists of areas of skin and hair with absence of melanocytes. These patches are present at birth and affected individuals may have an unpigmented forelock of hair.

Waardenburg syndrome is a condition associated with hypopigmented skin patches, iris color anomalies, hearing loss, and Hirshsprung's disease. Tietz syndrome involves diffuse hypopigmentation and sensorineural deafness. McCune–Albright syndrome is associated with café-au-lait spots, polyostotic fibrous dysplasia, and endocrine abnormalities (early puberty, hyperthyroidism). Linear and whorled nevoid hypermelanosis results in hyperpigmentation along Blaschko's lines and, similar to HI, can involve a variety of extracutaneous (mostly neurologic and skeletal) manifestations. Such complications occur in fewer than 20% of patients but can include seizures, developmental delay, skeletal anomalies, and other symptoms that can also be associated with HI (Di Lernia, 2007).

PATHOLOGY

Skin biopsy has shown focal areas with a reduced number of melanocytes in the epidermal basal layer and absence of extracellular melanin in the dermis (Tagawa et al., 1997). Additionally there are reduced numbers of melanosomes that have reduced melanin production (Stoebner and Grosshans, 1970). Unlike IP, there is no evidence of inflammation and in IP melanin can be found in macrophages.

Brain pathology can show abnormalities in neuronal migration. Hemimegalencephaly shows hamartomatous changes affecting the cerebral hemisphere with thickened cortex and loss of horizontal lamination, poor myelination, reactive astrocytes, and abnormal dendritic and axonal processes (Tagawa et al., 1997). White matter abnormalities suggest disorder of cortical lamination with resulting wallerian degeneration and altered myelination (Ruggieri et al., 1996). There can be heterotopic gray matter, astrocytic giant cells (Ross et al., 1982), periventricular cysts (Williams and Elster, 1990), and cerebellar hypoplasia (Pini and Faulkner, 1995).

MOLECULAR BASIS/PATHOPHYSIOLOGY

Clinical findings seen in HI are thought to be due to differential migration of genetically different – through chimerism or mosaicism – cell lines. Abnormalities are seen in the tissue or body part to which the aberrant cells migrate. The migration likely occurs during the second trimester as both melanoblast and cortical neurons migrate during this period (Ross et al., 1982). Hemihypertrophy can occur on one contiguous side of the body or may affect upper and lower extremities on different sides of the body (Donnai and Read, 1992). Similarly, skin lesions can occur contralateral or ipsilateral to hemimegalencephaly (Peserico et al., 1988; Battistella et al., 1990). Donnai hypothesized that two different cell lines in the brain may interfere with each other's normal migration, leading to migration deficits (Donnai and Read, 1992; Malherbe et al., 1993).

GENETICS

HI is typically a sporadic condition and data are strongest for mosaicism as the genetic etiology. Mosaicism occurs when there are two different populations of cells during embryologic development, often due to mutation or chromosomal nondisjunction in one cell line. The whorled pattern of hypopigmentation seen in HI is thought to be an expression of mosaicism in pigmentary cells derived from the neural crest. Clinically, mosaicism is more commonly detected in fibroblasts or lymphocytes from patients with both cutaneous and extracutaneous findings, possibly because genetic heterogeneity large enough to be seen on karyotype analysis is more likely to cause multisystemic findings than changes in patients with only cutaneous alterations. As a result, karyotype analysis of all patients with a combination of skin findings and an associated noncutaneous finding has been recommended (Orlow, 1995).

Patterned hypopigmentation can be due to mosaic disruption of melanin production at any point in the melanogenesis pathway. As a result, there is a vast amount of genetic heterogeneity in HI (Taibjee et al., 2004). It would be reasonable to propose that the different associated abnormalities (skeletal, brain, renal) are determined by which genes are affected in proximity to the pigmentary genes that result in the first obvious cutaneous signs of the condition. Karyotypic abnormalities reported in association with HI were reviewed by Taibjee et al. in 2004 and include trisomy 2 (Taibjee et al., 2004; Gupta et al., 2007; Ponti et al., 2014), trisomy 21 (Okanari et al., 2014), ring chromosome 20 (Cappanera et al., 2011), ring X chromosome (Fritz et al., 1998), and a translocation between X and chromosome 17. In this last case, X-inactivation results in mosaicism of genes residing on chromosome 17 that are silenced when the X chromosome to which they were translocated is inactivated (Hatchwell, 1996; Hatchwell et al., 1996).

While mosaicism is the most plausible explanation, multiple modes of genetic inheritance have been proposed and subsequently called into question (Ruggieri, 2000). The most likely explanation for the rare cases of familial HI is through inheritance of a balanced translocation or other chromosomal abnormality (Amon et al., 1990; Vormittag et al., 1992; Ruggieri and Pavone, 2000). As a result, recurrence risk of HI in subsequent children is exceedingly small, unless one of the parents has a chromosomal anomaly (Moss and Burn, 1988). But counseling should be done with caution for families with affected females due to the clinical overlap of HI and IP early in the course of the disease (Moss and Burn, 1988), and the possibility of balanced X-autosome translocation and inactivation (Hatchwell, 1996).

A study of patients with hemimegalencephaly found that 30% of such patients carried a mutation in the PI3K-AKT3-mTOR pathway; some of these also had skin findings consistent with HI (Lee et al., 2012) but mosaicism was not reported.

TREATMENT AND MANAGEMENT

A patient with skin and extracutaneous symptoms in whom the diagnosis of HI is considered should undergo karyotype analysis (Orlow, 1995). If an abnormality is found, parental samples should also be obtained to evaluate the risk of recurrence in subsequent children. IP should be considered in females presenting with lesions consistent with HI and lesions should be monitored closely for evolution typical of IP. All patients should be offered genetic counseling.

Consensus recommendations for screening of associated extracutaneous conditions do not exist and management is symptomatic. Patients with classic skin lesions should have a complete physical examination, including neurologic and ophthalmologic examination. Infants, particularly girls, should be evaluated at least every 3 months for the first year of life to monitor developmental progress, head growth, and evolution of skin lesions in the case of possible IP. Subsequent follow-up should be every 6–12 months to evaluate neurologic development, somatic growth, presence of precocious puberty, and musculoskeletal abnormalities, and to educate families regarding the risk of seizures. Given the high incidence of brain malformations, MRI should be performed in patients with abnormal neurologic exam, neurodevelopmental impairment or seizures. Musculoskeletal abnormalities may warrant appropriate imaging and orthopedic referral as well. Clinical scoliosis screening should be done yearly through adolescence. Ophthalmology evaluation should be done at least yearly until a reliable visual acuity can be obtained. Renal evaluation for function (creatinine clearance) and structure (ultrasound) for all patients with HI has also been recommended (Coward et al., 2001) and can be performed upon the initial visit as most genitourinary abnormalities are congenital. If precocious puberty is suspected patients should be referred to an endocrinology specialist to determine the cause and proper treatment. Dental evaluation should be undertaken beginning at the time of the first tooth eruption due to the risk of severe malposition and the associated risk of caries.

CONCLUSION

Hypomelanosis of Ito is a rare neurocutaneous disorder with possible multisystem involvement, with neurologic involvement being the most serious and most common noncutaneous manifestation.

References

Akefeldt A, Gillberg C (1991). Hypomelanosis of Ito in three cases with autism and autistic-like conditions. Dev Med Child Neurol 33: 737–743.

Amon M, Menapace R, Kirnbauer R (1990). Ocular symptomatology in familial hypomelanosis Ito. Incontinentia pigmenti achromians. Ophthalmologica 200: 1–6.

Ardinger HH, Bell WE (1986). Hypomelanosis of Ito. Wood's light and magnetic resonance imaging as diagnostic measures. Arch Neurol 43: 848–850.

Assogba K, Ferlazzo E, Striano P et al. (2010). Heterogeneous seizure manifestations in hypomelanosis of Ito: report of four new cases and review of the literature. Neurol Sci 31: 9–16.

Auriemma A, Agostinis C, Bianchi P et al. (2000). Hemimegalencephaly in hypomelanosis of Ito: early sonographic pattern and peculiar MR findings in a newborn. Eur J Ultrasound 12: 61–67.

Ballmer-Weber BK, Inaebnit D, Brand CU et al. (1996). Sporadic hypomelanosis of Ito with focal hypertrichosis in a 16-month-old girl. Dermatology 193: 63–64.

Bartholomew DW, Jabs EW, Levin LS et al. (1987). Single maxillary central incisor and coloboma in hypomelanosis of Ito. Clin Genet 31: 370–373.

Battistella PA, Peserico A, Bertoli P et al. (1990). Hypomelanosis of Ito and hemimegalencephaly. Childs Nerv Syst 6: 421–423.

Bhattacharya R (2014). A unique case of Ito syndrome presenting with juvenile myoclonic eplepsy and normal intelligence. AP J Psychol Med 15: 132–135.

Blaschko A (1901). Die Nervenverteilung in der Haut in ihre Beziehung zu den Erkrankungen der Haut. Beilage zu den Verhandlungen der Deutschen Dermatologischen Gesellschaf, VII. Congress zu Breslau, Mai.

Bloch B (1926). Krunkendemonstrationen aus der Dermatologischen Klinik Zürich. 1. Eigentümliche, bisher nicht beschriebene Pigmentaffektin (Incontinentia pigmenti). Schweitz Med Wochenschr 56: 404–405.

Bolognia JL, Orlow SJ, Glick SA (1994). Lines of Blaschko. J Am Acad Dermatol 31: 157–190, quiz 190–192.

Bonifazi E (2013). Differential Diagnosis in Pediatric Dermatology. Springer, Milan, Italy.

Browne RM, Byrne JP (1976). Dental dysplasia in incontinentia pigmenti achromians (Ito). An unusual form. Br Dent J 140: 211–214.

Cappanera S, Passamonti C, Zamponi N (2011). New association between ring chromosome 20 syndrome and hypomelanosis of Ito. Pediatr Neurol 45: 341–343.

Chevalier C, Colon S, Faraj G et al. (1997). Glomerulopathy and hypomelanosis of Ito. Nephrologie 18: 125–127.

Coward RJ, Risdon RA, Bingham C et al. (2001). Kidney disease in hypomelanosis of Ito. Nephrol Dial Transplant 16: 1267–1269.

Daubeney PE, Pal K, Stanhope R (1993). Hypomelanosis of Ito and precocious puberty. Eur J Pediatr 152: 715–716.

Di Lernia V (2007). Linear and whorled hypermelanosis. Pediatr Dermatol 24: 205–210.

Donnai D, Read AP (1992). Hypomelanosis of Ito. Lancet 339: 819–820.

Echenne BP, Leboucq N, Humbertclaude V (1995). Ito hypomelanosis and moyamoya disease. Pediatr Neurol 13: 169–171.

El-Sawy T, He L, Chiang MF et al. (2011). Retinoblastoma presenting in a child with hypomelanosis of Ito. Open Ophthalmol J 5: 55–58.

Fritz B, Küster W, Orstavik KH et al. (1998). Pigmentary mosaicism in hypomelanosis of Ito. Further evidence for functional disomy of Xp. Hum Genet 103: 441–449.

Gatter N, Hoppe B, Nützenadel F et al. (2007). A cutaneous disease with multisystem involvement: hypomelanosis of Ito may be associated with proteinuria, focal segmental glomerulosclerosis and end-stage renal disease. Nephrol Dial Transplant 22: 1796–1798.

Glover MT, Brett EM, Atherton DJ (1989). Hypomelanosis of Ito: spectrum of the disease. J Pediatr 115: 75–80.

Golden SE, Kaplan AM (1986). Hypomelanosis of Ito: neurologic complications. Pediatr Neurol 2: 170–174.

Gupta S, Shah S, McGaw A et al. (2007). Trisomy 2 mosaicism in hypomelanosis of Ito. Am J Med Genet A 143A: 2466–2468.

Happle R, Vakilzadeh F (1982). Hamartomatous dental cusps in hypomelanosis of Ito. Clin Genet 21: 65–68.

Hatchwell E (1996). Hypomelanosis of Ito and X;autosome translocations: a unifying hypothesis. J Med Genet 33: 177–183.

Hatchwell E, Robinson D, Crolla JA et al. (1996). X inactivation analysis in a female with hypomelanosis of Ito associated with a balanced X;17 translocation: evidence for functional disomy of Xp. J Med Genet 33: 216–220.

Ikeda T, Sato K, Miyaura T (1999). Fundus and fluorescein documentation of hypomelanosis of Ito. Arch Ophthalmol 117: 976–977.

Ishikawa T, Kanayama M, Sugiyama K et al. (1985). Hypomelanosis of Ito associated with benign tumors and chromosomal abnormalities: a neurocutaneous syndrome. Brain Dev 7: 45–49.

Ito M (1952). Studies on melanin. Tohoku J Exp Med 55: 1–104.

Iype M, Iype T, Geetha S et al. (2007). Hypomelanosis of Ito with cerebral malformation. Indian J Pediatr 74: 1044–1045.

Jelinek JE, Bart RS, Schiff SM (1973). Hypomelanosis of Ito ("incontinentia pigmenti achromians"). Report of three cases and review of the literature. Arch Dermatol 107: 596–601.

Khaku AS, Hedna VS, Nayak A (2013). Teaching NeuroImages: hypomelanosis of Ito. Neurology 80. e130.

Lee JH, Huynh M, Silhavy JL et al. (2012). De novo somatic mutations in components of the PI3K-AKT3-mTOR pathway cause hemimegalencephaly. Nat Genet 44: 941–945.

Malherbe V, Pariente D, Tardieu M et al. (1993). Central nervous system lesions in hypomelanosis of Ito: an MRI and pathological study. J Neurol 240: 302–304.

Manjila S, Miller BR, Goodman A et al. (2014). Pharmacoresistant epilepsy in hypomelanosis of Ito: palliative surgical treatment with modified anatomic posterior quadrantic resection. Clin Neurol Neurosurg 123: 15–17.

Moss C, Burn J (1988). Genetic counselling in hypomelanosis of Ito: case report and review. Clin Genet 34: 109–115.

Nehal KS, Pebenito R, Orlow SJ (1996). Analysis of 54 cases of hypopigmentation and hyperpigmentation along the lines of Blaschko. Arch Dermatol 132: 1167–1170.

Oguma E, Aihara T, Shimanuki Y et al. (1996). Hypomelanosis of Ito associated with neuroblastoma. Pediatr Radiol 26: 273–275.

Okanari K, Miyahara H, Itoh M et al. (2014). Hemimegalencephaly in a patient with coexisting trisomy 21 and hypomelanosis of Ito. J Child Neurol 29: 415–420.

Orlow SJ (1995). Congenital disorders of hypopigmentation. Semin Dermatol 14: 27–32.

Park JM, Kim HJ, Kim T et al. (2011). Sexual precocity in hypomelanosis of Ito: mosaicism-associated case report and literature review. Int J Dermatol 50: 168–174.

Pascual-Castroviejo I, López-Rodriguez L, De La Cruz Medina M et al. (1988). Hypomelanosis of Ito. Neurological complications in 34 cases. Can J Neurol Sci 15: 124–129.

Pascual-Castroviejo I, Roche C, Martinez-Bermejo A et al. (1998). Hypomelanosis of ITO. A study of 76 infantile cases. Brain Dev 20: 36–43.

Peserico A, Battistella PA, Bertoli P et al. (1988). Unilateral hypomelanosis of Ito with hemimegalencephaly. Acta Paediatr Scand 77: 446–447.

Pini G, Faulkner LB (1995). Cerebellar involvement in hypomelanosis of Ito. Neuropediatrics 26: 208–210.

Placantonakis DG, Ney G, Edgar M et al. (2005). Neurosurgical management of medically intractable epilepsy associated with hypomelanosis of Ito. Epilepsia 46: 329–331.

Ponti G, Pellacani G, Tomasi A et al. (2014). Hypomelanosis of Ito with a trisomy 2 mosaicism: a case report. J Med Case Rep 8: 333.

Rafay MF, Al-Futaisi A, Weiss S et al. (2005). Hypomelanosis of Ito and moyamoya disease. J Child Neurol 20: 924–926.

Ross DL, Liwnicz BH, Chun RW et al. (1982). Hypomelanosis of Ito (incontinentia pigmenti achromians) – a clinicopathologic study: macrocephaly and gray matter heterotopias. Neurology 32: 1013–1016.

Rott HD, Lang GE, Huk W et al. (1990). Hypomelanosis of Ito (incontinentia pigmenti achromians). Ophthalmological evidence for somatic mosaicism. Ophthalmic Paediatr Genet 11: 273–279.

Ruggieri M (2000). Familial hypomelanosis of Ito: implications for genetic counselling. Am J Med Genet 95: 82–84.

Ruggieri M, Pavone L (2000). Hypomelanosis of Ito: clinical syndrome or just phenotype? J Child Neurol 15: 635–644.

Ruggieri M, Tigano G, Mazzone D et al. (1996). Involvement of the white matter in hypomelanosis of Ito (incontinentia pigmenti achromiens). Neurology 46: 485–492.

Ruggieri M, Magro G, Polizzi A (2001). Tumors and hypomelanosis of Ito. Arch Pathol Lab Med 125: 599–601.

Ruiz-Maldonado R, Toussaint S, Tamayo L et al. (1992). Hypomelanosis of Ito: diagnostic criteria and report of 41 cases. Pediatr Dermatol 9: 1–10.

Selvaag E, Aas AM, Heide S (2000). Structural hair shaft abnormalities in hypomelanosis of Ito and other ectodermal dysplasias. Acta Paediatr 89: 610–612.

Silverberg NB (2010). Update on childhood vitiligo. Curr Opin Pediatr 22: 445–452.

Steiner J, Adamsbaum C, Desguerres I et al. (1996). Hypomelanosis of Ito and brain abnormalities: MRI findings and literature review. Pediatr Radiol 26: 763–768.

Stoebner P, Grosshans EM (1970). Incontinentia pigmenti achromians (ITO). An ultrastructural study. Arch Klin Exp Dermatol 239: 227–244.

Sun BK, Tsao H (2008). X-chromosome inactivation and skin disease. J Invest Dermatol 128: 2753–2759.

Sybert VP (1994). Hypomelanosis of Ito: a description, not a diagnosis. J Invest Dermatol 103: 141S–143S.

Tagawa T, Futagi Y, Arai H et al. (1997). Hypomelanosis of Ito associated with hemimegalencephaly: a clinicopathological study. Pediatr Neurol 17: 180–184.

Taibjee SM, Bennett DC, Moss C (2004). Abnormal pigmentation in hypomelanosis of Ito and pigmentary mosaicism: the role of pigmentary genes. Br J Dermatol 151: 269–282.

Tey HL (2010). A practical classification of childhood hypopigmentation disorders. Acta Derm Venereol 90: 6–11.

Tsutsumi T, Oguchi H (1991). Labial talon cusp in a child with incontinentia pigmenti achromians: case report. Pediatr Dent 13: 236–237.

Urgelles E, Pascual-Castroviejo I, Roche C et al. (1996). Arteriovenous malformation in hypomelanosis of Ito. Brain Dev 18: 78–80.

Vergine G, Mencarelli F, Diomedi-Camassei F et al. (2008). Glomerulocystic kidney disease in hypomelanosis of Ito. Pediatr Nephrol 23: 1183–1187.

Vivas D, García-Guereta L, Bret M et al. (2009). Images in cardiovascular medicine. Hypoplasia of the abdominal aorta and hypomelanosis of Ito: “pseudo-cauda equina” imaging. Circulation 120: 2025–2026.

Vormittag W, Ensinger C, Raff M (1992). Cytogenetic and dermatoglyphic findings in a familial case of hypomelanosis of Ito (incontinentia pigmenti achromians). Clin Genet 41: 309–314.

Weaver RG, Martin T, Zanolli MD (1991). The ocular changes of incontinentia pigmenti achromians (hypomelanosis of Ito). J Pediatr Ophthalmol Strabismus 28: 160–163.

Wierman ME, Beardsworth DE, Mansfield MJ et al. (1985). Puberty without gonadotropins. A unique mechanism of sexual development. N Engl J Med 312: 65–72.

Williams DW, Elster AD (1990). Cranial MR imaging in hypomelanosis of Ito. J Comput Assist Tomogr 14: 981–983.

Xu F, De Las Casas LE, Dobbs LJ (2000). Primary meningeal rhabdomyosarcoma in a child with hypomelanosis of Ito. Arch Pathol Lab Med 124: 762–765.

Zappella M (1992). Hypomelanosis of Ito is frequently associated with autism. Eur Child Adolesc Psychiatry 1: 170–177.

Zappella M (1993). Autism and hypomelanosis of Ito in twins. Dev Med Child Neurol 35: 826–832.

Handbook of Clinical Neurology, Vol. 132 (3rd series)
Neurocutaneous Syndromes
M.P. Islam and E.S. Roach, Editors

Chapter 22

Epidermal nevus syndromes

SARAH ASCH[1] AND JEFFREY L. SUGARMAN[2*]

[1]*Department of Dermatology, University of California San Francisco, San Francisco, CA, USA*

[2]*Departments of Dermatology and Community and Family Medicine, University of California San Francisco, Santa Rosa, CA, USA*

INTRODUCTION

Solomon et al. (1968) proposed the term epidermal nevus syndrome (ENS) to describe the association of epidermal hamartomas and extracutaneous abnormalities. Although many continue to use this term, it is now understood that we are not speaking of a single disease, but rather a heterogeneous group of disorders, with distinct genetic profiles defined by a common cutaneous phenotype: the presence of epidermal and adnexal hamartomas that are associated with other organ system involvement.

In 1895, Jadassohn coined the term "organ-naevus" to differentiate hamartomas of keratinocytes and epidermal appendages from nevocellular (melanocytic) nevi; this idea was built on by Mehregan and Pinkus (1965), and again by Solomon and Esterly in 1975 (Sugarman, 2004a). In the first comprehensive review of ENS, Solomon and Esterly (1975) emphasized that although some hamartomas have more sebaceous differentiation (i.e. nevus sebaceous (NS)) and others more epidermal differentiation (i.e. keratinocytic epidermal nevi (KEN)), many show differentiation toward several cutaneous appendages. In 1987, Diven highlighted the clinical importance of separating the extracutaneous manifestations associated with NS (i.e. NS syndrome) from those associated with keratinocytic or "verrucous" epidermal nevi (Diven et al., 1987). In 1995, Happle (1995) expanded this idea and has proposed that several subsets of ENS should be differentiated from one another. In a more recent review, Happle (2010a) outlined a total of nine well-defined syndromes, including both organoid (those EN containing appendageal hamartomas) and keratinocytic nevi. These include the NS syndrome (also known as Schimmelpenning–Feuerstein–Mims syndrome), Proteus syndrome, CHILD syndrome, Becker's nevus associated with extracutaneous involvement (so called "pigmented hairy EN" syndrome), nevus comedonicus syndrome, and phakomatosis pigmentokeratotica. It is likely that as the genetics of ENS are further elucidated, a new classification system based on the underlying genetic change (Fig. 22.1) and timing of the mutation in fetal development will emerge, similar to the new concept of "RASopathies" and "mosaic RASopathies" (Tidyman and Rauen, 2009; Happle, 2013).

Advances in molecular biology have revealed that the manifestations of ENS are due to genomic mosaicism. Varied clinical manifestations of ENS may be due in large part to the functional effects of specific genetic defects and the timing of the mutation in fetal development. The appreciation of genetic heterogeneity in epidermal nevi makes it easier to comprehend the wide variety of both cutaneous and extracutaneous manifestations. If a pluripotent cell acquires a mutation, its daughters will harbor and perpetuate the mutation even as they may differentiate into several distinct cell types. The idea of a postzygotic, but still pluripotent cell allows us to understand why there exist subtypes of EN with and without extracutaneous involvement. For example, some EN are caused by mutations in keratin 1 and 10. These mutations, when present in a nonmosaic, generalized form, cause a form of ichthyosis (epidermolytic hyperkeratosis) (Paller et al., 1994). Since the generalized form of this disease in not associated with extracutaneous abnormalities, one would not expect the EN caused by the mosaic form of these mutations to be associated with extracutaneous abnormalities.

*Correspondence to: Jeffrey L. Sugarman, MD, PhD, Associate Professor in Dermatology, Departments of Dermatology and Community and Family Medicine, University of California, San Francisco, 2725 Mendocino Ave, Santa Rosa, CA 95404, USA. Tel: +1-707-545-4537, E-mail: pediderm@yahoo.com

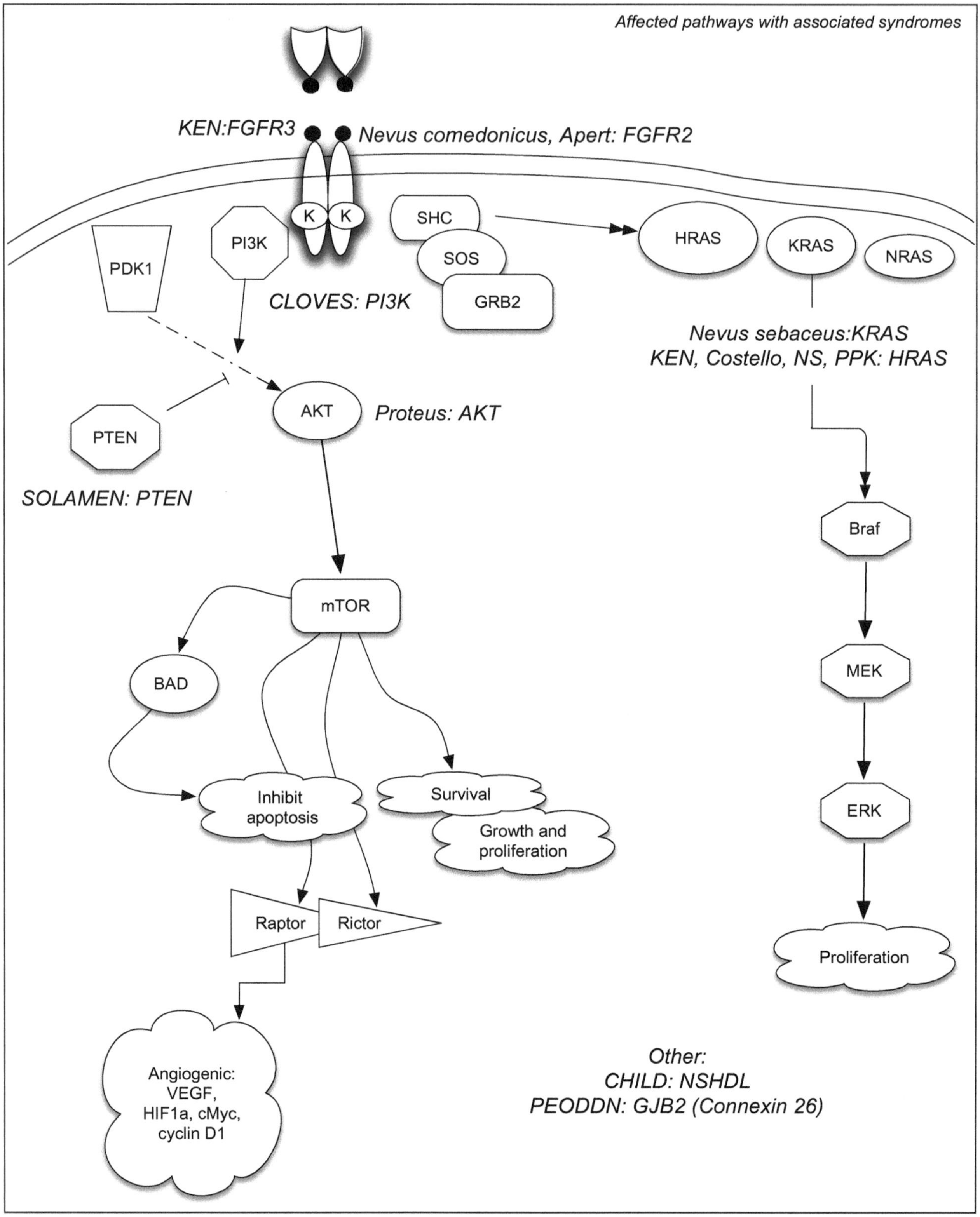

Fig. 22.1. Schematic of pathways with mutations and associated clinical diagnosis.
Apert, Apert Syndrome; CHILD, congenital hemidysplasia with ichthyosiform erythroderma and limb defects; CLOVES, Congenital lipomatous overgrowth, vascular malformations and epidermal nevus, also known as congenital lipomatous asymmetric overgrowth of the trunk, lymphatic, capillary, venous, and combined-type vascular malformations, epidermal nevi, skeletal and spinal anomalies; Costello, Costello Syndrome; KEN, Keratinocytic epidermal nevus; NS, Nevus Sebaceus; PEODDN, porokeratotic eccrine ostial and dermal duct nevus; PPK, phakomatosis pigmentokeratotica; SOLAMEN, Segmental overgrowth, Lipomatosis, Arteriovenous Malformation and Epidermal Nevus, also known as Type 2 segmental Cowden disease.

At this time, our categories are still largely based on clinical descriptions; however, new classification schemes are coming to light as we further our molecular genetic understanding of EN. For example, of the KEN, there are both those discussed above caused by keratin mutations (K1 and K10), leading to an epidermolytic phenotype, and another subset of nonepidermolytic epidermal nevi. At least one subset of the nonepidermolytic epidermal nevi is now known to be caused by mutations in fibroblast growth factor receptor 3 (FGFR3). In contrast to those with keratin mutations, these FGFR3 KEN have been shown to be associated in some instances with extracutaneous abnormalities, as this receptor is expressed in other tissues.

There is a rapidly expanding literature on the genetic basis of various epidermal nevi at this time, and as the genetic bases of different types of EN are more clearly delineated, the patterns of associated malformations are likely to be clarified more completely. Unfortunately, since our understanding of the genetic abnormalities causing EN is incomplete, comprehensive biological classification is impossible at the present time.

This chapter discusses various facets of the ENS, including clinical presentations, radiologic features, histopathology, genetic basis, pathogenesis, and management. All epidermal nevi can exist without extracutaneous involvement and this is by far the more common clinical scenario. We will try to delineate those ENS for which there is (1) cutaneous and well-documented central nervous system (CNS) involvement, including NS, PPK, CHILD and nevus comedonicus syndromes; (2) well-defined syndromes with associated EN, such as Proteus and CLOVES syndromes; (3) ENS without notable CNS involvement, but with varied systemic involvement; and (4) less well-defined ENS which are still under active investigation or noted by only a few scattered case reports. The delineations are not always possible, as there is often significant clinical overlap between subtypes, and many case series have lumped together or misclassified subtypes.

OVERVIEW OF CLINICAL MANIFESTATIONS

The incidence of epidermal nevi (EN) has been reported to range from 1 to 3 per 1000 live births affecting males and females equally (Solomon and Esterly, 1975; Alper et al., 1979). The percentage of individuals with EN who have extracutaneous abnormalities (i.e. ENS) is not precisely known, and many estimates in the literature are overstated due to ascertainment bias. The largest and least biased series is that of Rogers (1992), who reported on 233 cases, with 53 cases of epidermal nevi noted to have an associated anomaly. However, it is not clear from the data if each of these was an isolated anomaly, or if one child had multiple anomalies, thus complicating the estimation of prevalence of associated anomalies. The most common extracutaneous associations involve the CNS, ocular, and skeletal systems.

Cutaneous features

The cutaneous features of EN depend in part on the predominant cell type involved, the degree of cellular differentiation, the body site of involvement, and the age of the patient. Epidermal nevi follow several well-recognized patterns of cutaneous mosaicism, including "the lines of Blaschko." Blaschko's lines refer to the S-shaped or V-shaped whorled, streaked, and linear patterns that are recognized in many different cutaneous disorders. Blaschko, in 1901, meticulously recorded these lines that do not follow any known nervous, vascular, or lymphatic structures in the skin. Rather, they are felt by some to represent the dorsoventral migratory pathways of the neuroectoderm during embryogenesis. Happle described additional cutaneous patterns in 2002, some of which are seen in EN, including block-like (type 2, example Becker's nevus), phylloid (type 3), large patches without midline separation (type 4), and lateralization (type 5, example CHILD syndrome) (Happle, 2002; Molho-Pessach and Schaffer, 2011).

Various methods of classifying EN have been put forth over the years. The concept that nevi may show differentiation towards various skin structures, including pilosebaceous, apocrine, eccrine and keratinocyte, has been the basis of most clinical classifications to date. NS would show predominantly hamartomatous sebaceous glands, nevus comedonicus would be constructed of hamartomatous follicular units, KEN would involve only epidermal keratinocytes. Some (Happle, 2010a) have divided the "organoid" nevi from the "keratinocytic" nevi, separating the nevi that contain pilosebaceous and other adnexal dermal structures of the skin from the purely epidermal or keratinocyte-derived structures. Body site also seems to play a role in the clinical expression, as NS occurs primarily on the scalp and face, where there are greater numbers of sebaceous glands, and KEN are found more frequently on the torso and extremities.

Central nervous system

The association of EN with CNS abnormalities has been recognized for many years. In the mid 20th century, Schimmelpenning and subsequently Feuerstein and Mims were among the first to describe the association (Schimmelpenning, 1957, reviewed in Sugarman, 2004a; Feuerstein and Mims, 1962). Solomon and Esterly (1975) provided the first comprehensive review of the associated neurologic (and other organ system) abnormalities. Since then, many reports and reviews have provided more detail and insight into the spectrum of the neurologic abnormalities associated with ENS (Table 22.1; Paller, 1987; Eichler et al., 1989; Rogers et al., 1989, 1992; Hodge et al., 1991; Grebe et al., 1993). When other extracutaneous disease is also present, the incidence of CNS involvement becomes considerably higher, which may be a reflection of the inherent biological severity of the causative mutation, or more likely the timing of the mutation during development. Solomon and Esterly (1975), for example, found that 50% of patients with ENS have neurologic involvement. Estimates of the true incidence of CNS involvement in all patients with EN are probably lower than the 5–15% range cited in the literature, as true incidence studies have been hampered by ascertainment bias, the paucity of accompanying histologic data on the EN, and inconsistency in obtaining imaging studies documenting any structural CNS anomalies. In addition, definitions of clinical findings vary considerably in different reports. Interestingly, CNS involvement has been reported mainly in NS, but has been documented in association with KEN, as well as in Proteus syndrome, in nevus comedonicus, and in phacomatosis pigmentokeratotica, the details of which are discussed below with each syndrome, where known.

Both the site and type of EN appear to influence the risk of CNS disease, including, but not limited to seizures, intellectual disability, or structural anomalies. In Grebe's review of 74 patients with ENS, skin lesions on the face and scalp correlated strongly with CNS involvement, while lesions limited to the trunk and limbs were less associated with CNS involvement (Solomon and Esterly, 1975; Baker et al., 1987; Grebe et al., 1993). The NS subtype is more prevalent on the head and neck, likely reflecting the regional predominance of sebaceous glands in this area. CNS disease has been reported in a higher percentage of NS than keratinocytic nevi. Moreover, some authors have not found as striking a correlation between location and neurologic abnormalities.

Table 22.1

Associated neurologic anomalies in individuals with epidermal nevus syndrome (9)

Clinical	Structural
Seizures	Ventriculomegaly
Mental retardation/ developmental delays	Hemimegalencephaly
Hemi/quadriparesis	Cerebral vascular anomalies/ infarction
Hypotonia/hypertonia	Cortical atrophy
Cranial nerve palsy	Posterior fossa abnormalities
Cortical blindness	Gyral abnormalities
Deafness	Hydrocephalus
Segmental dysethesia	Calcifications
Macrocephaly/ microcephaly	Lipoma
	Neoplasms
	Isolated enlargement of the temporal lobe
	Agenesis of the corpus callosum
	Porencephalic cyst

Other extracutaneous features

As the genetic basis of the syndromes is elucidated, it has become clear that different ENS are likely to have distinct patterns of neurologic (as well as ocular and skeletal) involvement. Data on the incidence of extracutaneous features in some of the other subsets is sparse secondary to their relative rarity and the more recent characterization of some of these syndromes and where not clearly delineated will be discussed by system.

In the following sections we will address the well-defined syndromes and known associated manifestations with each syndrome. We will also provide separate sections of associated systemic manifestations that have not been clearly linked to one type of EN.

WELL-DEFINED EPIDERMAL NEVUS SYNDROME WITH CENTRAL NERVOUS SYSTEM INVOLVEMENT

Nevus sebaceus

Clinical features

Nevus sebaceus (NS) is relatively common, representing approximately one half of all EN (Rogers, 1992). NS is virtually always present at birth although some may not be noticed until later in childhood or until after puberty. Most have a salmon to yellow color and a characteristic smooth waxy surface (Fig. 22.2). Mehregan and Pinkus (1965) have described the natural history of NS, with an initial stage in infancy and childhood characterized by lesions that are relatively flat, due to the quiescence of sebaceous glands. At puberty, under hormonal influences, NS often thickens and develops papillomatous epidermal hyperplasia. Benign and rarely malignant neoplasms characterize a third stage, which may be present in as many as 10–15% of cases. In Rogers's 1992 series, nearly two-thirds of NS were localized to the scalp and one-third to the face; 5% are more extensive, involving scalp, face and neck and in one case, the anterior chest.

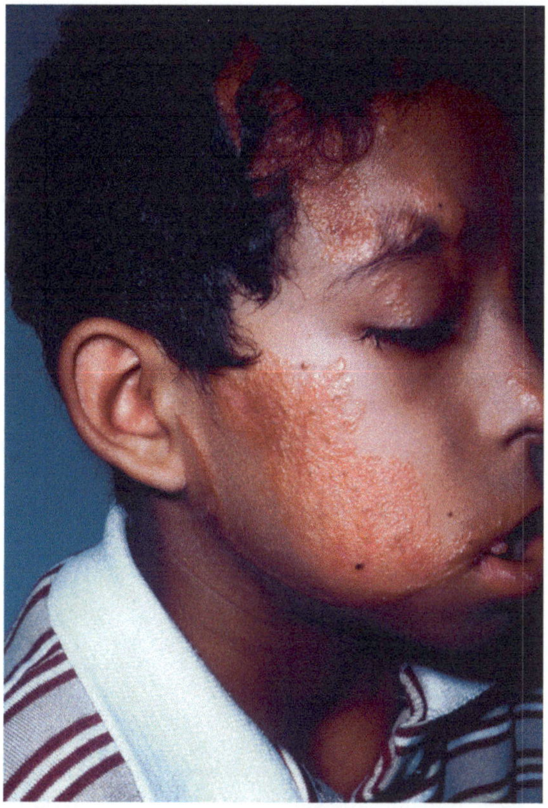

Fig. 22.2. Nevus sebaceus on the face of a young child with nevus sebaceus syndrome. From Sugarman (2004b) with permission from Cambridge University Press.

By definition, NS syndrome involves a NS coupled with extracutaneous malformations, usually involving the CNS, ocular, or skeletal systems. The major neurologic abnormalities include mental retardation and seizures. A review of 196 consecutive patients with NS, referred mainly by their pediatricians and studied over a period of 16 years, revealed 7% with neurologic manifestations (Davies and Rogers, 2002). The smaller incidence of associated neurologic anomalies in this study compared with the many other cited reports likely reflects the inclusion of all EN in a prospective manner in a dermatology clinic. Extensive nevi were four times as likely in the patients with neurologic manifestations compared to those without. Interestingly, a centrofacial location of the nevus was 10 times more likely in the patients with neurologic abnormalities (21% compared to 2% of patients with NS in other locations). Intellectual disability was found in 79% and seizures were present in 57% of neurologically affected patients. Neuroimaging was normal in the majority (75%) of those scanned with clinical neurologic abnormalities.

Skeletal, ocular, and oral anomalies associated with NS and other ENS have been documented without clear delineation of associated EN over the years. Thus, these anomalies are discussed under each organ system after the selected more well-delineated syndromes.

Histopathology

The histopathologic features of NS change with age, as do the clinical features. It is a complex lesion with abnormalities of components of the skin including hair follicles, sweat and sebaceous glands and the epidermis. Before puberty, the sebaceous glands are fairly small, but even at this stage the presence of incompletely differentiated hair structures may suggest the diagnosis. After puberty, one can find large numbers of mature sebaceous glands, often located high in the dermis and often communicating directly with the surface of the epidermis. The epidermis can be acanthotic or papillomatous, even in younger children (Fig. 22.3). In older lesions, there can be foci pale-staining glycogen-rich cells, suggesting outer root sheath differentiation. There is loss (sometimes a complete absence) of mature hair follicles, especially on the scalp. Ectopic apocrine glands are identifiable on the lower dermis in some patients. There are often buds of follicular germ that resemble foci of basal cell carcinoma but usually represent follicular induction of primitive hair germ. Many NS ultimately develop basaloid proliferations, which have been primarily shown to be benign growths (see below under Epidermal nevi and neoplasms), obviating the need for prophylactic excision, which has been the standard recommendation in the past (Calonje et al., 2012).

Molecular genetics and pathophysiology

It has been recently elucidated that NS are due to a postzygotic mutation in the Ras pathway (see Fig. 22.1). Several papers have recently identified specific activating mutations in HRAS and KRAS found in NS, but

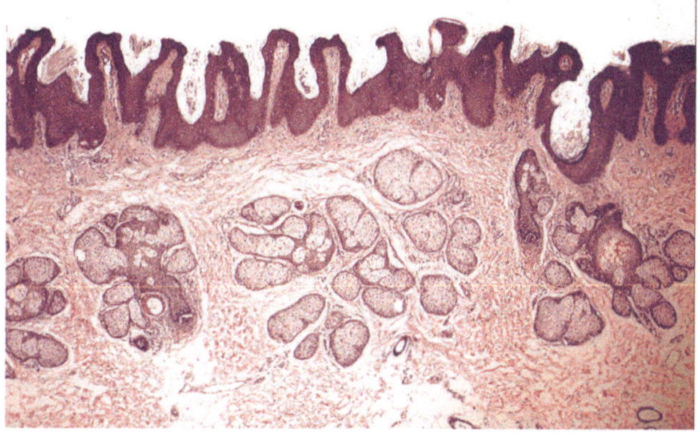

Fig. 22.3. Hematoxylin and eosin stained section of nevus sebaceus showing papillated epidermal hyperplasia and subjacent abnormal folliculosebaceous glands. From Sugarman (2004b) with permission from Cambridge University Press.

not in blood, buccal mucosa, uroepithelial cells or unaffected skin of affected individuals. Most striking is that fibroblasts underlying the NS, when cultured, did not harbor the mutation; however the cultured keratinocytes from the lesions did, thus confirming that the mutation is solely in the epidermal cells (Groesser et al., 2012; Levinsohn et al., 2012). It was also recently shown that expression of the same KRAS mutation within murine hair follicles reproduces features of NS including abortive hair follicles and epidermal and sebaceous gland hyperplasia (Mukhopadhyay et al., 2011).

Functional studies of the effects of HRAS revealed increased expression of phosphorylated ERK1/2, as well as AKT1, suggesting that gain of function mutation in a mosaic pattern is responsible for NS. When expressed in the germline, HRAS mutations are responsible for Costello syndrome (Aoki et al., 2005), which surprisingly is not associated with NS. This may be because the specific mutation is different (NS showing primarily HRAS c.37G > C, whereas Costello is HRAS c.34G > A). The significant difference in phenotype is likely due to the specific downstream effects of these mutations as well as which tissues are affected. KRAS mutations show a similar pattern, in that slightly different mutations in KRAS lead to Noonan syndrome and cardiofaciocutaneous syndrome, but the particular mutation noted in NS (KRAS c35G > A) has not yet been described in the germline. This may be because the germline KRAS mutations are so significant that they yield an embryonic lethal phenotype.

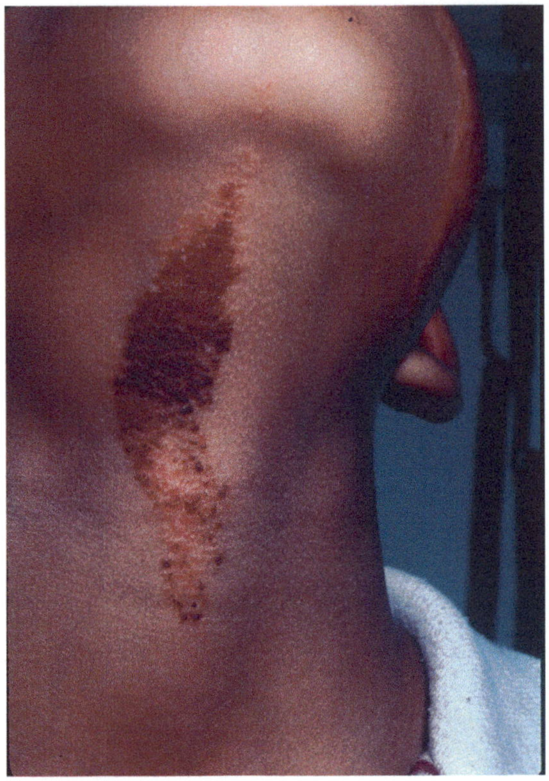

Fig. 22.4. Keratinocytic epidermal nevus on the neck. From Sugarman (2004b) with permission from Cambridge University Press.

Keratinocytic epidermal nevi

Clinical features

Keratinocytic epidermal nevi (KEN) are the other most common form of EN (Rogers, 1992). They appear as linear or whorled skin-colored to pink or slightly hyperpigmented plaques (Fig. 22.4). They are usually present at the time of birth but many have their onset during early infancy, sometimes extending over adjacent areas of skin for the first few months to years of life. Initially they may be flat, but over time they often become more elevated, verrucous, and darker in color. Acral lesions often have a more warty appearance (verrucous EN) (Solomon and Esterly, 1975). When the nail matrix is affected, the nail may be dystrophic. In the body folds, lesions are softer and less verrucous. The distribution and extent of the EN varies widely: they can be solitary, multiple, large or small, and are commonly found on the trunk or extremities. Lesions may be either unilateral (so-called "nevus unius lateris") or bilateral, usually stopping abruptly at the dorsal and ventral midline. When they are symmetric and bilateral they have been referred to as systematized EN or ichthyosis histrix. KEN are less common on the head and neck (Solomon and Esterly, 1975).

Historically, "keratinocytic EN syndrome" has been left out of the characterization of defined subsets of ENS. Clearly, KEN are associated with extracutaneous manifestations including oral and facial dysmorphisms, skeletal anomalies, as well as CNS abnormalities such as seizures and developmental delay (Collin et al., 2007; García-Vargas et al., 2008; Bygum et al., 2011; Ousager et al., 2012). However, as these have been grouped with other EN in the past, as with NS syndrome, these anomalies are also further discussed by organ system after the specific syndromes. Also, the cutaneous features of KEN are distinct from those of other ENS such as NS. The epidermolytic KEN (caused by mutations in K1 and 10) may be phenotypically distinct from other KEN, appearing as a thinner, softer and more scaly plaque (Fig. 22.5). These are not associated with extracutaneous manifestations. Thus, the extracutaneous manifestations seen in KEN may be distinct based on the underlying mutations. However, this "subset" probably does not represent a true single syndrome because it likely encompasses several unique syndromes. Thus, "keratinocytic EN syndrome" will surely be subdivided in the future as specific molecular defects are identified and correlated with extracutaneous clinical features.

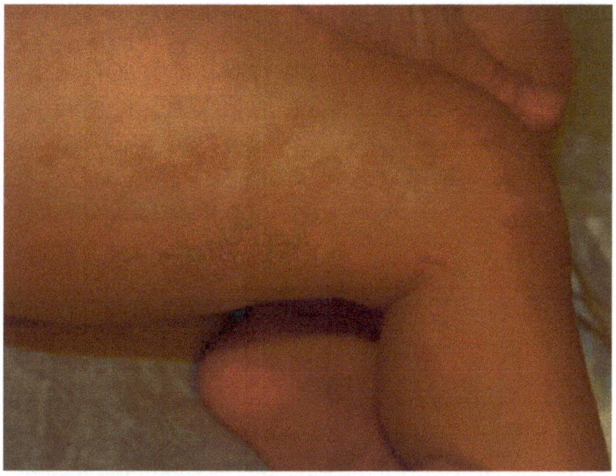

Fig. 22.5. Epidermolytic keratinocytic epidermal nevus on leg.

Histopathology

Histologically, keratinocytic or verrucous EN are characterized by acanthosis, orthohyperkeratosis, papillomatosis, and an expanded papillary dermis that is sharply demarcated from the surrounding normal skin. A total of 167 biopsy specimens were analyzed from 160 patients with EN, excluding NS. The most frequent type showed hyperkeratosis and epidermal hyperplasia (67%), followed by acrokeratosis verruciformis-like features with marked hyperkeratosis, hypergranulosis, and acanthosis (13%), epidermolytic hyperkeratosis (5%), seborrheic keratosis-like features (5%), psoriasiform hyperplasia (3.6%), verruca-like features (2%), and 1.2% each of porokeratosis-like features, Darier's-like changes, and acanthosis nigricans-like features (Su, 1982).

Molecular genetics and pathophysiology

Hafner et al. (2006) identified activating FGFR3 mutations, almost exclusively at codon 248 (R248C), in 11 of 33 (33%) patients with nonorganoid, nonepidermolytic epidermal nevi (nonepidermolytic KEN). In four of these cases, samples from adjacent histologically normal skin showed no mutation in FGFR3. Other case reports and series have followed, including a verrucous EN with the same mutation (Collin et al., 2007), and a series in which FGFR3 mutations were reported in 16/39 (41%) EN from 11/33 (33%) patients that were not found in normal adjacent skin or blood where tested (Hernández et al., 2007). Further, there is also a report of a case of a 5-year-old girl with a systematized nonepidermolytic EN and cerebral defects who was found to harbor the same mutation (R248C) in both the EN and in blood lymphocytes, whereas areas of normal skin showed wild-type FGFR3 only (García-Vargas et al., 2008). This same mutation has been well described in thanatophoric dysplasia when present in the germline, and likely represents here a postzygotic mutation leading to somatic mosaicism and a newly defined ENS (Happle, 2010a). Also, acanthosis nigricans (AN)-like epidermal nevi have been described in few case reports in the literature (de Waal et al., 2010), and AN also harbors an FGFR3 mutation, suggesting a possible association. In further investigations, Hafner et al. (2012) analyzed 72 KEN with DNA mutational analysis and found that 39% of KEN harbored RAS mutations (HRAS (29%) > NRAS (4%) > KRAS (1%)), 27% harbored FGFR3 or PIK3CA or a combination of mutations (see Fig. 22.1). Some 38% of KEN harbored no identified mutations in RAS, FGFR3, PIK3CA, and given the lack of abnormality, 10 of these were further screened for mutations in 19 other oncogenes and no mutations were uncovered. Finally, a single patient is reported with a mutation in HRAS p.G12S (c.34G → A) with systematized EN and multiple urothelial cancers (Hafner et al., 2011).

Nevus comedonicus

Clinical features

Nevus comedonicus (NC) is characterized by a localized collection of dilated follicles containing keratin. Some authors have split this entity into NC consisting of comedo-like dilated follicular orifices and "acne nevus" which produces sebum and may be complicated by inflammation and scarring. We take these entities to be overlapping enough to be considered together and refer to them collectively as NC. NC has been found on the face as well as the trunk and upper extremity (Guldbakke et al., 2007). NC may be complicated by superinfection with bacteria, chronic inflammation and scarring, and the development of large cysts with time (Guldbakke et al., 2007; Jeong et al., 2012). Some cases of NC may become "activated" during adolescence, presumably by the influence of androgen hormones.

Extracutaneous manifestations associated with NC include electroencephalographic (EEG) abnormalities and seizures, dysgenesis of corpus callosum, microcephaly, ipsilateral cataract, and corneal changes. Skeletal anomalies have also been associated with NC including hemivertebrae, scoliosis, polysyndactyly or clinodactyly, oligodontia, and absence of the fifth ray of a hand (Engber, 1978; Patrizi et al., 1998; Vidaurri-de la Cruz et al., 2004).

Histopathology

The microscopic characteristics of NC include poorly differentiated hamartomatous follicular structures (Cho et al., 2012) that may reflect the genetic basis of dysregulated growth and development. These changes consist of wide invaginations of the epidermis that resemble dilated infundibula that are filled with keratin. These changes, however, are not identical to acne. There is little if any inflammation in some cases, and closed comedones as seen in acne may not be present. Usually the hair itself is absent and the sebaceous glands are rudimentary. The epidermis may be thickened, with hyperkeratosis and acanthosis (Guldbakke et al., 2007; Cho et al., 2012). Some cases have revealed epidermolytic hyperkeratosis histologically (Schecter et al., 2004; Morillo et al., 2007). Follicular tumors, including trichofolliculoma and pilar sheath acanthoma, keratoacanthoma, syringocystadenoma papilliferum, hidradenoma papilliferum, and very rarely basal cell carcinoma or squamous cell carcinoma, have been documented within a NC (Dudley et al., 1986; Lee et al., 2002; Alpsoy et al., 2005; Kaliyadan et al., 2010; Zarkik et al., 2012; Tchernev et al., 2013).

Molecular genetics and pathophysiology

Fibroblast growth factor 2 (FGFR2) mutations have been implicated in NC (see Fig. 22.1). Munro and Wilkie (1998) reported on a 14-year-old boy with a nevus comedonicus who had a mutation in the fibroblast growth factor receptor 2 (FGFR2) identical to that found in Apert syndrome. The mutation was found in DNA analysis from the nevus comedonicus, but not from uninvolved skin or peripheral blood lymphocytes. Melnik et al. (2008) reported a similar case with the same mutation. This suggests that NC represents a mosaic condition for a mutation in FGFR2, which, if present in the germ line, would result in Apert syndrome. FGFR2 undergoes differential splicing, leading to FGFR2b and FGFR2c receptors that are expressed in the epithelium and mesenchyme respectively. The mutation seen in Apert syndrome and NC changes the affinity of the receptors for ligands that are differentially expressed in the epithelium and the mesenchyme. This disrupts the normal patterning that occurs as mesenchyme and epithelium interact reciprocally during development.

This concept also provides a framework for understanding the association of nevus comedonicus with skeletal anomalies. Apert syndrome is characterized by skeletal anomalies such as craniosynostosis, syndactyly, fusion of cervical vertebrae and severe acne. Individuals with NC and skeletal anomalies (i.e. nevus comedonicus syndrome) may be mosaic for FGFR2 in both skin and extracutaneous tissues secondary to a mutational event occurring earlier in development than seen in individuals without extracutaneous involvement. In an excellent review in 2009, Melnik outlines the indirect evidence derived from acne pathogenesis, Apert syndrome, and NC as to how the mutation of FGFR2 seen in all three conditions could lead to infundibular overexpression of IL-1α, which may be the driving force of comedo formation in nevus comedonicus and Apert syndrome (Melnik, 2009).

Phacomatosis pigmentokeratotica

Clinical features

Phacomatosis pigmentokeratotica (PPK) is defined by the association of speckled lentiginous nevus (also known as nevus spilus, a light brown patch with dark brown macules within) and NS. PPK is rare with approximately 30 reported cases in the literature. Interestingly, while the NS follows the lines of Blaschko, the melanocytic component usually occurs in a checkerboard (broad segmental areas respecting the midline) pattern (Tadini et al., 1998), and can even involve oral mucosa and conjunctiva (Taheri et al., 2011).

Neurologic, ocular, and skeletal anomalies are variably present. The neurologic defects reported in association PPK are segmental dysesthesia, mild mental retardation, seizures, and deafness, ptosis and strabismus (Tadini et al., 1998). Hemiatrophy of tissues underlying the nevi with associated muscular weakness appears to be a common finding. In addition, PPK has been associated with bony asymmetry including hypoplasia of the pelvis, kyphoscoliosis, and hemiatrophy (Happle et al., 1996). Scattered reports of tumors, including rhabdomyosarcoma and nephroblastoma, do exist, but it is not clear if these tumors were related to the underlying syndrome.

Molecular genetics and pathophysiology

Recently, Groesser et al. have identified identical HRAS mutations in both the sebaceous and the melanocytic nevi in six patients with PPK, suggesting that PPK is due to an activating postzygotic mutation in HRAS in a multipotent progenitor cell that then gives rise to both types of nevi (Groesser et al., 2012). The significance of this will be further discussed in the section on Genetics (see also Fig. 22.1).

Congenital hemidysplasia with ichthyosiform nevus and limb defects (CHILD) syndrome

Clinical features

The acronym CHILD syndrome (OMIM #308050) was introduced by Happle et al. (1980) and was reported as early as 1903 (Happle, 2010a). Earlier reports describing this entity used the term “congenital unilateral ichthyosiform erythroderma” (Rossman et al., 1963; Cullen et al.,

1969). The disorder is very rare, with fewer than 30 reported cases in the literature (Seeger and Paller, 2014). The syndrome is seen almost exclusively in females and is believed to be transmitted as an X-linked dominant trait. The cutaneous findings characterizing the syndrome are unilateral inflammatory erythematous patches often covered in dry, yellowish scales. They are often accentuated in body folds, such as the vulva, axilla, and gluteal fold (so-called "ptychotropism"), helping to distinguish CHILD syndrome from ILVEN (inflammatory linear verrucous EN). The nevus characteristically involves one side of the body with striking midline demarcation, with the majority of patients having right-sided involvement (Fink-Puches et al., 1997; Hummel et al., 2003). There may be thin bands of sparing following the lines of Blaschko. Interestingly, the cutaneous features of CHILD syndrome appear to undergo spontaneous partial regression during childhood.

The extracutaneous features of the reported cases to date (Seeger and Paller, 2014) include neurologic abnormalities, skeletal changes, cardiac anomalies, deafness, absence/hypoplasia of ipsilateral internal organs such as kidney and lung and liver. The skeletal anomalies can vary from minimal hypoplasia of the phalanges to complete absence of an extremity. Reported neurologic anomalies include ipsilateral hemispheric hypoplasia and EEG abnormalities. The skeletal changes are not necessarily limited to the limbs and have included the axial skeleton and skull. Life-threatening cardiovascular defects have also been reported (Happle et al., 1980; Seeger and Paller, 2014). CHILD syndrome may better be categorized as a form of localized ichthyosis, as opposed to an ENS. The primary defect in CHILD syndrome is one of barrier, whereas in an EN there is dysregulated growth of epidermal keratinocytes.

Histopathology

The histologic abnormalities in CHILD reflect the genetic underpinnings. There are alternating areas of ortho- and parakeratosis and numerous lipid-containing vesicles in the stratum corneum. There is a notable lack of well-defined layers of differentiated keratinocytes in the epidermis. There is also acanthosis and elongation of rete ridges characterized as a psoriasiform epidermal hyperplasia with variable reports of inflammatory infiltrates (Seeger and Paller, 2014).

Molecular genetics and pathophysiology

CHILD syndrome is caused by various loss of function mutations in the *NSDHL* gene that encodes a 3-β-hyroxysteroid dehydrogenase. This protein functions in the cholesterol biosynthetic pathway (König et al., 2000; Kim et al., 2005). The cutaneous pathogenesis appears to involve two mechanisms. The faulty cholesterol biosynthetic pathway results not only in deficient cholesterol synthesis but also in the accumulation of toxic metabolic intermediates (Elias et al., 2011; Paller et al., 2011). Recent reports show that the cutaneous findings in CHILD can be treated with a novel pathogenesis-based approach with topical lovastatin that inhibits the rate-limiting step in the cholesterol biosynthetic pathway, blunting the creation of the toxic intermediate compounds. Concomittant treatment with cholesterol to replace this essential component of the epidermal membrane was necessary. Topical treatment with lovastatin/cholesterol (but not cholesterol alone) virtually cleared skin lesions by 3 months, accompanied by histologic and ultrastructural normalization of epidermal structure and lipid secretion (Elias et al., 2011, 2012; Paller et al., 2011).

Additionally, a defective response to hedgehog signaling is now thought to play a role in disorders of cholesterol biosynthesis (Cooper et al., 2003; Stottmann et al., 2011; Morimoto et al., 2012) that share some of the skeletal and cardiovascular defects seen in CHILD syndrome.

OVERGROWTH SYNDROMES WITH EPIDERMAL NEVI

Historically, the following syndromes have been delineated as part of the family of ENS; however they are likely better understood as overgrowth syndromes with concomitant epidermal nevi. Of note, the epidermal nevi in the syndromes appear to harbor the same mutation as found in other affected tissues, implying that the epidermal nevi seen in these syndromes are best considered as a cutaneous manifestation of each type of overgrowth syndrome. While these syndromes will likely be considered in their own overgrowth category in the future, they are included here for completeness as these newer definitions are reaching maturity.

Proteus syndrome

Clinical features

Proteus syndrome was first described by Cohen and Hayden (1979) and since then the highly variable clinical features of the syndrome have been redefined, in a consensus conference in 1998 (Biesecker et al., 1999). The unifying theme for Proteus syndrome is irregular, relentless, progressive overgrowth of multiple tissues. The clinical criteria for diagnosis comprise two components: general (of which all three must be met) and specific (of which a well-delineated subset must be met). The general diagnostic criteria include mosaic distribution of lesions, progressive course, and sporadic occurrence. In addition, of the specific criteria, patients must have either a connective tissue nevus, or two of the following: EN, disproportionate overgrowth, ovarian or parotid tumors, or three of the following: lipomas or fat hypo/aplasia, vascular malformations, and characteristic facies. A cerebriform connective tissue nevus is considered nearly pathognomonic for the disease but is rarely present at birth. However, linear verrucous epidermal nevi are common and while they are subtle at birth, they increase and darken, often over the first year of life.

Extracutaneous findings include hemihypertrophy, asymmetric macrodactyly, skull and other skeletal anomalies, epidermal and connective tissue nevi, lipomas and lipoatrophy, and vascular (usually venous) malformations. These features may be present at birth, but more commonly develop over time. Cerebral involvement is not typical in Proteus syndrome but mental retardation with seizures and brain malformations have been reported and seem to be more common in those with characteristic facies. Children with Proteus syndrome can have early eruption of permanent teeth. Ophthalmologic abnormalities include strabismus, epibulbar tumors, nystagmus, myopia, and retinal pigmentary abnormalities (De Becker et al., 2000). Common complications seen in those with Proteus include deep vein thrombosis with possible pulmonary embolism (especially surrounding surgical procedures), aggressive overgrowth, and possible malignancy.

The histopathology reflects the particular tissue overgrowth involved.

Molecular genetics and pathophysiology

The underlying cause of Proteus syndrome is an activating mutation in AKT1, which is a kinase that mediates cell proliferation and apoptosis (see Fig. 22.1). This was demonstrated by Lindhurst et al. (2011), who performed whole exome sequencing of patients with known Proteus syndrome and unaffected controls. Of 29 patients with Proteus 26 had a somatic activating mutation in AKT1 in affected tissues, and did not harbor the mutation in unaffected tissues or peripheral blood. The unaffected controls did not carry an AKT mutation. Given AKT's role in control of cell proliferation, an activating mutation here can be easily understood to lead to a clinical phenotype of overgrowth.

CLOVES syndrome

Clinical features

CLOVES syndrome (congenital lipomatous overgrowth, vascular malformations and epidermal nevus; or congenital lipomatous asymmetric overgrowth of the trunk, lymphatic, capillary, venous, and combined-type vascular malformations, epidermal nevi, skeletal and spinal anomalies) describes patients who were thought formerly to have a form of Proteus syndrome. However, these patients were distinguished from Proteus patients owing to their nonprogressive and proportionate overgrowth that is described as "ballooning," as opposed to destructive. CLOVES syndrome is now described as a segmental overgrowth syndrome. An excellent review published by Mirzaa et al. (1993) describes CLOVES and related syndromes including fibroadipose hyperplasia (both of which harbor epidermal nevi), and delineates the expected comorbidities, recommendations for monitoring and extended differential diagnosis of these segmental overgrowth syndromes. Briefly, associated abnormalities include skeletal anomalies and orthopedic problems such as severe scoliosis; infiltrative lipomatous overgrowth, especially of the trunk; lymphatic malformations, cutaneous vesicles; vascular and hematologic abnormalities such as paraspinal high-flow lesions with spinal cord ischemia, central phlebectasias, and thromboembolism. These patients particularly require ongoing surgical care for the progressive overgrowth of truncal lipomatosus. Finally, these mutations have also been associated with various adult onset cancers, including gastrointestinal tract, brain, breast, bladder, lung, and thyroid. However, the role of this mutation may contribute more to the aggressive overgrowth behavior of the tumors, and not the transformation to tumor in and of itself. Wilms' tumor has been documented in two CLOVES patients; thus the role of tumorigenesis of this mutation in these patients remains unclear (Kurek et al., 2012).

Molecular genetics and pathophysiology

The underlying genetic cause of CLOVES syndrome is now known to be a mutation PIK3CA that constitutively activates the PI3K-AKT pathway, affecting a pathway similar to Proteus syndrome and Cowden syndrome (see Fig. 22.1). Kurek et al. (2012) showed this by analyzing samples from six affected individuals compared with unaffected controls. A custom enrichment assay was used to examine signaling pathways implicated in other overgrowth syndromes and PIK3CA mutations were identified using a massively parallel sequencing technique, and confirmed with specific analysis of each patient.

SOLAMEN syndrome

Clinical features

SOLAMEN syndrome (segmental overgrowth, lipomatosis, arteriovenous malformation and epidermal nevus; also known as type 2 segmental Cowden disease) has a distinctive EN. The associated EN has been dubbed a "PTEN nevus" in that is *p*apillomatous, *t*hick, *e*pidermal and *n*onorganoid, and to reflect its genetic underpinnings (Happle, 2007). As the name SOLAMEN encompasses, this syndrome clinically has segmental overgrowth including hemihypertrophy of limbs, macrocephaly, ballooning of toes; lipomatosis cutaneous and extracutaneous, cutis marmorata-like lesions, varicosities of involved leg, angiomas, extracutaneous vascular malformations, and CNS findings including hydrocephalus and seizures. Some reported cases of "Proteus-like syndrome" may ultimately be revealed to be SOLAMEN syndrome (Happle, 2010a). Associations also include thyroid adenomas, ovarian cystadenoma, and multiple breast tumors.

Histopathology

The specific histopathologic features of this type of EN remain to be defined.

Molecular genetics and pathophysiology

In SOLAMEN, PTEN shows a loss of function mutation in affected patients. SOLAMEN syndrome is probably best considered as a mosaic PTEN hamartoma syndrome or a mosaic PTEN-opathy (Mirzaa et al., 1993). This is analogous to the concept of Rasopathies and mosaic Rasopathies, as with Costello syndrome and KEN, or NS and NS syndrome. PTEN is considered to be a tumor suppressor gene, therefore loss of function leads to unregulated growth in affected tissues. There are some data to support the notion that the segmental pattern may be due to PTEN mosaic nullizygosity; however, specifically, nullizygosity as the mechanism of action has only been clearly demonstrated in one patient (Caux et al., 2007). This affects the same pathway growth regulatory pathway as seen in Proteus syndrome, Cowden disease and CLOVES (see Fig. 22.1).

EPIDERMAL NEVUS SYNDROMES WITHOUT ASSOCIATED CENTRAL NERVOUS SYSTEM INVOLVEMENT

Becker's nevus (pigmented hairy epidermal nevus)

The Becker's nevus syndrome does not have CNS abnormalities generally associated with it, although it can have involvement of other cutaneous, muscular, or skeletal defects that have been used to designate it as an EN syndrome (Danarti et al., 2004).

Clinical features

A circumscribed patch of hyperpigmentation with hypertrichosis and slight acanthosis characterizes Becker's nevus. When stroked or rubbed, Becker's nevi become more elevated and appear more infiltrated (pseudo-Darier's sign) secondary to piloerection of the hairs within the nevus. This stems from the increased numbers of smooth muscle fibers of the erector pili contained within Becker's nevi. The overlapping clinical features (hyperpigmentation, hypertrichosis, and increased smooth muscle bundles) between smooth muscle hamartoma and Becker's nevus have prompted some to consider smooth muscle hamartoma and Becker's nevus on a continuum (Karo and Gange, 1981). Several differences, however, may indicate that these are distinct entities. Smooth muscle hamartoma is congenital whereas Becker's nevus has its onset later in childhood and in adolescence. However, a recent case series of 188 cases from Italy in 2012 notes that 26% of Becker's nevi were observed at birth, not at puberty, and that similar numbers of males and females were affected. The Becker's nevi in the younger children were often hairless. In this series only 5% of the children had Becker's nevus syndrome and in these cases Becker's nevus was present from birth (Patrizi et al., 2012). The Becker's nevus is androgen-dependent, becoming more prominent and developing coarse hair after puberty in males. Becker's nevi are usually located in the shoulder region but up to 12% of cases have been noted on the lower extremities.

Pruritus, lichen planus, and acneiform eruptions with comedones, papules, pustules, and cysts have been reported in association with Becker's nevi (Terheyden et al., 1998; Danarti et al., 2004). Becker's nevi have been associated with ipsilateral hypoplasia of the breast, hypoplasia of underlying musculature, lipoatrophy, and underlying skeletal anomalies, including scoliosis, hemivertebrae, fused or accessory cervical ribs, pectus excavatum or carinatum, and internal tibial torsion (Happle and Koopman, 1997; Danarti et al., 2004). A report of a segmental Becker's nevi on the face with mucosal involvement has also been reported (Pahwa and Sethuraman, 2012). The genetics underlying Becker's nevus remain to be elucidated.

Histopathology

There is slight epidermal acanthosis and regular elongation of rete ridges, variable hyperkeratosis, and acanthosis. There is no increase in the number of melanocytes. The basal cell layer is hyperpigmented and melanophages are present in the papillary dermis. At times, the arrector pili muscles are increased in number and size, which can make these nevi difficult to differentiate from smooth muscle hamartoma (Danarti et al., 2004).

Inflammatory linear verrucous epidermal nevus (ILVEN)

Clinical features

Inflammatory linear verrucous epidermal nevus (ILVEN), described simultaneously by Altman and

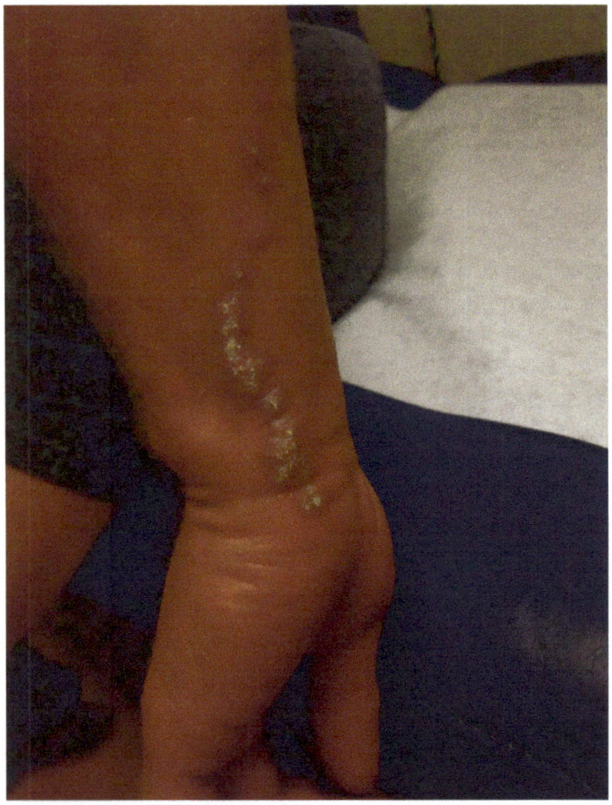

Fig. 22.6. Inflammatory linear epidermal nevus (ILVEN), following the lines of Blaschko.

Mehregan (1971) and by Kaidbey and Kurban (1971), represents approximately 6% of all EN (Rogers et al., 1989). It is characterized by linear pruritic, erythematous, and hyperkeratotic papules that often coalesce into plaques (Fig. 22.6). It is usually unilateral and often affects the lower half of the body, with the buttock being the most frequently affected site (Rogers et al., 1989; Lee and Rogers, 2001). It is often present at birth or develops with the first 6 months of life, which together with its resistance to standard treatment, and severe pruritus, may help to distinguish it clinically from psoriasis. A skin biopsy may be necessary to distinguish ILVEN from psoriasis.

There have been reports of ILVEN associated with ipsilateral skeletal anomalies (Golitz and Weston, 1979), although others have called into question the association of ILVEN with extracutaneous manifestations. Lee and Rogers reviewed 23 cases over a 13 year period and found two with skeletal anomalies but in both cases they were bilateral and thought to be coincidental. It has been suggested that the original reports associating ILVEN with severe skeletal defects were in fact incorrectly diagnosed cases of CHILD syndrome (Lee and Rogers, 2001; Happle, 2010a). To additionally cloud the picture, there are a number of reported cases of psoriasis and ILVEN coexisting in the same patients (Bondi, 1979), and there have been reports of arthritis associated with ILVEN (Al-Enezi et al., 2001). However, TNF-α-blocking biological agents have not shown clear success in treating ILVEN in patients who also have psoriasis, with some reports showing efficacy and others no significant change (Renner et al., 2008; Özdemir et al., 2012). Of note, a recent case reported two children with ILVEN of the upper extremity who were treated for epilepsy with valproic acid; both had resolution of their pruritus and subsequent resolution of their skin lesions. It is unclear whether the epilepsy in these children was related to the skin lesions, although upper extremity ILVEN is less common. However, the authors hypothesize that as valproic acid suppresses TNF-α *in vitro*, perhaps a similar anti-inflammatory process is occurring in the skin of these children (Diluvio et al., 2013). Finally, there has been some limited effectiveness of modern destructive laser therapies in a few patients (Alonso-Castro et al., 2012; Conti et al., 2013). Thus, the pathogenesis and treatment of ILVEN are an open area of debate. It remains unclear, although seemingly less likely, that ILVEN is truly associated with other systemic or anatomic abnormalities or defines a syndrome within itself. As with other EN, when the underlying mutation is identified it will likely guide our understanding of other possible associations; however, clinical observations to this point have not documented strong associations.

Histology

Histologically, ILVEN demonstrates orthohyperkeratosis alternating with parakeratosis (Dupre and Christol, 1977). Absent involucrin expression, differential staining of keratin 10, and differential staining of T cell subsets have been reported, but the histopathologic differentiation of ILVEN from psoriasis remains challenging at times (de Jong et al., 1991; Ginarte et al., 2000; Vissers et al., 2004).

Porokeratotic eccrine ostial and dermal duct nevus (PEODDN)

Clinical features

The nomenclature of this EN is still not finalized at this time. PEN (porokeratotic eccrine nevus) was originally described as "comedo nevus of the palm" (Marsden et al., 1979) and subsequently termed "porokeratotic eccrine ostial and dermal duct nevus" (PEODDN) (Abell and Read, 1980). In 2009, Goddard et al. proposed the name "porokeratotic adnexal ostial nevus" (PAON) to encompass both PEODDN (PEN) and porokeratotic eccrine and hair follicle nevus (PEHFN), a related entity. PEN and PEODDN are both used widely in the literature at this time. However, since the original description, there have been more than 40 reported cases of PEODDN with a male:female ratio of 1:1. PEODDN consists of verrucous, keratotic papules with keratin filled invaginations representing eccrine ducts located on the palms and soles. The majority of cases present at birth or soon after, and can present as neonatal erosions (Goddard et al., 2009). There have been several reports of widespread cutaneous involvement (Sassmannshausen et al., 2000; Dogra et al., 2002; Goddard et al., 2009; Easton et al., 2012). Of note, the cutaneous histopathology overlaps with that of KID syndrome (keratitis-ichythosis-deafness syndrome; due to a GJB2 mutation which encodes gap junction protein connexin 26); this has led to the hypothesis that this EN, as others, could be due to mosaicism. In 2012, Easton et al. reported two cases of patients referred for widespread PEODDN who were found to harbor a mutation in GJB2 in affected skin, but not in unaffected skin or blood; and Lazic et al. (2012) reported a case of KID syndrome with a concomitant PEODDN. This leads to the hypothesis that PEODDN could be a somatic mosaic of KID syndrome. This remains to be further elucidated. There have been reports of association of squamous cell carcinoma within the PEODDN, and these lesions should be clinically monitored (Goddard et al., 2009).

LESS WELL-DEFINED EPIDERMAL NEVI

There are scattered reports of other less well-defined EN that may or may not eventually be associated with syndromic patterns. Each of these has been reported with scattered associated anomalies, which will not be listed here; however, identification of these nevi should prompt a full physical examination for other possible physical findings and a brief literature search for current reports. Angora hair nevus (Schauder et al., 2000) is a broad, band-like area covered with long smooth white hair. There are only two case reports in the literature, both with associated extracutaneous anomalies. Woolly hair nevus (WHN) is a circumscribed area of curled unruly hair differing from the texture of the surrounding scalp hair (Reda et al., 1990). About 50% of patients with a WHN have an associated ipsilateral linear EN (Bovenmyer, 1979). Somatic HRAS c34.G > A, p.G12S mutations have been found in both the WHN and the associated KEN. This mutation is distinct from the more strongly activating HRASp.G13R mutation associated with NS. The WHN-associated weakly activating RAS mutation may allow abnormal hair growth (and thus the phenotype of woolly hair) while the more strongly activating RAS mutation seen in NS drives pilosebaceous progenitors toward sebocyte differentiation (and thus the phenotype of NS with its associated alopecia) (Levinsohn et al., 2014).

Apocrine nevi (nevus syringocystadenoma papilliferum) are composed of apocrine glands and are usually seen as a component of NS but have been reported rarely as the dominant or sole histologic feature. The clinical appearance is variable, with the nevus appearing gray to dark brown in color and papular, nodular, cystic, or verrucous (Solomon and Esterly, 1975). Eccrine nevi are composed of sweat ducts, and when present, are also generally found as a component of NS. There have been reports of apocrine nevi associated with skeletal anomalies (Vakilzadeh et al., 1976; Ando et al., 1991). Happle, in his 2010 two-part series on ENS (Happle, 2010a, b), outlines a few further epidermal nevi that are in the early stages of clinical definitions. These include: nevus trichilemmocysticus, which consists clinically of multiple comedo-like plugs, filiform hyperkeratosis, and trichilemmal cysts arranged along the lines of Blaschko (Ferreira et al., 2013); SCALP syndrome (sebaceous nevus, CNS malformations, aplasia cutis congenita, limbal dermoid, and pigmented nevus), described in four patients to date, NEVADA syndrome (nevus epidermicus verrucosus with angiodysplasia and aneurysms) (Happle, 2010a, b).

DISCUSSION OF EXTRACUTANEOUS MANIFESTATIONS

The majority of reports of EN and extracutaneous manifestations did not differentiate between the various clinical subtypes of EN. Additionally, these reports mostly occurred before molecular understanding was elucidated. This has made it more difficult to delineate which specific EN put patients at risk for specific associated anomalies, malformations, and systemic manifestations. We will review below by system and discuss (where possible) the associated specific ENs.

Central nervous system

Gurecki et al. (1996) reviewed 23 cases of biopsy proven EN with accompanying neurologic abnormalities. The authors also analyzed the results of CNS imaging studies including computed tomography (CT). Cases of keratinocytic nevi and NS were approximately equal in numbers. Nevi were located on the head and/or neck in 91% of the patients and 55% of the patients had seizures. All cases of neonatal seizures and infantile spasms had major hemispheric malformations; 53% had mental retardation or developmental delays. Of these cases, most were considered moderately or severely impaired. The most common CNS structural abnormalities were hemiatrophy (26%), vascular anomalies (26%), cranial bone deformities (26%), hemimegalencephaly (22%), gyral abnormalities (22%), and posterior fossa abnormalities (9%) (Fig. 22.7A, B). In Rogers' study (1992),

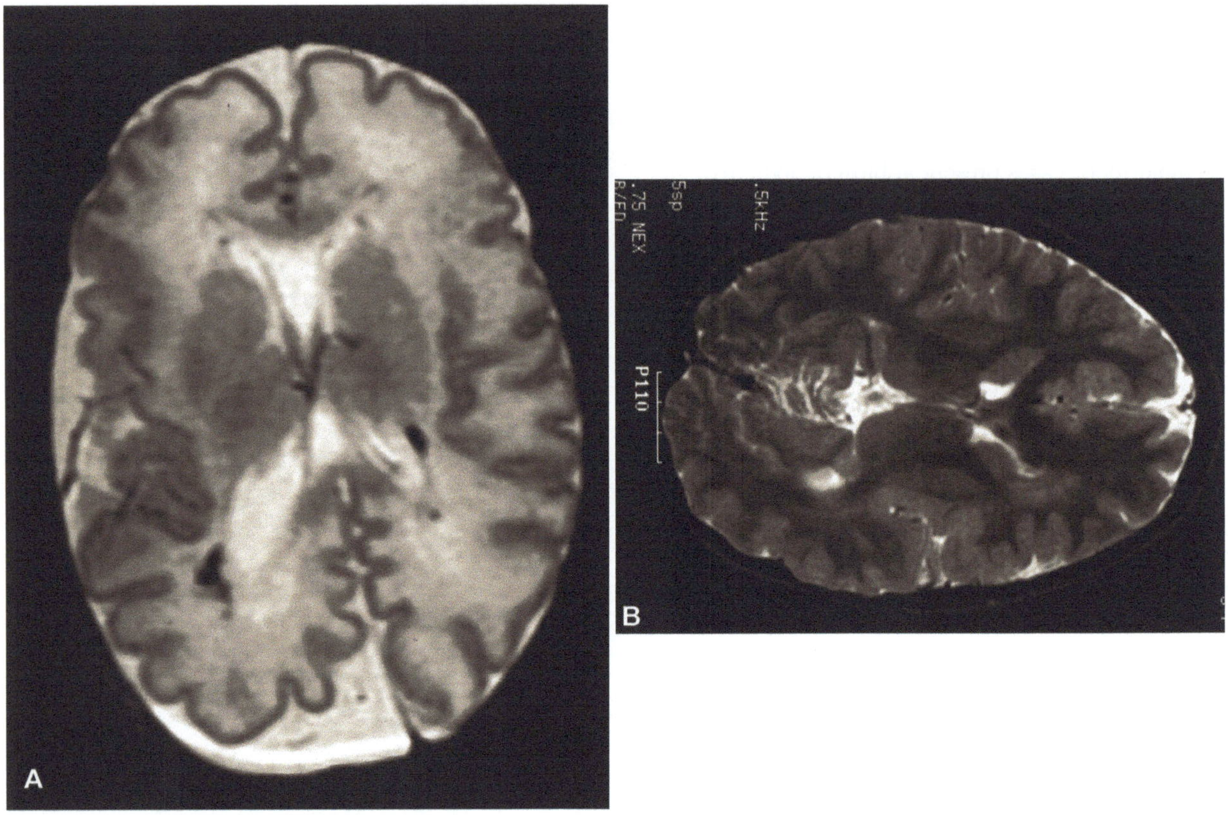

Fig. 22.7. (**A**) T2 weighted axial MRI image shows classic findings of hemimegalencephaly with enlarged left hemisphere and abnormal cortical gyral pattern, as well as dysplastic white matter. (**B**) T2 weighted axial MRI image shows small dark areas of hemorrhagic necrosis in the periventricular white matter, probably secondary to ischemia. From Sugarman (2004b) with permission from Cambridge University Press.

of the 106 cases of EN with a predominantly keratinocytic morphology, mental retardation was present in 7%, seizures in 5%.

In a review by Pavlidis et al. (2012) of hemimegalencephaly and ENS, 57 previously reported cases were identified in which the most frequent associated features were severe epilepsy, in about half of cases with neonatal onset, mental retardation/developmental delay, ocular/visual involvement, and facial abnormalities.

Other less commonly reported neurologic anomalies associated with EN include cortical atrophy, hydrocephalus, porencephaly, lateral ventricle enlargement, agenesis of the corpus callosum, Dandy–Walker malformation, hyperkinesia, cerebrovascular tumors and malformations, cerebrovascular accidents, hemiparesis and quadriplegia, cortical blindness, hemiparesis, quadriplegia, intracranial and intraspinal lipomas and CNS tumors (see Table 22.1) (Solomon and Esterly, 1975; Baker et al., 1987; Dodge and Dobyns, 1995; Mall et al., 2000).

The seizures in ENS are often associated with underlying structural brain abnormalities and are often difficult to control. The seizure types are variable and include generalized tonic, tonic-clonic, focal motor and infantile spasms. Major hemispheric malformations predispose patients to neonatal seizures, infantile spasms, and the Lennox–Gastaut variant (Gurecki et al., 1996).

While there is much analysis in the literature of the histopathologic features of cutaneous lesions in ENS, pathologic descriptions of CNS findings in such patients are rare. Prayson et al. (1999) examined the clinicopathologic features of three patients with ENS who underwent surgical resections for chronic epilepsy. Microscopic examination of resected cortical tissue demonstrated severe diffuse cortical dysplasia characterized by a disorganized cortical architectural pattern, a haphazard orientation of cortical neurons, and increased molecular layer neurons. There was also prominent cortical astrocytosis. Gyral fusion was seen in one patient. Pial glioneuronal hamartomas were observed in one patient. Neuronal heterotopia was observed in all three patients. Similarly, Pavone et al. (1991) noted a disturbed laminar pattern of the cerebral cortex on microscopic analysis of a patient with neurologic involvement. In addition, they noted atypical giant neurons, heterotopic neurons in both the white matter and the subarachnoid space, areas of marked astrocyte proliferation infiltrating into adjacent structures, and small angioma-like conglomerations of blood vessels. From an architectural standpoint, this may be analogous to the hamartomatous structures observed microscopically in EN.

Ocular

Many of the ocular abnormalities probably occur at 4–6 weeks gestational age (Gurecki et al., 1996). Ocular abnormalities in ENS may occur in several ways. First, primary dysplasia due to defects of ocular development may occur as a consequence of the mutational event. Other ocular abnormalities, for example, anterior chamber abnormalities, occur later in development and can be attributed to secondary failure. In addition, abnormal migration of CNS components can have downstream effects on the development of certain ocular structures. Last, direct extension of EN may involve the eyelid, lid margin, conjunctiva, and sclera.

Ocular choristomas and colobomas are the most common ocular findings associated with NS syndrome. Although many reports that discuss the association of ocular anomalies in ENS do not distinguish between the subtypes of EN, it appears that NS may be the most commonly associated. Rogers (1992) series found that 9% of 119 patients with EN had ocular anomalies, with strabismus being most common (6.7%). Grebe et al. (1993) reviewed 74 cases of ENS, selected because of the association with other multiple organ system abnormalities, and found ocular involvement in 39%. Lipodermoids were the most common (20%), followed by colobomas (9%), corneal opacities (7%), exo/esotropia (7%), retinal changes (scarring, degeneration, detachment, 7%), ptosis (5%), macroophthalmia (4%), and conjunctival growth disorders (3%).

Gurecki et al. (1996) reviewed the ophthalmologic involvement in 23 patients with biopsy-proven EN and accompanying abnormal neurologic examination. They found ophthalmologic involvement in nine out of 23 (39%) patients. Six of these (26%) had involvement of neuroectodermally derived ocular tissues, with anomalies such as colobomas of the optic disk and nerve, retinal dysplasia, and microphthalmia, with bilateral involvement in four patients. Similarly, Solomon and Esterly (1975) found ocular abnormalities in 33% of their group of 60 patients with ENS, with extension of the nevus to the eyelid in 11%, ocular dermolipoid in 9%, and pannus formation, nystagmus, coloboma, and cortical blindness each in one patient. Katz et al. (1987) described a case of optic nerve hypoplasia associated with NS. Although the ocular anomalies often occur in patients who also have neurologic anomalies, there are cases of EN with major ophthalmologic abnormalities in the absence of neurologic anomalies (Diven et al., 1987).

Skeletal

Solomon and Esterly (1975) attempted to classify the skeletal abnormalities by separating them into primary osseous changes, and secondary changes which may be attributed to another coexisting abnormality. Examples of primary skeletal abnormalities included incomplete formation of bony structures including ankle, foot, phalanges, and vertebrae, hypoplasia of pelvic bones and long bones, and bone cysts. Examples of secondary osseous changes included limb hypertrophy, kyphoscoliosis, asymmetry of the skull, short stature, spontaneous fractures, and rickets. Rogers et al. (1989) found that 15% of 119 patients with EN had abnormalities of the skeletal system. However, many of the associated anomalies were minor and common in the general population and may have been coincidental. Grebe et al. (1993) reviewed the clinical features of 74 cases of ENS and found skeletal anomalies in 50%. The high incidence of skeletal abnormalities in their series reflects the inclusion of patients with other organ system involvement. They found a diverse range of abnormalities, many of which could be attributed to tissue overgrowth. In Rogers' study (1992), of the 106 cases of EN with a predominantly keratinocytic morphology, 18% had skeletal abnormalities although less than half of these (7%) were thought to have a strong likelihood of association.

Oral

While only mentioned sporadically in the literature, it is clear that oral involvement can be part of ENS. In a recent report of ENS associated with ameloblastoma, Colletti et al. (2014) reviewed EN and associations with oral lesions. EN are often characterized by localized or diffuse verrucous growths on the lips, palate, gingiva, buccal mucosa, and tongue. They note that less common findings include tongue hypertrophy, cleft palate, high arched palate, and various dental anomalies, and can be associated with oral cavity neoplasms including ameloblastoma (three reported cases in the literature), odontoma, and giant cell granuloma (Colletti et al., 2014). Thus, especially in cases of EN on the head and neck, it is extremely important to complete a thorough oral examination and consider a referral for evaluation by a pediatric dentist or oral surgeon.

Other extracutaneous manifestations

There have been several case reports involving EN associated with various malformations of the genitourinary system, including hydronephrosis, horseshoe kidney, solitary kidney, duplicated renal collecting system, cryptorchidism, ureteropelvic junction obstruction, hypospadias, and ectopic ureter (Lansky et al., 1972; Mollica et al., 1974; Atherton, 1998). Reported cardiac anomalies include ventricular septal defect, coarctation of the aorta, pulmonic stenosis, tetralogy of Fallot, patent ductus arteriosus, dilated pulmonary artery, aortic stenosis, and hypoplastic left heart (Grebe et al., 1993). It is not clear whether a true association exists between these anomalies and EN or whether their association is merely by chance.

ENDOCRINE ABNORMALITIES ASSOCIATED WITH EPIDERMAL NERVOUS SYNDROMES

Sugarman and Reed (1969) were the first to report the association of ENS with hypophosphatemic rickets. Since then, there have been multiple reports of hypophosphatemic vitamin D-resistant rickets associated with NS syndrome (Olivares et al., 1999). CNS abnormalities were present in 50% of patients (7/14), specifically seizures (21%), ventriculomegaly (21%), mental retardation or developmental delay (36%) and spasticity, hypotonia, psychomotor retardation, congenital partial left III CN palsy, bilateral temporal lobe atrophy, right brain hemiatrophy, left temporal arachnoid cyst, and asymmetric lateral ventricles were also reported. Rickets, muscle weakness, and bone pain developed at an early age in many of the patients. There has also been a case of phacomatosis pigmentokeratotica associated with hypophosphatemic vitamin D-resistant rickets (Saraswat et al., 2003).

Yu et al. (2000) reported a case of ENS associated with the syndrome of inappropriate antidiuretic hormone (SIADH). Their case involved an infant with seizures, hyponatremia, and SIADH. There have been several reported cases of central precocious puberty associated with ENS (Tay et al., 1996).

The rare hypophosphatemic vitamin D-resistant rickets that is associated with NS is thought to result from abnormal phosphate excretion secondary to defective renal tubular reabsorption of phosphate. Some authors have proposed that this condition is analogous to the rare association of vitamin D-resistant hypophosphatemia and osteomalacia associated with mesenchyme-derived neoplasms (tumor-induced osteomalacia), in which the tumor produces a phosphaturic factor that leads to osteomalacia (Aschinberg et al., 1977; Carey et al., 1986). The phosphaturic factor responsible for tumor-induced osteomalacia (TIO) has been identified. These tumors secrete large amounts of FGF23 (Jonsson et al., 2003). Similarly, patients with autosomal dominant hypophosphatemic rickets have increased levels of circulating FGF23 due to the production of a mutant FGF 23 that makes it resistant to cleavage and degradation.

There have been reports of at least partial reversal of the hypophosphatemia with removal of a portion of an EN (Aschinberg et al., 1977; Ivker et al., 1997; Saraswat et al., 2003). In one compelling case report, a 12-year-old patient with large facial EN and vitamin D-resistant hypophosphatemic rickets improved following excision of several fibroangiomatous EN. Serum phosphate concentrations and tubular reabsorption of phosphorus both increased significantly after excision. These authors then homogenized the EN that had been initially removed, and infused them into the femoral vein of a dog. Two hours after the infusion, the tubular reabsorption of phosphorus decreased dramatically. Control infusions had no effect on phosphorus reabsorption. Based on these findings, they speculated that EN produce a phosphaturic factor that leads to osteomalacia (Aschinberg et al., 1977). These findings suggest that the hypophosphatemic rickets seen associated with NS is either related to overproduction of FGF23 (wild type or mutant) or involves another molecule regulating the phosphate homeostasis and skeletal mineralization axis. Recently, activating NRAS and KRAS mutations in bone and in the nevus have been identified in cases of NS associated with hypophosphatemic rickets providing the first evidence that elevated FGF23 levels, hypophosphatemia, and osteomalacia are associated with pathologic Ras activation and may lead to future insights into FGF23 regulation (Lim et al., 2014). This same report also included a case of giant melanocytic nevus associated with hypophosphatemic rickets which was also associated with RAS mutation. All five patients in this cohort had multiple extracutaneous findings, including foci of dysplastic bone, brainstem lipoma, thyroid nodules, eye, and cardiac abnormalities. The term cutaneous-skeletal hypophosphatemia syndrome has been designated for this condition given the broad multisystem systemic findings with a shared origin from an early somatic mutation (Lim et al., 2014).

EPIDERMAL NEVI AND NEOPLASMS

NS was once thought to have a 10–15% risk of malignant transformation, most commonly to basal cell carcinoma. Other cutaneous malignancies including squamous cell carcinoma and adnexal carcinomas have been described in association with EN; these are most commonly described with nevus sebaceus. Malignant transformation of EN may occur whether or not they are associated with other organ system abnormalities (Mehregan and Pinkus, 1965; Solomon and Esterly, 1975). Mehregan and Pinkus (1965) analyzed 150 cases of EN and found 52 tumors in 33 patients. These included basal cell carcinoma, syringocystadenoma papilliferum, syringoma, apocrine cystadenoma, apocrine carcinoma, hidradenoma, and sebaceous epithelioma. Rogers et al. (1989) found secondary tumors in only 7% of EN, including invasive squamous cell carcinoma, and a keratoacanthoma-like lesion.

Other mostly more recent reports, however, have modified our understanding of the biological behavior of NS. Domingo and Helwig (1979) described nine cases of aggressive malignant neoplasms associated with NS and reviewed the literature. In their study, malignant transformation occurred in older adults (median age 64.5 years) and was associated with aggressive neoplasms including squamous cell carcinoma, apocrine carcinoma, and other adnexal carcinoma. More recently, the magnitude of malignant potential of NS has been questioned. Cribier et al. (2000) retrospectively analyzed 596 cases of NS, 79% of which were located on the scalp. They found benign tumors in 14% of these, the most common of which included syringocystadenoma papilliferum (5%), trichoblastoma (5%), trichilemmoma (3%), and sebaceoma (2%), but found basal cell carcinoma in only 0.8% of cases. Interestingly, many of the cases of trichoblastoma (a benign neoplasm) had originally been classified as basal cell carcinoma but were rediagnosed using new criteria. Similarly, a retrospective analysis of 155 cases of NS by Jaqueti et al. (2000) found neoplasms in 21%, with trichoblastoma being most common (7.5%), followed by syringocystadenoma papilliferum (6%) and sebomatricoma (5%). Smaller numbers of apocrine hidrocystoma and apocrine poroma were also found. There were no cases of malignant neoplasm found. The decreasing incidence of tumors developing in NS could be the result of more frequent early excision for cosmetic reasons, but the data from Jaqueti et al. suggest otherwise; 61% of their cases were from adults. A very recent series of 707 cases of nevus sebaceus showed a rate of 2.5% with malignant tumors, all in adults, with basal cell carcinoma being the most common (Idriss and Elston, 2014). It seems likely that most of the tumors arising in NS that have in the past been interpreted as basal cell carcinoma are in fact examples of primitive follicular induction or trichoblastomas, and not authentic basal cell carcinoma. Yet still the debate rages on, with a new report in 2013 that used a follicular stem cell marker PHLDA1 (TDAG51) that showed that many of these tumors are possibly more related to basal cell carcinomas and not trichoblastomas (Sellheyer et al., 2013). In one report, loss of heterozygosity of the patched gene, which is known to be involved in the development of basal cell carcinoma, was found in 40% of the NS that were analyzed. It is unclear whether patched deletion in NS is associated with the propensity to develop basal cell carcinoma or other benign appendageal tumors (Xin et al., 1999). Current practice is generally to follow NS clinically (yearly in children) and to biopsy unusual growths or ulcerations, especially after adolescence. It is impossible to overemphasize the importance of clear communication with a knowledgeable dermatopathologist on any biopsied lesions within these nevi.

A recent report by Hafner et al. (2009) documents the occurrence of two basal cell carcinomas arising in a KEN, and the basal cell carcinoma portion was found to have a PIK3CA mutation that was not found in the EN cells. The report is significant because it suggests that not all cells within an EN will harbor the same mutation as a putative oncogenic cell. However, Bourdeaut et al. (2010) reported a case of co-occurrence of EN, rhabdomyosarcoma, polycystic kidneys, and growth retardation in an infant, and identified an oncogenic G12D KRAS mutation in both the epidermal component of the EN and in the rhabdomyosarcoma but not in the dermal component of the EN lesion or in unaffected tissues, including normal skin and blood.

There have also been many reports of various extracutaneous malignancies arising in patients with EN. These include astrocytoma, glioma, Wilms' tumor, nephroblastoma, salivary gland carcinoma, bladder cancer, rhabdomyosarcoma, and gastrointestinal carcinomas (Lansky et al., 1972; Solomon and Esterly, 1975; Dimond and Amon, 1976; Rosenthal and Fretzin, 1986; Atherton, 1998; Schulz and O'Leary, 2001). As data accumulate, it is most likely the types of tumors that will be found in relation to ENS will be better defined by the genetic mutations they harbor; however, the guiding principle for clinicians is to monitor clinically for concerning changes and biopsy or refer expeditiously for appropriate evaluation. Of note, while many EN harbor mutations in genes associated with tumor formation/suppression, often the specific mutations found in EN are different than those found in tumors (data surrounding this issue will be further discussed below with genetics). As we expand our knowledge of the complex interplay of tissue growth and overgrowth; our current "two-hit" hypothesis of tumors may take on new dimensions when separating dysregulated growth from malignant degeneration. Associations with other tumors specific to each ENS are discussed with each disease.

GENETIC BASIS AND PATHOGENESIS

Many lines of evidence suggest that the pathogenesis and clinical expression of ENS is based on genomic mosaicism. Mosaicism is the mixture of more than one genotypically distinct cell lineage within one organism. Epidermal cells are thought to originate in the neural crest and move to the periphery of the growing embryo by directional proliferation (Moss et al., 1993; Molho-Pessach and Schaffer, 2011). A somatic mutation occurring during the migration of embryonic ectoderm will only be clinically apparent if the mutation leads to a recognizable difference from the surrounding normal cells, which often follow Blaschko's lines. Thus, patterns of cutaneous findings that follow Blaschko's lines are believed to be a cutaneous expression of mosaicism. In this conceptual framework, mutations that occur earlier in development would lead to more extensive cutaneous involvement and a greater likelihood of other organ system involvement. In addition, the biological severity of a particular mutation is also likely to determine the extent and severity of clinical involvement.

The concept of lethal genes surviving by mosaicism has been proposed by Happle (1987) to explain the sporadic inheritance, asymmetric clinical distribution in multiple organs, the lack of diffuse involvement of entire organs, and the equal sex ratio of affected individuals that characterize the features of ENS. Happle postulated that these syndromes are due to the action of a gene product that if present in the germ line would be lethal, but is clinically manifested only when present in a subpopulation of cells, thereby surviving by mosaicism. However, mosaicism in ENS may involve a nonlethal gene defect, as is the case with epidermolytic hyperkeratosis. In the mosaic form, there is an EN of the epidermolytic type. If the gene defect is also present in the germ cells, this nonlethal mutation may be transmitted to the next generation as epidermolytic ichthyosis (formerly known as generalized bullous congenital ichthyosiform erythroderma or epidermolytic hyperkeratosis) (Happle, 1987).

Many specific examples have provided evidence that genetic mosaicism can cause the cutaneous phenotype of EN (Table 22.2). Paller (1987) found point mutations in 50% of the keratin 10 alleles of epidermal cells from patients with EN (of the epidermolytic type), while finding no mutations in adjacent clinically normal skin. They also found the same mutations in 50% of the keratin 10 alleles in all the cell types examined from their offspring. Similarly, Moss et al. (1995) has provided evidence of a keratin 10 mutation in affected cells of an individual with linear EN while showing that cells from unaffected adjacent epidermis had no mutation. Keratin 1 gene mosaicism has also been identified in an EN (of the epidermolytic type) in a woman whose son has epidermolytic hyperkeratosis and the identical genomic mutation. This is not surprising, as keratins 1 and 10 are obligate partners, providing function to the

Table 22.2

Genes identified in selected epidermal nevi

	Mosaic disorder (EN)	Associated mutation	Germline disorder (where known)
EN	Nevus sebaceus	KRAS	
	Keratinocytic EN	HRAS (c.37G > C)	Costello syndrome
	Nevus sebaceus		(c.34G > A)
	Phakomatosis pigmentokeratotica		
	Keratinocytic EN	FGFR3	Skeletal dysplasias, including thanatophoric dysplasia
	Epidermolytic EN	Keratins 1 and 10	Epidermolytic ichthyosis
	Nevus comedonicus	Fibroblast growth factor receptor 2	Apert syndrome
Overgrowth syndromes	CLOVES	PI3K (phosphotidyl inositol 3 kinase)	
	Proteus syndrome	AKT	
	SOLAMEN	PTEN	
Unclassified	NS associated with rickets	RAS	
	PEODDN	GJB2 (Connexin 26)	
	CHILD syndrome	3β-hydroxysteroid dehydrogenase	
	Wooly hair nevus	HRAS	

EN, epidermal nevus; NS, nevus sebaceus; CLOVES (Congenital lipomatous overgrowth, vascular malformations and epidermal nevus; also known as congenital lipomatous asymmetric overgrowth of the trunk, lymphatic, capillary, venous, and combined-type vascular malformations, epidermal nevi, skeletal and spinal anomalies); SOLAMEN (Segmental overgrowth, Lipomatosis, Arteriovenous Malformation and Epidermal Nevus; also known as Type 2 segmental Cowden disease); PEODDN (Porokeratotic Eccrine Ostial and Dermal Duct Nevus); CHILD (Congenital Hemidysplasia with ichthyosiform erythroderma and limb defects).

epidermis in a cell type-specific manner. The data now support this hypothesis for several subtypes of EN, as has been discussed throughout the chapter. There is mounting evidence, and it is now largely accepted, that postzygotic mosaicism is likely the mechanism that underlies many of the syndromes and nonsyndromic skin findings that have been described to date. Recently described examples include Sturge–Weber/isolated port-wine stains with GNAQ mutations (Shirley et al., 2013), and keratinocytic epidermal nevi/Costello syndrome with HRAS mutations (Hafner et al., 2012).

Even within a single nevus the same mutation with different functional endpoints has been noted. A report of a nevus marginatus which has both components of NS surrounded by a more KEN type lesion showed it to have the same mutation in HRAS in both portions of the nevus, and that nearby normal skin had only wild type HRAS (Groesser et al., 2013). This lends even further credence to the idea that a pluripotent progenitor cell acquires the mutation and gives rise to various more differentiated cells with distinct phenotypes. A similar phenomenon was seen with PPK, where an activating postzygotic mutation was noted with both the NS and nevus spilus in the same patient. This singular mutation for two nevi of distinct cell types highlights the importance of the embryonic timing of the mutational event in determining the clinical phenotype, but also raises further questions. For example, if HRAS mutations in ectodermal progenitor cells occur early enough (before neural crest differentiation) in embryogenesis to give rise to both sebaceous and melanocytic nevi (as proposed for PPK), why do we not see melanocytic nevi in cases of NS syndrome with widespread NS and neurologic complications (which suggests an earlier embryonic HRAS mutational event)? It is possible that the HRAS mutational events in NS syndrome occur after neural crest differentiation and thus are restricted to the epidermal lineage. However, this does not convincingly explain the involvement of multiple tissues and organ systems in some cases. Previous explanations accounting for the close relationship of a sebaceous and a melanocytic nevus had been proposed by Happle. According to his twin spotting hypothesis (also known as didymosis), somatic recombination of two nonallelic distinct heterozygous recessive mutations residing on a different chromosome of a homologous pair would be combined in a stochastically low probability event that has never been described outside *Drosophila* and the plant kingdom. The resulting homozygous daughter cells would give rise to two distinct nevi in close spatial proximity. However, the recent findings of Groesser et al. (2013) call into question the "twin spotting" (didymosis) hypothesis previously put forth by Happle.

To further highlight the possibility of multisystem involvement, FGF mutations provide an emerging example. It is well documented that FGFs play a vital role in embryonic development. They function as important signaling molecules between epithelial and mesenchymal boundaries (Ornitz and Itoh, 2001). Mutations in FGFs have been associated with developmental defects in neuronal, skeletal, ocular, and cardiovascular systems. Because the product of a single gene may be involved in the formation of different structures at distinct times during embryogenesis, a single genetic defect may cause anomalies of multiple structures in different organs during development. In addition, different genetic defects may give rise to similar phenotypes by disruption of embryogenesis at a common point in time. Specifically, FGFR3 mutations have been found in both benign skin lesions, such as seborrheic keratoses, and also in malignancies including urothelial carcinomas. In carcinomas these FGFR3 mutations have also been found in association with PIK3CA mutations, which are also known to be involved in a broad spectrum of malignant tumors. The exact type and location within the gene appears to have some effect on its malignant potential. Hafner et al. (2007) evaluated a number of EN and SKs at exons 9 and 20 of PIK3CA and FGFR3 (previously identified as hotspot mutations). While unaffected corresponding tissue was negative for mutations (where tested), 9/33 (27%) EN harbored PIK3CA mutations; all cases showed a substitution that is uncommon in cancers. By contrast, 10/62 (16%) SK revealed the typical cancer-associated PIK3CA mutations and displayed a wide range of FGFR3 mutations. In contrast to SKs, the EN harbored only the R248C FGFR3 mutation. Of note, only 5/95 total lesions evaluated showed mutations in both genes. Thus, the authors of the study argue that there may be a distinct pathogenesis based on the precise mutations keratinocytes develop and at what point the mutation is developed. This study cannot answer the question of whether the cells that ultimately became SKs harbored the mutation or acquired it at a later point. As the FGFR3 mutations have also been noted in KEN, there are many questions as the function and downstream effects of these mutations that lead to the phenotype. To evaluate the alterations in function from this mutation, Hafner et al. (2010) transfected human keratinocytes with R248C or wild type FGFR3 IIIb and studied growth effects *in vitro*. At subconfluence, the cells showed similar growth rates, but when the cells reached confluence, the FGFR3 mutants continued to grow, showed reduced apoptosis and decreased attachment to fibronectin when compared to wild type. The downstream mediators were examined, and ERK1/2 showed increased phosphorylation. This type of experiment begins to lay a possible mechanism where cells begin to overgrow, either because they carry a mutation in a mosaic form from birth (as in the case of an EN); or perhaps it is acquired later in life (as may be true with seborrheic keratoses).

While skin lesions often follow Blaschko's lines and mosaicism has largely been established, a direct link to the CNS anomalies seen has not been fully made. The cortical dysplasia seen in individuals with ENS likely represents derangement in neuronal migration, much of which occurs in the first two trimesters of gestation. It is well recognized that cortical dysplasia is associated with epilepsy and may be at least partly responsible for the increased incidence of seizures in ENS. In addition, cortical dysplasia has also been associated with certain low-grade neoplasms including ganglioneuromas, implicating a common underlying etiology for the abnormal neuronal migration in patients with ENS and the development of these tumors (Prayson et al., 1993, 1999). It appears that at least in a subset of patients with neurologic involvement, primary vascular anomalies may lead to secondary CNS pathology. For example, infarcts, atrophy, porencephaly, and calcifications are best explained in some cases by prior ischemia or hemorrhage (Pavone et al., 1991). Disruption of fibroblast growth factor activated Ras signaling has been associated with ocular coloboma as well as eye and brain development in general, providing a molecular bridge between the defects seen in the skin and in other organ systems (Cai et al., 2013). However, a recent paper by Minić et al. (2013) discusses the hypothesis that mosaicism and patterns analogous to lines of Blaschko may exist in the CNS. These findings are based on findings from brain imaging in incontinentia pigmenti patients and the patterns are based on the concept of radial unit cortical development. This has not yet been documented in ENS.

Despite the characterization of distinct ENS, classification of the majority of the developmental defects seen in ENS has not yet occurred. While many mutations in EN have been characterized, much is still unknown about the molecular pathways that lead from genotype to phenotype. The analysis of mutations from different EN will undoubtedly clarify the range of genetic defects and enhance our understanding of the role of genetic aberrations in the pathogenesis of the cutaneous and extracutaneous features seen in these disorders.

MANAGEMENT ISSUES

The management of the patient with ENS needs to be individualized, and should always include a thorough history including prenatal history, developmental history, family history and physical examination, and now genetic testing where appropriate. Careful and complete cutaneous examination should include the mucosa and areas covered by hair. A careful neurologic examination is critical, as is examination of the eyes, especially the conjunctiva, sclerae, and extraocular eye movements. Skeletal examination should include evaluation for kyphoscoliosis and evaluation of gait. Limb length and size should be measured to ascertain any asymmetry. Shoe wear patterns can serve as a clue to uneven weight distribution. Referral to appropriate subspecialists should be considered.

Generally, children with small isolated EN and a normal physical examination do not require further workup (except perhaps skin biopsy, see below). Diagnostic and screening laboratory investigations should be performed if abnormalities are suspected after history and physical examination.

The choice of imaging studies should be individualized depending on the nature of the suspected abnormalities and might include skull and chest radiographs, skeletal survey, CT scan of the head, magnetic resonance imaging (MRI), and EEG. Conventional X-ray and CT studies of the brain have demonstrated multiple abnormalities including cranial asymmetry, facial asymmetry, premature closure of the sphenofrontal suture, malformation of the sphenoid bone, dilated ventricles, cortical calcifications, cortical atrophy, and arachnoid cysts (Baker et al., 1987; Lazzeri et al., 1993). While head CT has been an extremely useful tool in delineating neurologic abnormalities in ENS, MRI is a better imaging modality for the demonstration of the abnormalities related to neuronal migration defects. This is due to the superior resolution of MRI in providing gray/white matter contrast (el-Shanti et al., 1992). Angiography has been used to delineate aneurysmal dilation in patients with neurovascular involvement. Skeletal abnormalities have generally been evaluated by X-ray studies, but there are cases involving soft tissue in addition to bony involvement that warrant CT or MRI to delineate the pathology. Skeletal survey may also be useful in excluding McCune–Albright syndrome in cases of ENS associated with precocious puberty, by the noting absence of polyostotic fibrous dysplasia. Other imaging modalities, such as cardiac echocardiography and intravenous pyelogram, have been employed to evaluate other organ system involvement.

Any child with an EN on the head or neck and developmental delay should have neurologic evaluation including brain imaging, even though the yield may be low. Seizures tend to occur early, if they are going to be manifested, and follow-up until school age is probably adequate for those in whom there are no neurologic abnormalities. Laboratories such as electrolytes including serum and urine calcium and phosphate, and liver and renal function tests should be considered when appropriate. An interdisciplinary approach may need to be adopted depending on the type and extent of other organ involvement and may require the coordination of different specialists including a dermatologist, pediatric neurologist, ophthalmologist, orthopedic surgeon, oral surgeon, plastic surgeon, and psychologist.

Skin biopsy or the newer technique of skin scrapings (see below) may be of benefit in some cases. It is certainly indicated to evaluate the possibility of malignancy if a particular area of an EN is changing, especially in NS, where there is a small but significant risk of the development of cutaneous malignancy. In the case of KEN, the microscopic findings of clumping of keratohyaline granules suggest a mutation in keratins 1 or 10. The sensitivity of skin biopsy for detecting these findings is not known, and their absence does not exclude the possibility of keratin mutations. Given the large number of mutations now known, it is possible that biopsy may be helpful to identify the specific mutation and perhaps identify potential future risks. Of note, a recent study was able to use superficial skin scrapings of affected skin to identify mutations of AKT1 that were found in other affected tissues as demonstration of a noninvasive method with which to obtain and analyze affected skin (Wieland et al., 2013). This technique may have great promise and utility, especially when addressing a pediatric population.

Referral to a medical geneticist may be warranted for further evaluation, including gene testing, regarding the possibility of transmission of a more generalized form of this disease (e.g. epidermolytic hyperkeratosis in the case of epidermolytic EN or Apert syndrome in the case of nevus comedonicus) to future offspring. The role of genetic counseling in other lesions is uncertain.

Management of patients with Proteus syndrome is complex. These patients all have relentless overgrowth that distorts and destroys normal tissue architecture, but each patient has unique specifics, thus each patient should have a strong multidisciplinary team, including dermatology, orthopedics, genetics, and general pediatric. An initial skeletal survey should be performed, with targeted follow-up thereafter. MRI should be obtained of clinically affected areas, as well detailed imaging (MRI or CT as clinically appropriate) of chest and abdomen, given the high rate of involvement and often lack of symptoms. It should be noted that anesthesia may be difficult where there are lung and skeletal malformations and an anesthesiologist should be made aware of the patient's diagnosis prior to procedures (Biesecker et al., 1999; Biesecker, 2006).

Prophylactic excision to prevent possible future malignancy may not be appropriate, especially if the surgery itself will be disfiguring. For NS, this approach was undertaken more often in the past when it was thought that the malignant potential was much greater than it is now known to be. A more practical approach may be to intervene with an incisional biopsy for any new growth in the nevus to precisely determine its nature. Excision of EN may also be warranted to ameliorate the disfigurement caused by the nevus itself. However, simple excision may not always be practical because of the location or size of the EN and because abnormalities in EN frequently involve not only the epidermis but also appendageal structures in the underlying dermis, removal of the epidermis alone may result in recurrence (Solomon and Esterly, 1975; Paller, 1987). Many alternative treatments of EN have been reported, and current therapies should be reviewed when treating any patient given the rapid development of this field at this time. Unfortunately, the efficacy of these alternative therapies is uncertain due to the lack of controlled trials. Many of the following therapeutic interventions are based on individual case reports where long-term follow-up is lacking.

When treating EN for cosmetic reasons, review of recent literature and consultation with dermatologic surgeons and/or plastic surgeons who are knowledgeable in the risks and benefits of treatment of these lesions is key. The risk with any of the treatments is disfigurement by scarring and/or recurrence of the lesions. There are many varied and recent reports of the various modalities used that cannot be fully summarized here, as techniques and modalities are in an active process of progress. Briefly, various lasers to treat vascular, inflammatory, and pigmented lesions have all been used with variable success. There have been reports of the successful use of carbon dioxide (CO_2) laser to remove linear EN without any obvious scarring (Hohenleutner et al., 1995; Boyce and Alster, 2002), but in many cases visible scars are evident, particularly if continuous wave CO_2 laser is used. The Q-switched ruby laser is occasionally effective in treating pigmented EN including Becker's nevus (Carpo et al., 1999).

Topical therapies are often less effective. Topical vitamin D_3 (calcipotriol 0.005%) has been tried in the treatment of ILVEN, where its most prominent effect was to relieve associated pruritus. Calcipotriol decreased the erythema and hyperkeratosis but the effects were only permanent in one patient (Mitsuhashi et al., 1997; Zvulunov et al., 1997). However, others have reported that calcipotriol is ineffective in treating the cutaneous features or the associated pruritus in ILVEN (Lee and Rogers, 2001). A large linear EN was successfully treated with a combination of topical tretinoin and 5% 5-fluorouracil (Efudex) with marked improvement. Unfortunately, the nevi recurred after 3–4 weeks, but the authors reported that reinitiation of therapy 2–3 times a month has had excellent results (Nelson et al., 1994). There is one report of shave excision using a dermatome followed by phenol peeling for the treatment of a verrucous EN with a reportedly good outcome (Bazex et al., 1995). NC may have comedo-like eruptions, which are primarily a cosmetic concern and have been treated accordingly with topicals, including tazarotene/ calcipotriene, tretinoin, 12% ammonium lactate and tacalcitol and pore strips (Inoue et al., 2000; Deliduka and Kwong, 2004; Guldbakke et al., 2007; Tchernev et al., 2013). However, NC may also harbor large cysts and have complications from recurrent infections including scarring, fistulas, and abscesses which may require significant surgical or ablative laser treatments (Guldbakke et al., 2007; Tchernev et al., 2013). Other approaches reported in the treatment of EN have included cryotherapy, keratolytics, and intralesional steroids, but none have been substantiated in controlled studies. Long-term follow-up is lacking from many of these case reports. As our understanding of the pathogenesis of EN expands, future management may be directed toward specific molecular defects.

References

Abell E, Read SI (1980). Porokeratotic eccrine ostial and dermal duct naevus. Br J Dermatol 103: 435–441.

Al-Enezi S, Huber AM, Krafchik BR et al. (2001). Inflammatory linear verrucous epidermal nevus and arthritis: a new association. J Pediatr 138: 602–604.

Alonso-Castro L, Boixeda P, Reig I et al. (2012). Carbon dioxide laser treatment of epidermal nevi: response and long-term follow-up. Actas Dermosifiliogr 103: 910–918.

Alper J, Holmes LB, Mihm Jr MC (1979). Birthmarks with serious medical significance: nevocullular nevi, sebaceous nevi, and multiple café au lait spots. J Pediatr 95: 696–700.

Alpsoy E, Durusoy C, Ozbilim G et al. (2005). Nevus comedonicus syndrome: a case associated with multiple basal cell carcinomas and a rudimentary toe. Int J Dermatol 44: 499–501.

Altman J, Mehregan AH (1971). Inflammatory linear verrucose epidermal nevus. Arch Dermatol 104: 385–389.

Ando K, Hashikawa Y, Nakashima M et al. (1991). Pure apocrine nevus. A study of light-microscopic and immunohistochemical features of a rare tumor. Am J Dermatopathol 13: 71–76.

Aoki Y, Niihori T, Kawame H et al. (2005). Germline mutations in HRAS proto-oncogene cause Costello syndrome. Nat Genet 37: 1038–1040.

Aschinberg LC, Solomon LM, Zeis PM et al. (1977). Vitamin D-resistant rickets associated with epidermal nevus syndrome: demonstration of a phosphaturic substance in the dermal lesions. J Pediatr 91: 56–60.

Atherton DJ (1998). Naevi and other developmental defects. In: RH Champion, JL Burton, DA Burns, SM Breathnach (Eds.), Textbook of Dermatology, 6th edn., Vol. 1. Blackwell Publishing, Inc., Malden, Massachusetts, USA, pp. 519–542.

Baker RS, Ross PA, Baumann RJ (1987). Neurologic complications of the epidermal nevus syndrome. Arch Neurol 44: 227–232.

Bazex J, el Sayed F, Sans B et al. (1995). Shave excision and phenol peeling of generalized verrucous epidermal nevus. Dermatol Surg 21: 719–722.

Biesecker L (2006). The challenges of Proteus syndrome: diagnosis and management. Eur J Hum Genet EJHG 14: 1151–1157.

Biesecker LG, Happle R, Mulliken JB et al. (1999). Proteus syndrome: diagnostic criteria, differential diagnosis, and patient evaluation. Am J Med Genet 84: 389–395.

Bondi EE (1979). Psoriasis overlying an epidermal nevus. Arch Dermatol 115: 624–625.

Bourdeaut F, Herault A, Gentien D et al. (2010). Mosaicism for oncogenic G12D KRAS mutation associated with epidermal nevus, polycystic kidneys and rhabdomyosarcoma. J Med Genet 47: 859–862.

Bovenmyer DA (1979). Woolly hair nevus. Cutis 24: 322–325.

Boyce S, Alster TS (2002). CO2 laser treatment of epidermal nevi: long-term success. Dermatol Surg 28: 611–614.

Bygum A, Fagerberg CR, Clemmensen OJ et al. (2011). Systemic epidermal nevus with involvement of the oral mucosa due to FGFR3 mutation. BMC Med Genet 12: 79.

Cai Z, Tao C, Li H et al. (2013). Deficient FGF signaling causes optic nerve dysgenesis and ocular coloboma. Dev Camb Engl 140: 2711–2723.

Calonje E, Brenn T, Lazar A et al. (2012). McKee's Pathology of the Skin with Clinical Correlations, Elsevier/Saunders, Edinburgh.

Carey DE, Drezner MK, Hamdan JA et al. (1986). Hypophosphatemic rickets/osteomalacia in linear sebaceous nevus syndrome: a variant of tumor-induced osteomalacia. J Pediatr 109: 994–1000.

Carpo BG, Grevelink JM, Grevelink SV (1999). Laser treatment of pigmented lesions in children. Semin Cutan Med Surg 18: 233–243.

Caux F, Plauchu H, Chibon F et al. (2007). Segmental overgrowth, lipomatosis, arteriovenous malformation and epidermal nevus (SOLAMEN) syndrome is related to mosaic PTEN nullizygosity. Eur J Hum Genet EJHG 15: 767–773.

Cho SB, Oh SH, Lee JH et al. (2012). Ultrastructural features of nevus comedonicus. Int J Dermatol 51: 626–628.

Cohen Jr MM, Hayden PW (1979). A newly recognized hamartomatous syndrome. Birth Defects Orig Artic Ser 15: 291–296.

Colletti G, Allevi F, Moneghini L et al. (2014). Epidermal nevus and ameloblastoma: a rare association. Oral Surg Oral Med Oral Pathol Oral Radiol 117: e275–e279.

Collin B, Taylor IB, Wilkie AOM et al. (2007). Fibroblast growth factor receptor 3 (FGFR3) mutation in a verrucous epidermal naevus associated with mild facial dysmorphism. Br J Dermatol 156: 1353–1356.

Conti R, Bruscino N, Campolmi P et al. (2013). Inflammatory linear verrucous epidermal nevus: why a combined laser therapy. J Cosmet Laser Ther 15: 242–245.

Cooper MK, Wassif CA, Krakowiak PA et al. (2003). A defective response to hedgehog signaling in disorders of cholesterol biosynthesis. Nat Genet 33: 508–513.

Cribier B, Scrivener Y, Grosshans E (2000). Tumors arising in nevus sebaceus: a study of 596 cases. J Am Acad Dermatol 42: 263–268.

Cullen SI, Harris DE, Carter CH et al. (1969). Congenital unilateral ichthyosiform erythroderma. Arch Dermatol 99: 724–729.

Danarti R, König A, Salhi A et al. (2004). Becker's nevus syndrome revisited. J Am Acad Dermatol 51: 965–969.

Davies D, Rogers M (2002). Review of neurological manifestations in 196 patients with sebaceous naevi. Australas J Dermatol 43: 20–23.

De Becker I, Gajda DJ, Gilbert-Barness E et al. (2000). Ocular manifestations in Proteus syndrome. Am J Med Genet 92: 350–352.

de Jong E, Rulo HF, van de Kerkhof PC (1991). Inflammatory linear verrucous epidermal naevus (ILVEN) versus linear psoriasis. A clinical, histological and immunohistochemical study. Acta Derm Venereol 71: 343–346.

de Waal AC, van Rossum MM, Bovenschen HJ (2010). Extensive segmental acanthosis nigricans form of epidermal nevus. Dermatol Online J 16: 7.

Deliduka SB, Kwong PC (2004). Treatment of nevus comedonicus with topical tazarotene and calcipotriene. J Drugs Dermatol JDD 3: 674–676.

Diluvio L, Campione E, Mordenti C et al. (2013). Inflammatory linear verrucous epidermal nevus and epilepsy: remission during valproic acid therapy. J Drugs Dermatol JDD 12: 18–19.

Dimond RL, Amon RB (1976). Epidermal nevus and rhabdomyosarcoma. Arch Dermatol 112: 1424–1426.

Diven DG, Solomon AR, McNeely MC et al. (1987). Nevus sebaceus associated with major ophthalmologic abnormalities. Arch Dermatol 123: 383–386.

Dodge NN, Dobyns WB (1995). Agenesis of the corpus callosum and Dandy–Walker malformation associated with hemimegalencephaly in the sebaceous nevus syndrome. Am J Med Genet 56: 147–150.

Dogra S, Jain R, Mohanty SK et al. (2002). Porokeratotic eccrine ostial and dermal duct nevus: unilateral systematized involvement. Pediatr Dermatol 19: 568–569.

Domingo J, Helwig EB (1979). Malignant neoplasms associated with nevus sebaceus of Jadassohn. J Am Acad Dermatol 1: 545–556.

Dudley K, Barr WG, Armin A et al. (1986). Nevus comedonicus in association with widespread, well-differentiated follicular tumors. J Am Acad Dermatol 15: 1123–1127.

Dupre A, Christol B (1977). Inflammatory linear verrucose epidermal nevus. A pathologic study. Arch Dermatol 113: 767–769.

Easton JA, Donnelly S, Kamps MAF et al. (2012). Porokeratotic eccrine nevus may be caused by somatic connexin26 mutations. J Invest Dermatol 132: 2184–2191.

Eichler C, Flowers FP, Ross J (1989). Epidermal nevus syndrome: case report and review of clinical manifestations. Pediatr Dermatol 6: 316–320.

Elias PM, Crumrine D, Paller A et al. (2011). Pathogenesis of the cutaneous phenotype in inherited disorders of cholesterol metabolism: therapeutic implications for topical treatment of these disorders. Dermatoendocrinol 3: 100–106.

Elias PM, Williams ML, Feingold KR (2012). Abnormal barrier function in the pathogenesis of ichthyosis: therapeutic implications for lipid metabolic disorders. Clin Dermatol 30: 311–322.

el-Shanti H, Bell WE, Waziri MH (1992). Epidermal nevus syndrome: subgroup with neuronal migration defects. J Child Neurol 7: 29–34.

Engber PB (1978). The nevus comedonicus syndrome: a case report with emphasis on associated internal manifestations. Int J Dermatol 17: 745–749.

Ferreira PS, Valente NYS, Nico MMS (2013). Multiple filiform keratoses and nodules in a 10-year-old girl. Pediatr Dermatol 30: 261–262.

Feuerstein RC, Mims LC (1962). Linear nevus sebaceus with convulsions and mental retardation. Am J Dis Child 104: 675–679.

Fink-Puches R, Soyer HP, Pierer G et al. (1997). Systematized inflammatory epidermal nevus with symmetrical involvement: an unusual case of CHILD syndrome? J Am Acad Dermatol 36: 823–826.

García-Vargas A, Hafner C, Pérez-Rodríguez AG et al. (2008). An epidermal nevus syndrome with cerebral involvement caused by a mosaic FGFR3 mutation. Am J Med Genet A 146A: 2275–2279.

Ginarte M, Fernández-Redondo V, Toribio J (2000). Unilateral psoriasis: a case individualized by means of involucrin. Cutis 65: 167–170.

Goddard DS, Rogers M, Frieden IJ et al. (2009). Widespread porokeratotic adnexal ostial nevus: clinical features and proposal of a new name unifying porokeratotic eccrine ostial and dermal duct nevus and porokeratotic eccrine and hair follicle nevus. J Am Acad Dermatol 61: 1060e1–1060e14.

Golitz LE, Weston WL (1979). Inflammatory linear verrucous epidermal nevus. Association with epidermal nevus syndrome. Arch Dermatol 115: 1208–1209.

Grebe TA, Rimsza ME, Richter SF et al. (1993). Further delineation of the epidermal nevus syndrome: two cases with new findings and literature review. Am J Med Genet 47: 24–30.

Groesser L, Herschberger E, Ruetten A et al. (2012). Postzygotic HRAS and KRAS mutations cause nevus sebaceous and Schimmelpenning syndrome. Nat Genet 44: 783–787.

Groesser L, Vogt T, Happle R et al. (2013). Naevus marginatus revisited: a combined organoid and nonorganoid epidermal naevus caused by HRAS mutation. Br J Dermatol 168: 892–894.

Guldbakke KK, Khachemoune A, Deng A et al. (2007). Naevus comedonicus: a spectrum of body involvement. Clin Exp Dermatol 32: 488–492.

Gurecki PJ, Holden KR, Sahn EE et al. (1996). Developmental neural abnormalities and seizures in epidermal nevus syndrome. Dev Med Child Neurol 38: 716–723.

Hafner C, López-Knowles E, Luis NM et al. (2007). Oncogenic PIK3CA mutations occur in epidermal nevi and seborrheic keratoses with a characteristic mutation pattern. Proc Natl Acad Sci U S A 104: 13450–13454.

Hafner C, Klein A, Landthaler M et al. (2009). Clonality of basal cell carcinoma arising in an epidermal nevus. New insights provided by molecular analysis. Dermatol Basel Switz 218: 278–281.

Hafner C, Di Martino E, Pitt E et al. (2010). FGFR3 mutation affects cell growth, apoptosis and attachment in keratinocytes. Exp Cell Res 316: 2008–2016.

Hafner C, Toll A, Real FX (2011). HRAS mutation mosaicism causing urothelial cancer and epidermal nevus. N Engl J Med 365: 1940–1942.

Hafner C, Toll A, Gantner S et al. (2012). Keratinocytic epidermal nevi are associated with mosaic RAS mutations. J Med Genet 49: 249–253.

Hafner C, van Oers JMM, Vogt T et al. (2006). Mosaicism of activating FGFR3 mutations in human skin causes epidermal nevi. J Clin Invest 116: 2201–2207.

Happle R (1987). Lethal genes surviving by mosaicism: a possible explanation for sporadic birth defects involving the skin. J Am Acad Dermatol 16: 899–906.

Happle R (1995). Epidermal nevus syndromes. Semin Dermatol 14: 111–121.

Happle R (2002). Dohi memorial lecture. New aspects of cutaneous mosaicism. J Dermatol 29: 681–692.

Happle R (2007). Linear Cowden nevus: a new distinct epidermal nevus. Eur J Dermatol EJD 17: 133–136.

Happle R (2010a). The group of epidermal nevus syndromes. Part I Well defined phenotypes. J Am Acad Dermatol 63: 1–22. quiz 23–24.

Happle R (2010b). The group of epidermal nevus syndromes. Part II Less well defined phenotypes. J Am Acad Dermatol 63: 25–30. quiz 31–32.

Happle R (2013). Nevus sebaceus is a mosaic RASopathy. J Invest Dermatol 133: 597–600.

Happle R, Koopman RJ (1997). Becker nevus syndrome. Am J Med Genet 68: 357–361.

Happle R, Koch H, Lenz W (1980). The CHILD syndrome. Congenital hemidysplasia with ichthyosiform erythroderma and limb defects. Eur J Pediatr 134: 27–33.

Happle R, Hoffmann R, Restano L et al. (1996). Phacomatosis pigmentokeratotica: a melanocytic-epidermal twin nevus syndrome. Am J Med Genet 65: 363–365.

Hernández S, Toll A, Baselga E et al. (2007). Fibroblast growth factor receptor 3 mutations in epidermal nevi and associated low grade bladder tumors. J Invest Dermatol 127: 1664–1666.

Hodge JA, Ray MC, Flynn KJ (1991). The epidermal nevus syndrome. Int J Dermatol 30: 91–98.

Hohenleutner U, Wlotzke U, Konz B et al. (1995). Carbon dioxide laser therapy of a widespread epidermal nevus. Lasers Surg Med 16: 288–291.

Hummel M, Cunningham D, Mullett CJ et al. (2003). Left-sided CHILD syndrome caused by a nonsense mutation in the NSDHL gene. Am J Med Genet A 122A: 246–251.

Idriss MH, Elston DM (2014). Secondary neoplasms associated with nevus sebaceus of Jadassohn: a study of 707 cases. J Am Acad Dermatol 70: 332–337.

Inoue Y, Miyamoto Y, Ono T (2000). Two cases of nevus comedonicus: successful treatment of keratin plugs with a pore strip. J Am Acad Dermatol 43: 927–929.

Ivker R, Resnick SD, Skidmore RA (1997). Hypophosphatemic vitamin D-resistant rickets, precocious puberty, and the epidermal nevus syndrome. Arch Dermatol 133: 1557–1561.

Jaqueti G, Requena L, Sánchez Yus E (2000). Trichoblastoma is the most common neoplasm developed in nevus sebaceus of Jadassohn: a clinicopathologic study of a series of 155 cases. Am J Dermatopathol 22: 108–118.

Jeong HS, Lee HK, Lee SH et al. (2012). Multiple large cysts arising from nevus comedonicus. Arch Plast Surg 39: 63–66.

Jonsson KB, Zahradnik R, Larsson T et al. (2003). Fibroblast growth factor 23 in oncogenic osteomalacia and X-linked hypophosphatemia. N Engl J Med 348: 1656–1663.

Kaidbey KH, Kurban AK (1971). Dermatitic epidermal nevus. Arch Dermatol 104: 166–171.

Kaliyadan F, Nampoothiri S, Sunitha V et al. (2010). Nevus comedonicus syndrome – nevus comedonicus associated with ipsilateral polysyndactyly and bilateral oligodontia. Pediatr Dermatol 27: 377–379.

Karo KR, Gange RW (1981). Smooth-muscle hamartoma. Possible congenital Becker's nevus. Arch Dermatol 117: 678–679.

Katz B, Wiley CA, Lee VW (1987). Optic nerve hypoplasia and the syndrome of nevus sebaceous of Jadassohn. A new association. Ophthalmology 94: 1570–1576.

Kim CA, Konig A, Bertola DR et al. (2005). CHILD syndrome caused by a deletion of exons 6-8 of the NSDHL gene. Dermatol Basel Switz 211: 155–158.

König A, Happle R, Bornholdt D et al. (2000). Mutations in the NSDHL gene, encoding a 3beta-hydroxysteroid dehydrogenase, cause CHILD syndrome. Am J Med Genet 90: 339–346.

Kurek KC, Luks VL, Ayturk UM et al. (2012). Somatic mosaic activating mutations in PIK3CA cause CLOVES syndrome. Am J Hum Genet 90: 1108–1115.

Lansky LL, Funderburk S, Cuppage FE et al. (1972). Linear sebaceous nevus syndrome. A hamartoma variant. Am J Dis Child 123: 587–590.

Lazic T, Li Q, Frank M et al. (2012). Extending the phenotypic spectrum of keratitis-ichthyosis-deafness syndrome: report of a patient with GJB2 (G12R) connexin 26 mutation and unusual clinical findings. Pediatr Dermatol 29: 349–357.

Lazzeri S, Mascalchi M, Cellerini M et al. (1993). Epidermal nevus syndrome: MR of intracranial involvement. AJNR Am J Neuroradiol 14: 1255–1257.

Lee SH, Rogers M (2001). Inflammatory linear verrucous epidermal naevi: a review of 23 cases. Australas J Dermatol 42: 252–256.

Lee HJ, Chun EY, Kim YC et al. (2002). Nevus comedonicus with hidradenoma papilliferum and syringocystadenoma papilliferum in the female genital area. Int J Dermatol 41: 933–936.

Levinsohn JL, Tian LC, Boyden LM et al. (2012). Whole-exome sequencing reveals somatic mutations in HRAS and KRAS, which cause nevus sebaceus. J Invest Dermatol 133: 827–830.

Levinsohn JL, Teng J, Craiglow BG et al. (2014). Somatic HRAS pG12S mutation causes woolly hair and epidermal nevi. J Invest Dermatol 34: 1149–1152.

Lim YH, Ovejero D, Sugarman JS et al. (2014). Multilineage somatic activating mutations in HRAS and NRAS cause mosaic cutaneous and skeletal lesions, elevated

FGF23, and hypophosphatemia. Hum Mol Genet 23: 397–407.

Lindhurst MJ, Sapp JC, Teer JK et al. (2011). A mosaic activating mutation in AKT1 associated with the Proteus syndrome. N Engl J Med 365: 611–619.

Mall V, Heinen F, Uhl M et al. (2000). CNS lipoma in patients with epidermal nevus syndrome. Neuropediatrics 31: 175–179.

Marsden RA, Fleming K, Dawber RP (1979). Comedo naevus of the palm – a sweat duct naevus? Br J Dermatol 101: 717–722.

Mehregan AH, Pinkus H (1965). Life history of organoid nevi special reference to nevus sebaceus of Jadassohn. Arch Dermatol 91: 574–588.

Melnik BC (2009). Role of FGFR2-signaling in the pathogenesis of acne. Dermatoendocrinol 1: 141–156.

Melnik BC, Vakilzadeh F, Aslanidis C et al. (2008). Unilateral segmental acneiform naevus: a model disorder towards understanding fibroblast growth factor receptor 2 function in acne? Br J Dermatol 158: 1397–1399.

Minić S, Trpinac D, Obradović M (2013). Blaschko line analogies in the central nervous system: a hypothesis. Med Hypotheses 81: 671–674.

Mirzaa G, Conway R, Graham JM, Dobyns WB (1993). PIK3CA-related segmental overgrowth. In: RA Pagon, MP Adam, TD Bird et al. (Eds.), GeneReviews™, University of Washington, Seattle, Seattle (WA).

Mitsuhashi Y, Katagiri Y, Kondo S (1997). Treatment of inflammatory linear verrucous epidermal naevus with topical vitamin D3. Br J Dermatol 136: 134–135.

Molho-Pessach V, Schaffer JV (2011). Blaschko lines and other patterns of cutaneous mosaicism. Clin Dermatol 29: 205–225.

Mollica F, Pavone L, Nuciforo G (1974). Linear sebaceous nevus syndrome in a newborn. Am J Dis Child 128: 868–871.

Morillo V, Manrique P, De Miguel E et al. (2007). Nevus comedonicus with epidermolytic hyperkeratosis. Eur J Dermatol EJD 17: 176–177.

Morimoto M, du Souich C, Trinh J et al. (2012). Expression profile of NSDHL in human peripheral tissues. J Mol Histol 43: 95–106.

Moss C, Larkins S, Stacey M et al. (1993). Epidermal mosaicism and Blaschko's lines. J Med Genet 30: 752–755.

Moss C, Jones DO, Blight A et al. (1995). Birthmark due to cutaneous mosaicism for keratin 10 mutation. Lancet 345: 596.

Mukhopadhyay A, Krishnaswami SR, Yu BDY (2011). Activated Kras alters epidermal homeostasis of mouse skin, resulting in redundant skin and defective hair cycling. J Invest Dermatol 131: 311–319.

Munro CS, Wilkie AO (1998). Epidermal mosaicism producing localised acne: somatic mutation in FGFR2. Lancet 352: 704–705.

Nelson BR, Kolansky G, Gillard M et al. (1994). Management of linear verrucous epidermal nevus with topical 5-fluorouracil and tretinoin. J Am Acad Dermatol 30: 287–288.

Olivares JL, Ramos FJ, Carapeto FJ et al. (1999). Epidermal naevus syndrome and hypophosphataemic rickets: description of a patient with central nervous system anomalies and review of the literature. Eur J Pediatr 158: 103–107.

Ornitz DM, Itoh N (2001). Fibroblast growth factors. Genome Biol 2. REVIEWS3005.

Ousager LB, Bygum A, Hafner C (2012). Identification of a novel S249C FGFR3 mutation in a keratinocytic epidermal naevus syndrome. Br J Dermatol 167: 202–204.

Özdemir M, Balevi A, Esen H (2012). An inflammatory verrucous epidermal nevus concomitant with psoriasis: treatment with adalimumab. Dermatol Online J 18: 11.

Pahwa P, Sethuraman G (2012). Segmental Becker's nevi with mucosal involvement. Pediatr Dermatol 29: 670–671.

Paller AS (1987). Epidermal nevus syndrome. Neurol Clin 5: 451–457.

Paller AS, Syder AJ, Chan YM et al. (1994). Genetic and clinical mosaicism in a type of epidermal nevus. N Engl J Med 331: 1408–1415.

Paller AS, van Steense lMAM, Rodriguez-Martín M et al. (2011). Pathogenesis-based therapy reverses cutaneous abnormalities in an inherited disorder of distal cholesterol metabolism. J Invest Dermatol 131: 2242–2248.

Patrizi A, Neri I, Fiorentini C et al. (1998). Nevus comedonicus syndrome: a new pediatric case. Pediatr Dermatol 15: 304–306.

Patrizi A, Medri M, Raone B et al. (2012). Clinical characteristics of Becker's nevus in children: report of 118 cases from Italy. Pediatr Dermatol 29: 571–574.

Pavlidis E, Cantalupo G, Boria S et al. (2012). Hemimegalencephalic variant of epidermal nevus syndrome: case report and literature review. Eur J Paediatr Neurol 16: 332–342.

Pavone L, Curatolo P, Rizzo R et al. (1991). Epidermal nevus syndrome: a neurologic variant with hemimegalencephaly, gyral malformation, mental retardation, seizures, and facial hemihypertrophy. Neurology 41: 266–271.

Prayson RA, Estes ML, Morris HH (1993). Coexistence of neoplasia and cortical dysplasia in patients presenting with seizures. Epilepsia 34: 609–615.

Prayson RA, Kotagal P, Wyllie E et al. (1999). Linear epidermal nevus and nevus sebaceus syndromes: a clinicopathologic study of 3 patients. Arch Pathol Lab Med 123: 301–305.

Reda AM, Rogers 3rd RS, Peters MS (1990). Woolly hair nevus. J Am Acad Dermatol 22: 377–380.

Renner R, Colsman A, Sticherling M (2008). ILVEN: is it psoriasis? Debate based on successful treatment with etanercept. Acta Derm Venereol 88: 631–632.

Rogers M (1992). Epidermal nevi and the epidermal nevus syndromes: a review of 233 cases. Pediatr Dermatol 9: 342–344.

Rogers M, McCrossin I, Commens C (1989). Epidermal nevi and the epidermal nevus syndrome. A review of 131 cases. J Am Acad Dermatol 20: 476–488.

Rosenthal D, Fretzin DF (1986). Epidermal nevus syndrome: report of association with transitional cell carcinoma of the bladder. Pediatr Dermatol 3: 455–458.

Rossman RE, Shapiro EM, Freeman RG (1963). Unilateral ichthyosiform erythroderma. Arch Dermatol 88: 567–571.

Saraswat A, Dogra S, Bansali A et al. (2003). Phakomatosis pigmentokeratotica associated with hypophosphataemic vitamin D-resistant rickets: improvement in phosphate homeostasis after partial laser ablation. Br J Dermatol 148: 1074–1076.

Sassmannshausen J, Bogomilsky J, Chaffins M (2000). Porokeratotic eccrine ostial and dermal duct nevus: a case report and review of the literature. J Am Acad Dermatol 43: 364–367.

Schauder S, Hanefeld F, Noske UM et al. (2000). Depigmented hypertrichosis following Blaschko's lines associated with cerebral and ocular malformations: a new neurocutaneous, autosomal lethal gene syndrome from the group of epidermal naevus syndromes? Br J Dermatol 142: 1204–1207.

Schecter AK, Lester B, Pan TD et al. (2004). Linear nevus comedonicus with epidermolytic hyperkeratosis. J Cutan Pathol 31: 502–505.

Schimmelpenning GW (1957). Clinical contribution to symptomatology of phacomatosis. Fortschritte Auf Dem Geb Röntgenstrahlen Nukl 87: 716–720.

Schulz U, O'Leary CP (2001). Spinal AVM, epidermal nevus, and rhabdomyosarcoma: a rare neurocutaneous syndrome? Neurology 56: 395–397.

Seeger MA, Paller AS (2014). The role of abnormalities in the distal pathway of cholesterol synthesis in the congenital hemidysplasia with ichthyosiform erythroderma and limb defects (CHILD) syndrome. Biochim Biophys Acta 1841: 345–352.

Sellheyer K, Cribier B, Nelson P et al. (2013). Basaloid tumors in nevus sebaceus revisited: the follicular stem cell marker PHLDA1 (TDAG51) indicates that most are basal cell carcinomas and not trichoblastomas. J Cutan Pathol 40: 455–462.

Shirley MD, Tang H, Gallione CJ et al. (2013). Sturge–Weber syndrome and port-wine stains caused by somatic mutation in GNAQ. N Engl J Med 368: 1971–1979.

Solomon LM, Esterly NB (1975). Epidermal and other congenital organoid nevi. Curr Probl Pediatr 6: 1–56.

Solomon LM, Fretzin DF, Dewald RL (1968). The epidermal nevus syndrome. Arch Dermatol 97: 273–285.

Stottmann RW, Turbe-Doan A, Tran P et al. (2011). Cholesterol metabolism is required for intracellular hedgehog signal transduction in vivo. PLoS Genet 7: e1002224.

Su WP (1982). Histopathologic varieties of epidermal nevus. A study of 160 cases. Am J Dermatopathol 4: 161–170.

Sugarman JL (2004a). Epidermal nevus syndromes. Semin Cutan Med Surg 23: 145–157.

Sugarman JL (2004b). Epidermal Nevus syndromes. In: ES Roach, VS Miller (Eds.), Neurocutaneous disorders, Cambridge University Press, Cambridge, pp. 88–104.

Sugarman GI, Reed WB (1969). Two unusual neurocutaneous disorders with facial cutaneous signs. Arch Neurol 21: 242–247.

Tadini G, Restano L, Gonzáles-Pérez R et al. (1998). Phacomatosis pigmentokeratotica: report of new cases and further delineation of the syndrome. Arch Dermatol 134: 333–337.

Taheri AR, Nikandish M, Mashayekhi V et al. (2011). Phacomatosis pigmentokeratotica associated with compound melanocytic nevus of the conjunctiva. Int J Dermatol 50: 994–998.

Tay YK, Weston WL, Ganong CA et al. (1996). Epidermal nevus syndrome: association with central precocious puberty and woolly hair nevus. J Am Acad Dermatol 35: 839–842.

Tchernev G, Ananiev J, Semkova K et al. (2013). Nevus comedonicus: an updated review. Dermatol Ther 3: 33–40.

Terheyden P, Hornschuh B, Karl S et al. (1998). Lichen planus associated with Becker's nevus. J Am Acad Dermatol 38: 770–772.

Tidyman WE, Rauen KA (2009). The RASopathies: developmental syndromes of Ras/MAPK pathway dysregulation. Curr Opin Genet Dev 19: 230–236.

Vakilzadeh F, Happle R, Peters P et al. (1976). Focal dermal hypoplasia with apocrine nevi and striation of bones (author's transl). Arch Dermatol Res Arch Für Dermatol Forsch 256: 189–195.

Vidaurri-de la Cruz H, Tamayo-Sánchez L, Durán-McKinster C et al. (2004). Epidermal nevus syndromes: clinical findings in 35 patients. Pediatr Dermatol 21: 432–439.

Vissers WH, Muys L, Erp PE et al. (2004). Immunohistochemical differentiation between inflammatory linear verrucous epidermal nevus (ILVEN) and psoriasis. Eur J Dermatol EJD 14: 216–220.

Wieland I, Tinschert S, Zenker M (2013). High-level somatic mosaicism of AKT1 c49G > A mutation in skin scrapings from epidermal nevi enables non-invasive molecular diagnosis in patients with Proteus syndrome. Am J Med Genet A 161: 889–891.

Xin H, Matt D, Qin JZ et al. (1999). The sebaceous nevus: a nevus with deletions of the PTCH gene. Cancer Res 59: 1834–1836.

Yu TW, Tsau YK, Young C et al. (2000). Epidermal nevus syndrome with hypermelanosis and chronic hyponatremia. Pediatr Neurol 22: 151–154.

Zarkik S, Bouhllab J, Methqal A et al. (2012). Keratoacanthoma arising in nevus comedonicus. Dermatol Online J 18: 4.

Zvulunov A, Grunwald MH, Halvy S (1997). Topical calcipotriol for treatment of inflammatory linear verrucous epidermal nevus. Arch Dermatol 133: 567–568.

Handbook of Clinical Neurology, Vol. 132 (3rd series)
Neurocutaneous Syndromes
M.P. Islam and E.S. Roach, Editors

Chapter 23

Lipoid proteinosis

JOHN A. McGRATH*
St John's Institute of Dermatology, King's College London (Guy's Campus), London, UK

INTRODUCTION

History

Lipoid proteinosis (LP; OMIM 247100) was first described in 1929 by a Viennese dermatologist (Urbach) and an otorhinolaryngologist (Wiethe) (Urbach and Wiethe, 1929). Approximately 400 cases have been reported in the world literature.

Origin of the name

The original term used by Urbach and Wiethe to describe the disease was "lipoidosis cutis et mucosae." The disease name lipoid proteinosis was derived from the histopathologic features on skin biopsy that showed abnormal protein and lipid inclusions. Lipoid proteinosis has also been known as lipoglycoproteinosis, or, due to the glass-like (hyaline) appearance of the dermis in affected skin, hyalinosis cutis et mucosae. Lipoid proteinosis is also sometimes referred to by the eponymn Urbach–Wiethe disease.

Epidemiology

Lipoid proteinosis is found throughout the world, although it occurs more commonly in some regions. Notably, it is particularly common in the Northern Cape province of South Africa, including Namaqualand, where propagation of a mutated common ancestral allele dating back to a mid-17th century settler from Germany has been identified (van Hougenhouck-Tulleken et al., 2004).

CLINICAL MANIFESTATIONS

Typical clinicopathologic features of lipoid proteinosis are illustrated in Figure 23.1. The disorder may not be evident at birth, in that there may be no obvious neurologic, skin, or mucous membrane features. Initial clinical signs are usually mucocutaneous: during infancy there may be some vocal abnormalities with a hoarse voice and subsequently some warty infiltration of the skin may occur. This skin thickening tends to progress over several years and is exacerbated by trauma. Neurologic abnormalities may become evident during early childhood or may not present until later life.

Hallmark signs and symptoms

In some patients, neurologic features, such as epilepsy and migraine, may predominate over the skin abnormalities and thus give the earliest clues to the underlying diagnosis (Claeys et al., 2007). Nevertheless, neurologic manifestations are highly variable and may only manifest or worsen with time. There may be bilateral circumscribed and symmetrical calcification in the medial temporal regions that can be associated with seizures. Neuropsychiatric associations include schizophrenia, anxiety, and variable abnormalities in mediating emotional responses, particularly related to perceived danger and threat. Headache is uncommon in lipoid proteinosis but is a well recognized feature (Arkadir et al., 2013). Other physical signs include dizziness, ataxia, slight psychomotor retardation, and amnesic impairment as well as transient brachiofacial paralysis (Maruani et al., 2007). Neurologic abnormalities in some families, however, may be very subtle or even nonexistent (Salih et al., 2011).

The first clinical sign affecting the skin and mucous membranes is usually hoarseness, caused by infiltration of the vocal cords (Hamada, 2002). The most typical skin infiltration occurs on the eyelids, where beaded papules occur; this is sometimes referred to as moniliform blepharosis. Other skin changes include waxy yellow

*Correspondence to: John A. McGrath, MD FRCP FMedSci, Professor of Molecular Dermatology, Dermatology Research Labs, Floor 9 Tower Wing, Guy's Hospital, Great Maze Pond, London SE1 9RT, UK. Tel: +44-20-71886409, Fax: +44-20-71888050, E-mail: john.mcgrath@kcl.ac.uk

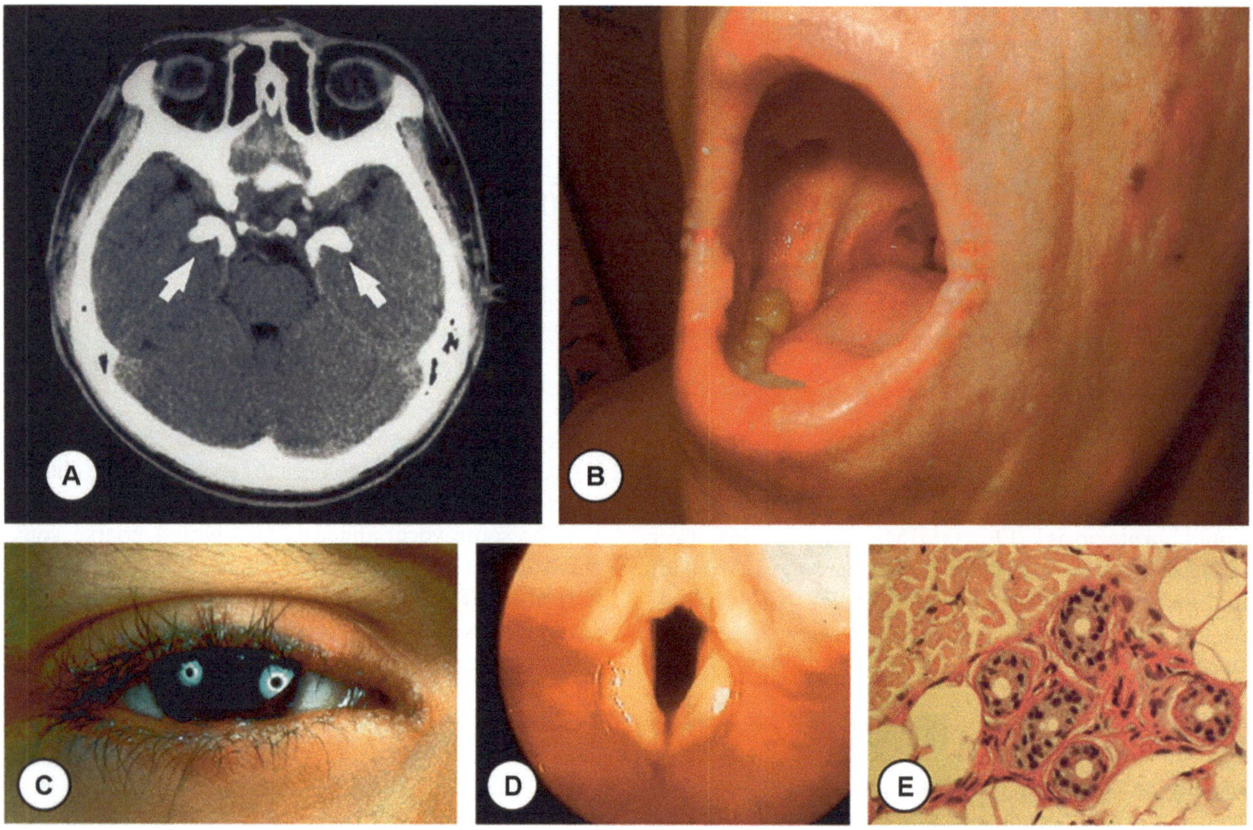

Fig. 23.1. Clinicopathologic features of lipoid proteinosis. (**A**) Bilateral bean shaped calcification in temporal lobes (arrows); (**B**) skin and mucous membrane thickening; (**C**) beaded papules on eyelids (moniliform blepharosis); (**D**) bilateral thickening of vocal cords; (**E**) skin biopsy showing thickened basement membrane around blood vessels, glands and fat (periodic acid–Schiff stain).

papules, nodules with generalized skin thickening. Friction to the skin can lead to changes in skin texture, particularly on the hands, elbows, knees, buttocks, and axillae (van Hougenhouck-Tulleken et al., 2004). In some patients, the skin infiltration can resemble viral warts. Before infiltration occurs the skin can be quite fragile, leading to trauma-induced blister formation and scarring (Hamada, 2002). Some of the scarring can be pock-like or acneiform, and is particularly evident on the face and extremities. Mucosae of the pharynx, tongue, soft palate, tonsils and lips can become infiltrated, leading to respiratory difficulty, often exacerbated by upper respiratory tract infection. Recurrent episodes of inflamed parotid and submandibular glands can occur. A useful physical sign in lipoid proteinosis is an inability to protrude the tongue due to a thickened sublingual frenulum. In combination with a hoarse voice, these features provide a useful clinical approach to making the diagnosis.

Involvement by systems

INTRACRANIAL CALCIFICATION

A classic feature of lipoid proteinosis is intracranial calcification. Calcium deposition can occur anywhere in the brain but typically occurs in the temporal lobes, and amygdala involvement is considered pathognomonic. Some patients show evidence of epilepsy and psychiatric symptoms, such as schizophrenia, mood disorders, and anxiety – sometimes in association with calcium deposits, although the clinicopathologic association is not always evident or predictable. A study in South Africa showed that five of 24 individuals with lipoid proteinosis had a history of epilepsy; two of these described classic features of temporal lobe epilepsy with déjà vu experiences and vague fugue states but any relationship between the epilepsy and intracranial calcification in these cases is not known (van Hougenhouck-Tulleken et al., 2004).

DYSFUNCTIONAL AMYGDALA

The focal nature of the neuroanatomic pathology to the amygdala in cases of lipoid proteinosis has led to studies on affected patients to derive information about the normal functioning of the amygdala. Human amygdala plays an important role in processing social signals such as facial expressions (including fear) and facilitates responses to them in face-sensitive cortical regions with relevance to social competence. Decision making may be impaired in lipoid proteinosis with evidence for impoverished emotional arousal rating and deviant punishment behavior (Scheele et al., 2012). Selective damage of the amygdala can lead to both decisions under ambiguity and decisions under risk being affected, particularly the latter being more reduced (Brand et al., 2007). Studies have also indicated that in the absence of a functional amygdala, prioritized visual processing may rely on alternative structures such as pulvinar and cortical visual areas, consistent with compensatory pathways (Bach et al., 2011). Patients with lipoid proteinosis may show impaired recognition of fearful faces but this is variable, suggesting that other brain regions might be able to compensate to support fear and social signal processing (Becker et al., 2012).

MEMORY

Studies on patients with lipoid proteinosis have indicated that an intact amygdala is necessary for generating retrograde and anterograde episodic memory changes in response to positive and negative emotions (Hurlemann et al., 2007). Patient studies have also indicated that the amygdala has a key role in relaying biological and social significance to autobiographical memory (Markowitsch and Staniloiu, 2001). Some patients have shown a selective memory impairment of autobiographic episodes with preserved memory for autobiographic facts, with other evidence that the amygdala is also critically involved in the processing of emotional memory. Patients with lipoid proteinosis may also have selective memory impairments which can be associated with marked effective emotional fluctuations. These findings have been used to suggest that the amygdala region confers an effective flavor to memories and thereby enhances the likelihood of their long-term storage (Markowitsch et al., 1994).

PANIC ATTACKS

Patients with lipoid proteinosis often report a sense of unprovoked anxiety, impending fear or panic, even in the presence of bilateral selective amygdala lesions (Wiest et al., 2006). Although a role for the amygdala in fear and panic responses has been implicated, such reactions can also be elicited in patients with lipoid proteinosis, notably following inhalation of 35% CO_2 (Feinstein et al., 2013). This finding indicates that the amygdala is not an absolute requirement for fear and panic reactions, and identifies a distinction between fear triggered by external threats from the environment versus fear triggered internally by CO_2. In patients with lipoid proteinosis the internal threat signaled by CO_2 is detected and interpreted as fear and panic despite the absence of an intact functional amygdala (Feinstein et al., 2013).

NEUROPSYCHOLOGICAL ABNORMALITIES

Individuals with lipoid proteinosis may show cognitively little deviation from unaffected subjects but they differ emotionally, particularly in their judgment of all emotions and facial expressions, and notably in both negative and positive processing (Siebert et al., 2003). Other assessments have revealed less correct recognition of emotional expressions such as fear, anger, disgust, surprise, sadness, happiness, and neutrality in lipoid proteinosis compared to control subjects, with worse performances on tests of memory, including digits forward and backward (Thornton et al., 2008). Individuals with lipoid proteinosis may appear remarkably dispassionate when relating highly emotional and traumatic life experiences with little sense of distrust or danger. What can appear as resilience in coping with life in some individuals may be a reflection that emotions such as fear and anger may be missing. Indeed, patients with lipoid proteinosis may have a truncated emotional life, in that they express very little negative emotion and are not good judges of character, being overly trusting of others, the implication being that they have an impoverished emotional landscape (Tranel et al., 2006).

Tumor/neoplastic potential

Lipoid proteinosis is not associated with neoplasia. The protein implicated in lipoid proteinosis, extracellular matrix protein 1 (ECM1), is overexpressed in certain malignant epithelial tumors, as well as other solid organ tumors with a role in epithelial stromal interactions but lack of ECM1, as occurs in lipoid proteinosis, is not associated with any increased risk of malignancy (Chan et al., 2007).

Imaging

Computed tomography (CT) of patients with lipoid proteinosis may show bilateral amygdala calcification with magnetic resonance imaging (MRI) revealing hypointense signals in these areas. Intracranial calcifications mostly occur in the amygdala, hippocampus, perihippocampal gyrus, or striatum (Gonçalves et al., 2010). Not all patients show intracranial calcification, however, and there may not always be a correlation between the neuroimaging abnormalities and clinical features. The calcific deposits tend to appear later in childhood and may become evident with age. Calcification rarely occurs at other sites, such as the skin.

Variants

Clinical heterogeneity is well recognized in lipoid proteinosis and the degree of neurologic or cutaneous pathology can show both intrafamilial and interfamilial variability; what underpins the heterogeneity is not known. Warty skin infiltration may be exacerbated by trauma and injury, and perhaps by sun exposure, but there is no known environmental or epigenetic modifier for the neurologic features.

Differential diagnoses

Lipoid proteinosis has characteristic clinicopathologic features that, when present, make the diagnosis straightforward, although in the presence of minor abnormalities or signs yet to emerge, the differential diagnosis can include variants of porphyria (including erythropoietic protoporphyria) and hyaline fibromatosis syndromes (previously known as infantile systemic hyalinosis and juvenile hyaline fibromatosis).

PATHOLOGY

The most common radiologic hallmark is the presence of bean- or comma-shaped intracranial calcifications in the temporal lobes and in the amygdala (Quirici and Rocha, 2013). This is usually more evident in patients who have had more longstanding lipoid proteinosis. In addition to medial temporal calcifications that are typical of lipoid proteinosis, additional calcification has been reported in the right posterior parietal region (Kachewar and Kulkarni, 2012). Bilateral amygdala calcification is more evident in patients with longer disease duration, and epilepsy, when present, maybe related to these calcifications (Appenzeller et al., 2006). Striatal calcifications may also be associated with generalized dystonia (Teive et al., 2004). Spontaneous intracerebral hemorrhage has also been reported in lipoid proteinosis (Messina et al., 2012). In addition to neurobiological models of anxiety, patients with lipoid proteinosis have also provided new insight at a neurochemical level. Notably, the amygdala damage is associated with abnormal serotonergic signaling via postsynaptic 5-HT_{2A} receptors, thus providing some insight into anxiety behavior. Nevertheless, in patients, the receptor signaling extends beyond the amygdala, thereby providing additional insight into human anxiety behaviors and identifying the 5-HT_{2A} receptor as a possible pharmacologic target for controlling pathological anxiety (Hurlemann et al., 2009).

With regards to tissue pathology, skin is the organ most commonly biopsied. This can show abnormalities in both the epidermis and the dermis. Typically, there is a homogenized appearance to the dermis with indistinct spaces between the collagen bundles; this is termed hyalinosis (Hamada, 2002). The epidermis may also show abnormalities with variable thickening and some hyperkeratosis (increased scaling).

MOLECULAR BASIS/PATHOPHYSIOLOGY

In 2002, lipoid proteinosis was mapped to lq21 and mutations were identified in *ECM1*, encoding extracellular matrix protein 1 (Hamada et al., 2002). There are at least four different isoforms of *ECM1* with known alternative splicing of exon 7. The majority of mutations are found in exons 6 and 7. The finding that there are several pathogenic mutations in exon 7 indicates that the *ECM1a* isoform (in which exon 7 is present) perhaps has the most important functional biological role. Most mutations are nonsense mutations or small insertion or deletion mutations, although a number of missense mutations have also been identified. To date, approximately 50 recessive mutations have been described in lipoid proteinosis (Chan et al., 2007). Heterozygous carriers (estimated to be approximately 1 in 400 of the general population) are asymptomatic. ECM1 is a multifunctional binding and scaffolding protein that interacts with a variety of extracellular and structural proteins that contribute to the maintenance and integrity of several tissues and organs (Sercu et al., 2008). At a protein level, ECM1 consists of several protein loops capable of protein–protein binding. Binding to various other extracellular matrix proteins, such as type IV collagen, laminin-332 and fibulin has been noted (Sercu et al., 2009). ECM1 also binds to enzymes such as matrix metalloproteinase-9 (MMP9), which may have relevance to some of the neuropathology, including epilepsy (Fujimoto et al., 2006). Lipoid proteinosis tissue also shows basement membrane thickening, including around several blood vessels (Hamada et al., 2002).

ANIMAL MODELS

There is no mouse model for lipoid proteinosis; attempts to knockout the *ECM1* gene resulted in embryonic lethality. The reason for the disparity between murine and human phenotypes is not known. Knockdown of ECM1 in a zebrafish model has resulted in evident developmental pathologies and may be a suitable model system for testing future therapies (Li and Uitto, 2013).

GENETICS

All cases of lipoid proteinosis are autosomal recessive. Several of the reported cases have been from consanguineous marriages and thus homozygous mutations are more common than compound heterozygous mutations in the published literature (Chan et al., 2007). Thus far, there is no evidence for genetic heterogeneity.

TREATMENT AND MANAGEMENT

The general approach

General measures include making an accurate diagnosis and dealing with acute clinical problems (epilepsy, migraine, anxiety, etc.). Direct Sanger sequencing of the *ECM1* gene is the most efficient way to establish the diagnosis, although lack of ECM1 immunostaining on a skin biopsy can also be used diagnostically (Chan et al., 2004). Life expectancy in individuals with lipoid proteinosis is normal, although premature death from acute-on-chronic respiratory failure in association with infiltration of the respiratory tract has been reported, particularly in affected individuals in rural South Africa (van Hougenhouck-Tulleken et al., 2004).

Therapeutic interventions

Attempts have been made to treat lipoid proteinosis with oral steroids, oral dimethylsulfoxide, oral retinoids, and intralesional heparin, but none of these has proved particularly helpful. Carbon dioxide laser surgery to the thickened vocal cords and the beaded eyelid papules may be beneficial in some individuals (Hamada, 2002).

References

Appenzeller S, Chaloult E, Velho P et al. (2006). Amygdalae calcifications associated with disease duration in lipoid proteinosis. J Neuroimaging 16: 154–156.

Arkadir D, Lerer I, Kiaphoiz L et al. (2013). Lipoid proteinosis with bilateral amygdalae calcifications, headache, and cognitive impairments. Neurology 81: 303–304.

Bach DR, Talmi D, Hurlemann R et al. (2011). Automatic relevance detection in the absence of a functional amygdala. Neuropsychologia 49: 1302–1305.

Becker B, Mihov Y, Scheele D et al. (2012). Fear processing and social networking in the absence of a functional amygdala. Biol Psychiatry 72: 70–77.

Brand M, Grabenhorst F, Starcke K et al. (2007). Role of the amygdala in decisions under ambiguity and decisions under risk: evidence from patients with Urbach–Wiethe disease. Neuropsychologia 45: 1305–1317.

Chan I, South AP, McGrath JA et al. (2004). Rapid diagnosis of lipoid proteinosis using an anti-extracellular matrix protein 1 (ECM1) antibody. J Dermatol Sci 35: 151–153.

Chan I, Liu L, Hamada T et al. (2007). The molecular basis of lipoid proteinosis: mutations in extracellular matrix protein 1. Exp Dermatol 16: 881–890.

Claeys G, Claes LR, Goethem JW et al. (2007). Epilepsy and migraine in a patient with Urbach–Wiethe disease. Seizure 16: 465–468.

Feinstein JS, Buzza C, Hurlemann R et al. (2013). Fear and panic in humans with bilateral amygdala damage. Nat Neurosci 16: 270–273.

Fujimoto N, Terlizzi J, Aho S et al. (2006). Extracellular matrix protein 1 inhibits the activity of matrix metalloproteinase 9 through high-affinity protein/protein interactions. Exp Dermatol 15: 300–307.

Gonçalves FG, de Melo MB, Matos VD et al. (2010). Amygdalae and striatum calcification in lipoid proteinosis. AJNR Am J Neuroradiol 31: 88–90.

Hamada T (2002). Lipoid proteinosis. Clin Exp Dermatol 27: 624–629.

Hamada T, McLean WHI, Ramsay M et al. (2002). Lipoid proteinosis maps to 1q21 and is caused by mutations in the extracellular matrix protein 1 gene (ECM1). Hum Mol Genet 11: 833–840.

Hurlemann R, Wagner M, Hawellek B (2007). Amygdala control of emotion-induced forgetting and remembering: evidence from Urbach–Wiethe disease. Neuropsychologia 45: 877–884.

Hurlemann R, Schlaepfer TE, Matusch A et al. (2009). Reduced 5-HT2A receptor signaling following selective bilateral amygdala damage. Soc Cogn Affect Neurosci 4: 79–84.

Kachewar SG, Kulkarni DS (2012). A novel association of the additional intracranial calcification in lipoid proteinosis: a case report. J Clin Diagn Res 6: 1579–1581.

Li Q, Uitto J (2013). Zebrafish as a model system to study heritable skin diseases. Methods Mol Biol 961: 411–424.

Markowitsch HJ, Staniloiu A (2001). Amygdala in action: relaying biological and social significance to autobiographical memory. Neuropsychologia 49: 718–733.

Markowitsch HJ, Calabrese P, Würker M et al. (1994). The amygdala's contribution to memory – a study on two patients with Urbach–Wiethe disease. Neuroreport 5: 1349–1352.

Maruani A, Djilali-Bouzina F, Abdallah-Lotf M et al. (2007). Urbach–Wiethe disease (lipoid proteinosis) with neurological involvement. Ann Dermatol Venereol 134: 62–64.

Messina MJ, Graziamaria N, Barbieri A et al. (2012). Spontaneous intracerebral hemorrhage in Urbach–Wiethe disease. Neurology 79: 1740–1741.

Quirici MB, Rocha AJ (2013). Teaching neuroimages: lipoid proteinosis (Urbach–Wiethe disease). Neurology 80: e93.

Salih MA, Abu-Amero KK, Alrasheed S et al. (2011). Molecular and neurological characterizations of three Saudi families with lipoid proteinosis. BMC Med Genet 12: 1–6.

Scheele D, Mihov Y, Kendrick KM et al. (2012). Amygdala lesion profoundly alters altruistic punishment. Biol Psychiatry 72: e5–e7.

Sercu S, Zhang M, Oyama N et al. (2008). Interaction of extracellular matrix protein 1 with extracellular matric components: ECM1 is a basement membrane protein of the skin. J Invest Dermatol 129: 1836–1837.

Sercu S, Lambeir AM, Steenackers E et al. (2009). ECM1 interacts with fibulin-3 and the beta chain of laminin 332 through its serum albumin subdomain-like 2 domain. Matrix Biol 28: 160–169.

Siebert M, Markowitsch HJ, Bartel P et al. (2003). Amygdala, affect and cognition: evidence from 10 patients with Urbach–Wiethe disease. Brain 126: 2627–2637.

Teive HA, Pereira ER, Zavala JA et al. (2004). Generalized dystonia and striatal calcification with lipoid proteinosis. Neurology 63: 2168–2169.

Thornton HB, Nel D, Thornton D et al. (2008). The neuropsychiatry and neuropsychology of lipoid proteinosis. J Neuropsychiatry Clin Neurosci 20: 86–92.

Tranel D, Gullickson G, Koch M et al. (2006). Altered experience of emotion following bilateral amygdala damage. Cogn Neuropsychiatry 11: 219–232.

Urbach E, Wiethe C (1929). Lipoidosis cutis et mucosae. Virchows Arch Path Anat 273: 285–319.

van Hougenhouck-Tulleken WV, Chan I, Hamada T et al. (2004). Clinical and molecular characterization of lipoid proteinosis in Namaqualand, South Africa. Br J Dermatol 151: 413–423.

Wiest G, Lehner-Baumgartner E, Baumgartner C (2006). Panic attacks in an individual with bilateral selective lesions of the amygdala. Arch Neurol 63: 1798–1801.

Index

NB: Page numbers in *italics* refer to figures and tables.

I

J

K

L

M

N

FSC
www.fsc.org
MIX
Papier | Fördert
gute Waldnutzung
FSC® C083411

Kolibri 360 GmbH
Ferdinand-Jühlke-Straße 7
99095 Erfurt, Deutschland
produktsicherheit@kolibri360.de